MICROBIOLOGY

ESSENTIALS AND APPLICATIONS

MICROBIOLOGY

ESSENTIALS AND APPLICATIONS

SECOND EDITION

LARRY McKANE

California State Polytechnic University

Pomona, California

JUDY KANDEL

California State University

Fullerton, California

McGRAW-HILL, INC.

NEW YORK ST. LOUIS SAN FRANCISCO AUCKLAND BOGOTÁ CARACAS

LISBON LONDON MADRID MEXICO CITY MILAN MONTREAL NEW DELHI

SAN JUAN SINGAPORE SYDNEY TOKYO TORONTO

MICROBIOLOGY
Essentials and Applications

This book is printed on acid-free paper.

1 2 3 4 5 6 7 8 9 0   VNH   VNH   9 0 9 8 7 6 5

ISBN 0-07-045154-0

ISBN 0-07-113513-8

9 780071 135139

This book was set in New Aster by York Graphic Services, Inc.
The editors were Kathi M. Prancan, Deena Cloud, and Eleanor Castellano;
the designer was Wanda Siedlecka;
cover illustration by Carlyn Iverson, Scientific & Medical Illustration
and Larry McKane;
the production supervisor was Richard A. Ausburn.
The photo editor was Kathy Bendo;
the photo researchers were Sue C. Howard and Ken Eward.
New drawings were done by Network Graphics.
Von Hoffmann Press, Inc., was printer and binder.

Part Openers: Carlyn Iverson, Scientific & Medical Illustration and Larry McKane

Library of Congress Cataloging-in-Publication Data

McKane, Larry.
    Microbiology: essentials and applications / Larry McKane, Judy
Kandel.—2nd ed.
        p.   cm.
    Includes index.
    ISBN 0-07-045154-0
    1. Microbiology.   I. Kandel, Judy.   II. Title.
QR41.2.M38   1996
576—dc20
                                        94-25301

Larry McKane, Ph.D. Larry McKane began his professional career as an assistant professor at Purdue University (Fort Wayne, Indiana). During his two years at Purdue, he taught general microbiology courses to biology and other science majors, and medical microbiology courses to nurses and physician assistants. While in Indiana, he developed the infectious disease component of the physician assistant program for Indiana University, a program that grew into a prototype used across the nation. In 1976 he transferred to California State Polytechnic University in Pomona, California, where he is currently a full professor of microbiology and Coordinator of the Microbiology Section. He has been selected by six graduating classes of his university to be inducted into Cal Poly's roll of outstanding professors. The broad range of microbiology classes he teaches includes students from all academic disciplines, not only microbiology majors. In addition, for eight years, he has offered workshops to other teachers on enhancing teaching strategies—how to create "the teachable moment" in classrooms and in academic literature. These workshops have attracted hundreds of participants from virtually all the states in the USA.

Within the last five years, Dr. McKane has expanded his activities to include consulting in the fields of bioremediation and environmental microbiology. He is still in the classroom, however, and his heart is still with the students, still learning new ways to activate student interest and imagination both in the classroom and in this textbook.

Harvesting the best of his microbiological expertise and blending it with the distillate of his other talents (and those of his coauthor), he feels that in writing this textbook they have indeed created interest and stimulated student imagination without sacrificing the essentials of sound, fundamental microbiology.

Dr. McKane currently lives in California with his wife, three children, and his microbes.

Judy Kandel, Ph.D., M.P.H. Judy Kandel lives in Los Angeles with her husband and son but is a native of Brooklyn, New York, where she received her undergraduate education and formative training as a biologist at Brooklyn College. She migrated west for graduate work in microbiology at UCLA. After receiving her Ph.D. in 1972, she began teaching at California State University, Fullerton (CSUF), where she is currently a professor of biological science. Over the years she has taught many different courses, but she consistently teaches general microbiology for biology majors. She also currently teaches three upper division electives—general virology, pathogenic bacteriology, and industrial microbiology.

As a faculty member, Dr. Kandel has been able to expand her activities and express her interests in microbiology in a variety of ways. These include earning a master of public health degree in epidemiology from UCLA in 1979, collaborating on the first and second editions of this textbook, establishing a consulting partnership that primarily serves the medical device and pharmaceutical industries, and supervising laboratory research, which always involves independent projects performed by graduate and undergraduate students at CSUF. She has published articles and made presentations at meetings on topics ranging from the structure and activity of diphtheria toxin and fungal killer toxins to microbial growth in liquid fuels and is currently examining the role of microbes in the intestinal tract of herbivorous fishes.

Dr. Kandel's concern for students is reflected in other recent activities. She has been the chair of the Undergraduate Advising Committee in her department for the past five years and was responsible for designing the mandatory advising program for biological science majors. She received an intramural grant to redesign the microbiology curriculum and created a more "brains-on," science-oriented lab experience and a more challenging but relevant lecture course. Out of this endeavor emerged opportunities to speak at American Society for Microbiology meetings and to present workshops at the national meetings. In 1993, Dr. Kandel was appointed for a two-year term as a Foundation of Microbiology Lecturer in the area of microbiology education. Her next challenge (but surely not her last) is to create enhanced pedagogical aids for the microbiology student using computer animation technology.

This textbook was created on personal time, family time. The sacrifices our families made were not only frequent and painful, but invisible, bringing them none of the professional satisfaction, recognition, or sense of accomplishment that reward the authors. We are not so naive as to believe that dedicating this book to them will compensate for the thousands of times we were "missing in action," but it might be a start.

To our families: Jan and David (our spouses) and Chris, Matthew, Julia, and Aaron (our children).

We dedicate this book to you.
We rededicate our lives to you.

Contents in Brief

Contents

PART FIVE

Infectious Processes and Host Responses 407

CHAPTER 17

Host–Parasite Interactions 408

CHAPTER 18

Acquired Immunity 440

CHAPTER 19

Immune Disorders 473

Preface

There's quicksand out there, waiting for introductory microbiology students. It's the trap of incidental facts, a morass of unconnected details that explain little but consume students in a soggy pool of material to memorize. In writing *Microbiology: Essentials and Applications,* one of our goals was to avoid creating such a quicksand book, concentrating instead on connections and concepts. This is not an encyclopedia of microbiology. It is the story of the science of microbiology and how it works, a book that students will read and learn from. In writing this text, we had one driving goal—to make sure that introductory microbiology students read the book and "get it." And we sincerely believe we have succeeded in this endeavor. In fact, one of our reviewers, who enthusiastically agreed that we accomplished this goal, suggested that we change the title of the book to *Manageable Microbiology.* We'll not be doing that, but we appreciate the compliment.

A Student-Oriented Book

Understanding and success are terrific motivators. Knowing where you are keeps you going. Our approach of making microbiology *accessible to students,* the people who are encountering this information for the first time, provides this essential encouragement. Without diluting basic microbiology, we have carefully weeded out unnecessary terminology and the disjointed little facts that cloud the student's ability to assemble a conceptual picture from all the parts.

Beyond this, what we are trying to accomplish in this second edition can be distilled to three comprehensive goals:

- To teach the broad spectrum of microbiology in a way that emphasizes understanding and insight. One of our underlying approaches is to emphasize the **significance** of each piece of information we present. We have worked hard to ensure that no reader leaves any section of this book with the question, "So what?"
- To create and maintain "the teachable moment," a time during which the mind is activated, alert, and absorbed in the topic. We engage the reader with clear, lively writing, crisp engaging chapter introductions, and illustrations that maintain student interest while informing, challenging, and stimulating critical thinking skills.
- To help create an informed citizenry, armed with the information, understanding, and cognitive skills needed to evaluate the complex sources of information and rumors that bombard today's public. We hope this book provides students with some of the tools they need to help realize the importance of microbiology to everyday life and to separate myths from facts. We also hope to help people make informed decisions, in their jobs, in their personal lives, and in the voting booth.

The following strategies have helped us accomplish these goals.

A Reader-Friendly Writing Style

The best teaching device a text can offer is a clear, well-organized, and interesting approach to writing. We have attempted to write "visually," helping students learn the material by creating vivid mental images from abstract concepts. Students appreciate the book's ease of reading, accessibility of information, reader friendliness, and ability to motivate by discussing why the principles and descriptions in the book are important or relevant.

Lively Chapter Openers

Every chapter begins with an invitation to the reader. Chapter introductions immediately engage the student with fascinating stories or perspectives, some with unexpected twists that reveal the real-life nature of microorganisms to be as fascinating as anything the most inventive novelist could create. In addition, many of the topics within a chapter get their own introductions, a way to allow students to "ease" into a complex subject with something familiar, then progress to the new, more challenging facts and concepts. This approach helps turn memorization into understanding.

The Illustration Program

One of the strongest features of the first edition was our innovative art program, which was praised as unparalleled in its effectiveness and appeal. We believe we have improved on that art in this edition, maintaining its excitement and effectiveness while making the images even more visually rich, stylish, and pedagogically innovative. These illustrations go far beyond their vibrant colors and engaging images—they simply do a better job of teaching. Almost every figure contains three-dimensional images, sparing the reader an unnecessary and often impossible mental leap from flat facsimiles to "real-life" representations. Colors are used as recognition guides that help steer students through each figure—when they encounter burnt orange, for example, they always know they are looking at a cell wall. Color is also used to show processes—for example, the flow of energy through metabolic pathways and ecosystems. (Hot colors symbolize energized structures whereas cool colors are used for objects in their low-energy states.) The bold sweeping arrows help transform static images into active ones, changing abstract concepts into more easily grasped physical images. These high-performance illustrations make microbiology more inviting and enjoyable to learn.

Special Features

We have invented the following special approaches that help provide the reader with direction, focus, and a bridge from abstract information to everyday life.

● **MYTH VS. FACT.** In every chapter we pinpoint and expose many of today's popular and sometimes dangerous myths, then replace them with facts. Like the erroneous beliefs that camels carry water in their humps and ostriches stick their heads in the sand to avoid danger, microbiological myths are very pervasive among the general public. (Does getting wet or chilled really cause a cold? Does vitamin C really help prevent the common cold and influenza? Of course not.) Students and instructors alike look forward to these spots. They are fun, fascinating, and pointedly revealing. Exposing the more dangerous myths can even be life-saving.

● **THE EXPLORERS.** This series reveals the real-life fascinating events that led to the unearthing of much of what we know about microbiology. These stories create a connection with the information and build on the reader's understanding of science. A few of these features are listed here to provide a sense of what they accomplish:

- Dedication Beyond the Call
- PCR Shatters a Technological Roadblock
- The DNA Detectives
- The Hazards of Laboratory Microbiology
- Microbiologists Investigate the "The Final Frontier"

● **BEYOND THE BASICS.** These essays take the reader beyond the mainstream chapter material. They present more challenging information that some instructors will assign but others will not, depending on the objectives of the class. By creating more flexibility in the text, these essays better enable the instructor to tailor the textbook to match his or her goals in class, as the following list reveals:

- Other Pathways for Catabolizing Glucose
- Bacterial Behavior
- The Generation of Antibody Diversity
- Bringing Order to the Chaotic Streptococci
- The Origin of the AIDS Virus

● **THE MICROBES.** These essays present topics that enliven the mainstream of microbiological information, topics that help bring the subject to life. Some are fascinating accounts of microbial strategies for overcoming seemingly insurmountable obstacles to survival; others tell of the impact of microbes in our personal lives. The following list of titles provides an indication of the interesting subjects discussed in this feature:

- Dangerous Ghosts of Dead Bacteria
- The R Factor—An Impending Disaster?
- Crippled Bacteria and Suicidal Microbes
- A Modern Disease for Modern Times
- A Vaccine against Nazis

● **CONCEPT CHECKS.** These pauses are found throughout the chapter at the end of each major section. Unlike the "review questions" found in several competing texts, these questions attempt to test whether the student has grasped the key concepts and significance of the material. They do not ask for names, definitions, or lists, but challenge a true understanding of the ideas. If students have mastered the concepts, they will be able to memorize the facts that apply to them. The reverse is not true.

End-of-Chapter Material

In addition to a list of key terms (with page numbers to indicate where the terminology is explained), we have departed from the standard end-of-chapter material that typifies most textbooks. This departure is designed to enhance the power of the following features to reinforce the chapter material and help build understanding.

● **REVIEWING CONCEPTS AND TESTING UNDERSTANDING.** Each chapter's material is revisited in two forms. The first of these, *Key Facts and Concepts,* provides the reader with a review of the critical points of the chapter. These facts and concepts enable students to see the same information from a different perspective. This summary information is grouped into packets that cluster together related information, helping students make connections that may otherwise have escaped them. This feature is followed by *Review and Problem-Solving Questions,* a battery of questions that allow self-assessment, reinforce the information, promote deeper understanding, stimulate critical thinking activities, and encourage problem solving. Students will exit each chapter with the encouragement of knowing they have mastered the topic. Again, success is a terrific motivator.

● **HEADLINES.** A recent article in the *ASM (American Society for Microbiology) News* reported that end-of-chapter references are rarely used by readers of introductory microbiology textbooks. Instead of hard-to-find additional readings, we have introduced "Headlines," a feature that summarizes a topical article about some recent scientific breakthrough in the field. Each of these readings is followed by thought-provoking questions that challenge the reader to critically examine the information and understand its significance. The Headlines feature provides students with additional reading experiences in a form that is readable.

Expanded Coverage

We have been very pleased by the large numbers of instructors who stayed loyal to this book over many years. They say they want a concept-oriented book that presents microbiology not only clearly but also with enthralling narrative and dynamic visuals, and that this book does that for them. To better appeal to these instructors and their students, we have expanded our coverage to provide a very balanced view of microbiology. Whereas the applications in the previous edition were predominantly medical, this edition discusses the influences and applications of microbiology in all aspects of life, from environmental microbiology to the role of microbes in space exploration. We have expanded the essentials and applications to have a greater breadth and depth. Recognizing that our microbe-vulnerable bodies are one of the common denominators shared by all students regardless of their major or interest, we decided not to sacrifice the medical examples but rather supplement them with discussions of the microbiology of soil, water, environment, industry, food, and agriculture. A brief look at our table of contents reveals the new coverage—a stronger emphasis on microbial interactions (Chapter 14), expanded coverage of host resistance and immunity (Chapters 17 through 19), stronger discussions of the revolution in microbial genetics (Chapters 8 and 9), and enhanced discussion of environmental and industrial microbiology (Chapters 26 and 27). This book also comes equipped with an epilogue, our views of some of the more promising areas of microbiology and what new information and applications await us as we enter the next century.

Acknowledgments

We wish to thank the many manuscript reviewers:

D. Andy Anderson, *Utah State University*

Patricia S. Astry, *Fredonia State University*

Gail F. Baker, *LaGuardia Community College*

Clinton L. Benjamin, *Lower Columbia College*

Robert J. Boettcher, *Lane Community College*

David Campbell, *St. Louis Community College at Meramec*

Garry Cartwright, *Northeastern Oklahoma State University*

Paul H. Demchick, *Barton College*

Donald S. Emmeluth, *Fulton—Montgomery Community College*

Bentley A. Fane, *University of Arkansas*

Paul Farnsworth, *University of Texas at San Antonio*

Charles Goldberg, *Borough of Manhattan Community College*

Diane S. Herson, *University of Delaware*

Fred D. Hinson, *Western Carolina University*

Anne Morris Hooke, *Miami University*

David Mack Ivey, *University of Arkansas*

Robert J. Janssen, *University of Arizona*

Alfred T. Mikell, Jr., *University of Mississippi*

Glendon R. Miller, *Wichita State University*

Edwin C. Roland, *Ohio University*

Mary J. Ruebush, *Montana State University*

Carl E. Sillman, *The Pennsylvania State University*

Warren S. Silver, *University of South Florida*

Richard T. St. John, *Widener University*

Pamela B. Tabery, *Northhampton County Community College*

James E. Urban, *Kansas State University*

William R. Wellnitz, *Augusta College*

We'd like to thank McGraw-Hill's senior editing supervisor, Eleanor Castellano, who smoothed out what would have been a very bumpy production road, and our development editor Deena Cloud, who helped even out our writing. Our job was also made easier by the effort of Hermann Strohbach, who made the design work. A heartfelt thanks also goes to Art Ciccone, our art "translator," for his keen eye and artistic expertise, both of which helped us transform our original illustrations and ideas into the highest quality art program of any book in the field. We also wish to thank all the beleaguered artists and photo editors who, to their enormous credit, endured our persistence and high expectations to ensure that each photo and illustration went well beyond the standard in beauty, interest, and teaching clarity.

Above all, we would like to thank Kathi Prancan, our biology editor, friend, and advocate, for her confidence in the talents and judgment of the authors, and for ensuring that we received what was needed for this book to be exactly what we envisioned it should be. Thanks for staying behind the wheel, Kathi. You steered us through some very rough terrain.

Larry McKane

Judy Kandel

Supplements Overview

For the Instructor

■ Overhead Transparencies

The expanded art program for the main text will provide the basis for an outstanding overhead transparency package. All of the 200 transparencies are in four colors, and label sizes have been increased for improved distance visibility.

■ Test Bank (by William Wellnitz, Augusta College)

With over 1300 questions, this printed test bank provides the professor with an extensive resource for developing exam material. Among the questions included for each chapter are multiple-choice, true/false, completion, matching, problem-solving, and conceptual. Each question is categorized by the type of learning being tested (recall, interpretation, analysis, or synthesis), and icons are used to denote questions with a higher level of difficulty.

■ Computerized Test Bank

The printed test bank is also available in the following computerized formats: IBM $5^1/_4$; IBM $3^1/_2$; and Macintosh.

For the Student

■ Student Study Companion (by Virginia Klair)

This innovative study tool is designed to provide students with a hands-on, interactive approach to learning. The *Student Study Companion* emphasizes active learning by employing exercises that allow students to create their own flow charts, concept maps, index card catalog, and diagrams. Text material is reinforced through nontraditional methods to help students apply their knowledge in different forms for greater learning. For each chapter, the *Student Study Companion* provides a chapter preview, extended outline, glossary, learning activities, and sample examination questions. The "Putting It All Together" section helps integrate concepts from each unit of the text. General study skills tips are also included throughout the guide to improve study habits.

MICROBIOLOGY

ESSENTIALS AND APPLICATIONS

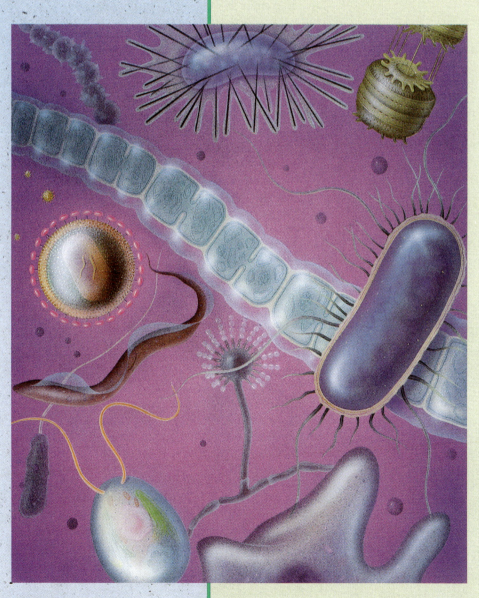

CHAPTER 1

Exploring the Microbial World

A few inhabitants of the invisible calm of microbes are depicted in this fanciful rendition of a "microbial landscape." Although unseen in everyday life, microorganisms are inescapable residents of our world. Living with microorganisms is impossible to avoid.

S ince the dawn of life some 3.5 billion years ago, our planet has collided with giant asteroids, erupted with explosive volcanoes at the rate of 20 Mount Saint Helen–sized blasts per month, accumulated in its atmosphere one of the most lethal chemicals in the history of life, endured three ice ages that extended glaciers well into what today are temperate zones, and suffered global episodes of disease. These and other natural disasters have claimed some of the earth's most magnificent species. The giants of the terrestrial earth, the dinosaurs, may have fallen victim to one of these episodes 65 million years ago in the most famous and perplexing vanishing act of all.

But perhaps the most profound episode of extinction, claiming almost every species on this planet, killed off not the earth's largest organisms, but its smallest. The little-publicized disaster occurred some 2 billion years ago, caused by organisms smaller than the period at the end of this sentence. These organisms produced a deadly new gas that accumulated in the atmosphere and poisoned the majority of life forms existing on the planet at that time. The earth has never been the same, for that deadly gas is still with us today. It is oxygen.

At the time, the earth was populated exclusively by microorganisms. **Microorganisms**, or **microbes**, are organisms that are too small to be seen with the unaided eye. Of course, many microbes escaped extinction, and their descendants are still with us. Among today's microorganisms are a group of pests we commonly refer to as "germs." Considering our problems with germs, extinction of all the earth's microbes might not seem like such a bad thing. We live in a society in which many people border on having phobias against microorganisms, believing them all to be germs. But suppose some global change did precipitate the selective loss of all the earth's microorganisms without directly killing any larger organisms. What would be the effect?

Life Without Microbes

Many people would celebrate the loss of all microbes as a wonderful event, and at first it might seem to be so. Colds, influenza, AIDS, cholera, tuberculosis, and all other infectious diseases would immediately vanish. It would no longer be necessary to spend billions of dollars on preservation techniques to prevent the microbial spoilage of food or decomposition of useful products. Termites in the home would no longer be a problem, for without wood-digesting microorganisms in their guts these insects would starve.

The celebration, however, would be woefully short-lived. You would immediately be stricken with diarrhea, having lost the trillions of bacteria that normally live in your large intestine and help stabilize its contents. Before long, you would begin running short of food. The first foods to disappear would be meats such as beef and lamb. Cattle and sheep could not digest their high-fiber diets without the microorganisms in their guts that do the job for them. Fish, shellfish, and other seafood would disappear about the same time, since the aquatic food chain is fueled by microscopic organisms that capture the sun's energy and convert it into forms that can be used for sustaining life. Without these photosynthetic microorganisms, virtually all freshwater and marine animals would starve to death.

In fact, every food chain on earth would eventually be extinguished. Natural decomposition of dead plants and animals would grind to an immediate

(a)

(b)

(c)

(d)

FIGURE 1-1 Microorganisms—could we do without them? Here are just a few of the countless assets that we would lose should microorganisms disappear. (*a*) This diverse picture of life would be transformed into a lifeless "moonscape" were it not for microorganisms (shown in inset) that decompose dead organisms and wastes, recycling their nutrients for reuse by plants and other food-producing organisms. (*b*) Many crops that provide food for the world's 5 billion people depend on microorganisms (see inset) that literally make fertilizer for the plants by transforming atmospheric nitrogen into a form the plants can use. These bacteria thrive in the roots of plants such as the alfalfa growing between the corn and wheat in this photo. (*c*) Animals such as cattle, sheep, or these mountain goats depend on the microorganisms in their stomach chambers (inset) to digest their high-cellulose diet of grass. Without microbial help, these animals would starve. (*d*) These beautiful orchids form a cooperative association with a fungus that assists in absorbing water and nutrients from the soil. The orchid cannot survive without the fungal roots (inset).

halt. The nutrients in the bodies of these dead organisms would not be returned to the soil to be reused by plants. Failing to find sufficient nutrients to grow, plants would disappear, and all plant eaters would soon follow. Without plants and animals to eat, humans would quickly be just another of the millions of species piling up in a thick layer of dead organisms on a dead planet.

Before starving, however, we would have other struggles—with breathing, for example. Without the photosynthetic microorganisms that live in the ocean and generate about half the world's oxygen, atmospheric oxygen would quickly decline. By then the loss of microbe-produced supplies such as wine, bread, many medicines, cheese, and soy sauce would probably seem inconsequential. Our sewage facilities would fail to remove environmentally disruptive organic materials from our wastewater, but that wouldn't matter anymore—we would all be dead.

Fortunately, this scenario is pure fantasy, one that would likely never happen. Even if some catastrophic chain of events led to the extinction of most of the world's organisms, microorganisms would very likely survive. Sharing our world with microorganisms is inevitable, and we benefit immeasurably from their presence and activities (Fig. 1-1).

CONCEPT CHECK

- What microbial activities are necessary to support human life on earth?

Life Without Microbiology

As essential as microbes are, those that cause damage pose a formidable force to be reckoned with. **Microbiology,** the study of microorganisms, has generated remarkably effective weapons in the battle against harmful microbes. Microbiological knowledge has doubled the human life span compared with that of people living before the "golden age of microbiology"—a period that spanned the last half of the nineteenth century, a period when microbiologists provided many of the advances that made life both safer and more tolerable. Before then, the connection between disease and microorganisms remained unproven. There was no way to prevent tetanus, polio, whooping cough, rabies, diphtheria, and other fatal diseases that took an enormous toll. Nor were antibiotics available to treat scarlet fever, epidemic typhus, plague, or dozens of other diseases that, although curable today, claimed millions of lives in the preantibiotic era. Before weapons against infectious disease were developed, people who lived to the age of 40 were considered old.

Go back another 200 years, in England, for example, and you would find conditions shockingly unhygienic. Bathing, for example, was considered dangerous, so people rarely washed. Human wastes were collected in chamber pots and discarded in the streets. These wastes spread cholera, dysentery, typhoid fever, and dozens of other diseases, but no one realized it in those days. People died of these diseases with no knowledge of their microbial origin. Even when people removed feces and urine from their community, the wastes were emptied into open cesspools. The pools would overflow with heavy rains and drain into wells, contaminating the water supply. Uncontrolled populations of rats carried the microorganisms responsible for the great bubonic plagues, the Black Death that killed one-fourth of the population of Europe, revisiting the continent in great epidemics for 300 years.

Not only were people helpless against microbial assaults on their health, they also lacked efficient ways to prevent microorganisms from quickly destroying their food supplies. Drying, salting, and pickling food were the most common ways to retard food spoilage. Without better preservation techniques, microorganisms ranked with crop-eating insects as the major competitors for human food supplies. Microbes not only spoiled food, they also attacked crop plants even before they could be harvested. The microbe that caused the Irish potato blight, for example, wiped out the major dietary staple of Ireland in the mid-1800s, leaving a million Irish dead of starvation in a single year.

Our modern times are unquestionably better than a world unimproved by microbiology. As we launch you on this exploration of the microbial world, notice all that microbes have to offer you, from maintaining life on the planet to helping fight disease. They have even provided the keys to unlocking some of life's most provocative secrets.

Microorganisms continue to be dangerous enemies, but they are even better allies. It's a good thing that most of them are on our side.

CONCEPT CHECK

- How has the study of microbiology improved the quality of human life?

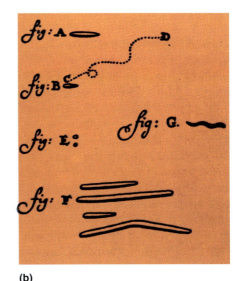

(a)   **(b)**

FIGURE 1-2 Father of microbiology. (*a*) Antoni van Leeuwenhoek with his simple microscope in hand. (*b*) Drawings from Leeuwenhoek's notebook of some of the microbes he observed.

A Science Changing the World

Despite their small size, microbes exert an immense influence on humans and on the world in general. Microbiologists explore these tiny organisms. They attempt to understand the activities of microorganisms as well as their impact on humans. Such knowledge helps us improve methods of controlling the growth of microorganisms, minimizing their harmful effects and maximizing their beneficial activities. Microbiology's benefits go beyond practical applications, however. Much of what we know about how living organisms work, and even about the origin of life itself, comes from microbiology. The cornerstone of countless basic and applied accomplishments that have enriched our lives and revolutionized the way we view the world has been microbiology, a science that was born—by accident—just over 300 years ago.

■ A New World's Discovery

In 1674 Antoni van Leeuwenhoek, an amateur lens grinder of extraordinary skill and patience, looked through his simple microscope and discovered a new world (Fig. 1-2). The existence of this universe of microscopic inhabitants had been suspected by a very few insightful scientists, but no one had ever seen them before. The "animalcules," as Leeuwenhoek called them in a series of letters to the Royal Society of London, were found in his mouth, in stagnant water, and in foods. Leeuwenhoek observed microbes in samples from nearly every environment he investigated. The microscopic organisms seemed to be everywhere. He reported that the human mouth contained more organisms than there were people in all of his United Netherlands. His discovery touched off a controversy that influenced scientific endeavor for the next 200 years. The controversy concerned the origin of microbes.

■ The Spontaneous Generation Controversy

Leeuwenhoek's discovery of microorganisms explained why many foods and drinks spoil, becoming distasteful and even dangerous. After sufficient time they simply become overgrown by microbes. Microbial proliferation is often evidenced by the cloudiness of a once-clear liquid. But scientists wanted to know where these organisms came from. Some people suggested that nonliving substances were converted into living organisms; in other words, they believed in the **spontaneous generation** of life from nonliving material (abiogenesis). Opponents of this explanation supported the theory of **biogenesis,** which stated that all organisms arise only from other living organisms, and insisted that overgrown food had been seeded with at least one viable (living) parent microbe. Although the biological origin of frogs and flies had already been proved, the idea of the spontaneous generation of microorganisms required two centuries to disprove.

Spontaneous generation appeared to be disproved when Lazzaro Spallanzani demonstrated in 1765 that beef broth that had been boiled to kill all the microbes in it remained **sterile** (free of all living organisms) as long as the container was plugged with a solid stopper. In response to criticism that air was necessary for spontaneous generation to occur, cotton plugs were substituted for the solid stoppers. (Cotton filters remove suspended particles from air passing through them, and microbes are among the trapped particles.) Broth that was protected from suspended particles remained sterile, demonstrating that the broth itself could not give rise to living organisms. Removing the cotton plug allowed microorganisms to enter and then overgrow the liquid within 18 hours. Thus, boiling did not destroy the ability of the broth to support microbial growth. Biogenesis seemed to be the only logical explanation for such a phenomenon.

Many proponents of spontaneous generation, however, remained unconvinced, arguing that passage through cotton altered the air and removed the "vital force" needed for spontaneous generation to occur. The controversy was finally resolved in 1861 by the powerful logic of Louis Pasteur, who was already on his way to becoming one of the most influential scientists of all time. He designed swan-necked flasks that allowed the introduction of fresh, unaltered air while trapping dust particles and microorganisms in the curved neck, thereby eliminating the need for the cotton plug (Fig. 1-3). Broth that had been boiled in these flasks remained sterile. Thus, Pasteur proved that neither broth nor air could spontaneously produce microorganisms, since sterility was preserved in the presence of both. Furthermore, when the flask was tipped, allowing the liquid to flow into the neck, visible microbial growth developed within hours (Fig. 1-3b). Pasteur

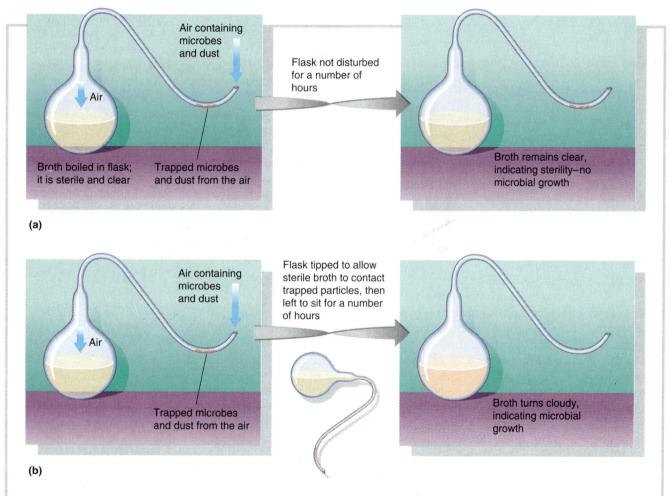

FIGURE 1-3 Swan-necked Pasteur flask. Dust settles in the neck of the flask, trapping particles and microorganisms while allowing unaltered air to reach the nutrient medium. (*a*) The broth remains clear and sterile. (*b*) If the flask is tilted so the broth flows into the neck and reaches the trapped particles, the liquid becomes cloudy within hours, indicating microbial proliferation.

thus demonstrated that the agents responsible for spoiling the broth were the microbes trapped in the flask's neck.

In spite of his convincing experiments, a few scientists continued to reject biogenesis, and for good reason—their own experiments failed to confirm Pasteur's claim. These investigators boiled and sealed liquid infusions of hay rather than beef broth. The hay infusions often became overgrown with bacteria, even when they were protected from all external sources of microbes. The conflicting results were due to certain bacteria that are found in hay but rarely in beef. These bacteria form **endospores,** protective structures that are among the most heat-resistant forms of life on earth. Many endospores can survive several hours of boiling and then, at lower, more hospitable temperatures, germinate to an actively growing stage. Thus, the boiled hay infusions were actually inoculated from within by the surviving endospores, resulting in overgrowth with no apparent source of microbial parents. With John Tyndall's discovery of heat-resistant endospores in 1877, all remaining arguments supporting spontaneous generation evaporated.

CONCEPT CHECK

- What is biogenesis?
- Why didn't microbes grow in Pasteur's swan-necked flasks?

The Microbes

Microbes include several major categories of organisms: (1) protozoa, (2) microscopic algae, (3) microscopic fungi (yeasts and molds), (4) bacteria, and (5) viruses. Although most are unicellular (single-celled) organisms, some, like the molds and many algae, are considered multicellular. Bacteria possess the least complex cellular structure. The cell structure of protozoa is similar to that of animal cells. The cells of algae are similar to those of plants, whereas fungi superficially resemble plant cells that lack chlorophyll and thus have lost the ability to photosynthesize. In spite of these resemblances, however, microorganisms are considered neither plant nor animal. Viruses lack all cellular structure and are referred to as noncellular particles. They are classically considered to be the simplest of all microbes.

Some typical members of each group of microorganisms are pictured in Figure 1-4.

CONCEPT CHECK

- What feature is common to all microbes?
- How do the various categories of microbes differ?

Microbes in the Environment

Microorganisms exist virtually everywhere in the *biosphere* (the thin envelope around the earth in which life exists). They are in our food, in the water we use for drinking and bathing, on our utensils, on our clothing and bed sheets, and on our bodies. The air we breathe carries a wide variety of microbes. Their varied nature allows some types to survive in even the most unlikely environments. Microbes have been found in the thin cold air miles above the earth. Other microbes thrive in natural hot springs at a temperature of 90°C, and at least one type grows well at 105°C. Some bacteria can grow in sulfuric acid concentrated enough to kill virtually all other types of organisms. Other microbes proliferate in distilled or deionized water, using minute amounts of nutrients dissolved from the air.

Microorganisms inhabit the surfaces of living human and animal bodies and grow abundantly in the mouth and the intestinal tract. There are ten times as many bacterial cells residing in your colon as there are human cells in your body. In fact, one-third of the dry weight of human feces is bacteria. Yet most persons are unaware of the presence of the trillions of microorganisms that inhabit their bodies because the microbes ordinarily stimulate no apparent physiological response and cause no disease. Such harmless microorganisms comprise the **normal flora,** those microorganisms that normally live on the human body in a harmonious relationship with their host.

MYTH: To remain healthy, all plants and animals must quickly eliminate all microbes from inside their bodies. Otherwise, disease will surely follow.

FACT: Some animals and many plants harbor inside their tissues beneficial microorganisms that supplement an animal's food supplies or assist a plant in its uptake of water and nutrients. ▶ (See Mutualism, Chap. 14, p. 335.)

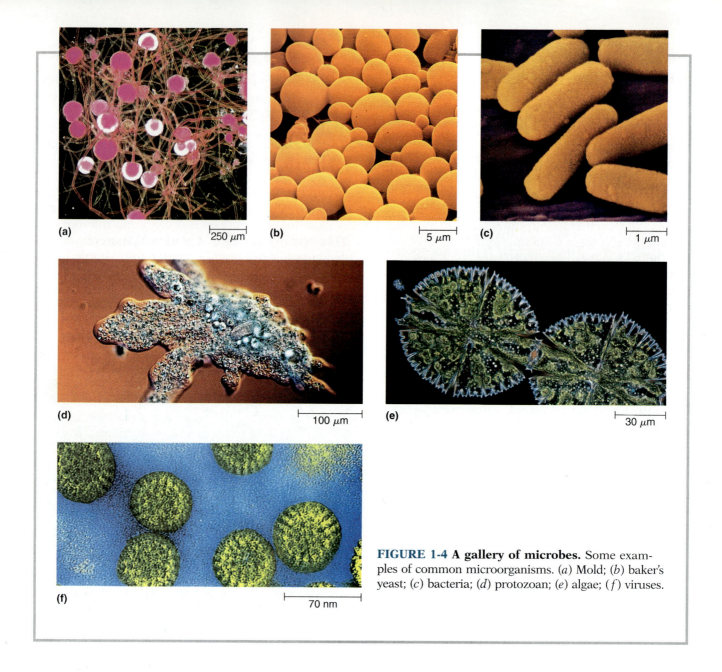

(a)   250 μm

(b)   5 μm

(c)   1 μm

(d)   100 μm

(e)   30 μm

(f)   70 nm

FIGURE 1-4 A gallery of microbes. Some examples of common microorganisms. (*a*) Mold; (*b*) baker's yeast; (*c*) bacteria; (*d*) protozoan; (*e*) algae; (*f*) viruses.

It is far easier to identify the types of environments that are devoid of microorganisms because only a few such places naturally exist on earth. In the vent of an erupting volcano, for example, all microbes would be incinerated. The interior of a healthy human body usually contains no microbes. (In humans and other animals, "interior" does not include the digestive tract or outermost regions of the respiratory tract, which are extensions of the external environment.) Your cerebrospinal fluid normally harbors no microorganisms. Similarly, the blood and tissue fluids of a healthy person are microbe-free, except for transient microorganisms introduced by trauma such as cutting the skin or accidentally biting the tongue. Urine in the bladder is also sterile, although it becomes contaminated with normal flora organisms during voiding. All the internal organs of a healthy person are free of microbial growth, thanks to a very effective complex of protective mechanisms that restrict microbes to the surface of the body and normally eliminate any cells that breach the surface. Because these protective mechanisms are integral to understanding the relationship between microorganisms and humans, they are generally considered a part of the microbiology curriculum.

CONCEPT CHECK

- How does the earth's distribution of microorganisms compare to that of humans?

Microbial Activities

Life requires biochemical activity. All living organisms must be able to change chemicals from their environment into forms that can be incorporated into cellular material. These and all the other chemical changes performed by an organism are known collectively as **metabolism.** The influences that microbes exert on their environment are direct reflections of their diverse metabolic activities. Considering the widespread distribution of microbes, it is fortunate that the metabolic processes of most do not harm, and in many cases actually benefit, humans or the global environment.

■ Beneficial Activities

Microorganisms contribute in countless ways to the environment, to our quality of life, and to our understanding of the way life works (Table 1-1). Although some of them were discussed earlier in the chapter, we'll close in on the details of a few of these beneficial activities.

● **ENVIRONMENTAL BENEFITS.** The health of the global environment is inescapably linked to its microbial inhabitants. In fact, without microorganisms, all life on earth would perish. Microorganisms decompose dead organic matter into simple nutrients that can be used by plants and other photosynthetic organisms. As decomposers, microbes are critical to the **biogeochemical cycle,** the flow of

TABLE 1-1   A SAMPLING OF BENEFITS DERIVED FROM MICROORGANISMS AND THEIR ACTIVITIES

In Natural Environments

Activity	Benefit
Decomposition of dead organic matter	Recycles nutrients throughout the biosphere.
Oxygen production	Photosynthetic aquatic microorganisms generate about half the atmosphere's oxygen.
Pest control	Insect diseases may help control crop-destroying pests.
Normal flora	Produce some nutrients for their host and compete with pathogens that could otherwise cause infectious disease in the host.
Soil fertility (nitrogen fixation)	A few types of bacteria convert atmospheric nitrogen to a form usable by plants.
Survival of cattle and other ruminants	Cellulose-digesting microbes in the guts of cattle, sheep, and other types of grazing or browsing animals allow animal to sustain itself on food that it cannot digest otherwise.
Aquatic food chains	Photosynthetic microorganisms in water provide the energy and nutrients to sustain themselves and feed virtually all aquatic consumers.
Terrestrial food chains	Microbial decomposition provides nutrients needed by the photosynthesizers that support dryland food chains. Some terrestrial animals live on aquatic organisms, thereby interlinking aquatic and terrestrial food chains.
Destruction of toxic compounds	Poisonous by-products of some organisms are naturally detoxified by microbial activity.

For Human Application

Activity	Benefit
Alcoholic fermentation	Beer, wine, and spirit production.
Antibiotic production	Source of many pharmaceutical products used to fight infectious diseases in humans and other animals.
Biopesticides	Specific insect-killing microorganisms are used to replace chemical pesticides to fight crop pests without killing useful animals or polluting the environment.
Bioremediation	Microorganisms can be used to clean up oil spills and destroy other environmentally disruptive substances, such as toxins at dump sites and chemical contaminants released as industrial waste.

(continued)

(a)

(b)   Isolated colony

FIGURE 1-11 From kitchen to laboratory. (*a*) Robert Koch with the basic tools of the microbiologist. The petri dishes contain a solid medium, and the flask contains broth, a liquid medium. The solid medium contains the solidifying agent agar, which was suggested to Koch by Ellie Hesse, who used the seaweed extract to solidify jelly. All other solids Koch had tried were digested to liquids by the microbes. This breakthrough allowed Koch to obtain pure cultures easily, enabling him to be the first to prove the germ theory of infectious disease, beating Louis Pasteur, who was working on virtually the same project in France. (*b*) Surface of solid medium (nutrient agar) inoculated with bacteria and incubated for 24 hours. An isolated colony can be selected to obtain pure cultures.

anthracis in the blood of people with anthrax. But Koch was the first to use a scientific approach to prove that the bacterium was the cause of the disease rather than the result of it. Such scientific approaches transformed microbiology from a collection of observations into a true science.

■ Immunization

The best approach to combating infectious disease is *prophylaxis,* that is, disease prevention. Among the most powerful prophylactic weapons against pathogens are those that enhance people's resistance to disease. During the course of some infectious diseases, the body develops **immunity,** resistance to further attacks by the same pathogen. Thus a person suffers from some illnesses such as measles only once. Immunity can be clinically induced by introducing harmless variants of pathogens into the body, which consequently develops resistance without acquiring the disease. This procedure, known as **vaccination** (*vacca*=cow), was first used successfully in 1798 by Edward Jenner, who used cowpox virus to protect against smallpox. Jenner had observed that people who contracted cowpox (a mild disease characterized by sores on the hands) never developed smallpox. He proved that inoculating people with pus from cowpox lesions safely protected them from smallpox (Fig. 1-12).

In spite of his success, Jenner never understood the mechanism by which his vaccine provided protection. We now know that the success of the vaccine was due to similarities in the physical and chemical structures of the smallpox and cowpox viruses. Immunity develops partially because the body produces specific blood substances, called **antibodies,** in response to the presence of foreign agents such as viruses or bacteria. Antibodies react with the specific agent that stimulated their production and prevent it from damaging the host (Fig. 1-13).

FIGURE 1-12 The first vaccination. Edward Jenner administers the first vaccine against infectious disease to 8-year-old James Phipps using pus from a milkmaid's cowpox lesion. His procedure protected this child from smallpox.

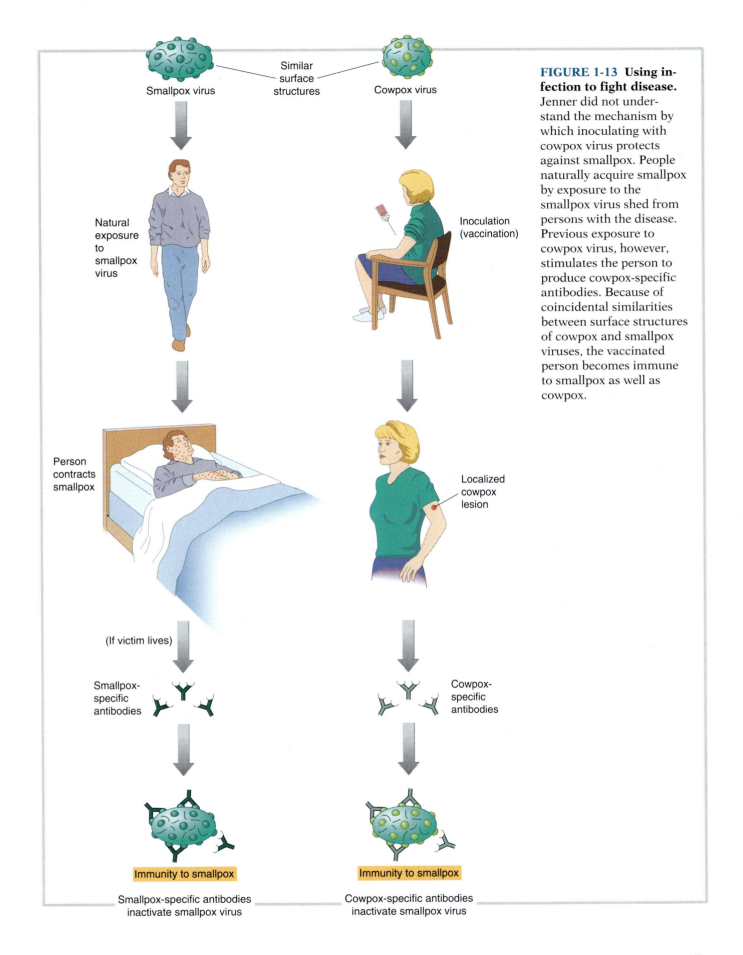

Smallpox virus

Similar surface structures

Cowpox virus

Natural exposure to smallpox virus

Inoculation (vaccination)

Person contracts smallpox

Localized cowpox lesion

(If victim lives)

Smallpox-specific antibodies

Cowpox-specific antibodies

Immunity to smallpox

Immunity to smallpox

Smallpox-specific antibodies inactivate smallpox virus

Cowpox-specific antibodies inactivate smallpox virus

FIGURE 1-13 Using infection to fight disease. Jenner did not understand the mechanism by which inoculating with cowpox virus protects against smallpox. People naturally acquire smallpox by exposure to the smallpox virus shed from persons with the disease. Previous exposure to cowpox virus, however, stimulates the person to produce cowpox-specific antibodies. Because of coincidental similarities between surface structures of cowpox and smallpox viruses, the vaccinated person becomes immune to smallpox as well as cowpox.

Vaccination became a useful tool for preventing other diseases when Louis Pasteur (who had his fingers in many scientific pies) developed methods for converting pathogens into harmless variants that could induce immunity. Using this method he developed successful vaccines against anthrax and against rabies. Vaccines developed by scientists since Pasteur have dramatically reduced the incidence of diphtheria, polio, whooping cough, tetanus, mumps, and measles.

■ Chemotherapy

Some diseases can strike the same person many times because infection fails to induce immunity. Malaria and some sexually transmitted diseases are examples of such recurrent illnesses. Fortunately, many diseases can be cured by **chemotherapy,** the use of chemicals that selectively inhibit or kill pathogens without killing the patient.

Many chemotherapeutic agents are **antibiotics,** chemicals produced by one type of microorganism that interfere with specific biological processes of other types of microbes. (**Caution:** Do not confuse antibiotics and antibodies.) Alexander Fleming's accidental discovery of penicillin in 1929 opened the door to the antibiotic era of medicine. Fleming observed that a contaminating mold growing on a solid medium somehow inhibited the growth of the bacteria that he had inoculated on the medium. The mold was *Penicillium notatum,* which inspired Fleming's name for the antibiotic. Years later other investigators purified this inhibitory chemical from mold cultures. Penicillin became the first of many effective antibiotics for treating infectious disease. Millions of human lives have been saved by antibiotic therapy, and the search continues for new and more effective antimicrobial agents.

■ Epidemiology

An important field within microbiology is **epidemiology,** an investigative science that provides information about the factors and conditions that contribute to the occurrence of diseases. Epidemiologists monitor the incidence of each illness in a population and the number of persons dying of each disease. This information helps authorities identify potential health hazards in the community so that appropriate precautions can be instituted. For example, an increase in the incidence of measles suggests the need for a more extensive immunization (vaccination) program. Epidemiologists also identify sources of disease outbreaks. Once identified, measures can be instituted to control the source of infection, thereby preventing the spread

of disease. ▶ (The details of epidemiology are discussed in Chapter 20.)

The development of medical microbiology has produced dramatic progress in the war against human diseases (Fig. 1-14, right-hand column). Citizens of developed countries no longer need fear the onslaught of crippling or killer epidemics of polio, diphtheria, whooping cough, typhoid fever, plague, yellow fever, or cholera. Yet outbreaks of these diseases can still occur if control measures become inadequate; thus these diseases are not completely defeated.

The first (and so far only) total microbiological victory over infectious disease culminated in 1980 when smallpox, a disease that scarred or killed as many as a fourth of the world's population before this century, became the first infectious disease declared by the World Health Organization to be successfully eradicated from the earth.

Despite these advances, infectious disease remains the most common reason people seek the care of a physician. Although most cases are mild and self-limiting, others are life-threatening. Infectious diseases such as tuberculosis, hepatitis, pneumonia, and AIDS (acquired immunodeficiency syndrome) continue to cause human suffering and death. In addition, as many as one in every ten patients still acquire diseases while hospitalized. Such hospital-acquired diseases are called **nosocomial infections**.

Control of nosocomial and other infectious diseases depends on an expanded knowledge of microorganisms and their activities as well as on the human body's response to them.

CONCEPT CHECK

- What is the germ theory of infectious disease?
- What findings must be demonstrated to prove the etiology of an infectious disease?
- What role in reducing infectious diseases is played by each of the following? Immunization; chemotherapy; epidemiological investigations.

Microbiology and the Process of Science

One of our newest sciences is devoted to studying the earth's oldest group of organisms. Microbiology is barely 300 years old, yet some of the organisms studied resemble those on earth 3.5 billion years

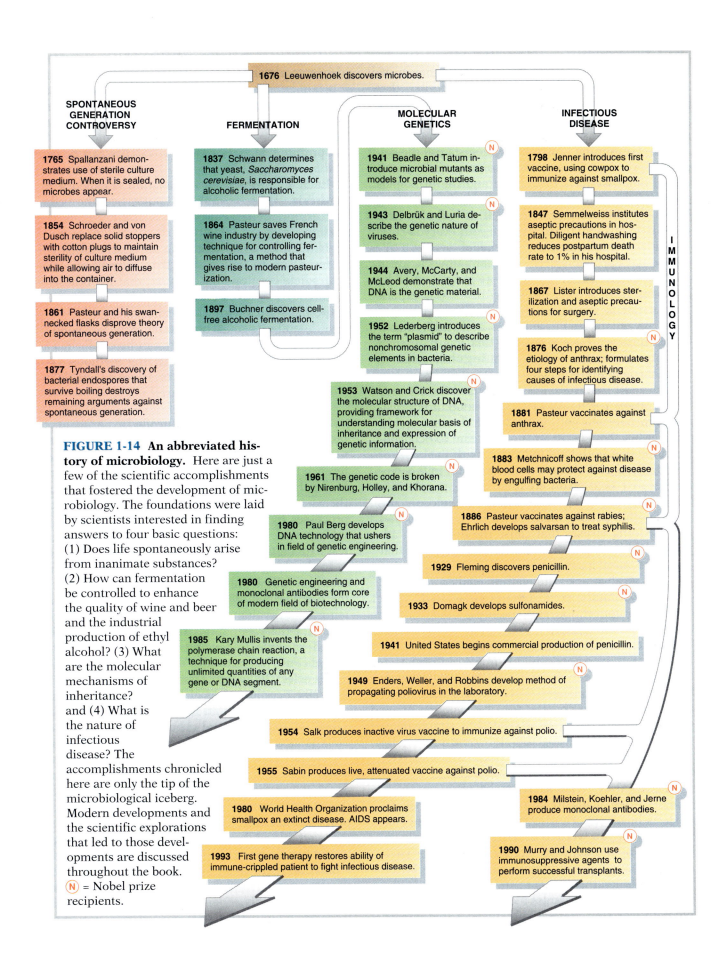

1676 Leeuwenhoek discovers microbes.

SPONTANEOUS GENERATION CONTROVERSY

1765 Spallanzani demonstrates use of sterile culture medium. When it is sealed, no microbes appear.

1854 Schroeder and von Dusch replace solid stoppers with cotton plugs to maintain sterility of culture medium while allowing air to diffuse into the container.

1861 Pasteur and his swan-necked flasks disprove theory of spontaneous generation.

1877 Tyndall's discovery of bacterial endospores that survive boiling destroys remaining arguments against spontaneous generation.

FERMENTATION

1837 Schwann determines that yeast, *Saccharomyces cerevisiae*, is responsible for alcoholic fermentation.

1864 Pasteur saves French wine industry by developing technique for controlling fermentation, a method that gives rise to modern pasteurization.

1897 Buchner discovers cell-free alcoholic fermentation.

MOLECULAR GENETICS

1941 Beadle and Tatum introduce microbial mutants as models for genetic studies.

1943 Delbrük and Luria describe the genetic nature of viruses.

1944 Avery, McCarty, and McLeod demonstrate that DNA is the genetic material.

1952 Lederberg introduces the term "plasmid" to describe nonchromosomal genetic elements in bacteria.

1953 Watson and Crick discover the molecular structure of DNA, providing framework for understanding molecular basis of inheritance and expression of genetic information.

1961 The genetic code is broken by Nirenburg, Holley, and Khorana.

1980 Paul Berg develops DNA technology that ushers in field of genetic engineering.

1980 Genetic engineering and monoclonal antibodies form core of modern field of biotechnology.

1985 Kary Mullis invents the polymerase chain reaction, a technique for producing unlimited quantities of any gene or DNA segment.

INFECTIOUS DISEASE

1798 Jenner introduces first vaccine, using cowpox to immunize against smallpox.

1847 Semmelweiss institutes aseptic precautions in hospital. Diligent handwashing reduces postpartum death rate to 1% in his hospital.

1867 Lister introduces sterilization and aseptic precautions for surgery.

1876 Koch proves the etiology of anthrax; formulates four steps for identifying causes of infectious disease.

1881 Pasteur vaccinates against anthrax.

1883 Metchnicoff shows that white blood cells may protect against disease by engulfing bacteria.

1886 Pasteur vaccinates against rabies; Ehrlich develops salvarsan to treat syphilis.

1929 Fleming discovers penicillin.

1933 Domagk develops sulfonamides.

1941 United States begins commercial production of penicillin.

1949 Enders, Weller, and Robbins develop method of propagating poliovirus in the laboratory.

1954 Salk produces inactive virus vaccine to immunize against polio.

1955 Sabin produces live, attenuated vaccine against polio.

1984 Milstein, Koehler, and Jerne produce monoclonal antibodies.

1980 World Health Organization proclaims smallpox an extinct disease. AIDS appears.

1990 Murry and Johnson use immunosuppressive agents to perform successful transplants.

1993 First gene therapy restores ability of immune-crippled patient to fight infectious disease.

IMMUNOLOGY

FIGURE 1-14 **An abbreviated history of microbiology.** Here are just a few of the scientific accomplishments that fostered the development of microbiology. The foundations were laid by scientists interested in finding answers to four basic questions: (1) Does life spontaneously arise from inanimate substances? (2) How can fermentation be controlled to enhance the quality of wine and beer and the industrial production of ethyl alcohol? (3) What are the molecular mechanisms of inheritance? and (4) What is the nature of infectious disease? The accomplishments chronicled here are only the tip of the microbiological iceberg. Modern developments and the scientific explorations that led to those developments are discussed throughout the book. Ⓝ = Nobel prize recipients.

Key Terms

microorganisms (p. 2)
microbes (p. 2)
microbiology (p. 4)
spontaneous generation (p. 5)
biogenesis (p. 5)
sterile (p. 6)
endospores (p. 7)
normal flora (p. 7)
metabolism (p. 9)
biogeochemical cycles (p. 9)
pathogens (p. 12)

enzymes (p. 12)
genes (p. 12)
asepsis (p. 15)
germ theory of infectious disease (p. 15)
Koch's postulates (p. 15)
pure culture (p. 15)
culture medium (p. 15)
colony (p. 15)
immunity (p. 16)
vaccination (p. 16)
antibodies (p. 16)

chemotherapy (p. 18)
antibiotics (p. 18)
epidemiology (p. 18)
nosocomial infections (p. 18)
anecdotal evidence (p. 20)
variables (p. 20)
hypothesis (p. 20)
controlled experiment (p. 20)
experimental group (p. 20)
control (p. 20)
placebo (p. 20)

Key Facts and Concepts

- **The study of microbes.** The world is populated by microorganisms. Microbiologists devote their professional lives to studying these microbes and their effects. Applications of their discoveries help in the fight against disease, spoilage, and deterioration of useful products and allow us to harness microbes and put them to work for the benefit of humans and the environment.
- **Evolution of microbiology.** The development of microbiology as a science began with the discovery of microbes and developed because of interest in the theory of spontaneous generation, in enhancing the proficiency of the powerful fermentation industry, in fighting infectious disease, and in understanding the molecular mechanisms responsible for genetic inheritance. Microbiology is expanding to other areas, such as environmental science, treatment of noninfectious disease, and crop improvement, and providing tools that help us understand evolution and life's subtle molecular and cellular processes.
- **Distribution of microorganisms.** Microorganisms inhabit virtually every niche in the biosphere. They are the most adaptable group of organisms, some of them living in conditions too hostile for any other form of life. They play an important role in determining the properties of every environment they inhabit.
- **Beneficial microbial activities.** Microorganisms sustain life on the planet. Decomposers recycle essential nutrients, returning complex organic molecules of dead organisms to their inorganic forms needed by photosynthesizers. Some fix molecular nitrogen, transforming this essential nutrient to a form usable by other organisms. Aquatic food

chains and webs depend on photosynthetic microorganisms as the primary food producers. Other benefits emerge from scientists learning how to use microbes to solve human problems or manufacture desirable products.
- **Detrimental microbial activities.** Infectious diseases constitute perhaps the most dreaded of microbial activities. Some of these diseases strike people directly; others affect us by killing livestock and crop plants. Microbes spoil our food, foul our water, and deteriorate useful products. In nature, some episodes of extinction may have been due to the introduction of new pathogens against which a species had no effective defense.
- **Medical microbiology.** Although a few scientists made some progress against infectious disease even before Koch proved the germ theory of infectious disease, most medical victories occurred in the "golden age" that followed. Antibiotics and other chemotherapeutic agents cure many types of diseases with their selective toxicity against microbes. Prophylactic measures have proven the most valuable against infectious diseases. Improved hygiene, sanitation, sewage disposal, and medical asepsis reduce the likelihood of disease transmission. Vaccination has produced a population of people who are immune to such diseases as tetanus, diphtheria, and polio.
- **Applied microbiology.** Microbes directly benefit humans by producing many of our foods, by producing industrial products and medicines, and by competing with pathogens for space on our bodies. Microbes play a central role in treatment of raw sewage so it can be safely released into the environment. We are using microbes to digest pollu-

tants and clean up oil spills. Each year we discover or invent new ways to harness the activities of microorganisms.

- **Environmental microbiology.** Microorganisms are among the most important influences in maintaining environmental stability. Excessive microbial proliferation is often one of the first indicators of environmental disturbance. Although microbes can be polluters, we are now using them as weapons to fight pollution.

- **Microbiology and the future.** Biotechnology and genetic engineering must be considered when projecting where the most important scientific breakthroughs will occur. Microbiologists will continue to enhance our understanding of life's origins and history and the mechanisms of elusive biological processes; these advances in basic understanding will be as valuable as the applications of that knowledge for improving our future and that of the organisms that share the earth with us.

Review and Problem-Solving Questions

1. Define: (a) Normal flora: (b) pathogen; (c) nosocomial infection.
2. What is the importance of microorganisms in (a) sewage treatment? (b) the pharmaceutical industry? (c) the preparation of vaccines?
3. List four ways in which microbial activities are (a) beneficial to humans; (b) detrimental to humans.
4. Describe a prophylactic method employed to decrease the occurrence of infectious disease.
5. Compare and contrast:
   (a) Spontaneous generation and biogenesis
   (b) Antibodies and antibiotics
   (c) Genes and enzymes
   (d) Decomposers and spoilage microbes
   (e) Normal flora and pathogens
6. Identify the major contribution(s) of each of the following people: (a) Antoni van Leeuwenhoek; (b) Louis Pasteur; (c) Robert Koch; (d) Joseph Lister; (e) Edward Jenner; (f) Alexander Fleming; (g) Ellie Hesse (see Fig. 1-11).
7. Microorganisms spoil many foods even if the foods are refrigerated. Develop a hypothesis that explains the source of the spoilage microbes.
8. What is the function of a control in a well-designed experiment?
9. What are two essential elements of an acceptable scientific hypothesis?
10. As the number of McDonald's restaurants in the United States has increased since 1980, so has the number of AIDS cases. Why is it incorrect to conclude that the AIDS epidemic is related to the growth of McDonald's?
11. Why is it important to publish results of scientific investigations?

CURABLE DISEASES THREATEN HALF A BILLION PEOPLE

While those of us living in developed countries are most threatened by heart disease and cancer, people in developing countries are still battling for their lives against microbial pathogens. Researchers interviewed by Ann Gibbons[*] point out that in 1990 the World Health Organization identified 12 microbial diseases (listed below) to be among the top 20 causes of death in the world, especially in tropical regions of Africa, Asia, and South America.

These scientists are particularly concerned with providing adequate medical protection against pathogens that have been virtually eliminated as killers in the developed countries. Such protection includes universal immunization of children and availability of drugs for chemotherapy.

[*]Ann Gibbons, in "News and Comment," *Science* 256:1135 (1992).

INFECTIOUS DISEASES: THE LEADING KILLERS IN 1990

Cause of Death	Estimated Number
Acute respiratory infections	6,900,000
Diarrheal diseases	4,200,000
Tuberculosis	3,300,000
Malaria	1,000,000-2,000,000
Hepatitis	1,000,000-2,000,000
Measles alone	220,000
Meningitis, bacterial	200,000
Schistosomiasis (parasitic tropical disease)	200,000
Pertussis alone (whooping cough)	100,000
Amoebiasis (parasitic infection)	40,000-60,000
Hookworm (parasitic infection)	50,000-60,000
Rabies	35,000
Yellow fever (epidemic)	30,000
African trypanosomiasis (sleeping sickness)	20,000 or more

SOURCE: Global health situation and projections, estimates 1992, World Health Organization, Geneva, 1992.

- What problems do you think might impede the delivery to susceptible populations of adequate protection from the diseases listed in the table?

- If money were the only obstacle, would you support allocating government funds for an international aid program? Justify your stand.

CHAPTER 2

Introduction to the Microbes

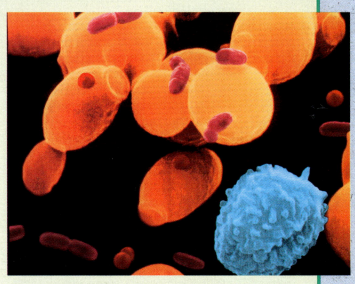

Although never seen by the unaided human eye, these residents of the microbial world each pack the enormous complexity needed for life into microscopically small packages.

It was the most important development in the history of earth. Yet when life emerged billions of years ago from the collection of inanimate molecules on the surface of the primitive planet, the original organism left no fossil evidence that reveals how life began (or if it did leave fossils, they remain undiscovered). No one knows what the first organisms were like or what harsh environmental conditions they had to endure in order to survive and produce progeny. The oldest known fossils date back about 3.5 billion years to a time when life had already been around for millions of years. We do know, however, that the first organisms were probably cells, for the fundamental unit of life for all organisms is the cell.

Even such a highly evolved, complex organism as a human being is actually an intricately coordinated complex of cells. Microorganisms, on the other hand, are generally **unicellular;** that is, the whole organism is one cell. In such cases a single microbial cell performs all the functions required to maintain itself and propagate. These biological functions are achieved by mechanisms strikingly similar to those of the individual cells that compose the bodies of plants and animals.

All living cells, in fact, share many biochemical similarities. They use many of the same chemical constituents to build their structural components and the same genetic code to transmit hereditary information from parent to offspring. They have similar, often identical, metabolic processes. All cells share many physical properties—a membrane that surrounds the cell and regulates the flow of chemicals into and out of the cell, genetic material that determines the cell's characteristics, and protein-synthesizing machinery. Like all cells, microorganisms reflect this unity of life at the cellular and subcellular levels. On the other hand, there are many properties among microbes that distinguish them morphologically, physiologically, and behaviorally from each other as well as from all other forms of life on earth. In this chapter we examine some of these similarities and differences. Keep in mind, however, that these similarities testify to the common ancestry of all organisms. You may be surprised at how much you have in common with bacteria, fungi, and the other organisms that inhabit the microbial world.

Eukaryotes and Prokaryotes

In spite of the similarities, microbes are not miniature versions of the larger cells of plants and animals. At one time it was commonly assumed that all cells contained the same internal structures and chemical components. In the 1950s, however, the electron microscope revealed striking differences between bacteria and the cells of other organisms. These physical differences formed the basis for classifying cells into two categories: eukaryotic cells and prokaryotic cells.[1] The most outstanding feature of **eukaryotic cells** (*eu* = true; *karyote* = nucleus) is a distinct nucleus surrounded by a membrane that separates it from the other contents of the cell (Fig. 2-1). Eukaryotic cell organization is common to the fungi, algae, and protozoa, as well as to more complex multicellular organisms, including humans.

[1]Sometimes written "eucaryotic" and "procaryotic." The different spellings reflect a preference for either the latinized form of the word (karyon) or the anglicized version (caryon).

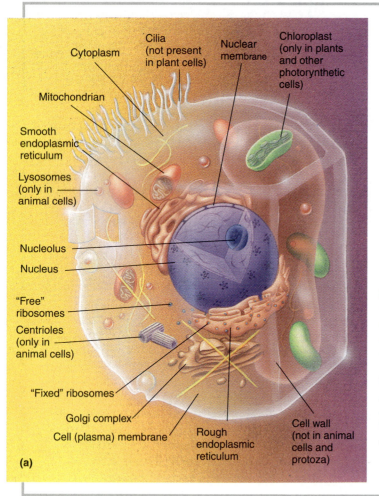

Cytoplasm
Cilia (not present in plant cells)
Nuclear membrane
Chloroplast (only in plants and other photorynthetic cells)
Mitochondrian
Smooth endoplasmic reticulum
Lysosomes (only in animal cells)
Nucleolus
Nucleus
"Free" ribosomes
Centrioles (only in animal cells)
"Fixed" ribosomes
Golgi complex
Cell (plasma) membrane
Rough endoplasmic reticulum
Cell wall (not in animal cells and protoza)

(a)

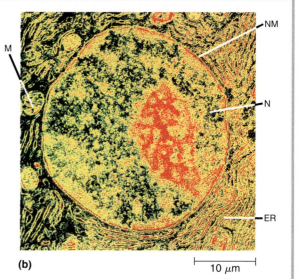

(b)

10 μm

NM
M
N
ER

FIGURE 2-1 Representative eukaryotic cell. (*a*) This stylized cell reveals the structures that characterize eukaryotic cells. Typical animal-like cells and protozoa are represented on the left part of the cell (yellow background); plant-like cells and algae are represented on the right (purple background). The division is somewhat artificial, since some cells have properties of both. (*b*) The eukaryotic cell contains a nucleus (N) surrounded by a membrane (NM). Mitochondria (M) and endoplasmic reticulum (ER) are also evident.

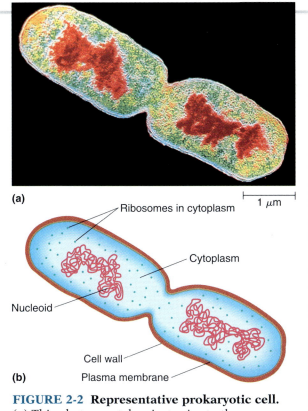

(a)

1 μm

Ribosomes in cytoplasm

Cytoplasm

Nucleoid

Cell wall

(b) Plasma membrane

FIGURE 2-2 Representative prokaryotic cell.
(*a*) This photo was taken just prior to the separation of the two cells following cell division. (*b*) This stylized cell shows a few of the structures characteristic of prokaryotic cells.

Prokaryotic cells (*pro* = before), on the other hand, do not contain a membrane-bound nucleus. Instead, their hereditary material is suspended in a portion of the cytoplasm called the nucleoid, or nuclear region (Fig. 2-2). The prokaryotic cell is characteristic of bacteria, which are divided into two groups, eubacteria and archaeobacteria.

The distribution of microorganisms among the world's organisms is shown in Figure 2-3. (**Caution:** Do not confuse eubacteria with eukaryotes.)

Beyond this nuclear distinction, eukaryotic and prokaryotic cells exhibit other important structural and chemical differences. Many of the structures and chemicals unique to prokaryotes determine the roles of these microorganisms in nature, from recycling nutrients to causing disease in humans. Differences between eukaryotic and prokaryotic cells also provide the basis for treating diseases caused by bacteria. Such practical considerations will be evident in the following discussion on cell structure and function.

CONCEPT CHECK

- What is the basis for classifying cells as prokaryotes or eukaryotes?
- Which groups of microorganisms possess prokaryotic cell structure? Which groups have eukaryotic cell structure?

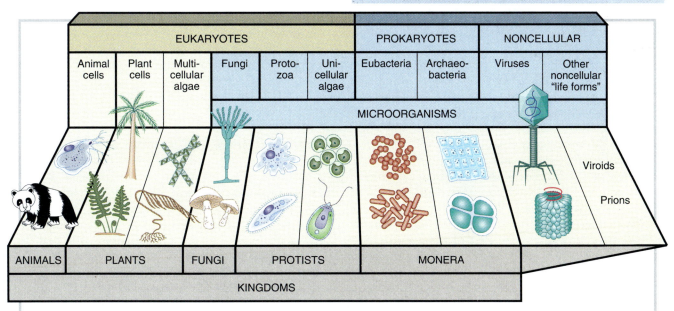

FIGURE 2-3 A spectrum of the world's organisms (the microorganisms are shown in the light blue shaded boxes). The eukaryotes include multicellular organisms such as plants and animals and some of the microorganisms (fungi, protozoa, algae). The prokaryotes are all microorganisms. Because of their noncellular structure, viruses are considered neither eukaryotic nor prokaryotic and are placed in a separate category. The kingdoms to which the various categories of cells belong are also shown.

The Eukaryotic Cell

The eukaryotic cell is characteristic of plants, animals, fungi, and protists (protozoa and algae). All cells, whether eukaryotic or prokaryotic, are surrounded by a *plasma membrane* and filled with a gelatinous fluid called the **cytoplasm.** The largest structure in the cytoplasm is the *nucleus.* The cytoplasm of all eukaryotic cells also contains **organelles,** intracellular structures enclosed by membranes that segregate their contents from the cytoplasm. The organelles carry out specialized tasks, such as generating energy for the cell or packaging substances. They include mitochondria, vacuoles, and, in photosynthetic cells, chloroplasts. Ribosomes and elaborate membrane systems called the endoplasmic reticulum and Golgi complex lace the cytoplasm of all eukaryotic cells. Also suspended throughout the cytoplasm is a microscopic network of filaments and tubules, the *cytoskeleton,* that provides a framework for the cell. The cytoplasm itself is in constant motion, a phenomenon termed *cytoplasmic streaming.* This movement helps transport suspended matter and dissolved nutrients throughout the cytoplasm.

■ Nucleus

The **nucleus,** which contains the genetic material, is separated from the rest of the cell by the *nuclear membrane.* The nucleus contains **DNA** (*deoxyribonucleic acid*), the genetic material that determines a cell's characteristics and transmits those properties to offspring. The DNA is found in **chromosomes** inside the nucleus. All eukaryotic cells contain two or more linear chromosomes (in contrast to prokaryotes, which have only one chromosome, a circular strand of DNA). The number of chromosomes in the nucleus is characteristic of a particular kind of organism (for example, human body cells have 46 chromosomes). In eukaryotic chromosomes, much of the DNA is compactly coiled around proteins called *histones.* Nonhistone proteins are also associated with chromosomes; they are thought to regulate gene expression. Part of the genetic material appears as a smaller body within the nucleus called a *nucleolus* (plural, *nucleoli*). The *RNA (ribonucleic acid)* of ribosomes is synthesized at this site. Differences in the appearance of the nucleus, especially in the chromosomes and nucleolus, are helpful in identifying certain pathogenic protozoa and are instrumental in the diagnosis of disease. For example, the protozoan that causes dysentery can be distinguished from harmless intestinal protozoa

by the appearance of the nucleus on microscopic examination.

When eukaryotic cells divide to form identical daughter cells, the chromosomes in the nucleus are duplicated so that there is a copy of the genetic instruction for each of the new cells. Such cell division occurs by the process of *mitosis* in eukaryotes. During mitosis, the parent cell synthesizes a complex apparatus that distributes the duplicated chromosomes accurately to the progeny cells.

■ Endoplasmic Reticulum

The **endoplasmic reticulum** is an internal membrane system that begins at the nuclear membrane and extends throughout the cytoplasm, almost to the plasma membrane. It increases the membrane surface area of the cell, enabling nutrients to reach all parts of the large eukaryotic cell while providing a passageway for the rapid export of waste products of metabolism. Without cytoplasmic streaming and the endoplasmic reticulum, essential materials would penetrate too slowly to support the growth of eukaryotic cells.

The membranes of much of the endoplasmic reticulum are covered with **ribosomes,** the sites of protein synthesis. The presence of ribosomes gives a bumpy appearance to the membrane, which is therefore referred to as the *rough endoplasmic reticulum (RER).* The RER often serves as a passageway for the export of the proteins synthesized by ribosomes on its surface. Ribosome-free portions of the endoplasmic reticulum are called the *smooth endoplasmic reticulum (SER).* These regions contain enzymes that perform essential reactions such as the synthesis of certain lipids.

■ Ribosomes

Cells also contain *free ribosomes* floating in the cytoplasm, not attached to any membrane. These free ribosomes produce most of the *intracellular* proteins, proteins that are retained within the cell, whereas the ribosomes attached to the endoplasmic reticulum generally synthesize *extracellular* proteins, proteins that are released from the cell. Eukaryotic ribosomes, whether free or attached to membranes, are 80 S particles. This S value is simply an indication of the particles' density as determined by their rate of sedimentation when they are subjected to thousands of times the force of gravity in a centrifuge. Heavier particles sediment faster and therefore are assigned a higher S value (called the *sedimentation coefficient*). For example, 80 S particles are heavier than 70 S particles.

Mitochondria

Most of the energy for cell functions is produced in organelles called **mitochondria** (singular, *mitochondrion*). Cellular energy, which is contained in an accessible molecule called *ATP,* is generated by **respiration,** a process that releases energy from food molecules. In eukaryotes, most ATP is generated in mitochondria, using molecular oxygen and special enzymes bound to the mitochondrion's membranes or dissolved in its fluid matrix. ▶ (The details of respiration are presented in Chapter 7.) Mitochondria contain their own DNA and ribosomes. Unlike the DNA in the cell's nucleus, the DNA in a mitochondrion is a single small circular piece. Mitochondrial ribosomes are 70 S in size, slightly smaller than the cytoplasmic ribosomes. Mitochondria use their DNA and ribosomes to manufacture many of their own protein constituents independently of the cell nucleus. They are capable of replicating within the cell and can produce more respiratory membranes in response to an increase in the cell's demand for energy.

Chloroplasts

Photosynthetic eukaryotic cells contain organelles called **chloroplasts.** Chloroplast membranes contain the green pigment chlorophyll, which is used in **photosynthesis,** the process by which cells use the energy of sunlight to convert carbon dioxide and water into organic compounds that contain chemical energy needed for biological activities. Like mitochondria, chloroplasts have their own single circular piece of DNA and 70 S ribosomes and are also self-replicating.

Vesicles

Many eukaryotic cells form intracellular membrane-bound sacs called **vesicles.** Some vesicles are formed by **endocytosis,** inward movement of the plasma membrane as it surrounds and engulfs particles outside the cell (Fig. 2-4). Protozoa, for example, create vesicles called food vacuoles that are formed when prey is engulfed by endocytosis. Other sacs in the cytoplasm are manufactured within the cell by internal membrane systems. **Lysosomes** are among the most important of these internally produced vesicles. Lysosomes are filled with *enzymes,* molecules that promote chemical reactions in the cell. These lysosomal enzymes are released into food vacuoles, where they digest engulfed particles.

Many of our white blood cells protect us by engulfing microbes that enter our body. The engulfed

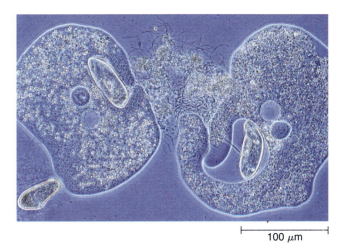

100 µm

FIGURE 2-4 Amoeba trapping a meal of bacteria by endocytosis.

prey is enzymatically destroyed after lysosomes empty their contents into the vacuole that contains the microbe.

Golgi Complex

Lysosomes and other membrane-bound vesicles are manufactured by the **Golgi complex,** an organelle composed of a series of flattened sacs (Fig. 2-5*a*). This organelle modifies proteins it receives from the ribosome-coated rough endoplasmic reticulum and packages them into vesicles. Lysosomes and certain other vacuoles remain in the cell as storage sacs for the chemicals they contain. Other vesicles (called secretory vesicles) migrate to the cell surface, where they fuse with the plasma membrane and discharge their chemical cargo outside the cell (Fig. 2-5*b*). Many hormones are secreted from cells by this mechanism. Another important category of secreted proteins are enzymes essential for extracellular (outside-the-cell) digestive processes. These enzymes are necessary if cells are to obtain soluble nutrients from particles or molecules that are too large to be transported into the cell.

Plasma Membrane

Directly surrounding the cytoplasm in all cells is the **plasma membrane** (also called the *cell membrane*). In addition to physically separating the cell from its environment, the plasma membrane determines which molecules are allowed to travel between the external medium and the interior of the cell, a characteristic known as **selective permeability.** This selectivity allows passage of needed molecules into the cell and allows wastes to leave.

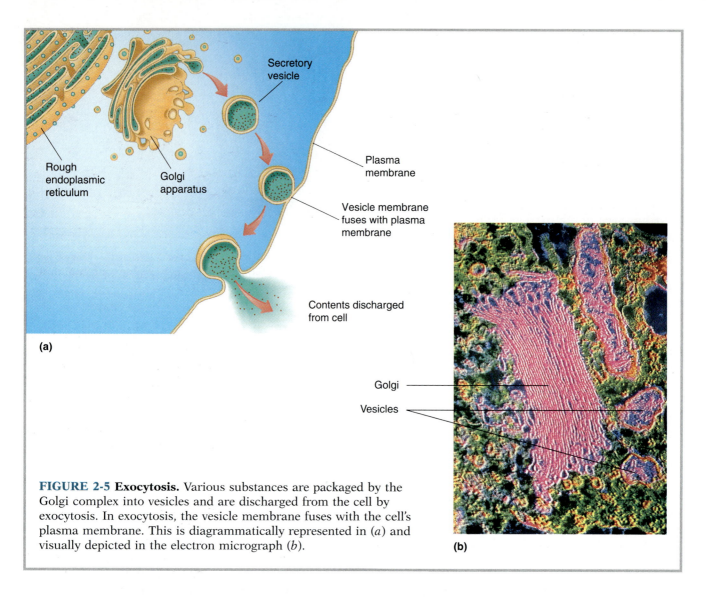

(a)

(b)

FIGURE 2-5 Exocytosis. Various substances are packaged by the Golgi complex into vesicles and are discharged from the cell by exocytosis. In exocytosis, the vesicle membrane fuses with the cell's plasma membrane. This is diagrammatically represented in (*a*) and visually depicted in the electron micrograph (*b*).

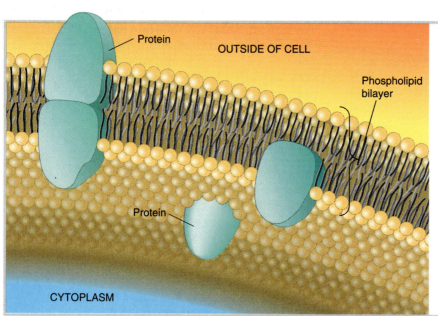

FIGURE 2-6 Plasma membrane structure. The plasma membrane is a semifluid bilayer of phospholipid molecules each with a hydrophilic (water-seeking) end and a hydrophobic (water-avoiding) end. The hydrophobic ends are always in the interior, where they are protected from water. The proteins dispersed in the membrane have multiple functions, including transport of molecules across this barrier. The semifluid bilayer and the embedded proteins create a structure that is often referred to as a "fluid mosaic," a useful term that reflects the ability of membrane proteins to move laterally in the membrane.

Membranes of eukaryotes and most prokaryotes have the same basic structure (Fig. 2-6). They are composed of a double layer (a *bilayer*) of *phospholipids* (a combination of lipid and phosphate), with proteins dispersed throughout the structure. These proteins generally mediate special membrane functions, notably the transport of substances into and out of the cell. Some membrane proteins are *receptor proteins* that recognize specific substances with which the cell can bind. A cell's activities change in response to the binding of surface receptors. For example, certain cells of your immune system launch an assault on an invading microbe when the foreign invader attaches to the corresponding receptor proteins on the host defense cells.

Eukaryotic membranes typically contain lipid compounds called *sterols,* with the type of sterol present varying from one group of organisms to another. For example, cholesterol is commonly found in human and other animal cell membranes, whereas ergosterol is predominant in the membranes of fungi. This difference is of practical importance because drugs that can selectively damage ergosterol-containing membranes without harming cholesterol-containing membranes are useful in treating human diseases caused by fungi.

■ Cell Wall

The plasma membrane is the outermost layer in most protozoa. In fungi and most algae, however, a **cell wall** surrounds the plasma membrane (Fig. 2-7). The cell wall gives the cell rigidity and shape. In addition, it protects the cell by enabling it to resist swelling and subsequent bursting. Consequently, under conditions where large volumes of water tend

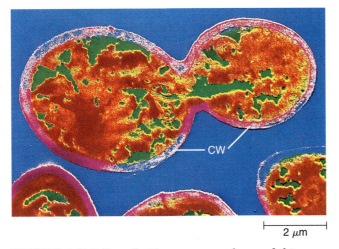

FIGURE 2-7 Cell wall. The outermost layer of this yeast cell (*Candida albicans*) is a rigid cell wall (CW).

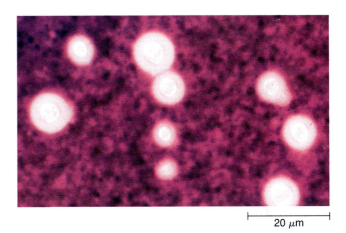

20 μm

FIGURE 2-8 Protective cover. Cells of the yeast *Cryptococcus neoformans* have been treated with india ink to show the surrounding capsule.

to enter the cell, such as immersion in pure water, *lysis* (bursting) of the cell is less likely.

Most eukaryotic cell walls are composed of some type of *polysaccharide*, a large molecule composed of repeating sugar subunits. The polysaccharide in the cell wall of plants and most algae is cellulose. Most fungi, on the other hand, have cell walls made of polysaccharides other than cellulose.

■ Capsule

In a very small number of eukaryotic species, the cell wall may be surrounded by an additional layer, the **capsule.** Under most environmental conditions, its presence is not essential for the cell's survival, but under some circumstances the capsule protects the microorganism from destruction. For example, the fungus *Cryptococcus neoformans* uses its capsule to escape the defenses of the immune system of an infected person. The capsule protects the fungal cell surface from the protective white blood cells that normally defend the body against infection. The encapsulated fungus may then become established in the lung and invade the central nervous system, causing a fatal brain infection. The large encapsulated cells can be seen by the microscopic examination of stained clinical specimens (Fig. 2-8). Nonencapsulated *Cryptococcus neoformans* cells are readily engulfed by white blood cells and are incapable of causing human disease.

■ Structures for Motility

Eukaryotic cells move by a number of different mechanisms. Amoebas (protozoans) move by sending out extensions of their surface called **pseudopods** (*pseudo* = false; *pod* = foot). The rest of the cell then flows toward the tip of the extended pseudopod. Human white blood cells also move in this manner. Other eukaryotic microbes depend on flexible appendages, flagella or cilia, for motility. **Flagella** (singular, *flagellum*) are long filaments that whip back and forth, propelling the cell forward. **Cilia** (singular, *cilium*) are shorter than flagella and generally more numerous but are otherwise identical in morphology and chemical composition. They function as tiny oars (Fig. 2-9*a*). A eukaryotic flagellum or cilium contains 10 pairs of hollow rods, called microtubules, that extend down the length of the appendage (Fig. 2-9*b*). Each microtubule slides back and forth next to the other tubule in the pair. This sliding motion bends the flagellum or cilium, producing the propulsive motion.

> **CONCEPT CHECK**
>
> - What structures are universal to all eukaryotic cells, and what are their functions?
> - What structures are present in only some (not all) eukaryotes, and what are their functions?

Eukaryotic Microbes

Distinctive features separate eukaryotic microorganisms into three categories: fungi, protozoa, and algae.

■ Fungi

The outstanding features that characterize most **fungi** (singular, *fungus*) are the presence of a cell wall, the lack of motility, and the absence of photosynthesis. Because fungi cannot derive energy from sunlight, they must depend on an external source of organic compounds to provide the energy and chemical building blocks necessary for growth and survival. Although these nutrients are usually obtained from dead organisms, some fungi can use living tissue as a food source, often causing disease in the process.

Some fungi develop as single cells called **yeasts** (Fig. 2-10*a*), whereas others produce networks of filaments characteristic of **molds** (Fig. 2-10*b*). A few fungi are *dimorphic* (*di* = two; *morph* = form), existing as either a yeast or a mold depending on en-

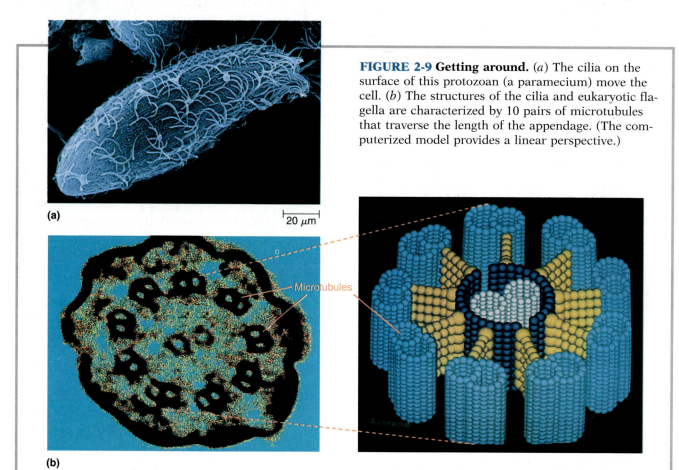

FIGURE 2-9 Getting around. (*a*) The cilia on the surface of this protozoan (a paramecium) move the cell. (*b*) The structures of the cilia and eukaryotic flagella are characterized by 10 pairs of microtubules that traverse the length of the appendage. (The computerized model provides a linear perspective.)

(a)

20 μm

(b)

Microtubules

Axoneme

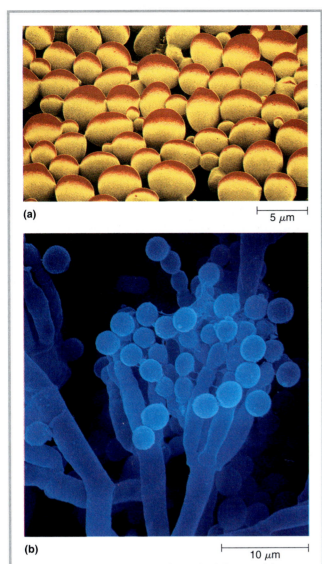

(a)   5 μm

(b)   10 μm

FIGURE 2-10 The two "faces" of fungi. (*a*)
Yeast. *Saccharomyces,* the yeast used in bread and
beer production. (*b*) Mold. *Penicillium,* a blue-
green mold that spoils millions of dollars worth of
food but also produces the life-saving antibiotic
penicillin. Both organisms have been false-colored.

Engulfed cells (food)

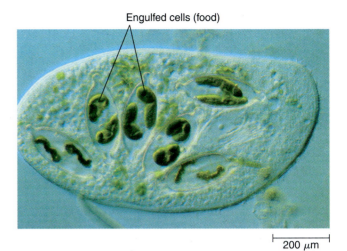

200 μm

FIGURE 2-11 Protozoan—predator and prey. This
ciliated protozoan is engorged with a recent meal of
green algae. The protozoan, in turn, will likely become
food for a larger predator, establishing a critical link in
the food chain. Even the largest animal on earth (the
blue whale) depends on such microscopic prey for its
entire food supply.

patients and are responsible for life-threatening
pneumonias. Fungi are also the sources of several
antibiotics, including penicillin, that have saved
countless lives threatened by devastating bacterial
infections.

Fungi are discussed in more detail in Chapter 11.

MYTH: Microbiologists study organisms that can be
viewed only with the aid of a microscope.

FACT: A few of the organisms that reside in the microbial kingdoms of Fungi and Protists are large
(macroscopic) organisms such as mushrooms and
some algae. Most of the members of these kingdoms,
however, are microscopically small and warrant
their inclusion in a microbiology course.

■ Protozoa

Protozoa (singular, *protozoan*) are unicellular eukaryotic organisms that are generally nonphotosynthetic[2] and lack a cell wall. Most protozoa are
motile, moving about by the action of pseudopods,
flagella, or cilia. Protozoa form an essential link in
many of the earth's food chains (Fig. 2-11). Some,
however, are responsible for widespread diseases.
For example, malaria is a protozoan disease that affects more people than any other infectious disease.

vironmental conditions. For example, a pathogenic
fungus may grow as a mold in the soil and as yeast
cells in the infected human body. Another group of
fungi are the *macrofungi,* filamentous organisms
that produce the large reproductive structures familiar to everyone as mushrooms.

Fungi cause some of the most persistent and disfiguring diseases still prevalent throughout the
world (including the United States). Common diseases caused by fungi include ringworm infections,
yeast infections of the mouth and vagina, and several serious systemic diseases. Some fungi take advantage of the depressed immune system of AIDS

[2] Some protozoa become photosynthetic in the presence of
visible light.

The agents of trypanosomiasis ("sleeping sickness") and amebic dysentery are also protozoa. This ubiquitous group of organisms are discussed in Chapter 12.

■ Algae

Algae (singular, *alga*) are a diverse group of organisms ranging from microscopic single cells (Fig. 2-12) to large multicellular seaweeds. They may be either motile or nonmotile, and most possess a cell wall. With one exception, algae are photosynthetic and are distinguished by the presence of chloroplasts within the cytoplasm. The chloroplasts are responsible for photosynthesis, the synthesis of organic material from carbon dioxide and water using sunlight as the sole energy source.

Algae are among the most important organisms on earth. They are not only the primary producers in aquatic food chains, they also generate about half of the earth's atmospheric oxygen (oxygen is a by-product of photosynthesis). Algae are discussed more fully in Chapter 12.

CONCEPT CHECK

- What features distinguish fungi, protozoa, and algae?
- What impact do these organisms have on the quality of human life?

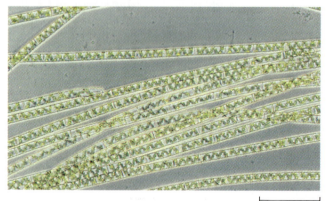

├────────────┤ 450 μm

FIGURE 2-12 Photosynthetic "microbes." Most algae are important food and oxygen producers, like this green alga, *Spirogyra*. Although algae are generally thought of as being of microscopic size, some are huge seaweeds. For example, the giant kelp that is found below the ocean surface or washed up on the beach can form 80-meter-long individuals. (These huge multicellular algae are now classified as plants rather than protists.)

The Prokaryotic Cell

The volume of a typical prokaryotic cell is approximately 1/125 that of a eukaryotic cell. The primary structural characteristic that distinguishes prokaryotes from eukaryotes is the absence of a nucleus, mitochondria, and other membrane-bound organelles or networks. Although physically different from the eukaryotic cells, prokaryotic cells must still perform many of the same functions. They must generate usable energy, eliminate wastes, and reproduce. Many prokaryotic cells achieve these ends using unique methods. For example, some prokaryotes can use chemical energy extracted from inorganic nitrogen or hydrogen sulfide, a process no eukaryote can duplicate. Other prokaryotic cells perform essential processes under extreme environmental conditions, such as high temperatures or low pH, that are intolerable to eukaryotes.

The prokaryotic cell is structurally less complex than the eukaryotic cell. All prokaryotes contain cytoplasm (but with no cytoplasmic streaming), a nucleoid, a plasma membrane, and ribosomes (review Fig. 2-2). A cell wall is also present in the majority of prokaryotic cells. ▶ (Although briefly described here for ease of comparison with eukaryotic cells, the details of these prokaryotic structures are discussed in Chapter 4.)

■ Nucleoid

The genetic material of a prokaryote is not separated from the cytoplasm by a nuclear membrane. It is seen in the electron microscope as the **nucleoid,** an area that appears lighter than the cytoplasmic contents (Fig. 2-2). The nucleoid contains a single circular chromosome of DNA about 400 times the length of the cell. Proteins help keep the prokaryotic chromosome compact. These proteins differ from the histones found in eukaryotic cells.

The nucleoid is duplicated just before a cell divides, usually by a process called *binary fission,* which results in the formation of two equivalent progeny cells. Unlike mitosis, however, binary fission does not require the same specialized apparatus for cell division.

■ Plasma Membrane

The prokaryote's plasma membrane, like the eukaryotic membrane, functions as a physical boundary and as a selective barrier in the passage of materials into and out of the cell. In addition, the prokaryote's plasma membrane assumes many of the biological functions performed by membrane-bound organelles in eukaryotic cells. For example, since there are no mitochondria or chloroplasts, enzymes for generating energy by respiration or photosynthesis are integrated into the plasma membrane. Some prokaryotes possess *mesosomes*, invaginations (infoldings) of the plasma membrane into the cytoplasm (Fig. 2-13). Many functions have been attributed to this structure, but recently the mesosome has created considerable controversy among scientists engaged in understanding prokaryotic structure and function (see THE EXPLORERS: The Bacterial Mesosome—Fact or Artifact?).

■ Ribosomes

As with eukaryotes, all protein synthesis in prokaryotes takes place at the ribosomes. Ribosomes are distributed throughout the cytoplasm; they are also

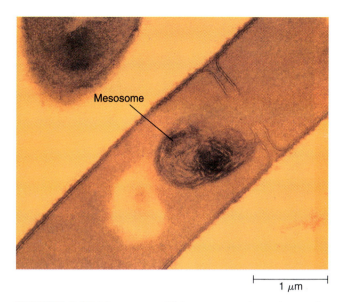

1 µm

FIGURE 2-13 Mesosome. This structure is an intracellular extension of the plasma membrane believed to function in cell division.

ity of eubacteria derive nutrients by decomposing organic materials. Some are so efficient at this that the low concentrations of nutrients in distilled water can support their growth.

A few eubacteria are photosynthetic. For example, **cyanobacteria** perform oxygen-evolving photosynthesis in a manner similar to eukaryotic algae and plants. Some cyanobacteria transform molecular nitrogen (N_2) into a form usable by plants, thereby introducing this essential nutrient into the food chain. Many of these microbes grow as filaments of cells, as illustrated by *Anabaena* in Figure 2-15. Although they are not pathogenic, some cyanobacteria grow in close association with a living host (see THE MICROBES: The Green Polar Bears). Other photosynthetic eubacteria perform a different type of photosynthesis that does not release oxygen.

Eubacteria have caused devastating epidemics of human suffering, perhaps none so dramatic as the bubonic plague (the Black Death) that killed one-

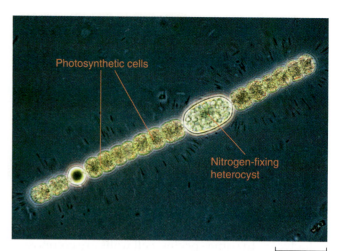

Photosynthetic cells

Nitrogen-fixing heterocyst

20 μm

FIGURE 2-15 Suppliers of usable nitrogen. The cyanobacterium *Anabaena* is typically arranged in chains of cells. Heterocysts are specialized structures in which nitrogen fixation occurs.

THE MICROBES

The Green Polar Bears

For several years visitors to a few zoos in the United States and Europe have observed an unusual group of animals—green polar bears. Although the bears were white in their arctic habitat, in captivity they acquired a greenish cast to their coats. The pigment is more noticeable in the warmer months and was at first believed to be algae growing on the hair surfaces. Microscopic inspection of the bears' hairs, however, revealed an unexpected finding.

The green color is indeed due to photosynthetic microbes, but they are inside the hair, not on its surface. The bears' outer "guard" hairs are hollow, and it is in these hollow shafts that the organisms have taken up residence. Microscopic examination and analysis of the prokaryote's pigments revealed that the photosynthetic freeloader belongs to the Chroococcales, a family of cyanobacteria. This blue-green bacterium apparently

gains access to the hollow space when the end of the hair breaks off and exposes the inner chamber.

The hairs of polar bears are especially susceptible to such colonization because they lack pigment. This permits light to penetrate into the hollow compartment, which also provides protection for the microbe. Such a relationship has been reported in no other animal and only in captive specimens of the polar bear. In

the wild the temperatures are too cold to encourage growth of the cyanobacteria.

Aside from cosmetic considerations, the microbes have no detrimental effects on the colonized bears. Zoo officials are nonetheless eager to display white polar bears and are experimenting with different environmental factors to discourage the microbe's growth. San Diego Zoo scientists have successfully controlled the problem by increasing water salinity.

Green polar bears. The color of these bears reflects the growth of cyanobacteria in the hollow guard hairs of the animals' fur.

fourth of the population of Europe in the fourteenth century. Although eubacteria continue to be a major cause of disease, their many beneficial activities more than offset their negative impact. Because of the importance of eubacteria, much of this book is devoted to discussing their biology, their impact on people, and methods for controlling their activities.

■ Archaeobacteria

Archaeobacteria comprise three groups of prokaryotes that thrive in environments that would be extremely hostile to other forms of life. These three groups are the extreme *halophiles* (salt lovers), *thermoacidophiles* (heat and acid lovers), and *methane-producing* bacteria. Some archaeobacteria grow near thermal vents in the ocean floor, at a depth of more than 1500 meters (about 1 mile) and in 250°C waters. Other remarkable archaeobacteria thrive in acidic solutions, such as acid runoff from mines. The unique construction of their plasma membrane and cell wall permits these organisms to survive where no others can.

CONCEPT CHECK

- What features distinguish archaeobacteria from eubacteria?
- What features distinguish prokaryotes from eukaryotes?

Viruses

A **virus,** unlike the microorganisms discussed so far, has no cell structure. Some viruses are composed solely of nucleic acid (DNA or RNA) surrounded by a protective protein shell. Others are surrounded by an additional membrane layer. They have no cytoplasm, no internal organelles, and no cellular machinery to synthesize their own protein or to produce energy. Viruses are **obligate intracellular parasites**—they require the biological machinery of a host cell to carry out the functions needed for their reproduction and survival.

Viruses cause infectious diseases of humans, plants, and animals. Bacteria, fungi, algae, and protozoa are also susceptible to viral infections. Viruses are so small they can be observed only with an elec-

tron microscope (Fig. 2-16). They are the subject of Chapter 13.

CONCEPT CHECK

- How do viruses differ from cellular organisms?
- Why are all viruses classified as obligate intracellular parasites?

Microbial Diversity and Antimicrobial Therapy

The biological differences between prokaryotes and eukaryotes provide a basis for treating diseases caused by bacteria. An essential property of chemical agents used for antimicrobial drug therapy inside the human body is **selective toxicity,** the ability of the drug to destroy invading prokaryotic cells while leaving the eukaryotic host cells unharmed. Drugs that selectively interfere with processes that occur solely in prokaryotes may be able to subdue a bacterial pathogen while having no effect on the eukaryotic cells of the human body. Such antimicrobial drugs are powerful weapons against infectious disease.

Selective toxicity is perhaps best represented by penicillins and cephalosporins. These antibiotics prevent the synthesis of the peptidoglycan layer of the eubacterial cell wall, a structure with no chemical equivalent in the eukaryotic human cell. Without their cell walls, most bacteria are so fragile they have little chance of survival. Some other therapeutically effective antibacterial agents selectively inhibit the activities of prokaryotic ribosomes. A few of these ribosome inhibitors have toxic side effects due to similarities between mitochondrial ribosomes and bacterial ribosomes, a similarity that has provided clues to the evolutionary origin of these organelles (see BEYOND THE BASICS: Bacteria and the Early History of Life).

Most antibacterial agents have no effect on eukaryotic pathogens. In addition, since eukaryotic pathogens, such as fungi and protozoa, have the same basic cell structure as human cells, few drugs can kill or inhibit them without having adverse effects on human cells. The few drugs that are used to treat diseases caused by fungi and protozoa are somewhat toxic to the human host as well. Some

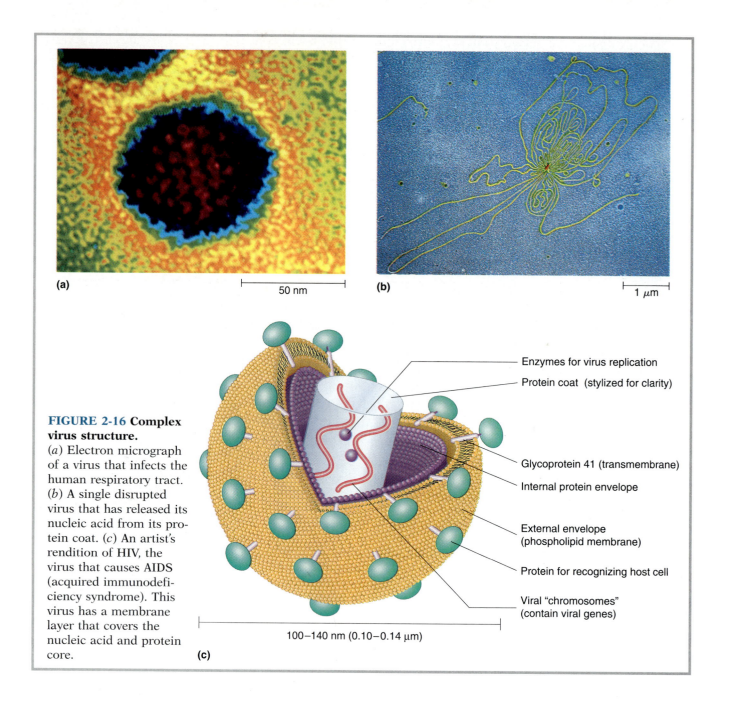

(a) 50 nm

(b) 1 μm

Enzymes for virus replication

Protein coat (stylized for clarity)

Glycoprotein 41 (transmembrane)

Internal protein envelope

External envelope (phospholipid membrane)

Protein for recognizing host cell

Viral "chromosomes" (contain viral genes)

100–140 nm (0.10–0.14 μm)

(c)

FIGURE 2-16 Complex virus structure.
(*a*) Electron micrograph of a virus that infects the human respiratory tract. (*b*) A single disrupted virus that has released its nucleic acid from its protein coat. (*c*) An artist's rendition of HIV, the virus that causes AIDS (acquired immunodeficiency syndrome). This virus has a membrane layer that covers the nucleic acid and protein core.

have very serious side effects and are employed only in life-threatening situations. Diseases caused by viruses are even more difficult to treat. Once a virus becomes an integral part of the infected cell, in most cases it is impossible to destroy it therapeutically without also killing the host cell.

Regardless of the microbe, however, many properties distinguish it from human cells. A more thorough knowledge of the structure and function of each organism increases the likelihood of developing safer, more effective drugs against a wider range of pathogens. Herein lies one of the great promises in the continuing battle against infectious disease.

CONCEPT CHECK

- Why is selective toxicity an important consideration in treating infectious disease?
- Why is it easier to find antimicrobial drugs against prokaryotic pathogens than against pathogenic eukaryotic microbes or viruses?

Bacteria and the Early History of Life

The early earth was an extraordinarily inhospitable place. The first atmosphere not only lacked the molecular oxygen we think of as conducive to life today, but most of the gases it did possess (methane, carbon monoxide, nitrogen, ammonia) would quickly poison most modern organisms. There was also no usable water. The earth was so hot, any liquid water that formed would have immediately exploded into vapor. The earth cooled, however, and torrential rains created seas that accumulated on the thin crust that was solidifying around the earth's surface. Vast numbers of volcanoes and enormous lightning storms shook the early earth. These were the conditions under which life began nearly 4 billion years ago.

The first organisms were single cells, much simpler than any cells living today. These first cells were the ancestors of all life on earth, from bacteria to dinosaurs to humans. Until recently, evolutionary theory proposed that the ancestor of all cells was in fact similar to some of today's bacteria, an ancient prokaryote that generated two lines of descent, one that gave rise to modern prokaryotic cells and one that produced eukaryotic cells. The discovery of archaeobacteria in the 1970s, however, forced us to reexamine this proposal. Most investigators now believe that the universal ancestor of all cells was a *progenote,* a cell much simpler than the simplest of present-day prokaryotes. From this progenote, *three* separate lines of cellular evolution developed: archaeobacteria, eubacteria, and an ancestral nucleated cell that lacked organelles. But where are modern eukaryotic cells in this scheme? Clearly this evolutionary puzzle lacks an essential piece.

The missing piece that links eukaryotic cells to the original progenote may be revealed by the remarkable similarities between bacteria (specifically eubacteria) and the mitochondria of modern eukaryotic cells. Mitochondria are not only the same size as typical bacteria, they also possess their own DNA (a circular strand, similar to that found in bacteria) and have their own ribosomes, which appear more prokaryotic than eukaryotic. These similarities suggest that mitochondria may have developed from an ancient eubacterium that was very efficient at generating usable energy by oxygen-dependent respiration. The original eubacterium entered and lived within the cytoplasm of a larger cell, establishing an **endosymbiotic** relationship, as shown in the figure below. After millions of years of such coexistence, the internal eubacteria lost their capacity for independent existence and developed into the organelle responsible for supplying the cell with energy. A similar interaction between a nucleated cell and a cyanobacterium is likely responsible for the development of chloroplasts in photosynthetic eukaryotes.

The similarity between true bacteria and mitochondria has practical significance in addition to being an interesting theory of evolution. It may explain why human cells are damaged by some antibacterial drugs. These adverse side effects may be due to inhibition of mitochondrial processes that are similar to their eubacterial counterparts. For example, drugs such as tetracycline and erythromycin, which interfere with bacterial protein synthesis, also inhibit mitochondrial ribosomes, although to a lesser degree. Similarities between mitochondria and eubacteria impose safety limitations that must be evaluated when choosing the most effective drug for treatment.

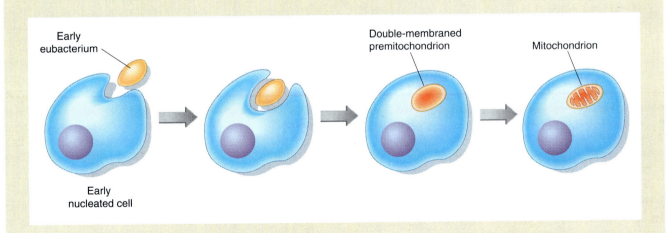

Early eubacterium

Double-membraned premitochondrion

Mitochondrion

Early nucleated cell

Key Terms

unicellular (p. 27)
eukaryotic cells (p. 28)
prokaryotic cells (p. 29)
cytoplasm (p. 30)
organelles (p. 30)
nucleus (p. 30)
DNA (p. 30)
chromosomes (p. 30)
endoplasmic reticulum (p. 30)
ribosomes (p. 30)
mitochondria (p. 31)
respiration (p. 31)
chloroplasts (p. 31)

photosynthesis (p. 31)
vesicles (p. 31)
endocytosis (p. 31)
lysosomes (p. 31)
Golgi complex (p. 31)
plasma membrane (p. 31)
selective permeability (p. 31)
cell wall (p. 33)
capsule (p. 33)
pseudopods (p. 33)
flagella (p. 34)
cilia (p. 34)
fungi (p. 34)

yeasts (p. 34)
molds (p. 34)
protozoa (p. 35)
algae (p. 36)
nucleoid (p. 36)
eubacteria (p. 38)
archaeobacteria (p. 38)
peptidoglycan (p. 38)
cyanobacteria (p. 40)
virus (p. 41)
obligate intracellular parasites (p. 41)
selective toxicity (p. 41)
endosymbiotic (p. 43)

Key Facts and Concepts

- **The unity of life.** All cells perform similar processes to survive and reproduce. They all have an outer plasma membrane that surrounds the cytoplasm and controls the selective movement of materials into and out of the cell. All possess DNA as the genetic material in chromosomes, and all require protein-synthesizing machinery.
- **Cellular dichotomy.** Cells are classified as either prokaryotic or eukaryotic depending on their microscopic anatomy.
- **Organelles.** Eukaryotes contain a membrane-bound nucleus and several different kinds of membrane-bound organelles, including the mitochondria, vesicles, endoplasmic reticulum, Golgi complex, and, in photosynthetic organisms, chloroplasts. Eukaryotic cells contain 80 S ribosomes, where protein synthesis occurs.
- **Eukaryotic microorganisms.** Algae, fungi (yeasts and molds), and protozoa are eukaryotic organisms. Algae are photosynthetic, whereas fungi and most protozoa are not. Algae and fungi contain cell walls, whereas protozoa do not. Many protozoa move about by means of pseudopods, flagella, or cilia. Some algal and fungal cells have flagella.
- **Prokaryotic anatomy.** Like all other cells, prokaryotes are surrounded by a plasma membrane, but, unlike eukaryotes, they lack membrane-bound organelles and internal membrane systems. In some prokaryotes, infoldings of the plasma membrane (mesosomes) extend into the cell. The prokaryotic cell contains a nucleoid rather than a membrane-bound nucleus. This nuclear region contains a single circular chromosome, which is suspended directly in the cytoplasm, not enclosed in a nuclear membrane. Prokaryotic ribosomes are smaller than those in the cytoplasm of eukaryotes.
- **Two types of prokaryotes.** Eubacteria and archaeobacteria are prokaryotic microorganisms. They differ in the composition of the plasma membrane, cell wall, and ribosomes. The plasma membrane of a eubacterium is similar to that of eukaryotes; archaeobacterial plasma membranes contain lipids with chemical bonds distinct from those of either eukaryotes or eubacteria. Eubacteria generally possess a cell wall that contains a unique compound, peptidoglycan; archaeobacterial cell walls do not contain peptidoglycan. Eubacterial ribosomes display a selective sensitivity to chemicals that are ineffective against archaeobacterial ribosomes or eukaryotic ribosomes. Chemicals that are effective against archaeobacterial and eukaryotic ribosomes are ineffective against eubacterial ribosomes.
- **Noncellular microbes.** Viruses are noncellular particles so small that they can be observed only with an electron microscope. They are composed of nucleic acid surrounded by a protective coat of protein and sometimes a lipid membrane.
- **Significance of differences between microbes.** In addition to adding to our understanding of life and its history, structural and biochemical differences among eukaryotes, prokaryotes, and viruses have countless applied benefits. These differences are important aids for identifying microbes in the laboratory. The differences also provide selective targets for antimicrobial drugs that interfere with the growth of pathogens while causing few adverse side effects for the host.

Review and Problem-Solving Questions

1. Describe the structure and functions of the following parts of a eukaryotic cell: (a) Nucleus; (b) mitochondria; (c) endoplasmic reticulum; (d) chloroplasts.
2. Compare the composition and functions of the plasma membranes of eukaryotic and prokaryotic cells.
3. List the identifying characteristics of fungi, algae, and protozoa.
4. In what ways are prokaryotic and eukaryotic cells similar?
5. List the properties that archaeobacteria have (a) in common with eubacteria; (b) in common with eukaryotes; (c) that are unique.
6. What characteristics of viruses differentiate them from bacteria and eukaryotic microbes?
7. Discuss the mechanism for selective toxicity of penicillin. What other sites in the prokaryotic cell may be selective targets? What sites in a fungal cell may be targets for selective toxicity?
8. What evidence exists to suggest that mitochondria evolved from a prokaryotic cell that formed an endo-symbiotic relationship with a primitive nucleated cell?
9. You discover a cell that contains a membrane-bound nucleus, DNA with histones, and peptidoglycan in the cell wall. How would you explain this seemingly paradoxical combination of traits?
10. A new microbe has recently been isolated from the sweat of microbiology students studying for exams. You have been given the job of examining and classifying this organism as either a eukaryote or a prokaryote.
    (a) What microscopic features can be used for identification?
    (b) Without the aid of a microscope, how might you be able to classify this organism?
11. Why is cytoplasmic streaming essential for the survival of eukaryotic cells? Propose a hypothesis that explains how prokaryotes survive without cytoplasmic streaming.

HEADLINES

A GIANT AMONG BACTERIA

In 1985, a new microorganism was isolated from the intestinal tract of tropical surgeonfishes. Because of its size (some are larger than 500 µm), the organism was thought to be a newly discovered protist. In 1991 evidence was presented[*] that revealed the gut microbe to be a giant prokaryote. What made this finding so astounding is that this unicellular organism, which is visible without a microscope, is the largest bacterium ever discovered and redefines the concept of size for prokaryotes.

- If you had isolated this new microbe, what evidence would you use to determine whether it was a prokaryote or a eukaryote?
- What problems might such a large prokaryote have that the typical smaller prokaryotes do not encounter?

[*]K. D. Clements and S. Bullivant, An unusual symbiont from the gut of surgeonfishes may be the largest known prokaryote, *Journal of Bacteriology,* 173:5359–5362 (1991).

CHAPTER 3

Tools for Investigating the Microbial World

Seeing bacteria and other microbes is much like trying to spot people on the ground in this satellite photo of the Grand Canyon. Not only must details on the ground be enlarged, but clarity of image must accompany the enlargement if individual organisms on the ground are to be discerned.

Because microorganisms are too small to be seen with the naked eye, we need tools to increase the power and accuracy of our observations and thus our knowledge of the microbial world. Microscopes have become indispensable as "eyes" for the discovery of information about the structure of cells. They are the instruments most often associated with microbiology. Microscopic observation, however, is only one method of detecting these organisms. Formation of **colonies** (visible accumulations of microorganisms) on laboratory culture media, for example, allows observation and characterization of microbes without the aid of any instrument. Other laboratory techniques enable us to detect and characterize microorganisms by demonstrating evidence of microbial biochemical activity or a specific immunologic reaction. Many laboratory procedures are needed to identify species of yeast or bacteria because their microscopic anatomy alone provides very little detail. Nonetheless, the microscope is usually the first tool the microbiologist uses in the systematic characterization of a microbe. Without this instrument, a microbe's anatomy, indeed the microbe itself, would remain hidden (Fig. 3-1).

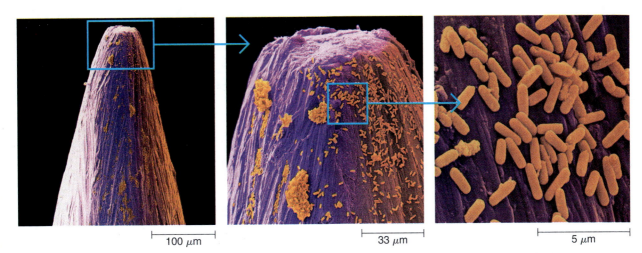

| 100 μm | 33 μm | 5 μm |

FIGURE 3-1 A culture on the point of a pin. The small size of bacteria can be appreciated when specimens on a pinpoint that has been dipped in *E. coli* are examined under greater and greater magnification.

Microscopic Observations

The earliest microscopes resembled magnifying glasses; they had a single lens, usually in a hand-held instrument. These *simple microscopes* (microscopes with one lens) magnified objects 200 to 300 times, thereby providing the opportunity to view objects with diameters below the lower limits of human vision (see THE EXPLORERS: The Father of Microscopy). With these instruments, however, only the gross form of cells could be discerned. Modern instruments are **compound microscopes;** that is, they contain at least two lenses. The second lens magnifies the image produced by the first lens, and consequently the size of the observed image depends on the magnification of both lenses. For example, combining a 10× lens with a 45× lens produces an image 10 × 45, or 450, times as large as the object being examined. Objects too small to be seen with a single lens are often observable through the compound microscope.

Although **magnification,** the process of enlarging an object's image, can theoretically be increased by using more powerful lenses or additional lenses, it will not be useful unless the details of the enlarged image are clearly visible. *Useful magnification* depends on the microscope's **resolving power** or *resolution*, its ability to distinguish two adjacent points as separate and distinct images rather than as a single larger image (Fig. 3-2). The lens system of the human eye, for example, has a resolving power of approximately 0.2 mm. This means that the eye cannot distinguish as separate two points or objects that are

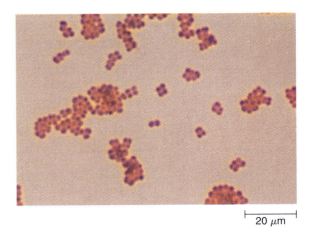

| 20 μm |

| 20 μm |

FIGURE 3-2 The importance of resolution is revealed when comparing these two photos, images of the same specimen at the same magnification, with the only difference being the higher resolution of the cells in the photo at the top.

The Father of Microscopy

Although Antoni van Leeuwenhoek is remembered as the first microbiologist, he was by trade a men's clothing merchant. More important, he was a man devoted to his hobby of grinding fine lenses. He did not invent the first microscope, but he was the first to report seeing microorganisms. He discovered a new world, that of the microbes.

Leeuwenhoek did not create his microscopes to search for this new world but rather to help him solve some scientific mysteries. His expertise at grinding lenses and his mastery of microscopic technique, however, allowed him to see bacteria, protozoa, and other microorganisms in such detail that we can identify many of his microbes from his recorded descriptions. These organisms were discovered quite by accident.

Leeuwenhoek was trying to determine "the cause of the pungency of pepper on the tongue." He hypothesized that there were tiny points on the pepper particles and that these tiny points irritated the tongue. To test his hypothesis, Leeuwenhoek soaked the pepper in water for several days to prepare it. What he didn't know was that pepper is a nutrient that promotes the growth of several types of microorganisms. When he finally mounted his pepper solution on his microscope, he failed to find the "pepper points," discovering instead the organisms that changed our view of the world forever. Many such accidents have opened new scientific frontiers; serendipitous discoveries by alert scientists are firmly embedded in the history of scientific exploration.

Leeuwenhoek found it easier to build a new microscope than to change specimens, since focusing the instrument was the most difficult aspect of using his microscope. Consequently, he produced hundreds of scopes, many of which still survive in museums, some containing the original specimen placed there by Leeuwenhoek.

Leeuwenhoek developed a very effective technique for using his instruments. Modern microbiologists are still not sure how he could have seen what he accurately described at the magnification attainable with his simple microscopes. It was proposed in 1976, 300 years after his initial observations, that Leeuwenhoek may have used an indirect source of illumination, creating an effect similar to that of the darkfield microscope.

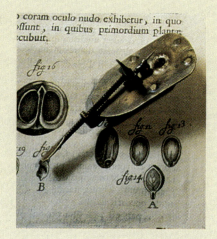

less than 0.2 mm apart (the approximate width of the point of a needle). Two objects that are each large enough to be visible will appear to be touching each other (or appear to be a single object) if the space separating them is less than 0.2 mm. Since the goal of microscopy is to increase the degree of detail that can be observed, both the magnification and resolution of a microscope must be maximized.

The useful magnification provided by a microscope is limited by its resolving power. Increases in magnification without corresponding increases in resolving power will produce a larger but blurred image. Thus a microscope with a resolving power of 0.2 μm has a maximum useful magnification of $1000\times$ (1000×0.2 μm $= 0.2$ mm), an image large enough for the human eye.

The resolving power of a microscope depends largely on the wavelength of the beam used for illumination. Shorter wavelengths give better resolution than longer wavelengths. Ultraviolet light and high-voltage electron beams have wavelengths shorter than those of visible light, and both are used in high-resolution instruments (Fig. 3-3). The resolving power of the light microscope depends not only on the wavelength but also on the amount of light that passes through the specimen and eventually enters the microscope through the objective lens. This property, called *numerical aperture*, is determined by the optical quality of the objective lens and the *refractive index* (degree of bending of light rays) of the material in the space between the specimen and the lens. The more light entering the lens, the greater the resolution.

■ Microscopic Dimensions

Since the average bacterial cell is approximately 0.001 mm in diameter, the dimensions of microorganisms and of their component parts are more conveniently expressed in units smaller than millimeters. The **micrometer** (μm) is one-millionth of

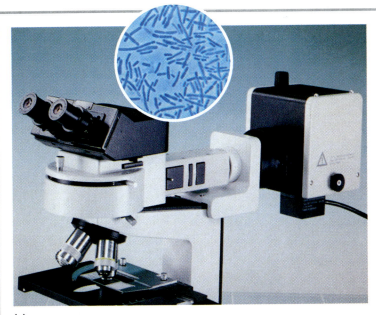

(a)

(b)

FIGURE 3-3 Ultraviolet and electron microscopes. The epifluorescence ultraviolet microscope (*a*) and the electron microscope (*b*) use radiation with shorter wavelengths than visible light to increase useful magnification.

a meter, or one-thousandth of a millimeter ($1 \mu m = 1 \times 10^{-6}$ m $= 1 \times 10^{-3}$ mm). This allows us to express the diameter of the average bacterial cell as $1 \mu m$. For even smaller structures such as viruses, the nanometer and angstrom are often employed. The **nanometer** (nm) is one-billionth (1×10^{-9}) of a meter, or one-millionth of a millimeter. The **angstrom** (Å) is one-tenth of a nanometer, or one ten-billionth (1×10^{-10}) of a meter. For ease of reference, these units are summarized in Table 3-1.

Tremendous size diversity exists among the microorganisms (Fig. 3-4). For example, the 27-nm poliovirus, one of the smallest viruses, is about one-thousandth the diameter of the average eukaryotic cell. This size difference is similar to the height difference between a housefly and an elephant. Size is one of the properties of a microorganism that is useful for determining its identity in the laboratory.

MYTH: The largest organisms on earth are the giant sequoias of California.

FACT: The earth's largest organism is a giant subterranean fungus that covers 37 acres (in northern Michigan), may weigh as much as 100 tons, and is at least 1500 years old. Ironic as it may seem, the earth's largest organism is being studied by microbiologists.

TABLE 3-1	UNITS OF MEASUREMENT	
Unit	**Fraction of a Meter**	**Example of Objects Measured**
Centimeter (cm)	10^{-2}	Whole chicken egg (~7 cm long)
Millimeter (mm)	10^{-3}	Flea (~1 mm long)
Micrometer (μm)	10^{-6}	Typical bacterium (1 μm diameter)
Nanometer (nm)	10^{-9}	Poliovirus (27 nm)
Angstrom (Å)	10^{-10}	Cell membrane thickness (10 Å)

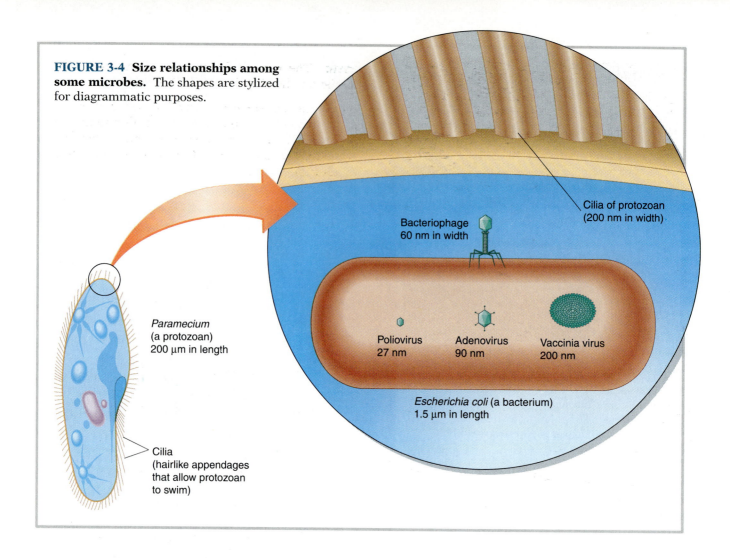

FIGURE 3-4 Size relationships among some microbes. The shapes are stylized for diagrammatic purposes.

Cilia of protozoan (200 nm in width)

Bacteriophage 60 nm in width

Poliovirus 27 nm

Adenovirus 90 nm

Vaccinia virus 200 nm

Escherichia coli (a bacterium) 1.5 μm in length

Paramecium (a protozoan) 200 μm in length

Cilia (hairlike appendages that allow protozoan to swim)

■ The Brightfield Microscope

The most commonly used microscope for general laboratory observations is the standard **brightfield microscope** (Fig. 3-5a). This instrument uses a *condenser*, a lens that focuses the light source directly on the specimen. Thus, the entire field of view is illuminated (Fig. 3-5b). The final magnified image is produced by a series of two additional lenses, the **objective lens,** which is located close to the object being examined (the specimen), and the **ocular lens** (eyepiece) through which the image is viewed. Most brightfield microscopes are equipped with at least three different objective lenses on a rotating nose-piece. Typically these provide magnifications of 10×, 45×, and 100×. The enlarged image produced by the objective lens is further magnified by the ocular lens, usually an additional 10×. The greatest magnification achievable with such a lens system is

1000× (10 × 100). The maximum resolving power of this microscope is approximately 0.27 μm, about one-thousandth the thickness of a human hair. To achieve this level of resolution, however, immersion oil must be placed between the specimen and the 100× objective lens. Immersion oil and glass have the same refractive index. Filling the air space between the glass lens and the glass slide with immersion oil directs the light rays into the lens, thereby increasing resolution (Fig. 3-6). Most microorganisms, except viruses and some very thin bacteria, can be clearly observed with a brightfield microscope.

The brightfield microscope is most often used as an aid in counting cells and in the laboratory identification of microorganisms. Microscopic observations of cell morphology and reactions of cellular constituents to various stains are discussed later in this chapter.

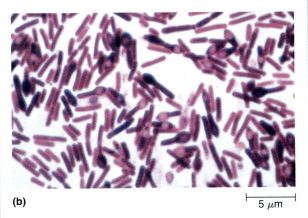

(a)

Occular lens

Objective lens

Specimen on slide

Stage

Condenser

Lamp (not shown)

(b)

5 μm

FIGURE 3-5 The brightfield microscope. (a) The path of light travels from the source of illumination (the lamp) through the condenser, which focuses available light on the specimen. The specimen is mounted on a slide on the stage. Light enters the objective lens, which magnifies the image of the specimen. The ocular lens further magnifies the image produced by the objective lens, usually about 10 times. Focusing knobs (not shown) provide coarse and fine adjustment. (b) Photomicrograph of *Clostridium botulinum* as it appears under the brightfield microscope.

MYTH: The greater the amount of available light, the better the visibility of the image.

FACT: Images are frequently "burned out" by too much light—lost in the glare of overillumination. In these cases, objects become visible only when light is reduced to the optimal (most satisfactory) level. In addition, too much light reduces the contrast between specimen and background to the point of invisibility.

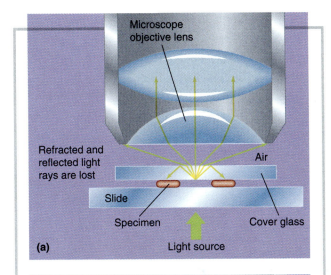

Microscope objective lens

Refracted and reflected light rays are lost

Air

Slide

Specimen

Cover glass

(a)

Light source

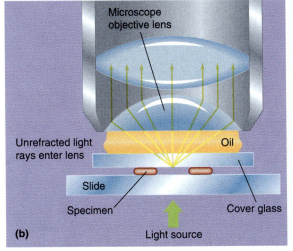

Microscope objective lens

Unrefracted light rays enter lens

Oil

Slide

Specimen

Cover glass

(b)

Light source

FIGURE 3-6 Why we use oil immersion lenses. Preventing light from refracting as it travels through the slide-specimen-lens pathway improves the quality of the image. (a) Light traveling from glass to air is bent because of the different light-refracting properties of air and glass. (b) Since immersion oil and glass have virtually identical light refracting properties, light is not bent as it goes from one into the other.

Contrast (sharp differences between light and dark areas) is as critical a factor as resolution when making microscopic observations. Although contrast can be artificially increased by staining the specimen, the staining procedure usually kills the specimen. To view living unstained cells, special optical devices that increase the visual contrast between specimen and background can be attached to a standard brightfield microscope. Two commonly used instruments that employ such devices are the darkfield microscope and the phase-contrast microscope.

■ The Darkfield Microscope

The **darkfield microscope** is equipped with a special condenser that uses a darkfield ring to block light from entering the objective lens directly from the light source. The result is a completely dark field

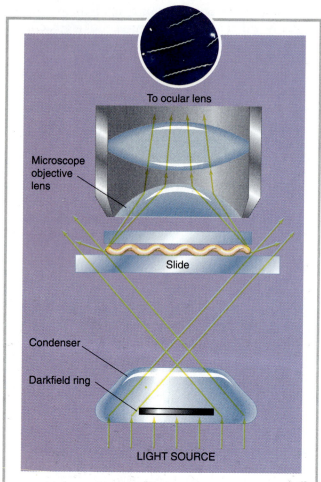

FIGURE 3-7 Darkfield microscopy. Because of the darkfield ring in the condenser, the only light that reaches the observer's eye is the light scattered by the specimen. The darkfield photo in the circle is *Treponema pallidum*, the bacterium that causes syphilis.

of view. Light that passes through the specimen is scattered, and some of it is deflected into the objective lens (Fig. 3-7), so the specimen appears brightly illuminated on a dark background. A bright object in the dark is much easier to see than a dark object in bright light. The darkfield microscope can therefore detect organisms that are too thin to be seen in the brightfield microscope, for example, the one that causes syphilis (Fig. 3-7). Because it cannot detect color, the darkfield microscope is not used for examining stained specimens. It is excellent, however, for viewing living unstained cells.

■ The Phase-Contrast Microscope

Living cells in liquid are best viewed with a **phase-contrast microscope.** This instrument highlights details between structures of different densities, such as a cell and its surrounding medium. The physical properties of each part of an illuminated specimen determine the speed and direction that light travels as it passes through that part. The optics of the phase-contrast microscope are designed to detect differences in transmitted light and translate them into patterns of shadows and light (Fig. 3-8). Different organelles within a cell also have different densities. The phase-contrast microscope therefore has the advantage of revealing subcellular anatomy without the use of stains, so it is not necessary to kill the specimen. Motility, phagocytosis, and cellular activities can be more clearly observed than with the standard brightfield microscope.

Both the darkfield and phase-contrast microscopes have limitations. Neither microscope provides greater resolution or magnification than the standard brightfield instrument. More important, however, neither instrument can be used for determining the color of a stained specimen. Because important diagnostic information is obtained from ascertaining the way a microorganism reacts to certain stains, neither darkfield nor phase-contrast can replace the standard brightfield microscope in the clinical or research laboratory.

■ The Fluorescence Microscope

A **fluorescence microscope** uses ultraviolet (UV) light as the source of illumination. Although invisible to the human eye, ultraviolet light causes some microbes to fluoresce (give off visible light). The fluorescence microscope is designed to provide a dark background so that the colorful fluorescent specimen appears to glow (Fig. 3-9). The most widespread uses of fluorescence microscopy employ *fluorochromes*, dyes that glow when exposed to ultraviolet light. For example, acridine orange is a flu-

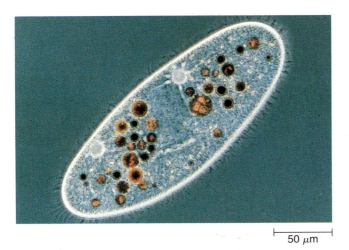

50 μm

FIGURE 3-8 Phase-contrast microscopy. Observation of these living protozoa using a phase-contrast microscope reveals organelles and movement, including the beating of cilia.

orescent dye that reacts with DNA and RNA. Cells stained with acridine orange can be easily seen using the fluorescence microscope, making it a useful tool for viewing bacteria.

Fluorescence microscopy is important in clinical diagnosis. *Mycobacterium tuberculosis* in sputum can be identified by its characteristic yellow fluorescence after staining with the fluorescent dye auramine. Since few other bacteria contain the waxy surface layer that combines with this dye, this examination is useful in the diagnosis of tuberculosis.

Fluorescence microscopy also aids in the rapid identification of other microorganisms when the **fluorescent antibody technique** is employed. In this technique a fluorescent dye is chemically attached to an antibody molecule that can recognize and combine with a specific infectious agent. For example, fluorescent-labeled antibodies can be mixed with a sample of tissue taken from a patient suspected of being infected with a particular microbe. If the microbe is present in the patient's tissue, the fluorescent antibodies will attach to it. This tissue will thus appear to glow when exposed to ultraviolet light. If the microorganism is absent, the antibodies fail to coat the tissue and the specimen appears dark. The fluorescent antibody technique is one of the most useful techniques available for the rapid diagnosis of legionnaires' disease, syphilis, rabies, herpesvirus infections, and other diseases caused by organisms that are difficult to culture.

■ The Electron Microscope

The electron microscope uses a beam of electrons instead of light. Because a beam of electrons has a much shorter wavelength than light, electron micro-

scopes have dramatically better resolution and therefore greater useful magnification than standard microscopes. Magnetic lenses focus the electron beam in much the same way that glass lenses focus light. The image formed by the electrons is projected onto a viewing screen (similar to a television screen) or onto photographic film to be permanently recorded as a photograph.

There are two types of electron microscopes: the transmission electron microscope and the scanning electron microscope.

● **THE TRANSMISSION ELECTRON MICROSCOPE (TEM).** The useful magnification of the **transmission electron microscope** is 1000 times greater than that possible with conventional brightfield microscopes. The TEM can produce an image 1 million times the actual size. The introduction of the TEM allowed the first observations of viruses and other particles too small to be detected with light microscopes. Similarly, the fine details of intracellular structures are revealed, providing new understanding of the fundamental life processes of all cells.

In the TEM, a thin specimen is bombarded with electrons. Depending on their composition, the components of the specimen either transmit, absorb, or deflect the electrons. Magnetic lenses collect and focus electrons that pass through the specimen. The image produced is a visual interpretation of this interaction of electrons with the specimen. Areas that are "electron-transparent" (allow electrons to pass through) appear light; areas that deflect electrons appear dark. To prevent air and dust from interfering with the electron path and causing random scattering, the specimen and magnetic lenses are sealed in a chamber under vacuum.

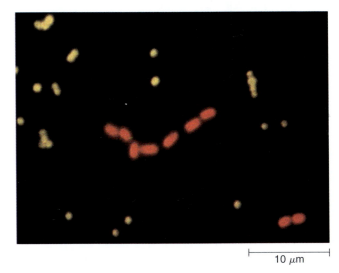

10 μm

FIGURE 3-9 Fluorescence microscopy. These marine bacteria glow when exposed to ultraviolet light.

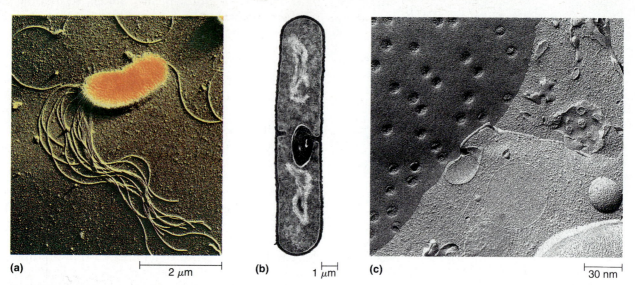

(a) 2 μm   **(b)** 1 μm   **(c)** 30 nm

FIGURE 3-10 Transmission electron microscopy. (*a*) Shadowcast. (*b*) Thin section through bacterial cell. (*c*) Freeze-etched interior of eukaryotic cell. All colors in photos are added by computer enhancement, as TEM generates only black-and-white images.

Most biological specimens appear transparent in the TEM and are virtually invisible unless stained to increase contrast between them and the background (Fig. 3-10a). Heavy metals that reflect electrons, most commonly tungsten, uranium, or lead, are used for this purpose. In addition, because electrons have poor penetrating power, internal cellular structures can best be viewed in a thin slice cut through a cell (Fig. 3-10b). Before it is sliced, the specimen is usually dehydrated and embedded in a plastic block. Using a special cutting instrument called an *ultramicrotome*, a single bacterial cell can be sliced into several hundred sections.

One method of preparing cells for transmission electron microscopy without sectioning is *freeze-etching*. The specimen is quickly frozen under vacuum. This makes it brittle. A special knife is then used to shatter portions of the cell away from underlying structures. The cell will break at its weakest parts, usually between the lipid bilayers of cellular membranes. In a eukaryotic cell, for example, breaks will occur at the cell membrane, nuclear membrane, organelle membranes, and membrane networks. The exposed surfaces of the fractured specimen reveal details of cellular architecture not available by other methods (Fig. 3-10c).

● **THE SCANNING ELECTRON MICROSCOPE (SEM).** The more recently developed **scanning electron microscope** reveals surface details of whole organisms in striking three-dimensional relief (Fig. 3-11). A moving electron beam scans the surface of a metal-coated specimen, causing the emission of secondary electrons from the specimen. The secondary electrons are used to create the image on the viewing screen. Because raised regions of the specimen emit more secondary electrons than depressed regions, the SEM generates pictures that reveal depth. Although only the surface can be observed, scanning electron microscopy has revealed unique information about the form and function of many microbes.

Neither SEM nor TEM allows the observation of living specimens. Furthermore, the harsh treatment of specimens in preparing them for viewing often introduces **artifacts** (artificial features not possessed

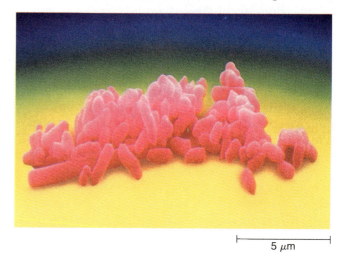

5 μm

FIGURE 3-11 Scanning electron microscopy. False-colored SEM of *Bacteroides*, one of the many bacteria that inhabit your intestines.

The Modern Microscope Revolution

A cell is a highly organized array of chemicals, complex enough to generate the remarkable property of life. Sugars, lipids, amino acids, and nucleic acids interact to form intricate cell structures whose details can usually be observed with an electron microscope. Until recently, simpler compounds suspended in the cytoplasmic matrix were invisible to the human eye. Several new technologies, however, have allowed us to peer into the cell at the level of individual molecules and atoms.

The distribution of specific molecules in a cell can be determined by coupling an *immunogold labeling technique* with electron microscopy. Antibodies that react with the molecules to be viewed are covalently bonded to gold particles about 5 to 20 nm in diameter. The labeled antibodies are incubated with the specimen, which is then examined with a transmission electron microscope. Although antibodies are too small to be detected with the electron microscope, the gold particles are not. They are visible as dark spheres and are deposited wherever an antibody has attached to its target antigen.

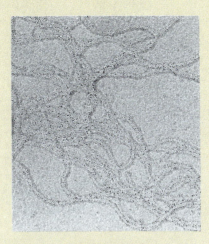

The *scanning tunneling microscope*, introduced in 1980, scans the specimen with such precision that it can produce images of atoms such as that of the DNA molecule in the photo to the right. With this instrument, a needle probe that has only a few atoms at its tip is placed at a fixed distance from the surface atoms of the specimen. The electrons circulating around the atoms at the tip of the probe just meet the electrons circulating around the atoms on the surface of the specimen. Voltage is applied, and electron flow produces a current called the *tunneling current.* The probe tip is then moved (much like a record needle) over the surface of the specimen.

The up-and-down movement of the probe as it travels over the contours of the specimen surface is monitored by a computer and translated into high-resolution images for viewing. The sensitivity of this system is striking. Movement of the probe through even the dimensions of an atom alters the tunneling current dramatically and is translated into a visual image.

by the living organism) due to shrinkage and distortion of the specimen. Skill and training are needed to distinguish artifacts from natural cellular structures in electron micrographs.

A comparison of the properties of each of the six commonly used kinds of microscopes is provided in Table 3-2. (See also BEYOND THE BASICS: The Modern Microscope Revolution.)

> **CONCEPT CHECK**
>
> - How do microscopes increase our ability to see small objects with clear detail?
> - What distinguishes standard brightfield microscopes from other types of compound microscopes and from electron microscopes?

Microbiological Stains

Most bacteria are transparent and difficult to see under the brightfield microscope. Therefore, specimens are routinely stained to increase visibility and to reveal additional information to help identify microbes. Staining is usually accomplished by fixing (attaching) the organisms to a slide prior to saturating them with a dye with which various parts of the cell react. Two fixation methods are commonly used, both of which kill the microbe. A specimen is *heat-fixed* by passing an air-dried smear gently through a flame and heating the glass until the cell's molecules change shape and adhere to the surface of the slide. The organism's shape is preserved, but

TABLE 3-2   COMPARISON OF SIX TYPES OF MICROSCOPES

Microscope	Maximum Useful Magnification	Resolving Power	Common Uses	Advantages	Disadvantages
Standard brightfield	1000×	0.2 μm	Observing stained specimens Counting microbes	Easy to use Relatively inexpensive Allows staining reactions to be interpreted Readily available	Lacks contrast; unable to resolve viruses and some very thin bacteria. Most intracellular structures must be stained to be observable. Artifacts may be introduced during staining procedures.
Darkfield	1000×	0.2 μm	Detecting unstained microbes not easily observed with brightfield microscopy	Allows living specimens to be viewed Reduces artifacts	Cannot be used to evaluate staining reactions. Subcellular detail not readily observable.
Phase-contrast	1000×	0.2 μm	Observing living cells Observing intracellular structures	Allows observation of intracellular biological activity Enhances subcellular details	Staining reactions cannot be evaluated.
Fluorescence	1000×	0.2 μm	Identifying micro-organisms Detecting specific infectious agents in tissue Detecting immunologic reactions	Allows very rapid identification of infectious agents	Limited to viewing specimens that naturally fluoresce or that are stained with a fluorescent dye.
Scanning electron microscope	100,000×	5 nm (0.005 μm)	Observing surface topography	Reveals three-dimensional shapes Resolution greater than brightfield	Observing internal detail requires freeze fracturing. Specimen must be killed. Expensive.
Transmission electron microscope	1,000,000×	0.2 nm (2 Å)	Observing fine structure of cells and viruses Diagnosing certain viral diseases and cancers Detecting certain large molecules	Permits viewing of objects too small to be detected by light microscopy	Specimen must be killed. Harsh preparation procedures increase the probability of artifacts in the specimen. Very expensive.

some cellular components may be damaged. *Chemical fixation* cements the cells to the slide, often without destroying their structure. Ethanol, formaldehyde, and glutaraldehyde are chemical fixatives used when it is important to maintain the integrity of the specimen.

Most dyes are charged molecules that react with cellular components of opposite charge. For example, methylene blue and crystal violet, which are positively charged, bind to the negatively charged nucleic acids and proteins inside cells and to polysaccharides on the bacterial cell surface.

There are two categories of stains, simple and differential. **Simple stains** employ a single dye, most commonly methylene blue, crystal violet, or fuchsin. Cells and the structures within them generally stain the same color. Simple stains, therefore, do little more than reveal characteristics of size, shape, and cell arrangement.

Differential stains, which require more than one dye, distinguish between structures within a cell or between different types of cells by staining them different colors. Some differential stains react with specific microbial structures such as flagella, capsules, and endospores. Another differential stain, the **Gram stain,** divides bacteria into two categories, the purple-staining gram-positive cells and the red-staining gram-negative cells. The Gram stain is the most important stain in the clinical microbiology laboratory. The *acid-fast* stain is another differential stain routinely performed whenever tuberculosis or leprosy is suspected.

■ The Gram Stain

The Gram stain was developed by Hans Christian Gram in 1884. The procedure requires four steps:

1. *Crystal violet* is added as a primary stain. It colors cytoplasm purple regardless of cell type.
2. *Iodine* is then used as a *mordant,* an agent that binds the dye to the cell and helps resist decolorization. It combines with crystal violet to form an insoluble complex inside the cell.
3. A *decolorizing agent* (alcohol or a mixture of acetone and alcohol) is then added to the specimen. The purple dye complex is retained by gram-positive organisms but is readily removed from gram-negative cells. Gram-negative cells will therefore become colorless at this stage, but gram-positive cells will remain purple.
4. The red dye *safranin* is applied as a counterstain. This stains the colorless gram-negative bacteria red while gram-positive cells remain purple (Fig. 3-12).

Examination of Gram-stained organisms usually provides a starting point for classifying, identifying, and characterizing bacteria. Such information helps to determine the source of microbes isolated as contaminants, for example, in the industrial production of foods and pharmaceuticals. The presence of gram-positive cocci (spherical cells) indicates that shedding of normal flora from humans is the likely source of contamination; the presence of gram-positive, endospore-producing bacilli (rod-shaped cells), on the other hand, suggests environmental sources. A Gram stain can steer decisions on how best to remove or destroy the contaminant and how to prevent future contamination by the intruding microbe.

This differential staining technique is also a fundamental step in diagnosis and treatment of disease. The Gram stain of clinical material taken directly from an infected patient can rapidly provide valuable information about the microorganism(s) causing the disease. This information helps guide the selection of subsequent tests needed to identify the bacteria. For example, identifying gram-positive cocci or gram-positive bacilli requires a different battery of tests than are used for identifying the gram-negative rods.

Gram stains of clinical specimens are especially important in determining the most effective antibiotic for critically ill patients who require immediate therapy. Penicillin, for example, is more effective against most gram-positive bacteria than against most gram-negative bacteria. Gram staining of clinical specimens, however, provides only a preliminary indication of the identity of the **etiologic agent**

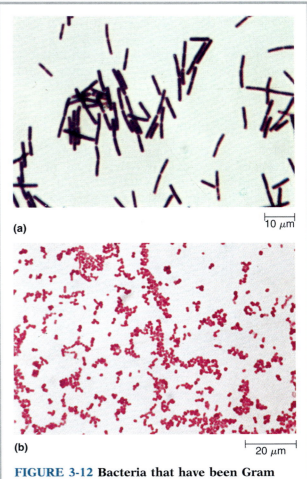

(a) 10 µm

(b) 20 µm

FIGURE 3-12 Bacteria that have been Gram stained. (*a*) Gram-positive bacteria (a species of *Bacillus cereus*). (*b*) Gram-negative bacteria (*Neisseria gonorrhoeae*).

(the organism causing the disease). Although it is an important first step, a Gram stain does not serve as a substitute for isolating the suspected microorganism and determining its biochemical characteristics and antibiotic susceptibilities.

■ The Acid-Fast Stain

The **acid-fast stain** is used to identify members of the genus *Mycobacterium*. The etiologic agents of tuberculosis and leprosy are both mycobacteria. Because mycobacteria contain high levels of lipid material that repel water-soluble dyes, they are difficult to stain by standard procedures. Hot carbol fuchsin, however, will penetrate the bacteria and stain them red. Unlike most other bacteria, the mycobacteria are difficult to decolorize once they are stained. Their resistance to decolorization by an acid-alcohol mixture explains the term "acid-fast." Following decolorization of non-acid-fast organisms, methylene blue is added as a counterstain to make the unstained bacteria visible. Mycobacteria remain red, whereas non-acid-fast bacteria stain blue (Fig. 3-13). The acid-fast stain is a rapid method for indicating the presence of mycobacteria in specimens such as sputum (material expectorated from a person's lower air passages), which may contain a large number of non-acid-fast normal flora.

■ Structural Stains

A variety of procedures can be used to selectively stain specific structures. In bacteria, such structures include endospores, flagella, cell walls, and capsules. Endospores are highly resistant to chemicals, heat, and irradiation and are primarily produced by members of the genera *Bacillus* and *Clostridium*. The

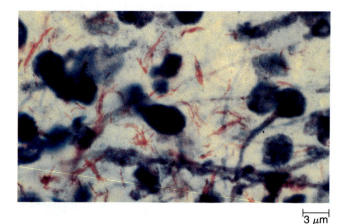

FIGURE 3-13 Results of the acid-fast stain. The red rods in this stained specimen of material from a lung are *Mycobacterium tuberculosis*, the organism that causes tuberculosis.

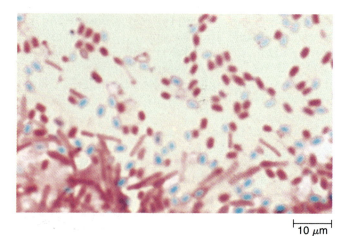

FIGURE 3-14 Endospore stain. (*Clostridium botulinum*). The green structures are endospores.

shape of the endospore and its location within the parent cell are important in identifying species within these genera. Endospores are not easily penetrated by most dyes, but hot malachite green will stain them. Decolorization with water washes the malachite green from the rest of the cell, but the endospores will retain the dye. Cells can then be counterstained with safranin. The green spores can readily be seen in striking contrast to the red cells (Fig. 3-14). ▶ (See Endospores, Chap. 4, p. 92.)

Flagella, structures that move a cell, require a special staining procedure to make them visible with the brightfield microscope. Bacterial flagella are about 20 nm in diameter, so thin that they are below the resolution of the light microscope. Before staining, they are treated with a mordant, for example, tannic acid, that coats the thin filament and increases its size. The dye then attaches to the mordant (Fig. 3-15). ▶ (See Flagella, Chap. 4, p. 88.)

Under certain conditions, some bacteria and yeasts produce capsules external to the cell wall. To a clinician, the presence of encapsulated microbes in a patient's tissues may indicate the nature of the etiologic agent and suggest a course of treatment. Capsules can be observed by combining simple staining with **negative staining,** a technique that dyes the background while leaving the cell colorless. India ink, nigrosin, or another dye that cannot penetrate the capsule is mixed with the microbe. The mixture is spread out on the slide to form a thin film, which is then dried. This leaves the whole cell and capsule unstained against a dark background. This preparation is heat-fixed and treated with a simple stain that stains only the cell (and not the capsule). The capsule is clearly visible as a colorless halo around the stained microbe (Fig. 3-16).

Table 3-3 summarizes the uses of some common bacteriologic stains.

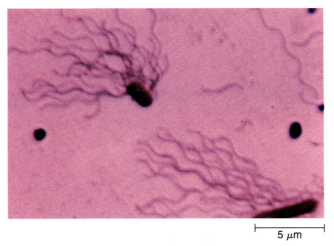

5 μm

FIGURE 3-15 Flagella stain. (*Salmonella typhi*)

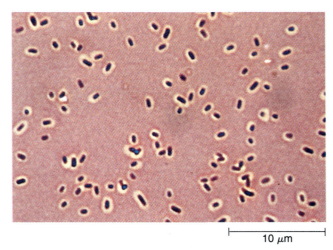

10 μm

FIGURE 3-16 Capsule stain. (*Klebsiella pneumoniae*)

TABLE 3-3	COMMON BIOLOGICAL STAINS USED IN BACTERIOLOGY		
Reagents	**Appearance of Stained Cells**		**Purpose**
Simple Stains			
Methylene blue	Blue		Simple stains are used for a variety of microorganisms. Most commonly used to
Crystal violet	Purple		determine the presence of microorganisms, for example, in urine or tissue specimens from patient; the Gram stain, however, is better for
Carbol fuchsin	Red		this purpose.
Negative Stain			
Nigrosine	Cells unstained; background dark		Nigrosine is used when cell is to be measured (prevents shrinkage since heat fixation is not used).
Gram Stain			
1. Crystal violet 2. Gram's iodine 3. Decolorizer 4. Counterstain	Gram-positive cells, purple; gram-negative cells, red (Occasionally in older cultures gram-positive cells will stain as gram-negative.)		A differential stain used as the first step in determining characteristic properties of any bacteria.
Acid-Fast Stain			
1. Hot carbol fuchsin 2. Acid-alcohol 3. Methylene blue	Acid-fast cells, red; non-acid-fast cells, blue		To help identify acid-fast bacilli, such as the agents of tuberculosis and leprosy.
Endospore Stain			
1. Hot malachite green 2. Water 3. Safranin	Endospores stain green; rest of the cell stains pink.		A differential stain used to detect and characterize bacterial endospores, structures resistant to staining by traditional techniques.

Culture Techniques

The morphologic simplicity of most microbes prevents their identification by microscopic examination alone since many of them look identical. Usually they must be isolated and grown in a **pure culture** (one that contains a single species). Pure cultures can then be used to determine biochemical and immunologic characteristics. Pure cultures are also needed for other microbiological applications, such as harvesting useful compounds produced by microorganisms. Isolation of microbes in pure culture requires aseptic technique and special methods of inoculation.

■ Aseptic Techniques

Aseptic techniques allow handling of materials without the introduction of microbial contaminants from air, water, hands, or other nonsterile sources. Once a culture is contaminated, it may be impossible to determine which microbes are contaminants and which are from the specimen. In addition, the contaminating microbes may multiply faster than the microbe of interest and may overgrow a sample. In such improperly handled clinical specimens, contaminants may obscure the diagnosis and result in incorrect treatment.

To prevent microbial contamination, exposure of the sample to nonsterile environments such as room air must be minimized, and contact between the sample and all nonsterile objects and surfaces must be avoided. For example, tubes and plates should be kept closed; they should be opened only when necessary. In addition, equipment for transfer, such as wire loops or forceps, should be sterilized by flaming. Strict adherence to aseptic technique prevents contamination and reduces the time required to achieve valid results. ▶ (See *aseptic technique*, Chap. 15, p. 350.)

■ Inoculation Techniques

Microbes can usually be isolated in pure cultures by inoculating solid media in a manner that results in the development of single colonies. Since all the cells in an isolated colony are descendants of a single organism, they are identical. Obtaining isolated colonies requires the use of a solid medium, since cells in a liquid cannot be kept separate. A solid medium is easily produced by adding the solidifying agent **agar** to a nutrient solution. Agar, an extract of seaweed, melts at 100°C and then solidifies the liquid in which it is dissolved when the temperature drops below 45°C. The melted medium is usually poured into *petri dishes*, specially designed shallow containers, where it is allowed to solidify.

The surface of agar media is ideal for the growth of isolated colonies. There are so many microbes in most specimens, however, that direct inoculation of a solid medium would result in overgrowth of the entire surface rather than isolated colonies. The sample must therefore be processed in such a way that cells are deposited on the medium far enough from each other that well-isolated colonies will develop. This can be accomplished by preparing streak plates, pour plates, or spread plates.

● **STREAK PLATES.** Organisms can be inoculated onto the surface of a plate with a wire loop or swab. A series of streaks spreads the inoculum over a large surface area until very few cells are left on the loop. Several patterns for streaking have proved effective for obtaining isolated colonies, but the most commonly employed technique is the *quadrant streak method* illustrated in Figure 3-17. **Streak plates** have four distinct areas of inoculation. The first quadrant can be inoculated with a loop or a swab. (Swabs are usually used to collect clinical specimens from the throat, eye, and vaginal tract and exudates from wounds or lesions. In industry, swabs are most commonly used to sample surfaces of large pieces of equipment.) The remaining three quadrants must each be streaked with a sterile instrument such as a flamed loop or fresh swab.

All streaking techniques depend on spreading the inoculum over a large surface area so that the organisms fall off the loop a single cell at a time. It is often difficult to isolate an organism in a single attempt, especially if there are many competing contaminants in the specimen. A second plate may be streaked using material from the area of the original streak plate that is richest in the desired organism.

● **POUR PLATES AND SPREAD PLATES.** Another way to obtain isolated colonies is to inoculate an agar medium with a sample that has been diluted in sterile liquid. In the **pour-plate technique,** a sample of the diluted culture, usually 1 mL, is added to melted agar at 45°C as it is poured into a petri dish. The pour-plate technique generates many subsurface

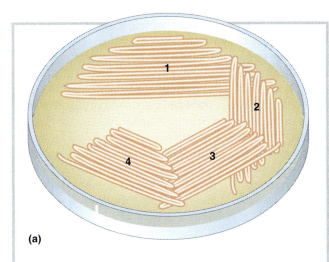

(a)

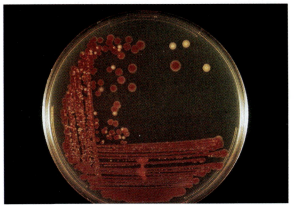

(b)

FIGURE 3-17 Streaking to obtain isolated colonies. The quadrant streak method is the most commonly employed technique for obtaining isolated colonies. (*a*) Quadrants are streaked successively through a portion of the previously streaked area (called a quadrant because it occupies about one-fourth the area of the surface). The first quadrant may be inoculated with a swab, after which a loop may be used. (*b*) A properly prepared streak plate reveals isolated colonies after a 24-hour incubation period.

colonies (colonies embedded in agar). **Spread plates** are prepared by spreading a small volume of the diluted sample across the surface of a solid medium by using a sterile bent glass rod. In this case, all colonies grow on the agar's surface. Pour plates and spread plates are generally employed for counting microbes.

Because of its convenience, the streak technique is usually employed for isolating bacteria from clinical specimens.

Identification Techniques

■ Cultural Characteristics

When growing on suitable solid media, most microbes form visible aggregates called *colonies*. The appearance of these colonies provides additional clues to the identity of the microorganisms. Cultural characteristics that contribute to identification are size, shape, pigmentation, and texture of the colonies (Fig. 3-18).

■ Biochemical Tests

A wide range of microbial activities can be determined by biochemical tests. These characteristics can be used to distinguish between microbes that appear morphologically identical. For example, different types of microbes can usually be distinguished by the characteristic pattern of nutrients each can utilize. In addition, microbes produce biochemicals such as acids, alcohols, gases, or specific enzymes, the detection of which may aid in identi-

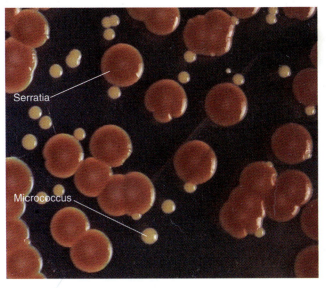

FIGURE 3-18 Cultural characteristics of bacteria growing on the surface of a solid medium. Notice the differences in appearence of colonies formed by the two species of bacteria (*Micrococcus* and *Serratia*) that are visible in this photo.

fying the organism. A series of biochemical tests can provide a microbial "fingerprint." ▶(Biochemical activities of microbes are discussed in more detail in Chapters 6 and 7.)

■ Immunologic Tests

Immunologic reactions can be used to identify microorganisms by determining whether the microbe can combine with specific antibodies. Antibodies can be obtained from the blood of animals that have been vaccinated (immunized) by injection with microbes or specific microbial components. These antibodies usually react only with the same type of organism or molecule that was used to stimulate their production; thus, reaction with a specific antibody is evidence of a microbe's identity.

Immunologic tests are particularly useful because they are rapid and can identify pathogens that cannot be grown on laboratory media. For example, *Treponema pallidum*, the bacterium that causes syphilis, is most often diagnosed by a fluorescent antibody test that uses fluorescence as the indicator of reaction (Fig. 3-19). ▶(Immunity and its applications for identifying microbes are discussed in Chapter 16.)

■ Genetic Tests

Every type of organism contains DNA that is unique to it alone, DNA that can be used for identification much like a fingerprint. Such genetic tests use labeled segments of the unique DNA from known microbes to probe the chromosome of the unknown organism. The probe will combine with the unknown organism's DNA only if its chromosome contains an identical segment. The label can then be detected. If the DNA probe is unique to another organism, however, it will not react, and no label will be detected. DNA probes are specific, and a positive reaction is proof of the microbe's identity. Biotechnological advances have made it possible to "grow" a microbe's DNA even when the organism is difficult to culture. Thus, sufficient DNA samples for identification of virtually any microbe can be obtained from a specimen even without culturing the organism. ▶(See Gene Amplification—Polymerase Chain Reaction, Chap. 9, p. 232.)

> **CONCEPT CHECK**
>
> • How are a microbe's colonies, biochemical activities, and immunologic reactions useful in identifying the microbe?
> • Why can antibodies or DNA probes be used to identify a single organism in a mixed culture?

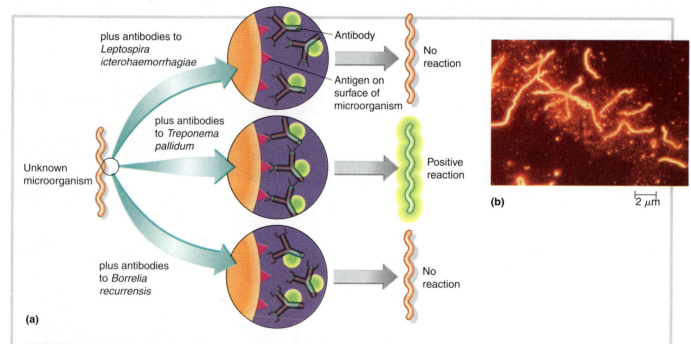

(a)

(b) 2 μm

FIGURE 3-19 Using immunofluorescence to identify a microbe. (*a*) The suspected organism is mixed with antibodies prepared against specific microbes. In this test, fluorescence identifies the organisms as *Treponema pallidum*, the bacterium that causes syphilis. (In actual testing, only the antibody against *T. pallidum* would be used.) (*b*) These bacterial cells (*Treponema pallidum*) glow when exposed to UV light because they have been coated with fluorescent-tagged antibody.

Key Terms

colonies (p. 46)
compound microscopes (p. 47)
magnification (p. 47)
resolving power (p. 47)
micrometer (p. 48)
nanometer (p. 49)
angstrom (p. 49)
brightfield microscope (p. 50)
objective lens (p. 50)
ocular lens (p. 50)

darkfield microscope (p. 52)
phase-contrast microscope (p. 52)
fluorescence microscope (p. 52)
fluorescent antibody technique (p. 53)
transmission electron microscope (p. 53)
scanning electron microscope (p. 54)
artifacts (p. 54)
simple stains (p. 57)
differential stains (p. 57)
Gram stain (p. 57)

etiologic agent (p. 57)
acid-fast stain (p. 58)
negative staining (p. 58)
pure culture (p. 60)
aseptic techniques (p. 60)
agar (p. 60)
streak plates (p. 60)
pour-plate technique (p. 60)
spread plates (p. 61)

Key Facts and Concepts

- **Microscopic observations of microorganisms.** The microscope is a fundamental tool for studying microorganisms. The useful magnifying capacity of a microscope is limited by its resolving power.
- **Modern microscopes.** Modern light microscopes are compound microscopes. The enlarged image of the specimen formed by the objective lens is further magnified by the ocular lens. The maximum resolution of the brightfield microscope using an oil immersion lens is 0.2 μm. Stain reactions can be interpreted with a standard brightfield microscope, where the color of the specimen is discernible.
- **Enhancing contrast.** Some variations on the brightfield light microscope are designed to enhance contrast rather than to increase useful magnification. With darkfield and phase-contrast microscopes, it is not necessary to stain the cells, so details of living microorganisms can be observed. The fluorescence microscope provides an especially rapid means for identifying microorganisms either by detecting natural fluorescence or by allowing the observation of specific immunologic reactions between a microbe and a fluorescent-labeled antibody.
- **Electron microscopes.** Electron microscopes use an electron beam focused by magnetic lenses, producing useful magnifications 1000 times greater than those achievable with light microscopes. In transmission electron microscopes, electrons pass through a thin specimen. In scanning electron microscopes, three-dimensional images of the specimen's surface are produced by electrons emitted from the specimen surface.
- **Microbiological stains.** Laboratory methods for detecting and characterizing microorganisms usually begin with microscopic examination of a fixed and stained specimen. Simple stains use a single dye and can help to identify the shape and size of an organism. Differential stains use more than one dye to distinguish between structures in a cell or different types of cells.
- **Some special stains.** The Gram stain procedure divides bacteria into gram-positive and gram-negative categories. Specialized stains, including the acid-fast, endospore, flagella, and capsule stains, are also useful in identification.
- **Pure cultures.** Prokaryotes lack the morphological complexity needed for microscopic identification. Further characterization usually requires isolating the microbe in pure culture. This can be accomplished by the streak-plate, pour-plate, or spread-plate techniques.
- **Other identification techniques.** Additional microbial properties can be determined by biochemical, immunologic, and genetic tests.

Review and Problem-Solving Questions

1. What advantages do compound microscopes have over simple microscopes? Why are light microscopes limited to two-lens systems?
2. Why is it impractical to increase magnification beyond the resolving power of a microscope?
3. What are the specific advantages and disadvantages of each of the following types of microscopes? (a) Brightfield; (b) phase-contrast; (c) fluorescence; (d) scanning electron; (e) transmission electron.
4. For what purpose would you use each of the following? (a) Simple stain; (b) negative stain; (c) acid-fast stain; (d) Gram stain.
5. Express the following measurements in micrometers: (a) 200 mm; (b) 17 m; (c) 75 nm; (d) 2 cm; (e) 155 Å.
6. Describe four laboratory procedures for identifying a bacterium. Under what circumstances is it important to identify bacteria?
7. What is a pure culture? Describe two situations in which a microbiologist must use a pure culture. How might you verify that you are working with a pure culture?
8. How could you show that endospores are present in a culture without using an endospore stain?
9. Describe the quadrant streak, spread-plate, and pour-plate techniques. What is the advantage of the quadrant streak method over the pour-plate method for isolating microbes in pure culture?
10. Name three common sources of microbial contaminants in the laboratory. How does the use of aseptic techniques prevent these microbes from contaminating pure cultures?
11. Why is agar medium held at 45°C before preparing a pour plate? Why isn't it poured directly when it is melted?

HEADLINES

COMPUTERS HELP MICROSCOPES "SEE" BETTER

Scientists are using computer-generated images of proteins and nucleic acids to help them "see" details of these structures that are not discerned even with microscopes of the highest resolution. Computer graphics are generated using information on chemical composition derived from scanning tunneling microscopy, X-ray crystallography, and other sophisticated techniques.

Once the three-dimensional molecular structure is completed, the computer can also be used to analyze how that molecule reacts to other chemicals. Such studies, for example, are being used to investigate enzyme activities and interactions of antibodies with the molecules to which they bind. They are also being used in computer-aided drug design where investigators simulate the interaction of drugs with specific sites on a target pathogen.

- How might the following computer-generated three-dimensional images be useful in designing molecules for disease prevention or treatment?
  - (a) Image of a surface molecule on *Neisseria gonorrhoeae* (the bacterium that causes gonorrhea) that the bacterium uses to attach to cells in the genital tract
  - (b) Image of the structure of an enzyme produced by the human immunodeficiency viruses (HIV) that is necessary for viral replication
  - (c) Image of the binding site on antibodies produced by some diabetic individuals that attack and destroy pancreatic cells

SOURCE: A. J. Olson and D. S. Goodsell, Visualizing biological molecules, *Scientific American,* 267: 76–81 (1992).

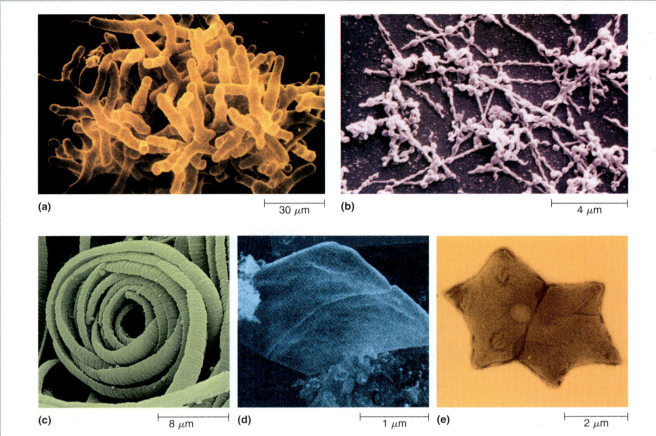

(a) 30 μm **(b)** 4 μm **(c)** 8 μm **(d)** 1 μm **(e)** 2 μm

FIGURE 4-4 Less common bacterial shapes. (*a*) Early growth of a filamentous bacterium, *Actinomyces naeslundii.* (*b*) *Mycoplasma* bacteria do not maintain a defined shape. (*c*) Complex arrangement of *Simonsiella muelleri.* Each "petal" (called a trichome) is an individual bacterial cell. (*d*) *Arcula,* the "square" bacteria, grow as thin sheets of cells. (*e*) *Stella,* unusual star-shaped bacteria.

organic material. Filamentous soil bacteria include *Streptomyces* species, producers of several important antibiotics.

A few bacteria lack rigid cell walls, and their flexible plasma membrane allows them to change shape. These bacteria, called *mycoplasmas,* are pleomorphic. Star-shaped *Stella* represent another uncommon bacterial morphology.

One particularly unusual microbe is the square bacterium *Arcula,* first described in 1981 after its discovery in a natural salt pond along the shore of the Red Sea. These bacteria are flat boxes with perfectly straight edges and sharp 90° angles at each corner (Fig. 4-4*d*). Although they are several micrometers in length and width, their thickness is a uniform 0.25 μm, just at the limits of resolution for the brightfield microscope. Each bacterium is a thin flexible sheet that can be bent and twisted by external forces, much like a thin sheet of flexible plastic. *Arcula* surfaces are unusually smooth. Smaller cells are usually perfectly square with dimensions of 2 × 2 μm. Larger cells are rectangles about twice

as long as they are wide (4 × 2 μm). Apparently these cells may remain attached to each other after division, producing sheets of squares. Their shape provides the basis for naming these microbes *Arcula,* from the Latin word *arca* for box.

■ Cell Arrangement

Arrangement of cells is a stable, genetically determined trait of a bacterial species, so it can be used to describe and identify a particular kind of bacterium. The grouping of the cells into a particular arrangement is determined by the plane or planes of division characteristic of the organism. For example, *Neisseria gonorrhoeae,* the causative agent of gonorrhea, divides along a single plane, producing a duplex of cells that adhere to each other. These cells then separate, divide, and form duplexes, generating pairs of cells called *diplococci* (Fig. 4-5*a*). Other cocci divide differently and form long chains called *streptococci* or aggregates determined by the incomplete separation of cells during reproduction (Fig. 4-5*b–e*).

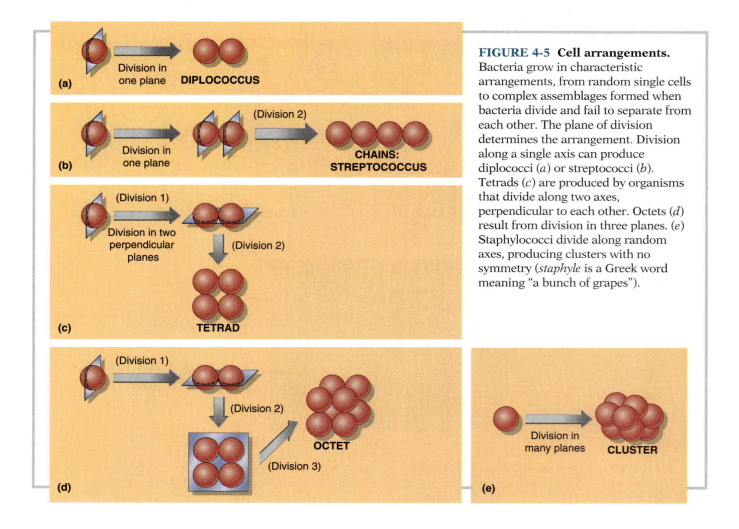

FIGURE 4-5 Cell arrangements.
Bacteria grow in characteristic arrangements, from random single cells to complex assemblages formed when bacteria divide and fail to separate from each other. The plane of division determines the arrangement. Division along a single axis can produce diplococci (*a*) or streptococci (*b*). Tetrads (*c*) are produced by organisms that divide along two axes, perpendicular to each other. Octets (*d*) result from division in three planes. (*e*) Staphylococci divide along random axes, producing clusters with no symmetry (*staphyle* is a Greek word meaning "a bunch of grapes").

Cell arrangement among the bacilli is much simpler than for cocci. Bacilli always divide through a single plane that is perpendicular to their longitudinal axis. They usually grow as single cells, although pairs (*diplobacilli*) or chains (*streptobacilli*) are occasionally observed. Some organisms—for example, *Corynebacterium diphtheriae*—bend at the point of division following reproduction, producing angular patterns that look like Chinese letters or resulting in a palisade arrangement that resembles a picket fence.

Spiral bacteria usually occur singly, although a few species form serpentine filaments with many cells attached end to end.

Cell arrangement can give some clues to the identity of an organism, but under certain conditions bacteria fail to grow in their characteristic patterns. For example, variations in culture medium, growth temperature, and specimen preparation often influence the arrangement observed. Thus, microscopic observations of cellular arrangement may be of limited diagnostic value.

■ Cell Structures

All prokaryotes have a nucleoid, ribosomes, and a plasma membrane. Most bacteria also have a cell wall, and some are further enveloped by a capsule or slime layer. Some types of bacteria also have cytoplasmic inclusions and various appendages (Fig. 4-6).

Unlike the larger features of eukaryotic cells, bacterial structures are difficult or impossible to distinguish using light microscopy. The finer details of subcellular structure are revealed best by electron microscopy. It is immediately evident that prokaryotes are structurally simpler than eukaryotes. How do bacteria accomplish the same life-sustaining processes that are carried out by the specialized organelles of eukaryotic cells?

Part of the answer has to do with their small size, which eliminates the need for extensive internal membrane networks. The other strategy involves the use of a single structure for multiple functions, functions that are performed in eukaryotes by several different membrane-bound organelles.

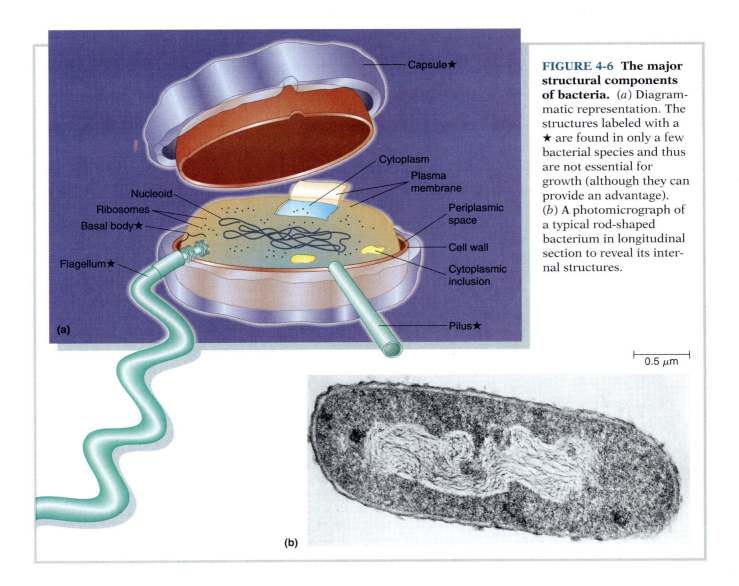

FIGURE 4-6 The major structural components of bacteria. (*a*) Diagrammatic representation. The structures labeled with a ★ are found in only a few bacterial species and thus are not essential for growth (although they can provide an advantage). (*b*) A photomicrograph of a typical rod-shaped bacterium in longitudinal section to reveal its internal structures.

Labels in figure (a): Capsule★, Cytoplasm, Plasma membrane, Periplasmic space, Cell wall, Cytoplasmic inclusion, Pilus★, Nucleoid, Ribosomes, Basal body★, Flagellum★

0.5 μm

(b)

CONCEPT CHECK

- What factors limit the size of bacterial cells?
- What bacterial shapes might you see when examining a specimen with a brightfield microscope?
- How does the plane of division determine the arrangement of bacterial cells?

Plasma Membrane

In prokaryotes, as in other cells, the cytoplasmic contents are separated from the environment by the **plasma membrane.** Major disruptions in this barrier result in the spilling of the cytoplasm from the cell and the death of the organism. The prokaryotic plasma membrane is the site of many functions that are accomplished by specialized internal organelles in eukaryotes. These include

1. Selective transport of molecules into and out of the cell
2. Secretion of extracellular enzymes
3. Respiration and photosynthesis
4. Regulation of reproduction
5. Cell wall synthesis
6. Monitoring of the environment

■ Structure of the Plasma Membrane

In eubacteria the plasma membrane is a phospholipid-protein bilayer similar to that in eukaryotic cells ▶ (see Plasma Membrane, Chap. 2, p. 31). Some of the proteins are anchored in the hydrophobic portion of the lipid bilayer, either inserted through a portion of the membrane (some facing the cell's interior, others on the exterior of the cell) or spanning the entire membrane. Different types of membrane proteins constitute one of the major differences between eukaryotic and eubacterial membranes, the other being the presence of sterols in the plasma membranes of eukaryotic cells but not in most eubacteria.

In archaeobacteria, lipids and protein also provide the framework for the plasma membrane, but their membranes are chemically unlike those of any other type of cell (Fig. 4-7). In archaeobacteria, the membrane lipids that make up the hydrophobic side chains extending into the middle of "typical" membrane bilayers are branched lipids, not fatty acids. In addition, ether bonds replace ester linkages and phosphate is often replaced by other hydrophilic compounds. Some archaeobacteria create their membranes from a single large molecule rather than as a true bilayer. Nonetheless, archaeobacterial membranes function like other plasma membranes. Their chemically unique properties, however, enable archaeobacterial membranes to withstand the high temperatures, salinity, or acidic conditions that characterize the extreme environments in which many archaeobacteria live.

■ Membrane Transport

Organisms must regulate the transport of nutrients into and waste products out of the cell. The plasma membrane acts as a *selectively permeable* barrier, determining which molecules move across the membrane. Certain types of molecules can freely penetrate membranes, whereas others are transported by specific protein carriers—the presence or absence of the corresponding carrier proteins in the plasma membrane determines whether a particular molecule can enter or leave a cell. In prokaryotes, molecules move across the plasma membrane by simple diffusion, facilitated diffusion, and active transport.

● **SIMPLE DIFFUSION.** Molecules that migrate freely through the lipid bilayer without the aid of protein carriers enter or exit the cell by **simple diffusion.** In this process, chemicals are distributed as a result of random molecular motion. The molecules move in response to a *concentration gradient*—they move from areas where they are in higher concentration to areas where they are in lower concentration (Fig. 4-8). (This is why the smell of baking bread quickly fills a room.) The cell spends no energy on transport by diffusion—substances automatically diffuse down a concentration gradient. Diffusion continues until *equilibrium* is reached. When the concentration of molecules is the same on both sides of the membrane, the gradient no longer exists.

The "holes" through which these substances diffuse across the membrane are created by a number of phenomena. Water, oxygen, carbon dioxide, and other small neutral compounds diffuse across temporary gaps created as the components of the fluid membrane are continuously rearranged. Small lipids need no holes; they simple dissolve in the lipid bilayer and move through.

● **OSMOSIS.** When the diffusing molecule is water, the process is termed **osmosis.** Since water readily passes through the membrane, its uptake or loss depends on its concentration in the environment relative to that in the cytoplasm and on the available space inside the cell. Because substances dissolved in water take up physical space, the higher the concentration of solutes (dissolved substances) in a so-

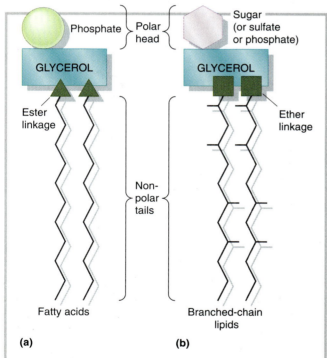

FIGURE 4-7 Prokaryotic plasma membranes. A comparison of the lipid components of eubacterial membranes (*a*) with the chemically unique archaeobacterial membranes (*b*). In archaeobacteria, the molecule comparable to the phospholipid of eubacteria and eukaryotic cells has the following characeristics: complex branched lipids replace straight-chain fatty acids, ether linkages replace ester linkages, and the charged phosphate group may be replaced by sulfate or a simple sugar.

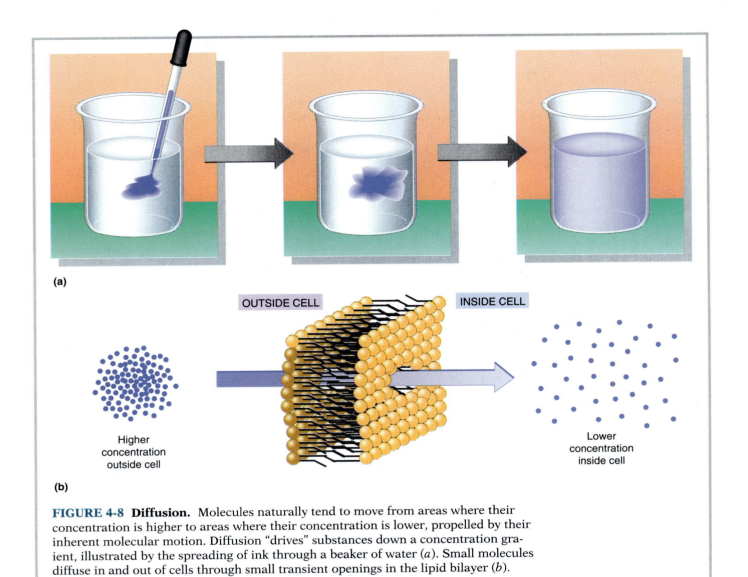

(a)

OUTSIDE CELL   INSIDE CELL

Higher
concentration
outside cell

Lower
concentration
inside cell

(b)

FIGURE 4-8 **Diffusion.** Molecules naturally tend to move from areas where their concentration is higher to areas where their concentration is lower, propelled by their inherent molecular motion. Diffusion "drives" substances down a concentration gradient, illustrated by the spreading of ink through a beaker of water (*a*). Small molecules diffuse in and out of cells through small transient openings in the lipid bilayer (*b*).

lution, the lower the water concentration. It is therefore traditional to say that *osmosis forces water to move in a direction from lower solute concentrations (greater water concentration) to higher solute concentrations (lower water concentration)*.

As a result of osmosis, a bacterial cell placed in pure water will accumulate water and tend to swell. Its rigid outer barrier (the cell wall), however, restricts that expansion and therefore limits the amount of water the cell can accumulate.

● **FACILITATED DIFFUSION.** Simple sugars and many small molecules that are not lipid-soluble pass through the membrane by the process of **facilitated diffusion** (Fig. 4-9). Each of these molecules binds at the membrane surface to a *permease*, a specific

protein receptor that escorts the molecule across the membrane. Permeases increase the rate of diffusion of molecules across the membrane. The protein binds with the material to be transported and changes shape to form a channel that opens on the opposite side of the membrane. The formation of a protein channel shields the molecule from the hydrophobic interior of the membrane, which would otherwise repel it much like oil repels water. Like simple diffusion, facilitated diffusion requires no energy expenditure by the cell, and net movement continues until equilibrium is reached.

Diffusion fails to provide bacteria with a solution to a crucial problem: how to build a stockpile of molecules on one side of the plasma membrane against a concentration gradient. In many natural

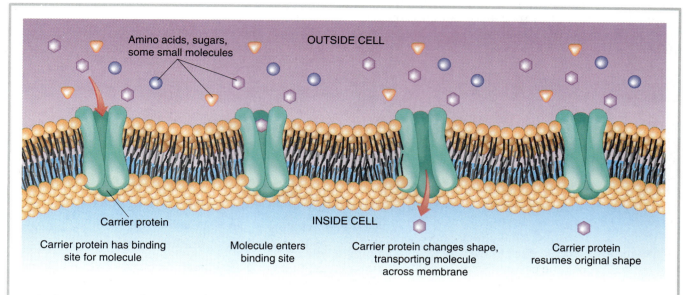

Amino acids, sugars, some small molecules

OUTSIDE CELL

Carrier protein

INSIDE CELL

Carrier protein has binding site for molecule

Molecule enters binding site

Carrier protein changes shape, transporting molecule across membrane

Carrier protein resumes original shape

FIGURE 4-9 **Facilitated diffusion.** Molecules migrate down a concentration gradient with the assistance of specific membrane carrier proteins. One possible mechanism for transport is that molecules bind to permease proteins, which change shape to form channels that open on the opposite side of the membrane.

environments, such as the ocean, nutrients are present in low concentrations. Bacteria in these environments must accumulate resources within their cytoplasm if they are to survive and reproduce. An inability to concentrate essential chemicals could restrict cell metabolism and growth. Conversely, a cell's survival depends on its ability to transport toxic and waste materials out of the cell, even against a concentration gradient. Trace amounts of some wastes can be fatal, so equilibrium might mean death for the cell.

● **ACTIVE TRANSPORT.** Unlike diffusion, moving molecules *against* a concentration gradient requires cells to spend energy. **Active transport** is the process used by prokaryotes and eukaryotes to transport materials against a concentration gradient. As in facilitated diffusion, the molecule to be moved across the membrane first binds to a specific carrier protein in the plasma membrane (Fig. 4-10). A subsequent change in the shape of the carrier protein moves the molecule to the opposite side of the membrane and releases it. In active transport, however, this configurational change requires energy, often provided by the release of chemical energy stored in ATP ▶ (this high-energy compound is discussed on p. 131). Each carrier protein provides a one-way transport mechanism. Some active transport proteins propel molecules into the cell, where they accumulate; others discharge materials into the surroundings.

● **GROUP TRANSLOCATION.** Many prokaryotes spend energy to transport nutrients by **group translocation,** an active transport process in which carrier proteins chemically modify molecules as they cross the membrane. In one such system, glucose reacts with phosphate as it enters the cytoplasm. The glucose phosphate compound, unlike plain glucose, cannot cross the membrane and is thus trapped inside the cell. Group translocation of glucose provides another advantage for bacteria. The glucose–phosphate bond contains considerable energy. This energy is used to prime the metabolic reactions that release the rest of the usable energy in the glucose molecule.

Simple diffusion, facilitated diffusion, and active transport are used by both prokaryotes and eukaryotes, whereas group translocation occurs only in prokaryotes. In many eukaryotes, membrane transport is preceded by endocytosis ▶ (see Vesicles, Chap. 2, p. 31). The engulfed materials are then digested, and the small molecules are moved through the vesicle membrane into the cytoplasm by one of the previously discussed mechanisms.

■ Secretion of Cell Products

Cells must often export proteins and other products into their environment. The cell wall, for example, is synthesized outside the plasma membrane. Molecules needed for constructing and repairing the cell wall are synthesized in the cytoplasm and transported across the plasma membrane. Specialized membrane proteins transport these subunits outside the plasma membrane and incorporate them into the growing cell wall.

Materials are also secreted from the cell to enable it to acquire nutrients. Large food molecules in the environment cannot be transported through the plasma membrane; they must be broken down into smaller molecules. Such *extracellular* (outside-the-cell) *digestion* is mediated by enzymes released from the bacterium through its plasma membrane. Other extracellular enzymes are used to destroy harmful chemicals such as antibiotics. For example, bacteria that produce penicillinases (penicillin-degrading enzymes) render the drug penicillin ineffective and are protected from its lethal effects. Another group of dangerous extracellular enzymes are the "spreading factors" that some pathogens secrete into the tissues of an infected host. These enzymes disassemble solid tissues that would otherwise wall off the pathogen.

Bacteria can also secrete proteins that kill other types of bacteria. These toxic compounds reduce the size of neighboring bacterial populations, thereby eliminating competition for nutrients. Many microbial pathogens release enzymes and toxins that attack their eukaryotic hosts. These proteins are necessary for successful invasion and are responsible for the specific damage to the host caused by the pathogen ▶ (see Microbial Virulence, Chap. 17, p. 414).

Proteins destined to be secreted by the cell are usually synthesized on *fixed ribosomes* that are attached to the plasma membrane. These proteins are passed through the membrane as they are assembled. Another function of the plasma membrane, therefore, is to provide a mechanism for conveying proteins to their extracellular site of action.

■ Respiration and Photosynthesis

In eukaryotes, mitochondria and chloroplasts carry out two of the most important processes on earth: respiration and photosynthesis. Bacteria perform both of these essential functions by using proteins embedded in the plasma membrane. Many of these proteins are similar or identical to those in mitochondria or chloroplasts.

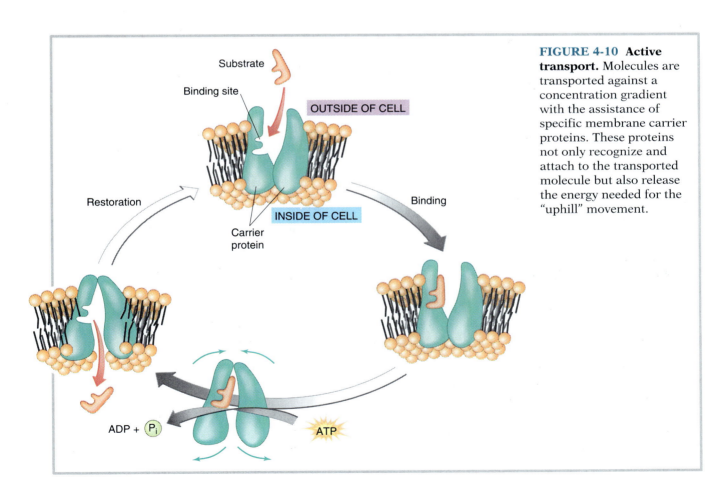

FIGURE 4-10 Active transport. Molecules are transported against a concentration gradient with the assistance of specific membrane carrier proteins. These proteins not only recognize and attach to the transported molecule but also release the energy needed for the "uphill" movement.

Substrate

Binding site

OUTSIDE OF CELL

INSIDE OF CELL

Carrier protein

Restoration

Binding

ADP + P$_i$

ATP

The greater its membrane area, the more rapidly the cell can carry out respiration or photosynthesis. In some bacteria the plasma membrane invaginates into the cytoplasm, significantly increasing the membrane surface area. Photosynthetic prokaryotes contain vesicles called *chromatophores* as well as extensive membrane infoldings called *thylakoids*. These structures harbor enzymes and pigments that perform functions similar to those of the eukaryotic chloroplast (Fig. 4-11).

■ Reproduction

Production of bacterial progeny is assisted by the plasma membrane. Specific proteins in the membrane attach to the replicating DNA and separate the duplicated chromosomes from each other. In addition, the cytoplasm of the two daughter cells is physically separated by the formation of a septum (cross wall), a structure assembled by the plasma membrane.

■ Monitoring the Environment

Some membrane proteins enable bacteria to respond to chemicals in their environment. Instead of transporting the chemicals across the membrane, these membrane proteins bind to specific molecules and send a signal to the cell's interior, indicating that the cell surface has encountered that specific chemical. The cell then responds to the surface message. For example, motile bacteria that detect glucose in the environment can direct their movement toward

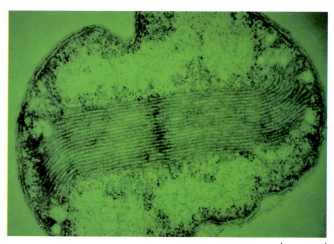

FIGURE 4-11 Internal membranes in a bacterium. *Nitrosococcus oceanus*, a photosynthetic prokaryote, contains abundant internal extensions of the plasma membrane. Enzymes and pigments used in photosynthesis are found in these structures.

the nutrient (see BEYOND THE BASICS: Bacterial Behavior and the Role of Membranes). Membrane proteins that act as sensory devices for the cell have been referred to as bacterial "cellular phones."

CONCEPT CHECK

- How does the plasma membrane accomplish many of the life-sustaining processes carried out by the specialized organelles of eukaryotic cells?
- In what ways does the archaeobacterial plasma membrane differ from the eubacterial plasma membrane?
- Compare the processes of facilitated diffusion, active transport, and group translocation.

Cell Wall

Although more than 2000 species of bacteria have been described, only about 100 are primarily human pathogens. The factors contributing to many activities, including the capacity of bacteria to initiate or promote disease, are often directly related to the exposed surface structures and appendages of the bacterial cell. The eukaryotic host often defends against microbial invasion by neutralizing these surface bacterial components.

The bacterial **cell wall** is the structure that immediately surrounds the plasma membrane. The presence of a cell wall creates a space that lies between the cell wall and the plasma membrane, a region called the **periplasm.** The shape of the cell wall molds the bacterium into its characteristic shape.

The most important function of the cell wall is to protect the cell physically. This protection is necessary because of the susceptibility of the plasma membrane to physical or osmotic *lysis* (disruption). In many aqueous environments bacteria accumulate water due to osmosis. Water influx increases cell size and would burst the cell (like an overinflated balloon) if the cell wall did not prevent the cell's expansion. In some bacteria the cell wall is strong enough to withstand 25 atmospheres (375 pounds per square inch) of internal osmotic pressure.

Beginning at the plasma membrane, the cell harbored inside the cell wall is called the **protoplast.** Removing the cell wall, usually by enzymatic digestion, releases the protoplast, now an unprotected cell whose outer barrier is the plasma membrane. These poorly protected cells will lyse in environ-

Bacterial Behavior and the Role of Membranes

Life for bacteria outside the laboratory is generally unpredictable. Nutrient availability and environmental conditions confronting the organism are rarely stable. Like other organisms, bacteria must respond quickly and appropriately to changes in environmental conditions. To detect such changes, cells constantly monitor their environment. Photosynthetic bacteria move or orient themselves toward sunlight; motile glucose-requiring organisms swim toward the sugar's source; cells adjust to changes in their osmotic environment to prevent fatal loss or explosive absorption of water. In other words, bacteria behave in response to their environment, behavior that allows them to move toward hospitable habitats and avoid dangerous ones. All this implies the presence of some type of sensory system to monitor their environment.

One type of sensory system for bacteria depends on proteins in the plasma membrane that detect changes in the environment. These proteins extend through the membrane, so one end is exposed to the environment outside the cell and the other is exposed to the cytoplasm inside the cell. The part of the molecule that protrudes from the outer cell surface performs functions different from those of the part inside the cell. The portion of the protein that extends into the periplasm (outside the membrane) is the *chemoreceptor*, which recognizes and binds with a particular molecular "signal" (usually a chemical in the environment). Binding of the signal molecule to the protein alters the intracellular portion of the membrane protein. When altered, this end of the protein activates another molecule, which then diffuses through the cytoplasm to the site in the cell where a response will be initiated.

One example of such sensory responsiveness is the ability of a bacterium to detect a chemical it "wants" and swim in the direction of that chemical. For example, if you place a capillary tube containing the nutrient glucose into a suspension of *E. coli,* the bacteria will swim toward the glucose diffusing from the tip of the tube and eventually move up the gradient into the tube.

Although the bacteria might vitally *need* the glucose, they don't really *want* it—this system works independently of thought processes, desire, and willful decisions, of which no individual cells are capable. The bacteria move toward the nutrient in response to molecular changes initiated by glucose binding to a flagellated cell. The glucose binds to the external end of a receptor protein, which causes a change in the cytoplasmic end of the protein. The altered inner end adds a phosphate molecule to the cytoplasmic molecules that regulate the direction in which the flagellum turns. *Increasing* concentrations of glucose in the environment bind more receptor proteins in the cell's membrane, which in turn phosphorylate more of the "messenger" molecules. These messengers diffuse to the flagellum and stimulate the basal body to continue turning in a counterclockwise direction. This keeps the bacterium traveling in the same direction (that of increasing nutrient concentration).

When there is less glucose in the direction in which a bacterium is moving, more receptor proteins remain empty, so less phosphorylated messenger is produced, and the flagellum spins in the *other* direction (clockwise, the direction that causes random tumbling). The resulting changes in direction of bacterial swimming continue until the cell finally goes (by chance) in the right direction, glucose concentrations begin to increase, and the flagellum continues to propel the cell in that direction.

---

ments that encourage an uncontrolled influx of water (Fig. 4-12). Protoplasts grown in the laboratory must be suspended in solutions containing high concentrations of salts or sugars. The use of such solutions prevents protoplast lysis by balancing the exchange of water across the plasma membrane. Protoplasts are also susceptible to lysis by physical trauma. For example, shaking a broth culture of protoplasts destroys the fragile cells.

All protoplasts assume a more or less spherical shape regardless of the shape of the bacteria from which they were derived. While still inside a cell wall, however, the protoplast's shape is rigidly fixed into a configuration that characterizes that paticular species, as shown in Fig. 4-13 (the shape of the cell wall is a genetically inherited trait). The empty cell wall fragment from which the protoplast escaped retains the shape characteristic of the original cell.

■ Structure of the Eubacterial Cell Wall

The cell wall of a eubacterium is a complex structure composed of several substances. Its impressive strength is due primarily to **peptidoglycan** (also known as murein or mucopeptide), a substance found only in these prokaryotes. Peptidoglycan is an enormous molecule composed of amino acids and sugars (*peptide* = chain of amino acids; *glycan* = sugar). The sugars *N*-acetylglucosamine (NAG) and *N*-acetylmuramic acid (NAM) alternate to form long parallel chains (Fig. 4-14*a*). NAM is a chemical that is unique to eubacteria. Attached to the NAM molecules are short side chains that are four or five

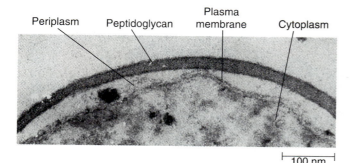

Periplasm   Peptidoglycan   Plasma membrane   Cytoplasm

100 nm

FIGURE 4-15 Gram-positive cell wall. The peptidoglycan lies outside the plasma membrane, with the periplasm in between.

wall impart properties to the outer surface of gram-positive cells that influence their ability to cause human disease. For example, the cell-wall-associated M protein of *Streptococcus pyogenes* effectively prevents its engulfment by white blood cells that protect the host from microbial invaders. In this way, the cell wall aids the ability of this bacterium to sur-vive host defenses. (*Streptococcus pyogenes* causes more types of disease than any other etiologic agent known. These diseases include strep throat, rheumatic fever, scarlet fever, impetigo, and serious kidney disorders.)

Teichoic acids and cell-wall-associated proteins are the major surface antigens of the gram-positive cell wall. **Antigens** are molecules that stimulate the host immune system, for example, to make antibodies. These antibodies specifically react with antigens on the bacterial surface, often assisting the host in eliminating the invading microbe. The antibodies recognize and react only with the type of antigen that stimulated their production. Because different types of cells have different surface antigens, antibodies against one type of bacterium do not protect against invasion by another species. Reactions of unknown bacteria with antibody preparations made against known bacteria are used in rapid identification.

In spite of these auxiliary compounds, the gram-positive cell wall is readily penetrated by penicillin

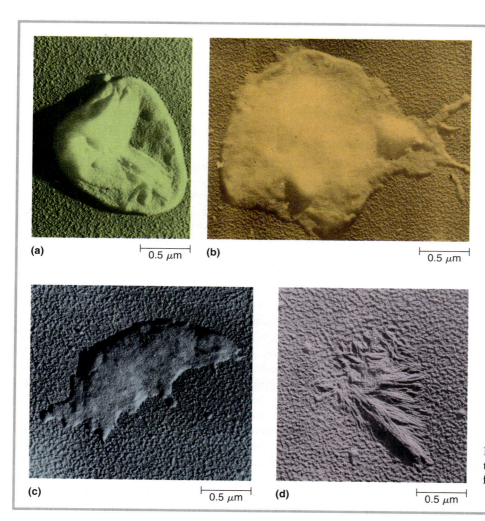

(a)   0.5 μm   (b)   0.5 μm

(c)   0.5 μm   (d)   0.5 μm

FIGURE 4-16 Lysozyme digestion of the cell wall isolated from a gram-positive cell.

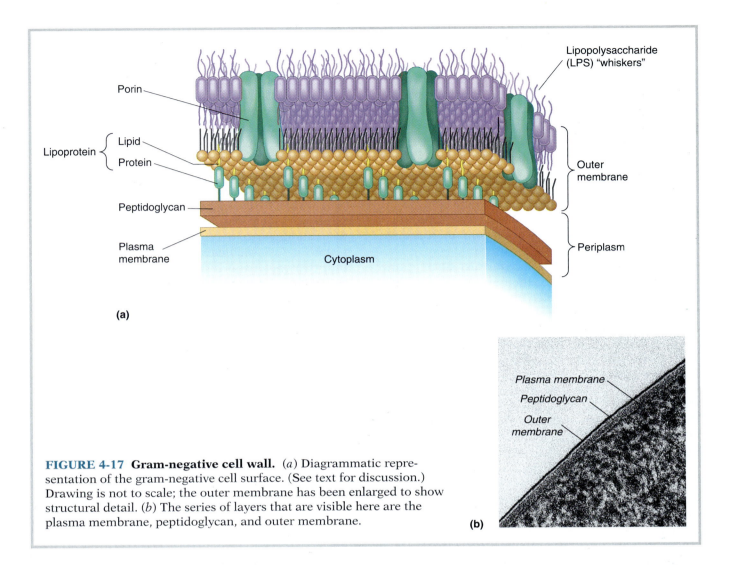

FIGURE 4-17 Gram-negative cell wall. (*a*) Diagrammatic representation of the gram-negative cell surface. (See text for discussion.) Drawing is not to scale; the outer membrane has been enlarged to show structural detail. (*b*) The series of layers that are visible here are the plasma membrane, peptidoglycan, and outer membrane.

and cephalosporin antibiotics (both of which prevent the synthesis of peptidoglycan) and *lysozyme* (an enzyme that digests the chemical bond between NAG and NAM). Lysozyme destroys the peptidoglycan, leaving cells with no protective cell walls (Fig. 4-16). Lysozyme is found in tears, saliva, and other secretions as a natural defense against bacterial diseases.

MYTH: Lysozyme in tears kills most microbes in the eye.

FACT: *Staphylococcus aureus*, an agent of pinkeye and other eye infections, is one of hundreds of lysozyme-resistant microorganisms that can live in the human eye. Fortunately, eyes are defended by additional resistance mechanisms such as mechanical flushing by tears (discussed in Chap. 17).

● **GRAM-NEGATIVE BACTERIA.** The cell wall of gram-negative bacteria is structurally more fragile than its gram-positive counterpart. Its peptidoglycan is much thinner, accounting for only 1 to 2 percent of the dry weight of the cell (compared to 20 percent in gram-positive bacteria). The exterior surface of the peptidoglycan layer is covered by the outermost portion of the cell wall, a series of layers called the **outer membrane** (Fig. 4-17). Together these outer layers provide a protective coating around the cell that resists penetration by some potentially toxic chemicals. Movement of materials out of the cell can also be impeded. Proteins released from the cell freely pass through the peptidoglycan layer, but their further movement is blocked by the outer membrane. These proteins become trapped in the periplasm. Gram-negative bacteria rely on periplasmic proteins for several important functions. Some serve as extracellular degradative enzymes, whereas others are receptors that help the bacteria recognize chemicals in their environment.

The outer membrane layer that is closest to the peptidoglycan is a lipid-protein composite, a **lipoprotein.** One end of lipoprotein is directly attached to peptidoglycan. The other end extends into a phospholipid-protein bilayer similar to the plasma membrane. The outer surface of the lipid bilayer, however, is composed of molecules of **lipopolysaccharide (LPS).** LPS consists of a molecule called *lipid A* that is chemically linked to a *polysaccharide* (a large molecule composed of repeating sugar subunits). The polysaccharides that extend outward from the lipid A are the outermost molecules of the cell wall and are major surface antigens of the gram-negative bacterial cell. Antibodies directed against polysaccharides of one kind of gram-negative bacterium usually do not protect against other gram-negative species.

In general, the structure of the outer membrane prevents passage of most hydrophobic molecules and large hydrophilic ones. Most small hydrophilic chemicals diffuse through channels in the lipid bilayer formed by proteins called **porins.** Porin channels are relatively nonspecific. Hydrophilic molecules smaller than the diameter of the channel will pass through. The size of the channel depends on the particular protein from which it is made. In *Escherichia coli*, molecules of molecular mass of up to 600 daltons[1] can penetrate. In certain other bacteria, the upper limit approaches 6000 daltons. Larger hydrophilic molecules are transported by facilitated diffusion in association with specific carrier proteins dispersed throughout the lipid bilayer. In the absence of a carrier, large hydrophilic molecules are excluded from the cell. Most hydrophobic compounds cannot penetrate beyond the outer surface layer of polysaccharides and are kept out of the cell. This makes gram-negative cells resistant to the effects of certain drugs, detergents, and dyes that kill gram-positive cells. Incorporating hydrophobic chemicals such as bile salts or eosin methylene blue into culture media favors the isolation and growth of gram-negative bacteria. The cell walls of gram-negative bacteria provide protection against these toxic compounds, whereas the growth of gram-positive cells is inhibited.

The properties of the outer membrane have medical importance. As a selectively permeable barrier, it protects gram-negative bacteria from the effects of many antipeptidoglycan chemicals such as penicillin and lysozyme. These antibacterial agents fail to penetrate the outer membrane and consequently are unable to either attack the peptidoglycan layer

(lysozyme) or interfere with its synthesis (penicillin). Determining the Gram reaction of a pathogen is therefore important for selecting the best antibiotic to treat serious infectious diseases when there is no time to wait for laboratory cultures and identification. The use of penicillin against gram-negative organisms will usually be of limited or no value. Important exceptions are *Neisseria gonorrhoeae* and *Neisseria meningitidis,* the gram-negative cocci that cause gonorrhea and meningitis. It is believed that unique protein components in the outer membrane of *Neisseria* allow penicillin to penetrate to its site of action and inhibit the synthesis of peptidoglycan.

LPS is also medically significant. When a gram-negative bacterium invades the human body, some of its progeny are engulfed by white blood cells in an attempt to prevent further spread of infection. Inside the phagocytic cell, however, the lipid A portion of LPS causes its host to produce compounds that are released into the bloodstream and may elicit toxic reactions. These reactions include fever, diarrhea, and potentially fatal shock. Because it is an integral part of the bacterial cell wall, lipid A is commonly referred to as **endotoxin** (see THE MICROBES: Dangerous Ghosts of Dead Bacteria). The presence of endotoxin enables gram-negative pathogens to produce symptoms of disease that are rarely provoked by infection with gram-positive bacteria. Every year an estimated 400,000 people suffer from symptoms of endotoxin release. Many of these individuals contracted systemic infections with gram-negative bacteria during surgery or as a result of trauma. In 1990, 70,000 people died from endotoxic shock in the United States alone.

■ Cell Walls and Gram Staining

The difference in staining behavior between gram-positive and gram-negative cells is due to differences in their cell walls. Although the exact mechanism of differentiation is still unclear, it may be due to dehydration of the thick peptidoglycan layer of gram-positive cells by alcohol, decreasing the porosity of the wall and trapping the crystal violet–iodine complex inside the cell so the cells do not decolorize when exposed to alcohol. The thinner peptidoglycan layer in gram-negative cell walls has larger pores that fail to retain the dye complex. The alcohol dissolves the outer layer, penetrates the cell, and washes away the crystal violet.

Any damage to the gram-positive wall that decreases its ability to retain the crystal violet makes a gram-positive cell *appear* to be gram-negative. For example, removal of the cell wall of a gram-positive

[1] A dalton is a unit of atomic mass. One dalton equals the atomic mass of a hydrogen atom.

bacterium produces a protoplast that stains gram-negative. Because cell walls deteriorate as cells age, "old" cultures of gram-positive bacteria will stain as gram-negative bacteria. Gram-negative cells, however, always remain gram-negative, regardless of age. These variations in gram-positive cells must be taken into account to interpret the results of Gram staining accurately.

The features that distinguish the cell walls of gram-positive and gram-negative bacteria are presented in Table 4-1.

● **ACID-FAST BACTERIA.** The peptidoglycan layer of mycobacteria is covered by a thick lipid layer that is responsible for the acid-fast stain reaction of mycobacteria. These lipids, composed primarily of waxes such as mycolic acid, interfere with penetration of Gram stain reagents but bind carbol fuchsin in such a way that the dye cannot be removed by an acid-alcohol rinse. The lipid layer creates a hydrophobic surface that may impede nutrient entry into the cell, perhaps explaining the slow growth rate of many mycobacteria. (For example, *M. leprae*, the mycobacterium that causes leprosy, divides once every 11 days, compared to the 20-minute optimal division time of many other types of bacteria.)

■ Wall-Deficient Variants

Although virtually all eubacteria have cell walls, one group, the **mycoplasmas,** exist naturally without these protective structures. Sterols in the plasma membrane provide mycoplasmas with some protection from osmotic lysis. Another protective factor is that mycoplasmas live as parasites inside eukaryotic hosts, where the fluids are osmotically favorable. Because they lack a rigid cell wall, mycoplasmas do not assume a defined and clearly recognizable shape. Instead they are highly pleomorphic organisms. Since the cell wall is not needed for their survival, these organisms are naturally resistant to cell wall inhibitors such as penicillin and cephalosporins. In the laboratory, mycoplasmas must be maintained on special media that protect them from osmotic lysis. Mycoplasma are more easily disrupted by physical and mechanical trauma than are typical "walled" bacteria.

Wall-defective or wall-deficient bacteria may also develop from cells that normally possess cell walls, for example, when they are exposed to penicillin or lysozyme. If the bacterium is situated in an osmotically favorable environment, the resulting protoplast may survive. Such an environment may be found within the human body, particularly in such areas as pus-filled wounds. These wall-less bacteria are called **L forms** ("L" for the Lister Institute in London, where they were first discovered). Most L forms will resynthesize their walls once the antibiotic is removed. Some, however, permanently lose the capacity to produce a cell wall. Because of their osmotic fragility, they are difficult to isolate or identify in most laboratories. Although their role in disease is still unclear, it is possible that L forms are responsible for some infectious diseases for which no organisms can be cultured. L forms may survive antibiotic therapy because of their lack of susceptibility to antipeptidoglycan agents. For example, relapses of infection after completion of penicillin therapy may result from the resumed growth of surviving L forms.

The major characteristics of the bacterial cell wall can be summarized as follows:

1. The cell wall provides protection from osmotic lysis.
2. It gives rigidity and shape to the cell.

TABLE 4-1 COMPARISON OF GRAM-POSITIVE AND GRAM-NEGATIVE CELL WALLS		
Property of Cell	**Gram-Positive Cells**	**Gram-Negative Cells**
Peptidoglycan	Thick layer, extensively cross-linked	Thin layer with few cross-links
Auxiliary compounds	Teichoic acids; proteins	Lipoprotein, outer membrane; lipopolysaccharide (endotoxin)
Resistance to osmotic pressure	More resistant (25 atm)	Less resistant (5 atm)
Penicillin sensitivity	Usually sensitive	Usually insensitive
Response to lysozyme	Digested cell wall	Resistant cell wall

Dangerous Ghosts of Dead Bacteria

Dead gram-negative bacteria retain their ability to injure or kill people. This dangerous "ghost effect" is caused by a residual endotoxin that retains its biological activity even after exposure to the conditions that killed the bacteria. This alarming fact is often evident when fever develops shortly after sterile solutions or objects such as needle catheters or prosthetic devices are introduced into the body. Even though such solutions and objects are sterilized to kill all microbial contaminants, some of the sterilization procedures fail to remove residual cell wall debris. For example, the activity of endotoxin survives steam heating at 120°C for 1 hour. This treatment far exceeds that required to kill any type of bacteria.

Even small amounts of residual endotoxin can elicit symptoms. Typical adults will develop fever within 3 hours after receiving as few as 1 million killed bacteria. This is about 300 bacteria per milliliter of blood, a tiny number considering that bacteria may typically achieve population sizes of 1 billion cells per milliliter.

To prevent such occurrences, all commercially prepared objects and substances to be introduced into the body are examined by the manufacturer for evidence of endotoxin contamination. One assay procedure detects the pyrogenic (fever-inducing) activity of endotoxin. Laboratory rabbits are fitted with rectal thermometers and injected with the test solution. (Solid objects must be rinsed in a pyrogen-free sterile solution, which is then tested for eluted endotoxin.) Three hours later, the rabbits' thermometers are examined for evidence of fever. Although the rabbit assay test (also called the *pyrogen test*) can detect endotoxin concentrations as low as 0.1 ng/mL, a large facility is required for housing the many rabbits needed.

A more convenient and sensitive assay procedure was developed after the 1968 discovery that amoebocyte blood cells from horseshoe crabs (*Limulus polyphemus*) contain a substance that visibly gels within 1 hour when mixed with very small amounts of endotoxin. This test, called the *limulus amoebocyte lysate* (*LAL*) assay, has replaced the pyrogen assay procedure in many labs (an advance hailed by rabbits). In addition to convenience, it is faster and is sensitive enough to detect the endotoxin from as few as 300 bacteria. Some chemicals such as albumin, a protein found in serum, interfere with the limulus assay. Samples containing these substances must be assayed by another method such as the pyrogen test.

---

3. In gram-negative bacteria, it provides protection from some antibiotics and destructive chemicals.
4. It determines a cell's characteristic Gram-stain reaction.
5. It houses periplasmic enzymes and proteins in gram-negative bacteria.
6. In most cells, surface antigens are components of the cell wall.
7. The cell wall may elicit certain toxic symptoms of diseases caused by gram-negative bacteria.
8. It provides a selective target for destruction by some antibiotics.
9. It anchors flagella so they can propel bacterial cells ▶ (see p. 88).

■ Structure of Archaeobacterial Cell Walls

The cell walls of archaeobacteria differ substantially from those of eubacteria. In fact, significant variability is found among the cell walls of archaeobacteria. One common feature, however, is their lack of peptidoglycan. Some archaeobacterial cell walls are composed of *pseudopeptidoglycan*. These walls are similar to peptidoglycan, but NAM is replaced with another sugar and D-amino acids are absent from the amino acid side chain. Other archaeobacteria possess walls composed solely of polysaccharide, protein, or glycoprotein. For example, *Halobacteria*, which are found in highly saline marine environments (8 to 36 percent salt concentrations), have cell walls composed of glycoprotein. For these bacteria, osmotic imbalance tends to draw water *out* of the cell, so there is no need for the rigidity provided by a peptidoglycan-rich wall. They protect themselves from osmotic dehydration by balancing the high sodium concentration in seawater with a high potassium concentration within the cell.

CONCEPT CHECK

- What advantages are conferred on bacteria by the presence of a cell wall?
- How does the structure of the gram-positive cell wall differ from the structure of the gram-negative cell wall?
- Describe three differences between gram-positive and gram-negative bacteria that are due to differences in their cell walls (other than the Gram stain reaction).

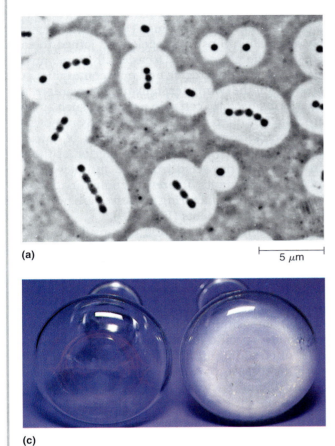

(a)

5 μm

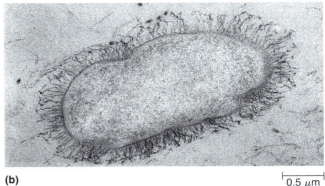

(b)

0.5 μm

FIGURE 4-18 Structures that surround the cell wall.
(*a*) *The bacterial capsule.* A thick polysaccharide capsule surrounds *Klebsiella pneumoniae*, a bacterium whose ability to infect humans is enhanced by its capsule.
(*b*) Glycocalyx cover, a mass of tangled fibrils extending from the cell surface, can be seen only with the electron microscope. A glycocalyx is present on most bacterial cells. Shown here is *Escherichia coli*. (*c*) The organism *Streptococcus mutans* was cultured in two flasks containing media that differed only in the sugar supplement. In the presence of sucrose (flask at right), a glycocalyx is formed, which cements the bacteria to the glass. Without sucrose (flask at left), the bacterium produces no glycocalyx and fails to attach to the surface of the flask.

(c)

Capsule, Slime Layer, and Glycocalyx

Most bacteria extrude some material that collects outside the cell wall to form an additional surface layer. A **capsule** is such a layer that adheres to the surface of the cell and forms a halo sharply delineated from the environment when differentially stained so that the cell is colored and the background is darker than the capsule itself (Fig. 4-18*a*). A surface layer that is loosely distributed around the cell and diffuses into the medium is referred to as a **slime layer**. A different structure, the **glycocalyx,**[2] is a tangled mass of thin polysaccharide fibers that extends from the bacterial surface (Fig. 4-18*b*). The capsule stain does not reveal the presence of the glycocalyx, which can only be seen by electron microscopy.

[2] "Glycocalyx" is often used to refer to any polysaccharide material outside the cell wall. Capsules and slime layers would then also be considered to be glycocalyxes.

In some bacteria, these layers help initiate infectious diseases ranging in severity from localized dental caries (the most costly disease in the United States) to life-threatening pneumonias. Host antibodies that bind to the surface layer often help eliminate these bacteria from an infected person.

The role of the capsule or glycocalyx in providing a bacterium with a selective advantage varies depending on the bacterium:

- A thick capsule protects some types of cells from dehydration (without preventing the inward passage of essential water-soluble nutrients and the outward passage of toxic waste products).
- The capsule protects some pathogens from being engulfed and destroyed by the body's white blood cells, thereby contributing to the organism's ability to cause disease. This is the critical factor in the ability of *Streptococcus pneumoniae* to survive the pulmonary defenses of an infected person and establish respiratory disease. Harmless strains of *Streptococcus pneumoniae* (such as the ones you are inhaling right now) differ from the dangerous strains solely by the presence or absence of the

■ Ribosomes

Proteins are assembled at the ribosomes, which occupy much of the cytoplasmic volume. Ribosomes consist of protein and nucleic acid. Prokaryotic ribosomes are 70 S, smaller than the 80 S ribosomes of eukaryotes. Ribosomes of eubacteria and archaeobacteria are of similar size but are chemically very different. Eubacterial and eukaryotic ribosomes also differ substantially, making bacterial ribosomes, like cell walls, selective targets for antibiotic action. Streptomycin, for example, binds to eubacterial ribosomes and alters their ability to function correctly in protein synthesis. The eukaryotic ribosomes of an infected person's cells continue their activity uninterrupted by the effects of the antibiotic. ▶ (The toxic side effects of streptomycin discussed on p. 388 are due to another property.)

■ Cytoplasmic Inclusions

A number of bacteria characteristically accumulate deposits of nutrient materials, usually phosphate, sulfur, carbohydrate, or fat, in structures called **cytoplasmic inclusions** or *cytoplasmic granules*. Energy reserves are usually stored as glycogen or starch (polymers of glucose) or poly-β-hydroxybutyrate (PHB), a lipid polymer formed only by bacteria. Most cells have the capacity to store only one kind of material, producing a characteristic inclusion that can be stained and microscopically observed. These stained inclusions may help to identify the organism. For example, *Corynebacterium diphtheriae* (which causes diphtheria) stores its reserves of inorganic phosphate in *metachromatic*, or *volutin*, *granules*. (When stained with methylene blue, the granules turn red.) Phosphate is used in the synthesis of ATP. *Yersinia pestis*, the bacillus that causes plague, also stores phosphate in observable inclusions.

Certain bacteria found in marine sediments contain chains of magnetite (Fe_3O_4) in deposits called **magnetosomes** (Fig. 4-20). The metal inclusion acts as a compass, allowing the bacteria to orient themselves along the earth's magnetic field.

Another intracellular component in some marine bacteria is the **gas vacuole.** Composed of a series of hollow protein cylinders that are permeable to atmospheric gases but not to water, the gas vacuole regulates the buoyancy of the cell. The concentration of gas in the vacuole is controlled by the number of protein cylinders. Aquatic photosynthetic bacteria need to adjust their buoyancy so that they reside at depths that provide adequate light, oxygen, and nutrients. They do this by increasing and decreasing the number of cylinders (and thereby the gas content) in the gas vacuole.

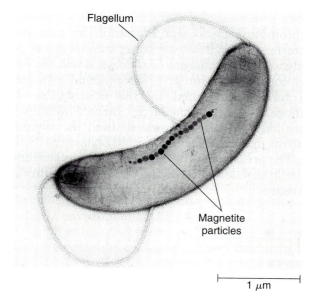

1 μm

FIGURE 4-20 Bacteria with compasses. Magnetosomes in this freshwater bacterium allow the organism to align along the earth's magnetic field. This orients the bacterium so it always swims downward toward the bottom sediment, which is its optimal environment.

Carboxysomes are found in bacteria that synthesize organic carbon from carbon dioxide. These cytoplasmic inclusions contain enzymes that direct key reactions in the conversion of carbon dioxide to organic carbon compounds. Cyanobacteria and other photosynthetic bacteria use energy derived from sunlight to convert carbon dioxide to glucose.

> **CONCEPT CHECK**
>
> - How do plasmids differ from the bacterial chromosome?
> - What properties are conferred on bacteria by the possession of gas vacuoles? carboxysomes? magnetosomes?

Appendages

Several structures project through the cell wall to form surface appendages. The most commonly observed bacterial appendages are flagella, axial filaments, and pili.

■ Flagella

Many genera of bacteria move by means of flagella. The location and number of flagella on a cell, as well

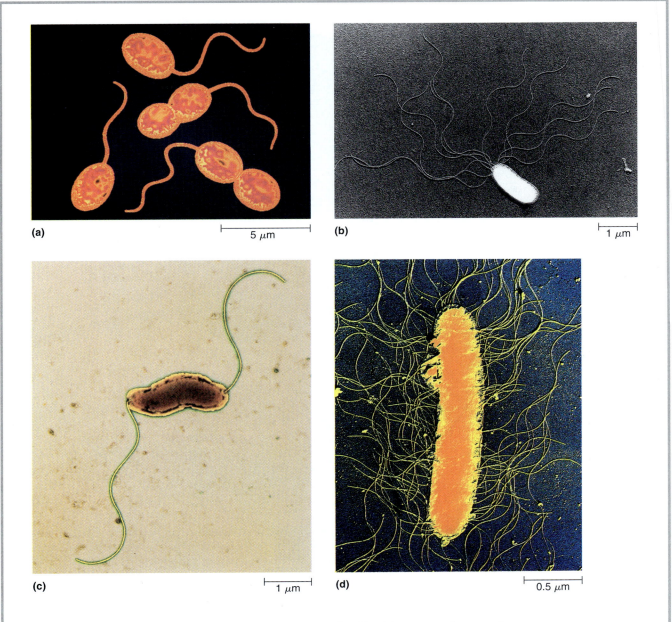

FIGURE 4-21 Arrangements of flagella on bacterial cells. (*a*) Monotrichous, polar; (*b*) lophotrichous, polar; (*c*) amphitrichous, polar; (*d*) peritrichous.

as the number of waves of an individual flagellum, vary according to bacterial species. These three factors, however, are constant for any given species.

Some organisms are *monotrichous* (*mono* = one; *trichous* = hair); they have a single flagellum (Fig. 4-21*a*). Some bacteria are *lophotrichous*, possessing many flagella arranged in tufts or clusters at one end (Fig. 4-21*b*). Other bacteria are *amphitrichous*, with flagella at both ends of the cell, either singly or in tufts (Fig. 4-21*c*). Since the flagella are on the ends (poles) of the cell in these three arrangements,

they are referred to as *polar flagella*. If flagella are distributed around the entire cell surface, the cell is said to be *peritrichous* (Fig. 4-21*d*).

Flagella are only about 20 nm in diameter, too thin to be detected by light microscopes unless special staining procedures are used. The flagella stain contains a chemical that precipitates onto the appendage and increases its diameter to resolvable dimensions.

The bacterial flagellum differs from its eukaryotic counterpart in a number of ways. Rather than

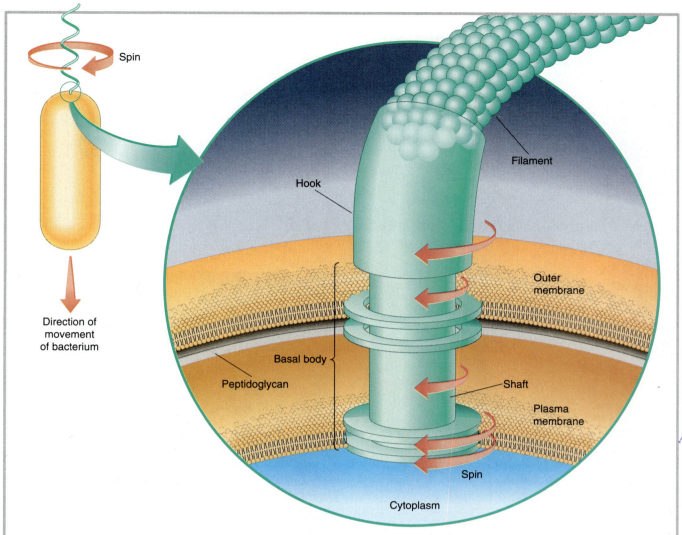

Spin

Direction of movement of bacterium

Hook

Filament

Outer membrane

Basal body

Peptidoglycan

Shaft

Plasma membrane

Spin

Cytoplasm

FIGURE 4-22 Bacterial flagellum acts like a propeller. The bacterial flagellum propels the cell by spinning like a propeller. The shaft of the basal body rotates in a series of rings anchored to the plasma membrane and the cell wall. The terminal ring is attached to the flagellum shaft and may be part of the "motor" that drives the flagellum. Rotation of the lower ring spins the flagellum.

a flexible whip that thrashes back and forth, it resembles a rigid corkscrew that spins, much like the propeller of a boat. The flagellum consists of a basal body, a hook, and a filament (Fig. 4-22). The basal body is the "motor." The ring embedded in the cytoplasmic membrane rotates, causing the filament to spin. The filament consists of a single type of protein called *flagellin* (the whole bacterial flagellum is about the diameter of one microtubule of a eukaryotic flagellum). The morphology of a flagellum is determined by the structure of its flagellin subunit.

The rotation rate of a flagellum has been measured to be as high as 300 revolutions per second. Flagellated bacteria are capable of very rapid move-

ment. *Pseudomonas aeruginosa*, for example, can travel 37 times its cell length in a second. This is equivalent to a 6-foot person swimming at approximately 150 miles per hour.

Bacteria are intermittent in their swimming activity. Motile bacteria alternate between short runs and tumbles. During a *run*, bacteria swim in a straight line. During a *tumble*, they turn in a random manner. When they resume their run, they move in whatever direction they were oriented in at the end of the tumble. For most flagellated bacteria, the run occurs when the basal body rings rotate counterclockwise, and the tumble occurs when the rings rotate in a clockwise direction. Although bac-

FIGURE 4-23 Axial filaments. The axial filament of this spirochete is composed of flagella fibrils that arise at each end of the cell and extend halfway down the organism, winding around its surface. The filaments can be clearly seen running down the middle of the corkscrew-shaped bacterium in this photo, even though they are covered by the cell wall.

terial motion during a tumble is random, the direction of movement can be oriented by **chemotaxis,** the ability to move toward chemical attractants such as nutrients and away from repellants such as poisons. (See BEYOND THE BASICS: Bacterial Behavior and the Role of Membranes, p. 77.)

■ Axial Filaments

Spirochetes move by means of specialized structures called **axial filaments** (Fig. 4-23). These structures are composed of two groups of fibers that originate within opposite ends of the cell and overlap in the middle. Structurally and chemically the fibers of the axial filament are similar to flagella, and they are sometimes called "endoflagella" (*endo* = within). Unlike true flagella, however, these fibers are covered by the spirochete's cell wall and are confined to the periplasmic space. When the cell moves, it rotates around its longitudinal axis and flexes and bends along its length.

■ Pili and Fimbriae

Pili (singular, *pilus*) are protein tubes that extend from the cell. Unlike flagella, pili play no role in motility. They are straighter, shorter, and thinner than flagella and can be observed only by electron microscopy (Fig. 4-24). They are found only on certain species of gram-negative bacteria, usually in large numbers. Their known functions include the following:

1. Conjugation between bacteria. In conjugation, bacteria join together temporarily and transfer genetic material (copies of chromosomes or plasmids) from one cell to the other. Conjugation requires a special long pilus called the *F,* or fertility, *pilus.* After attachment of the F pilus, DNA is transferred. ▶ (Genetic transfer by conjugation is discussed in detail in Chapter 8, p. 208.)
2. Attachment to surfaces such as the tissues of an infected person. Pili, like the capsule and glycocalyx, contribute to the establishment of infection by binding bacteria to cell surfaces in areas where they would otherwise be eliminated by the movement of body fluids. For example, certain pathogenic variants of *Escherichia coli* that produce a severe diarrhea have pili, whereas their nonpathogenic counterparts do not. The pathogens remain attached to the intestinal wall rather than being removed by the movement of the gastrointestinal contents. In males, *Neisseria gonorrhoeae* withstand the flushing action of urine by adhering to the urethral canal in a similar fashion; thus pili are critical in the development of gonorrhea in men.

Fimbriae (singular, *fimbria*) is another term for the short pili that occur in great numbers around the cell. They enable bacteria to attach to surfaces and to each other, so that the bacteria form clumps or films (called *pellicles*) on the surface of the liquid in which they are growing.

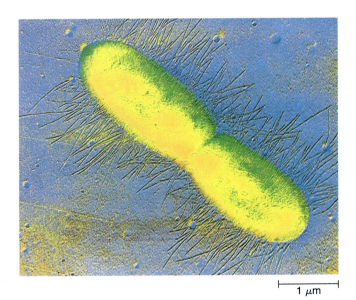

FIGURE 4-24 Pili. Electron micrograph of a dividing *Escherichia coli* cell showing numerous pili distributed over the cell surface.

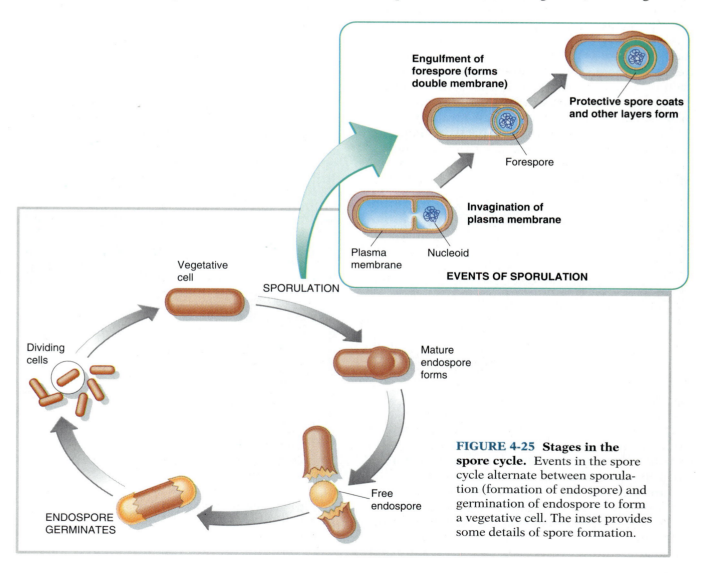

Endospores

A number of bacteria transform themselves into structures that do not grow or reproduce, exhibit absolute dormancy (no detectable metabolism), and are impervious to most of the adverse conditions that kill typical cells. These structures, called **endospores,** are remarkably resistant to heat, radiation, chemicals, and other typically lethal agents.

Endospore formation produces no new cells, so it is *not* a reproductive process. A single bacterium forms a single spore by a process called **sporulation.** A single *vegetative* (actively growing) cell emerges from the spore during **germination.** A bacterium may exist for indefinite periods as either a vegetative cell or an endospore, although endospores can survive for thousands of years (see THE MICROBES: Immortal Seeds of Death, p. 94).

Sporulation is triggered by the onset of unfavorable conditions, for example, the depletion of an essential nutrient in the bacterial environment. The spore requires 10 to 15 hours to form, and it must be completely developed before it encounters such adverse physical conditions as extreme heat or radiation. Otherwise it will not be prepared for the adversity and will be killed.

The name "endospore" was derived from the spore's development *within* the vegetative cell. Sporulation occurs in a series of stages (Fig. 4-25). The plasma membrane invaginates, enclosing a sec-

Engulfment of forespore (forms double membrane)

Protective spore coats and other layers form

Forespore

Invagination of plasma membrane

Plasma membrane

Nucleoid

EVENTS OF SPORULATION

Vegetative cell

SPORULATION

Dividing cells

Mature endospore forms

Free endospore

ENDOSPORE GERMINATES

FIGURE 4-25 Stages in the spore cycle. Events in the spore cycle alternate between sporulation (formation of endospore) and germination of endospore to form a vegetative cell. The inset provides some details of spore formation.

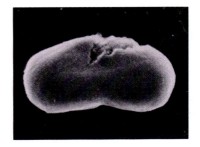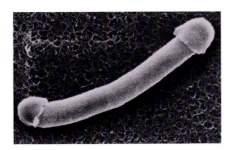

FIGURE 4-26 Germination of a spore. A new vegetative cell emerging from an endospore as seen in the scanning electron microscope.

tion of the cytoplasm that contains the bacterial chromosome, some ribosomes, and other cytoplasmic materials that will be needed for germination. Several impermeable layers develop around this core and protect it from damage by toxic chemicals. In addition, the developing spore synthesizes several spore-specific proteins and other molecules that help shield the endospore from environmental damage. For example, *dipicolinic acid* (a compound unique to endospores) in combination with calcium ions (Ca^{2+}) provides the spore with thermal resistance that greatly exceeds that of vegetative cells. (Some endospores can withstand several hours in boiling water.) Since proteins are less susceptible to thermal inactivation when dehydrated, the endospore's thermal resistance may be due in part to the loss of most of the water from the vegetative cell during sporulation.

Despite the fact that mature endospores are metabolically inert, they can respond quickly to changes in their environment, returning to the vegetative state within 15 minutes. During germination, the cell absorbs water and enlarges. At the same time the protective coats disintegrate and the vegetative cell emerges (Fig. 4-26). These changes are accompanied by the loss of the endospore's resistant properties.

The numerous layers surrounding the cytoplasm prevent dyes from penetrating the endospore. Endospores therefore remain uncolored by Gram staining and can be seen as clear areas within stained cells ▶ (see Fig. 3-14, p. 58). The spores can be differentially stained by using special procedures that help dyes penetrate the spore wall. The spore can then be observed as a colored structure when viewed with a standard light microscope.

The position of a spore within its parent cell is characteristic of the species and may aid in the microscopic identification of the bacterium. The position may be central, subterminal, or terminal. Additional diagnostic evidence can be obtained by observing the shape of the spore and whether it causes the parent cell to swell (Fig. 4-27).

Endospores are formed only by members of a few genera of bacteria. The majority of spore formers are rod-shaped organisms, members of the gram-positive genera *Bacillus* and *Clostridium*, most of which are nonpathogenic inhabitants of the soil. The pathogenic spore formers include the agents of tetanus (*Clostridium tetani*), gas gangrene (*Clostridium perfringens*), botulism (*Clostridium botulinum*),

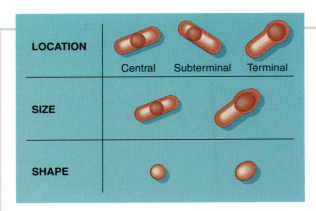

LOCATION Central Subterminal Terminal

SIZE

SHAPE

(a)

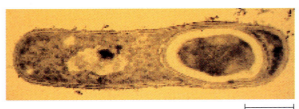

(b)   ⊢ 0.5 μm ⊣

FIGURE 4-27 The appearance of bacterial endospores within their parent cells. (*a*) Diagrammatic representation of the location, size, and shape of typical bacterial endospores. (*b*) Spores of *Bacillus subtilis* typically are subterminal and wider than the parent bacterium. Note the various layers found in the mature spore.

Immortal Seeds of Death

In addition to being one of earth's most heat-resistant forms of life, bacterial endospores may also be the oldest living microbes on this planet. Laboratory studies on endospore longevity have demonstrated that their viability is retained for at least 100 years. In the mid-1980s, however, these research findings were virtually eclipsed by discovery of an unrecorded disaster in the Soviet Union.

Evidence of the event was unearthed during an archaeological excavation at the site of a seventh-century settlement. The village was abandoned when it was apparently struck by an epidemic of anthrax 13 centuries ago, its homes burned and buried by the residents of the time to prevent spread of the disease. (Anthrax is a deadly disease of humans and other animals caused by *Bacillus anthracis,* an endospore-forming bacterium.) Microbiologists examining the site discovered viable anthrax spores in ceramic containers that had been closed since they were buried by the last inhabitants of the village. The endospores had lain dormant underground from A.D. 700 until they were eventually unearthed by modern-day archaeologists. These closed pots contained living organisms 1300 years old.

This finding would probably have been very disturbing to residents of England who found themselves haunted by an action that occurred during World War II. British authorities feared that the Germans were developing deadly bacteria into weapons of war, and in response they conducted a series of tests to determine if anthrax spores could survive when dropped and dispersed by bombs. The tests were carried out on Gruinard Island, an uninhabited strip of land 600 meters off the coast of Scotland.

Unfortunately, the experiments were a complete success, and the island was contaminated with deadly anthrax endospores, which were presumably harmless if left undisturbed. In October 1981, however, British officials received shocking news that the spores were being intentionally disturbed. They found a package containing 10 pounds of dirt removed from Gruinard Island that had been placed near the British laboratories that developed these biological weapons. The dirt, which contained viable anthrax spores, had been collected by members of Operation Dark Harvest, a group conducting a campaign of proenvironmental terrorism in retaliation against those who seeded the island. They claimed to have removed 300 pounds of contaminated soil from the island and threatened to distribute these "seeds of death" throughout Britain until the government decontaminated the island. Less than a decade after treating the soil with 5% formaldehyde in seawater, authorities declared that Gruinard Island was free of these seemingly immortal seeds of death.

and anthrax (*Bacillus anthracis*). Unfortunately, some pathogenic spore formers have been exploited as agents of biological warfare. (See THE MICROBES: Immortal Seeds of Death.)

Endospores are probably the most totally heat-resistant form of life. For example, many can survive in boiling water at 100°C for several hours. Exposure to temperatures above 120°C for 10 to 15 minutes is needed to kill the more heat-resistant types. Spores of the bacterium that causes botulism, a fatal food poisoning, can survive in foods that have been subjected to insufficient heating, such as boiling. They may later germinate into vegetative cells that secrete lethal botulinum toxin into the food. Another potentially fatal disease, tetanus, may occur when endospores are introduced into human tissue. *Clostridium tetani* spores, which remain dormant in soil, can germinate in the dead tissues of a deep wound. The vegetative cells then grow and secrete a toxin capable of causing paralysis and death.

CONCEPT CHECK

- What advantages are conferred on bacteria by the ability to form endospores?
- How does an endospore differ from a vegetative cell? How is it similar?

Key Terms

Key Facts and Concepts

- **Cell morphology.** Morphological properties, including cell size, cell shape, and cell arrangement, distinguish bacteria and are used to help categorize them. Most bacteria have one of three characteristic cell shapes: coccus, bacillus, or spiral. A few bacteria have other shapes, such as filamentous or square.
- **Prokaryotic cell components.** The components essential to all prokaryotic cells are the nucleoid, cytoplasm, ribosomes, and plasma membrane.
- **The bacterial membrane.** The plasma membrane of eubacteria is a phospholipid-protein bilayer constructed with fatty acids. Archaeobacterial membranes contain branched lipids. The plasma membrane serves as a boundary and a selectively permeable barrier that regulates transport of molecules into and out of the cell. Molecules move across the membrane by simple diffusion, facilitated diffusion, active transport, and group translocation. In prokaryotes, the plasma membrane performs many of the functions accomplished by the specialized organelles in eukaryotes, such as generating energy by respiration. Internal extensions of the plasma membrane are present in some bacteria. These extensions increase the surface area of the membrane, thereby enhancing its ability to function in respiration, reproduction, and photosynthesis. Some proteins in the membrane act as sensory devices to monitor the external environment.
- **The eubacterial cell wall.** With few exceptions, a rigid cell wall composed of a unique chemical, peptidoglycan, surrounds and protects the plasma membrane in eubacteria. The cell wall also deter-

mines the cell shape and the Gram stain reaction. In gram-positive bacteria, the cell wall contains teichoic acids in addition to a thick peptidoglycan layer. The cell wall of gram-negative bacteria contains a thinner peptidoglycan covered by an outer membrane. Porins in the outer membrane influence the permeability of the cell surface. Lipopolysaccharides confer endotoxin activity on gram-negative bacteria.
- **The archaeobacterial cell wall.** Most archaeobacteria also possess cell walls, but these walls lack peptidoglycan. They are composed of a peptidoglycan-like molecule (pseudopeptidoglycan), polysaccharide, protein, or glycoprotein.
- **Outside the wall.** Some bacterial cell walls are covered by capsules or an adhesive glycocalyx for attachment to surfaces. Capsules help some bacteria conserve water or nutrients. Some disease-causing bacteria use their capsule to escape host defenses.
- **Prokaryotic DNA.** The genetic information in prokaryotes is contained in a single circular molecule of DNA in a region called the nucleoid. Additional genetic information may be found on small extrachromosomal DNA circles called plasmids.
- **Prokaryotic ribosomes.** Eubacterial and archaeobacterial ribosomes are similar in size (both are 70 S) but are chemically distinct.
- **Storing food.** Some bacteria contain cytoplasmic inclusions in which excess nutrients are stored.
- **Moving about.** Flagella are the most common organelles of motility. They may be distributed in a polar or peritrichous arrangement. Prokaryotic

flagella consist of three parts: basal body, hook, and filament. Flagellated bacteria are often able to direct their movement through chemotaxis, in response to environmental signals.

- **Staying in place.** Pili are surface appendages that act as mediators of conjugation and attachment to surfaces.

- **Surviving in adverse environments.** The vegetative cells of several bacteria can develop into dormant endospores, which are much more resistant to heat, radiation, and chemicals than vegetative cells. Vegetative growth resumes when the spores germinate in a favorable environment.

Review and Problem-Solving Questions

1. Bacteria characteristically appear in specific shapes and arrangements. Draw and label a simple diagram of common bacterial shapes and arrangements.
2. Describe the structure and function of the following parts of a typical prokaryotic cell: (a) Plasma membrane; (b) nucleoid; (c) inclusions; (d) ribosomes; (e) flagella; (f) pili.
3. Compare and contrast the cell wall structures of gram-positive and gram-negative cells.
4. Describe the medical significance of (a) organisms that form endospores; (b) plasmids; (c) the lipopolysaccharide layer of the gram-negative cell wall.
5. What role does the glycocalyx, capsule, or slime layer play in certain disease processes?
6. Explain what happens when you treat the following organisms with penicillin: (a) Gram-positive bacteria; (b) mycoplasmas; (c) archaeobacteria.
7. How does the outer membrane differ from the plasma membrane?
8. Membranous structures play a vital role in eukaryotes. How do bacteria compensate for the lack of internal membranous organelles?

9. Bacteria and protozoa in need of energy are deposited in a medium containing glucose. Describe all the mechanisms available to each type of cell to acquire the sugar for metabolism.
10. Explain the following observations:
    (a) If you remove the cell wall from a flagellated bacterium, the organism loses the ability to move.
    (b) A diet devoid of sucrose does not lead to tooth decay.
    (c) Sporulation takes 10 to 15 hours, but germination may be completed within 15 minutes.
11. Motile bacteria are taken from a liquid culture and placed on a slide. A capillary tube containing a glucose solution is placed on the slide with one end extending into the culture. Describe what happens to the bacteria. What would happen if the tube contained a toxic compound?
12. Design an experiment to isolate (a) magnetotactic bacteria from sediment samples taken from swamps, marshes, rivers, or lakes; (b) endospore-forming bacteria from soil samples.

CHAPTER 5

Bacterial Growth and Laboratory Cultivation

Painted by prokaryotes. The colored bands in Yellowstone's Grand Prismatic Spring are due to the growth of pigmented bacteria and cyanobacteria. Each color corresponds to a different organism that can grow only in that particular temperature zone.

Cameras clicking, awestruck tourists flock by the millions to one of earth's most popular sites, drawn there by bacterial growth. As unlikely as that sounds, Yellowstone National Park has just such a place, one of the most unusual and beautiful natural sights on earth. The site is a hot sulfur pool framed in bands of striking color "painted" by bacteria growing in the waters. The temperature requirements for growth of the various bacteria that create these colored areas are so sensitive that there is no overlap between the adjacent rings of color. Not only must these bacteria be tolerant of the high sulfur concentration in the pool, but they must also be able to grow at temperatures much too hot for most other organisms. The heat-loving bacteria that proliferate in these hot waters are adapted to very narrow ranges of temperatures. Those that grow on the outer ring of the pond (the coolest part) cannot grow just a few feet inward. Those in this second ring cannot grow in the cooler or warmer temperatures of the adjacent rings, and so on until a series of different colored bacteria horizontally stratify to create the painted pool.

Understanding the dynamics of bacterial growth helps one fully appreciate the beauty of

■ The Process of Binary Fission

As a bacterial cell grows in size, it synthesizes and assembles the constituents of its cell wall and cell membrane. The cytoplasmic volume increases, rapidly filling with newly formed ribosomes and enzymes and with the replicating chromosome. The circular chromosome replicates into two identical molecules, one for each daughter cell. The cell grows until it reaches a critical size—double in length if it is a rod or double in diameter if it is a coccus. A septum develops that separates the two identical chromosomes and cytoplasmic contents into distinct compartments. Once the cell wall is completed, the daughter cells become independent.

■ The Potential of Exponential Growth

Each time a cell divides by binary fission it forms a new **generation** of cells (a generation is often called a "doubling" because the population doubles with each new generation). Each cell of the new generation can further divide by fission. A single dividing cell gives rise to two cells. The next division finds these two cells becoming four (2×2, or 2^2), then four cells become eight (4×2, or 2^3). After 10 doublings there are 2^{10} (more than 1000 cells), and after 20 generations there are over 1 million bacteria, all descendants of the one original cell. Because each generation has twice as many cells as the preceding generation, the population size increases at an exponential rate during active growth (Fig. 5-3). The bacterial population increases slowly at first, then explosively as more and more cells are added to the culture by cell division. The **growth rate,** the number of doublings per hour, remains the same throughout the active phase of growth, even though each doubling increases the population by billions of organisms at the end of the cycle. The growth rate does not vary whether one new cell is produced with each new generation or a million new cells are produced. Doubling a million cells, of course, has a more significant impact than doubling one cell.

The number of bacteria in a culture at any time depends on the size of the initial population and the number of generations that have been produced. Mathematically, this is expressed by the formula

$$B_f = B_i \times 2^n$$

in which B_f is the final number of bacteria, B_i is the initial population size, and n is the number of generations.[1] The number of generations can be determined by dividing the total amount of time the cells have been reproducing by the **generation time** (or **doubling time**), the amount of time it takes for the population to double. The generation time (GT) is the inverse of the growth rate (GT = 1/growth rate).

The doubling time is usually constant for each organism as long as physical and chemical conditions do not change. For example, under ideal conditions *Escherichia coli* has a generation time of less than 30 minutes. On the other hand, in the gut, where nutritional conditions fluctuate and competing microorganisms exist, the production of a new generation may require as much as 12 hours. Other microorganisms multiply even more slowly. *Mycobacterium tuberculosis* has a generation time of 12 hours on laboratory media and requires up to 6 weeks to produce a visible colony. *Treponema pallidum* has a generation time of 30 hours in rabbit testes; *Mycobacterium leprae* reproduces once every 13 days in the armadillo (Fig. 5-4). Neither of the last two microbes can be readily cultured on laboratory media and are usually propagated in living tissue.

The initial concentration of bacteria introduced into a culture medium influences the number of microbes present after a period of time. Larger amounts of inocula generate greater numbers of cells within a particular time period. To illustrate, suppose that 1000 of the potentially pathogenic organisms released in an unrestrained sneeze are transmitted into the respiratory tract of a susceptible host. This may be well above the *infectious dose,* the number of organisms needed to initiate infec-

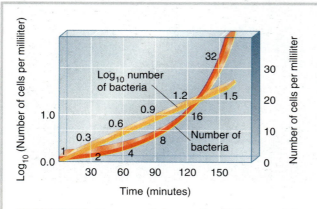

FIGURE 5-3 Exponential growth. This bacterial population has a doubling time of 30 minutes. Population growth is expressed as number of bacterial cells per milliliter and as logarithm of that number.

[1] This equation is usually written in logarithmic form (base 10) for mathematical convenience. $\log_{10} B_f = \log_{10} B_i + 0.3n$, where n = time/generation time. Thus $\log_{10} B_f = \log_{10} B_i + 0.3\, t/\text{GT}$. There are four unknowns in this equation: B_f, B_i, t, and GT. When three of the four are obtained from laboratory determinations, the final one can be determined mathematically.

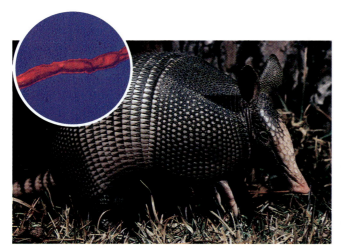

FIGURE 5-4 **Help from an armadillo.** Diagnosis of disease often depends on identifying the pathogen after it has been isolated and cultured. For the diagnosis and study of leprosy, however, this presents a problem, as the organism cannot yet be cultivated in vitro. In fact, *Mycobacterium leprae* (shown in inset) grows only in humans and armadillos. Diagnosing and studying the disease are therefore armadillo-dependent; without the animal, we would have no way to grow the organism for research and diagnosis.

tion in the host. Covering the mouth during sneezing may limit this number to 100, a difference of 900 organisms for the host defenses to eliminate. During the time it takes to mobilize host defenses, the invading pathogens may be actively reproducing. After one generation, sometimes only a half hour, the microbes from the unrestrained sneeze will have produced 2000 progeny, compared to 200 generated following the controlled sneeze. With more exponential growth, the difference between the numbers of organisms that the host defenses must eliminate becomes huge (after 20 generations, the difference is almost 50 million bacteria). The explosive reproductive capacity of microorganisms emphasizes the importance of rapid diagnosis and treatment of infectious diseases—delays allow pathogens to proliferate to sometimes uncontrollable concentrations.

The short generation times of most of the commonly encountered microorganisms facilitate the rapid production of colonies composed of billions of cells in less than a day after inoculation on a solid growth medium (Fig. 5-5). This burst of growth, however, falls far short of the growth *potential* of the organisms. A microbe with a doubling time of 30 minutes, two generations per hour, has the theoretical capacity to produce 2^{48}—or 281,474,976,710,656—progeny in 24 hours. The volume of this number of bacteria alone would be approximately 1 liter. Continued exponential growth for one or two more days would yield an astounding microbial population having a mass greater than

that of the earth. Obviously, microbial growth on agar or in liquid does not continue unchecked. Microbes do not come pouring out over the sides of the petri dish nor do they crowd together in a flask, creating a solid mass of cells. Despite their capacity to reproduce, there are some natural limitations on microbial growth. These limitations are illustrated by the kinetics of bacterial growth in a **batch culture,** a closed vessel or system containing a nutrient medium to which nothing is added and from which nothing is removed during the growth of the culture.

■ Bacterial Growth in Batch Culture

The dynamics of bacterial growth can be observed by inoculating bacteria into a liquid culture medium and measuring population sizes at regular time intervals. When these measurements are plotted on a graph, the resulting **growth curve** shows four distinct phases of growth: (1) lag phase, (2) logarithmic phase, (3) stationary phase, and (4) death phase. A typical growth curve is shown in Figure 5-6.

● **THE LAG PHASE.** When microbes encounter a new environment, they usually do not begin to multiply immediately but require a period of adjustment. During this **lag phase** there is no detectable increase in cell number. However, the microbes are metabolically active, adapting to their new environment before they begin to divide. If the cells are old or have been dormant, essential constituents that

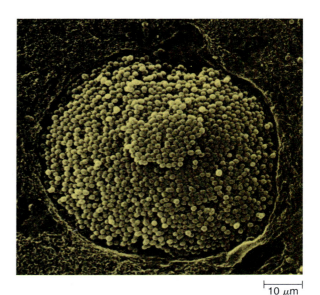

`10 μm`

FIGURE 5-5 **Colony development.** A young developing colony of *Staphylococcus aureus* as seen by scanning electron microscopy.

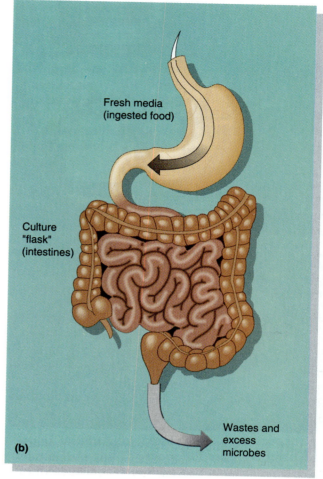

FIGURE 5-8 Continuous culture. (*a*) The chemostat maintains logarithmic growth by constantly replacing used medium with fresh medium. In nature, continuous cultures of bacteria are common. For example, the gastrointestinal tract (*b*) maintains organisms in the logarithmic phase of growth much as the chemostat does.

the organism itself, that is desired. Cultural conditions needed to encourage maximum product yield may differ from those promoting maximum growth rate. Understanding the growth requirements of microbes also provides us with a means to control their unwanted growth. Thus, in the laboratory and in industry, reproductive rates are increased or decreased by manipulating nutrients and physical conditions.

To survive, every cell on earth needs (1) water, (2) a source of energy to fuel the demands of the cell, and (3) chemical compounds to supply the building blocks from which the cell is composed. If an organism is to be successfully cultivated in the laboratory or in industry, these needs must be satisfied, in most cases by the culture medium.

Most microorganisms have simple nutritional needs; they can be cultivated with little difficulty on culture media containing sugars, water, and salts to provide essential elements. **Fastidious microbes,** on the other hand, have unusual nutritional requirements, making it more difficult and sometimes impossible to grow them in the laboratory. The isolation and identification of fastidious pathogens can therefore present a problem for the microbiologist.

■ Water

Most organisms have difficulty growing or even surviving in environments where water is scarce. Outside the cell, water carries dissolved particles that are to be transported into the cell. Within the cell, water serves as the solvent in which the cell's biochemical reactions occur and as the medium for elimination of soluble waste materials. Pathogenic organisms are generally bathed in an abundance of water within the tissues of their living host, and their survival outside the host often depends on their ability to resist drying. The bacteria that cause syphilis (*Treponema pallidum*) and gonorrhea (*Neisseria gonorrhoeae*), for example, are so sensitive that they die within 20 seconds after drying. Direct contact between infected and uninfected mucous membranes or skin is therefore necessary for these organisms to be transmitted from one host to another. In clinical specimens, microorganisms must be protected from drying by being placed in special liquid media for transport to the laboratory. Alternatively, the sample may be inoculated directly onto solid media at the time of collection. (The addition of agar to a liquid medium does not reduce its water content. It merely solidifies it.)

The amount of available water necessary for growth varies with the microbe, but most organisms become desiccated and metabolically inactive in dry environments. That does not mean that they are dead, however. Many types of bacteria survive for long periods in dry soil or dust, on inanimate objects, or in dehydrated foods. Dormant microbes remain viable and can resume growth when moisture is restored. In fact, desiccation is one of the most common methods of preserving bacterial cultures in the lab. The viability of dried cultures can be maintained for years.

■ Energy and Carbon Sources

Growth, movement, metabolism, protein synthesis, and many other essential cell processes demand energy. All organisms can be usefully divided into one of two categories according to their energy source. **Phototrophs** derive energy from sunlight through photosynthesis. **Chemotrophs,** on the other hand, depend on chemical energy harvested by the breaking of chemical bonds. Chemical energy sources may be organic (molecules containing carbon and hydrogen) or inorganic compounds. Most bacteria are *organotrophs*, organisms that use organic compounds, such as sugars and amino acids, for energy. Some bacteria are *lithotrophs* (*lith* = rock; *troph* = feeder), organisms that obtain energy from inorganic compounds, especially those containing nitrogen, iron, or sulfur. Lithotrophs are found only among prokaryotes.

But organisms need more than energy; they need construction materials for building more cell material, for cell growth and reproduction, and for repairing damage. The hub element for constructing cell materials is carbon, so important that this element is used to create another pair of categories for characterizing organisms according to their nutritional needs. **Autotrophs** use *inorganic* carbon in the form of carbon dioxide (CO_2) as their sole source of carbon, assimilating it into complex organic compounds of which the cell is made. **Heterotrophs,** on the other hand, require a supply of carbon in the form of organic molecules.

Bacteriologists usually combine these terms to characterize an organism's nutritional needs according to both its carbon and energy sources. This scheme produces four general categories: *photoautotrophs, photoheterotrophs, chemoautotrophs,* and *chemoheterotrophs* (Table 5-1). The majority of bacteria, including all pathogens, are chemo-

TABLE 5-1 NUTRITIONAL CATEGORIES AMONG MICROORGANISMS (BASED ON CARBON AND ENERGY SOURCES)

Category	Energy Source	Carbon Source	Representative Microbes
Photoautotrophs	Light	CO_2	Cyanobacteria, photosynthetic bacteria, algae
Photoheterotrophs	Light	Organic compounds	Photosynthetic bacteria
Chemoautotrophs (lithoautotrophs)	Inorganic compounds	CO_2	Sulfur-, iron-, and ammonia-oxidizing bacteria; several types of methane-producing bacteria
Chemoheterotrophs*	Organic compounds	Organic compounds	Protozoa; fungi; most bacteria

*Chemoheterotrophs usually use the same compounds for energy and carbon.

heterotrophs, obtaining both their energy and carbon from organic compounds. In nature, organic resources consumed by chemoheterotrophs are replenished by photoautotrophs and/or chemoautotrophs. Chemoautotrophic bacteria are the sole providers of organic material in environments with insufficient light to sustain photosynthetic organisms.

■ Essential Elements

In addition to carbon, all cells need hydrogen, oxygen, nitrogen, phosphorus, and sulfur. Hydrogen and oxygen, along with carbon, are essential for the synthesis of most organic compounds. (Some chemotrophs use the same compounds as sources not only of energy and carbon but also of these essential elements.) Phosphorus is needed for ATP and nucleic acids, sulfur for proteins, and nitrogen for nucleic acids and proteins.

These elements are usually supplied by the medium that is supporting the organism's growth, whether in the laboratory or in its natural environment. Most bacteria obtain their phosphorus from phosphate ions in the medium; sulfur from sulfate ions, hydrogen sulfide, or sulfur-containing amino acids; and nitrogen from ammonia. Ammonia, a molecule that can readily be absorbed by cells and incorporated into organic materials, is the most common source of nitrogen in nature. Other nitrogen sources include nitrates and atmospheric nitrogen gas (N_2). Only a select group of bacteria can "fix" N_2, that is, convert atmospheric nitrogen to ammonia.

Metals are also essential to the cell, but they are required in relatively small quantities. Potassium, magnesium, calcium, and iron are components of some critical cell structures, and they are needed for certain enzymatic reactions. Several other metals, called *trace elements*, are required in such minute concentrations that the levels normally found in the water used to prepare media in the laboratory are sufficient to meet the microbes' demands. Molybdenum, copper, cobalt, and zinc are examples of trace elements.

■ Organic Growth Factors

All cells also require *amino acids* for manufacturing proteins, *purines* and *pyrimidines* for making nucleic acids, and *vitamins*, which assist in many enzyme-mediated reactions. Organisms that cannot synthesize these organic compounds from the raw materials in their environment must be supplied with them. Some bacteria synthesize all these compo-

nents within the cell and do not need an external source. Such organisms have simple nutritional requirements and can grow on a medium that contains only a source of carbon, nitrogen, energy, water, and other essential elements. This is called a *minimal medium* because it supports only the growth of microbes that have minimal needs. Fastidious microbes fail to grow on minimal media.

> **CONCEPT CHECK**
>
> - Why are water, carbon, nitrogen, hydrogen, oxygen, and phosphorus essential for all cells?
> - How is energy acquired by phototrophs and chemotrophs?
> - How do autotrophs and heterotrophs obtain the diverse array of organic compounds needed for growth?

Physical Requirements for Bacterial Growth

In addition to nutrients, microorganisms need environmental conditions that are within a certain range to proliferate. Temperature, pH, oxygen, and osmotic pressure all influence survival and growth. Light is essential for phototrophs; bacteria that live deep in lakes and oceans require more pressure than those at lesser depths or on land. The range of tolerance for these conditions largely determines where an organism will be found in nature. The human body itself can be considered a series of microenvironments, each of which has the potential to harbor a limited spectrum of microorganisms. The physical and nutritional requirements of the more fastidious organisms, especially obligate parasites, are usually as complex as those of their hosts. These organisms are often difficult to culture in artificial environments.

■ Temperature

All organisms can be characterized by the range of temperatures within which they grow. For bacteria, the *minimum temperature* below which growth is impossible is usually 30 to 40°C lower than the *maximum temperature* that can support growth. Within this range is an *optimum temperature* at which the growth rate is maximal. Bacteria fall into one of three categories according to the temperature at which they grow best (Fig. 5-9):

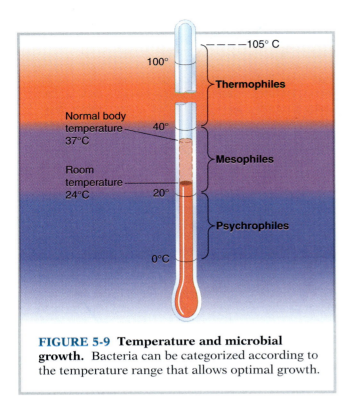

FIGURE 5-9 Temperature and microbial growth. Bacteria can be categorized according to the temperature range that allows optimal growth.

- **Thermophiles** are organisms whose optimum temperatures are above 40°C. Habitats from which thermophiles may be isolated include hot springs, tropical soils, compost piles, hot water heaters, hot tubs, and thermal vents in the ocean floor. (See THE EXPLORERS: Nature's Lawbreakers.) In general, most thermophiles are nonphotosynthetic prokaryotes. It has been suggested that the presence of heat-sensitive internal membrane systems or vesicles prevents most eukaryotes and photosynthetic prokaryotes from replicating at elevated temperatures.
- **Mesophiles** prefer temperatures between 20 and 40°C. Human pathogens are mesophiles, adapted to our 37°C body temperature (even with fever, human body temperature rarely exceeds 40°C). Although growth is less than optimal, some mesophiles can grow at the elevated temperatures found in foods kept warm on heating tables and in electric slow cookers. This phenomenon is responsible for many cases of foodborne illness associated with catered meals and cafeteria-style restaurants. Some mesophiles can even grow at very low temperatures, although somewhat slowly. Some of these organisms are responsible for food spoilage within the refrigerator.
- **Psychrophiles** grow best at temperatures below 20°C. They are all capable of growing at 0°C, and

some grow at –7°C. These organisms proliferate in deep ocean waters, in arctic and antarctic habitats, and in refrigerated or frozen foods.

MYTH: Most organisms that spoil food or drinks in the refrigerator are psychrophiles.

FACT: Although these food spoilers do grow (although slowly) at lower temperatures, most are mesophilic microbes that have optimal temperatures between 20 and 37°C. Consider the source of these contaminants. They are often cells shed from the human body or microbes that settle on food from room-temperature environments, such as the *Penicillium* growing on refrigerated cheddar cheese.

Low temperatures generally retard cell growth without killing microbes. Even freezing fails to kill most microorganisms in food or medical supplies. Once permissive temperatures are restored, growth resumes. In fact, rapid freezing, as in liquid nitrogen at temperatures of −196°C is a common mechanism for preserving viable bacterial cultures. Slow freezing, on the other hand, especially repeated freeze-thawing, generates large ice crystals that can cut cell membranes, killing even the hardiest bacteria. Some fragile microbes, such as *Neisseria* species, are cold-sensitive and die when merely refrigerated. Refrigerating specimens that are to be used for diagnosing gonorrhea or meningitis may therefore result in the failure to diagnose the disease accurately.

Low temperatures reduce membrane fluidity, restrict transport of essential nutrients, and slow down enzyme reactions. Below the minimum growth temperature, these processes are too slow to meet the demands of the cell. As temperatures increase, membrane transport occurs faster and enzymes mediate their reactions more quickly. Bacteria consequently grow more rapidly in warmer environments. At the optimum temperature, enzyme reaction rates are at their peak and growth rate is maximal. The organisms reproduce with the shortest doubling time. The optimal temperature for growth is usually about 5°C below the maximum temperature. Raising the temperature above the maximum usually kills the cell, by denaturing protein and irreversibly damaging molecules essential to the cell's survival. Excessive heat also destroys the cell membrane, and cells may burst. In these ways, heat actually kills cells (unlike cold, which induces protective dormancy). Fevers of 3 or 4°C above normal body temperatures are some-

Nature's Lawbreakers

The laws of nature place unavoidable restrictions on the activities of the earth's organisms. One of the "laws" formerly held to be most inflexible concerned heat: organisms do not grow at temperatures above 100°C. The boiling point of water was long believed to be the upper thermal limit for growth. Above this temperature, cytoplasm boils, proteins denature, DNA melts into separate strands, and lipids run into a nonfunctional mess. Even bacterial endospores, which survive boiling, cannot grow at such severe temperatures. Then, in 1982, remote-controlled submersibles equipped with cameras photographing the ocean floor off the Galapagos Islands made a remarkable discovery that led to modification of at least two of our "laws." Thermophilic bacteria (and several marine invertebrates) were found growing at 105°C in the hot waters next to superheated thermal vents in the ocean floor. Less than a year later, while biologists were still evaluating this remarkable discovery, another bacterium was brought up from the hot submarine floor, this one proliferating near vents with temperatures at 250°C (482°F)—well above the temperature that would ignite this page and burn your book.

How can this biological "outlaw" escape the consequences of its hostile habitat? Part of the answer is found in the depth at which the microbe lives. The enormous pressure exerted by 2500 meters of seawater protects cytoplasm from boiling. At this pressure (265 atm) water does not boil until it reaches 460°C (860°F). Proteins denature when heat

energy exceeds the intrinsic forces that maintain the bonds determining the molecules' shapes. A few extra chemical bonds could stabilize proteins to withstand tremendous heat. In addition, the proteins of extreme thermophiles contain at least five unusual amino acids not found in the proteins of other organisms. (However, it is not apparent whether these amino acids increase thermostability.) DNA might be stabilized by special proteins that maintain its configuration. The molecular structure of the plasma membrane of extreme thermophiles is substantially different from that of more cosmopolitan bacteria and eukaryotes. These microbes are archaeobacteria whose membranes are composed of long-chained ether-containing lipids that span the membrane, forming thermostable structures.

The presence of these extreme thermophiles deep in the lakes and oceans of the world provides the foundation for unique ecosystems, a whole community of organisms that do not depend even indirectly on sunlight as an energy source. Researchers exploring these totally dark underwater environments with remote control robotic submarines found a rich environment of crabs, sponges, and giant worms longer than you are tall. What is the source of energy and nutrients in these perpetually lightless habitats? The answer spews from fissures in the ocean floor, vents called *smokers* that gush out a rich soup of iron, sulfur, methane, and other inorganic chemicals. The hot material quickly mixes with ice-cold water, establishing a temperature gradient. Within this gradient, likely at temperatures no greater than 150°C, live bacteria that metabolize the abundant inorganic chemicals for energy while fixing dissolved carbon dioxide into organic molecules. These chemoautotrophic bacteria are the first link in the food chain that feeds a number of sea creatures—sponges,

worms, and clams. Although no doubt some organisms digest the bacteria, others have evolved a nonpredatory interaction in which both the animal and the bacterium benefit. For example, giant tube worms (see photo below) grow to 2 meters in length without ever eating anything. In fact, they don't even have a mouth. Instead they have teamed up with passenger bacteria that live in specialized intestinal tissues of the worm. The worm uses its blood-rich tentacles as gills to absorb oxygen and hydrogen sulfide spewing from the smokers. These chemicals then serve as substrates for bacterial metabolism within the worm. In return, organic compounds released from the internal bacteria (or cells that have died) are absorbed by the worm, providing it with all its essential nutrients.

Unraveling the mysteries of these light-independent, thermal rift ecosystems required the efforts of many types of scientists, from those controlling the remote submersible craft that made the remarkable discovery to microbiologists who characterized the chemoautotrophic bacteria found living inside the cells of the worms' blind digestive tracts.

times sufficient to inactivate an invading mesophilic pathogen. Medieval remedies based on this observation often included artificial inducement of fever and were occasionally successful if the pathogens were exceptionally heat-sensitive. Unfortunately, most pathogens are not killed by the heat of fever except at temperatures that are also high enough to cause similar heat-mediated damage to the infected person. Fevers above 41°C (105°F) can cause permanent brain damage and death.

■ pH

The measure of the relative acidity or alkalinity of a solution is called its *pH*. The pH value is expressed as a number from 0 to 14—the smaller the number, the greater the acidity. Acidic substances have pH values less than 7, whereas numbers above 7 indicate increasing alkalinity. Pure water is neither acidic nor basic and has a pH of 7 (neutral). The pH number is an expression of the concentration of hydrogen ions (H^+) in the solution—the higher the concentration, the more acidic the solution and the lower the pH. The value is the negative exponent of the H^+ concentration.[2] For example, when $[H^+] = 0.00001$ (10^{-5} M), the pH is 5. A more acidic solution at 0.01 (10^{-2} M) has a lower pH value (pH = 2).

A microbe's responses to pH are similar to those observed with heat; each organism has a minimum, optimum, and maximum pH for growth. Bacteria are classified into three categories based on their pH optimum. Most bacteria are *neutrophiles;* they grow best at a pH between 6 and 8. Most neutrophiles cannot survive at pH values below 4 or above 9, largely because hydrogen or hydroxyl ions (OH^-) inactivate proteins, disrupting transport in the plasma membrane and destroying enzymes in the cytoplasm. Many neutrophiles produce acidic or alkaline waste products of normal metabolism that can rapidly lower or elevate pH to intolerable levels. These cells survive in nature because the toxic end products become diluted or wash away in the environment. Laboratory culture media, however, often contain **buffers,** chemical substances that tend to resist changes in the pH even when acid or alkali is added.

A few bacteria prefer acidic or alkaline conditions. *Acidophiles* grow best below pH 5.5. These bacteria need hydrogen ions to stabilize the plasma membrane. Hydrogen ions do not accumulate within the cell, so the internal pH remains close to neutral. *Thiobacillus* species, for example, display pH optima between 2 and 3. They are found in soils acidified by volcanic runoff or by acidic drainage from mines (Fig. 5-10). Bacteria that thrive at pH greater than 9 are called *alkalophiles*. A few of these organisms can even grow above pH 11. Alkalophiles are most commonly found in alkaline lakes and soils. They can maintain a lower cytoplasmic pH than their surroundings because an active transport system helps them concentrate hydrogen ions from their environment. In addition, the lipids in the membrane

FIGURE 5-10 **Acid-lover's home.** In this mine area, the sulfuric acid used for dissolving metals has created conditions ideal for the growth of acidophiles such as *Thiobacillus*.

and amino acids in the proteins of alkalophiles are stable at high pH.

A variety of environments with broadly different pH levels exist within the human body. Most microbes cannot survive the extremely low pH in the human stomach. Pathogens that can tolerate prolonged exposure to strong acid can successfully establish infection after entering the body through the digestive tract. Most of these bacteria pass through the stomach to the safer neutral pH in the intestines, although some bacteria grow in the stomach. One such organism is the gram-negative bacterium *Helicobacter pylori*, which causes ulcers and possibly plays a role in the development of stomach cancer. Yet *H. pylori* is not an acidophile. It protects itself from the hostile gastric environment by attaching to mucus-secreting cells that line the stomach and raising the local pH. The bacteria produce an enzyme that generates high concentrations of ammonia around the cell. It is unknown exactly how the bacterial infection causes ulcers, but we do know that elimination of the bacteria before ulcers develop prevents ulceration.

---

[2]Chemical concentrations are expressed as the number of moles per liter, known as the *molarity* of the solution. A 10^{-2} M solution is one in which there is 0.01 mole of substance per liter of solution.

Key Terms

binary fission (p. 99)
generation (p. 100)
growth rate (p. 100)
generation time (p. 100)
doubling time (p. 100)
batch culture (p. 101)
growth curve (p. 101)
lag phase (p. 101)
logarithmic (log) phase (p. 102)
stationary phase (p. 102)
secondary metabolites (p. 102)
death phase (p. 103)
continuous cultures (p. 103)
fastidious microbes (p. 104)
phototrophs (p. 105)

chemotrophs (p. 105)
autotrophs (p. 105)
heterotrophs (p. 105)
thermophiles (p. 107)
mesophiles (p. 107)
psychrophiles (p. 107)
buffers (p. 109)
aerobes (p. 110)
facultative anaerobes (p. 110)
microaerophiles (p. 110)
anaerobes (p. 110)
osmotic pressure (p. 111)
halophiles (p. 113)
chemically defined medium (p. 115)
complex medium (p. 115)

transport media (p. 115)
enriched media (p. 115)
selective media (p. 116)
differential media (p. 116)
enrichment broths (p. 117)
slant (p. 119)
maintenance media (p. 119)
lyophilization (p. 119)
direct counts (p. 119)
indirect counts (p. 119)
plate count (p. 121)
colony-forming unit (CFU) (p. 121)
turbidity (p. 122)
standard curve (p. 122)
bioassays (p. 122)

Key Facts and Concepts

- **Reproduction by doubling.** Most bacteria multiply asexually by binary fission and have short generation times (rapid growth rates).
- **Phases of life in closed bacterial cultures.** When microbes encounter a new environment, they usually do not begin to multiply immediately but require a lag period. As long as nutrients are available and toxic waste products do not accumulate, microbes then continue to multiply in the log stage, as in continuous cultures. In batch cultures, however, nutrient exhaustion or toxic wastes eventually impede growth, leading to the stationary and death phases.
- **Nutritional characteristics.** To grow, cells need sources of energy and organic compounds. Organisms are characterized according to both their energy and carbon sources as either photoautotrophs, photoheterotrophs, chemoautotrophs, or chemoheterotrophs.
- **Environmental growth factors.** To proliferate, microorganisms also need environmental condi-

tions that are within a certain range. Temperature, pH, oxygen, and osmotic pressure all influence survival and growth. Light is an additional factor essential for phototrophs; bacteria that live deep in lakes and oceans require more pressure than those that normally grow close to the surface.
- **Culture media.** Pure cultures for identification or study can be obtained by using specialized media designed to enhance the growth of specific organisms, inhibit contaminants, or yield reactions characteristic of particular microbes. Cultures can be maintained by a number of methods that preserve microbial viability.
- **Counting microbes.** The concentration of microbes in a sample can be ascertained either by direct counts or by indirect methods such as plate counts or turbidity determinations. The choice of method usually depends on the availability of time and equipment and the need for an accurate determination of viable cells.

Review and Problem-Solving Questions

1. Draw and label a bacterial growth curve using the following data:

Time (hr)	0	1	2	4	8	12	16	20	24	28
Bacteria per mL	1×10^6	1×10^6	2×10^6	8×10^6	1×10^8	1.2×10^9	2×10^9	4×10^9	2×10^8	1×10^7

(a) What are the major events in each phase of microbial growth?

(b) During which phase in the growth cycle are organisms most susceptible to destruction by metabolic inhibitors?

(c) What two factors contribute to the termination of the log phase?

(d) Draw the growth curve you would expect if you transferred the bacteria from log phase into another flask containing the identical medium.

2. Distinguish between:
(a) Batch cultures and continuous cultures
(b) Synthetic and complex media
(c) Obligate aerobes, facultative anaerobes, obligate anaerobes, and microaerophiles
(d) Enrichment broths and enriched media
(e) Direct and indirect counts

3. Some pathogens are very sensitive to dehydration. How does this affect their mode of transmission?

4. Pathogenic organisms are chemoheterotrophs. Explain.

5. Are human pathogens thermophiles, mesophiles, or psychrophiles? Explain.

6. What is the function of each of the following types of media? (a) Transport media; (b) enriched media; (c) selective media; (d) differential media.

7. How would you isolate the following types of bacteria? (a) Cellulose-degrading bacteria; (b) nitrogen-fixing bacteria; (c) antibiotic-producing bacteria.

8. For determining microbial numbers, what are the advantages of the plate count over the direct cell count? What are the disadvantages?

9. What is a colony-forming unit? Why is this term used instead of cell number?

10. How are bioassays used to determine the concentration of various chemical compounds in products?

11. What types of physical conditions should be taken into consideration when trying to cultivate bacteria?

12. Why are some microbes more difficult to cultivate under laboratory conditions than others?

13. Products such as contact lens cleaning solutions contain preservatives to prevent the growth of any contaminating microbes. Design a bioassay to determine the effectiveness of the preservative in this product.

14. Suppose you were directing the industrial production of a product generated by E. coli during the stationary phase. What conditions would you create to optimize the output of the desired product?

15. You prepare a streak plate of a culture of E. coli on a medium labeled "trypticase soy agar." Nothing grows. Give at least three possible explanations for these results.

16. Growth of bacteria contaminating the surface of fresh vegetables leads to spoilage and is responsible for the short storage life of most vegetables. What physical factors might influence the growth rate of these bacteria? Design an experiment to test whether controlling these factors during storage would be an effective method for reducing spoilage.

HEADLINES

LIFE AFTER LOG

Some bacteria respond to nutrient depletion by forming protective endospores as they move into the stationary phase. A few survive prolonged starvation by assembling into aggregates and fruiting bodies. For most bacteria, however, the response to starvation is to enter into the stationary phase in a metabolically less active state. But evidence summarized by Siegele and Kolter* shows that in many bacteria, including E. coli, changes induced after logarithmic growth confer special properties that protect the bacteria from death. Starved bacteria are much smaller, almost spherical, and the chemistry of their plasma membrane and cell wall is modified. They become more resistant to heat, osmotic stress, and other environmental challenges.

- What advantages might small size and changes in surface chemistry provide a stationary-phase cell?
- The starved bacteria still maintain a low level of metabolic activity. Why is this activity important?
- Certain isolates of E. coli cannot survive starvation. Give a possible explanation.

*D. A. Siegele and R. Kolter, Life after log, Journal of Bacteriology, 174: 345–348 (1992).

CHAPTER 6

Metabolism— Universal Principles

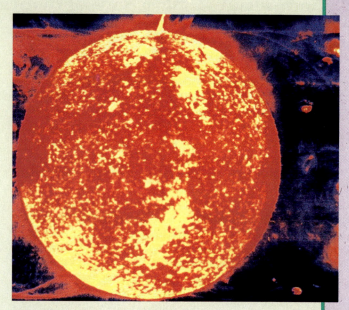

All life requires energy. For most living things this energy is supplied by sunlight, which is transformed by producers into forms usable by consumers and decomposers. Much of this energy ends up in complex molecules.

Christine Auburn stood tensely in the starting area, engulfed in a sea of excited, anxious runners. She awaited her first New York City marathon with a mixture of dread and euphoria, a feeling interrupted by the starter's pistol. The mass of people surged forward, soon crossing the Verrazano Narrows Bridge in a confluent stream of people that subjects the bridge to its most demanding structural challenge (Fig. 6-1). Christine's muscles were about to endure their most demanding challenge, as she would discover late in the race. Early in the race, however, her circulatory system carried ample energy (in the form of the sugar glucose) and oxygen to all her muscles. The oxygen would combine with the glucose, converting it to carbon dioxide and water and releasing the energy to power most of her 26-mile ordeal. She felt terrific.

Back in Christine's apartment, another organism, a bacterium growing in some leftover soup in a closed jar, was engaged in a similar effort. Like Christine, it was using glucose and oxygen to obtain the energy needed to fuel its life processes. In fact, Christine and the bacterium employed identical chemical strategies for obtaining this energy, breaking the six-carbon glucose in half, then completely disassembling each of these fragments until only six solo carbons remained. For a time, energy was abundant and its harvest was efficient.

Then both organisms "hit the wall." For Christine, the wall was at mile 18. For her bacterium and its progeny, it was at hour 18. For each, "the wall" was reached when there was no longer enough oxygen to burn glucose completely. Christine's sustained activity had exhausted the oxygen stored in her muscles, and her lungs and bloodstream could not keep up with the oxygen demand. With this oxygen debt, her muscle cells switched to fermentation, a less efficient mode of energy extraction that requires no oxygen. The muscle cells simply split the glucose into two fragments, releasing a small amount of energy, enough to sustain the muscles for a while in the absence of oxygen. The two molecular halves were then converted to lactic acid, a by-product that produces severe muscle aches and sensations of fatigue. At hour 18, Christine's bacterial culture had generated so many cells in the soup that they exhausted all the oxygen in the closed jar of liquid, although glucose was still abundant. The bacteria employed the same strategy as Christine did— they switched to fermentation to extract whatever energy they could from the sugar without the help of oxygen. The bacteria's rate of growth dropped to a fraction of what it was when oxygen was abundant. Lactic acid lowered the pH to inhibitory levels. They too had hit the wall.

As physically different as Christine and her bacteria appear to be, they are similar or identical in many of their resource-utilization strategies. In fact, many universal activities are employed by all organisms, from microbes to mammals. To explore these universal principles, we must venture into a realm of objects even smaller than microbes, the realm of chemicals and chemical reactions. All organisms, from microorganisms to humans, are composed entirely of chemicals. The survival of an organism depends on its ability to reorganize the chemicals from its environment so that usable energy is released or new molecules are formed, molecules that constitute the organism's own cellular materials. In other words, all organisms must acquire the raw materials and energy needed for *biosynthesis,* the construction of chemical components of growing cells.

The sum of all the cell-directed chemical reactions is called **metabolism.** Metabolic reactions determine where an organism can grow. They define how much energy it can generate from the nutrients in its surroundings. Meta-

bolic reactions distinguish one type of organism from all others. They do this by directing the synthesis of the large, highly organized biochemicals that perform the chemical activities needed to sustain life and form the structural fabric of the organism.

Studying microbial metabolism provides an understanding of life at its most subtle level. Knowledge gleaned from these studies has reduced human suffering and enhanced quality of life in many arenas (see THE EXPLORERS: The Rewards of Exploring Microbial Metabolism). This chapter is devoted to those metabolic principles that are shared by virtually all living organisms, such as the need for energy, use of small-step reactions producing the same fundamental classes of molecules, and the use of enzymes to direct the organism's chemical activities.

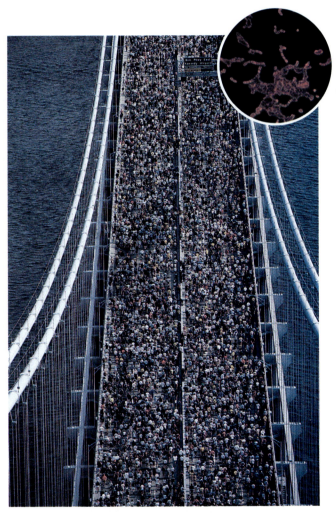

FIGURE 6-1 Different organisms, similar metabolic strategies. Such diverse organisms as humans and bacteria use many of the same strategies for generating energy and accomplishing other metabolic tasks.

The Rewards of Exploring Microbial Metabolism

The study of microbial metabolism contributes enormously to our understanding of life and our ability to apply basic biological principles to improving the quality of life on earth. Here are just a few of the areas in which our knowledge of the metabolism of microorganisms, mainly bacteria, contributes significantly to our well-being.

• *Understanding human metabolism.* The biochemical activities of all organisms are similar in many respects. Humans and microorganisms, for example, degrade glucose to the same three-carbon compound (pyruvic acid) using the same metabolic pathway. Scientists unraveling the nutritional requirements of microorganisms discovered the role of vitamins in human nutrition. The study of microbial metabolism has given us a window onto the essential metabolic activities of human cells.

• *Reduction of infectious disease.* Knowledge of a pathogen's metabolism helps direct the development of new media that better enable us to isolate and culture the microbe. Establishing the identity of an isolated organism depends largely on determining its biochemical characteristics, that is, its ability to metabolize certain compounds and to produce particular end products. Next the disease is treated with antimicrobial drugs, most of which operate by inhibiting specific metabolic reactions in microorganisms without interfering with human metabolism. Scientists continue to search for new drugs that target the pathogen and not the host. This search includes detailed investigations of microbial metabolism to identify reactions that are essential for microbial growth but are not found in humans.

• *Enhanced protection of the environment.* All natural substances are susceptible to microbial degradation. As decomposers and recyclers of natural organic materials, microbes not only maintain the earth's biogeochemical cycle, they also help dispose of the enormous quantity of wastes humans generate. They "detoxify" sewage and degrade much of the garbage in landfills. Microbes that perform unique types of metabolic reactions are often used in *bioremediation*—the removal of harmful substances by living organisms. Bacteria, for example, have been employed to "eat" oil released in tanker spills (as in the 1989 Alaskan oil spill) and detoxify industrial pollutants that threaten our water resources. The metabolism of these microorganisms helps keep our environment clean.

• *Retarding spoilage and deterioration.* Unfortunately, microbes do not distinguish between human waste materials and desired human goods. Biodeterioration, the destruction of useful substances by microbes, is accomplished by some or these same metabolic activities. Our knowledge of microbial metabolism helps us combat such unwanted, and often dangerous, microbial growth in products such as foods, pharmaceuticals, and implanted medical devices.

• *Production of commercial products.* Since the dawn of history, people have been putting microbes to work without really understanding that they were doing so. Microbial metabolism is basic to the production of beer, wine, cheeses, soy sauce, and other foods. Some microbes produce useful metabolic by-products such as antibiotics and drugs that help prevent rejection of transplanted organs. These microbes are grown in culture, and their products are harvested, purified, and sold. Microbes with new metabolic capacities are now being created in the laboratory. Some of these are bacteria that manufacture human insulin, growth factors, and enzymes in amounts and at a cost that makes these chemicals more commercially accessible than their counterparts produced by mammalian cells.

Microbiologists agree that they have yet to fully realize the natural metabolic versatility of microbes. In this chapter and the next, we examine the metabolic dynamics of microorganisms and probe the potential for directing their activities for our benefit.

Energy Concepts

Forming new cell material and maintaining order within an organism require *energy*—the ability to do work, to change things. Energy also drives the many mechanical activities of every organism, such as active transport of substances into and out of the cell, separation of duplicated genetic material during cell division, and movement of whole cells from one place to another. All organisms must harvest and use energy to maintain the processes needed for life. Some organisms acquire their energy directly from the sun, transforming it into a usable form stored in chemicals. These chemicals provide energy for other organisms that consume or decompose organisms or their wastes, breaking down energy-rich molecules to simpler states, releasing and using the stored energy in the process. Regardless of the source, the need for energy is universal; without a usable source of energy, an organism rapidly dies.

Potential and Kinetic Energy

The processes of life often require an organism to transform energy from one form to another. For example, when radiant energy from the sun is converted to a biologically usable form, it is stored in the chemical bonds that hold molecules together in an organized state. This state of energy "storage" converts sunlight to **potential energy** (energy in a state of "readiness" but doing no work). Releasing this potential energy converts it to **kinetic energy,** literally energy in motion, energy doing work. Chemical energy states may be easier to understand by using a mechanical analogy. An elevated object retains the energy required to raise it to its higher position. This potential energy can be converted to kinetic energy to do useful work if the object's fall is connected with some mechanism that directs the released energy to perform a particular task (Fig. 6-2). Just as the weight in the illustration drives the desired energy-requiring activity, cells have the "machinery" to couple energy-releasing chemical reactions with energy-consuming reactions or activities. For the next two chapters we will explore the nature of this "machinery" and how it works.

Chemical Energy

The potential energy stored in chemical bonds between atoms is referred to as **chemical energy.** All cells depend on releasing chemical energy to do the work of maintaining life. Even organisms that use sunlight as an energy source must first convert the kinetic energy of sunlight into chemical energy. In this form, energy can be released in a safe, controlled fashion to drive the cell's activities and growth. Cells may also store surplus energy in specialized chemicals that can later be "burned" when external energy sources are scarce.

Thermodynamics

Another area of energy dynamics must be discussed before moving on to the way organisms handle energy. This area has to do with two natural limitations on energy transformations, often called the *laws of* **thermodynamics** (thermodynamics is the study of energy transformations). The first of these principles states that energy can neither be created nor destroyed. Every organism, indeed every machine, requires an external source of energy if it is to work. Energy cannot be created from nothing.

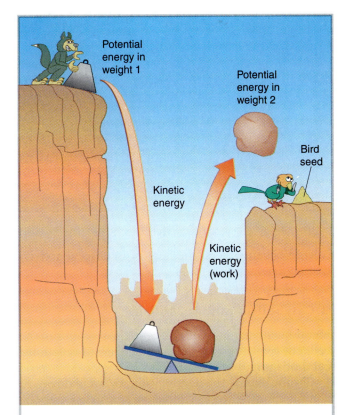

FIGURE 6-2 Energy transformations. The potential energy in one object (weight 1) is released as kinetic energy and then transformed back into potential energy stored in another object (weight 2). Although whimsical, cartoons are packed with lessons on energy transformations.

Even the sun, which provides the energy needed to fuel most biological activities on earth, creates no energy; rather it generates solar energy by a nuclear reaction that releases energy stored in the hydrogen atoms from which the sun is made.

The second law of thermodynamics concerns the natural inefficiency of energy transfers. It states that no energy transfer is 100 percent efficient—some of the energy will escape the transformation in an unusable form. When an automobile converts the chemical energy of gasoline into the mechanical energy of motion, about 85 percent of it is wasted as heat (which explains why cars need a cooling system.) Although living cells are much more efficient than cars, more than half the useful energy in the chemicals that fuel the life processes still escapes during chemical operations. This is the reason that life is a heat-generating process. From the heat released by microbes decomposing compost to the warmth of your body, the loss of energy is an inescapable consequence of biological activity. Even

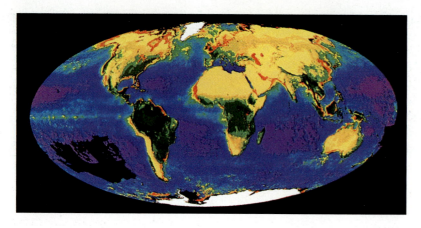

FIGURE 6-3 Using heat to detect life. This infrared satellite photograph of the world's oceans reveals the presence of algae in the frigid water by virtue of the heat generated by their biological activities. (Infrared film responds to differences in temperature rather than light.) The algae photograph red-orange.

in cold ocean waters, the heat of biological activity can be detected (Fig. 6-3).

■ Catabolism and Anabolism

Metabolism, the total of all chemical activities of the cell, consists of two general categories of reactions: catabolism and anabolism.

Catabolism is the degradation, or breakdown, of complex molecules to simpler molecules. Catabolic processes generally release energy; that is, they are *exergonic* (energy-yielding). Chemical energy stored in the bonds between atoms of complex molecules is released when chemical bonds are broken (or re-organized to a lower energy state). Some of this energy is trapped for use by the cell; the remainder is lost as heat. Some of the chemical products of catabolism are used by the cell to make new complex molecules, and the rest are expelled from the cell as waste products.

Anabolism is the biosynthesis of complex molecules from simpler compounds. The simple starting compounds may be transported into the cell, or they may be acquired from catabolic reactions. Molecular construction requires energy; anabolic reactions are therefore said to be *endergonic* (energy-consuming). The energy needed by these reactions is supplied by catabolic reactions and is absorbed into the chemical bonds created during the formation of complex molecules.

In general, the more complex a compound is, the more energy it contains, just as a house contains more energy than the stack of wood, bricks, and materials from which it was constructed. It contains the energy that construction workers spent assembling the randomly arranged building materials into a highly ordered complex structure. Complex chemicals contain the energy used to construct them from atoms and simpler molecules. This energy can be released by breaking up the complex structure into its simpler components (Fig. 6-4).

■ Oxidation and Reduction

Anabolism and catabolism cannot be completely understood in terms of changes in molecular com-

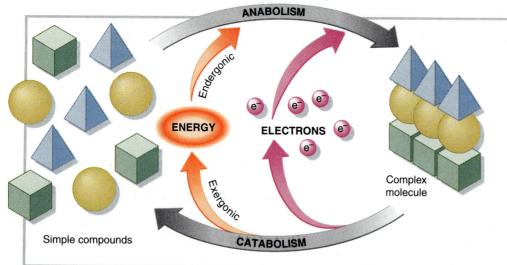

FIGURE 6-4 Catabolism and anabolism. The two types of metabolism are linked to each other by transfers of energy and, as discussed later in this chapter, electrons. Although not shown in this figure, all organisms require a fresh supply of nutrients and energy to replace the waste products and unused energy that are constantly being lost.

	Catabolic	Anabolic
Energy exchange	Exergonic	Endergonic
Type of reaction	Oxidative	Reductive
Nature of conversion	Complex → simple (degradative)	Simple → complex (biosynthetic)
Most energy found in	Substrates	Products

plexity (biosynthesis versus degradation) or energy transfers (endergonic versus exergonic). In all metabolic reactions, electrons are transferred as well (Fig. 6-4). Electron transfers determine whether a reaction is oxidative or reductive. **Oxidation** is the loss of electrons, whereas **reduction** is the opposite, a gain in electrons. Whenever a molecule is oxidized, its lost electrons are captured by another molecule, which is reduced in the process. *Every oxidation reaction, therefore, is coupled with a concurrent reduction reaction.*

Catabolism of a complex molecule releases energy and electrons. Biosynthesis, on the other hand, not only consumes energy but also requires a source of electrons. It can therefore be stated as a general rule that *highly reduced compounds are more energy-rich than highly oxidized compounds.* For example, the six carbon atoms of glucose are in a highly reduced state, and considerable energy is contained in this molecule. When an organism catabolizes glucose to six molecules of carbon dioxide (CO_2) and water, it transforms the carbon to a more oxidized state. This reaction is summarized in the equation:

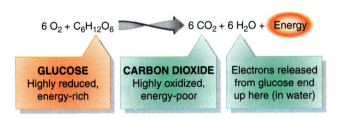

$$6\,O_2 + C_6H_{12}O_6 \longrightarrow 6\,CO_2 + 6\,H_2O + \text{Energy}$$

| GLUCOSE Highly reduced, energy-rich | CARBON DIOXIDE Highly oxidized, energy-poor | Electrons released from glucose end up here (in water) |

During this reaction, 12 electrons (carried by the 12 hydrogens) are transferred from glucose to oxygen, forming six molecules of water (since electrons are carried by hydrogens, *you can track the transfer of electrons by locating the H's*). Compared to glucose, carbon dioxide has little available energy because much of the chemical energy in glucose is released during oxidation (count its hydrogens). The same is true whenever we burn wood (cellulose). The highly reduced carbons in cellulose rapidly oxidize (burn), losing their electrons (and hydrogens) and reduc-

ing oxygen to water. In addition to water, the process liberates carbon dioxide and generates heat. Many cellulose-digesting microbes accomplish the same task but release the energy in gradual multistep metabolic pathways.

Table 6-1 compares some general properties of catabolic and anabolic reactions, showing the relationship of oxidation and reduction to catabolism and anabolism. This is why, in Figure 6-4, we showed the transfer of both energy and electrons.

■ ATP—The Energy "Currency" of All Organisms

Many cells obtain energy from complex molecules such as starch, glycogen, and lipids. They first degrade the large molecules to small molecular subunits, which are then oxidized by a series of reactions that sequentially release small amounts of energy that the cell can manage. This energy is lost, however, unless it is converted to a usable form that is readily available to the cell whenever it needs energy for life processes. Much of the energy from oxidation reactions is transferred to the chemical bonds of high-energy transfer compounds, the most common of which is **adenosine triphosphate (ATP)** (Fig. 6-5).

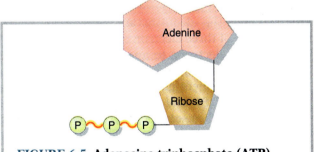

FIGURE 6-5 Adenosine triphosphate (ATP). Skeletal representation shows the two high-energy phosphate bonds (∼) of ATP, the common denominator in energy production and utilization. Adenosine diphosphate (ADP) has one less P∼P bond (thus less energy) but is otherwise identical to ATP.

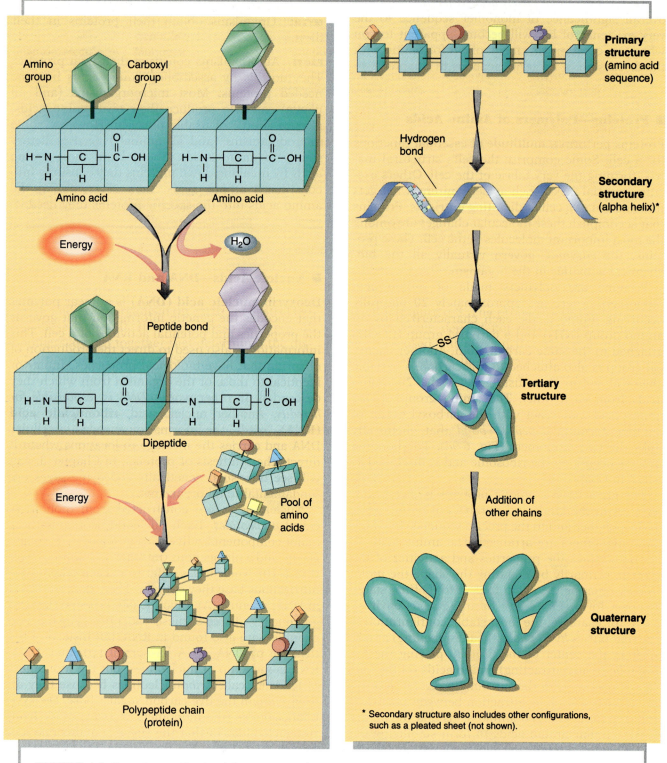

FIGURE 6-9 Protein synthesis. (a) Protein synthesis starts with the formation of a peptide bond between two amino acids, forming a dipeptide (two amino acids). This condensation reaction releases the H and OH as water, (Condensation reactions forming polymers always generate free water). The reaction continues until a polypeptide chain is formed. (b) The function of the resulting protein depends on the shape assumed by the finished protein. The primary structure (amino acid sequence) determines the subsequent levels of folding and therefore the ultimate function of the protein. The secondary structure may include other configurations, such as pleated sheets (not shown).

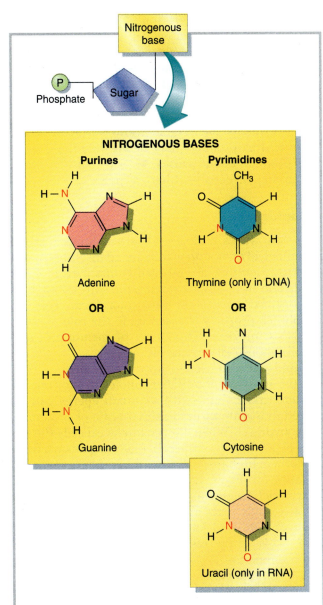

FIGURE 6-10 Nucleotides—the building blocks of nucleic acids. Each nucleotide, shown here in skeletal structure, consists of a sugar, a phosphate, and a nitrogenous base. The nature of the base determines which of five possible nucleotides the monomer is. These nitrogenous bases include single-ringed compounds called *pyrimidines* (cytosine, thymine, and uracil) and double-ringed compounds called *purines* (adenine and guanine). The oxygens and nitrogens (shown in red) can form hydrogen bonds with nitrogens and oxygens on different nucleotides.

of sequences are possible, creating the potential for an infinite number of unique types of organisms.

In cells, DNA is usually a double-stranded molecule. The order of nucleotides in one strand is complementary to the sequence in the other strand, since guanine (G) pairs only with cytosine (C), and adenine (A) pairs only with thymine (T) (Fig. 6-11). Double-stranded DNA resembles a twisted ladder in which the "rails" are linear phosphate–sugar linkages and the "rungs" are the A-T and G-C base pairs. Internal molecular forces twist the ladder into a helical configuration. For this reason DNA has been dubbed the "double helix." Unlike DNA, RNA molecules are usually single-stranded.

■ Polysaccharides—Polymers of Simple Sugars

Sugars, and all compounds constructed from sugars, are called **carbohydrates.** Carbohydrates are chemically defined as molecules that contain carbon, hydrogen, and oxygen, with a hydrogen-to-oxygen ratio of 2:1 (the same as that of water). In other words, carbohydrates are *hydrates of carbon* (*hydrate* = addition of water).

The simplest carbohydrates are simple sugars such as glucose. A simple sugar is called a *monosaccharide* (*mono* = one; *saccharide* = sugar). Sucrose (table sugar), on the other hand, is a *disaccharide;* it is composed of two simple sugars (glucose and fructose) bonded together. **Polysaccharides** are large polymers made up of simple sugars linked together; for example, starch is made up of chains of glucose molecules. From disaccharides to polysaccharides, the monomers are chemically linked to one another by *glycosidic bonds* (Fig. 6-12).

Although most polysaccharides are large molecules, they usually contain no more than two types of monomers. For example, cellulose, the cell wall material of most higher plants, is composed entirely of glucose subunits. Most other monosaccharides can be derived from glucose. Bacteria that do not use glucose as an energy source synthesize it from other compounds.

Many polysaccharides are manufactured by a cell for the sole purpose of storing excess energy (potential energy is stored in the chemical bonds of glucose and in the glycosidic bonds that bind the sugars together). These energy storage polysaccharides are easily digested, so the cell can quickly tap the stored energy whenever it needs it. Some bacteria store surplus energy as the polysaccharides glycogen or starch. These polysaccharides are more abundant in some cells during a time of energy surplus.

Polysaccharides are also major components of many bacterial structures. They form the backbone

chain). The linear sequence of nucleotides represents the information encoded in that molecule of DNA or RNA. Changing the nucleotide order alters the message, just as changing the order of letters in a word alters its meaning. With these four "letters" (the four different nucleotides), an infinite number

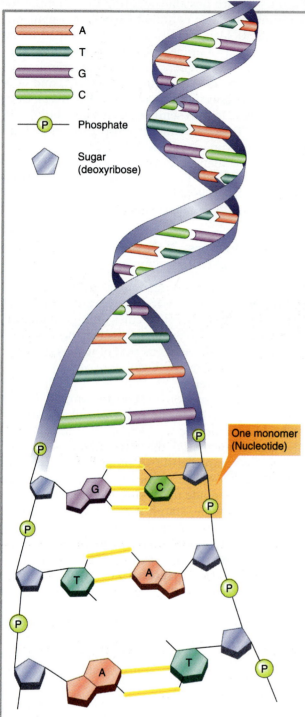

FIGURE 6-11 Structure of a DNA molecule. Nucleotides combine with each other to form a linear sequence of monomers. The nitrogenous bases of the nucleotides in one strand pair specifically with complementary bases in a second strand, forming a double strand with complementary nucleotide sequences. The ladder twists into a spiral, forming a double helix. Except for a few viruses, every organism stores its genetic information in the nucleotide sequence of such double helices.

of peptidoglycan and are found in the outer membrane of the gram-negative cell wall, bacterial capsules, slime layers, and glycocalyxes. Many of these surface polysaccharides are used for bacterial attachment and colonization. For example, dextran, a polysaccharide produced by the anaerobic oral streptococci, is a major component of dental plaque, which cements bacteria to tooth surfaces, where their acid by-products may deteriorate the enamel and produce dental caries. ▶ (See Capsule, Slime Layer, and Glycocalyx, Chap. 4, p. 85.)

■ Lipids

Fats, oils, waxes, and steroids are all lipids, a group of biological compounds that are all insoluble in water. However, they are all readily dissolved in other lipids and in hydrophobic organic solvents such as acetone, benzene, ether, and chloroform. This hydrophobic property is essential to the role of some lipids as integral components of membranes and membranous cell structures.

Lipids are also used to store surplus energy. They are the most weight-efficient molecules for energy storage (more than twice as much energy can be stored in a gram of lipid as in a gram of protein or carbohydrate). When energy intake exceeds the energy expenditures of the cell, many organisms, from bacteria to humans, will store the surplus as reserves of fat or oil (Fig. 6-13). They do this by assembling energy-rich fragments from glucose into long chains of fatty acids. Fats and oils consist of three molecules of fatty acid joined to a molecule of glycerol (Fig. 6-14). If the resulting *triglyceride* is liquid at room temperature (24°C), it is considered an oil; if it is solid, it is called a fat.

The lipids in cell membranes ▶ (described in Chap. 4, p. 72) are composed of only two fatty acid chains instead of three. The third carbon on the glycerol holds a negatively charged phosphate group (PO_4^{3-}) that, unlike hydrophobic fatty acids, is soluble in water. These important molecules are called **phospholipids** (Fig. 6-15). Because they have both hydrophobic and hydrophilic ends, phospholipids become spontaneously oriented so the fatty acid "tails" extend away from water, forming the double layer that is the fundamental structure of biological membranes.

Another group of important lipids is the *steroids*. These molecules contain no glycerol or fatty acid; instead they have a structure that consists of four fused rings (Fig. 6-16). Steroids help stabilize the membranes of all eukaryotic cells and at least one group of prokaryotes (the mycoplasmas). Cholesterol is a steroid found in the membranes of animal cells. Another steroid, ergosterol, is an important

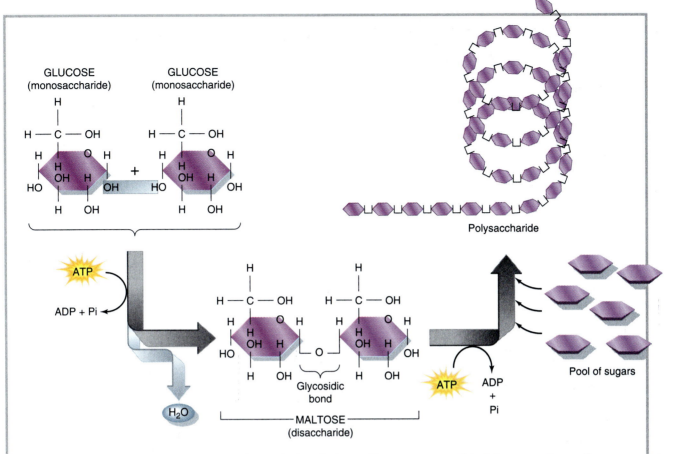

FIGURE 6-12 **Glycosidic bonds.** The glycosidic bonds formed between sugars bind them together to form a disaccharide sugar and subsequent longer polymers called polysaccharides. As with all polymers, the condensation reactions (dehydration synthesis) between monomers release free water molecules.

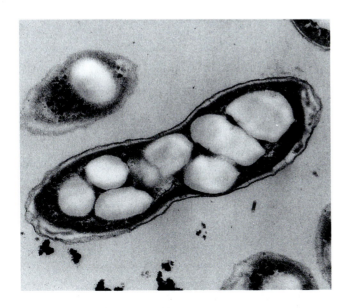

FIGURE 6-13 **Banking surplus energy.** This bacillus stores energy as globules of the lipid poly-β-hydroxy-butyric acid.

component of the membranes of fungi, but not animal cells, and provides a selective target for some antibiotics used for treating fungal infections. ▶ (See Target: Cell Membrane, Chap. 16, p. 385.)

CONCEPT CHECK

- What essential functions does each type of macromolecule perform in a cell?
- In what way do cells obtain the macromolecules necessary for their growth?

Metabolic Pathways— Molecular Modification Step by Step

The chemical processes of metabolism usually occur in a series of reactions rather than by single-step reactions. Each reaction slightly modifies the previ-

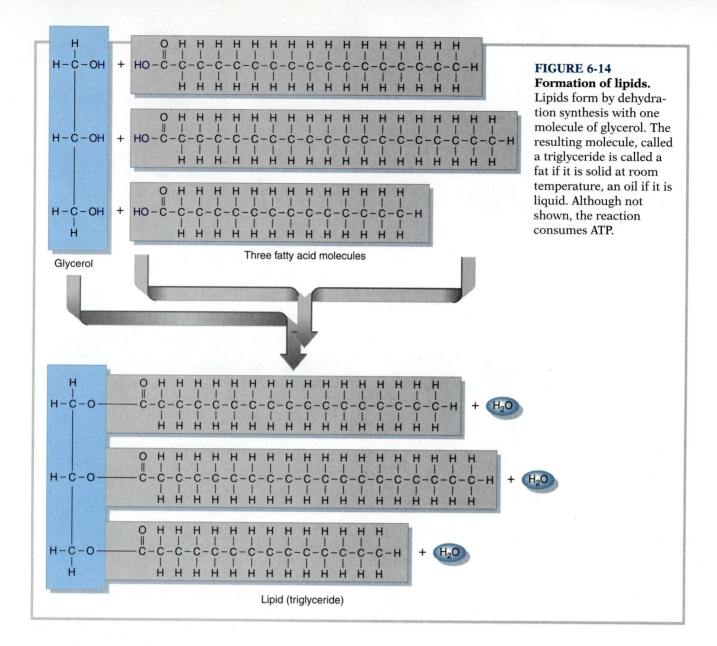

FIGURE 6-14
Formation of lipids.
Lipids form by dehydration synthesis with one molecule of glycerol. The resulting molecule, called a triglyceride is called a fat if it is solid at room temperature, an oil if it is liquid. Although not shown, the reaction consumes ATP.

Glycerol

Three fatty acid molecules

Lipid (triglyceride)

ous molecule until the entire sequence is completed. Each series of chemical changes is called a **metabolic pathway.** For example, cells break down glucose to carbon dioxide and water in a series of 19 reactions. You can accomplish the same process in a single reaction by simply igniting glucose in the presence of oxygen and letting it burn. But using a metabolic pathway to burn glucose provides two advantages over a single-step reaction. First, the ordered, gradual breakdown of glucose using metabolic pathways releases the energy gradually in easily managed bits rather than in a single explosive reaction. In this way, much of the released energy can be conserved for the cell to use. Second, metabolic pathways generate intermediate compounds in the process of producing end products. In the theoretical pathway.

$$[A] \longrightarrow [B] \longrightarrow [C] \longrightarrow [D] \longrightarrow [E]$$
$$\text{reaction 1} \quad \text{reaction 2} \quad \text{reaction 3} \quad \text{reaction 4}$$

A is the *substrate* (the starting compound) and E is the *final product;* B, C, and D are called the *metabolic intermediates.* These intermediates may be identical to compounds generated in pathways leading to other end products. For example, C may also be an intermediate in the formation of product Y,

$$[A] \longrightarrow [B] \longrightarrow [C] \longrightarrow [D] \longrightarrow [E]$$
$$\downarrow$$
$$[X] \longrightarrow [Y]$$

so C can be used by the cell to make either product E or product Y, depending on the organism's needs at that time. In this way, hundreds of different metabolic pathways in a cell are interconnected. Some

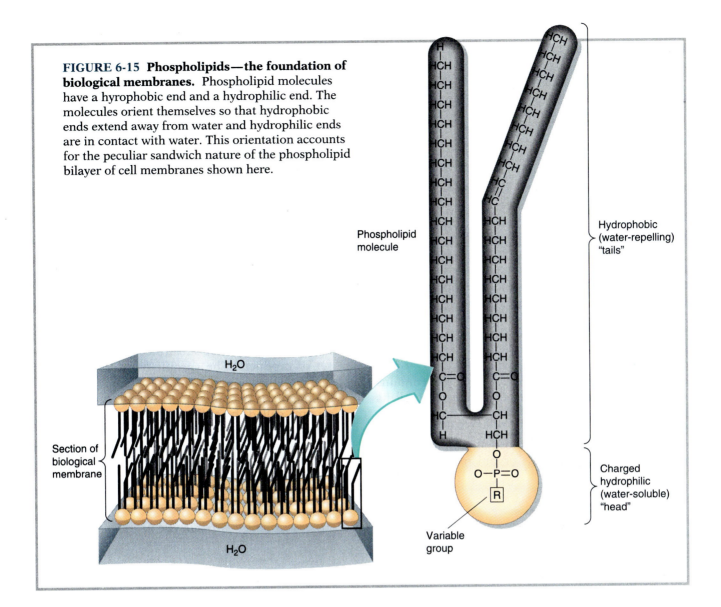

FIGURE 6-15 Phospholipids—the foundation of biological membranes. Phospholipid molecules have a hyrophobic end and a hydrophilic end. The molecules orient themselves so that hydrophobic ends extend away from water and hydrophilic ends are in contact with water. This orientation accounts for the peculiar sandwich nature of the phospholipid bilayer of cell membranes shown here.

Phospholipid molecule

Section of biological membrane

Hydrophobic (water-repelling) "tails"

Charged hydrophilic (water-soluble) "head"

Variable group

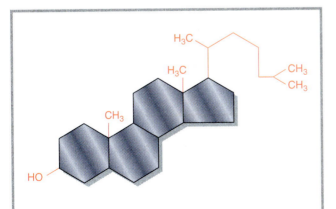

FIGURE 6-16 Cholesterol, a typical steroid. All steroids share the same four-ring skeleton shown in purple. The parts of the molecule shown in red are unique to this steroid (cholesterol). Although not shown, a carbon atom occupies each angle.

intermediates of glucose degradation, for example, are also substrates for the fat-producing pathway. Thus when glucose is abundant, these common intermediates are converted to fat, which is stored instead of being completely degraded to carbon dioxide and water. Linking pathways in this fashion provides a mechanism for diverting surplus intermediate compounds into the metabolic pathways that best meet the organism's constantly changing needs.

CONCEPT CHECK

- Why are metabolic pathways more advantageous to a cell than a single complete oxidation reaction?
- What is the significance of having metabolic intermediates shared by more than one pathway?

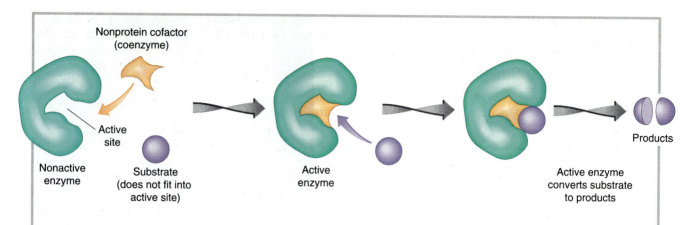

FIGURE 6-19 Role of accessory molecules in complex enzymes. The protein portion of the enzyme alone is nonfunctional. An accessory or helper molecule is required for activity. *Cofactors* or *coenzymes* are nonprotein helpers that bind temporarily to the enzyme to produce a better fit between the active site and the substrate or to neutralize repulsive charges between the enzyme and the substrate.

TABLE 6-3	COENZYME ACTIVITY OF B-COMPLEX VITAMINS	
Vitamin	**Coenzyme**	**Types of Reactions Assisted by Coenzyme**
Thiamin (B$_1$)	TTP	Removal of CO_2 from molecules (decarboxylation)
Riboflavin (B$_2$)	FAD	Hydrogen carrier in energy-generating reactions
Pyridoxine (B$_6$)	Pyridoxyl phosphate	NH$_2$ transfer (transamination); decarboxylation of amino acids; removal of H_2O (dehydration) from amino acids
Cobalamin (B$_{12}$)	Cobalamine	Protein and nucleic acid metabolism
Niacin (nicotinic acid)	NAD, NADP	Hydrogen carrier in energy-generating reactions and biosynthesis
Pantothenic acid (B$_5$)	Coenzyme A	Transfer of small organic molecular fragments in respiration and fatty acid metabolism
Folicin	Folic acid	Single-carbon transfer in nucleic acid and amino acid metabolism

TABLE 6-4	SUMMARY OF ENZYME PROPERTIES

1. Enzymes are composed of protein.
2. They have a high degree of specificity for substrate.
3. They are unchanged by the reactions they catalyze.
4. They have a high reaction efficiency.
5. They may require coenzymes for activity.
6. They are heat-sensitive.

■ Enzyme Nomenclature

The international system of naming enzymes calls for adding the suffix "ase" to the end of either the name of the substrate or the name of the reaction catalyzed. An enzyme that accelerates the breakdown of protein to amino acids is called a *protease*. The removal of a COOH (carboxyl) group, a common event in cellular metabolism, is mediated by a *decarboxylase*. (In decarboxylations, only CO_2 is removed, the H remaining with the original molecule.)

Other examples of enzyme nomenclature are provided in Table 6-5.

Enzymologists sometimes employ alternative systems of nomenclature when referring to certain degradative enzymes. Enzymes that break down their substrates are often described by adding the suffix "lytic" to the name of the corresponding substrate. For example, a protease is also called a *proteolytic enzyme* because it "lyses" (splits) proteins, and a lipase is a *lipolytic enzyme*. Finally, certain traditional labels for enzymes that were named before the international nomenclature scheme was generally adopted are still used. Some of the best known are trypsin and pepsin, which are both proteolytic enzymes (proteases) found in the mammalian stomach, and lysozyme, the peptidoglycan-lysing enzyme found in body secretions.

■ Conditions Affecting Enzyme Activity

Enzyme activity responds to changes in temperature, pH, and the relative concentrations of substrates and products.

● **TEMPERATURE.** Enzymes have optimum temperatures at which they operate most efficiently. For example, most human enzymes work best at 37°C, normal body temperature. Increases in temperature generally boost enzyme activity and accelerate reaction rates. This explains why many microorganisms grow more rapidly at higher temperatures. Above a critical temperature, however, growth stops. This is the temperature at which enzymes suffer heat damage. They change shape, often lose solubility, and coagulate (like the protein in a cooked egg). This protein damage, called **denaturation,** results in loss or alteration of enzyme activity. Once

an enzyme has been physically denatured, it usually cannot be repaired. Irreversible protein denaturation contributes to the effectiveness of superheated steam as an agent for killing microorganisms. Low temperatures approaching freezing, on the other hand, greatly reduce enzyme activity but rarely physically damage the enzyme. Refrigeration temporarily arrests microbial growth by slowing the rate of enzyme-controlled metabolism, but activity (and growth) is restored once the enzymes return to temperatures within their growth range. An organism's enzyme activity usually determines the optimal temperature at which the microbe grows.

● **pH.** An enzyme can be denatured by changes in pH as well as excessive heat. An enzyme that functions optimally at neutral pH (7.0) will usually be inactivated when its environment becomes too alkaline or too acidic. Microorganisms that live in extremely acidic or alkaline environments have developed mechanisms for maintaining a neutral pH in their cytoplasm ▶ (see pH, Chap. 5, p. 109).

● **SUBSTRATE, PRODUCT, AND ENZYME CONCENTRATIONS.** Relative concentrations of substrates and products also affect the rate of enzyme-catalyzed reactions. High substrate concentrations and low levels of product increase the reaction rate, whereas the opposite situation lowers it. Furthermore, the concentration of the enzyme itself plays a role in determining the reaction rate, with greater concentrations resulting in increased rates (up to a maximum). In fact, many metabolic reactions are at least partially regulated by either turning on or shutting off synthesis of the corresponding enzyme. For example, an organism might avoid wasting energy by turning off production of an enzyme when the corresponding substrate is unavailable. (The manner in which microbes turn enzyme synthesis on and off is discussed in Chapter 8.)

■ Chemical Inhibitors of Enzyme Activity

Many reactions essential to microorganisms, such as synthesis of bacterial cell walls, are not shared by humans. An agent that specifically inhibits the enzymes that catalyze these reactions can retard or completely arrest the growth of microorganisms without affecting human metabolism. In other words, these enzymes provide selective targets for antibiotics and chemotherapeutic agents. All effective antibiotics must possess such selective characteristics before they can be introduced safely into a diseased patient.

TABLE 6-5	EXAMPLES OF ENZYME NOMENCLATURE
Name Based on	**Enzyme**
Substrate	
Lipid	Lipase
Protein	Protease
Nucleic acid	Nuclease
Reaction Catalyzed	
Oxidation	Oxidase
Removal of hydrogen	Dehydrogenase
Transfer of amino group	Transaminase

Enzymes That Damage the Body

Many pathogenic microorganisms produce enzymes that contribute to their ability to invade the infected body, sometimes with destructive consequences. The potentially fatal symptoms of gas gangrene, for example, are largely due to lecithinase, an enzyme produced by the pathogen *Clostridium perfringens*. The substrate for this enzyme is lecithin, a component of human cell membranes. Dissolving cell membranes kills host tissue so it no longer presents an effective solid barrier to microbial invasion of surrounding tissues. The pathogen leaves a trail of enzymatically digested tissue as it invades new body sites. Lecithinase production, which can be detected by growing the pathogen on egg yolk media, has proved to be a valuable laboratory indicator in the diagnosis of gas gangrene (see photo).

Enzymes that contribute to the ability of a pathogen to injure people do so by one of two mechanisms: (1) direct destruction of host tissue, as seen in gas gangrene, or (2) interruption of essential metabolic processes. The pathogen that causes diphtheria, for example, produces a toxic enzyme that inactivates one of the components needed for protein synthesis by host cells. Unable to produce proteins, the host cells die, as does the infected owner of those cells if enough of them are impaired.

In addition to directly injuring the host, microbial enzymes may protect the pathogen by inactivating substances that would otherwise eliminate infection. For example, many species of bacteria produce the enzyme penicillinase, which destroys the antibiotic penicillin. The failure of this antibiotic in treating many *Staphylococcus aureus* infections is common, due to the emergence of penicillinase-producing strains of the bacterium.

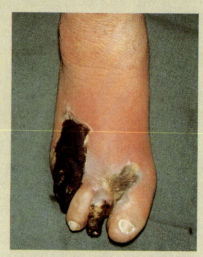

Deadly enzymes The effect of lecithinase in a patient with gas gangrene.

● **COMPETITIVE INHIBITION** Many selectively toxic inhibitors are *structural analogues* (molecules with similar shape) of the substrate of the enzyme inhibited. Because of this similarity in shape, the inhibitor competes with the substrate for the enzyme's active site (Fig. 6-20). Unlike the substrate, however, the inhibitor cannot be converted to products, and while it is in the active site it prevents normal substrate binding. In this tied-up state, the enzyme is inactive. Such **competitive inhibition** of enzymes is characterized by two important properties: (1) It is highly specific for a certain metabolic reaction, so only reactions with substrates similar to the inhibitor molecules will be affected; and (2) the effect is reversed once the inhibitor molecule is removed. The inhibitor is constantly dissociating from and rebinding with the enzyme. No structural damage is done to the enzyme. Therefore, inhibition is maintained only as long as the concentration of inhibitor is high enough for it to compete successfully with the substrate. As inhibitor concentrations decrease, the substrate overcomes the inhibitor, the inhibition is reversed, the enzyme resumes activity, and the organism begins to grow again.

This property of reversibility is an important consideration in the use of chemotherapeutic agents. Agents that work by competitive inhibition (sulfa drugs, for example) are *microbistatic;* that is, they do not kill microorganisms but merely inhibit cell growth. Most chemotherapeutic agents are used to stop the growth of the pathogenic microorganisms long enough for the body to eradicate the infection using its own defense mechanisms. Thus, if the antibiotic regimen is terminated too soon (such as when the symptoms disappear), the inhibited microbes may resume growth. The therapeutic role of competitive inhibitors is discussed in more detail in Chapter 15.

● **ENZYME INACTIVATION.** Enzymes can be inactivated by the heavy metals lead, silver, mercury, and arsenic. The inactivation is permanent and very broad in the proteins affected. These poisonous substances bear no resemblance to the substrate of the inactivated enzymes but attach nonspecifically to protein. Some of these inhibitors alter the enzyme's shape so that it is no longer active. Others bind in place of cofactors or prosthetic groups needed to ac-

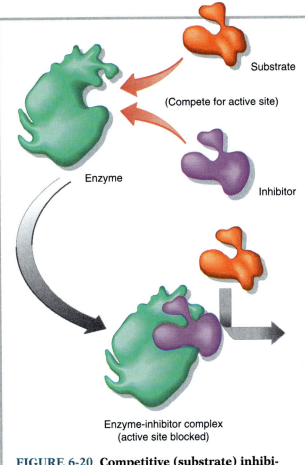

Substrate

(Compete for active site)

Enzyme

Inhibitor

Enzyme-inhibitor complex
(active site blocked)

FIGURE 6-20 Competitive (substrate) inhibition of enzyme activity. The inhibitor is a structural analogue of the substrate. The enzyme's active site recognizes both molecules, but the inhibitor cannot be converted to products.

tivate the enzyme. As a consequence, many critical metabolic functions are lost. These inhibitors affect virtually all enzymes, so inhibition is not selective for a specific reaction or a particular cell type. Human and bacterial cells are equally susceptible to the lethal effects of these agents. Any attempt to use heavy metal inhibitors systemically (throughout the entire body) as chemotherapeutic agents would eliminate the host along with the pathogen. Such nonselective enzyme inhibition accounts for the tragic consequences of lead poisoning in children who eat lead-containing paint peeling from walls. (These dangers are also the reason that lead-free gasoline was developed.) Heavy metal poisoning also follows ingestion of food that contains high concentrations of mercury. In Japan, mercury poisoning has been associated with consumption of fish caught in mercury-contaminated waters. Some heavy metals, however, such as those in the compounds mercuric chloride and silver nitrate, are used as topical agents to prevent or treat infections on the surface of the body, where they kill microbes by inactivating their enzymes.

CONCEPT CHECK

- Why does every trait of an organism depend on enzymes?
- How does an enzyme accelerate a chemical reaction, and why is it specific for that reaction?
- What factors of the cell environment affect enzyme activity? How does this affect cell growth?

Key Terms

metabolism (p. 127)
potential energy (p. 129)
kinetic energy (p. 129)
chemical energy (p. 129)
thermodynamics (p. 129)
catabolism (p. 130)
anabolism (p. 130)
oxidation (p. 131)
reduction (p. 131)
adenosine triphosphate (ATP) (p. 131)
phosphorylation (p. 132)

NAD (p. 132)
NADP (p. 132)
FAD (p. 132)
NADH (p. 132)
NADPH (p. 132)
$FADH_2$ (p. 132)
macromolecules (p. 133)
monomers (p. 133)
polymer (p. 133)
deoxyribonucleic acid (DNA) (p. 135)
ribonucleic acid (RNA) (p. 135)

carbohydrates (p. 137)
polysaccharides (p. 137)
phospholipids (p. 138)
metabolic pathway (p. 140)
enzyme (p. 142)
prosthetic groups (p. 142)
active site (p. 142)
coenzymes (p. 143)
cofactors (p. 143)
denaturation (p. 145)
competitive inhibition (p. 146)

Key Facts and Concepts

- **Fuel for life processes.** All organisms require a constant supply of energy to fuel the life processes. Energy conversions from kinetic to potential and back to kinetic occur as energy is trapped from its original source and transferred from organism to organism. An organism contains tremendous amounts of potential energy stored in its chemical bonds, energy that can be harvested by oxidizing these chemicals using catabolic pathways. Although most of the energy escapes as heat with each such transfer, enough of it is harvested to run the energy-consuming processes of life.

- **Metabolism and energy.** Metabolism is both energy-producing and energy-consuming. Catabolism is the degradation of complex molecules. Energy released by these exergonic reactions is used to form ATP, the universal energy currency in all organisms. Anabolism is the biosynthesis of complex molecules from simpler ones. Anabolic (biosynthetic) reactions consume ATP or other forms of energy.

- **Electron transfer.** Electrons are transferred in all metabolic reactions. Every oxidative reaction is coupled with a concurrent reductive reaction because electrons cannot be released by oxidation unless some other molecule is reduced, that is, accepts the electrons. Electrons are often temporarily transferred to coenzymes NAD, NADP, or FAD.

- **Metabolic pathways.** The capacity for life also depends on a cell's metabolic ability to extract building materials from the environment. The metabolic reactions that convert these materials into forms the cell uses for its structure and activities occur in series of reactions. These metabolic pathways allow the cell to control the release of energy and to generate intermediates that can be directed to other pathways.

- **Macromolecules.** Much of a cell's food supply and energy is used to construct macromolecules, polymers that are needed for necessary structures and for critical functions. Protein, nucleic acids, polysaccharides, and lipids are all macromolecules every organism must be able to manufacture in order to remain alive.

- **Enzymes.** Every organism depends on enzymes, proteins that catalyze virtually all metabolic reactions. The protein is often coupled with an essential nonprotein molecule such as a prosthetic group, coenzyme, or cofactor. Enzymes are highly specific in their catalytic action. Only molecules that can bind to an enzyme's active site can serve as its substrate. Factors such as temperature and pH can denature enzymes and kill cells by halting metabolic reactions. A cell can also regulate its own metabolism by controlling the concentration and activities of specific enzymes. Structural analogues can competitively inhibit an enzyme by preventing substrate binding to the active site. Heavy metals bind nonspecifically to protein and interfere with enzyme activity.

Review and Problem-Solving Questions

1. Distinguish between the following terms:
   (a) Oxidation and reduction
   (b) Exergonic and endergonic
   (c) Enzyme and coenzyme
   (d) Monomer and polymer
2. How do catabolic and anabolic pathways differ from each other? How are they similar?
3. What is the significance of the precise amino acid sequence in a protein?
4. Why may an organism get fat when it consumes large quantities of sugar?
5. List three ways in which RNA differs from DNA.
6. How do the chemical properties of phospholipids contribute to their role in membrane structure?
7. Do enzymes violate the law of thermodynamics that energy can neither be created nor destroyed? Explain.
8. The bacterium *Thermus aquaticus* has an optimum temperature for growth at about 71°C. How would you expect temperature to affect the activity of the enzymes from this organism compared to its effect on enzymes from *E. coli?*
9. Enzymes are protein, but not all proteins are enzymes. Explain.
10. What happens to the heat generated during metabolism?
11. Explain why most bacteria do not require a medium containing niacin, but some bacteria will not grow unless the medium contains niacin.
12. The following data were gathered by boiling an enzyme preparation or exposing it to sulfanilamide for 10 minutes and then inactivating the drug, thus returning the preparation to standard conditions.

Exposure	Relative Activity Level
None	100
Boiling	0
Sulfanilamide	100

Why is the effect of temperature on enzyme activity permanent but the effect or sulfanilamide is reversible?
13. What are the functions of the active site, prosthetic group, and coenzyme?
14. How does each of the following affect enzyme activity? (a) Temperature; (b) pH; (c) substrate concentration; (d) competitive inhibitors; (e) mercury salts

---

PROTEINS HELPING PROTEINS

The primary structure of a protein is determined by its amino acid sequences. However, the protein's activity depends upon folding the linear molecule into the correct three-dimensional shape. Each enzyme must form an active site capable of binding the substrate; structural proteins must also fold into their functional shape. Although the protein's primary structure dictates it shape, recent discoveries* have identified a class of protein's, called molecular chaperones, that play an essential role in protein folding. Chaperones are present in both eukaryotic and prokaryotic cells. They temporarily bind to unfolded proteins and couple ATP hydrolysis to the correct folding reactions.

- Would you classify these protein chaperones as enzymes?
- The concentration of many molecular chaperones increases when cells have been exposed to high temperatures. What advantage might it be to a heat-stressed cell to have these additional chaperones?
- Some chaperones simply cover certain amino acid sequences on a protein during the folding process. How might this affect the final shape of the protein (compared to folding in the absence of the chaperone)?

* E. Craig, Chaperones: helpers along the pathways to protein folding, *Science,* 260: 1902–1903 (1993).

CHAPTER 7

Dynamics of Microbial Metabolism

Harnessing microbial metabolism to dispose of waste. This "biocoil" reactor uses algae and other microbes to transform sewage into valuable protein that can be used as animal feed and other compounds that can be burned to generate electricity.

In the late 1980s, a huge barge loaded with tons of New York City garbage was launched on a trek that made front page news throughout the nation. No spectacular disaster befell the traveling trash; its newsworthy appeal was simply that it was on a long voyage to nowhere. Shipped to another state for disposal in a landfill, the garbage was refused by state officials, who simply had no space available for the extra burden on their facilities. The same fate awaited the barge as it steamed from state to state, and it eventually had to return to New York, where the rotting refuse remained until it disappeared in the middle of the night, probably to be dumped illegally at sea. The world watched with a mixture of amusement and alarm. Our waste disposal system was clearly failing us.

This incident might have been the first time that our crisis in waste management forced its way into the national consciousness. With the realization that landfills could not provide a perpetual solution to our waste disposal problems, strategists devised several plans to help forestall the impending exhaustion of landfill space. Recycling programs have been the most visible of these. But one of the first resources that officials turned to were microorganisms.

The diversity of microbial metabolism promises at least partial solutions to the waste disposal problem. Much of our refuse decomposes too slowly to keep up with the influx of new garbage. Microbiologists have successfully isolated some organisms and engineered others that can selectively speed up the decomposition process and generate valuable by-products, such as energy and animal feed, in the bargain. For example, 1 million tons of human and animal hair are discarded each year, plus another million tons of poultry feathers, all of which contain mostly keratin, a difficult-to-digest protein. When placed in biodigesters containing certain strains of *Bacillus licheniformis*, the material is fully decomposed within days (Fig. 7-1), generating marketable natural gas (methane) and a digested mass that can be used as livestock food. Scientists and entrepreneurs foresee a bright economic future and a healthier planet as we broaden our reliance on microbes to metabolize garbage to useful products. Before long, that wayward barge of trash might be considered a valuable resource rather than a migrating menace.

Harvesting such benefits requires a fundamental knowledge of the basics of microbial metabolism, which in many ways is identical to that of other organisms. In other ways, however, it is unique, making some bacteria capable of metabolic feats that no other organisms can accomplish, such as nitrogen fixation and the ability to thrive on energy extracted from inorganic chemicals. In this chapter, we examine the chemical activities of bacteria, earth's most metabolically diverse group of organisms.

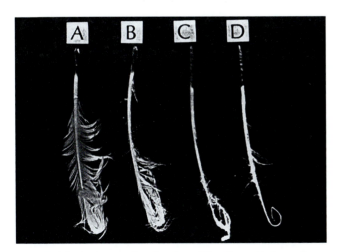

FIGURE 7-1 Bugs of a feather. The degradation of keratin in chicken feathers by *Bacillus licheniformis* PWD-1 (for "poultry waste disposal") is apparent in this sequence. The incubation times at A, B, C, and D are 0, 2, 7, and 10 days after inoculation with the bacterium.

Acquiring Energy for Life Processes

Energy acquisition is one of the fundamental properties of life, as discussed in the previous chapter. Depending on the organism, energy for ATP formation is usually acquired either from sunlight (by the process of photosynthesis) or from the catabolic oxidation of highly reduced compounds.

MYTH: All autotrophs are photosynthetic.

FACT: Among bacteria there are many chemotrophic autotrophs (or, more simply, chemoautotrophs), organisms that obtain energy by oxidizing inorganic chemicals and secure all their carbon by fixing inorganic carbon (carbon dioxide).

Nonphotosynthetic organisms are chemotrophs—they obtain their energy by oxidizing chemicals rather than from sunlight ▶ (see Energy and Carbon Sources, Chap. 5, p. 105). A few of these are *chemoautotrophs,* bacteria that harvest energy by oxidizing *inorganic* compounds and obtain carbon just as plants and other photoautotrophs do—by reducing CO_2 to organic compounds. Because they use no organic chemicals, chemoautotrophs are also known as *lithotrophs* (*litho* = rock or stone, suggesting inorganic material). Most nonphotosynthetic organisms, however, are *chemoheterotrophs* and obtain both energy and carbon by oxidizing highly reduced organic food molecules such as glucose.

■ Energy Acquisition by Chemotrophs

The basic energy-gathering strategy of most chemoheterotrophs is to convert food molecules to glucose (or a by-product of glucose metabolism). The chemical energy in glucose is then released by oxidizing the molecule, dismantling it into products that contain less energy. Some of this released energy is harvested to run the life processes of the cell. The metabolic fate of glucose depends on the extent to which the sugar is disassembled, as previewed in Fig. 7-2. The complete oxidation of glucose to CO_2 and H_2O by cells is a process in called *respiration.* Under certain conditions, however, some organisms cannot use their respiratory pathways; others lack them altogether. These cells partially disassemble glucose by a process called fermentation, generating some ATP and converting the remaining fragments

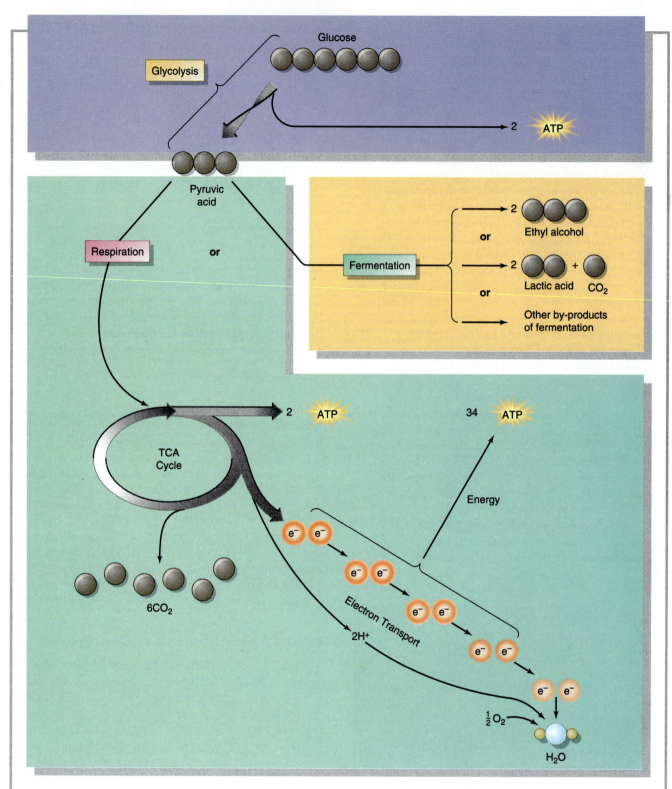

FIGURE 7-2 Preview of energy-acquiring strategies of chemoheterotrophs. Fermentation and respiration are two pathways for harvesting energy from glucose. Both strategies begin with glycolysis, the splitting of glucose into two molecules of pyruvic acid. In fermentation, the pyruvic acid is converted to organic by-products that still retain the carbon, electrons, and hydrogens—and therefore most of the energy—in the original glucose. In respiration, the pyruvic acid is completely oxidized. Respiration is a three-part operation: (1) glycolysis; (2) oxidation of pyruvic acid by the TCA cycle, which releases energized electrons; and (3) extraction of the energy in these electrons by the compounds of the electron transport system. The extracted energy is used to form ATP.

from glucose to a fermentation by-product. Notice that both respiration and fermentation begin with a pathway called *glycolysis,* the most common pathway for initiating glucose catabolism.

● **GLYCOLYSIS (EMBDEN-MEYERHOF PATHWAY).** The most universally used catabolite among microbial chemoheterotrophs is glucose. Some of the chemical energy in glucose can be liberated by **glycolysis,** a sequence of reactions in which the six-carbon sugar is split into two molecules of **pyruvic acid,** each containing three carbon atoms. Glycoly-

sis is accompanied by the production of ATP and NADH. The pathway consists of 10 metabolic reactions, each catalyzed by a specific enzyme. Each step of the pathway redistributes the energy contained in the original glucose molecule until some of it is concentrated in four high-energy phosphate bonds. This energy can then be used to form four ATP molecules. Because the phosphate is transferred directly from an organic substrate to ADP, this type of ATP formation is called *substrate level phosphorylation.* An abbreviated version of glycolysis is illustrated in Figure 7-3.

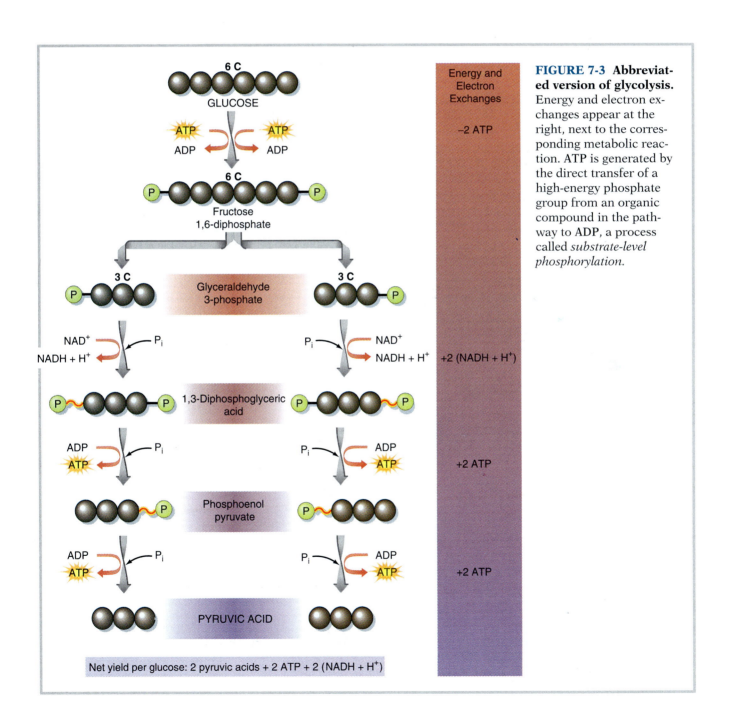

FIGURE 7-3 Abbreviated version of glycolysis. Energy and electron exchanges appear at the right, next to the corresponding metabolic reaction. ATP is generated by the direct transfer of a high-energy phosphate group from an organic compound in the pathway to ADP, a process called *substrate-level phosphorylation.*

The oxidation of glucose is initiated by activating the sugar with energy equivalent to two ATP molecules, a process analogous to priming the pump. In many bacteria, glucose is phosphorylated during active transport into the cell ▶ (see Group translocation, Chap. 4, p. 74). This makes the glucose molecule more reactive, and four ATP are eventually released. Because of this energy input, however, the net yield from glycolysis is two molecules of ATP per molecule of glucose, two pairs of electrons (carried by two NADH), and two molecules of pyruvic acid.

Production of the reduced coenzyme NADH during glycolysis depletes the supply of the electron acceptor NAD^+. Because NAD^+ is essential for oxidizing the intermediates of glycolysis, in its absence the pathway cannot proceed. The supply of NAD^+ is replenished by transferring the electrons in NADH to another molecule. The compound that ultimately acquires these electrons from NADH determines (1) the metabolic end products of glucose oxidation and (2) whether an organism is fermentative or respiratory. If an organic compound is both the electron donor and the final electron acceptor, the metabolic process is called **fermentation.** If an organic compound is the donor but an inorganic molecule, such as molecular oxygen, is the final electron acceptor, the process is termed **respiration.**

● **FERMENTATION.** Pyruvic acid or one of its derivatives is the final electron acceptor in most fermentations. Depending on the enzymes present, these compounds may be reduced by NADH to form ethyl alcohol, lactic acid, acetic acid, or a number of other metabolic by-products (Fig. 7-4). Regardless of the end product, all these processes are fermentations because an organic compound is the final electron acceptor. Fermentation requires no molecular oxygen, but some bacteria ferment glucose even when oxygen is abundant. Many obligate anaerobes and aerotolerant anaerobes generate all their energy through this metabolic pathway. (Aerotolerant anaerobes perform fermentation in the presence or absence of oxygen). For most facultative anaerobes, fermentation is the only method of obtaining energy in the absence of oxygen. Strict aerobes cannot obtain any of their energy by fermentation.

In most cases the only usable energy released from fermentation is generated during glycolysis. For example, the familiar alcoholic fermentation characteristic of brewer's yeast, *Saccharomyces cerevisiae*, converts pyruvic acid, the end product of glycolysis, to CO_2 plus a two-carbon compound, acetaldehyde, which is subsequently reduced by NADH to form ethyl alcohol. No additional energy is produced by these reactions. The release of car-

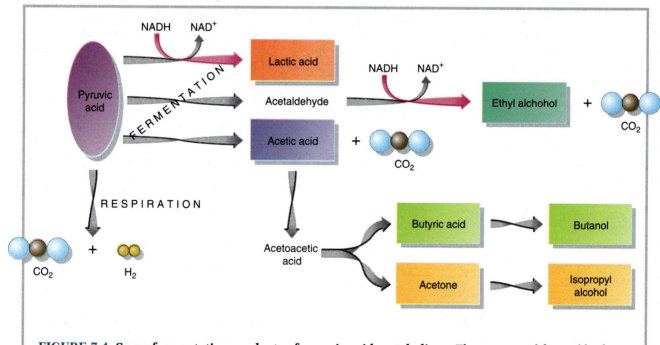

FIGURE 7-4 Some fermentation products of pyruvic acid metabolism. The compound formed by fermentation is characteristic of the organism and may be used to help identify the organism.

bon dioxide accounts for the familiar bubbling of alcoholic fermentation as well as the natural carbonation of beer and champagne. It is also the process by which yeast inflates bread dough to make it rise.

The organic waste products of glucose fermentation still contain considerable energy stored in their bonds. Some bacteria ferment these molecules discarded by other microbes. *Acetobacter*, for example, continues the degradation of ethyl alcohol into vinegar (French for "sour wine"). Propionibacteria extract energy from lactic acid by converting it to propionic acid, acetate, and CO_2. This process occurs naturally in Swiss cheese. The propionic acid contributes to the characteristic flavor of the cheese while the CO_2 creates bubbles that are trapped in the solidifying product, accounting for the holes in Swiss cheese.

Using the By-products of Fermentation. The end products of microbial fermentation have provided many medically and economically valuable products. In addition to producing various alcohols, industrial fermentations have been used to produce acetone, lactic acid, and formic acid. The final product depends on the microorganism employed and the medium in which it is grown.

Fermentation reactions generate organic end products that transform some edible substances into foods with more desirable characteristics. These altered properties usually help preserve foods and often enhance their flavor, texture, or digestibility. Dairy products, breads, soy sauce, pickled vegetables, and vinegar are produced by microbial fermentations. The role of microbes in food production is discussed in Chapter 27.

Cows, sheep, and certain other herbivorous animals possess a microbial flora that aids in the digestion of their plant diet. Bacteria and protozoa reside in an anaerobic chamber where they break down plant macromolecules into simple sugars, which they then ferment. The microbes obtain energy from the plant materials and produce fermentation end products such as acetic, proprionic, and butyric acids. These organic microbial wastes are absorbed by the animal, and the energy remaining in their chemical bonds is used to create ATP. The animal obtains additional nutrients by digesting the bacteria and protozoa that used the ingested plants for their own growth.

Fermentation as a Diagnostic Tool. The ability to ferment a particular sugar or other substrate depends on the genetic capacity of the organism to produce the necessary enzymes. Like other genetically determined characteristics (such as appearance,

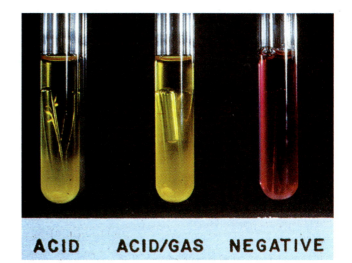

ACID   ACID/GAS   NEGATIVE

FIGURE 7-5 Fermentation "Fingerprint." Identifying an isolated bacterium, from an infected person's tissues, for example, is aided by fermentation tests. A particular species of bacteria produces enzymes that can ferment some sugars, but not all. By inoculating the organism into a battery of fermentation media that are identical except for the sugar present then, and watching for the pH indicator to change color (indicating an acid fermentation product), we can determine the pattern of sugar fermentation for that organism. In addition, some bacteria produce gas, which becomes trapped in the inverted tube. In an actual laboratory situation, many more sugars than the three shown here would be tested.

staining properties, motility, and antigens), fermentation reactions provide valuable clues for identifying microorganisms in the laboratory and are therefore important in the diagnosis of infectious disease (Fig. 7-5).

Protection of Humans by Bacterial Fermentations. The by-products of fermentation in normal bacterial flora may provide us some protection against infectious disease. For example, the lactic acid produced by the lactobacilli in the adult vagina opposes infection by lowering the pH below that which can be tolerated by many vaginal pathogens. A similar effect helps protect human skin from infection.

● **RESPIRATION.** The incomplete oxidation of glucose by glycolysis makes it a relatively inefficient process. At the end of glycolysis, approximately 95 percent of the chemical energy in glucose remains untapped in the chemical bonds of pyruvic acid. Fermentation provides no mechanism cells can use to directly harvest this energy. (Ethyl alcohol retains so much chemical energy it is used as an automobile fuel.) Respiration is an oxidative process that, unlike fermentation, efficiently transfers much of the energy remaining in pyruvic acid to form NADH and $FADH_2$. The energy in these electron carriers is then used to form ATP.

Electron Transport System.

The pair of electrons carried by NADH contains considerable energy. In respiratory organisms, some of the energy of these electrons can be cashed in for ATP by returning the electrons to their low-energy state and harnessing the energy released during the process. Many biologists compare these dynamics to the potential energy of water at the top of a waterfall. As water cascades over the rocks to its low-energy state, it releases this energy until it reaches the bottom. The energy of falling water can be used to perform useful work—to turn waterwheels that generate electricity, for example. Similarly, cells capable of respiration possess the equipment for harvesting the energy from electrons "falling" from their high-energy states.

During respiration the potential energy carried in electrons is extracted by passing the electrons through the **electron transport system** (ETS; also called the *respiratory chain*), which is a series of electron carriers attached to the cell membrane. In eukaryotes, these electron carriers are compartmentalized in the membranes of mitochondria; in prokaryotes, they are associated with the surface of the plasma membrane. Electron carriers are arranged according to their attraction for electrons, each member of the chain being more attractive to electrons than the previous molecule. Molecules with low electron affinities donate electrons to adjacent molecules with greater electron attraction. In this way the electrons flow through the respiratory chain, losing some energy with each successive transfer, in much the same way water cascading over a sequence of waterfalls loses energy as it pours down. The energy released by the flow of electrons through the ETS is harnessed for the synthesis of up to three molecules of ATP for every pair of electrons donated by NADH (Fig. 7-6). The production of ATP associated with the transport of electrons along the respiratory chain is called *oxidative phosphorylation.*

Three important types of molecules participate in most electron transport chains. They are flavoproteins, coenzyme Q, and cytochromes. *Flavoproteins* and *coenzyme Q* accept hydrogen and electron pairs. The electrons are donated to the next member of the chain, while the hydrogens are released as protons into the surrounding cell solution. Most of the carriers of the ETS are **cytochromes,** iron-containing molecules that accept high-energy electrons from the preceding member in the sequence and donate them to the next molecule in the chain.

Electron transport usually begins with a reduction of a flavoprotein by two hydrogen atoms. The pair of donated hydrogens consist of two electrons and a proton from NADH plus an H^+ (a proton) from solution. The reduced flavoprotein then transfers two electrons to the next member of the electron transport chain and releases the protons into solution. The electrons move through the ETS until they reach the last member of the chain. This final cytochrome has no subsequent member to accept its electrons, but it must be able to release them, since each cytochrome can hold only one pair of electrons. Only when a cytochrome transfers its electrons to the next member of the chain can it accept another electron pair. In most respiratory organisms, this is why molecular oxygen is needed. Molecular oxygen is the final electron acceptor.

Molecular oxygen has a greater affinity for electrons than any member of the electron transport chain. One atom of oxygen (one-half of an oxygen molecule) accepts one pair of electrons (and hydrogen ions) to form water. At this point the electrons have been completely stripped of their usable energy (they are at "ground level" state), and move out of the system in the water molecule. Thus in the presence of oxygen the chain remains open for passage of the next pair of electrons. This is the reason oxygen is essential to respiration in all aerobic organisms from bacteria to humans. In the absence of molecular oxygen, aerobic organisms die of energy starvation. The whole respiratory chain plugs up with electrons, the flow stops, and no ATP is produced. The respiratory poison cyanide has a similar effect. It binds irreversibly to the iron in a key cytochrome, preventing it from transporting electrons. The electron transport system is blocked, and therefore respiration is blocked. Any organism that relies strictly on respiration will be unable to obtain sufficient energy from glucose and will quickly die. When oxygen is used as the terminal electron acceptor, the process is called *aerobic respiration* (to distinguish it from the *anaerobic respiration* of bacteria that use an inorganic molecule other than oxygen as the terminal electron acceptor).

Membrane compartmentalization is very important to the generation of energy by respiration. Each member of the respiratory chain is anchored to a membrane in the proper sequence so each is physically separated from all but the immediately adjacent electron donors and acceptors. This prevents the cytochromes at the end of the chain from capturing the electrons directly from NADH and releasing all the energy in a single uncontrollable event. Mem-

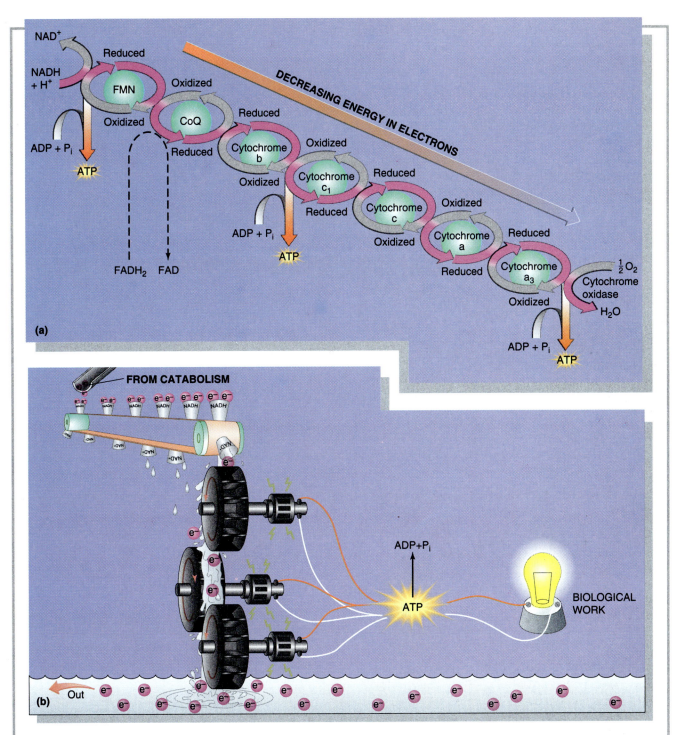

FIGURE 7-6 Electron transport. (*a*) Electrons are introduced into the electron transport chain by NADH (or FADH$_2$). Each molecule in the chain is reduced by electrons transferred from the previous molecule and is oxidized by losing electrons to the next member. (The molecules in the sequence vary according to the organism.) The electrons travel through the electron transport system, releasing energy as they move toward a lower, more stable energy state. The final oxidation rejoins a pair of protons from solution with oxygen and two low-energy electrons from the chain to form water. The energy released from the electrons is used to form molecules of ATP. (*b*) The energy dynamics of the ETS may be compared to a hydroelectric generating device in which water cascades over a series of water wheels. The water (electrons) flows into the system in a high-energy state and is carried in containers (NAD$^+$) to a series of waterwheels (cytochromes). The dynamo attached to each wheel generates energy (ATP) that can be used to run energy-requiring processes. Water (electrons) leaves the system in a low-energy state.

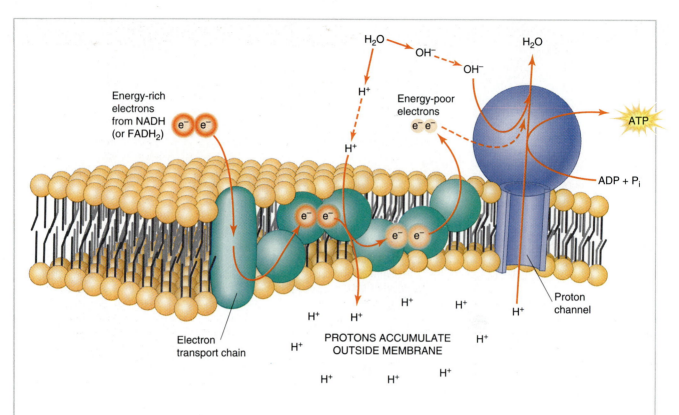

FIGURE 7-7 **Chemiosmotic generation of ATP.** Hydrogen ions accumulate on one side of the membrane, generating a protonmotive force. This gradient drives them through a channel equipped with the enzyme ATP synthetase, which uses the energy of proton flow to phosphorylate ADP to ATP.

brane compartments physically confine cytochromes to positions that ensure the transport of electrons in the proper sequence. NADH never contacts any chain member other than flavoprotein, which, in turn, has access only to the next member in the sequence.

Membranes are also important in providing a way of capturing the energy released from the "falling" electrons so it can be used to produce ATP. This process is called chemiosmosis.

Chemiosmosis—The Mechanism for Producing ATP.

The most accepted model that explains how ATP is formed during electron transport asserts that energy released from electrons is used to propel protons (H^+) across a membrane so they accumulate on one side of the barrier (much like water accumulates behind a dam). This process generates a proton gradient that represents a reservoir of energy called the **protonmotive force.** The protons eventually release this potential energy when they flow back across the membrane through proton channels. Sometimes the energy of the flowing protons is used directly by the cell, for example, to actively cotransport nutrients into the cell or to rotate fla-

gella. More often, these channels house an enzyme, *ATP synthetase,* that uses the energy of the proton flow to phosphorylate ADP to ATP. The synthesis of ATP using the energy created by a protonmotive force is called **chemiosmosis** (Fig. 7-7).

Aerobic Oxidation of Glucose.

The end products of glycolysis—two molecules of pyruvic acid— are the same for fermentative and respiratory organisms. Aerobic organisms, however, can harvest additional ATP from the two NADH produced during glycolysis. The electrons from the two reduced coenzymes can be cashed in for six ATP molecules at the electron transport system. Whereas fermentative organisms derive two ATP from each glucose oxidized by glycolysis, aerobes obtain eight ATP, two directly from glycolysis plus six from the electrons oxidized by the cytochromes. In addition, respiratory organisms harvest the energy remaining in pyruvic acid by catabolizing it to carbon dioxide, its most oxidized state, and water. The pyruvic acid's energy can be harvested regardless of where it came from— glycolysis or the end product of another pathway (see BEYOND THE BASICS: Other Pathways for Catabolizing Glucose).

Other Pathways for Catabolizing Glucose

Glycolysis is the most common pathway for metabolizing glucose to pyruvate, but it is not the only one available to microorganisms. A few bacteria, such as *Pseudomonas* and *Rhizobium*, use an alternative series of reactions, the *Entner-Douderoff pathway*. The end products of glucose degradation in this case are two molecules of pyruvate, one ATP, one NADH, and one NAPDH. The pyruvate can be further catabolized to CO_2 via the TCA cycle. The NADH generated throughout the process is recycled through the electron transport system. Thus the overall yield of ATP for this pathway is almost as great as that from glycolysis (see Appendix D for reaction sequences of these pathways).

The *pentose-phosphate pathway* often functions along with glycolysis or the Entner-Douderoff pathway. Pentoses (five-carbon sugars), such as ribose, are intermediate products in this pathway. Although the net yield of ATP from the complete breakdown of glucose is lower with this pathway than with glycolysis, the pentose-phosphate pathway serves other purposes for the cell.

1. It acts as an entry point for five-carbon sugars into metabolism. When pentoses are transported into the cell, they can be directed into this pathway and catabolized for energy.
2. The pentose-phosphate pathway is used to generate pentoses, such as ribose, from glucose for cell use. Ribose, for example, is needed to construct ATP, NAD^+ and $NADP^+$, and nucleic acids.
3. The pentose-phosphate pathway generates large quantities of NADPH reducing power that can be used for biosynthesis.

The amount of glucose shunted to glycolysis or to the pentose-phosphate pathway changes according to the cell's momentary needs for energy, intermediate compounds, or reducing power.

The following discussion describes how the complete oxidation of glucose by respiration releases 19 times as much usable energy (ATP) as fermentation does.

Aerobic Oxidation of Pyruvic Acid. Pyruvic acid, the end product of glycolysis, is oxidized by *decarboxylation*, removing one of the carbon atoms in the form of CO_2 (Fig. 7-8). Pyruvic acid decar-

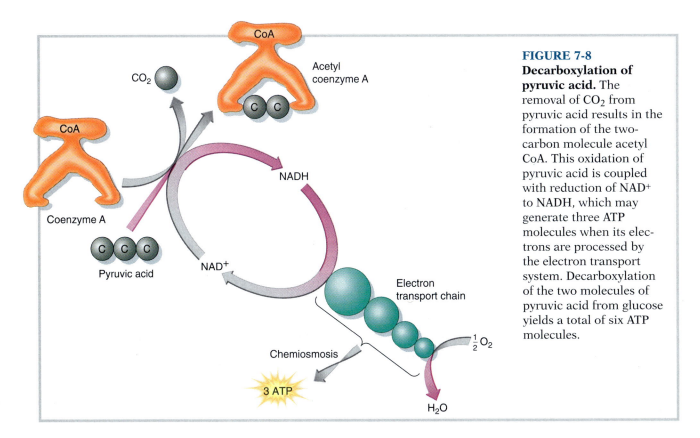

FIGURE 7-8

Decarboxylation of pyruvic acid. The removal of CO_2 from pyruvic acid results in the formation of the two-carbon molecule acetyl CoA. This oxidation of pyruvic acid is coupled with reduction of NAD^+ to NADH, which may generate three ATP molecules when its electrons are processed by the electron transport system. Decarboxylation of the two molecules of pyruvic acid from glucose yields a total of six ATP molecules.

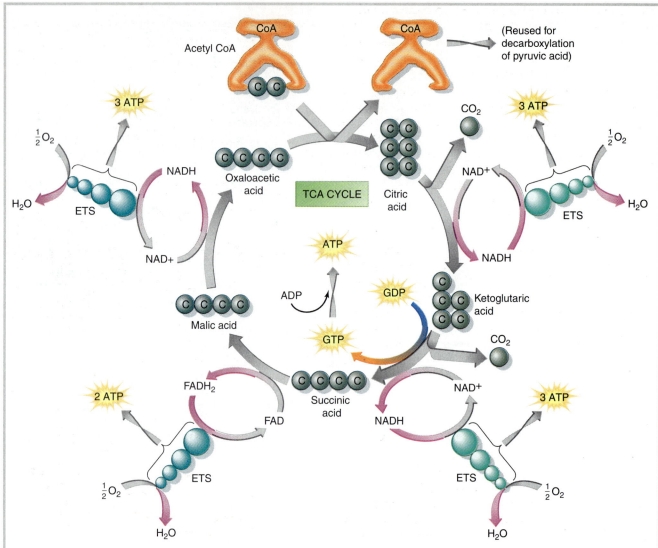

FIGURE 7-9 **Abbreviated version of the TCA cycle.** When coupled with the electron transport system (ETS), this cyclic pathway completes the total oxidation of glucose, to CO_2 and water. Each NADH may be sent through the ETS to yield three ATP molecules. The $FADH_2$ yields two ATP molecules when oxidized by the respiratory chain.

boxylation also produces a two-carbon fragment, called an *acetyl* group, that is temporarily hooked to the carrier molecule coenzyme A (CoA). The whole complex is called *acetyl coenzyme A*. The reaction also yields one NADH, which can generate three ATP molecules by passage of its electron pair through the electron transport system. Decarboxylation of the two molecules of pyruvic acid generated from each glucose molecule therefore yields two acetyl CoA and six ATP.

Tricarboxylic Acid Cycle (Krebs Cycle). Pyruvic acid decarboxylation is a metabolic bridge that transfers the carbon of glucose from the glycolytic pathway to the **tricarboxylic acid (TCA) cycle,** the pathway that completes the oxidation of glucose (Fig. 7-9). Acetyl CoA enters the tricarboxylic acid cycle by being enzymatically coupled to a four-carbon molecule (oxaloacetate). The resulting six-carbon compound is citric acid. (For this reason the TCA cycle is also called the *citric acid cycle.* Another common name is the *Krebs cycle,* for its discoverer.) During the successive steps, the two carbon atoms from acetyl CoA are oxidized to two molecules of carbon dioxide, leaving the four-carbon oxaloacetate to accept another acetyl CoA. In the course of the cycle, some of the energy released from the oxidation of citric acid is used directly to produce one ATP molecule. Most of the energy, however, is transferred by four electron pairs to three molecules of NAD^+ (forming three NADH) and one molecule of FAD (forming $FADH_2$). The energy from these electrons is then used to generate ATP at the electron transport system. In this way one acetyl CoA yields 12 ATP molecules when oxidized by the TCA cycle. Since two molecules of acetyl CoA are produced for each glucose oxidized, the final energy yield from the TCA cycle is 24 ATP molecules. Adding this to the 14 ATP derived from the previous oxidations, we find that aerobic oxidation of each glucose molecule yields enough energy to produce 38 molecules of ATP (compared to two ATP produced by anaerobic glycolysis). Thus, more than 90 percent of the ATP synthesized from glucose oxidation depends on the availability of oxygen as the final electron acceptor. The complete sequence of aerobic oxidation of glucose to water and CO_2 is distilled in Fig. 7-10.

Facultative anaerobes preferentially utilize the energy-efficient aerobic pathways when molecular oxygen is available, a phenomenon known as the *Pasteur effect*. Industrial fermentations that use facultative anaerobes are carefully controlled to exclude oxygen from the culture, since in the presence of molecular oxygen the useful by-products of fermentation either are not produced or are further oxidized to undesirable compounds.

In reality, most organisms do not produce 38 ATP from each glucose metabolized, since electrons are also needed for other reactions such as biosynthesis, cell motility, or, in a more unusual example, generating light (see Chap. 14, page 344). These electrons are therefore not cashed in for ATP, but their energy is not wasted since it is absorbed and stored in the newly synthesized molecules or used to drive other life-sustaining processes.

Anaerobic Respiration of Glucose. Some anaerobic organisms possess respiratory chains that convert the energy of electrons into ATP in the absence of molecular oxygen (Fig. 7-11). The terminal electron acceptor in such **anaerobic respiration** is an inorganic molecule other than oxygen, such as a sulfate, a nitrate, or carbon dioxide (Table 7-1). Anaerobic respiration is less efficient than aerobic respiration because the respiratory chains are shorter and they yield fewer ATP molecules per electron pair processed.

Anaerobic respiration is performed by some obligate anaerobes and facultative anaerobes. For example, metabolism by methane-producing obligate anaerobes (methanogens) occurs within digesters

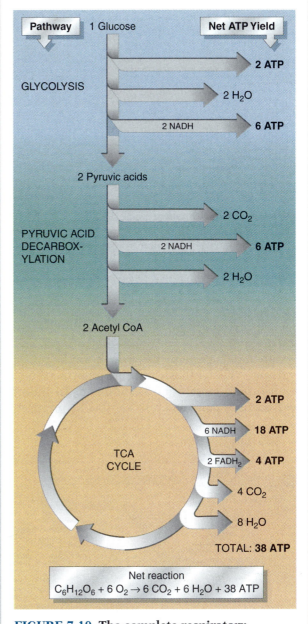

FIGURE 7-10 The complete respiratory sequence of glucose oxidation.

used to treat sewage and other organic wastes. In many places, these digesters are simultaneously used for treating wastes and for generating methane as a fuel. Other obligate anaerobes, the sulfate-reducing bacteria, corrode buried sewage pipes and equipment in the oil and shipping industries. Many facultative anaerobes use nitrate as an alternative terminal electron acceptor in respiration when oxygen is unavailable. Although less efficient than aerobic respiration, anaerobic respiration provides

FIGURE 7-11 **Anaerobes.** Bacteria living in this sulfur hot spring rely on anaerobic respiration, substituting sulfate for oxygen as the terminal electron acceptor.

Catabolism of Molecules Other Than Glucose.

Not all chemoheterotrophic organisms are limited to glucose as the sole source of energy. Many cells can tap energy stored in macromolecules by degrading them to their small molecular subunits (monomers). Because breaking the bonds between monomers consumes water, the splitting of polymers into monomers occurs by **hydrolysis** (*hydro* = water; *lysis* = split). Bacteria hydrolyze macromolecules to smaller fragments and monomers extracellularly by releasing specific enzymes into the sur-

more energy per molecule of glucose than fermentation. Nitrate, the final electron acceptor, must be present in sufficient amounts for this process to generate enough energy for growth. In the presence of oxygen, the synthesis of a key enzyme required for anaerobic respiration is inhibited. In this way, these bacteria can switch to aerobic respiration.

TABLE 7-1	EXAMPLES OF ANAEROBIC RESPIRATION	
Final Electron Acceptor	Final Product	Organism
SO_4^{2-}	H_2S, H_2O	*Desulfovibrio*
CO_2	CH_4	*Methanobacterium*
NO_3^-	NO_2^-	*E. coli;* *Pseudomonas*

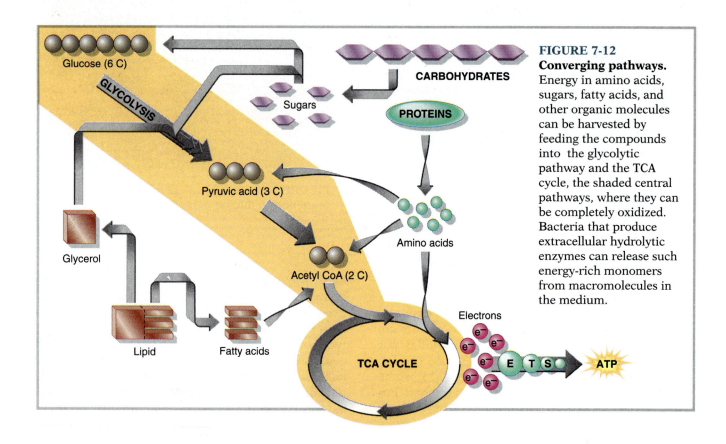

FIGURE 7-12
Converging pathways. Energy in amino acids, sugars, fatty acids, and other organic molecules can be harvested by feeding the compounds into the glycolytic pathway and the TCA cycle, the shaded central pathways, where they can be completely oxidized. Bacteria that produce extracellular hydrolytic enzymes can release such energy-rich monomers from macromolecules in the medium.

Cellulose—A Vast Untapped Food Resource

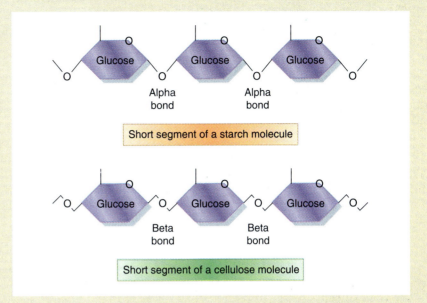

Short segment of a starch molecule

Short segment of a cellulose molecule

Cellulose is the most abundant carbon and energy source on earth. Found in trees, grasses, and other plants, cellulose provides chemical energy to countless chemoheterotrophic organisms but not to humans. People cannot digest cellulose; we have no molecular capacity to release the sugars that comprise the cellulose polymer. Ironically, cellulose is identical in its chemical content to starch, one of our most easily digested polymeric foods. Like cellulose, starch is a polymer composed entirely of glucose molecules. The significant difference is in the way the sugars are fastened to each other.

Starch is an "α-glucose polymer," which means that one glucose is linked to another by an alpha glycosidic bond (see figure). This bond is readily hydrolyzed by amylase, a digestive enzyme found in our saliva and intestinal tracts. Cellulose, however, is a "β-glucose polymer," with glucose molecules joined by beta glycosidic bonds. This linkage cannot be broken by animals, including humans, because we lack the necessary digestive proteins. When we eat plants, much of their energy and carbon remain locked in the glucose of undigested cellulose. The polysaccharide passes intact through our digestive tracts, providing needed fiber (bulk) that aids in formation of feces and in defecation but supplies us with no nutrients.

Cellulose's energy and carbon can be harvested by putting microbes to work. Cellulose-digesting bacteria and protozoa in the rumen of cattle, for example, allow the animal to survive in grasslands. Microbes in the hindgut of termites digest the wood in their diet. The wood-boring shipworm causes damage only because its resident cellulolytic bacteria allow it to digest the wood it eats. Other microbes, the free-living decomposers, recycle the carbon in cellulose of dead plants. These microbes produce the enzyme necessary to split glucose fragments from the cellulose plant wall, then metabolize the released sugars. They derive energy from the glucose molecules using pathways similar to those found in human cells. Our inability to produce cellulose-digesting enzymes, however, keeps the glucose in cellulose trapped in the molecule. This has spelled starvation and death for countless millions of people.

rounding medium. The monomers are transported into the cytoplasm, where they are metabolized to intermediates of common catabolic pathways (Fig. 7-12). A bacterium that lacks the appropriate extracellular hydrolytic enzyme to disassemble a particular polymer would starve in the presence of that polymer as the sole energy source (see THE MICROBES: Cellulose—A Vast Untapped Food Resource, p.164).

The ability of an isolated bacterium to hydrolyze a particular polymer (protein, lipid, or carbohydrate) provides valuable information for identifying the microbe. Some microorganisms derive energy from complex carbohydrates such as starch and cellulose, which are broken down into glucose. Proteins are digested to release amino acids, and lipids are degraded to glycerol (an intermediate in glycolysis) and a series of two-carbon compounds that are converted to acetyl CoA. All of these compounds can be completely oxidized to CO_2 and water through aerobic respiration. Other hydrocarbons, such as petroleum, can also be converted into intermediates of aerobic respiration.

Different sugars, amino acids, or other low molecular weight organic compounds may be the only available resources. We have already seen how five-carbon sugars are converted through the pentose-phosphate pathway to intermediates in glycolysis. Most other simple sugars also enter metabolism as intermediates in the glycolytic, Entner-Douderoff, or pentose-phosphate pathways.

Many bacteria can use amino acids as sole energy sources. Amino acids can be oxidized back to metabolic intermediates of glycolysis, pyruvic acid decarboxylation, and the TCA cycle by *deamination*, the removal of NH_2. In this way cells can completely convert amino acids to CO_2 and water whenever ATP and electrons are needed. These bacteria will grow on media containing proteins that have been partially digested into smaller chains of amino acids, such as peptones or tryptones. These are then broken up into individual amino acids by the bacteria, deaminated, and directed into one of the energy-generating pathways.

Some bacteria use natural gas (methane, the gas burned in gas ranges and gas heaters) as an energy source. These bacteria, called methylotrophs, oxidize methane or methanol (a one-carbon alcohol) for energy and carbon in the presence of oxygen. As with glucose metabolism, electrons released from oxidation of these compounds pass through an electron transport chain. A protonmotive force is generated, and ATP is synthesized. The single carbon of methane is linked with other carbons through a series of reactions to form sugars or amino acids. These molecules are then directed to biosynthesis according to the cells' needs.

● **CHEMOSYNTHESIS—AUTOTROPHY WITH A CHEMICAL ENERGY SOURCE.** Some chemotrophic bacteria derive energy solely from the oxidation of inorganic chemicals such as nitrates, sulfides, or sulfates. Electrons are removed from these metabolic substrates and transferred to an electron transport system. The transfer of electrons establishes a protonmotive force, which then generates ATP by chemiosmosis. In most cases, the final electron acceptor is oxygen, and these organisms are aerobes. However, their electron transport systems are less efficient than those of chemoheterotrophs, usually generating only one ATP per electron pair. Because they metabolize inorganic material for energy, these bacteria must get their carbon for building organic compounds from another source, usually CO_2. These organisms are therefore chemo*auto*trophs.

Chemoautotrophs require reducing power to convert CO_2 to organic carbon, but they cannot generate reducing power during energy production. Most chemoautotrophs apparently spend much of their ATP reserves in the production of reducing power. They use the ATP to drive a reverse electron transport system that transfers electrons from the

inorganic energy source to $NADP^+$. This not only consumes much of the ATP reservoir in the cell but also diverts some of the substrate from energy production. Chemoautotrophs, therefore, require high concentrations of their energy resource to grow efficiently.

Chemoautotrophs are the organisms that support the food chain in deep sea vents, an environment that is too dark for light energy to be harvested by phototrophs. Material escaping from the vents is rich in inorganic material, and seawater has ample CO_2, so this environment supplies all the components chemoautotrophic bacteria need to manufacture organic carbon. Some of the other organisms surrounding the vents live in symbiotic associations with chemoautotrophs (Fig. 7-13). They absorb organic carbon produced by the bacteria. The bacteria are the primary producers in a deep sea food chain that supports a variety of other vent dwellers. ▶ (See also THE EXPLORERS: Nature's Lawbreakers, p.108).

Some chemoautotrophs are important participants in the earth's biogeochemical cycles. Nitrifying bacteria, for example, convert ammonia to nitrates that plants can use more easily than the positively charged ammonium ion, which tends to bind tightly to clay particles in the soil. The activities of other chemoautotrophs are particularly useful to people. Some oxidize hydrogen sulfide to elemental sulfur, producing huge geological deposits of this element. Iron-oxidizing bacteria are used in metal mining. They generate metabolic products that react with copper or uranium trapped within rock, creating soluble forms that are easily recovered from ores. Similar techniques are being developed to use microbes to enhance gold and silver recovery. Iron- and

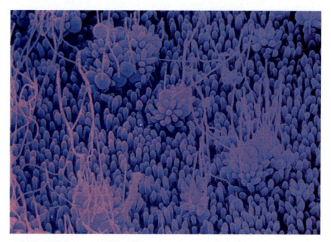

FIGURE 7-13 Microbial mat. This SEM shows sulfur-metabolizing bacteria from the tissues of a mussel at the Galapagos vent.

TABLE 7-2 SOME CHEMOAUTOTROPHIC PROCESSES

Compound Oxidized	Oxidized Product	Some Typical Genera	Natural Role or Significance of Process
Ammonium (NH_3)	Nitrite (NO_2^-)	*Nitrosomonas*	May reduce soil fertility by converting nitrogen to a more soluble form quickly washed away by rain. Important members of the nitrogen cycle in nature. Some species may be primary producers in deep ocean thermal vents.
Nitrite (NO_2^-)	Nitrate (NO_3^-)	*Nitrobacter*	Important members of the nitrogen cycle in nature. May reduce soil fertility by converting nitrogen to a more soluble form quickly washed away by rain. Some species may be primary producers in deep ocean thermal vents.
Ferrous iron (Fe^{2+})	Ferric iron (Fe^{3+})	*Gallionella, Leptothrix, Thiobacillus*	Important members of the iron cycle in nature. Some produce major geological iron deposits, cause corrosion of pipes and other products containing iron.
Hydrogen sulfide (H_2S)	Sulfate (SO_4^{2-})	*Thiobacillus, Thermothrix, Sulfolobus*	Important members of the sulfur cycle in nature. Some species may be primary producers in deep ocean thermal vents.
Sulfur (S^0)	Sulfate (SO_4^{2-})	*Thiobacillus, Thermothrix, Sulfolobus*	Important members of the sulfur cycle in nature. Some species may be primary producers in deep ocean thermal vents.
Hydrogen gas (H_2)	Water (H_2O)	*Alcaligenes, Aquaspirillum, Pseudomonas*	Poorly understood. Some species may be primary producers in deep ocean thermal vents.

sulfur-oxidizing bacteria are also helping to remove unwanted sulfur from coals. High-sulfur coals release health-threatening quantities of sulfur dioxide and other pollutants upon combustion. These wastes also contribute to the production of acid rain. The goal of coal desulfurization by microbial attack is to eventually reduce these toxic emissions.

Table 7-2 provides some examples of inorganic chemotrophic processes.

- How does an electron transport system coupled with chemiosmosis generate ATP?
- Why is aerobic respiration more efficient than fermentation or anaerobic respiration?
- How do chemoautotrophs generate ATP from inorganic molecules?

CONCEPT CHECK

- Why is glucose the preferred catabolite for most chemoheterotrophs?
- Why must NAD^+ be regenerated in the cell? How is it regenerated in fermentation? in aerobic respiration? in anaerobic respiration?
- What is the role of a terminal electron acceptor? Give examples of final electron acceptors in fermentation, aerobic respiration, and anaerobic respiration.

■ Energy Acquisition by Phototrophs

The chemical energy present in glucose and other reduced food molecules originates from the sun and is trapped by the photosynthetic machinery of phototrophs (Fig. 7-14). **Photosynthesis** is a process by which pigments harness the energy of light to form chemical energy. Since photosynthetic organisms require sunlight, they are only found in aquatic or terrestrial environments where sufficient light can penetrate.

The major pigments and overall process in photosynthetic eukaryotes and cyanobacteria are similar. Photosynthetic bacteria, on the other hand, use

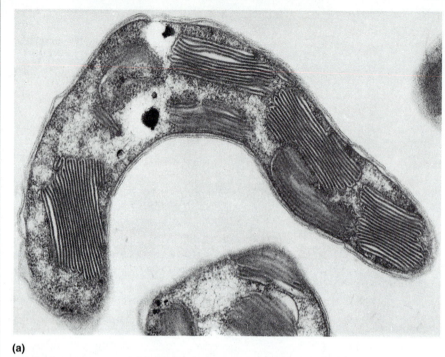

FIGURE 7-14 Photosynthetic machinery of prokaryotes. (*a*) Cyanobacteria are characterized by their membrane stacks of thylakoids. (*b*) The membrane components of photosynthetic bacteria are not free. These structures, called *chromatophores,* are extensions of the plasma membrane.

(a)

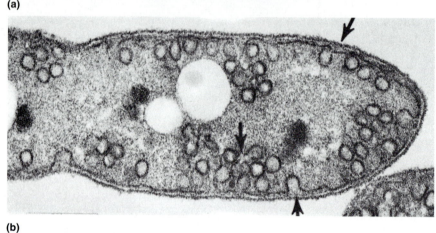

(b)

different pigments and catalyze a somewhat different series of reactions.

● **PHOTOSYNTHESIS IN EUKARYOTES AND CYANOBACTERIA.** The events of this type of photosynthesis are summarized in the following equation:

$$CO_2 + H_2O + \text{Light Energy} \xrightarrow[\text{Chlorophyll}]{} C_6H_{12}O_6 + O_2$$

Glucose

Chlorophyll is the primary pigment that absorbs light energy and releases *excited electrons* that temporarily contain some of the energy from sun-

light. Many biologists consider this to be Earth's most important reaction. The chemical energy harvested by glycolysis and respiration or by fermentation was originally introduced into the glucose molecules by light-excited electrons in photographs.

In oxygenic photosynthesis (shown in the equation) these excited electrons are used to produce NADPH and ATP, the reducing power and energy needed for converting CO_2 to glucose. Water is split to provide hydrogens (for reducing power), and oxygen is produced as a by-product. Oxygen-generating photosynthesis is referred to as *oxygenic.*

Oxygenic photosynthesis occurs in four steps as shown opposite (the flow of energy is shown in color):

1.

Light energy absorbed by a pigment called chlorophyll a excites two electrons to high-energy states. The loss of these two electrons from the chorophyll leaves a "hole." This electron hole is filled by electrons released from water by *photolysis* (the use of light energy to split water). Removal of these two electrons from water produces two protons and an atom of oxygen gas.

2.

The excited electrons pass through a membrane-bound electron transport system. As in respiration, a proton gradient is generated. The expelled hydrogen ions can recross the membrane only through channels containing ATP synthetase. Thus light energy is converted into the chemical energy of ATP, a process called **photophosphorylation.** Notice that a pair of unexcited electrons remain.

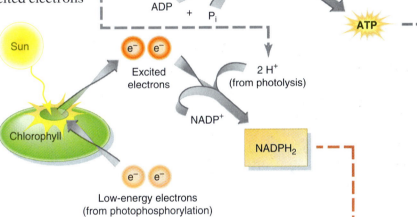

3.

Photoexcitement of a different chlorophyll a produces another pair of excited electrons. These electrons join with the protons released by the splitting of water (step 1) to reduce an $NADP^+$ to $NADPH + H^+$, generating a pool of reducing power. The electron "hole" in this chlorophyll is filled by the unexcited electrons left after ATP formation in step 2.

4.

No light is needed for the final synthesis events that manufacture sugar from CO_2. This series of reactions (shown in step 4) are called the light-independent reactions, the *Calvin-Benson cycle,* or simply *carbon fixation*. The sugar synthesized by this cycle serves as a reservoir of energy and building materials for the cell. When needed, the organic compounds produced by photosynthesis are metabolized for energy or used to synthesize more complex molecules.

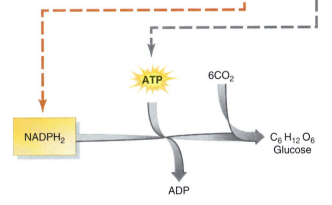

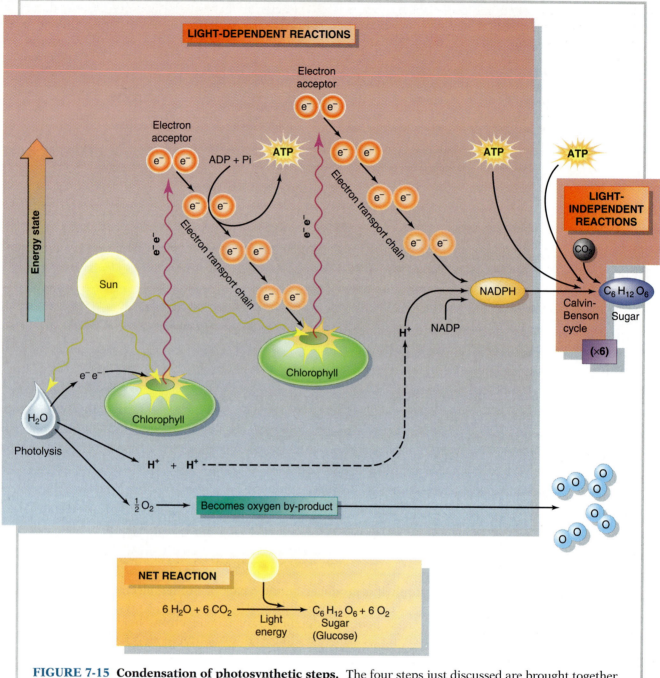

FIGURE 7-15 Condensation of photosynthetic steps. The four steps just discussed are brought together here in a single illustration, with increasing energy states located at the top of the diagram. The light-dependent reactions (steps 1 through 3 in the text) and light independent reactions (step 4 in the text) are separated into distinct boxes.

The oxygen generated by photolysis is eventually liberated as free molecular oxygen, the waste product of photosynthesis. Thus plants, algae, and cyanobacteria supply not only food for all consumers on earth but also virtually all of the molecular oxygen essential for the survival of aerobic organisms, including humans. A summary of photosynthesis is presented in Figure 7-15.

Some of the ATP for generating glucose during the light-independent reactions come from another photosynthetic process, *cyclic photophosphorylation.* In this process, electrons energized by light return to the same chlorophyll molecule that donated them. As these electrons cycle from low-energy to high-energy states and back again, they travel through an electron transport system that generates protonmotive force for producing ATP, as shown below:

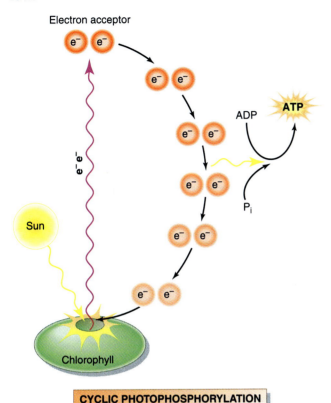

CYCLIC PHOTOPHOSPHORYLATION

In eukaryotes, pigments and electron transport systems are associated with the internal membranes of the chloroplast. In cyanobacteria, these components are bound to extensions of the plasma membrane. The similarity between photosynthesis in eukaryotes and cyanobacteria suggests an evolutionary relationship between the chloroplast and cyanobacteria. Current genetic evidence supports this idea. Life on earth likely originated in an anaerobic environment. Eventually an oxygenic photosynthesizing organism emerged. This ancient microbe gave rise to at least two lines of descendants: cyanobacteria and a prokaryotic ancestor of the chloroplasts (Fig. 7-16).

● **BACTERIAL PHOTOSYNTHESIS.** Photosynthetic bacteria (green and purple bacteria) perform photosynthesis that differs in several important ways from that in eukaryotes and cyanobacteria.

1. The major light-absorbing pigment in photosynthetic bacteria is **bacteriochlorophyll.**

2. Bacterial photosynthesis is *anoxygenic;* that is, it does not produce free oxygen. Light-activated electrons pass through an electron transport system and generate a proton gradient that drives the chemiosmotic synthesis of ATP. Low-energy electrons, however, return to the bacteriochlorophyll and refill the "hole" created by their removal (Fig. 7-17). Water is not split. Because the same electrons originally energized by light return to fill the hole they left, this process is cyclic photophosphorylation.

3. Photosynthetic bacteria generate reducing power from chemicals other than water. For example, some bacteria reduce $NADP^+$ directly with hydrogen (H_2). Others use hydrogen sulfide (H_2S) as the hydrogen donor. This leaves elemental

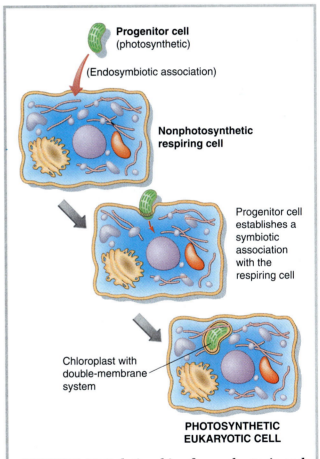

FIGURE 7-16 Relationship of cyanobacteria and chloroplasts. An oxygenic photosynthetic cell is believed to be the progenitor of both chloroplasts and cyanobacteria. A free-living descendant of this common ancestor is assumed to have established a symbiotic association with a respiring cell and evolved into the photosynthetic organelle of eukaryotes. The cell wall, if there was one, is not shown.

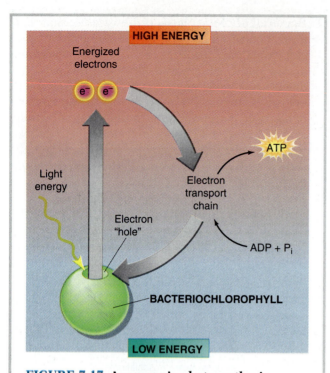

FIGURE 7-17 Anoxygenic photosynthesis.
Transport of light-activated electrons in anoxygenic photosynthesis is cyclic, the electrons returning to their original site in bacteriochlorophyll. This cyclic flow of electrons produces ATP only and generates no reducing power directly.

but convert light energy to ATP under anaerobic conditions. These bacteria are particularly interesting because their photosynthetic machinery does not require an electron transport system. Light stimulates a unique membrane-bound pigment system called **bacteriorhodopsin,** similar to the light-absorbing vision pigment (called rhodopsin) in the retina of vertebrate eyes. When light is absorbed by bacteriorhodopsin, protons are forced to the opposite side of the cell membrane. As in the other chemiosmotic systems, return of the protons drives the synthesis of ATP.

The methods by which organisms obtain energy are summarized in Table 7-3.

CONCEPT CHECK

- How do bacteria transform light energy into chemical energy?
- Compare cyanobacterial photosynthesis with bacterial photosynthesis.
- Explain how halophiles use the energy in sunlight to synthesize ATP even though they do not contain chlorophyll.

---

sulfur (S) as a waste product rather than oxygen:

$$12H_2S + 6\ CO_2 \xrightarrow[\text{Bacteriochlorophyll}]{} C_6H_{12}O_6 + 12\ S + 6H_2O$$

Other photosynthetic bacteria use thiosulfate or organic compounds such as succinate or malate as a source of reducing power. Most of these compounds have a higher affinity for electrons than do NAD^+ or $NADP^+$ and do not readily donate electrons to these carriers. The cells expend ATP to drive the reverse electron flow required to form reducing power.

4. Most photosynthetic bacteria, like cyanobacteria, use ATP and NADPH to reduce CO_2. However, some photosynthetic bacteria that metabolize organic molecules for reducing power can also use these compounds as a carbon source.

● **PHOTOSYNTHESIS IN ARCHAEOBACTERIA.** Some halophilic bacteria are facultative phototrophs. They perform respiration under aerobic conditions

Using Energy for Life Processes

Some of the energy generated by catabolic reactions is used for cell work such as membrane transport, motility, and luminescence. Most of the energy, however, drives anabolic reactions that create essential molecules for the cell. Most of these products of anabolism are proteins, nucleic acids, polysaccharides, and lipids. These four essential types of macromolecules are constructed by chemically linking the appropriate monomers ▶ (see Macromolecules—Giants of the Molecular Realm, Chap. 6, p. 133). Since biosynthesis is an endergonic reductive process, it requires both energy (ATP) and electrons (reducing power, usually in the form of NADPH). In addition, the cell requires a source of the monomers for constructing the organic components of the cell.

■ **Carbon Assimilation Among Autotrophs**

Photoautotrophs and most chemoautotrophs synthesize organic carbon compounds by reducing CO_2. The process is similar in both types of microbes,

TABLE 7-3 FIVE METABOLIC STRATEGIES FOR ACQUIRING ENERGY

Strategy	Type of Metabolism	Final e⁻ Acceptor	Energy Source
Photosynthesis	Anabolic	Chlorophyll and $NADP^+$	Light
Fermentation	Catabolic	Organic compound	Chemical (organic)
Aerobic respiration	Catabolic	O_2	Chemical (organic)
Anaerobic respiration	Catabolic	Inorganic compound other than O_2	Chemical (organic)
Lithotrophy	Catabolic	Usually O_2	Chemical (inorganic)

which fix six molecules of CO_2 into one molecule of glucose through a series of reactions called the Calvin-Benson cycle. Assembling a glucose molecule requires dozens of molecules of ATP for energy and NADPH to provide reducing power.

Phototrophs obtain the necessary ATP from the light-driven reactions of photosynthesis. Photosynthetic eukaryotes and cyanobacteria derive reducing power from the hydrolysis of water, whereas photosynthetic bacteria use other chemicals such as H_2S (hydrogen sulfide). Chemoautotrophs oxidize inorganic chemicals for both energy and reducing power.

Glucose generated from the Calvin-Benson cycle can be used by the cell for the production of organic compounds or stored for later use. The transformation of glucose to essential cellular monomers occurs through pathways similar to those found in heterotrophs.

Carbon Assimilation Among Heterotrophs

Chemoheterotrophs utilize preformed organic sources of carbon. When provided with abundant materials, they often transport into the cell small organic molecules and monomers that can be used directly in the synthesis of a macromolecule. When the environmental resources are restricted, the cell manufactures essential organic molecules by modifying intermediate compounds formed during catabolism (Fig. 7-18). These intermediates provide the cell with a method of directing metabolic compounds to the pathways that best satisfy the cell's changing needs. For example, when ample glucose (and therefore energy) is available to the cell, the cytoplasm is rich in all the intermediates of glucose catabolism, such as pyruvic acid, acetyl CoA, and the TCA

compounds. These catabolic intermediates are redirected to the biosynthesis of protein, lipid, polysaccharide, or nucleic acid.

Chemoheterotrophs use the energy and electrons released during catabolic reactions for biosynthesis. Before the electrons carried as NADH can be used for biosynthesis, however, they must be transferred to $NADP^+$. The following equation illustrates this transfer:

$$NADH + NADP^+ \longrightarrow NAD^+ + NADPH$$

This reaction is reversible; the direction is determined by an organism's energy needs. When energy is abundant, the production of NADPH is favored. (Biosynthesis of macromolecules is not only essential to growth but also an excellent method of storing energy.) Catabolic and biosynthetic reactions are therefore coupled not only by ATP (an energy carrier) but by electron carriers as well.

Virtually any type of molecule can be oxidized to a form that can enter the TCA cycle and be used for the production of ATP. Protein, for example, is metabolized to individual amino acids, many of which are then oxidized to acetyl CoA, ketoglutaric acid, or succinic acid. Each of these can then be directed into the TCA cycle for energy production. Conversely, when energy is abundant, these intermediates can be converted to amino acids for protein synthesis.

Lipid Metabolism

The fatty acids in lipids are constructed by sequentially attaching the two-carbon acetyl group of acetyl CoA to a growing chain. Fatty acids are attached to glycerol (also an intermediate of glycolysis) to pro-

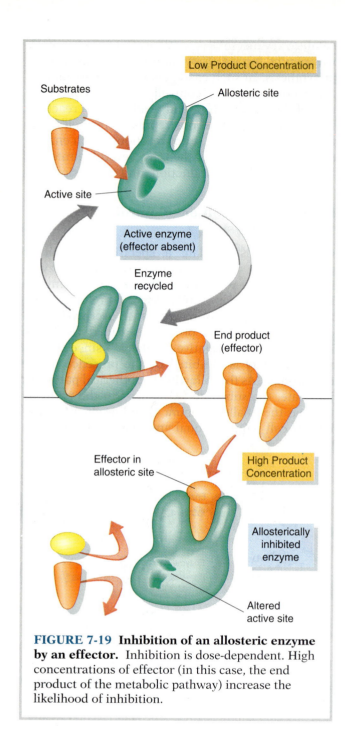

Low Product Concentration

Substrates

Allosteric site

Active site

Active enzyme
(effector absent)

Enzyme
recycled

End product
(effector)

Effector in
allosteric site

**High Product
Concentration**

Allosterically
inhibited
enzyme

Altered
active site

FIGURE 7-19 Inhibition of an allosteric enzyme by an effector. Inhibition is dose-dependent. High concentrations of effector (in this case, the end product of the metabolic pathway) increase the likelihood of inhibition.

with the substrate. In this way, the effector temporarily inactivates the enzyme (Fig. 7-19). The probability of an effector binding to the allosteric site increases as the concentration of the effector increases. Thus, the greater the concentration of the

end product of a pathway, the more the pathway will be inhibited. The inhibition is reversible, and the enzyme's activity is restored when the end product becomes scarce and dissociates from the allosteric site. The biosynthetic pathway operates only when its product is needed.

■ Induction and Repression—Control of Enzyme Synthesis

Allosteric inactivation of enzymes provides immediate control but wastes energy because it provides no mechanism for halting the synthesis of unneeded enzymes. Bacteria can also inhibit the synthesis of enzymes and turn it back on in response to changing conditions. In many metabolic pathways, for example, elevated concentrations of the pathway's product block enzyme synthesis. Such a system is referred to as *repressible* because enzyme synthesis is repressed by high product concentrations (these enzymes are not needed unless there is little of the pathway's product).

Many catabolic systems, on the other hand, are *inducible*. Enzyme synthesis occurs only when the corresponding substrate is available and increases according to the concentration of the substrate. *Escherichia coli*, for example, does not produce the enzymes necessary for utilizing lactose when lactose is not available. The presence of lactose, however, induces the synthesis of enzymes that catabolize it; induction is sustained until the supply of lactose is exhausted. Thus, many organisms with the genetic potential for utilizing several sugars will produce only the enzymes they need for metabolizing the sugars that are available to them. Mechanisms that regulate which enzymes are synthesized are part of the organism's genetic makeup; they are discussed in the next chapter, Principles of Bacterial Genetics.

CONCEPT CHECK

- What is an allosteric enzyme, and what is its role in feedback inhibition? Why is the first enzyme unique to the pathway often the only allosteric enzyme in that pathway?
- What advantage are feedback inhibition, induction, and repression to a bacterium?

Key Terms

glycolysis (p. 153)
pyruvic acid (p. 153)
fermentation (p. 154)
respiration (p. 154)
electron transport system (p. 156)
cytochromes (p. 156)

protonmotive force (p. 158)
chemiosmosis (p. 158)
tricarboxylic acid (TCA) cycle (p. 159)
anaerobic respiration (p. 160)
hydrolysis (p. 162)
photosynthesis (p. 166)

bacteriochlorophyll (p. 169)
bacteriorhodopsin (p. 170)
feedback inhibition (p. 173)
effector (p. 173)
allosteric enzymes (p. 173)

Key Facts and Concepts

- **Chemotrophy.** Chemotrophic organisms obtain energy by oxidizing chemical bonds. Both chemoheterotrophs and chemoautotrophs transfer electrons from their organic or inorganic substrates to an electron transport system. Through chemiosmosis, the electron transport system provides a means to extract the energy in the electrons to form ATP, greatly increasing the efficiency of the catabolic processes. As electrons pass down the respiratory chain, a proton gradient accumulates on one side of the membrane. This protonmotive force drives ATP synthesis as protons flow through membrane channels containing ATP synthetase.

- **Metabolizing glucose.** Chemoheterotrophs often use glucose as their energy source. Glucose catabolism usually begins with glycolysis, splitting the sugar into two molecules of pyruvic acid with a net yield of two ATP and two NADH. In fermentation, the final electron acceptor is an organic molecule, usually pyruvic acid or its derivative. No additional energy is provided to the cell, and the overall energy yield is two ATP.

- **Respiration.** In respiration, the final electron acceptor is an inorganic molecule. In aerobic respiration, oxygen is the final electron acceptor. In anaerobic respiration, the terminal electron acceptor is an inorganic molecule other than oxygen. Pyruvic acid is decarboxylated, and acetyl CoA is produced. This molecule is further degraded to CO_2 by the TCA cycle. Two additional ATP result directly from these reactions as well as 12 molecules of reduced NAD^+ and two molecules

of reduced FAD. The electrons are carried by the coenzymes to the electron transport system, where their energy is extracted and used to generate ATP. The overall yield from aerobic respiration is 38 ATP per glucose molecule.

- **Phototrophy.** Photosynthetic organisms acquire their energy by photophosphorylation, transforming light energy into chemical energy in ATP. Photosynthetic eukaryotes and cyanobacteria perform oxygen-generating photosynthesis. Bacterial photosynthesis releases no free oxygen.

- **Biosynthesis.** The intermediates of catabolism provide the raw materials for biosynthetic processes. Most of the monomers of protein, lipid, polysaccharide, and nucleic acid are manufactured from intermediate compounds generated from glycolysis, pyruvic acid decarboxylation, or the tricarboxylic acid (TCA) cycle. The intermediates may be diverted from catabolism to the synthesis of needed compounds.

- **Regulating metabolism.** Organisms can respond to changes in both their chemical environment and internal metabolic needs. Metabolism is carefully regulated by a cell so that necessary reactions are turned on and unnecessary reactions are inhibited. This is accomplished by allosteric inhibition of enzyme activity and by the processes of induction and repression of enzyme synthesis. These mechanisms for metabolic control are usually sensitive to the changing concentrations of the substrate or product of the pathway being regulated.

Review and Problem-Solving Questions

1. Fill in the following chart for organisms supplied with glucose in the medium.

	Number of ATP from Glycolysis (per Glucose)	Final Electron Acceptor	Representative End Product
Fermentation			
Aerobic respiration			
Anaerobic respiration			
Chemoauto-trophy			

2. A chemoheterotrophic bacterium is in the presence of an abundant supply of glucose but does not grow. List three possible reasons.

3. Explain why each of the following statements is false.
   (a) Oxygen is required only for organisms that can use aerobic respiration to acquire energy.
   (b) A facultative anaerobe will use the same amount of glucose in the presence or absence of oxygen.
   (c) A facultative anaerobe will produce more carbon dioxide from glucose in the absence of oxygen than in the presence of oxygen.
   (d) All enzymes unique to a biochemical pathway are allosteric enzymes.

4. On a single graph, draw the growth curves of a facultative anaerobe grown (a) under aerobic conditions and (b) under anaerobic conditions. Explain your graph.

5. What kinds of energy-generating metabolism can be found among strict anaerobes?

6. Explain how aerobic organisms obtain eight ATP molecules from glucose catabolism whereas fermentative organisms produce only two ATP molecules.

7. What happens to fermentative organisms in the presence of cyanide?

8. An anaerobe has the capacity to ferment or anaerobically respire. Explain which metabolic pathway is likely to be used.

9. Glucose can be catabolized by the Embden-Meyerhof, Entner-Douderoff, and pentose-phosphate pathways. What distinguishes these pathways from one another?

10. What advantage is it to a cell to use common pathways to catabolize sugars, fats, and amino acids rather than a series of unique pathways?

11. What would be some consequences of all chemoautotrophs disappearing from the planet? all cyanobacteria?

12. Why must cytochromes be bound to membranes in order to function correctly?

13. Complete the following table.

	Algae	Cyano-bacteria	Photosyn-thetic Bacteria
Site of photo-synthesis			
Light-absorbing pigment			
Oxygen generation			
Carbon source			

14. Why might an organism get fat when it consumes large quantities of sugar?

15. Why do chemoautotrophs require high concentrations of their energy resource to grow efficiently?

16. In a fermenting organism, if NADH transfers its electrons to $NADP^+$ for biosynthesis, what happens to pyruvic acid?

17. Distinguish between feedback inhibition, enzyme induction, and enzyme repression. Which responds the fastest to changes in the cell's environment? Which is least wasteful of the cell's resources?

18. How does an effector control the activity of an allosteric enzyme?

ANAEROBIC ADAPTATIONS

Diverse metabolic activities displayed by microorganisms have been used in the commercial production of foods, pharmaceuticals, and other industrial products. Because strict anaerobic bacteria are more difficult to work with than aerobes, they aren't as well studied. Even less is known about the anaerobes living in extreme environments such as temperatures above 60°C, concentrations of NaCl above 10 percent, or pH below 2 or above 9, and thus these bacteria represent largely unexplored groups of microbes. Susan Lowe et al.* review the biology of these anaerobes and suggest that they may possess useful metabolic adaptations to their environments, including the capacity to grow on toxic compounds or industrial waste products and the ability to produce novel organic compounds of industrial or pharmaceutical value. In addition, these anaerobes are a potential source of stable enzymes that can be used to catalyze reactions under extreme conditions.• Describe three useful products of anaerobic metabolism and the metabolic pathways through which they are produced.

• Outline the steps in the complete anaerobic metabolism of a polysaccharide. Where do lipids and proteins enter the energy-generating pathway in a bacterium that performs fermentation? anaerobic respiration?

• Facultative anaerobes exposed to oxygen switch from fermentative to a respiratory metabolism. What happens to strict anaerobes that are fermentative?

*S. E. Lowe, M. K. Jain, and J. G. Zeikus, Biology, ecology, and biotechnological applications of anaerobic bacteria adapted to environmental stresses in temperature, pH, salinity, or substrates, *Microbiological Reviews,* 57: 451-509 (1993).

CHAPTER 8

Principles of Bacterial Genetics

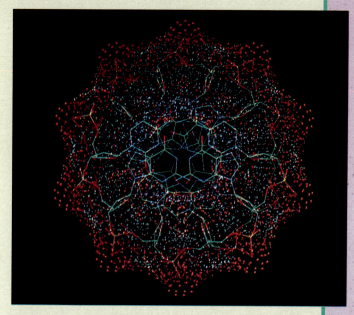

Peering into the secret of life. This unusual end view of the spiral DNA molecule is like gazing down the center of an enormously long spiral staircase. DNA contains the genetic instructions that direct an organism's self-construction and provides a mechanism for preserving those instructions from generation to generation.

The bacterial cell faced certain death. A simple chemical suddenly transformed its hospitable environment into a hostile setting from which it could not escape. The simple chemical was penicillin, injected into the bacterium's host by a nurse treating an infected patient. Soon the bacterium could no longer manufacture its cell wall, and the fragile cell burst like an overfilled balloon. Unfortunately for the patient, there was another bacterium, identical to the dead one except for one piece of genetic information, instructions that would save its life. The information instructed the cell's metabolic machinery to make an enzyme that cleaved the deadly penicillin into harmless chemical debris. Each time the bacterium divided, it supplied its daughter cells with a copy of the protective gene. The genetic information made its way from generation to generation, spelling the difference between life and death for all the bacterial cell's offspring. Each new daughter cell inherited the trait the same way you obtained your eye color and the texture of your hair from your parents.

Inheritance is not magic, nor is it beyond our comprehension. Within the past half century, scientists have demystified inheritance, replacing mystery with a fascinating and clearly understandable explanation of how each organism inherits the information that tells it whether to become a bacterium or a human.

The study of bacterial genetics has contributed enormously to our understanding of the genetics of all organisms. Many of the mechanisms discovered in relatively simple bacterial systems are very similar to corresponding mechanisms in humans. Bacteria are useful scientific models for studying the mechanisms of genetics for several reasons.

- They can be propagated so rapidly that dozens of generations can be studied in a short time.
- Large populations of essentially identical bacteria can be cultured from a single parental cell (essential for genetic homogeneity).
- Compared to eukaryotes, bacteria are genetically simple organisms. *Escherichia coli,* for example, possesses a single chromosome that contains almost 5000 genes. Human cells, with their 46 chromosomes and 100,000 genes, are much more complex and difficult to characterize genetically.
- Genetic material is readily transferred from one bacterial cell to another, so we can experimentally investigate the mechanisms of gene function.
- Bacteria require much less laboratory space than plants and animals.

Genes

The terms **genetics** is derived from the word "generation" and means the study of heredity and variation among generations of organisms. An organism's characteristics are acquired from its parents by means of the fundamental units of heredity, the **genes,** which are linearly arranged along the chromosomes. Genes direct the synthesis of all of an organisms's traits (Fig. 8-1), most of which are products essential to the organism's growth and survival. By passing copies of its genes to its offspring, an organism ensures that its progeny will inherit the parent's traits.

The entire complement of genes possessed by an organism is called its **genotype,** the organism's genetic potential. At any one time, however, not all the genes are *expressed;* that is, they are not all "turned on" in a way that allows the organism to follow the

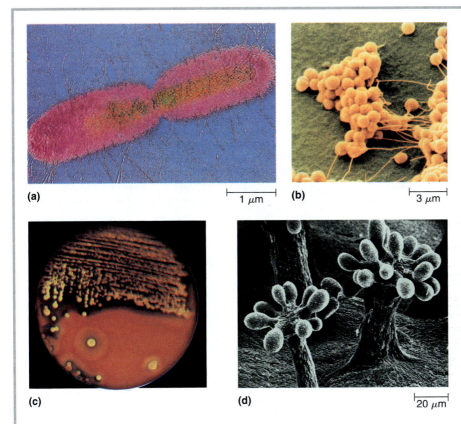

(a)　1 μm　**(b)**　3 μm

(c)　**(d)**　20 μm

FIGURE 8-1 Bacterial genetic traits. Just as the shape of your ears and the color of your skin are genetically determined, the microscopic morphology, culture characteristics, and metabolic activities of a bacterium are determined by its DNA. Here are just a few examples of genetic traits in prokaryotes: (*a*) The shape of this *Escherichia coli* and its ability to produce pili; (*b*) the arrangement of *Staphylococcus aureus* into grapelike clusters; (*c*) the colony characteristics of *Streptococcus pyogenes* growing on blood agar, including the production of an extracellular chemical that lyses the red blood cells suspended in the medium (producing a clear zone of beta hemolysis around each colony); (*d*) cell–cell interactions that form complex bacterial formations as this *Chondromyces crocatus* fruiting body.

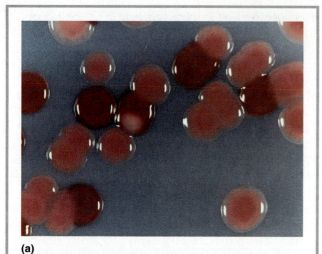

(a)

(b)

FIGURE 8-2 **Phenotype variations.** (*a*) *Serratia marcescens* produces a red pigment at 30°C. (*b*) At 37°C, pigment formation is inhibited and the colonies are white.

genetic instructions encoded in the gene. For example, many genes that control the production of enzymes necessary for digesting and using nutrients are turned on only in the presence of these nutrients. The genetic characteristics expressed at any given time constitute the organism's **phenotype.**

In bacteria, temperature, pH, age, and humidity are among the influences that dictate which genes are expressed and which are turned off as well as how the genes are expressed. For example, *Serratia marcescens* forms red colonies at 24°C and white colonies at 37°C (Fig. 8-2). The genotype is the same in both cultures, but the higher temperature inhibits production of functional pigment molecules. Thus, the phenotype has changed dramatically. Such cultural characteristics are important in laboratory identification of *Serratia marcescens,* an important

opportunistic pathogen responsible for many nosocomial (hospital-acquired) infections (see Chapter 25).

In contrast to alterations in phenotype, changes in genotype, called **mutations,** are relatively infrequent. The new genotype is stable and is inherited by all the descendants of the *mutant* organisms (the organisms containing the mutation).

CONCEPT CHECK

- How does a gene differ from a chromosome?
- Why is the genotype of an organism often different from its phenotype?

The Bacterial Chromosome and Plasmids

In prokaryotes, genetic information is contained in DNA arranged as a circular chromosome (Fig. 8-3). ► (See Nucleoid, Chap. 4, p. 87.) Bacteria are *haploid* organisms; that is, their cells contain a single copy of the chromosome (although during rapid growth, when DNA replication is faster than cell division, each cell may contain two identical copies of the chromosome). Because there is generally only one version of each gene in the cell, its genotype dictates the potential phenotype of that organism (unlike a diploid cell, in which any recessive gene is permanently masked by the presence of a dominant gene).

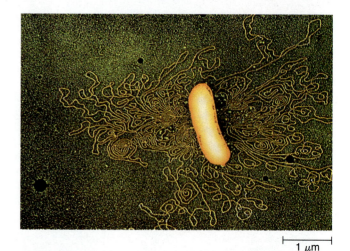

├─── 1 μm

FIGURE 8-3 **Genetic information in prokaryotes.** The DNA of the compacted chromosome released from this cell is many times the length of the cell.

The bacterial chromosome is approximately 1000 times as long as the cell containing it, but it occupies only about 10 percent of the cell volume. This long molecule is packed into the cell as a highly condensed *supercoil*. The DNA circle folds upon itself, and its twists and turns give it an extremely compact form. Supercoiling and untwisting of the DNA are controlled by enzymes called *DNA gyrases*.

Most of the information in the bacterial chromosome directs the synthesis of enzymes and structural proteins. For example, in *E. coli* about 90 percent of the chromosome carries information for the 3000 to 4000 proteins the cell can make. The other 10 percent manufactures transfer and ribosomal RNA (stable ribonucleic acids that participate in protein synthesis) or are regulatory segments that control the expression of the genetic material and allow the cell to turn genes on and off in response to environmental changes.

Most bacteria also contain smaller pieces of DNA that are not part of the chromosome. These extra-chromosomal circles of DNA are called **plasmids.** Although certain plasmids can integrate into the chromosome, most plasmids remain autonomous in the cytoplasm. The genes carried on plasmids specify different traits from those on the chromosome, traits that normally are not essential to the organism (Table 8-1). Plasmids do, however, contribute some advantages. Genes on the plasmid may encode information for such advantageous traits as an increase in metabolic options, resistance to antibiotics, or the capacity to synthesize compounds toxic to competing bacteria. But for cells grown under ideal conditions in the laboratory, plasmids are a burden. The bacteria must spend additional energy to maintain and replicate this genetic information, a handicap when their presence imparts no advantage.

In ideal conditions, plasmid-containing organisms are replaced by more efficient counterparts that lack the energy baggage. Billions of dollars depend on that not happening. As you will see in the discussion of genetic engineering in the following chapter, much of the biotechnology industry depends on making sure that plasmids carrying novel genetic information are maintained in the bacterial cells that are used as factories for the valuable prod-

ucts specified by introduced genes. Other plasmids are not so desirable, creating a serious threat to human health by protecting pathogenic bacteria from antibiotic treatment. These considerations are discussed more thoroughly later in this chapter and in Chapter 9.

CONCEPT CHECK

- What functions are performed by different regions on the bacterial chromosome?
- In what ways do bacterial plasmids differ from the chromosome?

DNA—The Genetic Material

Except for the genes of viruses that have only RNA, all genes are composed of the same genetic material: deoxyribonucleic acid (DNA). DNA stores specific genetic information that ultimately determines all the characteristics of an organism. Biological differences between organisms are due primarily to differences in the information encoded in their chromosomal DNA.

The biological tasks of DNA are threefold:

1. *Storage of genetic information.* DNA is the cell's blueprint and contains all the information needed to produce and maintain a unique organism.
2. *Inheritance.* Genetic information is precisely transmitted to all the organism's descendants.
3. *Expression of the genetic message.* Information stored in DNA is deciphered and used to direct what proteins are made by a cell, proteins that govern cellular activities and determine cellular characteristics.

The manner in which genetic information is stored, inherited, and expressed is basically the same in all organisms. The scientific detective work that revealed how cells accomplish these three tasks marks one of the most remarkable and exciting periods in the history of scientific achievement, a period that changed the way we view the living world (see THE EXPLORERS: The DNA Detectives).

■ Task 1: Storing Genetic Information

The unique structure of DNA accounts for its ability to store genetic information. DNA is a long double-stranded molecule made up of four kinds of

TABLE 8-1 REPRESENTATIVE PLASMID GENES	
• Antibiotic resistance	• Heavy metal resistance
• Antibiotic production	• Toxin production
• Catabolic enzymes	• Plant tumor induction
• Conjugation (plasmid transfer)	

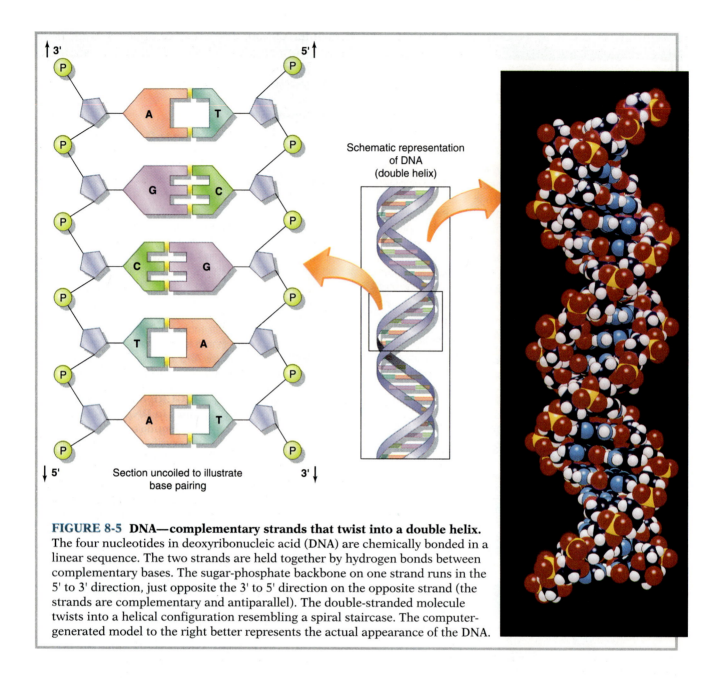

3' P

5' P

A ─ T

G ≡ C

C ≡ G

T ─ A

A ─ T

5' P

3' P

Section uncoiled to illustrate base pairing

Schematic representation of DNA (double helix)

FIGURE 8-5 DNA—complementary strands that twist into a double helix.
The four nucleotides in deoxyribonucleic acid (DNA) are chemically bonded in a linear sequence. The two strands are held together by hydrogen bonds between complementary bases. The sugar-phosphate backbone on one strand runs in the 5' to 3' direction, just opposite the 3' to 5' direction on the opposite strand (the strands are complementary and antiparallel). The double-stranded molecule twists into a helical configuration resembling a spiral staircase. The computer-generated model to the right better represents the actual appearance of the DNA.

pairs only with cytosine (C) in the other strand, and adenine (A) pairs only with thymine (T). In the double-stranded molecule, the two strands are *antiparallel*—their phosphate-sugar backbones run in opposite directions. Thus, the *5' end* (the end with the sugar's fifth carbon exposed) of one strand is opposite the *3' end* (the end with the sugar's third carbon exposed) of the complementary strand. DNA resembles a ladder in which the "rails" are linear phosphate-sugar linkages and the "rungs" are the A-T and G-C base pairs. Internal molecular forces twist the ladder into a helical configuration. For this reason DNA has been dubbed the *double helix* (Fig. 8-5).

● **THE "GENETIC LANGUAGE."**   The linear order of nucleotides can be arranged in an infinite number of sequences that carry messages written in a "genetic language." Just as verbal information is stored and transmitted by linear sequences of letters that form meaningful words and sentences, the genetic information necessary for determining every characteristic of an organism is encoded in the order of nucleotides in a DNA molecule. In the code we call written language, whenever the letters of a word are changed, the meaning of the word itself changes. Similarly, the meaning of the genetic message changes if the sequence of nucleotides is changed.

Mutation, therefore, may be defined more specifically as a permanent change in the nucleotide sequence in DNA. This may change the meaning of the stored information, thereby changing the characteristic that was specified by the gene. Different nucleotide sequences result in organisms with different characteristics, just as two books tell completely different stories by virtue of their unique sequences of letters.

MYTH: The products of mutation are generally "freaks" of nature, bizarre mutants with hideous features.

FACT: Mutation is quite a normal process that occurs to a small degree in all organisms, sometimes generating new, advantageous traits. Mutation is not only a common phenomenon, it is fundamental to evolutionary change and the formation of new species, such as humans. We are all mutants.

■ Task 2: Transmitting Genetic Information to Offspring

A cell ensures that all progeny cells get the necessary genetic information by producing exact copies of its chromosomal DNA and giving one copy to each of the daughter cells. This process, called **DNA replication,** is the way in which cells make identical copies of their DNA. Replication begins with the enzyme-catalyzed unwinding and separation of a portion of the two complementary strands of the DNA molecule (Fig. 8-6*a*). Specific DNA-binding proteins temporarily keep the unpaired single strands apart. This separation creates a Y-shaped structure called the *replication fork*, a site where DNA synthesis can begin. All bacterial cells maintain a reservoir of free nucleotides in their cytoplasm. In the presence of the enzyme **DNA polymerase** and single-stranded DNA, these free nucleotides combine with complementary unpaired nucleotides. Nearly as rapidly as the DNA strands separate, free nucleotides associate and bind with the unpaired regions, forming new complementary strands as the replication fork moves around the chromosome.

DNA replication proceeds only in a 5'-to-3' direction. But remember that the two strands of a DNA molecule run in opposite directions, one from 5' to 3' and the other from 3' to 5'. As a replication fork progresses along the chromosome, only one strand is continuously exposed in a 3'-to-5' orientation (Fig. 8-6*b*). This strand is synthesized with-out interruption. The other template strand, however, is oriented in the 5'-to-3' direction, so somehow the new complementary strand must be synthesized backward. This is accomplished by *discontinuous synthesis*. The polymerase attaches at a point "upstream" from the replication fork and synthesizes a short piece as it moves in a 3'-to-5' direction. It then pops off, waits for the replication fork to move, then reattaches and makes another segment. This produces a series of short strands with "nicks" in their backbones that are sealed into a continuous strand with an enzyme called **ligase.**

Because of the specificity in base pairing, two identical DNA molecules are produced, each containing one strand from the parent molecule and one newly synthesized complementary strand. Such replication is called *semiconservative* because each new double-stranded DNA molecule contains one strand conserved from the original molecule and one that is newly synthesized. Each new DNA molecule therefore encodes precisely the same genetic message that was stored in the original DNA molecule. This is the mechanism of heredity.

DNA synthesis occurs very rapidly. Under optimal conditions in bacteria, up to 1000 nucleotides can be polymerized in 1 second. Because the bacterial chromosome is a circle, strand separation creates two replication forks (Fig. 8-7). Replication of the circular bacterial chromosome is *bidirectional*—it proceeds from both ends at the same time. The replication forks meet, and the two chromosomes separate. In *E. coli*, replication of the chromosome takes approximately 20 minutes. The enzymes that catalyze DNA polymerization are not only fast, they are also highly accurate. The rate of mutation resulting from errors in base pairing is approximately 1 in 10^{10}. In *E. coli*, the rate of mutation in any particular gene is one in every million replications. Not only is DNA synthesis highly accurate, the enzyme "proofreads" the newly assembled strand. When an incorrect base is inserted, the enzyme usually removes it and replaces it with the correct one. The genetic stability of prokaryotes depends on this self-correcting activity.

■ Task 3: Deciphering the Genetic Message

Each gene on a chromosome is a message that directs the formation of one product, usually a protein (Fig. 8-8). This is called the "one gene, one polypeptide" theory of gene action. Originally it was called the "one gene, one protein" theory, but some complete enzymes and structural proteins contain more than one protein chain (polypeptide). In these

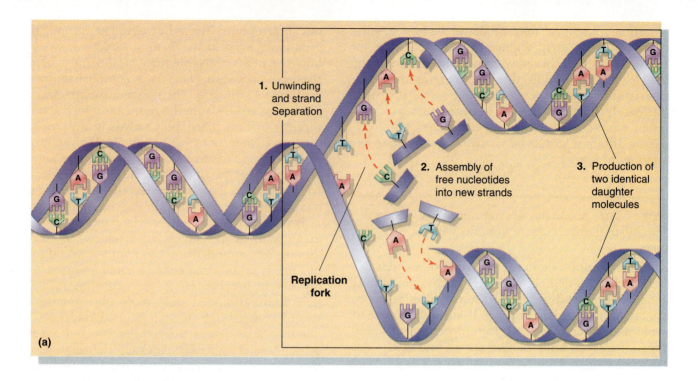

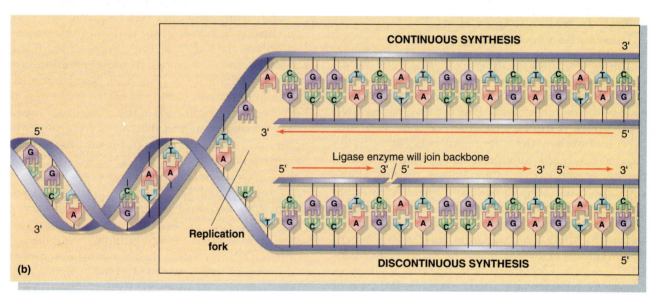

FIGURE 8-6 Replication of DNA. (*a*) *The overall result.* Strands separate, creating a replication fork. Energy for synthesis comes from the high-energy phosphate bonds in nucleotides themselves. Each of the new DNAs contains one strand from the original double helix and one newly synthesized strand. The two completed strands are identical in sequence to the original DNA molecule. Each daughter cell receives one of the copies. (*b*) *Coping with antiparallel strands.* The boxed portion of part (*a*) is examined in more detail here. As DNA strands separate, one strand is synthesized continuously, but the antiparallel strand must be copied in the opposite direction (away from the replication fork). This new strand is synthesized discontinuously, in short segments that are sealed into a single continuous strand with ligase.

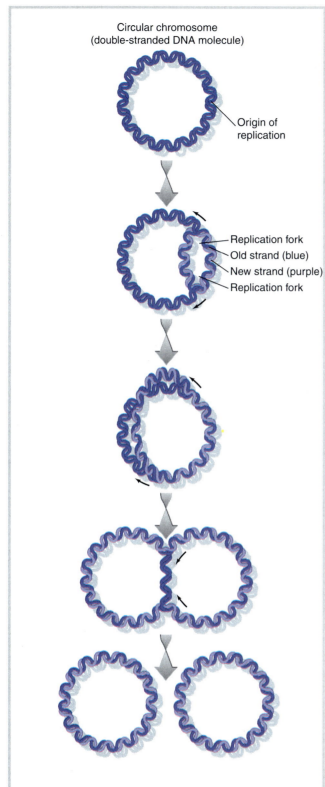

FIGURE 8-7 Replication of prokaryotic DNA. Because the bacterial chromosome is circular, there are two replication forks that travel away from each other. Synthesis continues in both directions until the replication forks meet.

Circular chromosome
(double-stranded DNA molecule)

Origin of replication

Replication fork
Old strand (blue)
New strand (purple)
Replication fork

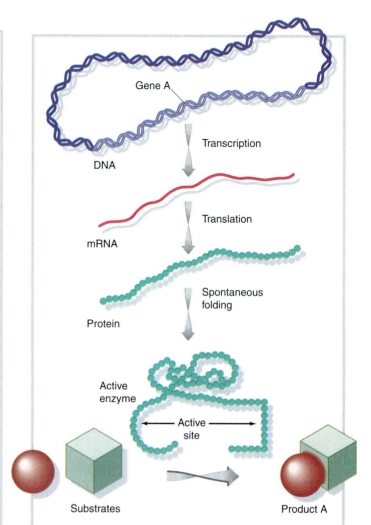

Gene A

DNA

Transcription

mRNA

Translation

Protein

Spontaneous folding

Active enzyme

Active site

Substrates

Product A

FIGURE 8-8 Flow of genetic information. The nucleotide sequence of gene A is encoded into a molecule of mRNA, which directs the synthesis of a specific protein. The protein gains its unique activity by folding into its functional three-dimensional shape. In this case the protein is an enzyme that catalyzes the synthesis of a particular substance (called product A here). Notice that DNA, mRNA, and protein are all linear polymers.

cases, more than one gene is involved in producing an active product. Because of this, most geneticists prefer to call this the "one gene, one polypeptide" principle. The gene designates the sequence of amino acids in the corresponding proteins. Thus the types of genes present in the DNA determine the types of structural proteins and enzymes made by the cell. The enzymes, in turn, catalyze the cell's metabolic reactions. *The ultimate control exerted by DNA is in dictating which enzymes and structural proteins are synthesized by the cell.*

Let's return to the importance of sequence. The

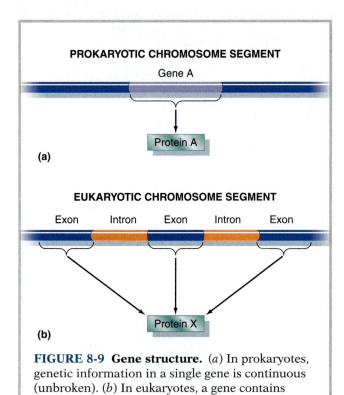

PROKARYOTIC CHROMOSOME SEGMENT

Gene A

Protein A

(a)

EUKARYOTIC CHROMOSOME SEGMENT

Exon   Intron   Exon   Intron   Exon

Protein X

(b)

FIGURE 8-9 Gene structure. (*a*) In prokaryotes, genetic information in a single gene is continuous (unbroken). (*b*) In eukaryotes, a gene contains protein-coding regions (called exons) that are interrupted by noncoding regions (introns).

sequence of nucleotides in the DNA of a gene determines the amino acid content and sequence of the corresponding protein, and the amino acid sequence ultimately determines the function of a protein. In prokaryotes, there is a direct correspondence between nucleotide sequence and amino acid sequence (Fig. 8-9*a*). By reading the DNA sequence, you can predict the amino acid sequence of the protein product. In eukaryotes, however, gene structure is more complex. In 1977, it was discovered that eukaryotic genes are interrupted by nucleotide sequences that do not code for amino acids. These functionally different regions are referred to as **exons,** *ex*pressed (protein coding) sequences, and **introns,**[1] *in*tervening (nonexpressed) sequences (Fig. 8-9*b*). The presence of the introns makes it impossible to predict the structure of a protein directly from the DNA sequence of a eukaryotic gene. As you will see next, intron segments must be removed from the message before the encoded information can be used to direct protein synthesis.

[1]Introns have been found in archaeobacteria and certain cyanobacteria in genes for transfer RNA (tRNA), but not in protein-coding genes. In cyanobacteria, the introns are closely related to those in chloroplast genes for tRNA. This is yet another indication that chloroplasts are derived from prokaryotic cells.

● **PROTEIN SYNTHESIS: TRANSCRIBING THE MESSAGE.** Proteins are synthesized at assembly sites called ribosomes, which are distributed throughout the cytoplasm of bacteria ▶(see Ribosomes, Chap. 2, p. 37, and Chap. 4, p. 88). The genetic message encoded in DNA must somehow be carried to the ribosomes, where these instructions direct the synthesis of proteins. Carrying the message to the ribosomes is the role of **messenger RNA (mRNA).**

Ribonucleic acid (RNA) is very similar to DNA in that it is a linear polymer of nucleotides. There are three primary differences between DNA and RNA: (1) The sugar in the nucleotides of RNA is ribose instead of deoxyribose; (2) unlike DNA, RNA usually consists of single strands; and (3) in RNA the nucleotide base uracil replaces thymine. Uracil pairs with adenine exactly as thymine does.

The first step in transmitting the genetic message from DNA to the ribosome is to precisely encode the information in DNA into a molecule of mRNA. Cells do this by assembling a linear molecule of mRNA along a temporarily single-stranded portion of the DNA molecule (Fig. 8-10). A chain of mRNA is synthesized using one DNA strand as a template to determine its nucleotide sequence (no mRNA synthesis occurs along the other DNA strand). The new mRNA is therefore complementary in sequence to this DNA "sense" strand and contains the same genetic information as DNA by virtue of its nucleotide sequence, which was determined by the sequence in DNA. The synthesis of mRNA molecules is called **transcription** because the genetic message from the DNA template is transcribed (copied) into a molecule of mRNA.

Transcription requires free RNA nucleotides, a DNA template, and **RNA polymerase,** the enzyme that catalyzes the assembly of ribonucleotide subunits into a polymer. This enzyme (which is also called *transcriptase*) binds to DNA at a gene-regulatory region called the **promoter.** RNA polymerase then moves along the DNA to the adjacent *coding region* (the part of the gene that directs the formation of the gene's product), and transcription continues until the enzyme encounters a series of nucleotides that signal termination of the process. At this site, the RNA polymerase and mRNA are released from the DNA, and transcription ceases.

Several differences characterize the RNA products of transcription in prokaryotes and eukaryotes:

1. In bacteria, a functional mRNA, one that can bind to ribosomes for translation into a particular protein, is produced. The 5′ end of the messenger contains the ribosome binding site, a sequence of nucleotides called the *leader region*.

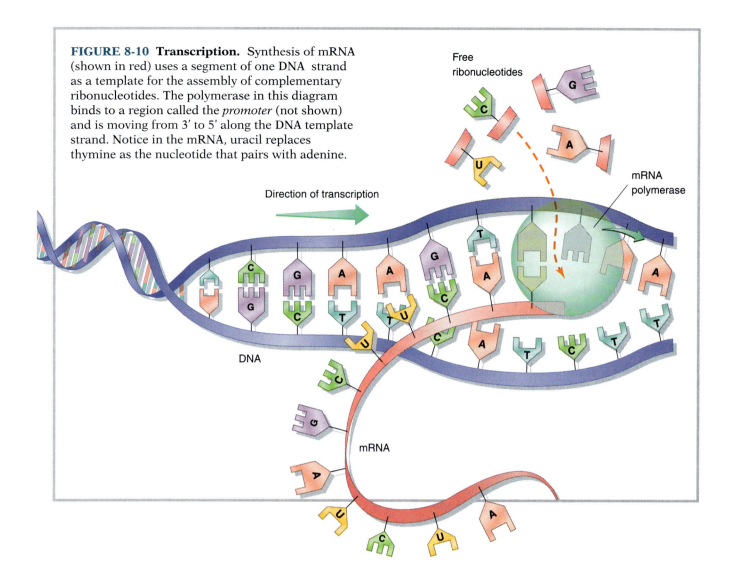

FIGURE 8-10 Transcription. Synthesis of mRNA (shown in red) uses a segment of one DNA strand as a template for the assembly of complementary ribonucleotides. The polymerase in this diagram binds to a region called the *promoter* (not shown) and is moving from 3' to 5' along the DNA template strand. Notice in the mRNA, uracil replaces thymine as the nucleotide that pairs with adenine.

Free ribonucleotides

Direction of transcription

mRNA polymerase

DNA

mRNA

2. Since there is no membrane separating the chromosome from the cytoplasm, the leader sequence of the bacterial mRNA usually combines with ribosomes as soon as it is produced. Protein synthesis in prokaryotes, therefore, occurs immediately as mRNA is assembled from the DNA template.

3. Many mRNAs in bacteria are *polygenic;* that is, they code for more than one protein (Fig. 8-11*a*). These proteins have related functions; for example, two enzymes may participate in the same metabolic pathway. Transcription of these genes is linked because they share a single promoter, but since each gene contains its own leader region for ribosome binding, each protein is synthesized separately. In eukaryotes, the RNA transcription product is usually *monogenic*, carrying the code for only one protein. In addition, the RNA must be modified before it becomes a func-

tional messenger because it contains introns (Fig. 8-11*b*). Posttranscriptional modification of eukaryotic RNA includes cutting and *splicing* the RNA to remove introns (see BEYOND THE BASICS: Ribozymes—Nonprotein "Enzymes"). In addition, specific nucleotides are added to the ends of the message, a guanine "cap" at the beginning of the transcript and a sequence of adenines that create a "poly A" tail. This is believed to protect the message from degradation and assist in its transport out of the nucleus and into the ribosome-filled cytoplasm, where it can direct the synthesis of protein. Thus, in eukaryotes, protein synthesis does not occur simultaneously with transcription.

● **PROTEIN SYNTHESIS: TRANSLATING THE MESSAGE.** The genetic message encoded in mRNA is translated by reading the code in groups of three nucleotides

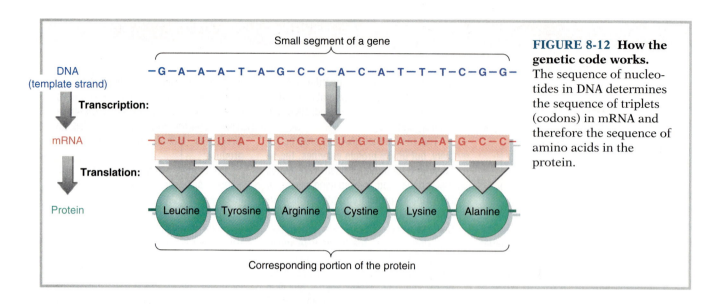

Small segment of a gene

DNA
(template strand)

−G−A−A−A−T−A−G−C−C−A−C−A−T−T−T−C−G−G−

Transcription:

mRNA

−C−U−U−U−A−U−C−G−G−U−G−U−A−A−A−G−C−C−

Translation:

Protein

Leucine — Tyrosine — Arginine — Cystine — Lysine — Alanine

Corresponding portion of the protein

FIGURE 8-12 How the genetic code works. The sequence of nucleotides in DNA determines the sequence of triplets (codons) in mRNA and therefore the sequence of amino acids in the protein.

serted into the protein whenever the appropriate codon occurs. Each tRNA therefore functions as a molecular decoder that translates the language of nucleic acids (nucleotide sequence) into the language of proteins (amino acid sequence).

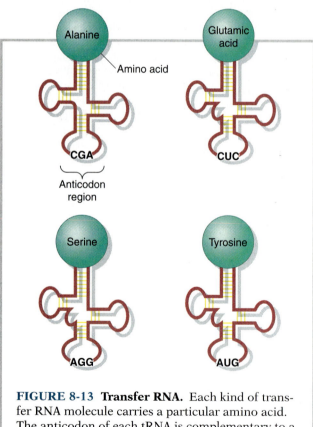

Alanine

Amino acid

CGA

Glutamic acid

CUC

Anticodon region

Serine

AGG

Tyrosine

AUG

FIGURE 8-13 Transfer RNA. Each kind of transfer RNA molecule carries a particular amino acid. The anticodon of each tRNA is complementary to a triplet sequence (codon) in mRNA. Only a few of the 61 different tRNAs are shown in this illustration. The representation of the tRNAs is stylized.

The cell maintains pools of different types of tRNAs, each specific for one kind of amino acid (Fig. 8-13). Each tRNA has a unique region of three nucleotides, called the **anticodon,** which is capable of binding to the complementary base triplet, the codon, on the mRNA. Each tRNA, therefore, recognizes one triplet on mRNA and inserts the correct amino acid when that triplet appears in the mRNA sequence. The tRNA with the anticodon CUC carries only the amino acid glutamic acid and recognizes only one codon (GAG) on mRNA. Thus, the codon GAG always specifies the insertion of glutamic acid into the growing protein chain.

Translation of the genetic code is made possible by the fact that each of the 20 amino acids has at least one unique codon (Fig. 8-14). Since the four nucleotides can be arranged in 64 possible triplet sequences, most amino acids have more than one codon, any one of which will specify its addition to the growing protein chain. There are three codons (UGA, UAG, and UAA), however, for which no amino acid is specified. These codons, known as *nonsense codons,* are used by a cell to mark the end of a protein chain and the termination of protein synthesis.

Protein synthesis takes place at ribosomes, structures composed of *ribosomal RNA (rRNA)* and protein. The cytoplasm of bacteria is densely packed with these structures (*E. coli* contains approximately 18,000 ribosomes per cell). Prokaryotic ribosomes are designated 70 S and are composed of two subunits, 30 S and 50 S. These subunits are joined only during the process of protein synthesis.

An abbreviated version of the events of translation is represented in Figure 8-15. (Many of the enzymes and cofactors have been omitted.) Each step in the figure (see p. 194) is numbered to correspond to the following description.

Step 1. The leader region at the 5' end of mRNA contains a nucleotide sequence that is complementary to a portion of the ribosomal RNA in the small subunit of the ribosome. This ensures that when the messenger RNA binds to the 30 S particle, it is aligned correctly, with the first codon (AUG) in position to accept the tRNA displaying the complementary anticodon (UAC). The larger subunit then associates to form a 70 S *initiation complex*. The mRNA shown is only a portion of the entire molecule, which would likely be at least 600 nucleotides in length. Notice the free tRNAs available in the cytoplasm. Each tRNA is carrying its specific amino acid.

Step 2. The tRNA with anticodon UAC is base-paired with the first codon (AUG) on the mRNA. The amino acid on the opposite end of the tRNA is aligned on the ribosome. The first codon on mRNA is usually AUG, the *initiator codon*. This triplet ensures that translation begins with the correct nucleotide. If the message were initiated at the second nucleotide, the remaining triplets would be read incorrectly. UGC would replace AUG as the first codon, the remaining triplets

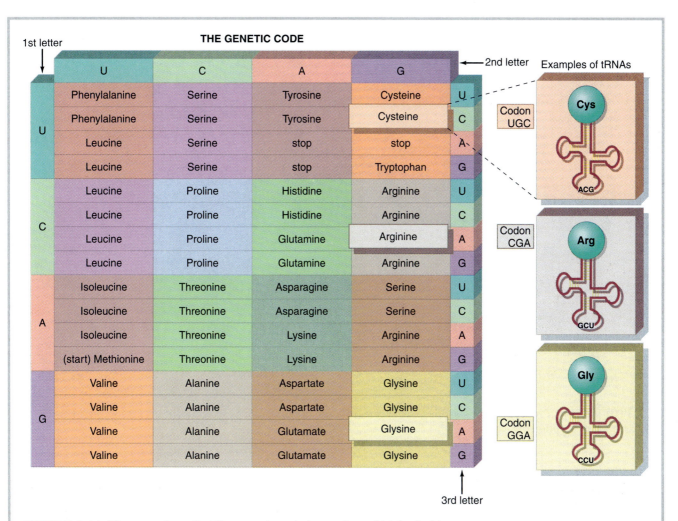

THE GENETIC CODE

FIGURE 8-14 The genetic code. The genetic code is a universal biological language. The correlation between the codon triplet and the amino acid it specifies is virtually the same in all organisms. To use the chart to determine which amino acid a codon specifies (for example, CGA), find the letter that corresponds to each of the three letters in the appropriate column or row. In this example, find the first letter (C) in the column on the left; follow the row to the right until you find the second letter (G); then find the amino acid in that square that matches the third letter (A). In this way, you've determined that CGA specifies the insertion of arginine. Cells have no such "decoder chart" to translate the message. tRNAs, a few of which are shown in this illustration, do the decoding.

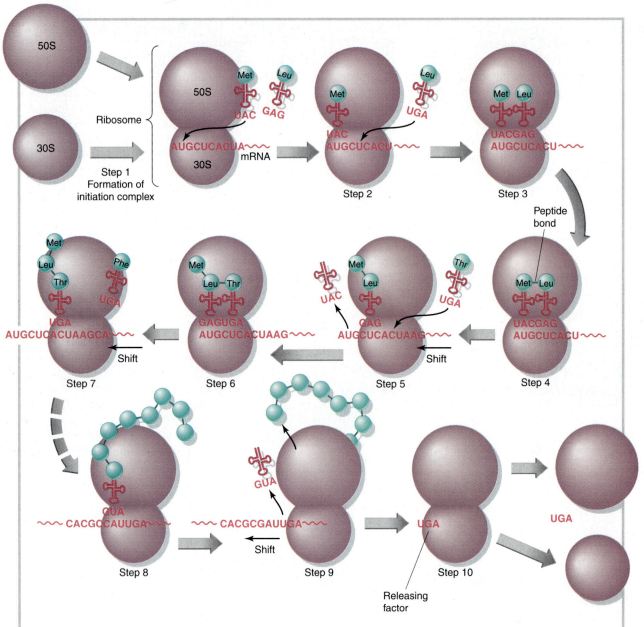

FIGURE 8-15 **Protein synthesis.** Simplified representation of translation as it occurs in the cytoplasm. (See text for description of steps.)

would be read out of synchrony, and the resulting protein would be a useless product of biological gibberish. The spatial order in which the triplets are read is called the *reading frame*.

Step 3. The second tRNA (in this case, the anticodon is GAG and the amino acid is leucine) base-pairs with the CUC codon. The two amino acids line up next to each other on the ribosome.

Step 4. The two amino acids are now enzymatically joined together by the formation of a peptide bond between them. The first amino acid (me-

thionine) dissociates from its tRNA. The growing polypeptide is now two amino acids long.

Step 5. The mRNA, tRNA, and growing amino acid chain shift three nucleotides to the left, bringing the next codon on the mRNA into position for accepting its specific tRNA. The first tRNA is released from the ribosome complex and returns to the cytoplasmic pool, where it can pick up another methionine. This *translocation* step requires energy derived from the hydrolysis of GTP (a high-energy compound similar to ATP).

Step 6. This time the codon is ACU, which binds

with the anticodon UGA of the tRNA carrying the amino acid threonine. Another peptide bond forms, and the growing polypeptide acquires another amino acid.

Step 7. Once again, as in step 5, the complex shifts to accommodate another tRNA. The elongation process continues in this fashion, the chain of amino acids growing in the proper sequence until the protein is completely synthesized.

Step 8. The process nears completion with *termination.* There are three chain-terminating codons that specify no amino acid and signal the end of the polypeptide chain. These are the *nonsense codons* UGA, UAA, and UAG. Since no tRNA binds with these codons, the amino acid inserted just before the termination codon becomes the terminal amino acid on the growing protein chain.

Step 9. The final step in the process is the release of the completed protein. When the entire complex is shifted, the finished protein is released from the ribosome.

The specific sequence of amino acids composing the protein polymer determines the protein's characteristic three-dimensional shape. This shape is essential to the function of the protein.

Protein synthesis occurs very rapidly in cells. In prokaryotic cells, ribosomes attach to mRNA and begin translation while the mRNA is still being transcribed from the DNA. Soon after a ribosome has translated the initial codons of the mRNA, the leader region is available for the attachment of another ribosome. Thus, the efficiency of protein synthesis is enhanced because many ribosomes can translate a single message at the same time, generating a chain of ribosomes held together by mRNA. This complex is called a *polysome.* Since each ribosome on the polysome is synthesizing a protein strand, a single molecule of mRNA may be simultaneously generating 25 to 35 identical protein molecules (Fig. 8-16).

We have seen how the genetic information stored in the specific nucleotide sequence of a DNA chromosome determines the characteristics of an organism by directing the synthesis of proteins that have unique amino acid sequences. The sequence of nucleotides in DNA determines the sequence of nucleotides in mRNA, which in turn dictates the sequence of the amino acids in the protein synthesized from this RNA molecule. This amino acid sequence ultimately determines the function of the protein, for example, its enzymatic activity, which in turn directs a specific metabolic reaction that creates a trait. The organism inherits this DNA sequence from parental cells and transmits the same genetic information to its progeny. All the life processes are ultimately determined by the sequence of nucleotides in DNA.

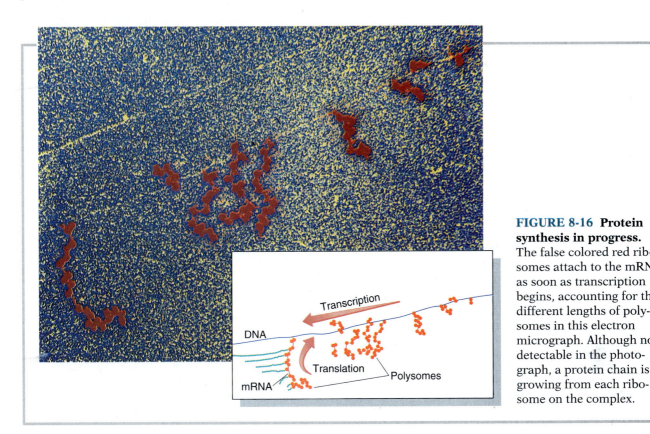

FIGURE 8-16 Protein synthesis in progress. The false colored red ribosomes attach to the mRNA as soon as transcription begins, accounting for the different lengths of polysomes in this electron micrograph. Although not detectable in the photograph, a protein chain is growing from each ribosome on the complex.

In other words, genetic information flows from DNA to RNA to protein. Some microbes, however (the virus that causes AIDS, for example), reverse part of the genetic flow, using RNA templates to create DNA copies. For more information on such exceptions, see THE EXPLORERS: Reversing the Genetic Flow.

CONCEPT CHECK

• Describe how and why exact copies of chromosomal DNA are produced in bacteria.
• How does the nucleotide sequence in DNA determine the amino acid sequence in a protein, and what is the significance of this process?
• What are the roles of mRNA, tRNA, and rRNA in protein synthesis?
• What does reverse transcriptase enable certain RNA viruses to do?

Regulation of Gene Expression

The expression of some genes cannot be regulated; these genes are transcribed and translated throughout the life of the cell. Such nonregulated genetic function is referred to as **constitutive protein synthesis.** Even among these genes, however, some are more active than others. Genes that encode enzymes for glucose metabolism, for example, bind RNA polymerase much more readily than do genes for most other proteins. Initiation of transcription occurs at a higher frequency for those genes with promoters that have the highest affinity for the polymerizing enzyme.

At any given time, many genes in a cell will be shut off and the production of the corresponding gene product temporarily halted. In other words, a cell rarely expresses its entire genetic potential. When *E. coli* is growing on glucose, for example, only 10 to 20 percent of its genes are active. *Resources are conserved by turning off genes for unneeded functions.* Many catabolic enzymes, for example, are not synthesized unless the substrate is available. In the absence of the substrate, transcription is turned off. Residual mRNA deteriorates within 2 minutes; after that the unneeded enzymes are no longer synthesized. The presence of the substrate turns on transcription of the appropriate gene, the enzyme is produced, and the substrate is utilized. This is control of gene expression by the process of gene **induction.** Inducible genes are turned on by the *inducer,* which is often the substrate of the corresponding catabolic pathway. Enzymes of biosynthetic pathways, on the other hand, are produced only when the end product of the pathway is needed. When this product is abundant, production of the enzymes is turned off by the process of gene **repression.** The product of the biosynthetic pathway must be present for enzyme synthesis to be repressed.

Induction and repression are methods of regulating enzyme synthesis by controlling transcription of mRNA from the regulated genes in response to varying concentrations of substrates and products.

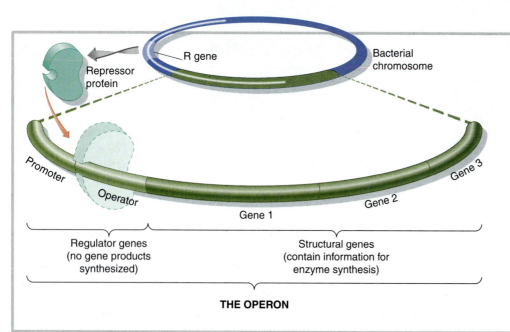

R gene

Repressor protein

Bacterial chromosome

Promoter

Operator

Gene 1

Gene 2

Gene 3

Regulator genes (no gene products synthesized)

Structural genes (contain information for enzyme synthesis)

THE OPERON

FIGURE 8-17 Clustered genetic elements of the classical bacterial operon. Genes 1, 2, and 3 synthesize enzymes in a single metabolic pathway. The promoter and operator segments are located on the end of the operon where transcription is initiated. Transcription of the adjacent genes is from left to right. The R gene produces a repressor protein that can bind to the operator region and turn off transcription.

Reversing the Genetic Flow

The flow of genetic information from DNA to RNA neatly explains the initial events in gene expression in most organisms, but some viruses use alternatives to this classical scheme. These viruses contain single-stranded RNA genomes packaged within their protective protein coats, and this is the only nucleic acid that directs the formation of progeny particles within host cells. Some of these single-stranded RNA viral genomes can act as messenger RNA the moment they are freed into the host cell's cytoplasm, where ribosomes attach and directly translate the message. Some others have a considerably more complex fate, one that contradicts the long-held belief that genetic information flows only from DNA to RNA. These viruses, in fact, reverse the flow of genetic information and use their RNA genome as a template to create a double-stranded DNA that directs the subsequent manufacture of messenger RNA. These *retroviruses* (*retro* = backward) include a group of important human pathogens, most notably the agents that cause AIDS and some forms of cancer. ▶ (See *retroviruses,* Chap. 13, p. 316.)

Even before it was discovered, this unique form of genetic replication was a topic of hot scientific pursuit. Chromosomes of cells infected with retroviruses were found to have DNA sequences complementary to the single-stranded RNA genome of the virus. In other words, a double-stranded DNA copy of the viral genome apparently integrates into the host cell's chromosome. This led to speculation that there may be another kind of polymerase, one that uses viral RNA as a template to create a complementary strand of DNA. The search for the new enzyme culminated in 1970 with the discovery by Temin and Baltimore of **reverse transcriptase** (RNA-dependent DNA polymerase) in retrovirus particles.

During replication of the retroviruses, the enzyme promotes the assembly of a single-stranded DNA molecule along the RNA viral genome. A second DNA strand complementary to the first is then synthesized as the RNA template is degraded. The result is a double-stranded DNA molecule that may integrate into the host chromosome. Integration of DNA replicas of retroviral RNA is an important component in the development of diseases caused by retroviruses.

But reverse transcriptase has its positives as well. The discovery of reverse transcriptase has provided an essential tool for molecular biologists and genetic engineers. It is now possible to use this enzyme to synthesize genes from isolated messenger RNA (instead of having to isolate the DNA itself). The advantages of this approach will be seen in the next chapter. For their discovery of reverse transcriptase, Temin and Baltimore received the 1975 Nobel prize.

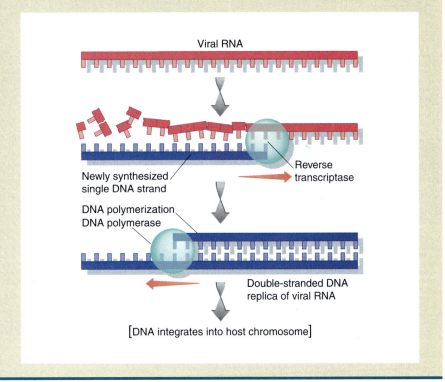

Viral RNA

Newly synthesized single DNA strand

Reverse transcriptase

DNA polymerization DNA polymerase

Double-stranded DNA replica of viral RNA

[DNA integrates into host chromosome]

The group of genes that provides a bacterial cell with this ability is called the *operon.*

■ The Operon

The **operon** is a segment on the bacterial chromosome that consists of the following elements (Fig. 8-17):

- *Structural genes.* These are genes that direct the synthesis of proteins with related functions, such as the enzymes of a single metabolic pathway. The genes are physically adjacent to one another and are regulated as a unit.
- *A regulator region consisting of a promoter and an operator.* Recall that the promoter is the site to

which mRNA polymerase binds to the DNA and initiates transcription—in this case, of a polygenic message. The **operator** region, which lies between the promoter and the structural genes, is the binding site for a specific *repressor protein*. When the operator site is occupied by repressor, RNA polymerase cannot bind to the promoter on DNA, transcription is blocked, and protein synthesis is turned off.

- *A **repressor gene** (R gene) located in another portion of the bacterial chromosome.* This gene directs the cell to produce the repressor protein that at-

taches to the operator region. The repressor protein has the additional ability to bind specifically with a small *effector,* the molecule that determines whether an operon will be turned on or turned off.

■ Inducible Operons

For many catabolic pathways, the substrate must be present before the pathway's enzymes can be produced. The operon is not expressed when the substrate (inducer) is not available to prevent the repressor from binding to the operator (Fig. 8-18a). In this state, transcription and enzyme synthesis are

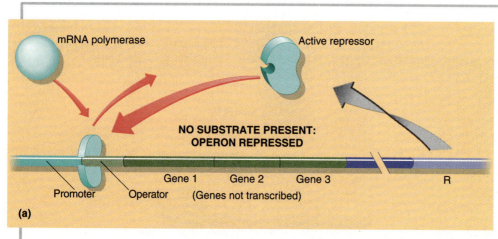

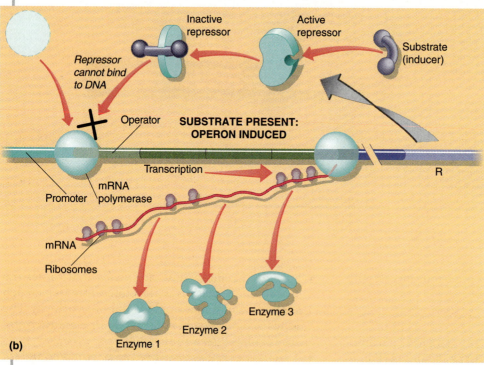

FIGURE 8-18 Regulation of an inducible operon. (a) In the absence of the inducer (the substrate S), the repressor protein binds to the operator and blocks transcription of genes by preventing binding of mRNA polymerase. The operon is turned off. (b) When the inducer is present, it binds to the repressor and prevents its attachment to the operator. Transcription of genes is unimpeded. The operon is induced, enzymes are produced, and the substrate is utilized.

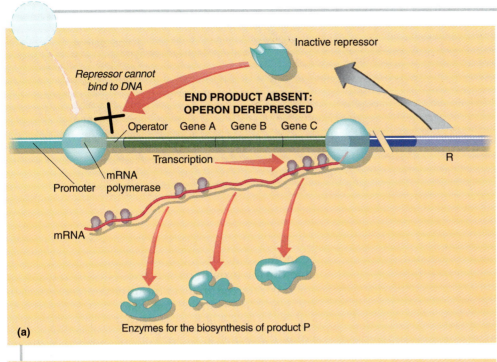

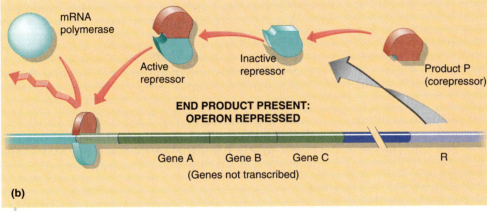

FIGURE 8-19 Regulation of a repressible operon. (*a*) In the absence of the corepressor (the product P), the inactive repressor protein fails to bind to the operator. Transcription proceeds, and the operon is actively expressed. (*b*) When the corepressor is present, it binds to and activates the repressor protein. The repressor-corepressor complex attaches to the operon and blocks transcription. The operon is repressed.

blocked. When present, the substrate induces the transcription of the operon by inactivating the repressor and preventing it from binding to the operator (Fig. 8-18*b*). Thus, if a bacterium is inoculated into a medium containing a new catabolite, such as lactose, the cell cannot use the substrate immediately. A lag period occurs until the sugar induces the formation of lactose-digesting enzymes.

■ Repressible Operons

Repressible operons differ from inducible operons in several ways:

1. They produce and regulate the enzymes of *biosynthetic* pathways instead of catabolic ones.

2. The effector is the *end product* of the pathway, not the substrate.
3. The repressor protein is inactive in the *absence* of the effector. The effector therefore functions as a corepressor that must be present before the operon can be turned off.

The enzymes of repressible pathways are thus produced whenever the pathway's final product (P), which is also the corepressor, is in low concentration and therefore needed by the cell (Fig. 8-19*a*). Under such conditions, the repressor protein cannot bind to the operator, and the operon is expressed. When the product is in ample supply, it activates the repressor, which binds to the operator and blocks transcription of the genes. This is a state

TABLE 8-3   TYPES OF MUTATIONS

Mutation	Description
Auxotrophic	Nutritional deficiency; loss of biosynthetic enzyme; requires end product in medium for growth.
Conditionally lethal	Essential gene product is functional only under certain conditions; the organism must be grown under conditions (temperature or osmotic pressure) that permit gene expression.
Antibiotic-resistant	Antibiotic fails to inhibit or kill the mutant; may be due to loss of binding site for drug, impermeability to drug, or synthesis of antibiotic-destroying enzyme.

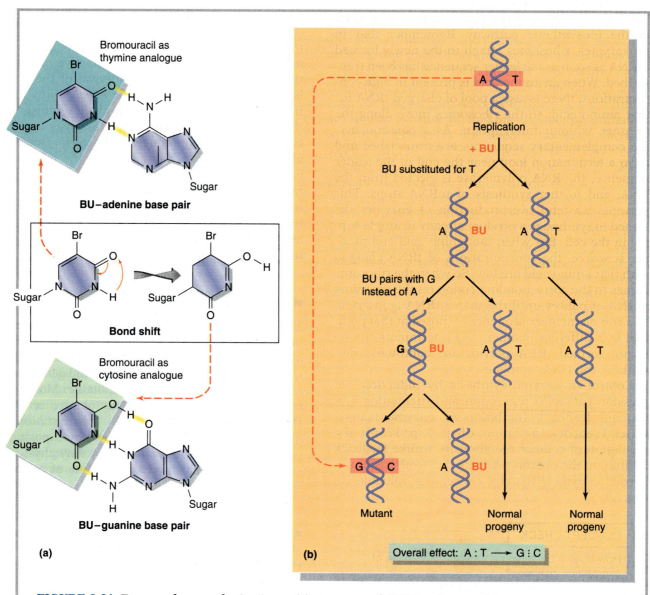

FIGURE 8-21 Base analogue substitutions. (*a*) Bromouracil (BU) is a base analogue of thymine and occasionally cytosine, depending on a shift in a double bond. (*b*) When BU is substituted for thymine, it forms a complementary pair with guanine, and a mutation occurs. Half of the subsequent progeny will carry a G:C base pair in place of the correct A:T base pair.

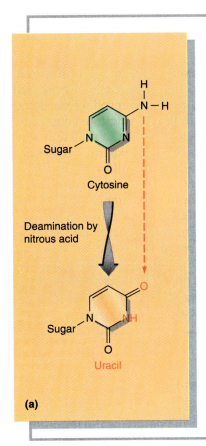

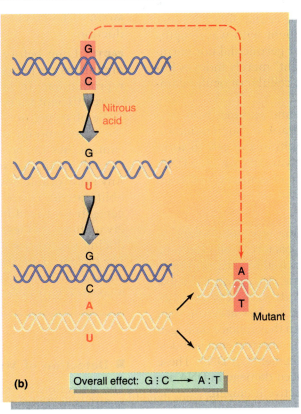

FIGURE 8-22 Base modification by nitrous acid. (*a*) Nitrous acid removes NH₂ from cytosine, converting it to uracil. (*b*) Deaminated cytosine (uracil) then forms a base pair with guanine, which then pairs with adenine, and a mutation occurs. Half of the subsequent progeny will carry a A:T base pair in place of the correct G:C base pair.

a substitute for thymine. It can pair correctly with adenine or incorrectly with guanine. When guanine becomes incorporated into DNA, it then pairs with its complement cytosine during the next replication cycle. Thus a thymine-adenine base pair is replaced by a guanine-cytosine base pair.

Some chemicals react directly with nucleotides already incorporated into DNA. These *DNA-modifying agents* cause the removal or addition of functional groups on the bases. Such changes often alter base-pairing properties (Fig. 8-22). For example, nitrous acid deaminates adenine, which then acts like guanine and bonds with cytosine instead of thymine.

Base analogues and DNA-modifying agents cause point mutations. They may create alternative codons that specify the same amino acid, so no change in phenotype occurs. Such *silent mutations* usually occur in the third base of the codon. Most point mutations are *missense mutations;* they change a codon to that of another amino acid. If the substituted amino acid does not negatively alter the protein's function, the cell will survive. Occasionally, point mutations change codons to termination codons. Protein synthesis stops prematurely when the ribosome reaches this signal, and a shortened nonfunc-

tional version of the protein is produced. Such mutations are called *nonsense mutations.*

Intercalating agents are flat molecules that insert themselves between base pairs and distort the structure of the DNA helix. As a result, bases are added or removed during replication, which results in frame-shift mutations. Messages produced from these regions of DNA will generally produce nonfunctional proteins.

● **RADIATION.** *Ultraviolet (UV) light* with wavelengths around 260 nm damages the DNA of microorganisms ▶ (see Ultraviolet Radiation, Chap. 15, p. 359). It triggers the formation of bonds between thymine bases that lie adjacent to each other in the same DNA strand. These *thymine dimers* acquire an abnormal configuration that impairs DNA replication and transcription. The amount of damage is proportional to dose and length of exposure. At high doses, UV light is lethal to cells and can be used as a germicidal agent.

Ultraviolet light with wavelengths around 260 nm induces mutations not only in bacteria but in the superficial cells of humans as well. Sunlight contains mutagenic ultraviolet light, and overexposure to sunlight is associated with an increased incidence

Key Terms

genetics (p. 179)
genes (p. 179))
genotype (p. 179)
phenotype (p. 179)
mutations (p. 180)
plasmids (p. 181))
DNA replication (p. 185)
DNA polymerase (p. 185)
ligase (p. 185)
exons (p. 187)
introns (p. 187)
messenger RNA (mRNA) (p. 187)
transcription (p. 188)

RNA polymerase (p. 188)
promoter (p. 188)
codons (p. 190)
genetic code (p. 190)
translation (p. 190)
transfer RNA (tRNA) (p. 190)
anticodon (p. 190)
ribozymes (p. 191)
constitutive protein synthesis (p. 196)
induction (p. 196)
repression (p. 196)
operon (p. 196)
reverse transcriptase (p. 197)

operator (p. 197)
repressor gene (p. 198)
catabolite repression (p. 200)
attenuation (p. 201)
mutants (p. 201)
wild type (p. 201)
mutagen (p. 201)
transposable elements (p. 204)
recombination (p. 205)
transformation (p. 207)
transduction (p. 207)
conjugation (p. 208)

Key Facts and Concepts

- **DNA.** The characteristics of an organism are determined by genetic information encoded in its DNA. The two strands of this double-stranded helix are linked by complementary pairs of nitrogenous bases. Genetic instructions are encoded in the linear sequence of nucleotides that make up an organism's DNA.

- **Prokaryotic DNA.** Genetic information in prokaryotes is encoded in the DNA of the circular chromosome and plasmids. Plasmids may provide bacteria with the ability to conjugate, to survive exposure to antibiotics, or to produce toxins.

- **Passing on the genes.** Precise replication of DNA ensures that, except for mutations, progeny bacteria inherit the same genes their parent cells possess. Each progeny DNA molecule consists of one strand from the parent cell and one newly synthesized strand.

- **Following instructions.** Genetic instructions in DNA direct the synthesis of proteins. In this way, encoded information becomes expressed as cellular characteristics. These proteins can be structural components of the cell or enzymes that control the cell's metabolism. Each physical or chemical property of an organism develops under the direction of at least one gene.

- **Protein synthesis.** The encoded information in DNA is transcribed into a complementary molecule of mRNA. At the ribosomes the mRNA serves as a template that determines the order in which

amino acids assemble into a growing protein chain. This amino acid sequence determines the activities of the protein.

- **Regulation of gene expression.** Cells regulate gene expression in response to their changing needs or environmental conditions. The bacterial operon provides a mechanism by which whole blocks of related genetic functions can be induced or repressed. Constitutive proteins, on the other hand, are constantly synthesized.

- **Mutations.** Mutations are permanent changes in the genetic code, often producing new or altered properties in the mutant. The frequency of mutation may be accelerated by mutagens.

- **Genetic transfer in bacteria.** Bacteria can transfer genetic information from a donor to a recipient. The transferred genes can be stably incorporated into the recipient's chromosome by recombination. The DNA may be transferred by bacteriophages (transduction), by physical contact between cells (conjugation), or as free nucleic acid liberated into the surrounding medium and subsequently absorbed by recipient cells (transformation). The transferred genetic material is a copy of either a portion of the donor's chromosome or a plasmid.

- **Genetic engineering.** Plasmids provide vehicles used by genetic engineers for the gene transfers central to the field of recombinant DNA technology.

Review and Problem-Solving Questions

1. Distinguish between:
   (a) Genotype and phenotype
   (b) Transcription and translation
   (c) Induction and repression
   (d) Transduction and transformation
2. If the nucleotide sequence in one strand of DNA is

   A A T C C G A T C G A T T C G

   (a) What is the nucleotide sequence in the complementary strand?
   (b) What is the nucleotide sequence in the corresponding mRNA (assuming this is a segment from a coding region of the gene)?
   (c) What is the amino acid sequence in the protein encoded by this information? (Use the chart in Fig. 8-14.)
   (d) What is the function of the nucleotides in the noncoding regions of the gene?
3. Explain the role of tRNA in protein synthesis. Why can it be considered a "molecular decoder"?
4. What would be the likely effect of a mutation that creates (a) a nonsense codon in the middle of a gene? (b) a missense codon in the middle of a gene? (c) a one-nucleotide addition in the promoter region?
5. Compare conjugation when the donor is an Hfr cell versus an F$^+$ cell.
6. What might be the effect of mutations in each of the following parts of the bacterial operon? (a) Operator; (b) promoter; (c) repressor protein; (d) structural gene (the coding region of the gene).
7. Why wouldn't attenuation be an effective means of regulating all genes?
8. How does a corepressor differ from an inducer?
9. Why does mixing dead encapsulated bacteria with living nonencapsulated bacteria produce living encapsulated bacteria, as discovered by Griffith?
10. What would you expect to happen to a DNA fragment containing the human gene for insulin production if it were transported into the cytoplasm of E. coli?
11. What problems might be encountered in translating the human gene for insulin by a bacterial cell?
12. What conclusions can you draw about the mechanism of genetic transfer in each of the following cases?
    (a) The donor and recipient are separated by a semipermeable membrane (water, molecules, and viruses can pass through, but cells cannot), and transfer continues.

(b) DNase (a DNA-destroying enzyme) is added to the medium, and gene transfer continues.
(c) Constant shaking of the medium prevents transfer.

13. Compare the structure of a gene and its corresponding RNA transcript in eukaryotes and eubacteria by filling in this table, following the example in row a:

Characteristic	Eukaryotes	Prokaryotes
(a) Monogenic	Yes	No
(b) Polygenic		
(c) Contains introns		
(d) Simultaneous transcription and translation		
(e) 5′ end of mRNA is the initiation codon		

14. All bacteria need a source of iron. Many bacteria can produce iron-binding proteins when grown in low concentrations of the metal. These proteins are not expressed when iron is plentiful in the cell's environment. Suggest a possible mechanism by which iron controls expression of these genes.
15. You wish to create an E. coli strain that can overproduce the amino acid glutamic acid, which is used for making the common flavoring agent monosodium glutamate (MSG). The enzymes in the pathway for glutamic acid production are under the control of a repressible operon and are also subject to feedback inhibition (see p. 173). How would you proceed?
16. In bacteria, what advantages does recombination have over mutations as a souce of genetic variability?
17. A cell is exposed to bromouracil (BU), a base analogue of thymine that often combines with guanine instead of adenine. The cell incorporates one molecule of BU into the strand used to make mRNA in the coding region of the gene for DNA polymerase.
    (a) What are the possible effects of this event?
    (b) What would be different if the BU were incorporated into the promoter region of the gene? in the leader region?
    (c) What would be different if the BU were only incorporated into the complementary strand?
18. Is an operon's repressor protein an inducible, repressible, or constitutive protein? Explain your answer.

BACTERIAL SEX IN OCEAN WATERS MAY POSE DANGERS FOR PEOPLE

Goodman and associates* recently reported that bacteria normally found in ocean waters conjugate with bacteria from human digestive tracts. Enteric bacteria are frequently introduced into the marine environment in wastewater from sewage treatment plants and in waters used in agriculture, food production, and other industrial processes. These enteric bacteria often find themselves in an ocean environment that can provide little dissolved organic carbon for growth. Under such sparse conditions, the starving microorganisms survive but do not multiply. These authors discovered that starved E. coli and typical marine bacteria of the genus Vibrio can conjugate and transfer plasmids when grown in laboratory conditions that mimic the ocean environment.

- Design your own experiment to determine whether conjugation between E. coli and Vibrio can occur in nutrient-sparse marine conditions.
- What are some dangers posed by gene transfer between human enteric bacteria and microorganisms indigenous to the environment?

*A. E. Goodman, E. Hild, K. C. Marshall, and M. Hermansson, Plasmid transfer between bacteria under simulated marine oligotrophic conditions, Applied and Environmental Microbiology, 59: 1035–1040 (1993).

CHAPTER 9

Genetic Applications and Biotechnology

Giant fermentation tank for growing gene-spliced bacteria that produce interferon. The area is bathed in UV light and the technician is clothed in protective clothing, precautions against contaminating the culture.

An eerie violet light bathes the white suits of these scientists and reflects off their glass masks, obscuring their faces. The light is used to trigger invisible bands of DNA to glow brightly in the dark. In this way, these modern molecular geneticists can locate and extract DNA, isolating plasmids away from bacterial chromosomes. What they are doing seems to many people to be as futuristic as they look (their suits protect them from the mutagenic effects of the UV light as well as protecting the culture from contamination). These scientists are genetically manipulating bacteria and other organisms until they have an organism with a novel combination of traits never assembled in nature. How can these "genetic designers" create new genetic combinations, in some cases producing new kinds of organisms? Furthermore, why would they want to?

These are some of the questions we answer in this chapter, somewhat broadly at first, then providing details as the chapter unfolds. By studying the intricate details of bacterial genetics, scientists have discovered the essential tools for manipulating genes. In less than 25 years, this technology has allowed us to produce new gene combinations and create genetically unique organisms. The most common of the new organisms are bacteria that contain genes for products never before synthesized by bacte-

cellulose filters (Fig. 9-9). The cells are lysed, cellular proteins are degraded, and DNA is denatured into single strands, which remain bound to the filter. A radioactively labeled nucleic acid probe is then added, and DNA hybrids containing the radioactive probe form wherever the foreign gene is present. All unbound probe molecules are washed away. The only radioactivity that remains is bound to the foreign genes. When X-ray film is exposed to the filter, dark spots appear where radioactive emissions exposed the film. These dark areas identify which bacteria successfully acquired the foreign gene.

A similar procedure is used to identify which cells are successfully transformed by detecting the products of foreign genes instead of the genes themselves. In this method, cells grown on agar are lysed and treated with a radioactively labeled antibody specific for the foreign protein. The antibody binds only to those cells containing the corresponding protein. Photographic film is then exposed to the agar, and the dark spots indicate which colonies contain the new gene. This technique not only identifies the cells that contain the foreign gene, it specifically spots those cells that are expressing the gene, since it is the gene product that is being detected. Of course, if the gene product is directly observable, such elaborate detection procedures are not necessary. For example, some assays depend on the introduction of the firefly luciferase gene into bacteria. The successfully engineered bacteria are easily identified—they are the ones that form luminescent (glowing) colonies. Bacteria engineered to digest cellulose to help reduce our mountains of paper waste are also relatively easily located. Only those cells that have successfully accepted and expressed the gene for digesting cellulose will grow on media in which cellulose is the sole energy source.

■ Expressing the Foreign Gene

Even if gene probes reveal that a cell has acquired the foreign gene, the desired product must be constructed if the procedure is to be successful. A few tests for products were discussed in the previous subsection, but gene expression doesn't always follow gene insertion; sometimes the engineered bacteria do not produce a functional protein. This may occur if posttranslational modifications are necessary for an active molecule—for example, the removal of introns from the RNA or the addition of sugar or lipid groups to the amino acid chain. These reactions, which are commonplace in eukaryotic cells, are rare in bacterial cells. In bacteria-to-bacteria gene transfers, as long as the appropriate control regions are present, the newly acquired DNA is transcribed and translated into protein.

MYTH: Once a microbe is successfully engineered to express a foreign gene in the laboratory, the major part of the battle is won. It is then a simple matter of locating the financial support for building the needed equipment.

FACT: Another set of problems is posed by "scaling up," engineering the systems for manufacturing the huge cultures needed to yield marketable amounts of product. These tanks, often 100,000 liters (25,000 gallons) in volume, must be designed to distribute oxygen and all other needed substances equally to all parts of the culture in spite of its size. The need to prevent microbial contamination, mutation that results in functional loss of the gene, and the breaking open of cells that produce their proteins intracellularly present additional scaling-up problems, as does purification of the final product.

■ Downstream Processing

Once the engineered cells have created the protein, *downstream processing*—the isolation and purification of an active protein—begins. Proteins are either stored within cells or secreted into the culture fluid. In either case, the first step in downstream processing is to separate the cells from the broth. If the proteins are intracellular, they are released by disruption of the cells. Purification procedures use traditional biochemical methods: precipitation, extraction, and chromatography (separation of the desired substance by passing the solution through a solid matrix). The final step before approval for sale is to demonstrate that the product is both safe and effective (see BEYOND THE BASICS: Protecting the Public's Safety—Testing Recombinant Products).

CONCEPT CHECK

- How are restriction endonucleases and reverse transcriptase used to obtain genes for genetic engineering?
- Why are plasmids typically used as cloning vectors in bacteria?
- What is the purpose of inserting marker genes, such as those for antibiotic resistance, into plasmids?
- How are probes used in the detection of recombinant bacteria?

Many genetically engineered products are used to prevent or treat disease. Vaccines, for example, are made from surface antigens of virulent pathogens. Transplanting a viral gene that encodes information for making the virus's coat protein may create a microbe that pumps out large quantities of a single viral antigen. The immunogen may then be readily harvested from the growth medium. Before it is used as a vaccine, however, the manufacturer must show that it will actually protect people from disease and that it carries no risk of deleterious side effects. Hormones, enzymes, and other products used in or on humans must be approved for use by the Food and Drug Administration (FDA). Their approval is based on data from independent laboratory test results and clinical trials as well as data provided by the manufacturer.

Among the first tests to be performed for safety are those that test the product on cell cultures. These tests may also include the Ames test to detect potential carcinogenicity if the material has never been administered to humans. One of the tests that must be performed on all products derived from bacterial culture is an *endotoxin test*. This test is designed to detect any residual cell wall debris that was not removed during the purification of the product. ▶ (See THE MICROBES: Dangerous Ghosts of Dead Bacteria, Chap. 4, p. 84.) Products that are nontoxic in cell culture are then administered to laboratory animals, usually mice and rats. When these preclinical trials are successfully completed, the manufacturer submits the data in an application for an investigational new drug (IND). Approval of this application is permission to administer the product to humans. The first clinical tests (Phase I clinical trials) determine the safety of the product in small sample populations. Those products that are nontoxic or whose side effects are considered acceptable are then tested for efficacy, first in small numbers of individuals (Phase II clinical trials) and then in larger populations (Phase III clinical trials). Such studies usually require that some people (called a study group) receive the drug and others (the control group) do not. Final approval for marketing the product depends on the data derived from Phase II/III clinical trials. It often takes more than 7 years and as much as $100 million to take a product through the entire process.

Environmental Release of Engineered Organisms

The manipulation of bacteria and viruses has created several kinds of useful products. Some bacteria are factories for the manufacture of proteins, most of which are pharmaceutical products or enzymes. Other bacteria acquire unique combinations of genes and are genetically engineered specifically for release into the environment, where they may be used to degrade pollutants, enrich soils, or protect plants. For example, the first microbe to be patented was created to digest oil.

Genetically engineered microbes (GEMs) cannot be tested outside the laboratory without permission from the Environmental Protection Agency (EPA) or the U.S. Department of Agriculture (USDA) or both. These government agencies are assigned the responsibility of determining the safety of releasing genetically altered organisms. Potential hazards from the release of newly created organisms are related to possible gene transfer to other organisms and the effect of engineered microbes on local ecology. Any new microbe may affect the plants, insects, animals, and humans in the community that it enters. For example, what if antibiotic resistance markers used in the development of the GEM were transferred to other soil bacteria, from there into bacteria in cattle, and ultimately into bacteria that inhabit or infect humans? To solve this problem, alternative markers, such as genes that direct the cell to produce a pigment, are used to track the fate of the GEMs (Fig. 9-10). Organisms that are patho-

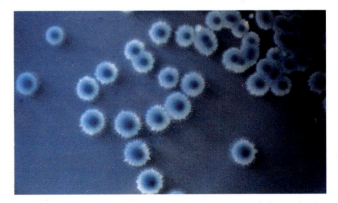

FIGURE 9-10 Signals from an engineered microbe. One way to monitor the environmental movement of microbes is to insert a marker gene for a color reaction. The *Pseudomonas* in this photo was given an *E. coli* gene for breaking down a chemical pollutant. It was also given a gene that made it produce and release a bright blue compound.

genic or that can still establish themselves as normal human microflora are not used unless the bacterium (such as *E. coli*) has been modified so it cannot survive in humans (for example, by equipping it with a conditionally lethal "suicide gene" or one of the other safeguards described in the box THE MICROBES earlier in this chapter).

CONCEPT CHECK

- What are the potential risks of releasing modified microbes into the environment?
- What are suicide genes? How can they limit the spread of engineered bacteria?
- How can bacteria be modified to monitor their fate in the environment?

Using Microbes to Transfer Genes into Plants

One goal of genetic engineering in plants is to provide healthier, more nutritious crops, crops with improved disease resistance, crops with greater yields, or crops that can grow on land that is of marginal value for the cultivation of traditional crop plants. Some of the genes transferred to plants would increase resistance to herbicides, pests, and disease, improve tolerance to environmental conditions, enable the plant itself to fix nitrogen directly from atmospheric nitrogen, and even increase the amino acid composition of the edible parts, making the engineered plant a better source of nutrients.

As with bacteria, plant cells most readily accept and express foreign genes that are incorporated into specific carrier vehicles. One such vehicle is the plant pathogen *Agrobacterium tumefaciens*. Normally this bacterium infects plants through damaged tissue and causes crown gall disease, the development of a tumorlike mass (Fig. 9-11a). The bacterium contains a large plasmid called the **Ti** (tumor-inducing) **plasmid.** A copy of a portion of this plasmid is transferred into plant cells, this portion then inserting itself into the plant's chromosome. This integrated plasmid segment, called **transfer DNA** (**tDNA**), triggers rapid growth of the infected cell, forming the crown gall tumor. In fact, once the DNA is incorporated into the plant, tumor production proceeds in the absence of *A. tumefaciens*. Transfer DNA is transcribed by the plant cell. Its products include enzymes needed for the biosynthesis of plant growth hormones that are responsible for increased cell multiplication.

The natural gene transfer system of the Ti plasmid has been adapted for genetic engineering (Fig. 9-11b). Using recombinant DNA technology, foreign genes are inserted into modified forms of the Ti plasmid of *A. tumefaciens*. The modifications include removal of tumor-inducing genes from the tDNA (so the resulting plants are normal) and addition of antibiotic resistance genes (as markers to track the plasmid). Often promoters from a plant virus are inserted as control elements of the foreign genes. The recombinant plasmid can be readily incorporated into *A. tumefaciens*, which transfers it into plant cells. Cells that contain the integrated plasmid and the foreign genes can be identified by growth on a medium containing the appropriate antibiotic. An entire plant can be regenerated from one of these genetically modified cells. Succeeding generations would acquire the new trait by normal inheritance.

Several nonmicrobial techniques have also been used to transfer foreign genes into plant cells. Protoplasts are obtained by removing the cell wall from plant cells, then treated with electric pulses, which allows DNA to pass through pores in the cell membrane. Alternatively, microprojectiles are "shot" into intact cells. These are nucleic acid–coated particles that are propelled through the plasma membrane.

The release of engineered plants is carefully controlled to prevent potentially harmful organisms (or their products) from being introduced into the environment. Because the entire biosphere depends on plants as food resources, new plant proteins may indirectly influence all forms of life, not only humans. In addition, human food resources that have been genetically engineered must be proven safe for consumption before they can be marketed. Another concern in the release of genetically engineered plants is the potential transfer of engineered genes to other plants, particularly weeds. For example, the establishment of the genes for herbicide resistance within the weed population would make control of these pests extremely difficult. Several thousand field trials of bioengineered crops have already occurred with no apparent problem.

CONCEPT CHECK

- Why is *Agrobacterium tumefaciens* a successful vehicle for gene transfer in plants?
- What modifications are made to the natural Ti plasmid when it is used in genetic engineering?

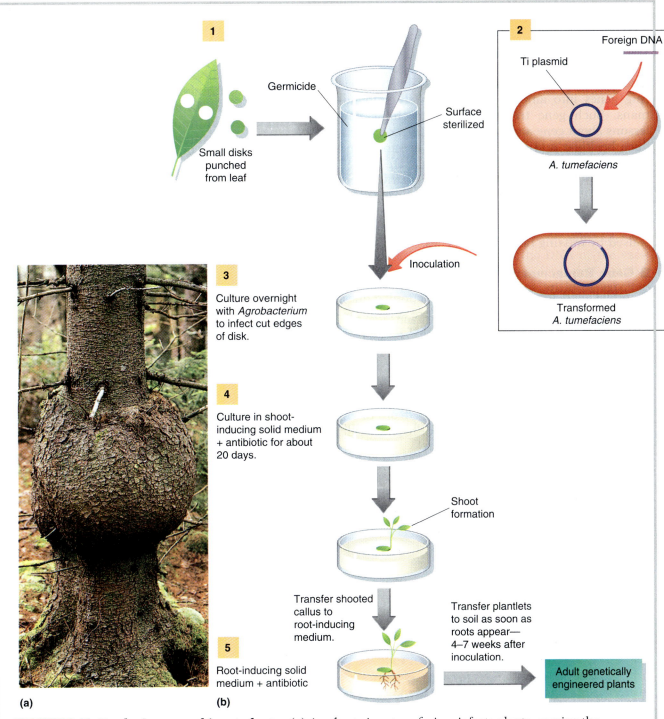

FIGURE 9-11 Producing recombinant plants. (*a*) *Agrobacterium tumefaciens* infects plants, causing the formation of ugly tumors called galls. These pathogenic bacteria contain the tumor-inducing (Ti) plasmid that, when transferred by conjugation to plant cells, is incorporated into the plant chromosome. (*b*) The Ti plasmid is used to carry foreign genes. (1) Disks are cut from plant leaves. (2) Foreign DNA is inserted into the transfer DNA (tDNA) of the Ti plasmid, which has been modified so that it cannot cause tumors and so it contains an antibiotic-resistance marker gene. The modified Ti plasmid is used to transform *A. tumefaciens*. (3) Leaf disk cells are then infected with the modified *A. tumefaciens*, which transfers its plasmids to plant cells. (4) The leaf disks are grown in media containing the antibiotic. Only cells containing the plasmid will grow; the antibiotic kills the cells that do not have the resistance gene carried on the plasmid. (5) The transformed cells generate plants containing the foreign genes.

International rules of nomenclature have been developed to facilitate communication between scientists of different nations. These rules require that all names be latinized. The genus name appears first, and the first letter is always capitalized; the species name is never capitalized. Both names are always italicized or underlined. The *Genus species* convention of writing the name instantly allows you to recognize that this pair of words denotes an organism. This helps distinguish it from other pairs of words, such as "tinea pedis," the latinized term that identifies the disease athlete's foot. Ideally, an organism's name conveys some information about the organism's properties (see THE MICROBES: What's in a Name?). For example, *Staphylococcus aureus* was named for its shape (*coccus* = seed, a spherical structure) and arrangement (*staphylo* = grapelike clusters) and for the gold color of its colony (*aureus* = gold) (Fig. 10-1). Another member of the genus, *Staphylococcus epidermidis,* is named for its natural habitat, the skin.

CONCEPT CHECK

- Why are the classification and naming of bacteria necessary for the existence of microbiology as a science?
- What is the advantage of using a binomial nomenclature rather than common names for microorganisms?

Classification of Prokaryotes

Classification of prokaryotes is complicated by the very nature of the organisms. Unlike eukaryotes, prokaryotes are all single-cell organisms that show very little distinguishing detail. There are hundreds of different species of bacteria that have exactly the same appearance, so morphological criteria must be supplemented with other types of tests to distin-

FIGURE 10-1 Biological nomenclature at its best. The name of this organism (*Staphylococcus aureus*) is packed with information about the microbe, revealing its shape, arrangement, and color of colony.

THE MICROBES

What's in a Name?

Most bacteria were discovered and named before strict rules for bacterial nomenclature were developed. Many names honor the microbiologist who first described the organism. For example, the genera *Escherichia, Salmonella, Shigella,* and *Neisseria* derive their names from the microbiologists Escherich, Salmon, Shiga, and Neisser, respectively. Simi-

larly, *Rickettsia prowazekii,* the cause of epidemic typhus, is named for two microbiologists, Ricketts and Prowazek, both of whom became infected with this pathogen and died while studying it. Other names describe properties characteristic of the bacterium. *Bacillus* (rod), *Treponema* (spiral), *Clostridium* (spindle-shaped), and *Mycobacterium* (funguslike bacterium), for example, refer to the morphology of the organism. *Proteus,* a motile bacterium that spreads over the surface of solid media and gives rise to a variety of colonial forms, is named after a Greek god who could change his shape. *Clostridium botulinum,* the etiologic agent of botulism,

derives its species name from the Latin word *botulus,* meaning sausage. Early outbreaks of this foodborne illness in Europe were traced to the ingestion of homemade sausages contaminated with the toxin-producing bacteria.

In 1976, a new bacterium was isolated as the agent of a sometimes fatal pulmonary infection. The first recognized victims were people attending a convention of the American Legion in Philadelphia. The organism was named *Legionella pneumophila* (*pneumo* = lung; *phile* = lover), commemorating the first known victims and their disease.

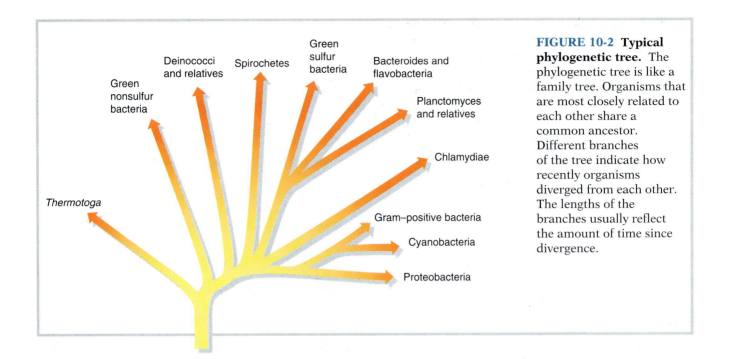

FIGURE 10-2 Typical phylogenetic tree. The phylogenetic tree is like a family tree. Organisms that are most closely related to each other share a common ancestor. Different branches of the tree indicate how recently organisms diverged from each other. The lengths of the branches usually reflect the amount of time since divergence.

guish between apparently identical organisms. A second complication interferes with our ability to determine the members of a particular species, a problem that arises from the absence of sexual reproduction in bacteria. Classification of eukaryotes (especially animals) into species is generally based on the ability of two organisms to produce fertile offspring by sexual reproduction with each other and not with members of other species. This reflection of genetic relatedness is not applicable to prokaryotes; some can attach to other bacteria and transfer genetic information, but they do not reproduce in this way. We must therefore turn to other criteria for determining what constitutes a particular species. In other words, how do we determine if a bacterium is genetically similar enough to the members of a particular species to justify being considered the same type of organism?

There is currently no way to answer this question for all prokaryotes. In fact, there is no single official scheme for classifying prokaryotes. The lack of agreement stems from the fact that the most stable taxonomic classification scheme is one based on similarity in chromosomal DNA, the ultimate indicator of genetic relatedness. Such information not only helps categorize organisms but also can be used to construct a phylogenetic tree, much like your family history can be documented in a family tree (Fig. 10-2). Closely related bacteria descended from a common ancestor more recently than did more distantly related bacteria, which have a more ancient ancestor in common. This type of classification helps us understand the evolutionary relationship among organisms but requires extensive information about the genetic composition of all known bacteria. Although advances in molecular biology have provided us with the means for comparing the genetic makeup of prokaryotic cells, it will take many years of further research to genetically examine all the known species of bacteria and establish an accurate and comprehensive system for revealing genetic (and thus evolutionary) relatedness. Such a system is called *phylogenetic* classification.

It is important to remember, however, that one major purpose of classifying bacteria is to provide a practical system for identifying organisms. Such a practical classification scheme (called *determinative* classification) need not reflect the true genetic or evolutionary relatedness of organisms. Determinative classification systems are usually based on observable properties.

MYTH: The animal kingdom is the most diverse of the five kingdoms; that is, it has the most species.

FACT: Although the animal kingdom has the most species discovered so far, the Monera kingdom is likely the most diverse. Every animal (including perhaps 30 million yet-to-be-discovered insects) will have a complement of its own microbial residents. On every type of plant and animal, many unique species of bacteria reside as normal flora or infect it as pathogens.

■ Genetic Classification

There are currently about 4500 known species of prokaryotic organisms, three times as many as the 1500 recognized in 1985. Where did these new species come from? Some of the increase is due to the discovery of new organisms, but many of these bacteria were "created" when genetic studies revealed that some bacteria thought to be variants of a single organism were distinct enough to be given separate names. The genetic criterion for classifying prokaryotes into species relies on directly comparing the nucleic acids of different bacteria. Prokaryotic cells are easiest to analyze for DNA similarities because they have only one chromosome. Even with prokaryotes, however, determining DNA relatedness requires specialized (and expensive) equipment and training, so genetic analysis is performed most often in research laboratories.

● **DNA BASE COMPOSITION.** A linear molecule of DNA is composed of nucleotides, each of which contains one of four bases—adenine, cytosine, guanine, or thymine ▶ (see Nucleic Acids—DNA and RNA, Chap. 6, p. 135). In *base composition studies*, the amounts of two DNA components, guanine (G) and cytosine (C), are measured, yielding a value called the **percent G + C.** The sum of the guanine and cy-

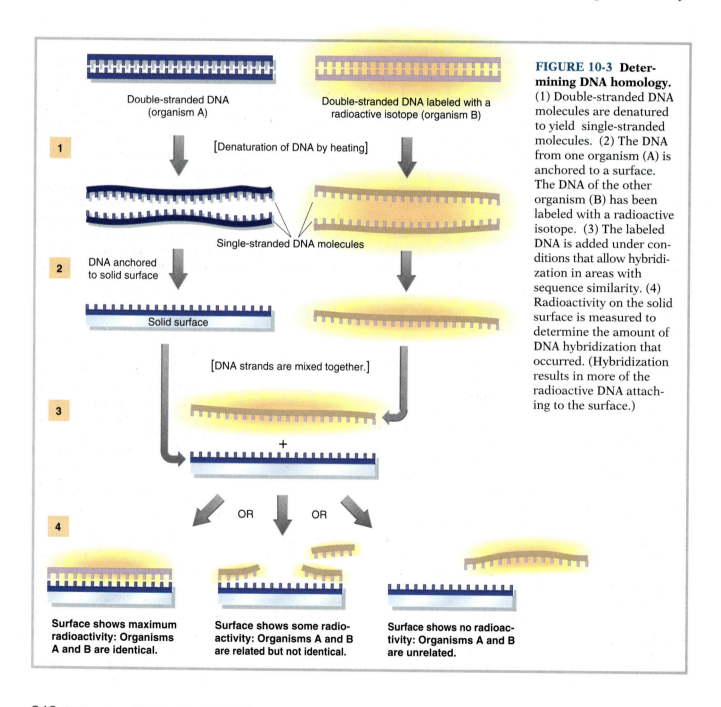

FIGURE 10-3 Determining DNA homology. (1) Double-stranded DNA molecules are denatured to yield single-stranded molecules. (2) The DNA from one organism (A) is anchored to a surface. The DNA of the other organism (B) has been labeled with a radioactive isotope. (3) The labeled DNA is added under conditions that allow hybridization in areas with sequence similarity. (4) Radioactivity on the solid surface is measured to determine the amount of DNA hybridization that occurred. (Hybridization results in more of the radioactive DNA attaching to the surface.)

Double-stranded DNA (organism A)

Double-stranded DNA labeled with a radioactive isotope (organism B)

1 [Denaturation of DNA by heating]

Single-stranded DNA molecules

2 DNA anchored to solid surface

Solid surface

[DNA strands are mixed together.]

3

+

4 OR   OR

Surface shows maximum radioactivity: Organisms A and B are identical.

Surface shows some radioactivity: Organisms A and B are related but not identical.

Surface shows no radioactivity: Organisms A and B are unrelated.

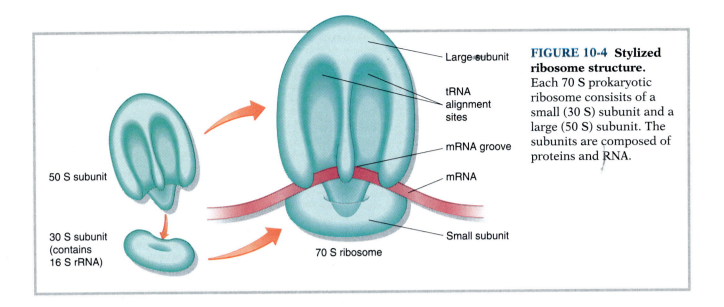

FIGURE 10-4 Stylized ribosome structure. Each 70 S prokaryotic ribosome consisits of a small (30 S) subunit and a large (50 S) subunit. The subunits are composed of proteins and RNA.

Labels on figure: Large subunit; tRNA alignment sites; mRNA groove; mRNA; Small subunit; 50 S subunit; 30 S subunit (contains 16 S rRNA); 70 S ribosome

tosine present in an organism's DNA is stable from generation to generation. The ranges reported among different types of bacteria vary from about 25 to 75 percent G + C, and organisms that are related have similar or identical G + C content. Not all organisms with similar G + C content, however, are related, since the nucleotide bases may be arranged in a completely different order (just as two sentences with the same number of the letter "e" might be completely different if the letters are arranged in a different sequence). This analysis, therefore, can determine that two organisms are different but cannot be used to ensure that they are related. To rule out coincidence, DNAs with the same percent G + C must be subjected to a stricter test for genetic relatedness, such as a test for DNA homology.

● **DNA HOMOLOGY.** Percent G + C determinations provide an indication of the base composition of an organism's DNA. An organism's properties, however, are determined by the *sequence* of the bases in DNA, a sequence that "spells out" the organism's characteristics just as the sequence of letters spells out the meaning of this sentence ▶ (see "The Genetic Language," Chap. 8, p. 184). Thus, if two bacteria are the same type of organism, their DNA base sequences will be very similar. **DNA homology** techniques measure similarities in the sequential order in which the molecular building blocks of DNA are arranged (Fig. 10-3). Nucleic acids from two bacteria (with similar percent G + C) are incubated together under conditions that allow *hybridization*, the formation of bonds between the nucleic acid from one bacterium and the nucleic acid of the other bacterium in regions with similar sequences. The percent of hybridization between the two DNAs determines the degree of DNA homology.

● **RIBOSOMAL RNA ANALYSIS.** Knowledge of the evolutionary history of prokaryotes, especially those that are not closely related, requires a direct comparison of DNA sequences. The actual order of the nucleotides in an entire bacterial chromosome is not yet available, but shorter segments of nucleic acids can be analyzed. Microbiologists have turned to gene coding for a small nucleic acid, **16 S ribosomal RNA (16 S rRNA)**, a component of prokaryotic ribosomes, for information on evolutionary relatedness (Fig. 10-4). Of all the genes that are available, why was the 16 S rRNA gene selected for most genetic analyses?

- It is found in all cells. (In eukaryotic cells, its equivalent gene codes for a comparable but slightly larger molecule, 18 S rRNA.)
- It is a manageable size for sequence analysis, as it is composed of approximately 1500 nucleotides.
- It codes for a molecule with a long history. Every living cell in every kingdom on today's earth contains these ribosomal molecules. The most logical explanation for this is that they have been around since the earliest ancestral cells and are essential molecules for protein synthesis. Organisms whose evolutionary paths may have diverged from each other millions of years ago may still retain some of the same base sequences in these RNAs, which were inherited from their common ancestor.

Analysis of the 16 S rRNA genes begins by amplifying the gene using the polymerase chain reaction. The DNA is then enzymatically cleaved into smaller fragments. The nucleotide sequences of these pieces are determined, and a computer is then used to compare sequences from different organisms. The emerging data support the theory that a primitive progenitor cell gave rise to three distinct cell types, establishing the ancestors of the eukary-

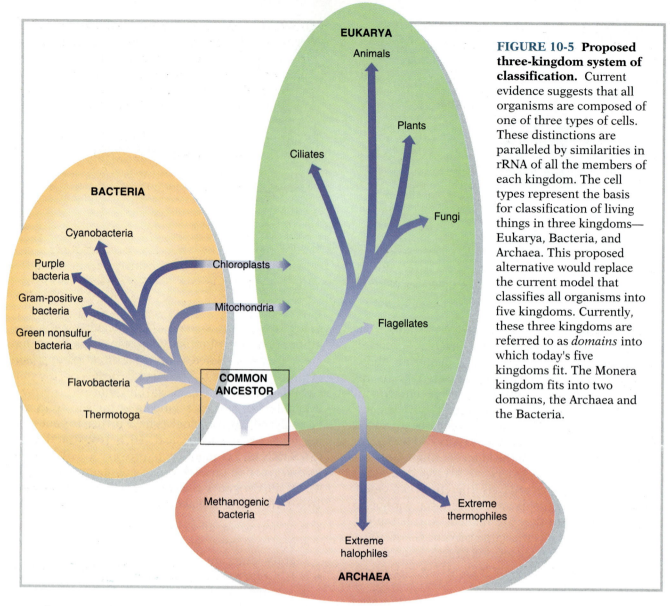

EUKARYA

Animals

Plants

Ciliates

Fungi

Chloroplasts

Mitochondria

Flagellates

BACTERIA

Cyanobacteria

Purple bacteria

Gram-positive bacteria

Green nonsulfur bacteria

Flavobacteria

Thermotoga

COMMON ANCESTOR

Methanogenic bacteria

Extreme thermophiles

Extreme halophiles

ARCHAEA

FIGURE 10-5 **Proposed three-kingdom system of classification.** Current evidence suggests that all organisms are composed of one of three types of cells. These distinctions are paralleled by similarities in rRNA of all the members of each kingdom. The cell types represent the basis for classification of living things in three kingdoms—Eukarya, Bacteria, and Archaea. This proposed alternative would replace the current model that classifies all organisms into five kingdoms. Currently, these three kingdoms are referred to as *domains* into which today's five kingdoms fit. The Monera kingdom fits into two domains, the Archaea and the Bacteria.

otes, the eubacteria, and the archaeobacteria. Each of these cell types is characterized by unique 16 S rRNA nucleotide sequences, called *signature sequences*, that define that cell type. So compelling is the evidence for these major taxonomic divisions that it has been recently proposed that the current five-kingdom system of classification be revised and all living things placed into *three* kingdoms based on these three lines of cellular evolution (Fig. 10-5). Ribosomal RNA analysis has also verified another, more widely accepted hypothesis that chloroplasts and mitochondria descended from eubacteria that established an intracellular existence in eukaryotic ancestor cells millions of years ago.

The evolutionary history within the eukaryotes, eubacteria, and archaeobacteria is also recorded in their rRNA compositions. For example, despite the wide morphological and physiological diversity among eubacteria, the entire group can be traced

back to the 11 distinct evolutionary groups diagrammed in Fig. 10-2. Members of each group share characteristic rRNA sequences that distinguish them from members of other groups.

The 11 groups of eubacteria are distinguished from each other by the following characteristics:

- *Proteobacteria* are a large and diverse group of gram-negative bacteria derived from a common ancestral photosynthetic bacterium. Most members have lost their photosynthetic machinery and are chemotrophic. The proteobacteria include enteric gram-negative rods, pseudomonads, sulfate- and sulfur-reducing bacteria, and organisms such as *Agrobacterium, Neisseria, Legionella, Vibrio, Haemophilus,* and the obligate intracellular parasitic *Rickettsia.*

- *Cyanobacteria* contain chlorophylls and accessory

pigments similar to those in chloroplasts of algae and green plants (the photosynthetic eukaryotes) and perform similar oxygenic photosynthetic processes. Genetic analysis clearly indicates that cyanobacteria and the chloroplasts of algae and green plants derive from a common ancestor.

- *Gram-positive bacteria* include rod-shaped endospore-forming species of *Bacillus* and *Clostridium* and non-spore-forming rods and cocci such as lactobacilli and staphylococci. The filamentous and irregularly shaped actinomycetes and the acid-fast mycobacteria also belong to this group. The mycoplasmas appear to be wall-less forms related to clostridia.
- *Chlamydia* are all members of a single genus and are obligate intracellular parasites that cannot grow outside a host cell. They lack critical enzymes for generating compounds needed for energy transfers.
- *Planctomyces* group members are budding bacteria with cell walls that lack peptidoglycan.
- The anaerobic *Bacteroides* and aerobic *Flavobacterium* group members are related gram-negative eubacteria with opposite oxygen requirements.
- *Green sulfur bacteria* perform anoxygenic photosynthesis using sulfur or sulfide as an electron source.
- *Spirochetes* are corkscrew-shaped gram-negative bacteria that possess an axial filament.
- *Deinococcus*, a radiation-resistant gram-positive coccus, is genetically related to *Thermus,* a gram-negative thermophile.

- *Green nonsulfur bacteria* perform anoxygenic photosynthesis but are found to be evolutionarily distinct from green sulfur bacteria.
- *Thermotoga* and *Thermosipho* were isolated from geothermally heated marine sediments and are among the most thermophilic eubacteria.

Ribosomal RNA analysis of the archaeobacteria separates them into major groups characteristic of methanogens, extreme halophiles, and thermoacidophiles. ▶ (See Archaeobacteria, Chap. 2, p. 41.)

● **PHYLOGENETIC ASSIGNMENTS.** Ribosomal RNA analysis helps define whether a prokaryote is a eubacterium or archaeobacterium and the specific group to which it is most closely related. As we begin to explore environments previously unavailable to microbiologists, we will undoubtedly discover organisms that have never before been described or even encountered by humans. (See THE EXPLORERS: "Growing" Unculturable Microbes.) Finding a signature sequence will be especially useful when trying to identify microbes with unusual combinations of properties that don't fit tidily into any group according to determinative characteristics. But even organisms that are already classified may be moved into different taxonomic categories as new facts about their molecular makeup are discovered.

Information on the details of microbial relatedness at the molecular level is evaluated by an international commission. One of the more difficult tasks

THE EXPLORERS

"Growing" Unculturable Microbes

Marine microbiologists have long been aware of the discrepancy between the numbers of bacteria they see in a sample of seawater and the number of bacteria that grow when the sample is plated. Even using a variety of media and varying incubation conditions, the best that these workers usually do is to culture a small percentage of the ocean's diverse prokaryotic inhabitants. Recently,

the polymerase chain reaction has been used to help identify natural bacterial populations.

Reproducing an organism's DNA and not the entire cell is the key to this approach. First, DNA is isolated by lysing all the microbes in a sample. PCR is then performed to amplify portions of the DNAs, in essence "growing" the genomes of the bacteria in the sample. The amplified DNAs are transferred into cultures of *E. coli*, and those *E. coli* that have acquired a foreign gene coding for 16 S ribo-somal RNA are identified. These *E. coli* can be cultured and thus provide sufficient foreign 16 S rRNA genes for analysis.

Several surprising results have been

found in the analysis of ocean waters. For example, among the bacteria in greatest abundance in a sample from the Sargasso Sea, a region in the Atlantic Ocean, are ones that have never yet been cultured. Even more interesting perhaps is the fact that some of these bacteria are also predominant in Pacific Ocean waters. Why are these organisms so abundant, and how did they get to be so widespread? Why hasn't anyone been able to grow them? Such unanswered questions will undoubtedly increase in number as molecular techniques make it easier to uncover the diversity of microbial life, not just in seawater but from wherever on our planet microbiologists can obtain samples.

of the commission is to define how genetically similar two organisms must be for both to be assigned to the same genus and species. Organisms that are about 20 to 60 percent related according to DNA homology are usually considered members of a single genus. Organisms that are more than 60 percent related are assigned to a single species. There is substantial natural variability within a species (organisms may be 40 percent unrelated and still belong to the same species). This variability may be expressed as differences in immunological, structural, and physiological characteristics. For example, one organism may have the ability to produce a capsule while another does not, or one organism may be able to use a particular sugar as a nutrient whereas another cannot. These differences are often used to further subdivide the members of a species into subgroups called **strains.** One strain for each species is designated as the *type strain*. Its characteristics are used to define the species, and all organisms similar to the type strain are included in that species. Different strains within the species may differ from each other in biochemical properties, morphological characteristics, ability to infect different hosts, or reactivity with different antibodies.

If an organism is too different to be assigned to any existing genus or species, it is declared a newly discovered organism and given its own name. Measurements of DNA, for example, were used to show that the agent of legionnaires' disease was a newly discovered bacterium genetically unrelated to any previously known. Decisions of the commission are published in the *International Journal of Systematic Bacteriology*.

■ "Practical" Classification

For most known bacteria, their percent G + C has been determined, but we know few details about other aspects of their genetic makeup. The lack of genetic data makes it currently impossible to assemble a classification scheme that is both phylogenetic in its ability to reveal evolutionary relationships and practical in its usefulness as a tool for routine identification of unknown bacteria. Identification of bacteria isolated from clinical and environmental sources relies primarily on a combination of morphological and biochemical characteristics that are relatively easy to determine in the laboratory. Using data from such determinative tests, the investigator turns to a classification scheme to find a species whose characteristics match those of the unknown bacterium. Perhaps the most widely used classification scheme for identifying prokaryotes is described in ***Bergey's Manual***.

■ Bergey's Manuals

Bergey's Manual for Determinative Bacteriology has been a practical guide for the identification of bacteria since it was first published in 1923. Its classification scheme relies primarily on observable characteristics to divide bacteria into taxonomic groups. The 1994 (ninth) edition separates the prokaryotes into 35 groups; 30 groups encompass the eubacteria, and the remaining five form the archaeobacteria. The groups are distinguished by such properties as Gram stain reaction, morphology, the ability to produce endospores, and mechanisms of metabolism, motility, and reproduction (Table 10-1). A few key tests performed on an isolated unknown bacterium yield sufficient information to determine which section of *Bergey's Manual* the organism belongs in (see Identification of Prokaryotes, below).

A four-volume edition of a companion text, *Bergey's Manual for Systematic Bacteriology*, was released, one volume at a time, between 1984 and 1989. This manual attempts to bridge the transition period created by the introduction of genetic analysis into the classification of bacteria. It partly classifies bacteria on the basis of available genetic data and relates the genetic data to those morphological and biochemical tests that are simple to perform in the laboratory. For convenience, however, morphological distinctions form the basis for assigning all the members of the prokaryotes to 33 sections, all of which are listed in Appendix B. (Figure 10-6 shows some of the more unusual features that separate bacteria into their own section.) One of the unfortunate consequences of this dual function (genetic taxonomy and identification) is that evolutionarily related organisms are occasionally assigned to different sections. For example, *Mycoplasma* is not even included in the same volume as the gram-positive bacteria to which members of this genus are related.

CONCEPT CHECK

- Why can't prokaryotic species be determined in the same way eukaryotic species are?
- What techniques are most useful in determining the information necessary for phylogenetic classification?
- What are some advantages and disadvantages of a phylogenetic classification scheme compared to one based on observable properties?
- Why has there been a need for nine editions (so far) of *Bergey's Manual of Determinative Bacteriology*?

Group	Distinguishing Features	Some Important Genera	Importance	Topic Discussed in Chapter:
Eubacteria				
1. Spirochetes	Flexible, coiled morphology; motility by axial filament	*Treponema* *Borrelia* *Leptospira*	Syphilis Lyme disease Leptospirosis	23 24 24
2. Aerobic/ microaerophilic, helical/vibroid gram-negative bacteria	Capable of growth only in presence of oxygen; rigid, curved or coiled morphology; motility by flagella	*Azospirillum* *Bdellovibrio* *Helicobacter* *Campylobacter* *Spirillum*	Nitrogen fixer in soil Parasite of bacteria Stomach ulcers Gastroenteritis Rat bite fever	26 14 22 22 24
3. Nonmotile (or rarely motile) gram-negative curved bacteria	Capable of growth only in the presence of oxygen	*Spirosoma*	Soil microbe	26
4. Gram-negative aerobic/micro-aerophilic rods and cocci	Capable of growth only in the presence of oxygen; straight rods or cocci	*Acetobacter* *Agrobacterium* *Bordetella* *Legionella* *Methylomonas* *Neisseria* *Pseudomonas* *Rhizobium* *Thermus* *Xanthomonas*	Vinegar production Plant pathogen; used in genetic engineering of plants Pertussis Legionnaires' disease Methane oxidizer Gonorrhea Pathogen of burns, septicemia; water/soil microbe Symbiotic nitrogen fixer Source of polymerase for PCR Xanthan production	27 9,27 21 21 26 23 21,25 26 14,26 9 27
5. Facultative anaerobic gram-negative rods	Capable of growth in the presence or absence of oxygen	*Escherichia* *Haemophilus* *Photobacterium* *Shigella* *Salmonella* *Vibrio* *Yersinia*	Genetic engineering; intestinal and urinary tract infections Meningitis Luminescent symbiont Dysentery Food poisoning; typhoid fever Cholera Plague	22 21 14 22 22 22 24
6. Gram-negative; anaerobic; straight, curved, and helical bacteria	Capable of growth only in the absence of oxygen	*Bacteroides* *Selenomonas* *Succinomonas* *Thermotoga*	Normal flora in intestinal tracts; causes some abscesses Rumen bacteria Rumen bacteria Geothermal marine sediments	22,24 14 14 5

Continued

FIGURE 10-6 **Some unusual prokaryotes.** Some properties of bacteria are so unique they warrant placing their owners in their own category. Examples of such traits are (*a*) gliding motility (these fruiting bodies of *Chondromyces crocatus* are formed by gliding cells aggregating and differentiating into complex sporangioles), (*b*) appendaged bacteria (the bacterial stalks of the *Caulobacter* shown here form a "rosette"), and (*c*) sheathed bacteria (these *Sphaerotilus natans* cells can be seen distinctly inside a tubular sheath that holds them together as filaments).

Identification of Prokaryotes

For identification of bacteria that have been previously characterized and classified, determinative tests are performed on a pure culture of the organism to be identified. It usually takes 10 to 30 tests to identify a bacterium. The "average" bacterium, however, contains more than 3000 genes, so these tests are actually examining a very tiny portion of the organism's properties. The most important criteria relate to Gram stain reaction, shape, endospore formation, and response to oxygen. For example, the vast majority of medically important bacteria are found in 18 of the sections in *Bergey's Manual*. Identification of a pathogen from an infected patient usually begins by determining its Gram stain reaction and shape. This information reduces the possible choices of identity to members of one to three of these 18 groups. Within each of these groups, the family, genus, and species are determined by more

specific morphological and biochemical tests. For example, clinical laboratories spend considerable time and effort identifying bacteria described in section 5 (gram-negative facultative rods). Many of these gram-negative bacilli are important pathogens; others are members of the normal human flora or are environmental contaminants. They are all facultative anaerobes that can grow in the presence or absence of oxygen. Members in this section are divided into three families, Enterobacteriaceae, Vibrionaceae, and Pasteurellaceae, which are easily distinguished. Only Enterobacteriaceae possess peritrichous flagella and are negative when tested for oxidase activity. (Oxidase is a protein used by some bacteria in respiration.) Based on DNA homology studies, the Enterobacteriaceae are subdivided into more than a dozen genera. Among these are *Salmonella*, *Klebsiella*, *Escherichia*, and *Yersinia*. Because they are all virtually identical in morphology, identification of Enterobacteriaceae members is based primarily on biochemical tests. *Bergey's*

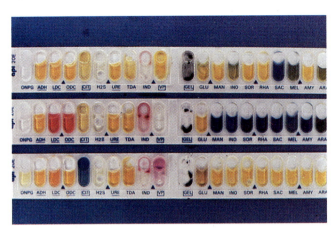

FIGURE 10-7 What's in a test. Commercial identification systems such as the one shown are composed of a limited series of biochemical tests. Isolates with non-characteristic responses often are identified with limited certainty, and additional tests may have to be performed to confirm their identification.

Manual lists the characteristic responses of species within the family. Enterobacteriaceae to more than 40 biochemical tests. Commercially available systems incorporate many of these tests in kits designed to facilitate identification of clinically relevant bacteria. The manufacturer usually supplies documents that describe the typical responses of the specific organisms for which the system is designed (Fig. 10-7).

One major problem arises when identification is based solely on observable traits. Since members of a species may vary by as much as 40 percent in DNA homology, what precisely are the characteristics that must be assayed to ensure that an organism is identified correctly? Such disturbing questions still plague microbiologists. For example, what is an *E. coli? Bergey's Manual* says it ferments lactose. Then what is an organism that does not ferment lactose but has all the other characteristics of *E. coli*? Fortunately, some of these problems have been resolved through the use of automated identification technologies and DNA (gene) probes.

■ Automated Identification Systems

Several methods relying on computer analysis have been developed to aid in the rapid and accurate identification of organisms that are routinely identified from clinical and environmental samples. These systems collect data on specific characteristics of the unknown isolate and compare this information to a computerized database of known bacteria. One system, for example, tests the substrate-using versatility of bacteria by measuring growth of the organism in 96 different media contained in small wells in a single plastic plate (Fig. 10-8). Another commercially available system measures the profile of the lipids extracted from the bacterial plasma membrane. The results are fed directly into a computer, which then determines the most likely identity of the unknown. Sometimes the characteristics of the isolate do not correspond to any bacterium in the database. This may occur because the databases are limited and do not contain information on all known bacteria. Alternatively, the organism may be a newly discovered bacterium. (See BEYOND THE BASICS: What's in a Number?)

BEYOND THE BASICS

What's in a Number?

The search for new sources of antibiotics has lured scientists into previously unexplored environments. Hundreds of microorganisms may be isolated from a single sample from a virgin habitat, and many may be bacteria that have never before been identified and therefore are not among the organisms listed in any of the available classification schemes and computerized databases.

Investigators wishing to test the drug-producing abilities of new bacteria want to know how many different organisms they have isolated. For example, among the hundreds of colonies that were formed from the original sample, there may be only 10 different species. Genetic analysis may give them the answer but is often not a practical alternative. Fortunately, comprehensive testing of microbial characteristics can be combined with computer technology to determine the variety of bacteria present in a population.

Numerical taxonomy is the comparison of organisms based on testing a large number of characteristics, each given equal weight in the comparison. Usually 50 to 100 characteristics are measured to give a valid result. Most often these include morphology, biochemical activities, and growth characteristics. The greater the similarity in test results between organisms, the more closely related the organisms are believed to be (organisms that have an 80 percent similarity rating or greater are usually the same species). This method reveals which colonies truly represent different organisms.

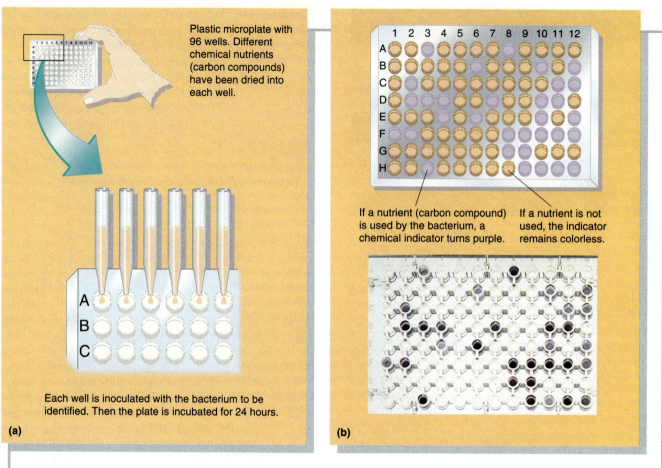

Plastic microplate with 96 wells. Different chemical nutrients (carbon compounds) have been dried into each well.

If a nutrient (carbon compound) is used by the bacterium, a chemical indicator turns purple.

If a nutrient is not used, the indicator remains colorless.

Each well is inoculated with the bacterium to be identified. Then the plate is incubated for 24 hours.

(a)

(b)

FIGURE 10-8 Food for (computer) thought. By inoculating an organism to be identified into 96 different media (*a*), we can determine which substrates the organism uses (purple indicates growth). (*b*) The computer then interprets the utilization patterns and identifies the organism.

■ Gene Probes

DNA probes are single-stranded fragments of DNA composed of nucleotide sequences found only in a particular bacterial species ▶ (see Fig. 9-9, p. 225). The probes are labeled with radioactive isotopes or fluorescent dyes and added to the DNA of the organism to be identified. The probe hybridizes to the unknown DNA if the two sequences are complementary. The detection of the label therefore indicates the identity of the unknown bacterium. Testing for the presence of definitive genetic sequences is now possible for many bacteria through the development and commercialization of DNA probes.

> **CONCEPT CHECK**
>
> - Why are many biochemical tests needed to accurately identify a bacterium while only a single DNA probe may be required for the same identification?
> - How have computers aided in the rapid and accurate identification of bacteria?

Key Terms

Key Facts and Concepts

- **Taxonomy and nomenclature.** Systematic classification of prokaryotes establishes a practical method to identify microbes and to assign a meaningful binomial name to each organism.
- **Phylogeny.** Phylogenetic classification provides indications of genetic (and evolutionary) relatedness of organisms. Phylogenetic classification of prokaryotes is based on nucleic acid studies. DNA homology indicates relatedness at the genus and species level. Ribosomal RNA analysis is most useful for uncovering evolutionary linkages.

- **Going by the book.** Routine bacterial identification is usually based on morphological and biochemical tests that reflect the genetic composition of the organism. Such information is compiled in a practical guide called *Bergey's Manual*.
- **Automated systems for identifying microorganisms.** Computers digest and interpret dozens, sometimes hundreds, of character traits. Some of these systems use numerical taxonomy for either identifying an organism or determining how closely related different organisms are.

Review and Problem-Solving Questions

1. List five major characteristics currently used in the classification of bacteria.
2. Describe how each of the following methods measures genetic (DNA) relatedness: (a) Percent G + C; (b) 16 S ribosomal RNA analysis; (c) DNA homology.
3. How does *Bergey's Manual of Determinative Bacteriology* differ from *Bergey's Manual of Systematic Bacteriology*?
4. Why is the Gram stain an important initial step in the diagnosis of infectious disease?
5. You have isolated an organism from the intestinal tract of a fish. Outline the tests you would perform in trying to determine whether this organism has ever been isolated and described before.
6. How are microbes distributed in the five-kingdom classification system compared to the three-kingdom system? What evidence suggests that there are only three kingdoms (or domains)?
7. Should traits controlled by genes on a plasmid be used in determining the taxonomic identity of an organism? Defend your position.
8. Clostridia and mycoplasmas have chromosomes that differ greatly in size, but these organisms appear to be genetically related. Why isn't chromosome size a

good indicator of genetic relatedness?
9. Describe what techniques you might use to identify bacteria in each of the following situations.
   - (a) You have been given a sputum sample to test for *Mycobacterium tuberculosis*.
   - (b) You have been given a water sample to test for the presence of a lactose-negative *E. coli*.
   - (c) You have a blood sample from a patient with a new infectious disease. No one has yet been able to culture this organism from blood.
10. Mitochondria probably descended from an ancestral proteobacterium. Design an experiment to determine which of the currently existing proteobacteria is the closest relative to human mitochondria.
11. Genetic data indicate that members of the genus *Shigella* (the agents of bacterial dysentery) have greater than 70 percent genetic homology with *Escherichia coli*. What kind of information has been used to traditionally classify these organisms into different genera? Despite their genetic similarity, they have not been reclassified as different strains of the same species in the current edition of *Bergey's Manual of Determinative Bacteriology*. What possible reasons may have prompted this decision?

COUNTING CREATURES GREAT AND SMALL*

At a workshop sponsored by the National Science Foundation (NSF), researchers discussed how the world's biological diversity could be cataloged, beginning with the identification of every species living within a single defined locale. This All Taxa Biodiversity Inventory (ATBI) will hopefully reveal information that will lead to better conservation, management, and utilization of biological resources. At this point, the problems facing researchers in global diversity, especially those studying microorganisms, seem staggering. However, scientists participating in the workshop had no doubt about the value of the information that could be harvested from a successful exploration.

- What specimens would you collect to isolate *all microbes*

existing in an acre of tropical rainforest?
- What can you do to prevent the introduction of nonindigenous (contaminant) microbes shed by the researchers at a site?
- How can the masses of data generated by this study be stored and organized so that they will be accessible to a variety of users?
- If you were a member of Congress and were asked to support this project with a $25 million budget, would you vote to allocate these funds? Support your decision.

*Carol Kaesuk Yoon, Counting creatures great and small, *Science*, 260: 620–622 (1993).

CHAPTER 11

Fungi

False-colored SEM of *Aspergillus,* or common bread mold. The fungus is a decomposer, a common air-borne contaminant, and a life-threatening pathogen in some persons with lowered defenses.

They saw the disaster coming but could do nothing to avert it. The first subtle signs of catastrophe were noticed during the mid-1800s. A few potato plants had developed brown flecks on their leaves and stems. But this transformation was terrifying for the Irish, whose potato crops were their primary source of food. They helplessly watched as virtually every potato vine in the country rotted and died, the affected fields looking as though they had been burned. During the 14 years of the great potato blight, a million people starved to death, and millions more desperately bolted from their country, confronted with the heartbreaking choice of staying and starving or emigrating to a new land.

The whole world felt the effects of the Irish potato blight, which forever altered the history of the countries that received the newly arriving families. Many of the politicians, scientists, and industrial captains who have since directed the course of events in the United States, for example, descended from Irish immigrants who would never have left their homeland were it not for this microscopic killer.

Even today, fungi remain among our major enemies, causing some of the most persistent infectious diseases known and robbing us of the resources on which our lives often depend. Other fungi, in contrast, are essential allies to humans and other life forms. Whether friends or foes, the fungi (like all groups of microbes) continue to shape human history.

Fungi were among the first microorganisms to be investigated scientifically, making **mycology**, the study of fungi, one of the first microbiological sciences. These early investigators soon discovered that **fungi** (singular, *fungus*) include molds, yeasts, and a group of macroscopic organisms often called the fleshy fungi. They are all eukaryotic, nonphotosynthetic organisms, usually enclosed by cell walls that are composed of *chitin*, a polysaccharide of *N*-acetylglucosamine subunits (unlike plant cells, which have cell walls of cellulose, a polysaccharide composed of glucose subunits). Many fungi are familiar to all of us—molds that grow on bread, fruit, and cheese; mildew in damp textiles; yeast used in baking and brewing; mushrooms and toadstools. Some fungi produce antibiotics that we use therapeutically against many bacterial infections. Among the fungi are organisms that excrete a wide range of degradative enzymes that attack virtually any organic material. Such degradative activities make fungi essential participants in recycling natural wastes in our environment, decomposers in the biogeochemical cycle ▶ (see Environmental Benefits, Chap. 1, p. 9). Unfortunately, their degradative proficiency also results in the unwanted growth of fungi that destroy useful materials. Learning ways to minimize their negative impact without compromising their positive contributions is one of the tangible benefits of studying the biology of the fungi.

Morphology of the Fungi

Fungi are generally larger than bacteria, with individual cell diameters ranging from 1 to 30 μm. Microscopic fungi exist as either molds or yeasts or both. The **molds** form large multicellular aggregates of long branching filaments called **hyphae** (Fig. 11-1*a*). These tubelike hyphae are responsible for the fluffy appearance of the macroscopic mold colony (Fig. 11-1*b*). **Yeasts,** on the other hand, are single cells that rarely form filaments (Fig. 11-2*a*). Yeast colonies are usually characterized by a smooth surface similar to that of many bacteria (Fig.11-2*b*).

Fleshy fungi produce large, macroscopic reproductive structures, the best known examples of which are mushrooms and toadstools. The aboveground fleshy structure, however, represents only part of the organism, most of which grows beneath the soil as microscopic filaments—hyphae. Because the fleshy fungi form hyphae, they are discussed with the molds.

The reproductive structures of fungi are called **spores.** Each spore is capable of generating a new colony in a favorable environment. Fungi may produce spores that are either asexual or sexual (Fig. 11-3). **Asexual spores** are produced by the processes of mitosis and cell division. Each spore is genetically identical to the parent. **Sexual spores,** on the other hand, are products of a sexual cycle, an alternation between the *diploid* state (in which each nucleus contains two sets of chromosomes) and the *haploid* state (each nucleus containing one set of chromosomes). One of the characteristic events in

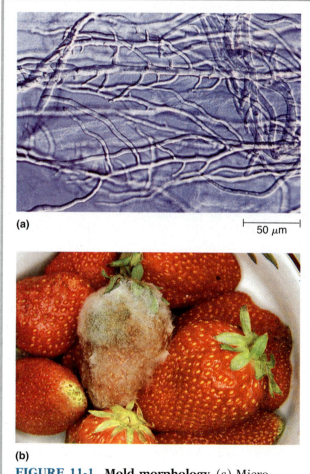

(a)

50 μm

(b)

FIGURE 11-1   Mold morphology. (*a*) Microscopic and (*b*) macroscopic appearance of typical molds.

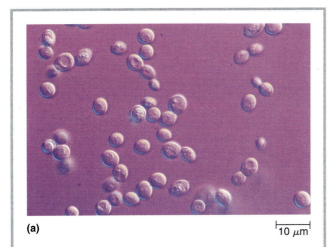

(a)

10 μm

(b)

FIGURE 11-2 Yeast morphology. (*a*) Microscopic and (*b*) macroscopic appearance of typical yeasts.

sexual reproduction is *fertilization,* the fusion of two haploid nuclei to form a diploid nucleus. This requires the presence of two sexually compatible haploid nuclei. Some fungi produce only one type of nucleus, and reproduction occurs only when the fungus comes in contact with a strain that is a different genetic *mating type. Self-fertilizing* molds, on the other hand, can produce nuclei with different mating types in the same organism.

The number of chromosomes in a diploid nucleus is reduced in half by *meiosis.* During this process, there is exchange and reassortment of parental chromosomes; the resulting haploid cells contain some genetic material from each parent. Sexual reproduction, therefore, provides a mechanism for the creation of new combinations of genetic information.

Many fungi can produce both sexual and asexual spores. The type of reproduction that occurs depends on environmental conditions, the availability

of compatible mating types, and, in the case of parasites, the nature of the host. Some fungi require more than one host to complete their reproductive cycle. (See THE MICROBES: Fungal Survival Strategies—Simple to Ingeniously Complex.)

■ Molds

● **HYPHAE.** The hyphae of a mold colony grow as an intertwined mass of filaments collectively called a **mycelium.** Molds are classified and identified partially on the basis of whether the hyphae are septate or aseptate. **Septate hyphae** are filaments with cross walls, or septa, that partition the hyphae into individual cellular compartments (Fig. 11-4). Most septa have pores that allow the migration of cytoplasm and many organelles. In **aseptate hyphae** there are no physical boundaries to distinguish individual cells in the hyphae.

Most hyphae are **vegetative**; that is, they are actively growing and form the main body of the colony. **Aerial hyphae** support specialized reproductive structures and further contribute to the fluffy appearance of the mold colony.

● **MOLD SPORES.** Most mold spores are produced by specialized regions at the ends of aerial hyphae, although some spores are formed from vegetative hyphae. Mold spores have the following characteristics:

- They are generally produced in large numbers.
- They are easily disseminated.
- Some are resistant to conditions that would kill the vegetative cell.

Thus, the collective functions of mold spores are threefold: reproduction, dissemination, and protection of the species against adverse environmental conditions.

The types of asexual and sexual spores produced by fungi vary in structure and in their mechanism of formation.

Asexual Spores. Many of the fungal spores and structures responsible for spore production are illustrated in Figure 11-5. Perhaps the most common kind of spore is the **conidium** (plural, *conidia*), which forms in clusters at the tips of specialized aerial hyphae (Fig. 11-5*a*). Conidia are not particularly resistant to adverse conditions, but they are lightweight and easily disseminated through the air, enhancing spread of the fungus. The conidia of many fungi can infect people and cause disease.

Sporangiospores are asexual spores contained in

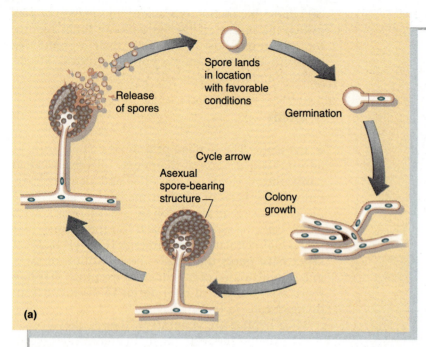

(a)

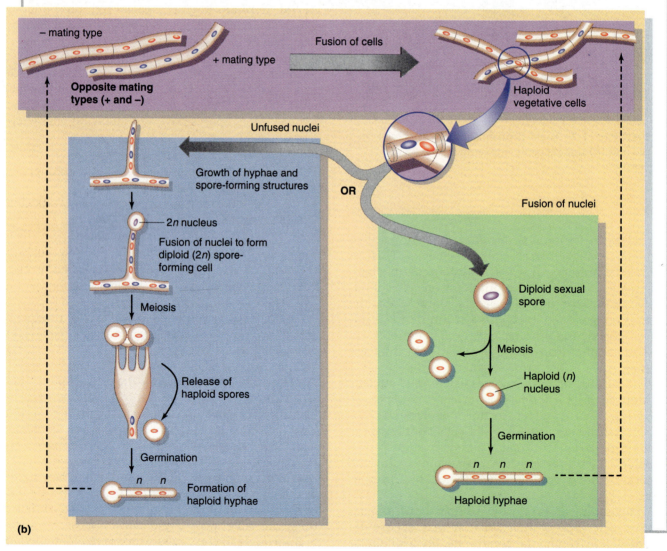

FIGURE 11-3 **Generalized life cycle of a hypothetical fungus.** Most fungi can reproduce asexually and sexually. (a) Asexual spore formation and germination of one type of mold. (b) Sexual reproduction. During the sexual stage the haploid (n) mycelium of two compat-ible hyphae fuse, resulting in a cell that contains two haploid nuclei (n + n). Depending on the species, the nuclei may immediately fuse to form a diploid cell (2n) (green background) or fuse only after the growth of the fungal colony and sexual structures (blue background). As a result of meiosis, the diploid cell produces haploid nuclei (n) with new combinations of genetic information. The haploid cells germinate to form vegetative mycelium. With a few exceptions, mature fungi are not diploid.

Fungal Survival Strategies—Simple to Ingeniously Complex

A variety of "lifestyles" contribute to the success of the 150,000 known species of fungi. The simplest organisms reproduce asexually and have only a single mature morphology. More complex life cycles are often associated with fungi that undergo sexual reproduction. Many of these organisms exhibit interesting adaptations to their environment.

One asexual strategy, however, is both complicated and "ingenious." This reproductive strategy, displayed by a species of *Pilobolus,* helps ensure the organism's success in its natural habitat—animal feces. The fungus requires an iron-containing organic compound provided by microorganisms specifically found in the feces of plant-eating animals. To survive, *Pilobolus* must transfer its spores from drying and decaying substrate to fresh material. *Pilobolus* produces a phototropic spore-bearing structure that violently discharges its spores, shooting them up to 25 cm away. Before shooting, however, the fungus orients its sporulating structures in the direction of the sun, so the ballistic release of progeny spores not only disseminates offspring from the original colony, but also propels the spores in a direction that maximizes their being deposited in the surrounding vegetation. The spores travel away from shadows (and the nearby obstructions that cast them). This increases the likelihood that they will be ingested by another plant eater and released in the animal's dung after passage through the gastrointestinal tract, providing it with a fresh source

of needed nutrients. The photosensitive spore-bearing structure of *Pilobolus,* therefore, improves the fungus's ability to survive. But how does the fungus detect and orient itself toward the light?

The sporangium of *Pilobolus* is packed with thousands of spores and sits on top of a swollen base (see photo). The

base contains a material that functions as a lens, focusing light rays on the base walls. The structure rotates until the light rays fall upon an area rich in carotene. This signals that it is oriented toward the light (and thus open space). Meanwhile, the base has been accumulating fluid under pressure. Ultimately this fluid squirts out, forcing the sporangium to be released at speeds averaging 10 meters (about 30 feet) per second. This mechanism of forcible ejection of the sporangium is responsible for the fungus's name; *Pilobolus* is Greek for "hat thrower."

Some of the most complex fungal life cycles are characteristic of plant pathogens that require two hosts. For example, *Puccinia graminis,* which causes stem rusts of cereals (wheat, oats, barley), alternates between cereal grain and barberry plants. Both hosts are necessary, and in the absence of either the organism does not survive. Between the two hosts, the fungus forms five different types of spores. The first is a sexual

spore produced when the fungus grows on the cereal plant. In the spring, these spores are released from the grain and invade the barberry plant, the only other known host. Here they germinate and develop into a second spore type, haploid asexual spores. Compatible asexual spores fuse to form a single cell with two separate nuclei. In late spring, wind transports these spores from the barberry plant to new grain plants, where they germinate, forming hyphae that invade the new host. The infected cereal grain becomes covered with the fourth type of spore, the red or orange spores that give rusts their name.

Fungal fruiting bodies.

Cells of host

These spores also infect new cereal plants and spread the infection. This is the only stage that reproduces itself, producing new cells by mitosis.

In the autumn, the fifth type of spore develops. This dikaryotic spore has a thick wall and is usually the form that survives the winter months. When spring arrives, the two nuclei in this spore fuse to form a diploid cell. This diploid cell produces the first type of spore, and the sexual cycle is completed. These sexual spores germinate when they infect barberry plants, thereby initiating another complex round of replication.

saclike structures called **sporangia** that are borne at the end of aerial hyphae (Fig. 11-5*b*). The sac erupts, discharging the mature spores. Aquatic fungi produce sporangiospores with flagella. These motile sporangiospores are called **zoospores**. Sporangiospores are normally harmless to people but occasionally infect those with impaired resistance and cause a fatal disease. **Chlamydospores** (Fig. 11-5*c*) are formed by differentiation of vegetative hyphae rather than by special structures. They are resistant to heat, drying, and freezing and often contain glycogen or lipid energy reserves. Chlamydospores are

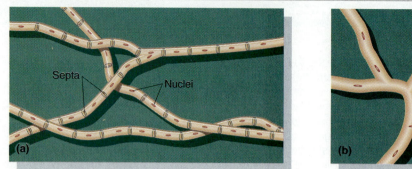

FIGURE 11-4 Characteristic hyphal structures. (*a*) Septate hyphae; (*b*) aseptate hyphae. (Note the absence of septa between nuclei.)

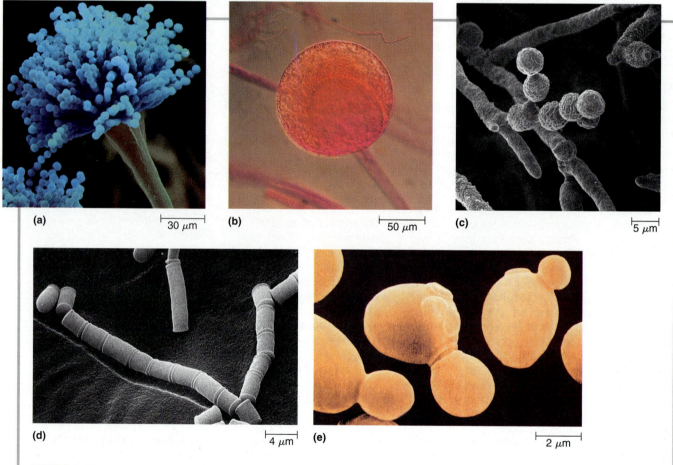

FIGURE 11-5 Various sporulating structures and the corresponding asexual spores.
(*a*) Clusters of conidia; (*b*) sporangiospores packaged inside a sporangium, a sac that releases spores when it ruptures; (*c*) spherical chlamydospores attached to vegetative cells; (*d*) arthrospores; (*e*) blastospores.

produced primarily for surviving adverse conditions and long intervals of time without nutrients.

Individual cells within the hyphae occasionally undergo thickening of their cell walls and fragment away from the parent filament, forming barrel-shaped **arthrospores** (Fig. 11-5*d*). Like conidia,

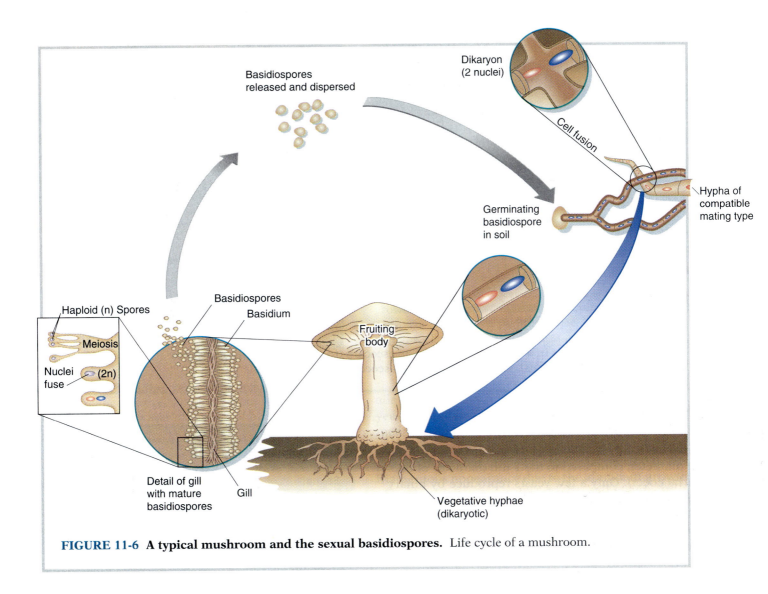

FIGURE 11-6 A typical mushroom and the sexual basidiospores. Life cycle of a mushroom.

arthrospores may be dangerous infectious agents for humans. A few molds produce asexual spores by budding. These buds are referred to as **blastospores** (Fig. 11-5*e*).

Sexual Spores. Sexual fungal spores are rarely agents of human disease. Structures that bear sexual spores in fungi are called **fruiting bodies.** The mushroom, for example, is a common fungal fruiting body that produces **basidiospores,** spores formed at the tip of a clublike structure called a **basidium** (plural, *basidia*). The underside of the mushroom looks like compartments separated by vertical walls called gills. These gills are covered with basidia (Fig. 11-6). Beneath the mushroom, in the soil, is the mold colony itself, consisting of a mat of intertwined hyphae, sometimes several feet in diameter.

Another kind of sexual spore, the **ascospore,** is found in a saclike structure called an **ascus** (plural, *asci*). Several asci are usually contained in a single larger fruiting body (Fig. 11-7). **Zygospores** and **oospores** are two kinds of sexual spores that are not associated with well-developed fruiting bodies. Zygospores are encased in a thick, darkly pigmented wall that makes them somewhat resistant to extremes in temperature and desiccation (Fig. 11-8). Oospores are large flagellated spores produced by certain aquatic fungi. Common types of sexual and asexual spores of fungi are listed in Table 11-1.

Vegetative Reproduction. New mold colonies can also be formed from fragments of hyphae. This process is called *vegetative reproduction* because it requires no spores. The hyphae fragments simply continue the same growth activities that were going on at the time of fragmentation. The fragmented cells produce a new colony that is identical to the parent.

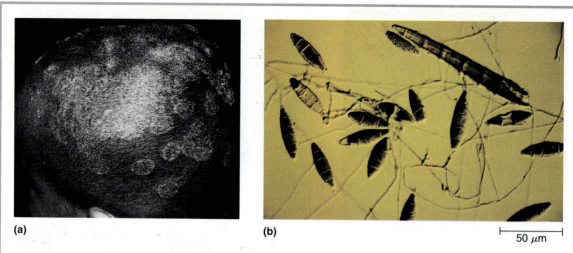

(a) **(b)**

50 μm

FIGURE 11-18 **Dermatophytoses.** (*a*) Tinea capitis. The dermatophytes often grow in a radial pattern on the skin with an elevated margin that gives the appearance of a circular worm beneath the cutaneous layer. This feature is responsible for the common name, "ringworm," although the pathogen is a fungus, not a worm. (*b*) Microscopic examination of cultured material from lesions may reveal hyphae and spores characteristic of the responsible dermatophyte.

• *Cutaneous mycoses* affect only the skin, hair, and nails. Most cutaneous mycoses are caused by a group of fungi called **dermatophytes** (*dermis* = skin; *phyte* = plant[1]). They cause diseases collectively known as *dermatophytoses* or "ringworm,"

[1]Clearly, fungi are not plants. This terminology is a holdover from the original classification of these microbes in the plant kingdom. Unfortunately, some medical terminology still reflects its inaccurate origins. "Ringworm" is another example of this, the disease caused by fungi and not a worm at all (the swelling at the advancing edges of the circular lesion suggested the presence of a worm beneath the skin).

which are among the most common and persistent of all human infections (Fig. 11-18).

• *Subcutaneous mycoses* affect the subcutaneous tissue below the skin and occasionally affect bone. They are slowly progressing but extremely persistent diseases. The organisms are usually introduced by a puncture wound. The most common of these diseases is *sporotrichosis*, which is caused by a dimorphic saprophytic fungus that grows as a mold in soil and as a yeast in the human body (Fig. 11-19).

• *Systemic* ("deep") *mycoses* infect the internal or-

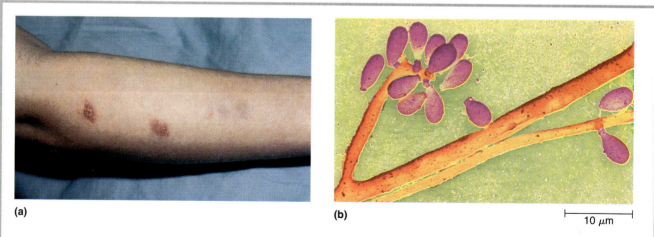

(a) **(b)**

10 μm

FIGURE 11-19 **Sporotrichosis.** (*a*) Clinical picture showing progression up the lymphatics; (*b*) microscopic appearance of *Sporothrix schenkii* in culture.

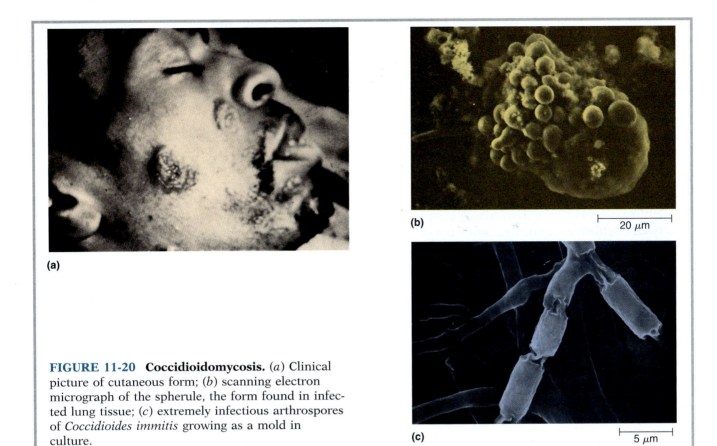

FIGURE 11-20 Coccidioidomycosis. (*a*) Clinical picture of cutaneous form; (*b*) scanning electron micrograph of the spherule, the form found in infected lung tissue; (*c*) extremely infectious arthrospores of *Coccidioides immitis* growing as a mold in culture.

(a)

(b)   20 μm

(c)   5 μm

gans and may spread throughout the host (Fig. 11-20). They are usually difficult to eradicate, and many patients with deep mycosis die in spite of extensive medical care (especially when the central nervous system is involved). Three of the four major systemic mycoses in North America are caused by dimorphic fungi (Table 11-4).

Since many fungi that cause superficial or cutaneous mycoses are skin parasites by nature, the sources of infection are usually humans or animals. The disease is transmitted directly from an infected person or animal to a susceptible host. The agents of subcutaneous and systemic mycoses, however, are normally saprophytic fungi growing in the soil. These fungi do not need human or animal hosts for survival. Because the primary source of infection is the soil, humans acquire these mycoses only when

TABLE 11-4   MAJOR SYSTEMIC MYCOSES OF NORTH AMERICA

Disease	Fungus	Appearance in Infected Tissue	Appearance In Culture[*]	Dimorphic
Coccidioidomycosis	*Coccidioides immitis*	Spherule[†]	Mold	Yes
Histoplasmosis	*Histoplasma capsulatum*	Yeast	Mold	Yes
Cryptococcosis	*Cryptococcus neoformans*	Yeast	Yeast	Rarely
Blastomycosis	*Blastomyces dermatitidis*	Yeast	Mold	Yes

[*]Cultures grown at room temperature (25°C).
[†]A nonfilamentous form of the fungus characterized by a sac filled with spores. When liberated, each spore matures into a spherule.

the spores of soil organisms are either inhaled or introduced into the body through a break in the skin.

Opportunistic Fungi. Some fungi that cannot infect healthy humans can cause serious and often fatal mycoses in people whose resistance has been lowered. Diseases caused by these **opportunistic fungi** may be either cutaneous or systemic. Many diabetic people are susceptible to pulmonary infection by the common saprophytic molds of the genera *Rhizopus* and *Mucor* (both zygomycetes) and *Aspergillus*. Other high-risk groups susceptible to opportunistic fungal infections are AIDS patients, alcoholics, leukemia patients, and persons on chemotherapy for treatment of cancer. Perhaps the most common opportunistic mycosis, especially among hospitalized and AIDS patients, is candidiasis, which is caused by the yeast *Candida albicans*. Unless the predisposing condition can be corrected, many opportunistic mycoses can be fatal, even with antibiotic treatment. ▶ (See Fungi, Chap. 25, p. 691.)

Mycotoxicoses. Some fungi produce toxic substances that poison a person who ingests them. These poisonous substances are collectively called **mycotoxins** (*myco* = fungus; *toxin* = poison). The most common mycotoxicoses follow ingestion of poisonous mushrooms, such as those in the genus *Amanita*, the notorious "death angel." These mushrooms contain lethal substances that destroy liver cells and excite the nervous system. Other poisonous species include the phosphorescent mushrooms in Fig. 11-21.

Other poisonous substances may be produced by fungi growing on grain, nuts, and other agricultural products. For example, *Claviceps purpurea* growing on rye produces a variety of substances, some of which cause the ergot poisoning discussed earlier. Beer and peanut butter made from grains and nuts contaminated with common species of *Aspergillus* and *Penicillium* can contain aflatoxin levels high enough to poison the consumer. The toxin is not destroyed by cooking. Fortunately, in the United States federal law requires that peanuts be tested for evidence of aflatoxin contamination. Other fungal toxins have milder effects. "Drunken bread eaters' disease," for example, is a condition in persons who consume bread made from flour contaminated with a fungus that produces an intoxicating substance.

Mycotoxicosis occurs more frequently among

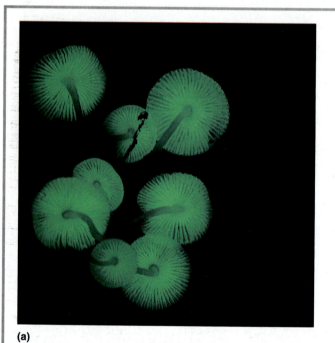

(a)

(b)

FIGURE 11-21 **Beautiful killers.** (*a*) These glowing mushrooms, photographed by their own light, contain more than phosphorescent pigments. Their powerful neurotoxin can kill a person within hours of ingestion. Not all poisonous mushrooms are so strikingly identifiable. Some are almost identical to edible species and can only be distinguished by microscopic examination. (*b*) Same mushroom photographed with a flashbulb.

domestic animals than among humans because animals are more likely to ingest the contaminated food. This often causes substantial losses of livestock. In 1960, 100,000 turkeys died after eating ground peanut meal in which *Aspergillus flavus* was growing. It was this outbreak of "turkey X disease" that led to the discovery of aflatoxins.

Most fungal toxins are only produced when the fungus grows in moist environments at relatively high temperatures. Unfortunately, these conditions are often found in silos and grain storage facilities. Most of these substances are toxic in extremely minute quantities, and lethal concentrations are produced long before the grain shows visible evidence of mold growth or spoilage.

● **CONTROL AND TREATMENT OF FUNGAL DISEASE.** Since fungi are eukaryotic, antibacterial agents are useless in eradicating mycoses. The first antibiotics that could inhibit eukaryotic fungal pathogens without having a similar effect on the eukaryotic human host cells became available in 1959. Before that time, disseminated systemic infections by *Coccidioides* and *Cryptococcus* were usually fatal. Antibiotics called *polyenes* are now used with reasonable success for treating deep mycoses, depending on the specific disease. Another group of antifungal agents are proving even more effective against many fungal diseases, reducing their duration and mortality. These agents, the *imidazoles*, are also targeted against the fungal cell membrane. Some of the imidazoles are used in topical preparations for localized cutaneous and mucosal infections. Others are designed for treating systemic mycoses. ▶ (Antifungal therapy is discussed in greater detail in Chapter 16.)

● **FUNGAL DISEASES OF PLANTS.** Plants provide a nutritional feast for many organisms, but the poor water availability on plant surfaces keeps bacteria from becoming successful plant parasites. The absence of competing bacteria, however, creates an easier opportunity for fungal invasion, and consequently virtually every plant is susceptible to fungal destruction. Rusts, smuts, rots, blights, mildews, galls, and wilts are among the array of agriculturally important plant diseases caused by fungi (Table 11-5).

Some plant pathogens are obligate parasites and cannot grow in the absence of a particular host. Many of these fungi can infect only a single plant species. Sometimes only some of the strains within a species are susceptible due to the development of genetic resistance to fungal pathogens in other strains. When their host dies, these obligate parasites must either find another susceptible host plant

TABLE 11-5 FUNGI PATHOGENIC FOR PLANTS		
Fungus	**Host Plant**	**Disease**
Phytophthera infestans	Potato	Late blight
Ceratocystis ulmi	Dutch elm tree	Dutch elm disease
Puccinia graminis	Cereals	Stem rust
Rhizopus spp.	Fruits	Soft rot
Claviceps purpurea	Cereals	Ergot
Erysiphe graminis	Cereals	Powdery mildew
Ustilago maydis	Corn	Smut
Endothia parasitica	Chestnut tree	Blight

or produce dormant resistant forms to survive. Facultative parasites, on the other hand, can also exist as saprophytes in the soil. Even when their host is absent, they can continue to grow.

The devastating loss of crop plants due to fungal disease may result from a number of different mechanisms. Invasion is the first step in all infections. A pathogenic fungus may invade a plant through openings such as stomata (the holes in leaves through which CO_2 enters the plant for photosynthesis), through wounds, or by direct penetration of plant tissue by fungal hyphae (Fig. 11-22). In this case, the

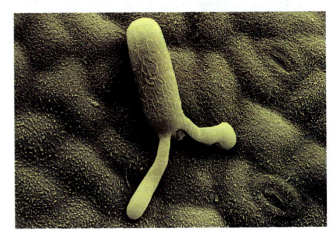

FIGURE 11-22 Plant invasion. This spore from a plant pathogen (*Erisyphe pisi*) is germinating on a leaf of its host. It will soon penetrate the plant's tissues with hyphae.

Key Terms

mycology (p. 256)
fungi (p. 256)
molds (p. 256)
hyphae (p. 256)
yeasts (p. 256)
spores (p. 256)
asexual spores (p. 256)
sexual spores (p. 256)
mycelium (p. 257)
septate hyphae (p. 257)
aseptate hyphae (p. 257)
vegetative hyphae (p. 257)
aerial hyphae (p. 257)
conidium (p. 257)
sporangiospores (p. 257)

sporangia (p. 259)
zoospores (p. 259)
chlamydospores (p. 259)
arthrospores (p. 260)
blastospores (p. 261)
fruiting bodies (p. 261)
basidiospores (p. 261)
basidium (p. 261)
ascospore (p. 261)
ascus (p. 261)
zygospores (p. 261)
oospores (p. 261)
budding (p. 262)
dimorphic fungi (p. 263)
oomycetes (p. 264)

zygomycetes (p. 265)
ascomycetes (p. 265)
aflatoxins (p. 265)
ergot (p. 265)
basidiomycetes (p. 266)
deuteromycetes (p. 266)
Sabouraud's agar (p. 267)
lichen (p. 269)
mycorrhizae (p. 269)
mycoses (p. 270)
dermatophytes (p. 272)
opportunistic fungi (p. 274)
mycotoxins (p. 274)

Key Facts and Concepts

- **Fungal morphology.** The fungi are nonphotosynthetic eukaryotic organisms with cell walls. They exist as unicellular yeasts or as molds, which are aggregates of filamentous hyphae. Hyphae may be septate or aseptate. A few fungi are dimorphic, growing as molds under certain conditions and as yeasts under other conditions.
- **Fungal reproduction.** Molds and yeasts are capable of both asexual and sexual reproduction. The reproductive structures in both cases are called spores. Asexual spores include conidia, sporangiospores, zoospores, chlamydospores, arthrospores, and blastospores. Sexual spores include zygospores, oospores, ascospores, and basidiospores.
- **Classifying the fungi.** The types of sexual spores produced provide the basis for taxonomic assignment of a fungus to one of five groups—the oomycetes, zygomycetes, ascomycetes, basidiomycetes, or deuteromycetes. Of these groups, only the deuteromycetes have no known means of sexual reproduction.
- **Cultivating the fungi.** Fungi grow best at pH between 5 and 6 and at higher sugar concentrations than bacteria. Many can be grown on artificial media in the laboratory.
- **Saprophytes and parasites.** Most fungi are saprophytes and are essential decomposers in biogeochemical cycles. A few species are human

pathogens capable of causing unusually persistent diseases. Many are important plant pathogens.
- **Saprophytic fungi.** Among the most common organisms on earth, they are our only sources of many valuable products—leavened bread and pastries, some cheeses, soy sauce, wine, beer, and pharmaceuticals such as penicillin. On the other hand, they are major sources of deterioration of commercially valuable products.
- **Mutualistic associations.** Fungi have formed mutually beneficial associations with photosynthesizing microbes to form lichens, with plants to form mycorrhizae, and with several wood-using insects.
- **Mycoses.** Human fungal pathogens cause diseases ranging from superficial infections to life-threatening systemic disease. The opportunistic mycoses invade people whose resistance is compromised. Control of mycoses depends largely on correcting predisposing factors, reducing exposure to infected people or to environments in which infectious spores commonly reside, and using effective antifungal antibiotics in patients suffering from fungal disease.
- **Fungi and plant disease.** Pathogenic fungi destroy virtually any kind of plant. They are responsible for rusts, smuts, blights, and other devastating infections of trees and crops.

Review and Problem-Solving Questions

1. Compare fungi and eubacteria by filling in the following table.

	Fungi	Eubacteria
Spores produced		
Steroids in cell membane		
Cell wall composition		
Heterotrophic/ autotrophic		
Sensitivity to penicillin		
Sensitivity to polyenes		

2. What is a fungal spore? List and describe five types of asexual fungal spores.
3. What is the advantage of a sexual life cycle for a fungus?
4. Why do mycotoxicoses occur more frequently among animals than among humans?
5. List the four medically important groups of fungi and their characteristics.
6. What medium is usually employed for the primary isolation of fungi? What characteristics make it suitable for that task?
7. Why are dermatophytes restricted to growth on the cutaneous layers of the body?
8. Identify some factors that contribute to the development of an opportunistic mycosis.
9. Why are mycoses more difficult to treat than bacterial diseases?
10. Why are most plant pathogens fungi rather than bacteria?
11. How do plant pathogens survive over the winter when their hosts are unavailable?
12. Certain fungi can degrade jet fuel when small quantities of water contaminate the fuel during storage. Commercially prepared fungicides are designed to inhibit the growth of these fungi and protect the integrity of the fuel.
    (a) Design an experiment to isolate a jet fuel–degrading fungus.
    (b) Design an experiment to test the effectiveness of a commercial fungicide for jet fuel.
13. Complete the following table.

	Sexual Spores	Asexual Spores
Oomycetes		
Zygomycetes		
Ascomycetes		
Basidiomycetes		
Deuteromycetes		

14. (a) How does a dikaryotic cell differ from a diploid cell?
    (b) What is the role of dikaryotic cells in the life cycle of basidiomycetes?
15. You are president of a new biotechnology company, and one of your staff scientists suggests that you develop a product that uses insect-attacking fungi as pesticides against household roaches. What do you think of this idea? What problems do you anticipate you would encounter? Describe a feasible procedure for tackling this assignment, creating a product that is ready for the market.

HEADLINES

AN EVOLUTIONARY LINK BETWEEN ANIMALS AND FUNGI

By comparing ribosomal RNAs, three scientists[*] have discovered an unexpected relationship between animals and fungi. Fungi and animals apparently share a unique evolutionary history distinct from that of plants. Our last common ancestor is thought to have been a flagellated protist, which may explain why some present-day water molds and flagellated protozoa are similar in structure.

- Fungi have been classified as plants, as protists, or given their own kingdom. What characteristics of fungi support these classifications? What characteristics are inconsistent with these classifications?

- What are the implications of the animal–fungus relationship to the development of selectively toxic antifungal drugs?

[*]P. O. Wainright, G. Hinkle, M. L. Sogin, and S. K. Stickel, Monophyletic origins of the metazoa: an evolutionary link with fungi. *Science,* 260:340–342 (1993).

CHAPTER 12

Protists—Protozoa, Algae, and Slime Molds

The intricacy of this foraminiferan shell is captured by the microscope and enhanced by polarized light. The protozoan that produced the shell has long since died. Enormous numbers of foraminiferan shells settled on ocean floors millions of years ago, producing thick sediments. Geological upheaval exposed these sediments, creating the famous White Cliffs of Dover in England. Protozoa are one of the three groups in the Kingdom Protista.

With alarming speed, the "phantom of the ocean" materializes, kills its victims by the thousands, then disappears back into its sanctuary. Although it might sound like some maritime myth, this story is disturbingly factual. The "phantom" is a real killer. Individuals consist of only one cell; however, swarms of these organisms lurk dormant on the seabed, showing no signs of life until "dinner" arrives. Sometimes within minutes after the arrival of live fish, the dormant killer transforms into motile cells that swim toward their victims and release a powerful nerve toxin that paralyzes and kills the fish. Sometimes a million fish die in a single episode. The dead fish then serve as dinner for the killer, which sucks nutrients from fragments of the carcasses (Fig. 12-1). Then, as quickly as they appeared, the motile cells are gone, once again transformed into their dormant states, which settle back to the seafloor where they await the arrival of another set of hapless victims. Within just a few hours after its lethal emergence from the ocean depths, the phantom has disappeared.

This deadly species is one of the many fascinating members of the protist kingdom, a group that contains some of the world's most beneficial microbes as well as some of the most deadly. As important and interesting as these organisms are, however, there are more than a few taxonomists that believe that the protist kingdom shouldn't exist at all.

"Taxonomic misfits that have no business sharing a kingdom" is the sentiment echoed by some biologists frustrated with trying to definitively describe the protists. Protists are a collection of eukaryotes that have been placed in the same kingdom more or less by default rather than because of any broad taxonomic similarities. The **protists** are all single-cell or colonial eukaryotes that do not fit into the plant, animal, or fungus kingdoms. Historically, they include three fundamentally different groups of organisms: protozoa, algae, and slime molds (Table 12-1). Although genetic studies show that these three categories are not valid taxonomic distinctions today, they are still useful groupings for studying the diverse organisms contained in this kingdom.

The protist kingdom embodies more than 100,000 microbial species, many of which have profound world significance. More than half the people on earth will suffer from a proto-

TABLE 12-1	THE PROTISTS
Group	**Major Characteristics**
Protozoa	Unicellular, heterotrophic, no cell wall; usually motile.
Algae	Unicellular or colonial, usually photosynthetic; possess cell walls.
Slime molds	Complex life cycle; feeding forms include unicellular wall-less organisms that aggregate to produce spore-forming reproductive structures.

zoan infection at some time in their lives, and many of these infections are deadly. For example, no single bacterial pathogen causes more human death and suffering than the protozoa that cause malaria. Algae, on the other hand, which rarely cause human disease, benefit all our lives through their photosynthetic, oxygen-generating activities. That is not to say they have no negative consequences. Some algae, such as the "phantom" microbe just described, kill people and animals with their toxins. Others form huge mats on the surfaces of lakes, part of a chain of events that can suffocate many of the organisms in an aquatic ecosystem. These are just a few of the dynamic effects of the protists, a group of organisms that have helped sculpt the character of the biosphere for almost 1.5 billion years.

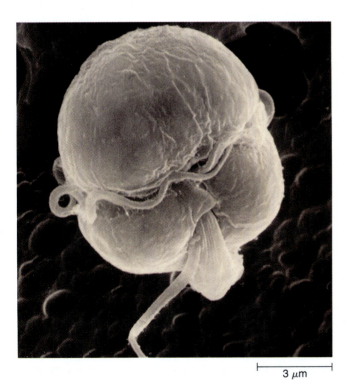

3 μm

FIGURE 12-1 Phantom of the ocean. This member of the protist kingdom, the alga *Gymnodinium*, kills passing fish with a powerful toxin, then attaches to the dead meat with its tonguelike appendage (shorter structure at bottom of the cell) and sucks nutrients from the carcasses, the cells make a phantom-like exit, disappearing into the ocean's depths.

Evolution of the Protists

Two billion years ago, the living world consisted solely of prokaryotic bacteria and the first nucleated cells. Many of these early eukaryotes were not well adapted to their changing environment and did not survive. The others gave rise to eukaryotic cells that were the ancestors of modern protists. For example, the few protists that are anaerobic, lacking mitochondria, are believed to have evolved from the earliest anaerobic nucleated cells. Some of these early anaerobic cells formed endosymbiotic relationships with various respiring bacteria, which very likely led to the evolution of mitochondria. ▶ (See BEYOND THE BASICS: Bacteria and the Early History of Life, Chap. 2, p. 43.) The vast majority of protists descended from these aerobic organisms.

Photosynthetic protists not only acquired their

mitochondria from oxygen-respiring bacteria but also received their chloroplasts in a similar fashion, by endosymbiosis with ancient oxygen-producing photosynthetic bacteria. The diversity among protists therefore reflects their distinct origins. Many present-day algae and protozoa are structurally similar to fossil protists trapped in Triassic amber 230 million years ago. Other early aerobic protists, however, abandoned their unicellular lifestyle. Their experimentation with evolution led to the development of multicellular algae and the "higher" eukaryotes—the fungi, plants, and animals.

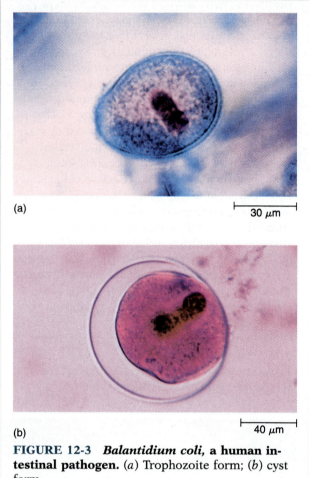

(a) ⊢————⊣ 30 μm

(b) ⊢————⊣ 40 μm

FIGURE 12-3 *Balantidium coli,* **a human intestinal pathogen.** (*a*) Trophozoite form; (*b*) cyst form.

> **CONCEPT CHECK**
>
> - How are anaerobic protozoa, aerobic protozoa, and algae related to each other and to bacteria?

Introduction to the Protozoa

■ Morphology

Protozoa (singular, *protozoan*) are nonphotosynthetic unicellular organisms with a eukaryotic cell structure. They are further characterized by the absence of cell walls, which distinguishes them from the fungi and nonphotosynthetic algae. Some protozoa possess a *pellicle,* a protein layer that increases the rigidity of their cell membrane. The pellicle provides some protection from osmotic stress and is responsible for the characteristic shape of those pro-

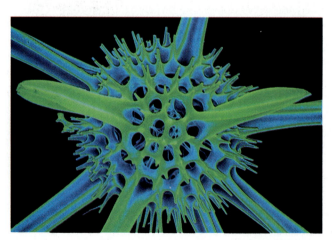

FIGURE 12-2 **Radiolaria.** These protozoa enclose themselves in a protective silica shell with holes through which the cell extends cytoplasmic projections for movement and food procurement. Only the shell is shown here.

tozoa that have stable morphologies. Protozoa range in length from about 2 μm (the smallest stage in the life cycle of the human *Leishmania* parasite) to 20,000 μm (2 cm, or about 4/5 inch). Thus, the smallest protozoa are similar in size to many bacteria; the largest are large enough to be seen without the aid of a microscope. Some protozoa, the radiolaria and foraminifera, possess intricate shells that are several centimeters in diameter (Fig. 12-2).

Most protozoa are **polymorphic** (*poly* = many; *morph* = form); they undergo morphological changes during different phases of their life cycles. Some of these protozoa alternate between two forms, the actively feeding trophozoite and the dormant cyst (Fig. 12-3). **Trophozoites** feed and reproduce as long as environmental conditions are favorable. The onset of adverse conditions that would kill the trophozoite may trigger its transformation into a cyst, thereby increasing the chances for species survival. The **cyst** is a dehydrated, thick-walled, protective form of the organism similar in function to the bacterial endospore. Cysts survive in both dry and aquatic environments, which allows them to be disseminated over large distances. Cysts form in response to environmental changes such as nutritional deficiency, desiccation, increased temperature, or reduced pH. For example, the trophozoite of the soil organism *Naegleria* transforms itself into a cyst when the soil gets too dry. When moisture returns to the soil, the trophozoite emerges from the cyst. Some cysts are also reproductive structures. Nuclear replication occurs as these cysts form, and many organisms eventually emerge when the cyst germinates.

Cyst formation is common among protozoa and is essential to the survival of some pathogenic species. For example, the cyst forms of intestinal parasites survive for long periods in water, soil, or food as well as in the extremely acid environment of the human stomach. Trophozoite forms of the same organism survive poorly outside the body and if ingested are quickly destroyed by stomach acid. Without cysts, the protozoan could not establish an infection in the gastrointestinal tract. The cyst is therefore the only infectious stage of intestinal protozoa; trophozoites play no role in transmission of intestinal illness. Once in the intestine, however, the fragile trophozoites emerge from the cysts, and re-

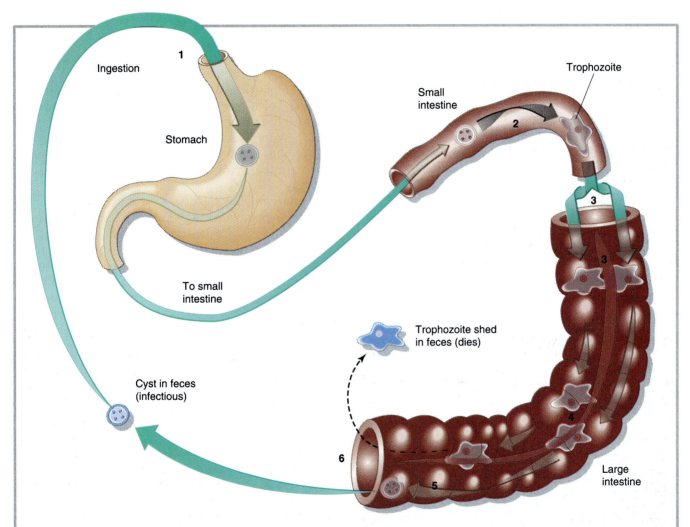

FIGURE 12-4 Life cycle of an intestinal protozoan. (1) Cyst carried in food, water, or on fingers enters the digestive tract. (2) Trophozoite emerges from the cyst in the small intestine. (3) Trophozoites colonize the large intestine, where they multiply by binary fission and (4) may invade the intestinal wall. (5) Encystment of the organisms is triggered by decreasing concentrations of water as the pathogens move into the more dehydrated regions of the large intestine. (6) Cysts and trophozoites are shed into the environment in feces. The formation of cysts is essential to the survival of the organisms, since trophozoites die when they pass into the environment in the feces.

production resumes in the new host (Fig. 12-4).

Not all protozoa form cysts. Protective cysts are unnecessary if the organisms are transmitted directly between living hosts without exposure to hostile environments. *Trichomonas,* for example, a genital pathogen that is transmitted from one person to another by direct sexual contact, forms no cyst. Similarly, no cysts are formed by protozoa that are transmitted by vectors, living organisms (usually a biting arthropod) that transfer the protozoa from one vertebrate host to another. Vectors protect the parasites from extreme changes in temperature, moisture, nutrient availability, and other factors in the external environment. Since arthropods are the only vehicles for transmitting these pathogens, the diseases they cause are geographically restricted to regions that favor the survival and proliferation of their respective vectors. Malaria, for example, is prevalent in humans only in areas populated by its vectors, the *Anopheles* mosquitoes, and human sleeping sickness only where there are tsetse flies. Control of these diseases depends on eliminating their vectors from these regions.

Some protozoa are **pleomorphic.** This means that they have several different trophozoite forms, often depending on their environment. (Caution: The term "pleomorphic" should not be confused with "polymorphic," which denotes the capability of producing both trophozoite and cyst forms.) Organisms with multiple hosts are often pleomorphic, with different trophozoite morphologies in the different host species. The morphology of the trophozoite may also vary in different tissues within the same host. For example, an organism may bear flagella for swimming through body fluids but lose the appendages when it invades heart tissue. Identification of characteristic forms within infected tissues is important in the microscopic diagnosis of protozoan illness.

■ Nutrition and Metabolism

Since protozoa are nonphotosynthetic chemoheterotrophs, they require an external source of food for energy and materials for biosynthesis. They obtain their nutrients in one of two ways. Some are *osmotrophs,* absorbing dissolved nutrients directly through their cell membrane. For example, parasitic protozoa that absorb their nutrients from the blood and tissues of their host are osmotrophs. Others, the *phagotrophs,* feed by engulfing soluble organic material or solid food particles, forming intracytoplasmic vesicles called *food vacuoles.* Amoebas, for example, can take up particulate matter anywhere along their surface by **phagocytosis** (Fig. 12-5). In this process, the cell membrane extends around the extracellular particle until the particle is completely surrounded and engulfed, forming a food vacuole within the cell.

Ciliated organisms such as the paramecia create currents to sweep food particles into a specialized mouthlike structure on the membrane, the *cytostome,* where they are trapped in food vacuoles. Enzymes stored in lysosomes are discharged into the vacuole, where the ingested particle is digested. The end products of digestion are absorbed into the cytoplasm, where they are either metabolized by the cell for energy or used to synthesize structural components. Any waste products remaining in the food vacuole are usually released from the cell by *exocytosis,* a process in which the vacuole membrane fuses with the plasma membrane, spilling the vacuole contents out of the cell.

Without a cell wall for protection from osmotic lysis, many protozoa are threatened by the problem of excess water accumulating within their cytoplasm. The osmotic environment of freshwater protozoa, in particular, "forces" water into the cell. Most of these protozoa solve the problem with osmoregulatory organelles called **contractile vacuoles** (Fig. 12-6). Water accumulates within these vesicles. When they are full, the vacuoles contract and expel their contents from the cell, "bailing out" the excess water.

Free-living protozoa are usually aerobic. Some parasitic protozoa, however, live in environments where oxygen may be limited; indeed, some intestinal protozoa require low oxygen concentrations or even anaerobic conditions. Resident aerobic bacteria help reduce the oxygen level in the intestine and may contribute to host susceptibility to protozoan infection. In rare instances, antibacterial chemotherapeutic agents may cure a person of an intestinal protozoan infection by temporarily suppressing the bacteria on which the protozoa depend.

■ Reproduction

Most protozoa reproduce by both asexual and sexual processes, although some organisms have no sexual cycle. Others can exhibit sexual cycles but do not require them for survival. A third group reproduces asexually in one host and sexually in another.

The principal mode of asexual reproduction is **fission,** cell division producing two or more progeny cells (Fig. 12-7). **Binary fission** is the most common. Flagellated protozoa always divide in a longitudinal direction (along their length), whereas protozoa with cilia always divide in a transverse direction. Some protozoa reproduce asexually by the process of **multiple fission.** In this case the nucleus divides many times before partitioning of

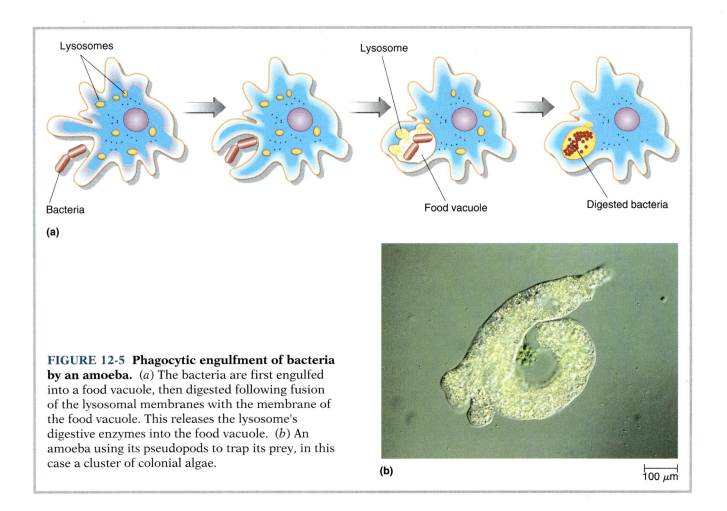

(a)

Lysosomes

Bacteria

Lysosome

Food vacuole

Digested bacteria

FIGURE 12-5 Phagocytic engulfment of bacteria by an amoeba. (*a*) The bacteria are first engulfed into a food vacuole, then digested following fusion of the lysosomal membranes with the membrane of the food vacuole. This releases the lysosome's digestive enzymes into the food vacuole. (*b*) An amoeba using its pseudopods to trap its prey, in this case a cluster of colonial algae.

(b)

100 µm

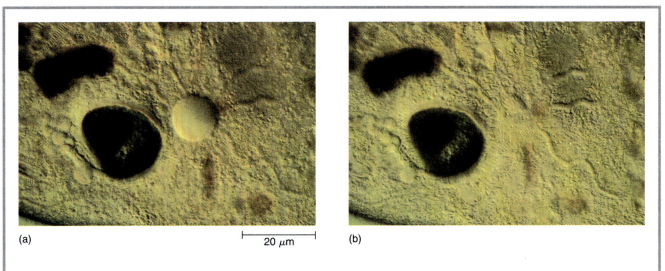

(a)

20 µm

(b)

FIGURE 12-6 Getting rid of excess water. The contractile vacuole expels water from the cell, protecting it from osmotic lysis. In this paramecium, radiating canals surrounding the vacuole collect water and fill the vacuole (*a*). When full, the vacuole and canals contract, squeezing the water from the cell and leaving the vacuole empty (*b*). The canals and vacuole immediately begin refilling with water that enters the cell from its hypotonic environment.

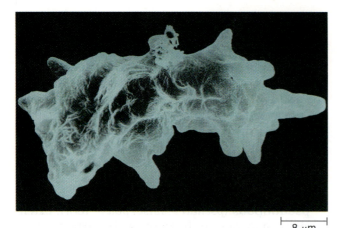

FIGURE 12-9 **Sarcodina.** A typical amoeba, this protozoan changes morphology as it moves or engulfs food.

8 μm

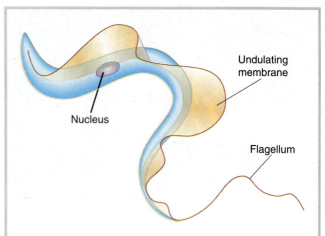

FIGURE 12-10 **An undulating membrane.** The flagellum of this blood and brain parasite (a typical trypanosome that causes sleeping sickness) is attached to a membrane that helps it move through body fluids.

forming pseudopods (Fig. 12-9). The pseudopods extend out from the cell, and the rest of the cytoplasm flows to the furthest point of extension. The formation of pseudopods allows the amoebas to move and to feed by phagocytosis.

Foraminifera and *radiolaria* are amoebas that have rigid internal or external skeletal structures made of calcium or silica (such as those shown on p. 280 and in Fig. 12-2). These organisms move and feed by extending their pseudopods through pores in their shells. The famous White Cliffs of Dover in England are composed of these protozoan skeletons that, over millions of years, settled to the ocean floor. Eventually, geological activity thrust the massive deposit above sea level.

Amoebic trophozoites multiply by binary fission. During cyst formation, nuclear multiplication may occur, resulting in several nuclei in a single cyst. Each nucleus becomes surrounded by its own membrane within the cyst wall. The number of nuclei found in a cyst is often characteristic of the species and is diagnostically important for distinguishing harmless intestinal protozoa from *Entamoeba histolytica*, the causative agent of amebic dysentery and the major pathogen among sarcodina.

■ Mastigophora

Mastigophora (*mastigo* = whip; *phora* = bearer) possess flagella in the adult stage of their life cycle. Some flagella propel the cell by pushing or pulling it; others may be used for steering. In some mastigophora, a flagellum is attached to a thin, loose membrane along most of the length of the cell, forming a structure called an *undulating membrane* (Fig. 12-10). This structure is characteristic of some dangerous blood and tissue parasites and is believed to facilitate movement of these flagellates through

viscous body fluids by amplifying the propulsive forces generated by the flagellum.

Some flagellated organisms produce cysts; others, such as those that are transmitted by direct contact between hosts or by vectors, have no cyst stage. Flagellates reproduce only by longitudinal fission, an asexual process.

Most species of mastigophora are free-living. Some, however, cause human diseases, the most serious of which are African sleeping sickness (caused by *Trypanosoma gambiense* and *Trypanosoma rhodensiense*) and Chagas's disease (*Trypanosoma cruzi*). Two other parasitic mastigophora cause less severe but very prevalent diseases. They are *Giardia lamblia*, an intestinal pathogen, and *Trichomonas vaginalis*, an agent of genital infections.

■ Ciliophora

Ciliophora are protozoa characterized by the presence of cilia on the cell surface. Cilia are structurally identical to eukaryotic flagella but are shorter and more numerous. Often, as in paramecia (Fig. 12-11), the entire ciliate body is covered with thousands of these hairlike projections that beat like the oars of a boat in a coordinated fashion to propel the organism. In some organisms cilia are restricted to certain regions or are fused together to form tufts called *cirri*. These structures can function as legs that the protozoa use to creep along surfaces. Cilia may also aid in the feeding process, sweeping food particles toward the cell and into the oral opening, the cytostome. Some ciliophora attach to solid sur-

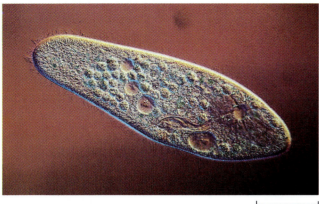

FIGURE 12-11 A typical paramecium. This protozoan is covered with cilia used for motility and for sweeping food into oral stoma.

60 μm

faces such as rocks and use their cilia exclusively for obtaining food.

The ciliates are unique among the protozoa in that each cell possesses two types of nuclei. The *macronucleus* contains multiple copies of each gene. It is responsible for directing cell growth and asexual reproduction by transverse binary fission. The smaller *micronucleus* contains a set of the same genetic information, but that information is not transcribed or used in controlling the routine activities of the cell. The information in the micronucleus is used for genetic recombination during conjugation.

Ciliates play host to a variety of smaller microorganisms. Certain algae, for example, live in cytoplasmic vacuoles of some ciliates, converting the heterotrophic host into a functional autotroph. Other ciliates house bacteria that confer the ability to produce toxins that kill sensitive protozoa, thereby eliminating potential competitors or predators. The killer protozoan is resistant to the toxin.

Balantidium coli, which causes diarrhea, is the only species of ciliophora known to cause human illness.

■ Sporozoa

All **sporozoa** have three things in common: (1) they lack motility in their replicative forms; (2) they are intracellular parasites; and (3) they alternate asexual and sexual reproductive cycles. Some sporozoa carry out their sexual and asexual cycles in the same host; others require multiple hosts. For example, *Toxoplasma gondii* multiplies sexually only in members of the cat family but develops asexually in humans and other mammals, where it often causes serious disease.

Coccidia is another genus of sporozoa whose members are animal parasites, particularly of chickens. Entire flocks consisting of tens of thousands of birds have been destroyed by these intestinal pathogens. Soil is contaminated by cysts of the organism released in the birds' feces. Young chickens picking food from the ground swallow the cysts and subsequently develop fatal coccidiosis (not to be confused with the human fungal disease, coccidioidomycosis). To prevent large outbreaks, commercial chicken feed often contains anticoccidial chemicals.

The four species of sporozoa of the genus *Plasmodium* are the agents that cause malaria. With an estimated 100 million cases annually, malaria is the most common serious infectious disease of humans.

> **CONCEPT CHECK**
>
> - Explain the origin of morphological diversity among organisms classified as protozoa.
> - What criteria are used to classify pleomorphic protozoa? polymorphic protozoa?

Significance of the Protozoa

Most of the 40,000 known species of protozoa are found in oceans, fresh water, or damp soil. Free-living protozoa are critical links in the food chain, feeding on other microorganisms and subsequently serving as a nutrient source for higher organisms. Other protozoa reside in the tissues and fluids of animals and humans. Fortunately, these internal residents are often harmless or even beneficial to the host. Cattle, for example, consume many soil protozoa while eating grass as their major food source. These microbes become established in the rumen, the first chamber of the bovine stomach. There they partially digest cellulose, the main constituent of grass and hay (the process is assisted by cellulolytic bacteria that share the rumen). As the protozoa multiply, they produce nutrients that the animal could not otherwise extract from its diet. These nutrients are transported along with many of the protozoa through the other stomach chambers and the intestinal tract. Along the route, most of the organisms are digested by the cow and thus serve as a source of protein. The cells that survive are shed in the feces and reseed the soil.

Cellulolytic protozoa are also responsible for the ability of certain termites to utilize wood as food. Soon after they hatch, young termites are infected

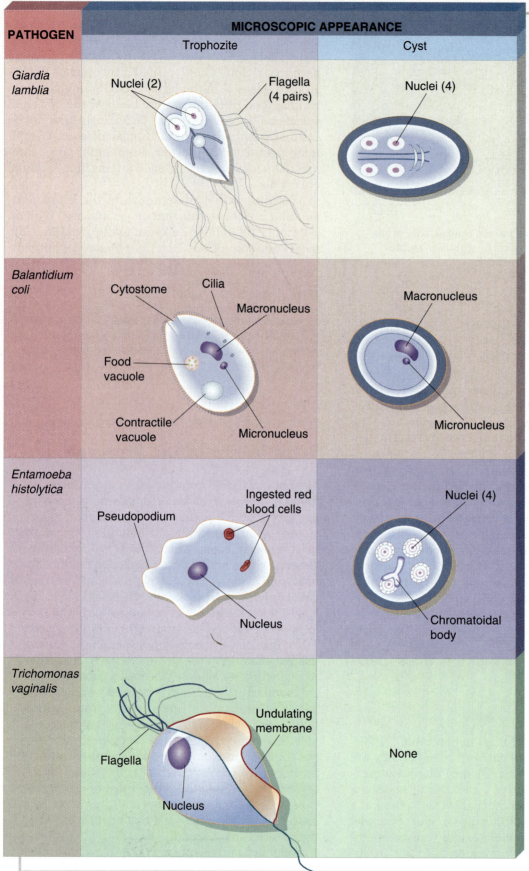

PATHOGEN	MICROSCOPIC APPEARANCE	
	Trophozite	Cyst
Giardia lamblia	Nuclei (2) — Flagella (4 pairs)	Nuclei (4)
Balantidium coli	Cytostome — Cilia — Macronucleus — Food vacuole — Contractile vacuole — Micronucleus	Macronucleus — Micronucleus
Entamoeba histolytica	Pseudopodium — Ingested red blood cells — Nucleus	Nuclei (4) — Chromatoidal body
Trichomonas vaginalis	Undulating membrane — Flagella — Nucleus	None

FIGURE 12-12

Disease-producing protozoa. *Giardia lamblia, Balantidium coli,* and *Entamoeba histolytica* are responsible for many human intestinal infections. *Trichomonas vaginalis* is a common vaginal infection.

with protozoa by eating a mixture of cellulose and protozoa regurgitated directly from the alimentary tracts of their parents. These protozoa increase to populations exceeding 10 million organisms per milliliter of intestinal content. Without these intestinal protozoa, your house would just sit undigested in the termite's stomach, and the insect would starve.

Both invertebrates and vertebrates may host parasitic protozoa. The relationship between parasite and host is usually fairly specific. Although the majority of protozoa cannot survive in the human body, several cause human diseases, of which three are included among the World Health Organization's list of top 20 causes of death worldwide. These are malaria, amebic infection, and trypanosomiasis.

Diseases Caused by Protozoa

Generally, pathogenic protozoa initiate infection in one of four primary sites in humans: the intestinal tract, the genital tract, the tissues and bloodstream, or the central nervous system. Of these sites, the intestinal tract is the only one in the human body normally populated by harmless protozoa. Isolation of protozoa from any other anatomical location is indicative of disease.

■ Intestinal Infections

Three protozoa, the flagellate *Giardia lamblia*, the ciliate *Balantidium coli*, and the amoeba *Entamoeba histolytica*, are the most common protozoa to cause human intestinal infections (Fig. 12-12). All three organisms have similar life cycles (typified by that shown in Fig. 12-4). The diseases, which are acquired by consuming contaminated food or water, are most prevalent in areas where sanitation and health education are poor. Since 1972 *Giardia lamblia* has caused more outbreaks of waterborne disease in the United States than any other pathogen. These outbreaks have usually been associated with deficiencies in water treatment systems (see THE MICROBES: Enjoy America—But Don't Drink the Water).

Of the three human intestinal pathogens, *Entamoeba histolytica* causes the most serious disease, amebic dysentery (bloody diarrhea). The pathogen releases enzymes that break down intestinal tissues (*histolytica* = tissue-lytic). The pathogen can spread and cause fatal abscesses of the liver, lungs, heart, and brain. About 10 percent of the world's population may be carriers of this protozoan, each infected person capable of shedding 3 million cysts per day. In the United States, a million people may be infected with *Entamoeba histolytica*. ▶ (See *Amebic Dysentery*, Chap. 22, p. 590.)

THE MICROBES

Enjoy America—But Don't Drink the Water

Protozoan infections are often considered to be characteristic of developing nations where sanitary conditions are often poor and insect populations are uncontrolled. Indeed, fatalities from protozoan infections are highest in those countries. Within the United States, however, at least one protozoan has alarmed public health authorities with its frequency of attack. The pathogen is *Giardia lamblia*, a common cause of waterborne outbreaks of diarrhea. Victims are often residents of rural environments or visitors whose vacations are abruptly terminated by the disease. *Giardia* infections also spread quickly through child care facilities, especially those in which diaper changes are not followed by thorough hand washing with soap and hot water.

Outbreaks of giardiasis can usually be traced to contaminated drinking water that has been inadequately treated. Routine chlorination procedures, for example, are ineffective against *Giardia* cysts. However, *Giardia* cysts are usually removed from municipal water supplies by filtration. Defective filtration systems that fail to remove the cysts mechanically are responsible for many incidents of giardiasis.

Giardia cysts may also be found in untreated waters, such as in streams, rivers, or lakes that are contaminated with animal or human feces. Even those cool, clear waters that seem so pure and refreshing to hikers may be harboring the infectious form of the protozoa and cause "hiker's diarrhea." Vacationers, campers, and hikers who drink from untreated sources can prevent their vacations from being ruined by boiling their drinking water to kill *Giardia* cysts. Chlorine tablets, however, provide no protection.

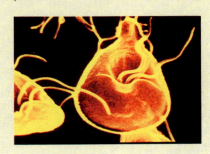

■ Genital Infections

The flagellated protozoan *Trichomonas vaginalis* is one of the world's most common causes of genital infections (Fig. 12-12). In some areas of the United States, 50 percent of the women examined show microscopic evidence of *Trichomonas* infection. The disease, *trichomoniasis* (commonly called a "tric infection"), is localized, usually causing vaginitis—a chronic, often irritating inflammation of the vagina. Males are also infected with the organism but usually show no symptoms. Because he has no indication of infection, a male may unknowingly reinfect his sex partner after she has been successfully treated with the antiprotozoan drug Flagyl ▶ (see Trichomoniasis, Chap. 23, p. 623).

■ Blood Infections

Human blood parasites are nearly always acquired by the bite of an infected arthropod vector and are generally limited in their geographic distribution to areas where the vectors are prevalent. Occasionally, however, they are transmitted by blood transfusions. The blood is the primary site of infection, but subsequent invasion of the brain, viscera, and other secondary sites may occur. Among the most debilitating protozoan blood infections are trypanosomiasis and malaria.

African trypanosomiasis is caused by *Trypanosoma gambiense* or *Trypanosoma rhodesiense* (Fig. 12-13). These trypanosomes invade the central nervous system, eventually causing mental deterioration, coma, and death. Both organisms are transmitted by the bite of the tsetse fly, a vector found only in Africa and southern Arabia. A single infected insect may shed the parasite in its saliva for 3 months, spreading the disease to large populations. Each year in Africa there are 250,000 cases of human trypanosomiasis.

American trypanosomiasis, also known as Chagas's disease, is caused by *Trypanosoma cruzi*, commonly found among domestic and wild animals in Central and South America. The vectors are triatomine bugs, blood-sucking insects that have an affinity to areas of the body, particularly the head, where blood vessels are close to the body surface. The insects' attraction to the lips accounts for their nickname, the "kissing bug." Another common site of attack is the eyelid (Fig. 12-14). The kissing bug does not directly inoculate its victims but deposits feces contaminated with trypanosomes on the skin

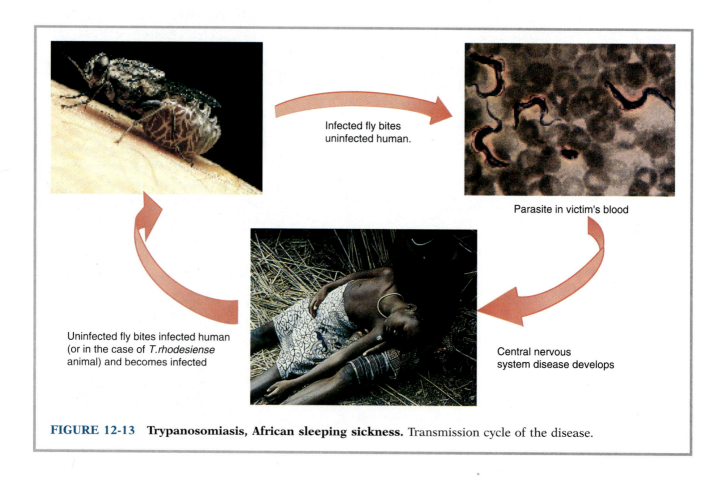

Infected fly bites uninfected human.

Parasite in victim's blood

Uninfected fly bites infected human (or in the case of *T.rhodesiense* animal) and becomes infected

Central nervous system disease develops

FIGURE 12-13 **Trypanosomiasis, African sleeping sickness.** Transmission cycle of the disease.

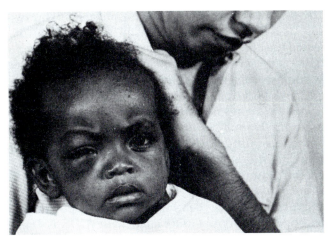

FIGURE 12-14   Chagas's disease. A doctor examines a young child suffering from an acute form of Chagas's disease. Swollen eyelids are a typical manifestation of acute infection.

surface while biting. Scratching the wound facilitates the entry of the protozoa through the bite and into the bloodstream. From the bloodstream, the organisms localize in the heart and central nervous system. In Central and South America, Chagas's dis-

ease is the major cause of heart disease for those under 40 years old. The damage they cause is often fatal.

More than a million people die of malaria every year. Virtually all malaria deaths occur in areas inhabited by the female *Anopheles* mosquito. This vector has been successfully controlled in most of the United States and other developed countries. However, a large part of the populated world is still at risk of contracting malaria (Fig. 12-15). Before mosquito eradication programs, however, malaria was common in the southeastern United States, where 600,000 cases were reported in 1914. Since 1950, only 13 cases within the United States have been caused by insect inoculation. (Cases of malaria that are acquired in the United States are from transfusion of blood harvested from a *Plasmodium*-infected donor.)

Four members of the genus *Plasmodium* cause malaria in humans: *Plasmodium malariae, P. vivax, P. ovale,* and *P. falciparum*. The disease occurs when any of the four species is present in the saliva of an anopheles mosquito and is injected into the human bloodstream by the insect's bite. The parasites infect the liver and ultimately release progeny back

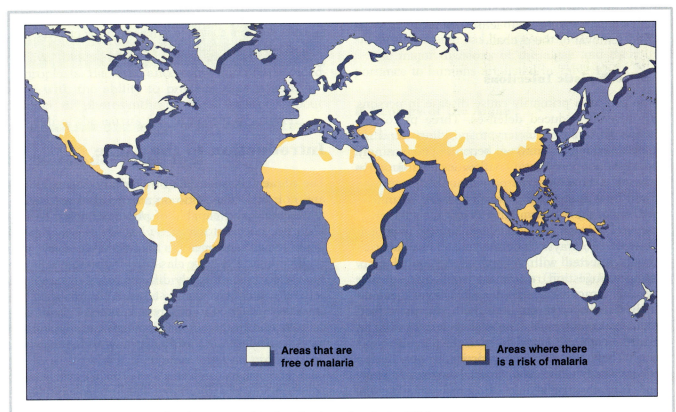

Areas that are free of malaria

Areas where there is a risk of malaria

FIGURE 12-15   Areas of risk. Geographic distribution of malaria, 1994.

TABLE 13-2   COMPARISON OF ENVELOPED AND NONENVELOPED VIRUSES

	Enveloped	Nonenveloped
Virion	Nucleic acid, capsid, and envelope	Nucleic acid and capsid
Location of attachment sites	On membrane envelope	On protein capsid
Infectivity of nucleocapsid alone	Noninfectious (requires presence of envelope)	Infectious
Ether-sensitive?	Yes	No

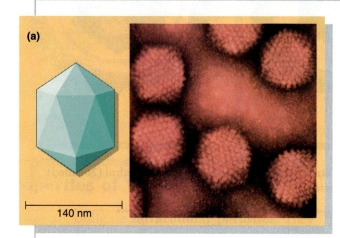

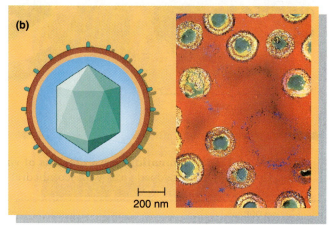

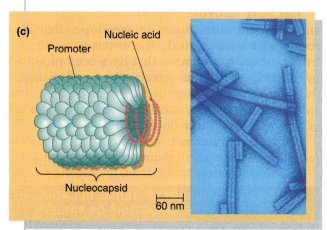

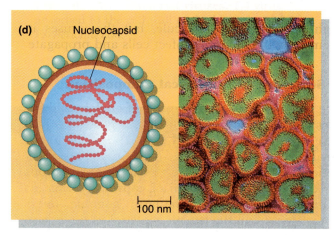

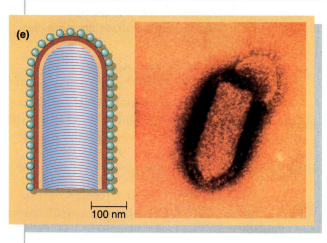

FIGURE 13-4 Morphology of viruses. (*a*) Cubic symmetry—naked (adenovirus); (*b*) cubic symmetry—enveloped (herpesvirus); (*c*) helical symmetry (tobacco mosaic virus); (*d*) enveloped helix—pleomorphic (influenza virus); (*e*) enveloped helix—bullet-shaped (rabies virus).

For nonenveloped viruses, host recognition sites are on the nucleocapsid. Because nothing but protein and nucleic acid is needed to form an infectious unit, the nucleocapsid is the virion. In enveloped viruses, the recognition sites are found on the envelope; thus their virion consists of the nucleocapsid and the additional membrane envelope. If this envelope is dissolved with a lipid solvent such as ether, the remaining nucleocapsid is incapable of infecting the host cell. Enveloped viruses, therefore, are said to be ether-sensitive. Nonenveloped viruses, on the other hand, are not inactivated by ether since lipid is not part of the virion and the protein capsid is not altered by such solvents. Table 13-2 presents a comparison of the enveloped and nonenveloped viruses.

In addition to structural proteins, the nucleocapsids of several types of viruses contain proteins with enzyme activities essential for viral infectivity. Some enzymes alter the host cell surface to facilitate virus entry. Most virion enzymes, however, are required for viral nucleic acid replication. For example, vaccinia virus, a DNA virus that reproduces in the cytoplasm of its host cell, is physically prevented from using host DNA-replicating enzymes and transcriptases, which are confined to the nucleus. Thus, vaccinia viruses must supply their own enzymes for these tasks so they can direct the production of progeny viruses.

The shape of nonenveloped viruses is determined by the manner in which the protomers are arranged around the nucleic acid core. Viruses with *cubic symmetry* are icosahedra, sphere-like structures with 20 triangular sides (Fig. 13-4*a* and *b*). Other viruses have *helical symmetry*. They resemble long rods with the protomers arranged around a spiraled coil of nucleic acid (Fig. 13-4*c*). The length of the helix is determined by the size of the nucleic acid. Envelopes alter the appearance of the virion, often resulting in pleomorphism (no stable shape) because the outer membrane is somewhat fluid. The shape of pleomorphic viruses may be determined by the physical forces exerted on the envelope, although the symmetry of the nucleocapsid inside remains unaffected by these influences (Fig. 13-4*d*). A few enveloped viruses, such as the peculiar bullet-shaped rabies virus (Fig. 13-4*e*), have more stable shapes.

Some viruses possess *complex* viral morphology, perhaps best represented by some bacterial viruses (bacterial viruses are called **bacteriophages,** or simply *phages*) (Fig.13-5). The nucleic acid is stored in the virus *head,* analogous to the capsid of viruses with cubic symmetry. The head is attached to the *tail,* a hollow tube covered with a sheath of contractile protein. The tail terminates in a *base plate.*

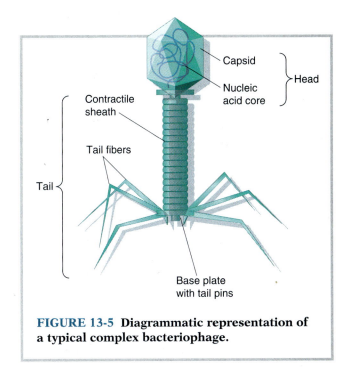

FIGURE 13-5 Diagrammatic representation of a typical complex bacteriophage.

Fibers that emanate from the base plate are used by the virus to recognize and attach to its specific host bacterium. If any of these structures is removed or damaged, the virus can no longer attach to and infect its host bacterium.

■ Genetic Composition

Viruses are unique in that they possess only a single type of nucleic acid, either DNA or RNA but never both. In some viruses, RNA is the genetic material instead of DNA. All microbes other than viruses have DNA for storing genetic information and RNA for translating the stored information. The nucleic acid of some viruses is single-stranded, whereas that of others is double-stranded. Thus, four possible configurations exist: (1) double-stranded DNA, (2) single-stranded DNA, (3) double-stranded RNA, and (4) single-stranded RNA. In addition, the nucleic acid is in the form of a linear chromosome in some viruses and a circle in others.

The amount of genetic information contained in viral nucleic acids ranges from a mere four genes for the smallest viruses to several hundred genes for the larger ones. Generally, these genes are housed on a single chromosome; however, a few viruses have genomes consisting of separate segments of nucleic acid. For example, influenza virus has a segmented genome composed of eight separate pieces of RNA.

■ Eclipse

The **eclipse phase** is an event unique to the life cycle of viruses. After entering a host cell, the virus particle disintegrates into its structural constituents. This is the beginning of viral eclipse. Eclipse continues until nucleocapsids reappear in the infected cell, progeny that are identical to the parent virus (the nucleocapsids of most enveloped viruses do not acquire their envelopes until they leave the cell, at which time eclipse ends). As remarkable as such an event seems, the processes by which viruses fall apart and reassemble are fairly well understood. They are discussed with viral replication later in this chapter.

The properties that distinguish viruses from all other organisms are summarized in Table 13-3.

■ Host Specificity

Every organism on earth is believed to be susceptible to viral infection. Animals, plants, algae, protozoa, fungi, and all prokaryotes are attacked and killed by viruses. Each virus can infect only specific cells, and many viruses are restricted to a single host species or to a single type of tissue within that species. Some viruses are so specific they infect only one or two strains within a species. For example, a bacteriophage that infects one strain of the bacterium *Staphylococcus aureus* may be incapable of attacking another strain that is identical except for a single molecule on the cell's surface. A few viruses, on the other hand, infect a broad range of natural hosts. Rabies virus, for example, can be transmitted among many different animal hosts including humans. Host specificity determines the types of organisms in which a pathogenic virus can cause infection and disease.

■ Classification of the Viruses

Viruses are divided into major categories according to whether they infect animal, plant, or prokaryotic cells. Further classification is based primarily on physical and chemical properties of the virus and on the range of hosts that can be infected. These properties are usually considered in the following order of importance:

1. Type of nucleic acid (DNA or RNA).
2. Single- or double-strandedness of the nucleic acid.
3. Capsid morphology (size, symmetry, and number of capsomeres).
4. Presence or absence of envelope (thus, sensitivity to inactivation by physical or chemical agents, especially ether).
5. Host range.

Although viruses have been assigned names according to the classical binomial system of nomenclature, they are still referred to by common names. The taxonomy of some viruses that infect animal cells is provided in Table 13-4.

CONCEPT CHECK

- How do viruses differ from cells? from plasmids? How are they similar?
- What problems must viruses confront because they are obligate intracellular parasites?
- Describe the variations in structure and chemistry that contribute to the diversity among viruses.
- Are viruses living entities? Defend your answer.

Viral Replication

Unlike cellular organisms, viruses cannot reproduce by fission. Because they have no independent metabolism, they must use a host cell to perform all the functions necessary for producing new infectious virions. Sometimes the virus will replicate in the host cell, followed by the release of progeny virions. This is called a *productive infection*. These viruses efficiently redirect host cell metabolic machinery to favor the production of new virus particles rather than new host cell material. New progeny viruses are then released at the expense of the cell itself. Often the progeny escape from the host cell by lysing it. Other viruses are less damaging, infecting and producing progeny without killing the host. Still other types of viruses produce no virions when they infect. Instead, they integrate their genome into the host chromosome and replicate

TABLE 13-4 CLASSIFICATION OF ANIMAL VIRUSES

Type of Nucleic Acid	Strandedness of Nucleic Acid	Morphology of Nucleocapsid	Envelope	Group*	Representative Human Disease
DNA	Single	Icosahedral	No	Parvovirus	Fifth disease
	Double	Icosahedral	No	Papovavirus	Warts
			No	Adenovirus	Respiratory infections
			Yes	Herpesvirus	Chickenpox; cold sores; genital herpes; infectious mononucleosis
		Complex	No	Poxvirus	Smallpox
RNA	Double	Icosahedral	No	Reovirus	Diarrhea
	Single	Icosahedral	No	Picornavirus	Polio: common cold
			Yes	Togavirus	Yellow fever; encephalitis
		Helical	Yes	Orthomyxovirus	Influenza
				Paramyxovirus	Measles; mumps
				Rhabdovirus	Rabies
				Coronavirus	Common cold
				Retrovirus	AIDS; possible role in tumor induction
		Unknown	Yes	Arenavirus	Lassa fever

*Additional properties, such as shape and size of the virion, are used to differentiate between groups.

along with the cell's chromosomes. Even these viruses, however, are capable of productive infection.

Productive infections of all viruses consist of five steps: (1) attachment, (2) penetration and uncoating, (3) synthesis of viral components, (4) assembly of viral components, and (5) release of progeny virions. The way these tasks are accomplished depends on the type of virus and on the type of cell infected (animal, plant, or bacterial).

■ Animal Viruses

The general scheme of viral replication presented in Figure 13-6 is typical of many nonenveloped animal viruses that replicate in the cytoplasm (as opposed to the nucleus).

● ATTACHMENT. The first event in viral infection is attachment of the virus to the surface of a susceptible cell. Viruses are not motile, and attachment follows random collision between the virus and host cell. This step requires the specific interaction of two complementary molecules, the **attachment site** on the surface of the virus and the **receptor site** on the cell's surface. Attachment sites are located in the capsid of nonenveloped viruses and on the specific proteins in the membrane of enveloped viruses. Receptor sites on animal cells are membrane proteins, polysaccharides, or lipids. The number of receptor sites may vary from 500 to 500,000 per cell. Since attachment and receptor sites interact in much the same way that two pieces of a puzzle fit together, this step is highly specific. A proper "fit" must exist before attachment can occur. Most viruses, therefore, show an extremely high degree of specificity for their host cells. In many cases, viral infectivity can be neutralized by molecules such as antibodies that cover the attachment sites before the virus can adhere to the cell's receptor sites. Formation of such antibodies is one way we develop immunity to viral diseases.

It may seem antievolutionary for cells to possess receptor sites that make them sensitive to viral infection. Most of these surface molecules, however, are not suicide sites; that is, their cellular function is not to welcome viruses. These receptor sites normally perform important functions for the cell—transporting substances across the cell membrane, for example. Viruses, however, have acquired the ability to use these complex surface molecules for attachment, as ports for initiating infection. The cell is faced with a dilemma. Eliminating complex surface molecules would prevent it from performing critical functions and would spell cell death much

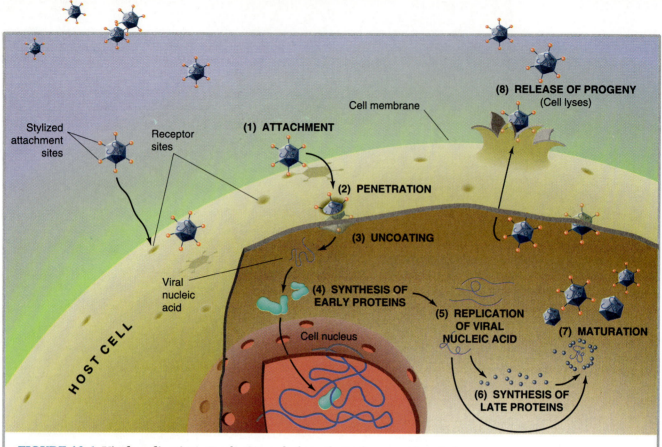

FIGURE 13-6 Viral replication. Replication of a hypothetical nonenveloped virus within the cytoplasm of a susceptible animal cell. See text for discussion of each step.

like eliminating the doors and windows of a house would render the structure useless. To be alive, therefore, is to be vulnerable to viral infection.

● **PENETRATION AND UNCOATING.** The second step in viral infection is penetration of the host cell. Some enveloped viruses enter the host cell by fusion of the viral envelope and the host cell membrane (Fig. 13-7a). These viruses contain an envelope "fusion" protein that rearranges the lipid membrane components of host and virus. Eclipse begins as the naked nucleocapsid is released into the cytoplasm. Most nonenveloped and enveloped viruses are engulfed by **receptor-mediated endocytosis,** a type of phagocytosis. The viral particles enter the cytoplasm in a membrane vesicle. The nucleocapsids are released from this intracellular vesicle by one of the mechanisms depicted in Figure 13-7b and c.

Before most viruses can reproduce, viral nucleic acid must be released from the capsid. Some viruses are uncoated as they escape from the vesicle. (In most cases, the **uncoating** process is poorly understood.) For nonenveloped viruses, eclipse begins with the uncoating step. The virus is now naked nu-

cleic acid (DNA or RNA) floating freely in the host cell's cytoplasm. No infectious virus can be found in the cell during eclipse.

● **SYNTHESIS OF VIRAL COMPONENTS.** Once free of the protein coat, the viral nucleic acid competes with the cell's chromosomes for control of the biological machinery. For production of progeny, the virus must (1) replicate its genome and (2) produce the structural proteins for its capsid and envelope (if it has one).

The first virus-directed act is the synthesis of viral mRNA. Some DNA-containing viruses use host enzymes to produce viral mRNA; others carry their own enzymes incorporated in the virion. The viral mRNA uses cellular ribosomes to direct the cell in the synthesis of **early proteins,** so named because they are produced soon after infection. Most early proteins are enzymes. Certain types of viruses produce enzymes that interfere with the expression of host cell genetic information, leaving the viral nucleic acid as the primary director of metabolism. Almost all viruses, however, produce early enzymes that replicate the viral genome, although the smaller

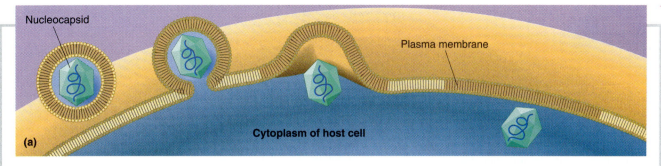

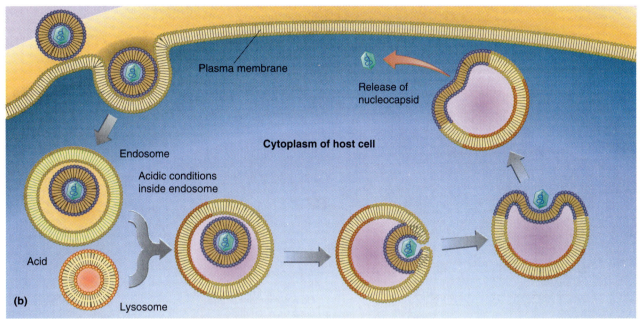

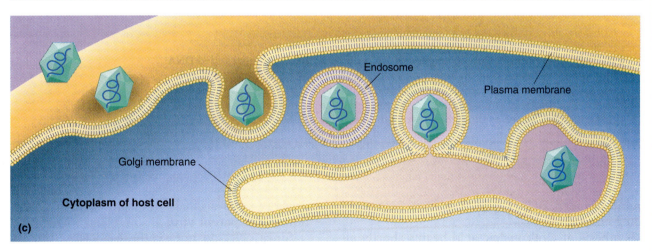

FIGURE 13-7 Gaining entrance to the host cell. Some mechanisms by which viral nucleocapsids enter host cells. (*a*) Fusion of viral envelope and cell membrane. The nucleocapsid is released directly into the cytoplasm. (*b*) Penetration of enveloped virus by endocytosis. The virus's envelope adsorbs to the cell membrane, and the entire virion is engulfed into a vacuole called an endosome. Acidic conditions in the vacuole cause the viral envelope to fuse with the vesicle membrane, releasing the nucleocapsid into the cytoplasm. (*c*) Penetration of nonenveloped virus by endocytosis. The endosome membrane fuses with an internal membrane system (Golgi complex or endoplasmic reticulum), releasing the free nucleocapsid.

complementary positive strand is synthesized using an enzyme contained in the viral capsid. The newly formed positive strands direct the synthesis of early proteins.

The virus that causes AIDS represents still another class of single-stranded RNA viruses, the **retroviruses** (*retro* = backwards). These viruses manufacture a *double-stranded* DNA copy of their RNA genome. This DNA copy integrates into the host cell chromosome and from there produces more viral genomes (RNA). This inverted flow of genetic information from RNA to DNA requires a unique enzyme (contained within the capsid) called *reverse transcriptase.* ▶ (See THE EXPLORERS: Reversing the Genetic Flow. Chap. 8, p. 197.)

After the viral chromosome has been repeatedly duplicated, cellular metabolism is redirected toward the synthesis of **late proteins.** These are the structural components that assemble into the capsid. Some late proteins are incorporated into the envelope; still others are triggers for host cell lysis.

● **ASSEMBLY.** The assembly of viral nucleic acid and structural proteins into nucleocapsids occurs in one of two ways. Either (1) the nucleic acid is packaged in a preassembled capsid, as shown in step 7 of Figure 13-6, or (2) the nucleic acid associates with the capsid proteins (protomers) during the capsid assembly process, as shown below.

For nonenveloped viruses, eclipse ends with the formation of the first virion, that is, the first intracellular nucleocapsid. Since nucleocapsids of enveloped viruses are noninfectious, enveloped viruses remain in eclipse until they acquire their membrane coverings.

● **RELEASE.** Some host cells burst during viral release. Virus-infected cells may literally explode, discharging infectious viruses capable of initiating another round of infection. Large numbers of noninfectious particles—empty capsids or capsids filled with incomplete genomes—are also released. The number of virions produced depends on the virus and the host cell.

MYTH: Virus-infected cells burst because they become engorged with so many viruses the overdistended plasma membrane pops.

FACT: Developing viruses that lyse their host cells damage the cell surface. Some direct the cell to produce a protein that disrupts the integrity of the membrane. Others instruct the cell to produce an enzyme that breaks down the protective cell wall so the cell will lyse by osmotic influx of water.

Most enveloped viruses leave the host cell intact, escaping by **budding** from the plasma membrane. This process, called *exocytosis,* resembles a reversed version of viral penetration by membrane fusion

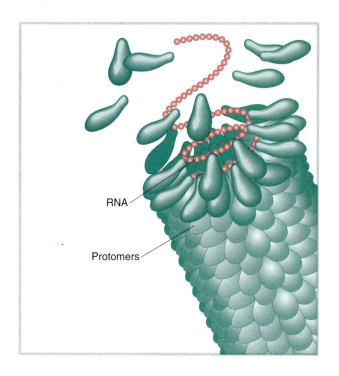

RNA

Protomers

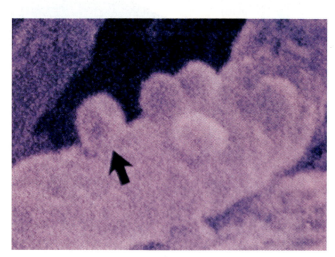

FIGURE 13-10 **Budding.** Enveloped viruses escape the host cell by budding, acquiring their membrane covering in the process. This shows numerous viruses budding from the surface of an infected cell.

TABLE 13-5 SOME IMPORTANT PLANT VIRUSES

Virus	Nucleic Acid*	Morphology	Enveloped?	Plant Host	Insect Host	Effect on Plant
Cauliflower mosaic	ds DNA	Cubic	No	Cauliflower	Aphid	Spots (altered pigments) on leaves
Tobacco mosaic	ss RNA	Helical	No	Tobacco	None†	Spots on leaves
Potato yellow dwarf	ss RNA	Helical	Yes	Potato	Leafhopper, aphid	Reduced rate of growth
Rice dwarf	ds RNA	Cubic	No	Rice	Leafhopper	Reduced rate of growth
Maize streak	ss DNA	Cubic	No	Corn	White fly, leafhopper	Elongated necrotic (dying) lesions on leaves
Tomato bushy stunt	ss RNA	Cubic	No	Tomato	Beetle	Abnormally small

*ds = double-stranded; ss = single-stranded.
†Entry through wounds.

(Fig. 13-10). Some late viral proteins are incorporated into the host cell membrane. Each nucleocapsid becomes surrounded by a portion of the modified membrane, forming a bud that pinches off as the virion escapes. This event also seals the holes in the host cell's membrane. The acquisition of the envelope is the final stage in maturation of these viruses. Some types of virus (herpesviruses, for example) acquire their envelopes by budding from internal cell membranes such as the nuclear membrane or the Golgi apparatus. The intracellular virion is then packaged in a vacuole. In some cases, the enveloped virus exits the host cell by fusion of the vacuole membrane with the plasma membrane. Because the host cell survives the release of enveloped viruses, more viruses, in some cases several hundred thousand, can be produced than in a lytic infection.

■ Plant Viruses

Viruses that infect plant cells must get through the protective cell wall to enter (or escape) their host cell. Plant viruses may enter through previously damaged cell walls. Most viruses, for example, are introduced by insects and other arthropods that feed on the plant. Progeny virions are then transmitted by direct cytoplasmic transfer to other cells throughout the plant. Infection usually results in death of the host cell and damage to or death of the plant. The characteristics of several viruses that attack commercially important crops are presented in Table 13-5.

■ Bacterial Viruses

Bacteriophages demonstrate some interesting variations in their replication cycles, differing from other kinds of viruses primarily in their mode of penetration and release. Figure 13-11 shows the attachment and penetration of one type of bacteriophage. Attachment sites on the tail fibers recognize and fasten to receptor sites on the bacterial cell wall, bringing pins on the base plate of the viral tail into contact with the cell wall. Enzymes in the tail digest the adjacent cell wall, allowing molecules to be released through the hole in the cell wall. When these molecules reach the virus, they trigger a change in the shape of the base plate. This initiates a spring-like contraction of the bacteriophage tail, thrusting the hollow core tube through the cell wall, thereby directly injecting the nucleic acid. The empty capsid remains outside the cell, so intracellular uncoating is unnecessary. Inside the cell, the free viral nucleic acid is used immediately to synthesize messenger RNA. Phages that are highly virulent produce early proteins that completely take control from the host cell. Cellular DNA, for example, is degraded within minutes into nucleotides that are later reused to synthesize viral DNA.

Assembly of these viruses occurs after many copies of viral DNA and structural proteins have ac-

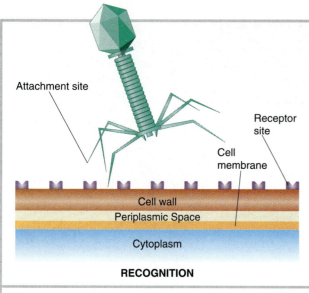

Attachment site

Receptor site

Cell membrane

Cell wall

Periplasmic Space

Cytoplasm

RECOGNITION

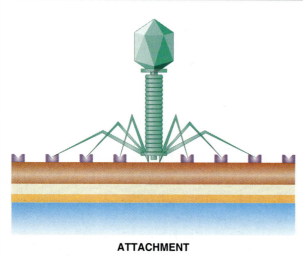

ATTACHMENT

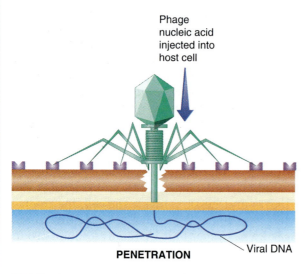

Phage nucleic acid injected into host cell

Viral DNA

PENETRATION

FIGURE 13-11 T-2 Bacteriophage. Steps in adsorption to the cell wall of *Escherichia coli* and subsequent penetration.

cumulated in the cell. The head, tail, and tail fibers are assembled independently and joined with each other in a specific sequence. First tail fibers join with the tail, then the DNA-filled head attaches to the tail. A late protein damages the host cell membrane, releasing an abundant supply of viral enzymes that digest the peptidoglycan layer of the cell wall. Progeny bacteriophages escape from the osmotically fragile cell in an explosive lytic event. The entire lytic cycle takes about 20 minutes, and each virus-infected cell releases an average of 100 virus progeny.

■ One-Step Growth

The events that characterize infection by any lytic virus can be demonstrated by a **one-step growth experiment.** This experiment is performed by simultaneously infecting every host cell in a culture with virus. The infecting viruses replicate once, but the absence of uninfected cells in the culture prevents progeny viruses from initiating a second infection cycle. Figure 13-12 illustrates the viral growth curve generated by such an experiment using a lytic bacteriophage. Immediately after penetration of viral nucleic acid, the eclipse phase begins. No infectious particles can yet be detected in the culture. Production of intracellular viruses continues until the host cell lyses, releasing all particles from that cell in a single burst.

The *rise period* is the time required for all infected cells to release progeny viruses. Usually the number of extracellular virions increases rapidly once the rise period begins. The final yield of progeny viruses can be used to determine the **burst size,** the average number of progeny produced by a single infected cell.

■ Lysogeny—An Alternative to Lysis

Some viruses can infect a cell without producing progeny or damaging the host. The virus-infected cells may reproduce for generations, each cell containing a copy of the viral chromosome. Because the viral chromosome replicates only when the cell's chromosome replicates, these cells usually show no evidence of infection. In bacterial cells, this state is called **lysogeny,** and such virus-containing bacterial cells are said to be *lysogenic.*

Bacteriophages capable of lysogeny are called **temperate phages.** The establishment of lysogeny by temperate phages occurs by the following mechanism (see Fig. 13-13):

1. After penetration, the viral chromosome directs

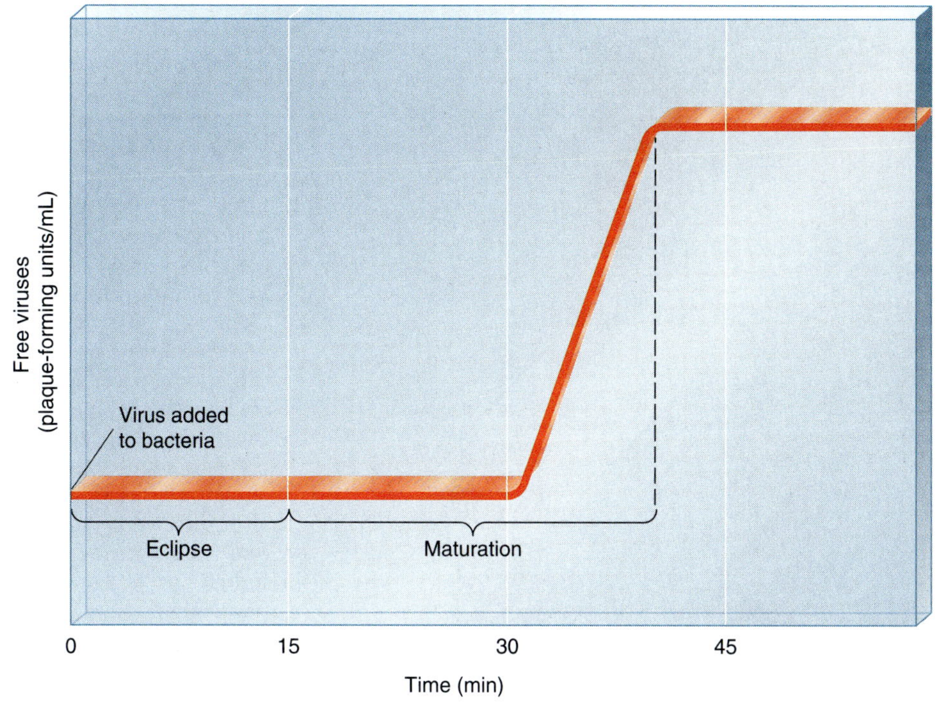
Free viruses
(plaque-forming units/mL)
Virus added
to bacteria
Eclipse
Maturation
0
15
30
45
Time (min)

FIGURE 13-12 One-step growth curve. T-2 bacteriophage infection of *Escherichia coli.* Immediately after infection the virus enters the eclipse phase. Progeny virions begin to appear in cells after 15 minutes. The rise period begins when lysis releases viruses from a host bacterium and continues until the last bacterium has lysed.

production of repressor proteins that specifically bind to the viral chromosome and turn off replication of viral DNA and synthesis of most viral proteins.

2. For most lysogenic phages, the repressed viral DNA then integrates into and becomes a physical part of the host chromosome. The integrated viral DNA is now called a **prophage.**
3. The viral DNA replicates along with the bacterial chromosome, so all progeny cells inherit one copy of the prophage in the chromosome and thus carry the potential for producing the temperate bacteriophage.

The integration site in step 2 is usually between two specific genes, and a particular phage will insert only at that site. Thus only one copy of the prophage will usually be present in the bacterial chromosome. One unusual bacteriophage, however, can insert anywhere in the host genome, even in the middle of a gene. This phage, called Mu, derives its name from its tendency to cause *mu*tations, alterations in the structure of DNA, often manifested as a change in the genetically acquired property that corresponds to the gene that is altered by viral insertion.

The presence of a prophage prevents subsequent infections by other viruses of the same type, as repressor protein in the lysogenic cell quickly binds to the infecting virus's newly injected DNA. Lysogenic cells therefore become *immune* to infection by the same type of virus.

An occasional lysogenic cell undergoes **induction,** the onset of the productive lytic cycle. Induction occurs whenever the prophage is no longer repressed and produces progeny viruses, eventually lysing the host cell. Induction occurs spontaneously, but its frequency is enhanced by irradiation with ultraviolet light or exposure to agents that interfere with DNA replication. Under these conditions, the repressor protein is destroyed by an activated bacterial enzyme. This enzyme is part of the bacteria's "SOS response," which rescues the cell by repairing damaged DNA ▶ (see Error-Prone Repair, Chap. 8, p. 204). In lysogenic bacteria, activation of the repair mechanism also eliminates the repressor protein, thereby allowing the virus to escape a dying host cell. Like the parent bacteriophage, viruses produced following induction are also temperate; most will establish lysogeny in the cells they subsequently infect.

The advantage of lysogeny to the virus is reflected by a biological principle: the most successfully adapted parasite is one that does not harm its host. To kill the host would destroy the parasite's life-support system. The bacterium also benefits from the

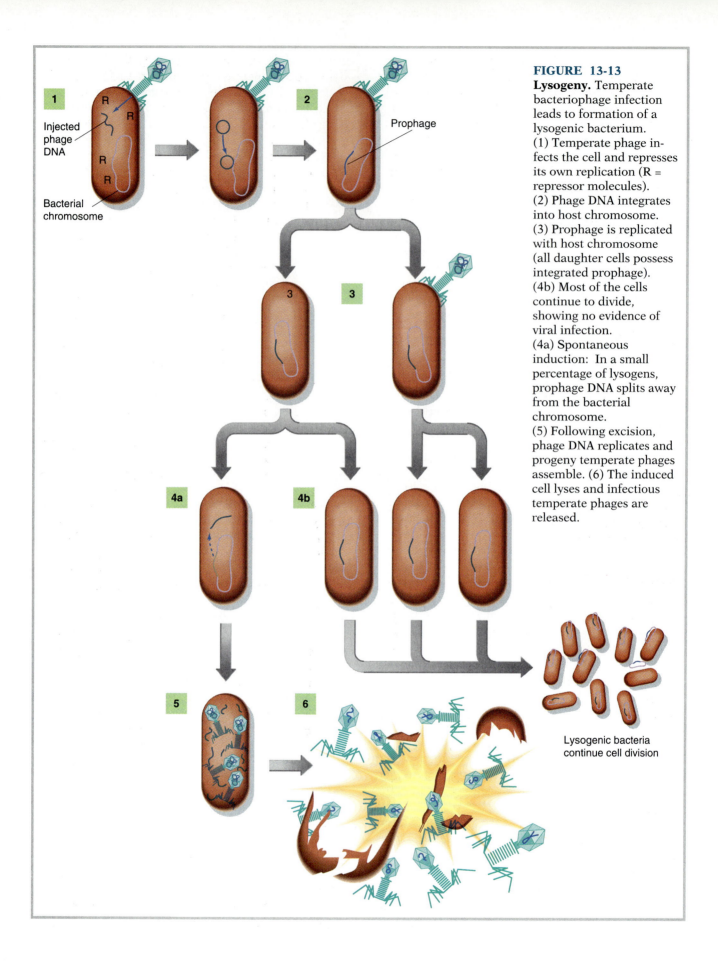

FIGURE 13-13

Lysogeny. Temperate bacteriophage infection leads to formation of a lysogenic bacterium.
(1) Temperate phage infects the cell and represses its own replication (R = repressor molecules).
(2) Phage DNA integrates into host chromosome.
(3) Prophage is replicated with host chromosome (all daughter cells possess integrated prophage).
(4b) Most of the cells continue to divide, showing no evidence of viral infection.
(4a) Spontaneous induction: In a small percentage of lysogens, prophage DNA splits away from the bacterial chromosome.
(5) Following excision, phage DNA replicates and progeny temperate phages assemble. (6) The induced cell lyses and infectious temperate phages are released.

Injected phage DNA

Bacterial chromosome

Prophage

Lysogenic bacteria continue cell division

lysogenic association in that the virus's repressor protein protects the cell from lytic infection by viruses of the same type as the prophage. In addition, bacteriophages may bring in new genes to the cell either by transduction ▶ (see Transduction, Chap. 8, p. 207) or by lysogenic conversion (below). These new genes may increase the bacterium's ability to survive—by providing it with resistance to antibiotics, for example.

● **LYSOGENIC CONVERSION.** Some lysogenic cells have properties not shared by their nonlysogenic counterparts. The acquisition of new properties following lysogeny is called **lysogenic conversion.** These changes are directed by genes on the phage chromosome that are not repressed during lysogeny, genes that are expressed as long as the host cell retains the prophage.

The medical significance of lysogenic conversion is illustrated by its role in the pathogenesis of diphtheria. The potentially fatal symptoms of this disease are elicited by a potent exotoxin, a soluble poison released from the bacteria growing in the victim's throat. Only lysogenic strains of this pathogen can produce the exotoxin and elicit symptoms of diphtheria. The prophage, not the bacterial chromosome, carries the toxin-encoding genes. Diseases such as scarlet fever and botulism, both caused by bacterial exotoxins, are also believed to be caused exclusively by lysogenic strains of their respective pathogens.

■ Oncogenic Viruses

Lysogeny is a phenomenon found only in bacteria, but an analogous system exists in animal cells infected with viruses that integrate their DNA into the host cell's chromosomes. The ability to integrate may be related to *oncogenesis* (the production of cancer) by some of these viruses.

"Cancer" is a single name applied to a number of diseases characterized by uncontrolled growth of the body's own cells. These malignant cells may arise from several types of normal tissue cells. Cancer cells have lost their sensitivity to the signals that inhibit excessive reproduction of normal cells and may metastasize (spread) throughout the body from their original sites of formation.

For some animals, such as chickens and mice, it has been clearly demonstrated that certain cancers are caused by **oncogenic viruses.** Viruses are also associated with several types of cancer in humans. The *papillomaviruses,* for example, are capable of maintaining their DNA in susceptible host cells in both plasmid and integrated forms. These viruses cause warts in all parts of the body. Genital wart lesions are among one of the most prevalent sexually transmitted diseases in America. Women with genital lesions containing certain papillomaviruses appear more likely to develop cancer of the cervix. *Herpesviruses* are another type of DNA virus associated with tumors in humans. Some persons infected with a herpesvirus called the Epstein-Barr (EB) virus (the agent of infectious mononucleosis) seem more prone to develop certain cancers of the nasopharynx. In addition, Burkitt's lymphoma, a malignant tumor of lymphoid tissue in the facial region, is linked to EB virus infection in black African children. Evidence also indicates that the hepatitis B virus is an etiologic agent of hepatocarcinoma, a form of liver cancer that is a major cause of death worldwide. Nucleic acid from this virus is found integrated in the tumor cells.

Other viruses that may induce human cancers are members of a group of RNA retroviruses that are frequently isolated from patients with leukemia. In their host cells, retroviruses produce DNA from their own RNA. The viral DNA then integrates into the host cell's chromosomes. Among this newly introduced viral nucleic acid may be a genetic segment called an **oncogene,** which is responsible for transforming normal cells to cancer cells. Normal human cells also have regions on their chromosomes comparable to oncogenes. These genes, called *proto-oncogenes,* normally direct the synthesis of products needed for cell growth and development. Excessive production of these gene products, however, can lead to uncontrolled growth and transformation to a tumor cell. When viral nucleic acid containing an oncogene is introduced into a cell, the cell may be unable to regulate synthesis from the virus's oncogene. The excess product may trigger the unrestrained cell growth associated with cancer. Alternatively, some viruses do not bring in oncogenes but promote overactivity of the host's proto-oncogene when they integrate into the host chromosome. Even in the absence of a virus, however, the cell's proto-oncogene may be activated when *carcinogens,* environmental factors such as ultraviolet light and certain chemicals, interfere with control of its expression.

The magnitude of the role of viruses in human cancer is currently a hotly debated matter. Carcinogens are certainly major causes of human cancer, but some oncogenic viruses may work in concert with carcinogens to produce tumors. Some oncologists believe that viruses may play a role in causing half of all types of human cancers. Whoever is right, one fact remains optimistically clear—viruses have helped build our understanding of the genetic basis for cancer induction, just one of the benefits provided by this generally maligned group of microbes (see BEYOND THE BASICS: Viruses—More Than Mere Microscopic Menaces).

■ **Viroids and Prions**

Viroids are unusual infectious entities associated with diseases of plants. They consist solely of a small circle of single-stranded RNA unprotected by a cap-

---

BEYOND THE BASICS

Viruses—More Than Mere Microscopic Menaces

Since their discovery, viruses have acquired a reputation as the world's nastiest menaces. They are guilty of causing dozens of human diseases, from mild cases of the common cold to fatal assaults by the AIDS virus. They threaten our food supplies by attacking livestock and crop plants. *Time* magazine reflected the popular fear of viruses by calling them "mankind's deadliest enemy." And many biologists embrace this fear; one Nobel prize–winning scientist describes viruses as "a piece of very bad news wrapped in protein." Even their name (*virus* is Latin for poison) suggests that they are thoroughly detrimental influences. Could such a prevalent "villain" have any possible human benefits? The answer is an unqualified yes.

The simplicity of viruses has provided keys to unlocking many of the mysteries of life. Our current understanding of how cells work, for example, would have been impossible without viruses to simplify the overwhelming complexity of the cell's molecular landscape. It would have been impossible to learn how cells follow genetic instructions to make protein, for example, by studying a normal animal cell with its thousands of proteins. After viral infection, the same cell makes as few as six proteins—those dictated by viral genes. In fact, much of what we know about the storage of genetic information in genes and the expression of genetic traits required viruses as the pivotal research tool. Without viruses, contemporary molecular biology would be unimaginable.

We are also using viruses as tools in our ongoing battle against human disease. Genetic deficiency diseases could be permanently cured if the defective gene were replaced with a working copy. For example, in people with faulty dystrophin genes, such "gene therapy" could correct the protein deficiency that leads to muscular dystrophy. One of the major obstacles to gene therapy, however, is the need to develop a mechanism to introduce the replacement gene into the cells of a patient. Viruses are currently the most promising vectors for ferrying replacement genes into people. During their development, viruses can be "tricked" into enclosing purified dystrophin genes in their capsids in place of viral genes. These gene containers still attach to their specific host cells but release the replacement genes into the cell. Once the newly acquired genes are integrated into the host cell's chromosomes, the cell will manufacture normal levels of dystrophin and the disease will be cured. Although currently experimental, most scientists believe that gene therapy is a near reality for virtually any disease caused by defective genes.

Viruses are also being used to create a new generation of vaccines that protect against a broad variety of infectious diseases. These "piggyback" vaccines use a primary virus such as vaccinia that infects and safely replicates in the human host (vaccinia is the relatively harmless virus used to stimulate immunity against smallpox). Introducing genes from other viral or even bacterial pathogens into the vaccinia virus's genome creates a gene carrier that, in human host cells, directs infected cells to produce proteins characteristic of the piggyback pathogens as well as its own genes. The vaccinated person then responds by launching an immune response against all the foreign proteins, conferring protection against the corresponding pathogens. We may soon be able to use such a modified vaccinia virus to simultaneously stimulate immunity against smallpox (although no longer needed), hepatitis, herpesvirus, whooping cough, and polio. ▶ (See Genetically Engineered "Piggyback" Vaccines, Chap. 9, p. 231.)

Viruses help propel the young field of genetic engineering by providing scientists with a source of the specific genes to be introduced into another organism. Although the gene-donating cell normally has only one copy of a particular gene, its cytoplasm may contain thousands of mRNA molecules transcribed from that gene. So while isolating a particular gene from a cell is a "needle-in-a-haystack" proposition, obtaining the corresponding mRNA is much easier. The enzyme reverse transcriptase can then be used to assemble ample quantities of the desired gene (using mRNA as a template). Retroviruses provide the only source of reverse transcriptase. Once the genes are available, other viruses are used to carry these genes and introduce the foreign DNA into recipient cells.

Viruses have also provided us a more enduring gift, that of helping us evolve into what we are. The natural relationship between the genes of virus and host has always been far more significant than a simple matter of genetic conquest. During intracellular development, a small percentage of developing viral particles "accidentally" enclose host genes rather than viral genes in the capsid. When these "viruses" infect a cell, they introduce new genes (but no lytic virus genes) into the recipient. The new genes may confer upon the recipient a new, and perhaps advantageous, combination of traits. In this way, viruses may have hastened the evolution of every species on earth, including humans. Without these "microscopic menaces," the human species may not have ever made an appearance on Planet Earth.

sid. Viroid RNA is replicated by host enzymes, but it is not translated within the infected plant cell. Viroid-infected plants are stunted in growth and abnormal in development. For example, viroids are the agent of cadang-cadang disease of coconut palms (the disease derives its name from the sound of falling coconuts). A major epidemic has destroyed most of the coconut palm population in the Philippines (see Fig. 13-14).

Although no viroids have been isolated from animal cells, the hepatitis delta agent is a possible candidate. It is composed of single-stranded RNA that has a high degree of similarity to some viroids. Although an envelope surrounds the capsid, the envelope is not encoded by the hepatitis delta agent RNA. It is produced by a coinfecting virus, the hepatitis B virus. Hepatitis delta agent is a *defective virus* that cannot successfully infect without the aid of its coinfecting helper virus.

Several slowly progressive neurological diseases are caused by another group of infectious agents called **prions.** Kuru and Creutzfeld-Jakob syndrome, two human prion diseases, are characterized by incubation periods of months to years (see THE

FIGURE 13-14 Barren harvest. This Philippine plantation of coconut palms has been stripped of all fruit by the cadang-cadang viroid. Cadang-cadang is one of the most devastating plant diseases known, killing about a million coconut palms each year. The epidemic is also ravaging the Philippine economy, half of which relies on coconut products.

EXPLORERS: The Laughing Death). Once symptoms appear, however, the disease progresses rapidly, and death is inevitable within a year after the victim

THE EXPLORERS

The Laughing Death

In the mid-1950s an unusual and deadly phenomenon threatened to extinguish a small tribe of people in New Guinea. Most of the women and many of the children were being attacked by a fatal neurological disorder that began by causing its victims to giggle uncontrollably. The syndrome baffled American epidemiologist Carleton Gajdusek, who analyzed soil, drinking water, food, and even the ashes in the fires in search of the etiologic agent and its mode of transmission. After months of inquiry, Gajdusek discovered that the tribe was cannibalistic. As an expression of respect for their dead relatives, the survivors would consume portions of the corpses, including the brain. Years after preparing the brains for cooking, the women and chil-

dren would begin the fatal giggles.

Kuru, as the disease was named, was subsequently shown by Gajdusek to be caused by a previously undiscovered type of pathogen, originally called a *slow virus* because of its 2-to-20-year incubation period. The agent, now recognized as a prion, is transmitted by eating the infected neurological tissue of someone who has died from the laughing death or by cutaneous inoculation of the virus while preparing the brain. The tribe's extinction was avoided when they were persuaded to abandon their cannibalistic tribute to their dead. Gajdusek's detective work on slow viruses won him the 1976 Nobel prize in medicine.

Although kuru has been virtually controlled, another prion-mediated disease, Creutzfeld-Jakob disease (CJD) continues to cause invariably fatal neurological infections throughout the world. The transmission of this rare disease is still an enigma, but two cases were caused when contaminated electrodes were inserted into the brain during neurosurgery. Another case followed a corneal

transplant from a patient with undiagnosed CJD. CJD is also an occupational hazard among neurosurgeons and neuropathologists. Each year 200 people in the United States die of the disease.

Several animal diseases are also caused by prions. The oldest documented animal disease is scrapie, an infection of sheep. The striking similarities between the symptoms and agents of scrapie and those of certain human diseases have led some scientists to hypothesize that prions originally entered the human population in scrapie-contaminated food. (This presumably does not happen any longer because sheep are monitored and infected animals are destroyed.) Some evidence to support this idea comes from our understanding of a prion disease of cattle called mad cow disease. The appearance of mad cow disease in England coincided with a change in cattle feed, a change that included the addition of tissue from sheep (although the sheep were not symptomatic with scrapie).

shows signs of illness. Although their infectious natures and small size suggest similarities to conventional viruses, no viruslike particle has been observed, even by electron microscopy. In addition, prions are resistant to physical and chemical treatments that inactivate conventional viruses. Nucleic acid has yet to be detected in preparations of these infectious agents. This has led to the speculation that prions are infectious agents that consist solely of protein. Tissues of infected people contain high concentrations of a protein that resembles a normal cell protein, suggesting that prions enter cells and stimulate the gene responsible for their synthesis. This gene produces the normal protein, which is then modified to become a prion.

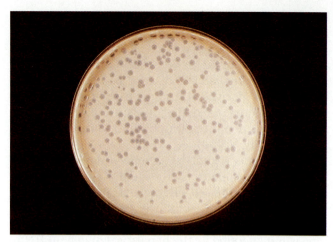

FIGURE 13-15 **Bacteriophage plaques.** The number of viruses in the volume applied to the plate is ascertained by counting the plaques in this confluent lawn of susceptible bacteria.

> **CONCEPT CHECK**
>
> - What host factors do viruses use for a productive infection? What kinds of proteins are encoded by viral genomes?
> - How do viruses harm their host cells?
> - Describe what is happening to viral nucleic acid in the bacterial host during lysogeny. What advantage is lysogeny to bacteriophages? to bacteria?
> - How may viruses carrying oncogenes cause cancer? In what ways may other viruses cause cancer?
> - What distinguishes viroids and prions from each other and from viruses?

Detection and Cultivation of Viruses

Although viruses are too small to be seen with the light microscope, they may be detected by electron microscopy. However, it is often easier or more reliable to employ an indirect method for detecting viruses. Many of these indirect techniques depend on our ability to cultivate viruses *in vitro* (in laboratory glassware such as petri dishes, test tubes, or flasks). Because they are obligate intracellular parasites, however, viruses cannot be cultivated on inanimate media such as nutrient agar. They require a living cell system. Bacteriophages, for example, can be propagated on susceptible bacterial cultures. In addition, their effects on host cell cultures are often observable, providing a means of detecting the presence of viruses. Just as a single bacterium generates an isolated colony on a suitable nutrient medium, the progeny of a single virus may create a macroscopically visible effect on the surface of sus-

ceptible host cells. When a suspension of virus-sensitive bacteria is spread over the surface of a solid nutrient medium and incubated, a confluent "lawn" of bacterial growth covers the surface of the agar. If bacteriophages are present in the bacterial suspension when it is spread over the surface, each virus that infects and lyses a cell produces progeny viruses that infect and destroy surrounding cells. The process continues until, by the time the bacterial lawn begins to appear, a small zone of clearing will have developed around the original virus-infected cell. Each such zone of clearing, called a **plaque** (Fig. 13-15), represents one bacteriophage applied to the plate.

If too many phages are applied to the seeded medium, all the bacteria will be infected, resulting in confluent lysis and no bacterial growth on the surface. If the virus suspension is diluted, however, the number of isolated plaques that appear can be used for determining the concentration of infectious bacteriophages in the original suspension. This is accomplished, as with the bacterial viable count, by multiplying the dilution factor by the number of plaques on the plate ▶ (see Plate Counts, Chap. 5, p. 121). In this respect, a bacteriophage plaque is analogous to a bacterial colony.

MYTH: If a bacterial lawn containing a viral plaque is incubated long enough, the plaque will enlarge until all the cells on the plate are lysed.

FACT: Plaques remain fairly small because viruses require growing host cells to produce viral progeny. The lawn of bacteria eventually exhausts the nutrients and stop growing. At that point viral replication stops, no additional cells can be lysed, and the plaque stops enlarging.

Viruses that infect animal cells can be detected in a similar fashion, using susceptible animal cells as the indicator. **Cell culture,** the growth of animal cells on artificial media, is the most common system for detecting and cultivating animal viruses. Animal cells are grown in a **monolayer,** a uniform layer one cell thick on the inner surface of a bottle or test tube (Fig. 13-16a). A virus that infects a single cell may ultimately form a plaque by a process similar to plaque formation by bacteriophages. Viral infection of cell cultures may induce several other morphological changes—fusion of cells into larger cells with multiple nuclei (called "giant cells" or *polykaryotes*), clumping of cells, or the development of **inclusion bodies** (intracellular aggregates of developing viruses). These virus-induced changes in cell cultures are referred to as **cytopathic effects (CPEs)** and are easily observed with a light microscope.

The appearance of CPEs is often characteristic of the type of virus and is consequently a useful diagnostic aid for identifying viruses isolated from an infected patient and grown on cell culture (Fig. 13-16b). Cytopathic effects also develop in a patient's cells during natural infection. Thus many viral diseases are rapidly diagnosed by microscopically observing stained material taken directly from the patient's lesions. For example, intranuclear inclusion bodies (called Cowdry bodies) in epithelial cells are characteristic of herpesvirus infection. In the absence of CPEs, some enveloped viruses may be detected by *hemadsorption,* the attachment of red blood cells to the surface of virus-infected cells. This phenomenon is due to the affinity of some viral envelope proteins for red blood cells. These envelope proteins are present in the surface membranes of cells infected with the viruses.

Although cell culture is the preferred method of viral cultivation, some viruses can only be grown in living experimental animals. Other viruses are cultivated on embryonated chicken eggs and are detected when pocks develop on membranes surrounding the embryo. Some viruses, such as influenza virus, can be detected by their tendency to cause clumping of red blood cells in test tubes, a phenomenon known as *hemagglutination.*

One of the most useful methods of viral detection is the serological reaction between the antigens of the virus and an antibody of known viral specificity. For example, if a virus obtained from an infected patient reacts with a herpes-specific antibody, the identity of the virus is confirmed. Most viruses can be detected (and identified) by such immunologic techniques. Indicators of positive antibody–virus reactions are discussed later in this chapter ▶ (see Diagnosis of Viral Diseases, p. 326).

Viruses can also be rapidly identified by their reaction with DNA probes. The specificity of these reactions eliminates the need to isolate the virus. If a virus is present in low concentrations or if its DNA is integrated into the host chromosome, the amount of viral DNA can first be amplified using the PCR reaction. This technique has been used to detect papillomavirus DNA in genital lesions and HIV in newborns.

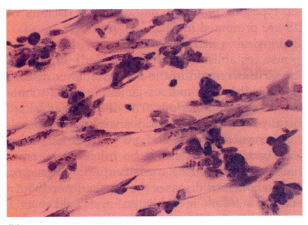

(a)　　　　　　　　　　　　　　　　**(b)**

FIGURE 13-16 **Cell culture.** (*a*) A monolayer typical of healthy cells; (*b*) cytopathic effect on the same cells infected with herpesvirus. Both photos at a magnification of 100×.

Key Terms

obligate intracellular parasites (p. 306)
capsid (p. 307)
protomers (p. 307)
capsomeres (p. 307)
nucleocapsid (p. 307)
envelope (p. 307)
virion (p. 307)
bacteriophages (p. 309)
eclipse phase (p. 309)
attachment site (p. 311)
receptor site (p. 311)
receptor-mediated endocytosis (p. 312)
uncoating (p. 312)

early proteins (p. 312)
positive-strand viruses (p. 314)
negative-strand viruses (p. 314)
retroviruses (p. 316)
late proteins (p. 316)
budding (p. 316)
one-step growth experiment (p. 318)
burst size (p. 318)
lysogeny (p. 318)
temperate phages (p. 318)
prophage (p. 319)
induction (p. 319)
lysogenic conversion (p. 321)

oncogenic viruses (p. 321)
oncogene (p. 321)
viroids (p. 322)
prions (p. 323)
plaque (p. 324)
cell culture (p. 325)
monolayer (p. 325)
inclusion bodies (p. 325)
cytopathic effects (p. 325)
latent infection (p. 326)
interferons (p. 328)

Key Facts and Concepts

- **Properties of viruses.** Viruses are noncellular bio-logical entities, obligate intracellular parasites that possess no enzymes for independent metabolism. A single type of nucleic acid (DNA or RNA) is present in each type of virus. Viral nucleic acid is surrounded by a protective protein coat called a capsid that is composed of repeating protomer subunits. For some types of viruses, the nucleocapsid is surrounded by a membranous envelope. Virus classification is primarily based on type of nucleic acid, capsid morphology, virion morphology and host range.

- **Viruses and host cells.** Viruses depend entirely on the cell's metabolic machinery to produce progeny viruses. The host cell is often killed by such infections, but some viruses allow the cell to live and continue to release new virions.

- **Events in viral replication.** Viral infection begins when the attachment site on the surface of the virion to a specific receptor site on the host cell membrane. Animal viruses penetrate into the host cell by fusion or receptor-mediated endocytosis. Uncoating releases viral nucleic acid and marks the beginning of eclipse. Bacteriophages release their nucleic acid directly into the bacterial cytoplasm. During eclipse, infectious particles are not present.

  The viral nucleic acid directs synthesis of early proteins, most of which are enzymes needed for controlling the host or for the replication of viral nucleic acid. Most DNA viruses use the host RNA polymerase to produce mRNA. Positive-strand RNA viruses serve as mRNA. Negative-strand RNA viruses carry a transcriptase in the virion

that, upon infection, makes a positive strand using the negative strand. Retroviruses carry reverse transcriptase that manufactures a DNA copy of the RNA genome. The capsid and envelope proteins are late proteins that are assembled with nucleic acids into nucleocapsids. Lytic viruses burst the host cell during release. Most enveloped viruses leave the host cell by budding.

- **Lysogeny.** The DNA of temperate bacteriophages integrates as a prophage into the host chromosome, resulting in lysogeny, which is maintained by a repressor protein produced by the prophage. Some bacterial cells acquire new properties following lysogeny. Lysogenic conversion explains why some diseases, such as diphtheria, are caused only by lysogenic bacteria.

- **Oncogenic viruses.** Oncogenic viruses cause the production of tumors in animals. Several types of DNA viruses (papillomavirus, herpesvirus, hepatitis B virus) are associated with cancer in humans. The only oncogenic RNA viruses are found among the retroviruses.

- **Viroids and prions.** Viroids and prions are unusual infectious agents. Viroids consist of naked RNA and cause several diseases in plants. Prions contain protein only and are agents of several slowly progressing neurological diseases.

- **Detecting viruses.** Lytic viruses are detected by the development of plaques on a lawn of sensitive host cells. Inclusion bodies and cytopathic effects are virus-induced changes that are useful diagnostic aids.

- **Pathogenic viruses.** Most pathogenic viruses produce acute or asymptomatic infections that rapidly run their course and stimulate permanent immunity. Recovery is probably due to a combination of neutralizing antibodies, interferon, and protective host cells that can recognize and destroy virus-infected cells. Some viruses establish latent infections which may be periodically reactivated.

• **Prevention and control of viral diseases.** Only a few chemotherapeutic agents are effective in the treatment of viral diseases. Many viral diseases have been successfully controlled by preventive measures—vaccination, control of vector populations, isolation of infected individuals, improved sanitary conditions, and control of animal reservoirs of infection.

Review and Problem-Solving Questions

1. Fill in each space in the following table with a "yes" or "no."

	E. coli	Rickettsias	Viruses	Viroids	Prions
Intracellular parasite	_____	_____	_____	_____	_____
Reproduce by fission	_____	_____	_____	_____	_____
Display eclipse phase	_____	_____	_____	_____	_____
Contain DNA and RNA	_____	_____	_____	_____	_____
Filterable at 0.22 μm	_____	_____	_____	_____	_____
Contain ribosomes	_____	_____	_____	_____	_____
Possess enzyme activity	_____	_____	_____	_____	_____
Sensitive to antibiotics	_____	_____	_____	_____	_____

2. Distinguish between (a) envelope and capsid; (b) capsid and capsule; (c) plaque and colony; (d) protomer and prophage.
3. Describe the events of infection, replication, and release for (a) nonenveloped animal viruses; (b) enveloped viruses; (c) lytic bacteriophages; (d) lysogenic bacteriophages.
4. Why is it unlikely that a bacteriophage would be enveloped?
5. Give one reason most viruses are specific for one type of host cell. How does this relate to disease?
6. Some viruses escape the host cell by reversing the penetration process. Explain.
7. How would each of the following aid in the detection of viruses? (a) Electron microscopy; (b) plaque formation; (c) cell culture; (d) light microscopy; (e) serological reactions; (f) hemagglutination.
8. Why is lysogeny believed to be more common than lytic phage infection?
9. List five similarities between animal viruses and bacteriophages.
10. Describe how you would count the number of bacteriophages capable of infecting E. coli in a sample of raw sewage.
11. *Transfection* is a process of transformation in which isolated viral nucleic acid rather than the intact particle is used for infection of the host cell. In many cases, uptake of viral nucleic acid results in the synthesis of viral progeny. Based on your knowledge of viral structure, predict whether transfection experiments using the following viral nucleic acids will yield a productive infection. Explain your answers. (a) HIV (retrovirus); (b) influenza (negative-strand virus); (c) poliovirus (positive-strand virus); (d) hepatitis delta virus (defective RNA virus).
12. Refer to the graph for each of the following.
   (a) The burst size (viral yield) is _____.
   (b) Draw, on the same graph, the expected one-step growth curve for this virus if lysis is delayed.
   (c) What is occurring during each phase of the curve?

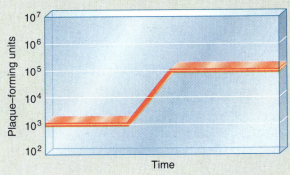

13. Indicate why each of the following statements is false.
   (a) A lysogenic bacterium contains a viral genome but is otherwise no different from the bacterium without the virus.
   (b) Interferon protects people primarily from a second exposure to a virus.
   (c) A plaque is formed like a colony except that it is made of viruses.
   (d) Enveloped viruses that enter by fusion are released from the cell by lysis.
14. Some viruses typically yield larger plaques than other viruses. What factors might determine the size of plaques formed by bacteriophage infection?
15. How does viral replication differ from cell division? Identify at least five potential targets for antiviral therapy in the lytic cycle of an animal virus.
16. Bacterial pneumonia is a common sequel in patients recovering from influenza. One hypothesis suggests that bacteria can more readily attach to the surface of virus-infected cells. Design an experiment to test this hypothesis.
17. Single-stranded RNA bacteriophages are the smallest viruses known. They contain only four genes and undergo a lytic cycle in *E. coli* and related gram-negative bacteria. Suggest a possible life cycle for these viruses, and indicate the products of the four viral genes.
18. Because of its broad antiviral activity, interferon is thought to represent a possible therapeutic agent against the common cold.
   (a) Design an experiment to test the usefulness of interferon as a chemotherapeutic agent against the common cold.
   (b) What problems might be associated with interferon treatment for the common cold?
19. In the spring of 1993, an unusual and fatal respiratory disease claimed the lives of dozens of people in Arizona, Utah, Colorado, and New Mexico. Investigators from the Centers for Disease Control and Prevention and local public health agencies identified a virus (called hantavirus) from some of the victims. How might you prove that this was the cause of the epidemic of respiratory infections?

HEADLINES

"STEALTH" VIRUSES EVADE HOST DEFENSES

The strategies by which several animal viruses "cleverly" avoid detection and destruction by their hosts have been uncovered in the past few years. It should not be surprising to learn that these parasites have once again resorted to stealing host genes to elude the immune system. For example, certain poxviruses contain a gene similar to a host gene that codes for a membrane protein that binds tumor necrosis factor (TNF), one of the host molecules that signals immune cells to attack virus-infected cells.[*] The virus gene produces a soluble molecule (rather than a membrane-associated protein) that binds TNF before it can reach its target on immune cells. Other poxviruses have been found to produce and release soluble forms of the receptor for interferon. Epstein-Barr virus, a herpesvirus that causes infectious mononucleosis, counters interferon in another way. It has stolen a gene for a protein whose function is to suppress production of interferon.

- How would synthesis of the soluble receptor for interferon help protect the poxvirus from an infected host's interferon?
- How might the poxviruses and herpesviruses have acquired ("stolen") these genes?
- Scientists are trying to outwit viruses at their own game. They are thinking of using a soluble form of the surface receptor for a virus in antiviral therapy. How might this protect a host from infection? For what possible reason might this approach fail?

[*]M. Barinaga, Viruses launch their own "Star Wars," *Science*, 258: 1730–1731 (1992).

CHAPTER 14

Microbial Interactions

Fatal attraction. This deep-sea anglerfish is life's final image for any fish attracted to its "fishing lure." A devoured fish is as much a victim of bioluminescent bacteria as of the predator's jaws. Fish attracted to the light generated by the glowing bacteria in the lure quickly disappear into the angler's formidable mouth.

Commensalism

Mutualism

Mutualism Between Microbes
Mutualism Between Microbes and Plants
Mutualism Between Microbes and Animals

BEYOND THE BASICS: A Glowing Success Story

Antagonism

Antagonism Between Microbes
Plant Parasites
Animal Parasites

The sleek fish deftly navigated a perilous course through dangerous ocean waters teeming with predators. He survived because of quick reflexes and good luck, managing to make victims of some less fortunate creatures while avoiding predators. What appeared to be another unwary victim lay straight ahead, its glowing body wiggling enticingly in the dark ocean depths. The sleek fish didn't have to know what it was—anything smaller than him that moved was a candidate for dinner. He approached it with the disarming glide that rarely failed to produce juicy results. A moment later, the confident hunter seemed to simply disappear. The glowing "prey," however, was still there, just above the formidable mouth of a deep-sea angler fish whose lethal strike is faster than the human eye can see. The angler fish earns its name from its predatory ability to fish for other fish, using a luminous lure that protrudes from its head to attract its catch.

The unfortunate fish lost its life as much to bacteria as to another fish. The deep-sea angler, like so many other organisms, depends on an obligate association with bacteria for its survival. If not for its interaction with the luminescent bacteria residing in its lure, it would be unable to attract dinner, and its daggerlike teeth would go unused. The angler would starve. The

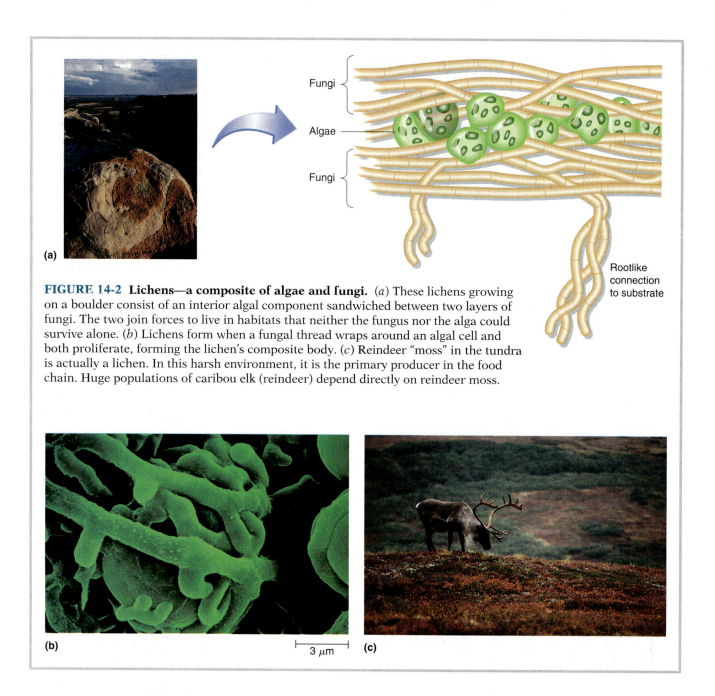

FIGURE 14-2 Lichens—a composite of algae and fungi. (*a*) These lichens growing on a boulder consist of an interior algal component sandwiched between two layers of fungi. The two join forces to live in habitats that neither the fungus nor the alga could survive alone. (*b*) Lichens form when a fungal thread wraps around an algal cell and both proliferate, forming the lichen's composite body. (*c*) Reindeer "moss" in the tundra is actually a lichen. In this harsh environment, it is the primary producer in the food chain. Huge populations of caribou elk (reindeer) depend directly on reindeer moss.

Labels on figure: Fungi, Algae, Fungi, Rootlike connection to substrate, 3 μm

mecium they inhabit in exchange for protection from ingestion by other predators (Fig. 14-3). Another mutualistic passenger within some paramecium cells is a bacterium that confers a killer trait on its host. The bacterium produces a protein toxic to strains of paramecia that lack the endosymbiont. This gives the killer strain a competitive edge over its noninfected neighbors vying for the same habitat and nutrients. The relationship is obligate for the bacterium, which depends on its host for food.

Microbial endosymbionts also live inside much larger organisms such as plants and animals. These mutualistic interactions are discussed later in this chapter.

● **LYSOGENY.** Lysogeny is a mutualistic relationship in which a host bacterium contains viral DNA integrated into its chromosome, an interaction that benefits both virus and bacterium ▶ (see Lysogeny—An Alternative to Lysis, Chap. 13, p. 318). The host houses and replicates the viral nucleic acid while viral repressor proteins prevent infection of the host by superinfecting viruses. In some cases, lysogenic bacteria have an enhanced ability to cause human disease such as scarlet fever or diphtheria (although such heightened virulence is not necessarily an advantage to either the virus or the bacterium). In these cases, the viral genome contains the gene for production of toxin. In other cases, the integrated

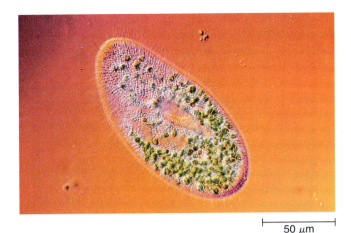

50 μm

FIGURE 14-3 **A working passenger.** The photosynthetic microbes that live inside this paramecium supplement its food and oxygen supply. The mutualistic endosymbiont inside the larger cell receives shelter, protection, and simple nutrients needed for photosynthesis.

viral genome contains genes that alter the bacterial surface antigens, making the bacterium more resistant to host defenses, a clear advantage for both the bacterium and the virus that relies on it for life support.

● **MICROBIAL AGGREGATES AND BIOFILMS.** Individual microorganisms may also team up with other members of their own species and with other species, creating complex communities that grow more successfully than individual microbes. Many of these aggregates grow in the form of mats or as **biofilms,** embedded in thin sheets of slime on surfaces. Some aggregates, for example, the microbial mats that cover ponds (Fig. 14-4a), are familiar to practically everyone. Others are more subtle but no less important (Fig. 14-4b). In fact, more than 99 percent of all bacteria in natural and industrial environments exist in biofilms and not as the free-floating organisms usually studied in pure cultures within microbiology laboratories. Growth within aggregates has a stabilizing influence on the environment and is significant in the recycling of nutrients in the biogeochemical cycles that regenerate materials needed to sustain life on our planet ▶ (see *Environmental Microbiology,* Chap. 26, p. 706).

Biofilms are also associated with microbial destruction of many useful products, with dangerous contamination of health delivery equipment (such as catheters and intravenous fluid systems), and with survival of human pathogens in sewage and other environments (even in the body) where individual cells would be eliminated. These advantages include protection of the cells in the interior of the aggregate from predation, from harmful changes in conditions, and from toxic chemicals in the environment. Even when lethal conditions kill the outer cells, interior microbes form a reservoir of new cells that may flourish when the environment becomes more favorable. It may be that some failures in an-

(a)

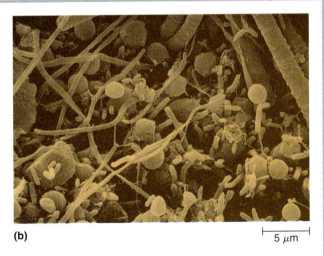

(b)

5 μm

FIGURE 14-4 **New respect for "pond scum."** (a) Microbial mats in natural ecosystems, such as these aggregates of photosynthetic algae, bacteria, and other microbes floating on the surface of a freshwater pond, help stabilize the environment in ways that scientists are just now beginning to understand. Living in mats enables them to survive many perils that might kill them if they lived individually. (b) A "microbe's eye" view of a biofilm. This natural biofilm is composed of many species of hetero-trophic bacteria plus a few thick filaments of cyanobacteria.

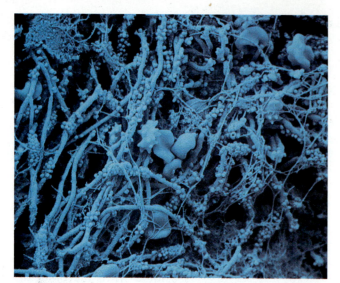

FIGURE 14-5 **Dental plaque—a destructive team.** The many organisms that live in dental plaque are necessary to translate their presence into tooth decay. None of these organisms could damage teeth without the help of the other members of the team.

timicrobial therapy against antibiotic-sensitive bacteria are due to the protection afforded by biofilms.

Perhaps the mutualistic aggregates that are most familiar to people are the biofilms on their teeth, aggregates known as dental plaque. Dental plaque is a prerequisite for dental caries (tooth decay) and periodontal (gum) disease, the leading cause of tooth loss. Plaque consists of many types of bacteria that are essential to the formation of plaque and the subsequent production of caries (Fig. 14-5). Some bacteria in plaque produce the sticky matrix (dextran) that holds the biofilm together and cements the aggregate to the tooth surface. Other organisms consume oxygen, creating the anaerobic conditions in which the dextran-producing bacteria thrive. Still other species produce the acid byproducts that erode dental enamel, leading to tooth decay. (Dental caries is discussed in Chapter 22, Diseases Acquired Through the Alimentary Tract.)

■ Mutualism Between Microbes and Plants

The association of microbes with plants is often beneficial to both participants. The most common forms of mutualism between plants and microbes are the associations that enable some plants to use molecular nitrogen and the root-enhancing partnerships that form between plants and fungi.

● **NITROGEN FIXATION.** Bacteria in the genus *Rhizobium* infect the roots of leguminous plants such

as clover, peas, soybeans, and alfalfa. Infection triggers genetic interactions that transform both the plant and bacteria. ▶ (See Nitrogen Fixation, Chap.

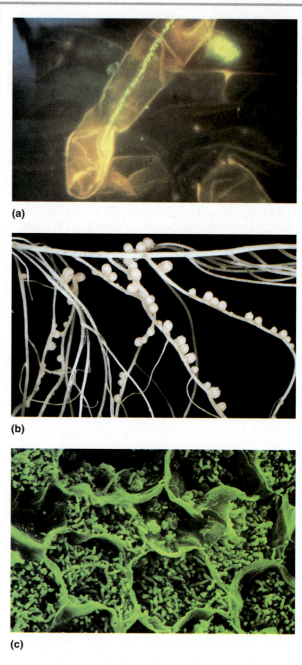

(a)

(b)

(c)

FIGURE 14-6 **One of life's most important associations.** (*a*) An infection thread forms when nitrogen-fixing bacteria invade root hairs of susceptible leguminous plants such as alfalfa or soybeans. (*b*) The bacteria eventually reside in a root nodule formed by the plant to harbor its fertilizer-producing passengers. (*c*) The *Rhizobium* bacteroids are revealed in this micrograph of material extracted from the root nodule.

26, p. 716.) The plant develops a root nodule that houses the gram-negative bacteria. Within this nodule, the rod-shaped *Rhizobium* changes into a *bacteroid*, a form that lacks a cell wall and is irregular in shape. These bacteroids perform *nitrogen fixation*, converting atmospheric molecular nitrogen (N_2) into ammonia (NH_3) for both plant and bacterial use (Fig. 14-6). The plant shares its photosynthetic products with the bacteria in the nodules. Together they assimilate inorganic nitrogen and carbon into useful organic forms. Enough usable nitrogen compounds dissolve into the soil from the root nodules to fertilize the plot for subsequent crops of nitrogen-requiring plants. This is the reason farmers rotate crops, periodically planting a leguminous crop to recharge the soil with nitrogen.

● **MYCORRHIZAE.** Fungi form stable associations, called *mycorrhizae*, with plant root systems ▶ (see Fig. 11-14, p. 270). Virtually every crop plant is associated with these fungal mycelia, which may cover the root surface or actually penetrate the root tissue. The fungus boosts the efficiency of root function by increasing the absorptive surface area of the roots. The fungus then relays its generous harvest of water, nitrogen compounds, and dissolved minerals to the plant's roots. The plant in turn supplies the heterotrophic fungus with sugars and other organic nutrients. The association is so important that many ornamental and crop plants simply won't grow without fungi that form mycorrhizae on their roots.

■ Mutualism Between Microbes and Animals

Although before birth or hatching, animals may be devoid of microbes, they become exposed to microorganisms the moment they are born (or hatch). As they grow and develop, they acquire normal flora organisms on their external surfaces and within their digestive tracts. Both vertebrates and invertebrates establish mutualistic relationships with these microbes, frequently depending on them for survival. For example, virtually all strict herbivores rely on gut microbes to digest plant material. In turn, animals who depend on these herbivores for food depend indirectly on the herbivore gut microbes. Although most microbe–animal mutualistic relationships are limited to the digestive tract, there are a few important exceptions. For example, protective lactobacilli reside in the human vagina, where they maintain an environment that inhibits the growth of many pathogens. Other exceptions include some luminous bacteria that grow on fish, squid, and ctenophores and mutualistic bacteria that coat the eggs of fish and marine animals, protecting them from fungal infection in the open ocean.

● **GUT MICROBES.** In ruminant herbivores such as cattle and goats, digestion depends on bacteria and protozoa in the *rumen*, a special digestive chamber that functions as a fermentation tank. The microbes digest complex organic constituents of grass, most notably the structural polysaccharides cellulose and lignin, into carbohydrate monomers. They then ferment these compounds into fatty acid end products, primarily acetic, propionic, and butyric acids. These organic acids are absorbed through the walls of the rumen into the animal's bloodstream, where they are used for energy-generating metabolism.

The rumen cavity is well suited for its function and can be compared to a continuous culture chamber. It is anaerobic, the temperature and pH are constant, and the microbial population is large. Over 30 species of bacteria plus several types of protozoa reach concentrations of 10^{11} organisms per gram of rumen contents. The microbes multiply, digesting the animal's food and converting it to the sugars, amino acids, and other products needed for their own growth within the rumen. The animal literally cultures a crop of food in its rumen. The crop is harvested by transferring the microbes from the rumen to another stomach chamber, where the bacteria and protozoa are digested by the animal's enzymes, providing essential amino acids to the host. Not only do grasses and other plants have high concentrations of fiber that cannot be digested without the help of the ruminant's microbes, these plants are also poor sources of protein. Ruminants would need protein supplements were it not for microbial metabolism.

The normal flora microbes that reside in the gut provide at least two other benefits for virtually all mammals. They produce vitamins that are used by the host animal, and they compete with pathogens introduced into the area, often preventing the establishment of infectious disease.

Many invertebrates also exist on a diet of cellulose, even though they can't digest it any more easily than a cow can. Wood-boring termites, for example, depend on intestinal microbes to extract nutrients from the cellulose they eat. They survive on a wood diet only because of these microbes. Cellulose digestion by wood-boring shipworms is also performed by symbiotic bacteria, in this case, bacteria that live within a gland connected to the worm's digestive system. These cellulolytic bacteria extend the metabolic repertoire of their hosts even further by fixing nitrogen, thereby helping the shipworm meet its nitrogen requirements even though it grows exclusively on wood. Both complex physiological processes are performed by the same bacterium.

Blood leeches obtain protein and fat from the red blood cells of their prey. These nutrients, however, would remain locked in erythrocytes that would pass

7. The development of symbiotic light emission in the squid *Euprymna scolopes* was studied by exposing juvenile squids to several seawater suspensions. The graph to the right represents the results of the experiment. What conclusions can you draw from this graph?

8. What advantages do organisms that live in biofilms have over those that are free-floating? What are the disadvantages?

9. Without microbes, all multicellular life on earth would cease. Discuss three reasons why plants and animals would die.

10. A cattle rancher is concerned with the health of his livestock and decides to treat his cows with antibiotics to prevent them from getting sick. His friend the farmer thinks that this is a great idea and decides to spray his crops with a fungicide, a chemical that he hopes will kill a fungal parasite that might infect his plants. They come to you for your expert opinion. What is your advice?

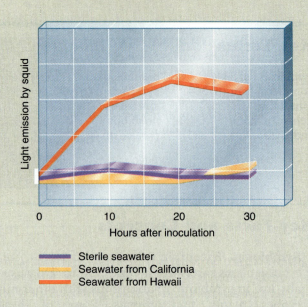

Sterile seawater
Seawater from California
Seawater from Hawaii

HEADLINES

BACTERIA AND INSECT REPRODUCTION

A team of biologists* recently discovered endosymbionts resembling bacteria present in the cytoplasm of several insects, including wasps, butterflies, mosquitoes, and houseflies. The strains harbored by these insects alter mitosis in their eggs and affect reproduction of their host. Some insect eggs contain bacteria that are responsible for parthenogenesis, the ability of unfertilized eggs to develop into female progeny. Endosymbionts in other insect eggs prevent the development of eggs that have been fertilized by sperm carrying the same endosymbiont. Analysis of the sequences encoding 16 S ribosomal RNA from endosymbionts from diverse insect taxa show that the bacteria are closely related and belong to the proteobacteria. Among the closest known relatives are members of the rickettsias.

- Would you characterize these relationships as commensal-

ism, mutualism, or parasitism? Defend your answer by showing what advantage or disadvantage is conferred on each participant in the interaction.

- Discuss the implications of the finding that endosymbionts from diverse insect taxa seem to be related. What characteristics do these bacteria have in common with rickettsias?

- Because they cannot be cultured outside of the host cell, would you consider these endosymbionts to be organelles of the host's cells rather than bacteria? How do they differ from mitochondria?

* R. Stouthamer, J. A. J. Breeuwer, R. F. Luck, and J. H. Werren, Molecular identification of microorganisms associated with parthenogenesis, *Nature*, 361: 66–68 (1993).

PART FOUR

Antimicrobial Methods

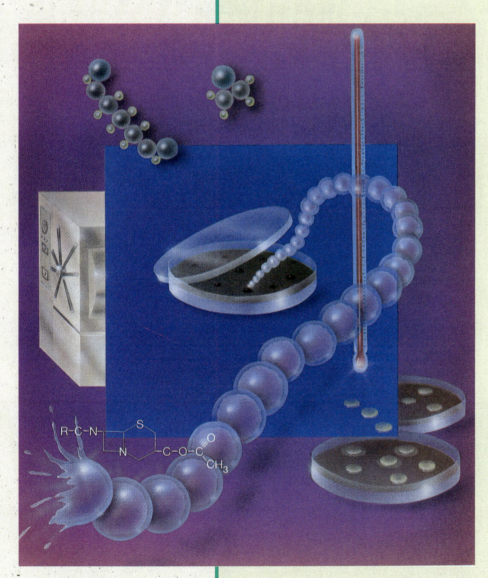

CHAPTER 15

Control of Microorganisms

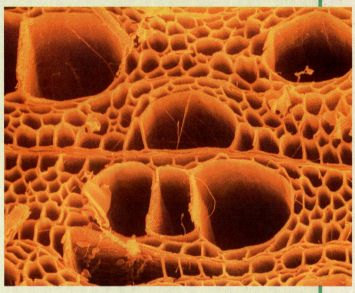

These cell walls, the constituents of wood, are about to be decomposed by a fungus (the filament growing in the open space). Such unwanted decomposition of wood and other valuable products is just one of the targets in our fight to control the growth of microorganisms.

The government closed its six-story Social Security Administration building and sent 1200 workers home. One of their colleagues had just died of an infectious disease acquired on the job, and several others were gravely ill. In September 1991, the Sacramento, California, building was declared unsafe, contaminated with a bacterium that threatened the health of anyone who entered. The menacing microbe was *Legionella pneumophila*, the agent of legionnaires' disease that was first discovered when it killed 29 people attending an American Legion convention in 1976. This same bacterium had created other "sick buildings" by establishing itself in cooling towers and spreading through air-conditioning systems. The government building remained closed until the microbe was destroyed and changes made to prevent its return.

But what changes had to be made to prevent the return of such outbreaks? In fact, how do we protect ourselves from detrimental microbes, from the pathogens that threaten our health to the environmental organisms that damage our products and foul our water supplies? Fortunately, physical or chemical control procedures can significantly reduce, and in some cases eliminate, the microorganisms in a designated environment or in an infected person. These protective measures are particularly important in hospitals, where there are many sources of pathogenic organisms. Microbial controls are also essential to the safe preparation of food, water, pharmaceutical agents, cos- metics, and other products for use in or on the human body (Fig. 15-1). In addition, virtually all organic-based materials, from wood to paints, will deteriorate unless protected from microbial destruction.

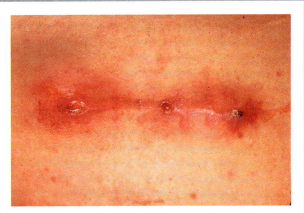

(a)

(b)

(c)

FIGURE 15-1 Some detrimental effects of microbes that can be prevented or retarded. (*a*) Infected surgical wound; (*b*) swimming pool overgrown with algae; (*c*) a potato supporting the growth of mold.

Antimicrobial Effects

The control of microorganisms often depends on establishing conditions that cannot be tolerated by microbes. Antimicrobial conditions are created by microbicidal or microbistatic agents (Fig. 15-2). **Microbicidal** (*cide* = kill) agents kill microorganisms and therefore have an irreversible and permanent effect. **Microbistatic** (*static* = standstill) agents inhibit microbial growth and multiplication, thereby preventing an increase in the number of microorganisms. Microbistatic agents do not kill or eliminate microorganisms. The microbe persists and can resume growth once the agent is removed. Therefore, microbicidal agents are generally preferred over microbistatic ones. **Germicidal** is another general term that refers to the destruction of microorganisms. Agents that specifically kill (or inhibit) bacteria, fungi, or viruses are referred to as *bactericidal* (or *bacteristatic*), *fungicidal* (or *fungistatic*), or *virucidal* (or *virustatic*). *Sporicidal* agents kill bacterial endospores, the most resistant forms of microbes.

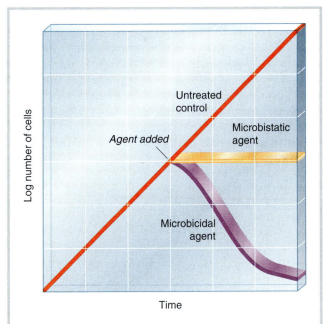

FIGURE 15-2 Effects of microbistatic and microbicidal agents. Organisms may be inhibited or killed by physical or chemical antimicrobial agents.

Antimicrobial agents perform one or more of the following processes:

- **Sterilization** eliminates all forms of life, including vegetative cells, spores, and viruses. Sterilization treatments also destroy potentially infectious nucleic acids such as viroids. During sterilization, either organisms are killed or they are physically removed from the objects or substances being treated. "Sterile" is an absolute term—there is no such thing as "almost sterile." An object or environment is not sterile as long as it contains even a single viable microbe.

  Although sterilization kills all microorganisms, it does not necessarily eliminate all harmful microbial effects. Endotoxins and other bacterial products often remain on objects and may consequently be introduced into the body. Washing and rinsing remove microorganisms and reduce endotoxin contamination.

- **Disinfection** eliminates the vegetative forms of most potentially hazardous and pathogenic organisms but does not ensure the elimination of all microbes. Bacterial spores, tubercle bacilli, and many viruses are particularly resistant to common *disinfectants* (disinfecting agents). Like sterilizing agents, disinfectants are used only on inanimate objects and never on body surfaces. Disinfection is generally employed if sterilization is either impossible or unnecessary. The purpose of disinfection is to minimize the risk of infection or product spoilage by reducing the number of microbes, especially pathogens, in the inanimate environment.

- **Sanitization** supplements disinfection with cleaning. This ensures the absence of dirt or organic debris as well as infectious microbes. Sanitization is typically employed for equipment used in food preparation and for reusable instruments in hospitals. Sanitization requirements are usually determined by public health agencies.

- **Antisepsis** is the inhibition or destruction of microorganisms on the surface of living tissue in an attempt to prevent infection. *Antiseptics*, the chemicals used in this process, must not harm the tissues on which they are used and therefore are milder (and possess less antimicrobial activity) than disinfectants.

These four means of eliminating potentially harmful microbes—sterilization, disinfection, sanitization, and antisepsis—are **decontamination** procedures.

Antimicrobial agents are also used to prevent microbial damage of various materials and to treat infectious disease in humans and other animal species.

- **Preservation** prevents the deterioration of products from microbial activities either by the addition of a chemical *preservative* or by establishing a physical environment inhospitable for microbial growth. For example, materials can be stored in cold or frozen states, oxygen can be removed during vacuum packing, or water can be withdrawn from the cells by lyophilization. ▶ (See Chap. 5, p. 122, for a discussion of factors affecting microbial growth.) These techniques retard spoilage and the growth of pathogens in foods, pharmaceutical preparations, and biological products for use in or on the human body. Preservatives are also essential in preventing destruction of "nutrient-rich" products that will be continuously exposed to the environment. Wood posts, for example, are treated to reduce rotting by soil microbes, paints contain preservatives so that they do not discolor, and jet fuels contain chemicals that prevent microbes from fouling fuel lines and from digesting the fuel during storage.

- **Chemotherapy** is treatment of disease by introducing chemicals into the human (or animal) body. Chemotherapeutic chemicals for treating infectious disease inhibit or kill microorganisms. Most of these agents are antibiotics, chemicals synthesized by microorganisms, usually a bacterium or fungus, which at very low concentrations (micrograms per milliliter) inhibit or kill other microbes. Many drugs similar to antibiotics are chemically synthesized in the laboratory. Since chemotherapeutic agents are used inside the body, they must exhibit selective toxicity for the target microorganisms, with little or no toxicity for human tissues. The unique characteristics of these antimicrobial agents are the subject of Chapter 16.

The most appropriate antimicrobial strategy is one that limits the microbial population without damaging the person, animal, plant, or object being treated. When possible, the best way to do this is to avoid microbial contamination. **Aseptic techniques** are precautions that help prevent contamination of culture materials, equipment, personnel, or the environment. Because aseptic procedures prevent the accidental introduction of microorganisms, they constitute a major element in combating disease and decomposition.

MYTH: Electric dishwashers that clean with very hot water and detergent sterilize dishes and utensils.

FACT: Dishwashing machines kill or remove most microbes (thereby sanitizing eating and cooking utensils), but many microbes remain afterwards. Nonetheless, commercial dishwashers are more effective at eliminating contaminating microorganisms than washing dishes by hand.

CONCEPT CHECK

- How do microbicidal agents differ from microbistatic agents?
- For what purposes is sterilization the appropriate control process? When is disinfection sufficient?
- How does preservation differ from disinfection? How does antisepsis differ from chemotherapy?

Factors Affecting Antimicrobial Activity

Some antimicrobial agents are microbicidal under one set of conditions and microbistatic under others. They may lose effectiveness as concentrations decrease or as conditions for use become suboptimal. Factors that influence the activity of antimicrobial agents are (1) the susceptibility of the microorganism, (2) the number of microorganisms, (3) the concentration or dose of the agent, (4) the length of exposure, and (5) environmental conditions.

● **MICROBIAL SUSCEPTIBILITY.** Microbes vary in their response to different antimicrobial agents (Table 15-1). Vegetative bacteria, fungi, and enveloped viruses are usually most susceptible to destruction. Vegetative cells of the mycobacteria that cause tuberculosis and leprosy, however, are covered by a waxy coating that protects them from many antimicrobial chemicals. In addition, the hepatitis B virus and several types of fungal spores are resistant to some disinfectants and are persistent problems in hospitals. *Bacillus* and *Clostridium* endospores are especially difficult to eliminate. The most difficult infectious materials to inactivate contain prions, such as the agent of the neurologically degenerative Creutzfeld-Jakob disease.

● **NUMBER OF MICROORGANISMS.** Microorganisms die when physical or chemical conditions irreversibly damage essential cell components. All organisms present, however, do not die simultaneously when a critical exposure is achieved. Death occurs logarithmically—a fixed percentage of the popula-

| TABLE 15-1 | DESCENDING ORDER OF RESISTANCE TO GERMICIDAL CHEMICALS |

Most resistant

Bacterial Endospores
Bacillus subtilis
Clostridium sporogenes

Mycobacteria
Mycobacterium tuberculosis var. *bovis*

Nonlipid or Small Viruses
Poliovirus
Coxsackie virus
Rhinovirus

Fungi
Trichophyton spp.
Cryptococcus spp.
Candida spp.

Vegetative Bacteria
Pseudomonas aeruginosa
Staphylococcus aureus
Salmonella choleraesuis

Enveloped or Medium-Sized Viruses
Herpes simplex virus
Cytomegalovirus
Respiratory syncytial virus
Hepatitis B virus
Human immunodeficiency virus (HIV)

Least resistant

tion will die during each minute of exposure to the agent. Antimicrobial effectiveness therefore depends on the initial concentration of the microbial population. Dust-covered objects, for example, are usually heavily contaminated (each gram of dust contains about 1 million organisms). The number of organisms in fecal matter and pus is even greater. Removing microbes by washing objects in a detergent and rinsing with water dramatically reduces microbial contamination and increases the likelihood that subsequent antimicrobial treatment will be adequate.

● **CONCENTRATION OR DOSE OF THE AGENT.** Diluting microbicidal chemicals usually weakens their antimicrobial activity. At lower concentrations they become microbistatic or lose antimicrobial activity completely. The antimicrobial effects of temperature or radiation also depend on the intensity of exposure. Low doses may inhibit growth, whereas high doses may sterilize. With a few important exceptions, the more concentrated or intense the exposure to any germicidal agent, the more likely it is that target organisms will be destroyed.

● **LENGTH OF EXPOSURE.** Because microbial death is a function of time, the longer microbes are exposed to potentially lethal conditions, the greater the number that will be killed (Fig. 15-3). For sterilizing agents, the exposure time must be long enough to ensure that the probability of even a single cell surviving is less than 1 in a million. This requires a knowledge of the initial population size and the death rate of the population. (See BEYOND THE BASICS: Dangerous Assumptions?) In contrast, microbistatic agents are effective only as long as they are present and must be used during the entire time inhibition is to be maintained.

● **ENVIRONMENTAL CONDITIONS.** Temperature, pH, and moisture affect the efficiency of most antimicrobial agents. In addition, some chemical agents are absorbed by blood, mucus, feces, tissue, and other organic materials that sharply reduce antimicrobial activity and therefore eliminate them as effective antiseptics. Objects can be rinsed prior to disinfection to prevent interference by organic debris. One reason antimicrobial procedures fail is the presence of biofilms ▶ (see Microbial Aggregates and Biofilms, Chap. 14, p. 337). Many types of microorganisms adhere to surfaces within biofilms, which are not removed without vigorous scrubbing.

This is extremely important when treating medical or dental instruments. Some antimicrobial agents are impeded by soaps and detergents that remain as thin films on skin or object surfaces. This difficulty can be minimized by thorough rinsing prior to disinfection or antisepsis.

CONCEPT CHECK

● How do population size, bacterial composition, and local environmental conditions influence the activity of antimicrobial agents?

Physical Agents for Controlling Microbes

The most common physical methods of control use moist heat, dry heat, radiation, or filtration.

■ Moist Heat

Although any organism can be killed by excessive heat, the lethal temperature depends on the heat re-

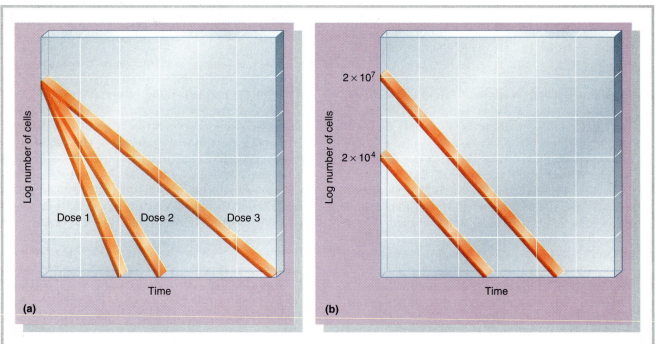

FIGURE 15-3 Cell death following exposure to a microbicidal agent. (*a*) The time required to kill identical concentrations of cells depends on the dose of the germicidal agent. In this diagram, dose 1 is the most concentrated and dose 3 is the least concentrated. (*b*) When the dose is constant, the time required to kill increases as the starting number of cells increases.

Dangerous Assumptions?

Most consumers assume that the canned and packaged foods and medicines we obtain from stores and the injections we receive from our doctors are "germ-free." We put our faith in commercial sterilization procedures performed to protect us from microbes that may infect our bodies or spoil products we wish to use. For a product to really be "germ-free," the control procedure to which it is subjected must eliminate all microbes present in and on the treated material. Is consumer confidence in these steriliza- tion procedures justified, or are you risk- ing your health whenever you use a "ster- ilized" product?

Sterilization of products intended for injection or implantation into humans is regulated by the federal government. All manufacturers must submit data that prove that their sterilization protocols are valid, that is, that they destroy all mi- crobes in and on the material. To do this, the manufacturer must determine the product's *bioburden,* the number of con- taminants that are typically present prior to sterilization. Knowing the bioburden allows the manufacturer to design a process that will be severe enough to kill all contaminants. In fact, the process is designed to kill millions more microbes than are expected.

To determine sterilization conditions, the manufacturer must also determine the rate of death of resistant spores ex- posed to the sterilizing agent. Such in- formation is obtained by inoculating a test material with a known number of spores and counting the survivors after different periods of exposure to the agent. From these data a graph can be generated that represents the dynamics of cell death. This provides information on the **D value** (decimal reduction time), the time required to kill 90 percent of the initial organisms, reducing the popula- tion to one-tenth its original size. Once the bioburden and D value have been de- termined, the minimum time required to kill the expected population at a defined,

fixed temperature is known. The steril- ization cycles designed for medical de- vices, for example, are increased beyond the minimum so that there is virtually no possibility of any organism surviving treatment.

Unique problems are presented by products derived from mammalian tis- sues or cells in culture. Tissues of hu- man origin are considered to be of high- est risk for transmission of viruses or prions. For example, in one study it was shown that recipients of human growth hormone isolated from cadaver pituitary glands contracted Creutzfeld-Jakob dis- ease. Most tissue-derived products can- not be exposed to heat or chemical sterilization treatments. Although high- resolution filters can eliminate most viruses from protein preparations, they cannot be used for blood or other cell- containing biological products and are not guaranteed to remove prions. A va- riety of treatments may be necessary to ensure the inactivation or removal of pri- ons. Because of the difficulties of re- moving prions, product suppliers con- centrate on deriving their materials from uncontaminated sources.

---

sistance of the organism and the amount of water in the environment. Moist heat, especially steam, ef- fectively kills cells by coagulating their proteins (critical enzymes, for example). In the absence of water, heat does not coagulate protein. Dry heat kills cells by oxidizing essential constituents, a process that requires much higher temperatures than can be achieved with moist heat. Three of our most com- mon antimicrobial processes rely on moist heat: pasteurization, boiling, and autoclaving (saturated steam under pressure). Of these, only autoclaving ensures sterilization.

● **PASTEURIZATION.** In order to prevent undesirable microbes from spoiling wine, Louis Pasteur gently heated the juice to kill contaminants before inocu- lating it with yeast to start fermentation. This process became known as **pasteurization**. Similar processes are used today by dairies, breweries, and other food industries to prevent spoilage and the transmission of disease. The selected combination of temperature and duration of heating is one that

kills the most heat-resistant pathogens commonly transmitted by that medium without damaging product quality. The target pathogen for pasteur- ization of milk, for example, is *Coxiella burnetti,* the rickettsia that causes Q fever ▶ (see Chap. 21, p. 545). It is destroyed by *low temperature holding* (LTH) *pasteurization,* exposure of the product for 30 minutes to a temperature of 62.8°C (144°F). More commonly, milk is treated by *flash* [high tempera- ture, short time (HTST)] *pasteurization*—heating for 15 seconds at 71.6°C (161°F) followed by rapid cool- ing. A newer process, *ultra high temperature* (UHT) *pasteurization,* is even faster, requiring only a 3-sec- ond exposure to superheated steam at 141°C (311°F). Although none of these procedures kills all microbes, they all eliminate the common pathogens. Any bacteria that are less heat-resistant than *Cox- iella burnetti,* such as the pathogens that cause the milkborne diseases tuberculosis and brucellosis, are readily destroyed by pasteurization.

Occasional outbreaks of milkborne disease still occur. These outbreaks are usually caused by failures in the pasteurization process, or they can be traced to the consumption of raw (unpasteurized) milk. A 1985 outbreak of *Listeria* infections in southern California, for example, was traced to contaminated cheese. *Listeria* infections are particularly dangerous in pregnant women, since the fetus is especially susceptible to the pathogen. (More than 100 women infected by the contaminated cheese lost their fetuses during the California outbreak.) An investigation by public health authorities showed that pasteurized and unpasteurized milk had been stored in adjacent containers. Small pinhole leaks in the container walls allowed microbes in the raw milk to contaminate the pasteurized product. Consumers of raw cow's milk are at high risk for milkborne diseases, particularly *Salmonella* and *Campylobacter* infections ▶ (see Chap. 22, pp. 588 and 589). Although certification requires that the total bacterial count in raw milk be determined, the absence of these pathogens from product purchased by the consumer is not guaranteed.

MYTH: Unpasteurized milk is more nutritious and health-promoting than pasteurized dairy products, because pasteurization destroys vitamins and nutrients.

FACT: This dangerous myth has taken a high toll in human lives. Pasteurized milk is no less nutritious than raw milk—pasteurization destroys only microbes, not nutrients. The chances of contracting a potentially fatal *Salmonella* infection is 200 times greater among people who drink raw milk than among those who drink pasteurized milk.[1] Hundreds of people who have died of milkborne diseases would be alive today had they chosen pasteurized instead of raw milk.

● **BOILING.** Heating increases the temperature of water until it reaches 100°C, the temperature at which water boils at normal atmospheric pressure at sea level. Further heating fails to raise the temperature of water because the additional heat energy escapes from the liquid in steam.

Although most vegetative cells are eliminated by

[1]The incidence of *Salmonella dublin* infection among people who drink certified raw milk (CRM) is 500 cases per million people in the United States compared with 3 cases per million among people who do not drink CRM. Of the patients who were hospitalized for *Salmonella* infections, 26% died. [From *Morbidity and Mortality Weekly Report*, 33:14 (April 13, 1984).]

10 minutes of boiling, some endospores, notably those that cause botulism, survive 5 hours of continuous boiling. This process is therefore considered a procedure for disinfection rather than for sterilization. Boiling is used when small instruments are to be disinfected or when there is no access to equipment needed for sterilization (for example, in emergency deliveries of babies at home). Surgical instruments, however, should always be sterilized and not merely disinfected.

In some cases, intermittent exposure to steam can be used in *fractional sterilization* (also called *tyndallization*, after the person who developed it). This process requires 3 days and is useful only for materials that can support microbial growth. On the first day, the material is steam-heated for 30 minutes, cooled, and incubated at 37°C. The high temperature and moist heat kill all vegetative cells and stimulate the germination of heat-resistant endospores. A similar treatment on the second day destroys the germinated cells and triggers the outgrowth of any remaining endospores. These bacteria die when the material is once again heated for 30 minutes on the third day. The most common uses of fractional sterilization are for media that cannot withstand the higher temperatures of the autoclave and that cannot easily be sterilized by other methods.

Although the higher temperature of boiling water kills more effectively than pasteurization, it is not routinely used in food industries because such high temperatures often damage the product.

● **AUTOCLAVING (USE OF STEAM UNDER PRESSURE).** The most common instrument for sterilizing heat-

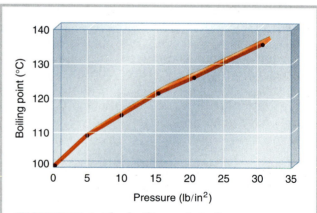

FIGURE 15-4 The boiling point of water as a function of pressure. Each point on the line indicates the pressure needed to change the boiling point of water to the temperature indicated by the same point.

Can You Can or Can't You Can?

In the early nineteenth century, the French military suffered major troop losses because of famine. To prevent these noncombat losses, French officials conducted a contest to develop a successful method for preserving foods. The winner, Nicholas Appert, heated food in sealed glass jars, thereby inventing the process of canning. Heating kills spoilage microbes, and sealing keeps out contaminants. Appert, however, had no knowledge of microbes or why his process worked. He believed that heat destroyed "ferments," nonliving factors in air that spoiled food.

Today, the canning method of food preservation is routinely used in industry and in the home. Two processes are generally employed, the boiling-water bath and the high-temperature, high-pressure method. Boiling easily eliminates the major spoilage bacteria, yeasts, and molds. Some foods contain only these contaminants and are sterilized after a few minutes at 100°C. Fruits and other high-acid foods with pH below 4.5 are too acidic for the survival of heat-resistant pathogens, notably *Clostridium botulinum*. Tomatoes, for example, are safe to eat after boiling (although mixing tomatoes with onions or peppers renders the combination susceptible to *C. botulinum* growth). Some nonacid foods can also be safely preserved by boiling and sealing. Such foods as jams, syrups, and sweetened condensed milk contain high sugar concentrations that inhibit the growth of most spoilage microbes and pathogenic bacteria, including endospore formers.

Most low-acid foods require sterilization and must be exposed to temperatures that exceed 100°C. The only practical canning methods that sterilize these foods without destroying them depend on pressure cookers and autoclaves. "Pressure canning" (actually, high-temperature canning) is essential for ensuring the safety of meats and vegetables. Most botulism outbreaks can be traced to nonacid products that have been boiled or otherwise processed at too low a temperature.

stable materials is the **autoclave.** This instrument sterilizes with saturated steam under pressure. The pressure increases the boiling point of water, thereby increasing the temperature to which water can be heated (Fig. 15-4). Pressure cookers, for example, build pressures within the closed vessel, so heated water vaporizes to steam at temperatures above 100°C (see THE EXPLORERS: Can You Can or Can't You Can?). The autoclave, like a pressure cooker, uses increased pressure to raise the temperature required to produce steam. It is these higher temperatures that destroy cells; the effects of pressure are not lethal. Standard autoclaves are usually operated at 15 lb/in² above atmospheric pressure, allowing the temperature to reach 121°C. Spores are killed after 15 minutes at these high temperatures. Other autoclaves use even higher temperatures (132 to 136°C) and pressures (27 to 33 lb/in²) but for shorter periods of time (3 to 10 minutes). This is an advantage in emergencies and for sterilizing rubber and other materials that may deteriorate with prolonged exposure to heat.

All air in the autoclave chamber is replaced by steam (Fig. 15-5a). When the chamber is filled with saturated steam, everything in it is killed. *Saturated steam* is water vapor that readily condenses on the surface of a cooler object and in doing so transfers tremendous amounts of heat energy to the object. *Saturated steam heats an object about 2500 times more efficiently than hot air at the same temperature.* In addition, dry articles (dressing, linens, etc.) acquire the moisture needed for sterilization at these temperatures. As steam condenses, it creates a partial vacuum that draws more steam into the autoclave. This process continues until the entire load has been penetrated by heat and thus brought to the temperature of the chamber. Once the sterilization temperature has been reached throughout the chamber, timing of the sterilization cycle begins. The temperature is maintained for the duration of the cycle needed to ensure sterilization.

Longer periods of time are needed for heat to completely penetrate large volumes and bulky items. Loads that completely fill the autoclave also present problems because steam may fail to penetrate all the material. Thus it may be better to run two smaller packages than a single large one.

After liquids have been autoclaved, pressure should be returned to normal by exhausting the steam *very slowly* from the chamber. Otherwise, the temperature of the liquid will exceed the boiling point at the reduced pressure. Slow exhausting prevents boiling over of loosely capped liquids and the explosion of liquids packed in airtight containers. After dry objects have been autoclaved, however, the steam can be rapidly exhausted. Most modern autoclaves are equipped with separate cycle controls for liquids and dry materials so that the exhaust

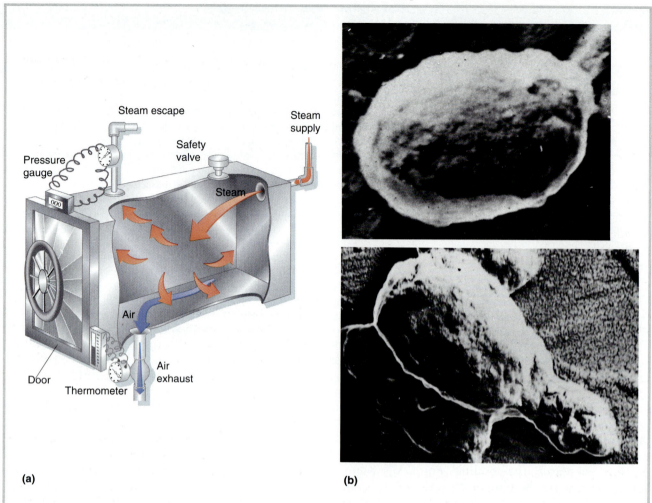

FIGURE 15-5 The autoclave. (*a*) Steam enters the top of the chamber, expelling cooler, heavier air out the exit at the bottom. The exit is sealed automatically when the chamber is filled with steam at the appropriate temperature. In some autoclaves, the air is removed with a vacuum pump before steam is introduced. These exhaust systems have a valve that automatically closes when pure steam begins to escape. (*b*) The effectiveness of autoclaving is revealed by these "before and after" electron micrographs of a bacterial endospore.

process can be tailored to the articles being autoclaved.

All items should be packaged to prevent their recontamination when they are removed from the autoclave. Paper, cotton, and cloth packaging become wet during autoclaving and are permeable to contaminants unless they are allowed to dry thoroughly before being removed from the autoclave. Most modern autoclaves create a vacuum at the end of the cycle to draw off residual moisture.

Materials not suitable for autoclaving include powders, heat-sensitive compounds, and water-insoluble substances (such as mineral oil) that steam cannot penetrate. Microbes suspended in oil are exposed only to dry heat during autoclaving.

The only infectious agents known to survive 4 hours of autoclaving are prions. Instruments contaminated with tissues that can harbor prions are sterilized by a combination of extensive autoclaving and treatment with 1 N sodium hydroxide.

Autoclaves are also used to decontaminate biological wastes prior to disposal. Most biotechnology companies, for example, autoclave their unwanted fermentation liquids. After heat treatment, the liquid can be safely dumped down drains destined for the sewers. Much of the infectious medical waste generated in hospitals is also inactivated by steam sterilization.

Dry Heat

Dry heat is generally used in three ways: flaming, incineration, and baking. A flame is commonly used to sterilize loops and needles so that microorganisms can be transferred without contamination. The mouths of test tubes and containers are also routinely flamed, although the heat is not sufficient to sterilize these surfaces. Scalpels and other sharp metal instruments are damaged by repeated flaming and are usually packaged and autoclaved. Incineration is useful for destroying heavily contaminated materials or tissues in the hospital and cultures of pathogenic microbes discarded in the laboratory. If the entire article is completely burned and there is no unburned material expelled in the exhaust, the elimination of pathogenic agents is ensured.

Baking in hot-air ovens requires prolonged exposure to temperatures between 150 and 180°C. The length of exposure depends on how readily heat can penetrate the material, since all parts to be sterilized must reach critical temperatures. Higher temperatures require shorter periods of exposure (Table 15-2). Temperatures below 150°C require too much time to be practical. Temperatures above 180°C often damage materials. Hot-air ovens are used for sterilizing materials such as glassware or metal instruments that can tolerate prolonged heat exposure. Articles made of rubber, plastic, or fabric may be destroyed by intense heat and should never be exposed to baking. Powders, oils, and waxes that are either destroyed or not effectively sterilized by the moist heat of the autoclave are also candidates for dry-heat sterilization. Water-based liquids, however,

TABLE 15-2 CONDITIONS FOR DRY HEAT STERILIZATION

Operating Temperature		Sterilization Time,
°C	°F	hours
121	250	>6
150–160	302–320	>3
160–170	320–338	2–3
170–180	338–356	1–2

can be heated only to 100°C and cannot be sterilized in hot-air ovens.

Radiation

Cosmic rays, gamma rays, X-rays, ultraviolet light, and visible light are all forms of radiation. When these rays strike an organism, energy may be absorbed by cellular constituents, causing cell damage or death. Radiation with the shortest wavelengths (wavelengths at or below those of ultraviolet light) has the greatest energy and is therefore the most lethal (Fig. 15-6). High amounts of such radiation are not common in the environment.

Ionizing radiation and ultraviolet radiation are two types used in microbial control. Sterilization with radiation does not rely on heat for killing and is often referred to as "cold sterilization." Radiation is commonly employed for sterilizing heat-sensitive materials such as disposable plastic products and materials that cannot withstand moisture.

Microwave radiation, which has been tested for germicidal potential, has much lower energy. Microwaves are antimicrobial only if the material reaches temperatures high enough to kill contami-

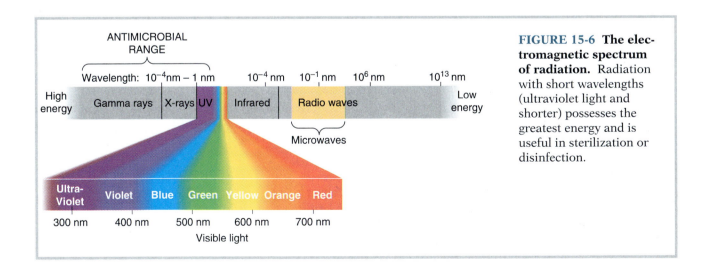

FIGURE 15-6 The electromagnetic spectrum of radiation. Radiation with short wavelengths (ultraviolet light and shorter) possesses the greatest energy and is useful in sterilization or disinfection.

nating organisms. Most microwave ovens are not designed as sterilization chambers and do not reliably destroy pathogenic organisms within foods.

● **IONIZING RADIATION.** Some rays have such high energy that they cause biologically active molecules to lose electrons. This ionizes the molecules so they can no longer perform critical cellular functions. High doses of such **ionizing radiation** kill every organism they strike, resulting in sterilization. Ionizing radiation is therefore a practical means for sterilizing many types of plastic products. Some materials, however, are damaged by radiation. For example, many plastics cannot be sterilized by radiation. Among the more recent applications of ra-

diation are the sterilization of vaccines, antibiotics, and other biological preparations. It is also used to reduce the number of contaminating insects and microbes in certain foods, thereby retarding spoilage. In the United States, the law requires that all irradiated foods be labeled.

MYTH: Irradiated food is dangerous because lingering radiation in the food makes the product "radioactive."

FACT: Irradiated food is actually safer than nonirradiated food because potential microbes are killed by the radiation, which does not make the food in

THE MICROBES

A Modern Disease for Modern Times

Fast food restaurants, one of modern society's symbols, have compounded an already serious infectious disease dilemma, that of widespread foodborne illness. In 1992, a major outbreak of a deadly illness erupted in four western American states when hundreds of people who had eaten hamburgers at a fast food chain suffered bloody diarrhea, after which many experienced serious kidney damage. At least four people died. The microbe responsible for the outbreak was a pathogenic strain of *Escherichia coli*, a species found in virtually every mammal's normal intestinal flora. This strain (called *E. coli* 0157:H7), unknown only a decade before, was introduced into ground beef during processing, when feces from slaughtered dairy cows was inadvertently mixed with the meat. Public health investigators discovered that 100,000 pounds of the contaminated meat was ground and formed into patties. The ground beef contained viable bacteria that could infect a person and cause illness unless the meat was thoroughly cooked. Compared to *Salmonella* and other foodborne pathogens, which

require ingestion of thousands of bacteria to produce symptoms, the presence of just a few *E. coli* 0157:H7 organisms can cause overt disease.

It is now estimated that 20,000 cases of *E. coli* 0157:H7 infection occur every year in the United States, and Canadian health authorities claim that the number may be 10 times that high. The deadly *E. coli* strain has also invaded home kitchens, where improperly cooked ground meats (including pork, poultry, and lamb as well as beef) lead to sporadic outbreaks of disease. Measures that once controlled such outbreaks by minimizing contamination and by setting cooking temperatures to ensure that dangerous microbes were killed are no longer adequate, according to Morris Potter of the Centers for Disease Control. "Something has changed, and these measures are no longer able to destroy [the bacteria]."*Until the problem can be identified and corrective measures instituted, eating medium rare ground meat will continue to be risky business.

Why have these outbreaks been associated with hamburger, but not steaks and other unground meats? Because microbes contaminate only the surface of meats, steaks, roasts, and other unground meats are virtually sterilized by cooking since their surfaces are next to the heat and achieve adequate temperatures. Grinding the meat, however, introduces the microbes at the surface into

the interior of the meat, essentially turning the entire mass of the meat into surface area. The bacteria in the interior are often protected from being killed by cooking because the center of the meat is cooler than the outside surfaces.

In light of the recent *E. coli* outbreaks, the Food and Drug Administration (FDA) recommends that hamburger (indeed all ground meats) be cooked until *the center* reaches 86.1°C (155°F) or until the interior is no longer pink and the juices run clear. These recommendations are especially important to follow in light of new discoveries on the prevalence of the pathogen. *E. coli* 0157:H7 is more common than *Shigella* (the prototype agent of bacterial dysentery) and frequently lives in the intestine of healthy cattle.

As microorganisms evolve, new strains and species will continue to create new health problems. Such changes in populations are inevitable, natural selection favoring the species best adapted for surviving and reproducing in constantly changing environments. As we change, so will the microbes that interact with us, and so must our procedures for controlling them. Perhaps optimism will soon return without over-confidence tagging along.

*Daniel P. Puzo, *Los Angeles Times,* March 4, 1993, p. H 28.

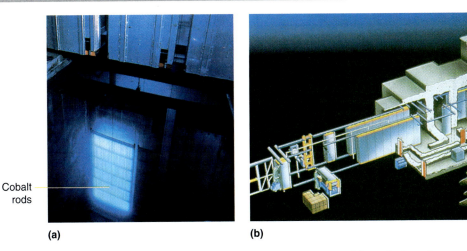

FIGURE 15-7 **Ionizing radiation.** (*a*) Gamma rays are generated by cobalt rods, which are immersed in water within a protected region of the radiation facility. (*b*) Materials to be sterilized are loaded onto a monorail conveyor system and automatically transported past the radiation source.

the least bit radioactive. Had the ground beef in the 1992 outbreaks of *E. coli* foodborne illness in the western United States been irradiated, the outbreak would not have occurred. (See THE MICROBES: A Modern Disease for Modern Times.)

Two types of ionizing radiation are commonly used for sterilization: *gamma rays,* which are emitted by radioactive elements (cobalt-60, for example), and *high-energy electrons,* which are produced by electron accelerators. The penetrating power of gamma radiation makes it useful for sterilizing large loads or bulk items. For example, cases of petri dishes, packaged and wrapped for shipping, can be sterilized in one large batch. Strict safety precautions are required to shield people from the radiation, since it is equally damaging to human cells (Fig. 15-7). High-energy electrons penetrate less efficiently and are less hazardous than gamma rays. They are consequently most useful for sterilizing smaller, individually wrapped articles.

● **ULTRAVIOLET RADIATION** Cellular DNA absorbs the energy of radiation at wavelengths between 250 and 260 nm and forms aberrant chemical bonds between adjacent thymine nucleotide bases. This is the basis of the antimicrobial effects of **ultraviolet radiation.** These UV-induced *thymine dimers* distort the DNA strands, impair chromosome replication prior to cell division, and interfere with the transcription and expression of genes. Thymine dimers

are lethal when they occur in genes for essential functions or when DNA replication is blocked. ▶ (See Radiation, Chap. 8, p. 203.)

The germicidal effects of ultraviolet radiation are dose-dependent—longer exposures and greater doses (higher wattage or less distance from the source of radiation) increase the number of vegetative cells killed. Some bacterial endospores are protected from UV radiation by proteins formed during sporulation. These spore proteins bind to DNA and alter its configuration, making it difficult for thymine dimers to form. Ultraviolet light is therefore a poor sterilizing agent. Another major limitation of ultraviolet light is its lack of penetrating power. Although UV rays pass readily through dust-free air and clear water, they fail to penetrate ordinary glass or many plastics, turbid solutions, thin films of grease, or milk. Therefore, their germicidal activity is limited to exposed surfaces or to liquids that are circulated through very thin tubes made of clear quartz (which UV light can penetrate). Effectiveness also decreases with increased distance from the radiation source. In addition, UV light can severely damage the retina of someone who looks directly at the bulb, and prolonged skin exposure contributes to the development of skin cancer. Given these limitations, the major use for ultraviolet light has been as a disinfecting agent for air and surfaces in surgical rooms and laboratory safety cabinets. These facilities are exposed to UV light prior to their use, and, of course, all people are restricted from the area during radiation unless they are wearing protective clothing and face shields.

drinking, swimming pools, whirlpool baths, and hot tubs (although bromine is often used in hot tubs). It reacts with the water to produce hypochlorous acid, the bactericidal agent that effectively eliminates such common waterborne pathogens as *E. coli*, *Salmonella typhi*, and *Entamoeba histolytica*. The cells are killed when chlorine oxidizes and thus inactivates critical enzymes. Continued disinfection, however, requires maintaining adequate chlorine levels. High temperatures and rapid circulation increase the rate of chlorine evaporation, quickly reducing concentrations to ineffective levels. These circumstances are associated with outbreaks of skin infections caused by *Pseudomonas aeruginosa* in swimming pools, hot tubs, and whirlpools. Automatic chlorinators monitor and maintain free chlorine levels at adequate concentrations. Some municipalities use chlorine to treat wastewater and sewage, a controversial procedure that requires much higher concentrations because chlorine is inactivated by organic materials. ▶ (See Sewage Disposal, Chap. 26, p. 724.)

Hypochlorites are chlorine-containing inorganic compounds such as bleach. Hypochlorites also react with water to produce hypochlorous acid. They are commonly used as disinfectants, especially by the food and dairy industry, to decontaminate equipment. These compounds, however, may damage fabrics and corrode metals. If the material cannot withstand treatment, an alternative method must be employed. Strong hypochlorite solutions are effective against hepatitis B virus and the HIV virus. In addition, hypochlorites are the only chemicals that can inactivate prions, the agents that cause Creutzfeld-Jakob disease. The CDC recommends using household bleach (1:100 dilution) to disinfect surfaces contaminated with blood. They also advise intravenous drug abusers who share needles and syringes to decontaminate them using full strength bleach. Hypochlorites are not used as antiseptics because they irritate skin and tissues. However, several organic chlorine compounds, such as chloramines, are nonirritants and are used for treating skin and wounds, where they slowly release hypochlorous acid.

Another long-employed antimicrobial agent is *iodine*. In dilute solutions of water or alcohol (tincture of iodine), iodine rapidly inactivates microbes by irreversibly combining with proteins. Iodine is useful for disinfecting thermometers, for reducing microbes on skin sites selected for surgery or needle puncture, and for treating cuts and wounds. Unfortunately, iodine stains skin and fabric and stimulates nerve endings in the skin. (Who can forget the sensation of iodine on a scraped knee?) Pain can be minimized by using **iodophors,** complexes of io-

dine and surfactants that slowly release the iodine as a potent nonirritating antiseptic. Povidone-iodine, the most widely used iodophor, is commonly employed as a presurgical hand wash and to prepare skin sites for surgery. Other iodophor preparations are designed for disinfecting small medical implements.

● **CHLORHEXIDINE.** The popular antiseptic **chlorhexidine** is active against both gram-positive and gram-negative bacteria and some fungi but is useless against mycobacteria, bacterial spores, and viruses. It kills cells by interfering with plasma membrane permeability and essential membrane-associated enzymes. It is also active against enveloped viruses. Chlorhexidine is nontoxic to humans and retains antimicrobial activity for several hours, even in the presence of soaps and organic matter. Unlike hexachlorophene, it is not absorbed through the skin, so there is no danger to the fetus of a mother using chlorhexidine. Chlorhexidine has been used extensively as a component of antiseptic lotions for surgical scrubs, in cleaning the skin and mucous membranes, and in decontaminating wounds.

● **HEAVY METALS.** Ions of heavy metals readily bind with and inactivate proteins, even in very low concentrations (Fig. 15-12). Their effect is not selective for microbes, so, as antiseptics, they must be used in dilute concentrations and only topically. The most common of these antiseptics are the mercurials (compounds containing the heavy metal mercury) mercurochrome and merthiolate and the silver-con-

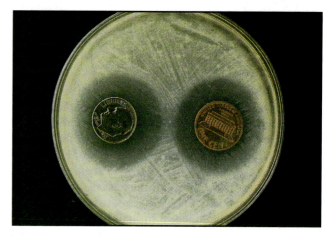

FIGURE 15-12 Toxic activity of heavy metals. The clear zones surrounding the dime and the penny is due to cell death from the few silver and copper atoms that dissolve and diffuse into the medium (an effect called oligodynamic action). Heavy metals kill human cells as effectively as bacterial cells and are therefore limited to topical use.

TABLE 15-5 COMMON USES FOR CHEMICAL ANTISEPTICS AND DISINFECTANTS

Use	Agent(s) Employed
Handscrub	
Routine washing	Soap, detergent, or iodophor; water; and "elbow grease"
Preoperative scrub; high-risk situations	Iodophor, chlorhexidine, hexachlorophene
Skin Preparation	
Routine injections	Alcohol (often used with an iodophor)
Surgical	Iodophor, chlorhexidine, hexachlorophene
Instrument Decontamination*	
Lensed instruments for internal examination of the body	Ethylene oxide, glutaraldehyde, vaporized hydrogen peroxide
Thermometers	Alcohol and iodine
Other small medical instruments	Alcohols, quaternary ammoniums, iodophors, phenolics, chlorinated solutions, formaldehyde, or glutaraldehyde soak
Environmental Control	
Linens and clothing	Ethylene oxide or chlorines
Floors, walls, and other surfaces	Phenolics, chlorines, iodophors, quaternary ammoniums
Utensils	Chlorinated solutions, quaternary ammoniums
Water	Chlorinated solutions

* Any instrument that will be introduced into the tissues or bloodstream *must* be sterilized. The preferred method for heat-stable items is autoclaving.

---

taining agents silver nitrate and silver chloride. Silver nitrate solutions are used to irrigate infected urinary bladders and to prevent eye infections. Silver nitrate drops (or antibiotics) are added to the eyes of all newborn children in the United States to prevent blinding *Neisseria gonorrhoeae* and chlamydia infections acquired during the birth process.

● **OZONE.** Ozone (O_3), a form of oxygen, is a gas used in water and wastewater treatment. It effectively kills bacteria, viruses, fungi, and protozoa. Ozone gas, however, is unstable in solution and decomposes quickly. Continuous ozone generation is therefore necessary during treatment.

The common uses of chemical antiseptics and disinfectants are summarized in Table 15-5.

■ Chemical Preservatives

Chemical preservatives are added to foods, pharmaceuticals, cosmetics, and other products to prevent the growth of microbes in the product while it is stored on the shelf or after it has been opened for use. Even sterilized products contain preservatives if they are to be used many times after opening. Many chemical disinfectants and antiseptics are also useful as preservatives. Some chemicals, how-

ever, are designed solely for use as preservatives. Some of the common chemical preservatives used in consumer products are indicated in Table 15-6.

● **FOOD PRESERVATION.** Food preservation is an ancient art. With no scientific data, our ancestors effectively retarded food decomposition with salts,

TABLE 15-6 PRESERVATIVES IN COMMON PRODUCTS

Product	Preservative Employed
Vaccines	Phenols; formalin (formaldehyde diluted in water)
Injectable drugs (multiple-use containers)	Phenols
Ophthalmic solutions	Chlorhexidine; alcohol
Cosmetics	Parabens
Foods	Benzoic acid/benzoates; sorbic acid/sorbates; parabens; sulfites

sugars, and acids. ▶(See THE MICROBES: Are You Worth Your Salt?, Chap. 5, p. 113.) Today chemical preservatives are used to prolong the shelf life of foods that would otherwise provide a hospitable environment for microbial growth. Such foods generally contain water, are at or near neutral pH, and are stored at room temperature. Chemical preservatives used in foods must be safe to consume because they are eaten with the food.

Most common food preservatives are acids or derivatives of acids. They reduce the microbes' intracellular pH, interfere with enzyme activity, and alter membrane permeability. Acetic acid, sorbic acid, benzoic acid, and citric acid are added to a wide variety of products. Other kinds of preservatives include sulfur dioxide (added to wines), sulfites (used to treat dehydrated fruits and vegetables), and nitrates and nitrites (used in meat processing). ▶(See discussion of microorganisms and food spoilage, Chap. 27, p. 743.)

● **PRESERVATION OF PHARMACEUTICALS, COSMETICS, AND OTHER INDUSTRIAL PRODUCTS.** Modern technology has created new pharmaceutical products and new problems. For example, contact lens cleaning solutions must remain free of microorganisms for the duration of their use or they will serve as reservoirs of organisms that cause serious eye infections. Medicines and vaccines have to be maintained in a sterile condition after they are opened. The failure to preserve pharmaceuticals appropriately threatens the user. Contaminated oral medications may cause gastrointestinal disease; contaminated lotions or creams may result in skin infections; contaminated ophthalmic solutions may lead to blindness.

Cosmetics also contain preservatives to prevent possible infection of the user. The preservative also retards deterioration of the cosmetic. Other industries that depend on preservatives include those manufacturing natural glues and adhesives; wool, cotton, and leather; paints and inks; resins and polishes; and wood. The wood industry is the largest consumer of industrial chemical preservatives. Creosote, the primary wood preservative, is used on the bottom of fenceposts so they won't rot. Creosote effectively protects wood from decay, but it contains carcinogens and must be used correctly to prevent endangering the handler.

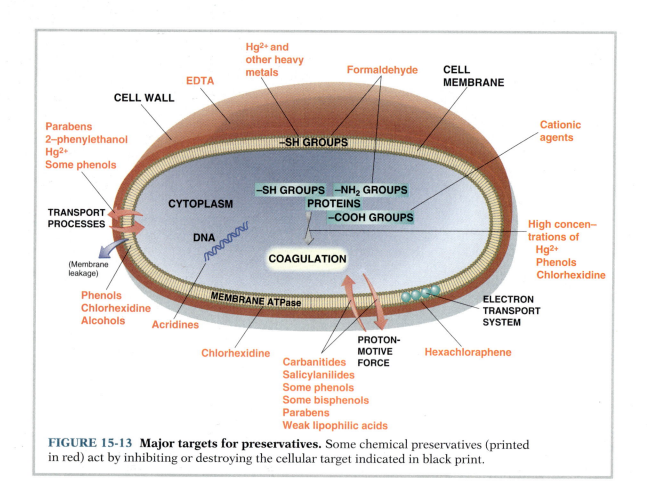

FIGURE 15-13 Major targets for preservatives. Some chemical preservatives (printed in red) act by inhibiting or destroying the cellular target indicated in black print.

A summary of the major targets of some preservatives is provided in Figure 15-13.

Evaluation Tests for Sterility

Although autoclaving, baking, ionizing radiation, and some chemical treatments theoretically kill all life forms, they fail to sterilize when improperly used. Undetected failures can have serious, even fatal, consequences (as when, for example, a contaminated solution is injected directly into a person). All sterilization procedures must be carefully and consistently monitored to detect failures and ensure sterility.

Several approaches can be adopted to evaluate the sterility of treated items:

- *Recording devices* on sterilizing equipment are used to measure operating conditions—temperature, moisture, pressure, gas content, and length of exposure to the sterilizing condition. The readings indicate any deviations from standard conditions that may have interrupted sterilization. Even ideal readings, however, do not guarantee that the entire load has been subjected to identical sterilizing conditions.
- *Random sampling* can provide an indicator of sterilization. Since it is not possible to monitor every article individually for contamination, items can be randomly selected for evaluation. This approach, however, depends on accurately predicting which media and culture conditions will most likely detect any pathogens present. It also requires a statistical analysis to determine how many items should be sampled to provide confidence in the results.
- *Chemical indicators* can be placed throughout the load to provide evidence of local sterilizing conditions. Tapes and strips impregnated with sensitive chemicals change color when the conditions

necessary for sterilization are achieved. Some indicators are sensitive to temperature, others to steam, gas, or radiation (Fig. 15-14).

- *Biological indicators* provide the most accurate test to determine whether the load is sterile. The most commonly used **biological indicators** are *spore strips*, pieces of filter paper impregnated with highly resistant bacterial spores. They are placed in areas where sterilizing conditions are most difficult to achieve—the lower front of the autoclave where air pockets may reside, the coolest areas in a hot air oven, or the interior of bulky items. When the sterilization cycle is completed, the strips are aseptically removed, placed in broth, and incubated. The absence of growth indicates that sterilization was achieved. The choice of bacterial spores depends on the sterilization process being monitored. The most resistant organism is selected because its destruction indicates that more sensitive microbes have also

FIGURE 15-14 Indicators of sterility. Chemical indicators can be used to label each item. The indicator changes color at sterilization temperatures.

bicidal (p. 349)	chemotherapy (p. 350)	phenolics (p. 365)
obistatic (p. 349)	aseptic techniques (p. 350)	quaternary ammonium compounds (p. 365)
micidal (p. 349)	D value (p. 353)	iodophors (p. 366)
sterilization (p. 350)	pasteurization (p. 353)	chlorhexidine (p. 366)
disinfection (p. 350)	autoclave (p. 355)	biological indicators (p. 369)
sanitization (p. 350)	ionizing radiation (p. 358)	phenol coefficient (PC) (p. 370)
antisepsis (p. 350)	ultraviolet radiation (p. 359)	use-dilution method (p. 370)
decontamination (p. 350)	filtration (p. 360)	in-use tests (p. 370)
preservation (p. 350)	ethylene oxide (p. 363)	

Key Facts and Concepts

- **Antimicrobial agents.** Antimicrobial agents are employed to destroy potentially harmful microbes or inhibit their growth. Microbicidal agents kill microbes by causing irreparable damage to essential components. Microbistatic agents only inhibit the growth of microorganisms and do so only while in contact with the microbe.

- **Decontamination.** Many physical and chemical agents are used to eliminate microbial contaminants from inanimate articles. Some chemicals are nontoxic enough for use as antiseptics on the surface of the human body.

- **Selecting the agent.** The selection of an agent depends on the ability of the object to withstand the treatment. Success of the treatment depends on length of exposure and concentration of the agent as well as on the nature and number of the contaminating microbes. A greater number of microbes requires more exposure time or increased doses of the antimicrobial agent.

- **Sterilization.** Sterilization, the complete elimination of all life, is achieved by moist heat (autoclaving), dry heat, radiation, filtration, and four germicidal chemicals—ethylene oxide, hydrogen peroxide, glutaraldehyde, and formaldehyde. The success of sterilization procedures may be evaluated with chemical or biological indicators. Chem-

ical indicators determine if the conditions in the apparatus were adequate for sterilization. Biological indicators determine whether the sterilization process has actually killed an especially resistant test organism, usually bacterial endospores.

- **Disinfection.** Disinfection does not destroy all microbes but eliminates potentially harmful contaminants. Boiling, pasteurization, and most antimicrobial chemicals are disinfectants. Bacterial endospores are particularly resistant to these agents.

- **Antisepsis.** Low-level disinfectants can be used as antiseptics on body surfaces. These chemicals include alcohols, quaternary ammonium compounds, iodine and iodophors, mercurials, silver-containing compounds, and chlorhexidine. They are most effective if the microbial populations on the body surface are first reduced by mechanical methods such as washing. This preparatory step also removes organic debris that interferes with the antimicrobial activity of many disinfectants.

- **Evaluating antimicrobial activity.** Disinfectants are evaluated in the laboratory by determining the highest dilution effective against a test organism (use-dilution method). In-use tests evaluate both disinfectants and antiseptics under actual conditions of use.

Review and Problem-Solving Questions

1. How do the concentration or dose of an antimicrobial agent and the duration of treatment influence the agent's activity?
2. Why is it advisable to wash highly contaminated objects with soap and water before treating them with antimicrobial agents?

3. Distinguish between the following:
   (a) Sterilization and pasteurization
   (b) Antisepsis and disinfection
   (c) Use-dilution and in-use tests
   (d) Chemical indicator and biological indicator
4. What are the advantages and limitations of the fol-

lowing processes for sterilization? (a) Boiling; (b) autoclaving; (c) irradiation; (d) dry heat; (e) ethylene oxide; (f) filtration.

5. How would you test an autoclave's performance?
6. Which chemical agent would you use for each of the following?
   (a) Prepare skin for an injection.
   (b) Prepare skin for abdominal surgery.
   (c) Clean a thermometer for the next patient.
   (d) Clean an electrode used in brain scans prior to reuse on another patient.
   (e) Wash the floors in the hospital emergency room.
   (f) Decontaminate the lab counter after a blood spill.
7. You are performing a plate count of commercial preparations of pasteurized and raw (unpasteurized) milk. To do this, you dilute your samples in water and plate aliquots into plate count agar (PCA). For each of the following items used in your project, select the most appropriate sterilization method and justify your choice. (a) Plastic petri dishes; (b) plate count agar; (c) glass pipettes; (d) water for dilutions; (e) inoculating loop; (f) contaminated petri dishes (after the plate count).
8. Why not autoclave dry wrapped surgical instruments in the same load as liquid media?
9. The president of the local drug company has hired you to inspect its facilities because shipments of its drug product have been found to be contaminated with bacteria and returned by its customers. You find that after manufacture the drug is filtered and a technician pipettes it into bottles that have been autoclaved.
   (a) On the basis of this information, what are the possible sources of contamination?
   (b) What further tests would you do to determine the actual nature of the problem?
   (c) What recommendations would you make to the president to reduce contamination?
10. Design an experiment to test the effectiveness of the preservative in solutions used for rinsing contact lenses.
11. A cosmetic company routinely performs bioburden analyses of its products to determine the level of microbial contamination. (Cosmetics do not have to be sterile.) Recently the company hired a new microbiologist. Since this person started working, no microbes have been detected in any product assayed by bioburden analyses. What are some possible explanations for these findings? How would you test your hypotheses?
12. Why is much more time needed to sterilize materials at 121°C with dry heat than with moist heat?
13. (a) Examine the labels of three or four pharmaceuticals and foods in your home, and identify any preservatives present in each product.
   (b) How are unpreserved pharmaceutical and food products protected against destruction by microorganisms?
14. You are given two bottles. One bottle contains a bactericidal agent, and the other contains a bacteristatic

agent. How would you determine which chemical is in each bottle?
15. Cylinders onto which *B. subtilis* endospores had been dried were soaked in various dilutions of a new disinfectant. Ten cylinders were tested at each dilution. After 10 minutes, the cylinders were removed, rinsed, and placed into tubes of trypticase soy broth. The results are tabulated as follows.

Dilution of Disinfectant	Number of Tubes Showing Growth
1:10	0
1:100	1
1:1000	3
1:10,000	10

   (a) Interpret the results of the test and indicate your recommendations for use of this disinfectant.
   (b) How would you determine if this chemical could also be used as an antiseptic? as a food preservative?
16. Claims that garlic has antimicrobial activity are common. Your entrepreneurial friend has decided to manufacture a product consisting primarily of diluted garlic juice and market it as a surface disinfectant and wants you to become his partner.
   (a) Based on the following table, which represents inoculations with *S. aureus*, what is the phenol coefficient of this product? ($-$ = no growth; $+$ = growth.)

Length of Exposure	Dilution of Phenol			
	1:70	1:80	1:90	1:100
5 minutes	$-$	$+$ $+$	$+$	
10 minutes	$-$	$-$ $+$	$+$	

Length of Exposure	Dilution of Garlic Juice			
	1:100	1:200	1:300	1:400
5 minutes	$-$	$-$ $-$	$+$	
10 minutes	$-$	$-$ $-$	$-$	

   (b) Design an experiment to help you determine whether this product will work as a disinfectant for the work surfaces in the microbiology labs at your school.
17. For each of the following agents, indicate situations in which it would be effective and situations in which it would be ineffective. (a) Surfactant; (b) alcohol; (c) iodine; (d) chlorhexidine.
18. In 1993, a well-publicized study reported that hot-air-blowing hand dryers were less sanitary than paper towels, which left the hands 40 percent more microbe-free. Having noticed that the study was funded by the paper towel industry, you wonder just how reliable these results are. Design an experiment to test the reliability of these findings, making sure that adequate controls are established.

HIV FOUND "HIDING" IN DENTAL EQUIPMENT

The danger of dentist-mediated transmission of AIDS from patient to patient came into sharper focus as David Lewis and coworkers* examined the most likely routes of transmission. Two common pieces of dental equipment, the drill and the angle (an instrument used in cleaning and polishing teeth), were used on HIV-infected people and then tested for the presence of the virus that causes AIDS. In all cases, these hand-held dental instruments were found to harbor the human immunodeficiency virus. The virus was still detected after the pieces were soaked in a chemical germicide, a standard procedure in dentistry. Apparently, the virus is able to avoid destruction by "hiding" in inaccessible joints or crevices or within the greases and oils used to lubricate the equipment. On reuse, the viruses may be released in a spray of materials. Although the risk to the patient is believed to be very low, the Food and Drug Administration wants dentists to change the way they recycle these instruments.

- What methods would you recommend dentists use to eliminate HIV from their instruments or to reduce the risk of infection for their patients?
- Up until 1988, the hand pieces of these instruments could not withstand high temperatures. What should dentists do with these older instruments?
- Many dentists and dental hygienists wear masks and gloves to protect themselves (and their patients) while performing treatments. How do these barriers work?
- A company has designed a new instrument with a configuration that they believe can be cold-sterilized. If you were head of the FDA, what kinds of tests would they have to perform before you would license their device?

* David L. Lewis, M. Arens, S. S. Appleton, K. Nakashima, J. R. Yu, R. K. Boe, J. B. Patrick, D. T. Watanabe, and M. Suzuki, Cross-contamination potential with dental equipment, *Lancet,* 340: 1252–1254 (1992).

CHAPTER 16

Antibiotics and Chemotherapy

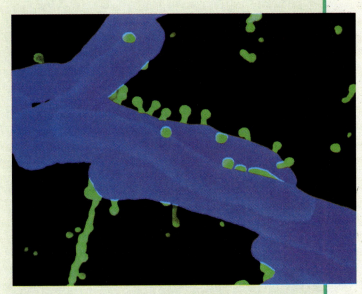

Introducing genes into tuberculosis bacteria using bacteriophage helps us select effective antimicrobial agents to treat patients suffering from drug-resistant disease. The introduced genes make the bacteria glow unless inhibited by an antibiotic. This is just one of many modern strategies for improving the effectiveness of antimicrobial treatment of the disease.

I n 1969, the U.S. Surgeon General announced to a large congressional audience that it was time to "close the book on infectious disease." Our modern weapons against these diseases were about to win the war—antisepsis, vaccination, hygiene, and antibiotics were proclaimed the imminent victors. Even our old adversary tuberculosis was steadily decreasing and appeared to be on its way to being controlled before the turn of the century. The proclamation turned out to be tragically premature. Today not only have infectious diseases persevered in the face of our modern arsenal, but as a group they remain the leading cause of death in the world—not cardiovascular disease, not cancer, not even accidents, but infectious diseases. The emergence of dangerous new afflictions and the resurgence of old adversaries such as tuberculosis prompted one epidemiologist to proclaim, "The microbes are back, with a vengeance."

One dangerous aspect of this microbial resurgence resides in the failure of traditional drugs to successfully eradicate pathogens in infected patients. Microbes are rapidly developing resistance to our most reliable antibiotics, drugs whose previous effectiveness helped create the false sense of security that still exists in many quarters of industrialized society. People speak of the current "antibiotic era of medicine" as a time when either there is a drug to cure any disease or there soon will be. Unfortunately, this state of overconfidence is undermining our efforts against the pathogens. It perpetuates carelessness in hygiene and aseptic technique. People are more likely to take a risk if they believe that a simple dose of antibiotics will erase the consequences of any mistakes.

The news is not all bad, however. Antibiotics are still potent weapons against many pathogens. Millions of people are alive today because of the effectiveness of antibiotics and antimicrobial chemotherapeutic agents, and the outlook for developing even more effective agents for treating infectious disease is good. The "shot in the dark" approach in searching for new drugs is being replaced with a more targeted approach, using the computer to design chemical compounds based on our knowledge of a particular pathogen and how antimicrobial agents work.

The battle between pathogens and people will very likely continue to seesaw, with no clear victor. But the outlook for us winning the major battles is very good, as long as we diligently monitor new microbial countermeasures against our chemical weapons and respond accordingly. This chapter is about those chemical weapons, the countermeasures the microbes take to neutralize their curative effects, and the ways we might respond to fight back.

The Early Days of Disease Treatment

The search for agents to cure infectious disease began long before people were aware of the existence of microbes. These early attempts used natural substances, usually native plants or their extracts, and many of these herbal remedies proved successful. Eating the bark of the cinchona tree, for example, prevented and frequently cured malaria. Only in modern times was quinine shown to be the active ingredient in the bark. Similarly, emetine, a treatment for amebic dysentery, was originally derived from ipecacuanha root. Ancient Egyptians discovered that eating bread overgrown with blue-green mold helped persons afflicted with certain diseases recover. Until 50 years ago, however, such successes were all too rare, and millions of lives were lost to diseases that today are treated with **chemotherapeutic agents.** These are chemicals administered to people (or animals) for treating disease.

The scientific search for effective chemotherapeutic agents began about 100 years ago in the laboratory of Paul Ehrlich (Fig. 16-1). Often called the founder of modern chemotherapy, Ehrlich conducted exhaustive experimental trials to find what he called a "magic bullet," a compound that would kill the pathogen without harming the patient. His efforts focused on the antimicrobial activity of metallic compounds and chemical dyes. He believed that metallic compounds, especially mercury, arsenic, and antimony, would prove useful because forms of these metals had been used to treat disease since the Middle Ages. Often however, the metal compound caused severe side effects or even killed the person receiving it. Ehrlich hoped to find derivatives of these compounds that were less toxic but retained their therapeutic effectiveness. The first "magic bullet" proved to be compound 606, an arsenical used to treat syphilis. Ehrlich's most important contribution to chemotherapy, however, was an extension of his experiences staining infected tissues with chemical dyes. He observed that certain stains selectively reacted with microbes and not with the host tissue. Ehrlich hypothesized that some of these dyes might also be toxic for microbes. This ability to differentiate between the host and the pathogen would make these dyes ideal chemotherapeutic agents. Following up on Ehrlich's idea, investigators screened thousands of chemical dyes, ultimately leading to the 1932 discovery of the sulfa drugs.

The accidental discovery of penicillin in 1929 ushered in the antibiotic era of chemotherapy (see THE EXPLORERS: Miracle Drugs from Moldy Cantaloupes, Sewage, and Dirt). **Antibiotics** are chemicals produced by microorganisms that, in very low concentrations, selectively kill or inhibit the growth of other microbes. Antibiotics that are nontoxic to human cells can usually be safely introduced into an infected person to combat pathogens. The introduc-

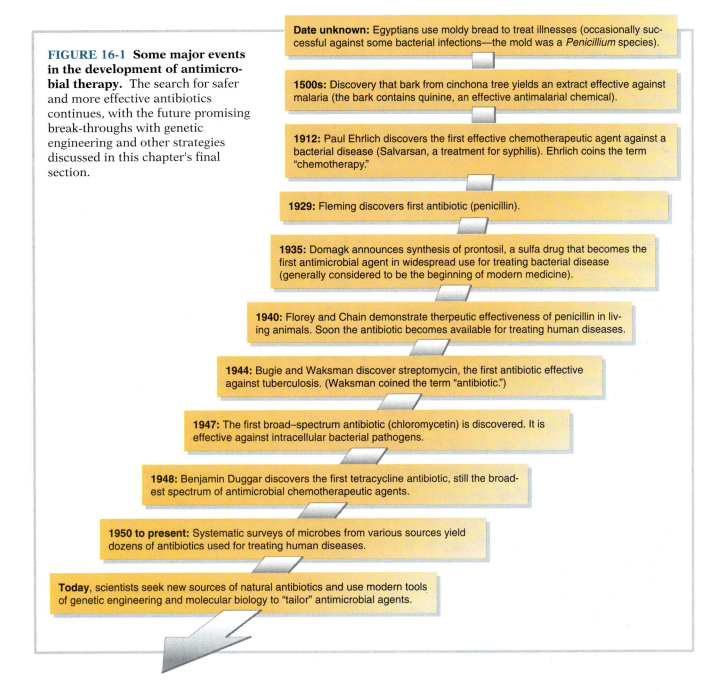

FIGURE 16-1 Some major events in the development of antimicrobial therapy. The search for safer and more effective antibiotics continues, with the future promising break-throughs with genetic engineering and other strategies discussed in this chapter's final section.

Date unknown: Egyptians use moldy bread to treat illnesses (occasionally successful against some bacterial infections—the mold was a *Penicillium* species).

1500s: Discovery that bark from cinchona tree yields an extract effective against malaria (the bark contains quinine, an effective antimalarial chemical).

1912: Paul Ehrlich discovers the first effective chemotherapeutic agent against a bacterial disease (Salvarsan, a treatment for syphilis). Ehrlich coins the term "chemotherapy."

1929: Fleming discovers first antibiotic (penicillin).

1935: Domagk announces synthesis of prontosil, a sulfa drug that becomes the first antimicrobial agent in widespread use for treating bacterial disease (generally considered to be the beginning of modern medicine).

1940: Florey and Chain demonstrate therpeutic effectiveness of penicillin in living animals. Soon the antibiotic becomes available for treating human diseases.

1944: Bugie and Waksman discover streptomycin, the first antibiotic effective against tuberculosis. (Waksman coined the term "antibiotic.")

1947: The first broad–spectrum antibiotic (chloromycetin) is discovered. It is effective against intracellular bacterial pathogens.

1948: Benjamin Duggar discovers the first tetracycline antibiotic, still the broadest spectrum of antimicrobial chemotherapeutic agents.

1950 to present: Systematic surveys of microbes from various sources yield dozens of antibiotics used for treating human diseases.

Today, scientists seek new sources of natural antibiotics and use modern tools of genetic engineering and molecular biology to "tailor" antimicrobial agents.

tion of penicillin, for example, saved the lives of countless individuals from diseases such as scarlet fever, pneumococcal pneumonia, puerperal (child-bed) fever, and gram-positive bacterial infections. Most of the medically important antibiotics in use today were discovered through systematic laboratory surveys between 1939 and 1963 (Table 16-1). The most effective agents were isolated from the molds *Penicillium* and *Cephalosporium* and from members of the bacterial genera *Streptomyces* and *Bacillus*. All these organisms are common contaminants in the soil and air.

> **CONCEPT CHECK**
>
> • Distinguish between an antibiotic and a chemotherapeutic agent.

Choosing the Best Chemotherapeutic Agent

No single antimicrobial drug is safe in all patients or effective against every infectious disease. The fol-

Miracle Drugs from Moldy Cantaloupes, Sewage, and Dirt

One of the most famous and fortunate accidents in history spawned a new medical era. The accident occurred in the laboratory of an alert microbiologist, Alexander Fleming. On a day in 1928 Fleming discovered that one of his cultures of *Staphylococcus aureus* was contaminated with a common airborne fungus. Before discarding the accidentally contaminated petri dish, Fleming noticed that the blue-green mold colony was surrounded by a zone in which no bacteria were growing (Fleming's original plate is shown in the photo). Outside this zone, bacterial colonies were numerous. The English scientist reasoned that the fungus produced some soluble compounds that diffused into the medium and prevented bacterial growth. He subsequently isolated the fungus *Penicillium notatum* and found that broth in which the mold had grown inhibited the growth of several species of bacteria. Fleming had discovered the first antibiotic. He called it penicillin.

For more than 10 years, Fleming's discovery was ignored by the pharmaceutical industry because the fungus did not produce enough antibiotic to be of practical use for treating bacterial diseases. In 1939, Howard Florey and Ernst Chain organized a group of scientists in Oxford to further investigate this promising drug. Although they conclusively demonstrated the therapeutic power of the drug in vivo, it required many months to get enough penicillin to treat even a few patients. What little penicillin could be harvested was so scarce that the urine of treated patients was collected and the drug reisolated for use in other patients.

One problem was the stingy amount of penicillin produced by Fleming's mold, so the team began an aggressive search for a more generous species of *Penicillium*. One tenacious assistant, nicknamed Moldy Mary, made daily trips to the markets of Peoria, Illinois, scrounging through rotting food and bringing back those with mold growth. The day Mary trotted in holding a moldy cantaloupe was the day that ushered in the antibiotic era of medicine in earnest. The rotting cantaloupe provided another species of mold, *Penicillium chrysogenum*, this one producing large quantities of the antibiotic. This mold produced penicillin in quantities great enough to allow its large-scale commercial production.

It is impossible to know how many people owe their lives to that rotting cantaloupe. Penicillin saved the lives of countless World War II soldiers who would otherwise have died of wound infections. It has been saving lives ever since and remains one of our safest, most effective antibiotics. For their contributions, Fleming, Florey, and Chain received the Nobel prize in 1945.

Since Fleming's fortunate accident, other therapeutically valuable antibiotics have been isolated from microbes growing in such places as sewage and soil. These discoveries were part of investigations specifically designed to isolate antibiotic-producing microbes. The first systematic searches for antibiotics in soil were performed in the United States under the direction of Selman Waksman. In 1943, he discovered streptomycin, the first of many therapeutically useful antibiotics. A few years later, Italian scientists isolated a *Cephalosporium* mold with antimicrobial activity from a sewage outflow. This organism was sent to the research group in England, who extracted a minor product of this mold. The cephalosporin drugs so widely used today are derived from this product.

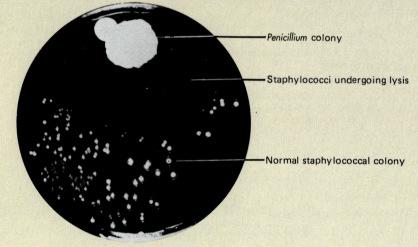

Penicillium colony

Staphylococci undergoing lysis

Normal staphylococcal colony

lowing factors influence the therapeutic value of antimicrobial drugs and must be considered if the patient is to receive the most effective chemotherapy.

- The selective toxicity of the drug
- The susceptibility of the pathogen to the chemotherapeutic agent
- The drug's spectrum of activity
- Possible adverse reactions to the drug
- The site of infection and the drug's ability to reach those tissues
- Metabolism of the drug in the body
- Duration of treatment
- Interaction with other drugs the patient may be taking

TABLE 16-1   SOME MAJOR ANTIMICROBIAL CHEMOTHERAPEUTIC AGENTS*

Antibiotic	Year of Discovery	Source	Spectrum of Activity
Isoniazid[†]	1912	Chemical synthesis	Mycobacteria
Penicillin G	1928	*Penicillium*	Gram-positive bacteria and *Neisseria*
Sulfa drugs[†]	1935	Chemical synthesis	Gram-positive and gram-negative bacteria
Griseofulvin	1939	*Penicillium*	Fungi
Chloroquine[†]	1941	Chemical synthesis	*Plasmodium* sp.
Streptomycin	1943	*Streptomyces*	Gram-negative bacteria; mycobacteria
Bacitracin	1945	*Bacillus*	Gram-positive bacteria
Chloramphenicol	1947	*Streptomyces*	Gram-positive and gram-negative bacteria
Polymyxin	1947	*Bacillus*	Gram-negative bacteria
Tetracycline	1948	*Streptomyces*	Gram-positive and gram-negative bacteria
Cephalosporin	1948	*Cephalosporium*	Gram-positive bacteria
Neomycin	1949	*Streptomyces*	Gram-negative bacteria
Nystatin	1950	*Streptomyces*	Fungi
Erythromycin	1952	*Streptomyces*	Gram-positive bacteria
Cycloserine	1954	*Streptomyces*	Gram-positive bacteria
Amphotericin B	1956	*Streptomyces*	Fungi
Vancomycin	1956	*Streptomyces*	Gram-positive bacteria
Metronidazole[†]	1957	Chemical synthesis	Protozoa; anaerobic bacteria
Kanamycin	1957	*Streptomyces*	Gram-negative bacteria
Rifamycin	1957	*Streptomyces*	Mycobacteria
Gentamicin	1963	*Micromonospora*	Gram-negative bacteria
Zidovudine[†]	1964	Chemical synthesis	Retroviruses
Acyclovir[†]	1974	Chemical synthesis	Herpes viruses
Ketaconazole[†]	1978	Chemical synthesis	Fungi
Fluoroquinoline[†]	1983	Chemical synthesis	Gram-negative bacteria

*Synthetic derivatives based on these drugs are not listed separately.
[†]Because these drugs are not produced by microbes, they are not considered antibiotics. But like antibiotics, their inhibiting effects are selective against microorganisms, and they are consequently effective chemotherapeutic agents.

■ Selective Toxicity

Antimicrobial agents are chemotherapeutically valuable only if they have **selective toxicity,** that is, if they inhibit or kill the microbe without producing undesirable side effects in the person being treated. The safest drugs are those that interfere with metabolic processes unique to prokaryotes. Penicillins and cephalosporins, for example, inhibit the synthesis of the peptidoglycan required for a functional bacterial cell wall. This target is not present in eukaryotic cells, which provides the basis for the drugs' selective toxicity.

Drugs that are toxic for eukaryotic pathogens (fungi or protozoa) are generally less selective. The host and the pathogen share so many cellular similarities that an agent affecting the microbe will likely have some adverse effect on human cells. Many drugs that effectively eliminate eukaryotic pathogens have such toxic side effects that they are used only when the danger of the patient dying of the disease exceeds the danger of toxic reactions to the drug.

Drugs that are selectively toxic against viruses are even more scarce. Because viruses become integral parts of their host cells, it is difficult to interfere with viral processes without also damaging the host. Consequently, most viral infections cannot be cured by most antimicrobial drugs. Fortunately, the natural defenses of the host limit the course and consequences of most viral infections.

■ Susceptibility of the Pathogen

For an antimicrobial drug to be of value it must kill or inhibit the growth of the pathogen within the body. In other words, the pathogen must be *susceptible* to the drug. The discovery of antibiotics fostered hope that infectious disease would soon be eradicated. Dangerous staphylococcal infections, for example, seemed universally curable by penicillin. However, within a year after the commercial introduction of penicillin for wide-scale use against staphylococcal infections, most of the target pathogens were *resistant* to the antibiotic—susceptible staphylococci had been replaced by strains of the bacteria that were not killed by the antibiotic. Penicillin resistance occurred in many other pathogens as well.

In fact, antibiotic-resistant pathogens appear soon after the introduction of any antibiotic, and the more an antibiotic is used, the more likely it is that drug-resistant organisms will emerge. Isolation and identification of the pathogen may suggest an appropriate drug, but the emergence of drug-resistant microbes makes it impossible to predict effectiveness solely on the basis of the microbe's identity. The isolated pathogen should therefore be tested for susceptibility to various antimicrobial chemicals to ensure that it can be killed or inhibited by the agent selected. The methods of testing for antimicrobial susceptibility and drug resistance are discussed later in this chapter.

■ Spectrum of Activity

Antimicrobial agents are either narrow-spectrum or broad-spectrum drugs according to the range of microorganisms they are usually effective against. **Broad-spectrum drugs** affect a wide number of microorganisms, whereas **narrow-spectrum drugs** are more limited in the types of cells they affect. Some narrow-spectrum drugs, for example, inhibit only gram-positive bacteria. Usually the ideal agent has the narrowest spectrum that is effective against the identified pathogen, since broad-spectrum agents disrupt the normal microbial flora that con-

tribute to the ecological balance and health of the patient. However, broad-spectrum drugs such as tetracyclines, chloramphenicol, and sulfa drugs are often used to treat mixed infections caused by several pathogens or for emergency situations in which there is no time to wait for laboratory results. In most cases such "shotgun therapy" is unwarranted.

■ Adverse Drug Reactions

Chemotherapeutic agents may have mild to fatal side effects. These may include such general symptoms as chills, fever, headache, nausea, or rash. More severe toxic reactions may damage the liver, kidney, or nervous system. For example, drugs that accumulate in the kidney cause renal damage and should not be used in older people or people with previous renal disease. Antimicrobial agents that pass through the placenta and cause fetal damage should not be used in a pregnant woman even if the drug is harmless to the mother. Similarly, antibiotics that are secreted in breast milk should not be used by mothers who are breast-feeding.

Perhaps the most common adverse effects of antibiotic therapy are the secondary infections that develop because the normal flora have been disrupted by broad-spectrum antibiotics. These **superinfections** are caused by fungi or bacteria whose growth is usually controlled by competition from the normal flora. If these opportunistic pathogens are resistant to the antimicrobial agent used, they may rapidly replace the disrupted flora and cause diarrhea, vaginitis, severe inflammation of the colon, or pneumonia. Sometimes superinfections can be controlled by withdrawing the antibiotic and allowing the normal flora to repopulate. Many times, however, another antibiotic is used to eliminate the opportunistic organisms. Development of opportunistic superinfections can usually be avoided by initially using a drug with the narrowest spectrum of activity against the target pathogen, thereby sparing most normal flora bacteria.

■ Site of Infection and Drug Distribution Within the Body

Antimicrobial agents have no therapeutic effect unless they reach the site of infection in concentrations high enough to incapacitate the pathogen. Antibiotics that cannot cross the barrier between the blood and the central nervous system, for example, are useless for treating meningitis unless injected directly into the cerebrospinal fluid. Abscesses, which are walled-off, localized accumulations of microorganisms, are protected against chemicals that fail to penetrate the abscess walls. Similarly, the gall bladder protects the typhoid bacillus from the effects of antimicrobial agents, often creating "healthy" carriers who continually shed typhoid bacteria in their feces in spite of antibiotic therapy. Intracellular pathogens are protected from penicillin, streptomycin, and other antibiotics that penetrate poorly into human cells. Tetracyclines easily enter host cells and are therefore effective against rickettsial and chlamydial diseases. Sulfonamides are ineffective against pathogens in necrotic (dead) tissue, which contains compounds that compete with the drug.

The amount of antimicrobial agent found at the site of infection also depends on the route of administration. Localized infections are best treated with drugs that accumulate at the site of infection. Gastrointestinal infections, for example, respond best when treated with agents that are taken orally and are poorly absorbed from the intestine so that they remain at the site of infection. Local infections of the urinary bladder are best treated with agents that concentrate in the urine even though serum drug levels may be too low to be antimicrobial.

For effective treatment of many diseases, the antibiotic must attain elevated concentrations in the patient's blood. Intravenous injections of penicillin, for example, ensure high concentrations throughout the bloodstream within 30 minutes. The intramuscular route is not as rapid, and somewhat lower concentrations are achieved. Oral administration is even slower and yields a substantially lower blood concentration (Fig. 16-2).

Drugs that are too toxic for internal use may be valuable as topical agents if the infection is localized on the body's surface.

■ Metabolism of the Drug

Many drugs are metabolized by the body. In some cases these changes increase their antimicrobial effectiveness. Prontosil, for example, is inactive until converted by the body to sulfanilamide (see THE EXPLORERS: Prontosil—A Cure in Camouflage). Such enhancement is rare, however, with most metabolic changes diminishing antimicrobial effectiveness. For example, many chemotherapeutic agents are destroyed by the low pH of the stomach. Others are bound and inactivated by serum proteins. Therefore, the amount of effective antibiotic in the body may be considerably lower than the amount administered.

■ Duration of Treatment

Most drugs are metabolized or excreted before the infectious agents are eliminated. These drugs must be periodically readministered to maintain thera-

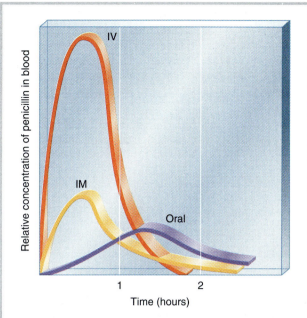

FIGURE 16-2 The influence of route of administration on blood levels of penicillin G. Both oral and intramuscular (IM) routes yield blood concentrations below that needed for therapeutic effectiveness, limiting use of penicillin G to intravenous (IV) delivery.

peutic levels. The schedule of administration is critical, and even a delay of one-half hour may reduce the effectiveness of therapy. This problem is partially solved by including another chemical that prolongs the drug's effect. For example, procaine and benzathine delay absorption of penicillin from an intramuscular injection site so there are therapeutic concentrations in the bloodstream for longer durations. Probenecid administered with penicillin competes with its removal from the blood by the kidneys and therefore delays excretion of the drug. It is routinely included in the single large dose of penicillin used for treating gonorrhea.

MYTH: Persons taking antibiotics can stop taking them when symptoms disappear.

FACT: Infectious diseases are usually not cured as soon as symptoms disappear, which often occurs before all pathogens are eliminated. The unkilled cells often resume growth and cause a recurrence of the disease unless therapeutic concentrations of the chemotherapeutic drug are maintained until all pathogens are eradicated from the infected patient.

Prontosil—A Cure in Camouflage

Theoretically it shouldn't have worked. Yet this miraculous new drug, the first of its kind, saved thousands of lives in the 1930s. The wonder drug, called prontosil, was the product of a German scientist, Gerhard Domagk. He had synthesized the simple sulfur-containing dye with hopes that its tendency to stain bacteria but not human cells would translate into selective toxicity when given to persons with bacterial infections. From the start prontosil showed evidence of doing just that. The drug cured mice of the streptococcal infections Domagk had experimentally given them. Without prontosil, they died. People suffering from similar infections were also cured with prontosil. Prontosil was accepted for widespread use, and soon the number of women dying of uterine infections following childbirth was reduced by more than half. In one of the most dramatic confirmations of the drug's effectiveness, prontosil saved the life of Domagk's own daughter. Its ability to cure infectious diseases was unquestionable.

Results of attempts to determine how the drug worked, however, baffled scientists, who found that prontosil does not kill or even inhibit bacteria growing on artificial media although it readily cures many diseases caused by these pathogens. This apparent paradox was resolved by the discovery that prontosil is metabolized in the body to sulfanilamide, a compound that inhibits bacterial growth both in the body and on artificial media. Sulfanilamide specifically interferes with a metabolic reaction that is essential for bacterial growth but not needed by eukaryotic cells, thereby accounting for its selective toxicity. Domagk was fortunate to have used animals for studying prontosil; had he used bacterial inhibition tests he might never have discovered it. Sulfanilamide was the first of the sulfa drugs, which even today are among our most valuable antimicrobial chemicals for treating infectious disease.

Although awarded the Nobel prize in 1939, Domagk was forbidden by Adolph Hitler to accept the prize. Germany was officially opposed to the Nobel award since a journalist who was an open critic of militarism and the Nazi party had received the Nobel peace prize in 1935.

■ Drug Interactions

Several antimicrobial agents are sometimes administered together to increase the spectrum of activity against mixed infection or when treatment of undiagnosed disease is urgent. Drug combinations are best employed when they result in *synergism*, an enhancement of activity greater than the sum of the two agents when used alone. For example, carbenicillin (a penicillin derivative) weakens the bacterial wall structure and enhances the penetration of gentamicin into the interior of the cell, where this second drug impairs protein synthesis. Synergistic combinations are sometimes effective against infections that normally do not respond to either drug alone. Synergistic combinations allow some drugs to be used in lower concentrations, thereby reducing negative side effects. In some cases, combination therapy reduces the likelihood that drug resistance will develop; mutants resistant to one antibiotic are eliminated by the other drug. There is at least one major disadvantage of combination therapy—it favors the emergence of multiple-resistant strains that contain R factors. ▶ (See Conjugation, Chap. 8, p. 208.) These strains can transfer antibiotic resistance to susceptible pathogens.

Some antimicrobial agents *antagonize* each other and should not be used in combination. For example, bacteristatic agents such as tetracycline should not be used with penicillin, which kills only actively growing cells. Other drugs may inactivate or precipitate one another and cannot be administered simultaneously in a single solution. Antibiotics may also react with food. For example, tetracycline should not be taken with milk, which reduces its antimicrobial effectiveness. Sometimes the antimicrobial agent reacts with commonly consumed substances and produces toxic reactions in the patient. Perhaps the most common example of such a danger is when alcohol is consumed by a person taking metronidazole (Flagyl), an antibiotic commonly used for treating *Trichomonas* vaginal infections. The antimicrobial drug interferes with the metabolism of ethyl alcohol, which may accumulate to toxic concentrations if alcoholic beverages are consumed during the course of metronidazole therapy.

CONCEPT CHECK

- Why is it important that chemotherapeutic agents are selectively toxic?
- Briefly explain how the effectiveness of chemotherapeutic agents is influenced by the target microorganism, site of infection, treatment duration, and other drugs the patient is taking.

Antimicrobial Mechanisms

Microbistatic chemotherapeutic agents impede microbial growth until the host defense mechanisms can eventually destroy the pathogens. Most microbicidal agents can kill pathogens with no assistance

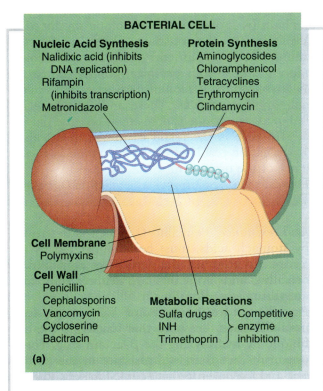

BACTERIAL CELL

Nucleic Acid Synthesis
Nalidixic acid (inhibits DNA replication)
Rifampin (inhibits transcription)
Metronidazole

Protein Synthesis
Aminoglycosides
Chloramphenicol
Tetracyclines
Erythromycin
Clindamycin

Cell Membrane
Polymyxins

Cell Wall
Penicillin
Cephalosporins
Vancomycin
Cycloserine
Bacitracin

Metabolic Reactions
Sulfa drugs ⎫
INH ⎬ Competitive enzyme inhibition
Trimethoprin ⎭

(a)

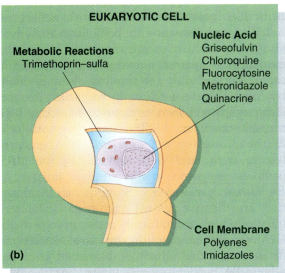

EUKARYOTIC CELL

Metabolic Reactions
Trimethoprin–sulfa

Nucleic Acid
Griseofulvin
Chloroquine
Fluorocytosine
Metronidazole
Quinacrine

Cell Membrane
Polyenes
Imidazoles

(b)

FIGURE 16-3 Microbial targets of chemotherapeutic agents. (*a*) Antibacterial drugs; (*b*) drugs used against eukaryotic pathogens.

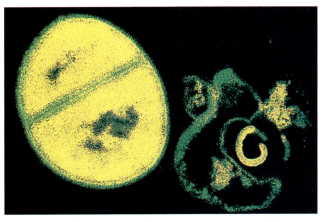

FIGURE 16-4 Bacterial target—the cell wall. The lysed remains of *Staphylococcus aureus* following treatment with antibiotics (cell wall inhibitors) contrast dramatically with the bacterium's appearance before exposure to antibiotics (the intact dividing cell at left).

from the host defenses. These agents function by selectively disrupting (1) cell wall synthesis, (2) cell membrane function, (3) protein synthesis, (4) nucleic acid synthesis, or (5) other specific metabolic reactions. The microbial targets of the major drugs are identified in Figure 16-3.

■ Target: Bacterial Cell Walls

Agents that prevent the synthesis of peptidoglycan produce osmotically fragile bacterial cells that lyse unless kept in a medium that prevents the influx of water (Fig. 16-4). Bacteria with defective cell walls survive poorly in the human body. Antibiotics that inhibit cell wall synthesis include bacitracin, vancomycin, cycloserine, and antibiotics such as penicillins and cephalosporins. Because these agents cannot destroy existing peptidoglycan, they are effective only against actively growing bacteria. Gram-positive bacteria are sensitive to most of these agents; gram-negative bacteria usually are not. The outer membrane of most gram-negative cell walls prevents the antibiotic from penetrating to its site of action. Eukaryotic cells lack peptidoglycan and are completely resistant to penicillins and cephalosporins, which are among the least toxic of all antibiotics. Cycloserine, bacitracin, and vancomycin, however, have toxic effects unrelated to their antimicrobial activity that limit their value as chemotherapeutic agents.

● β-**LACTAM ANTIBIOTICS—PENICILLIN AND CEPHALO-SPORINS.** Penicillins and cephalosporins are the most common of the β-lactam antibiotics, which are similar in structure and activity (Fig. 16-5). Each of

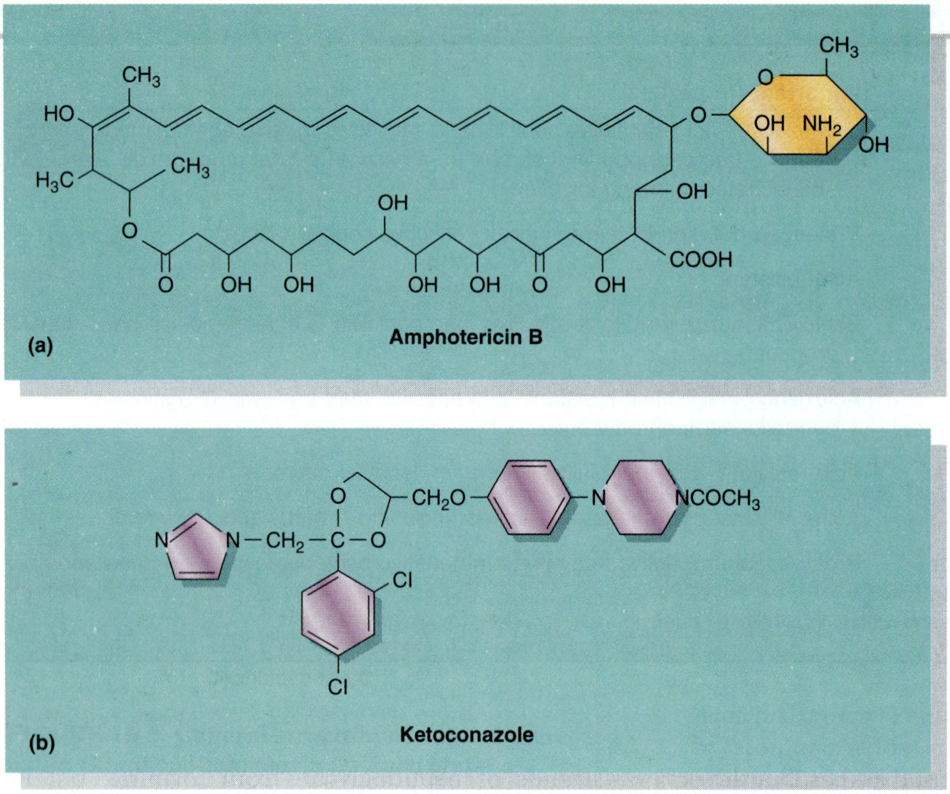

FIGURE 16-6 Some antibiotics targeted against the eukaryotic cell membrane. (*a*) A typical polyene; (*b*) a representative azole.

terial infections in burn patients. Their toxic side effects, however, prevent their widespread systemic use.

● **POLYENES.** The most therapeutically valuable of the *polyenes* are those that bind with ergosterol, the sterol in fungal membranes, creating membrane pores and causing leakage of cell metabolites. Human cells contain cholesterol instead of ergosterol and are therefore not as susceptible as fungi to these agents. The sterols in human cells, however, are somewhat affected by the polyenes. Toxicity for the patient, therefore, should be considered when using these antibiotics.

Polyenes are among the few antimicrobials available for the treatment of cryptococcosis, coccidiodomycosis, aspergillosis, mycormycosis, candidiasis, and other potentially fatal systemic fungal infections. The drug most often used is amphotericin B, an antibiotic whose therapeutic and toxic doses are so close that full therapeutic doses are used only in life-threatening situations. The tendency of amphotericin B to increase the porosity of the cell membrane, however, is used to facilitate the entry of other drugs into the fungal cell. Lower, less toxic doses of amphotericin B are often administered in synergistic combination with otherwise ineffective drugs that alone cannot reach their intracellular sites of action.

Another polyene, nystatin (named for New York State, where it was discovered), is too toxic for systemic therapy and is used primarily as a topical agent to eliminate *Candida* infections from oral or vaginal mucous membranes. Systemic use of nystatin would eliminate the patient along with the pathogen. It can, however, be taken orally to eliminate *Candida* from the bowels. It is so poorly absorbed through the intestinal epithelium that there is no systemic exposure to the drug when it is administered by this route. Women who frequently have vaginal candidiasis are sometimes treated with oral doses of nystatin to lower the likelihood of autoinoculation of the vagina with yeast from the anus.

● **AZOLES.** The *azoles* are another group of synthetic compounds that interfere with the cell membrane of fungi. Unlike the polyenes, these compounds inhibit the *synthesis* of ergosterol. One group of azoles, the imidazoles, are effective broad-spectrum antifungal drugs with few or no serious side effects. Clotrimazole and miconazole, for example, are found in topical ointments such as those used for vaginal yeast infections. Ketoconazole has been used successfully for treating systemic infections caused by several pathogenic molds and yeasts (Fig. 16-7). Ketoconazole is also effective against the dermatophytes. Another group of azoles are the tri-

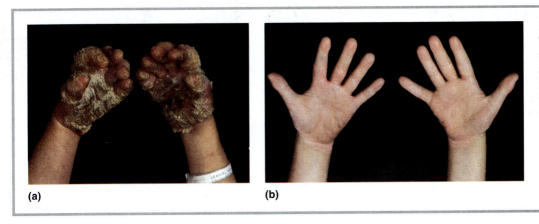

FIGURE 16-7
The effect of ke-toconazole on a chronic yeast infection. (*a*) Before treatment; (*b*) after treatment.

azole derivatives, such as fluconazole and itraconazole, which have even fewer side effects than the imidazoles. For *Cryptococcus* and oral *Candida* infections in AIDS patients, fluconazole is the drug of choice. (Itraconazole is an experimental drug in the United States. It has not yet been approved by the FDA for general use.)

■ Target: Protein Synthesis

Some antibacterial chemicals can selectively inhibit bacterial protein synthesis by disabling prokaryotic ribosomes while causing little deleterious effect on eukaryotic ribosomes (Fig. 16-8). The selective toxicity of these agents is often enhanced by their abil-

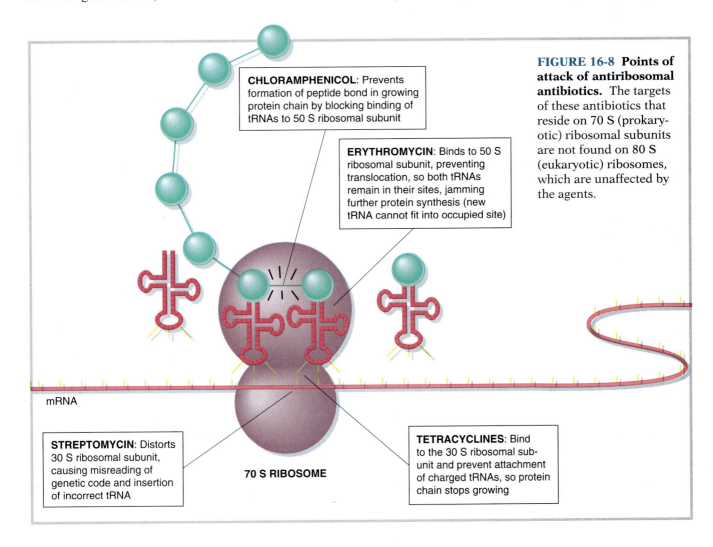

CHLORAMPHENICOL: Prevents formation of peptide bond in growing protein chain by blocking binding of tRNAs to 50 S ribosomal subunit

ERYTHROMYCIN: Binds to 50 S ribosomal subunit, preventing translocation, so both tRNAs remain in their sites, jamming further protein synthesis (new tRNA cannot fit into occupied site)

STREPTOMYCIN: Distorts 30 S ribosomal subunit, causing misreading of genetic code and insertion of incorrect tRNA

TETRACYCLINES: Bind to the 30 S ribosomal subunit and prevent attachment of charged tRNAs, so protein chain stops growing

mRNA

70 S RIBOSOME

FIGURE 16-8 Points of attack of antiribosomal antibiotics. The targets of these antibiotics that reside on 70 S (prokaryotic) ribosomal subunits are not found on 80 S (eukaryotic) ribosomes, which are unaffected by the agents.

ity to pass through bacterial membranes more readily than through eukaryotic membranes.

● **AMINOGLYCOSIDES.** The *aminoglycosides* are a group of structurally similar antibiotics that attach to the bacterial ribosome and interfere with accurate translation of the genetic code. Unfortunately, they also affect the small ribosomes in the mitochondria of eukaryotic cells ▶ (see BEYOND THE BASICS: Bacteria and the Early History of Life, Chap. 2, p. 43), which perhaps accounts for their severe side effects. Prolonged use or high dosages of any aminoglycoside can cause deafness, and most are toxic to the kidneys as well. Aminoglycosides are therefore used systemically only against serious gram-negative infections. They are poorly absorbed from the gastrointestinal tract and are usually administered by injection. The aminoglycosides include streptomycin, gentamicin, neomycin, tobramycin, and kanamycin (and its semisynthetic derivative amikacin). With the exception of gentamicin, all aminoglycosides are produced by *Streptomyces* species. (Gentamicin is spelled "micin" rather than "mycin" to denote that its origin is *Micromonospora*, not *Streptomyces*.)

● **TETRACYCLINES.** The *tetracyclines* display the broadest spectrum of all antibiotics. They are active against both gram-positive and gram-negative bacteria. Because tetracyclines readily penetrate cell membranes, they are the drug of choice for treating intracellular bacterial infections caused by chlamydias, rickettsias, and *Brucella*. Tetracyclines prevent the binding of transfer RNA to 70 S ribosomes. Since 80 S ribosomes are not affected, the inhibition is specific for prokaryotes.

Tetracyclines are acid-stable and are readily absorbed from the gastrointestinal tract, so they are administered orally. Milk products and iron-containing foods reduce antimicrobial effectiveness and should be avoided when tetracyclines are taken. The antimicrobial spectrum of tetracyclines is so broad that their extended use disrupts the normal flora and encourages secondary infections by tetracycline-resistant staphylococci or the yeast *Candida albicans*. In addition, these antibiotics discolor developing teeth (Fig. 16-9) and may retard normal growth of bones. Tetracyclines should therefore not be prescribed for pregnant women (they cross the placenta) or for young children.

● **ERYTHROMYCIN, CLINDAMYCIN, AND CHLORAMPHENICOL.** Several ribosome inhibitors prevent peptide bond formation during bacterial protein synthesis. They do no structural damage to ribosomes, so they are all bacteristatic agents. Ery-

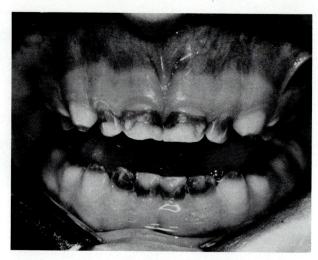

FIGURE 16-9 **A side effect of tetracycline therapy.** These teeth are permanently discolored as a result of tetracycline therapy during childhood.

thromycin is primarily active against gram-positive bacteria and commonly used as an alternative treatment for infections in people who are allergic to penicillins. It is the drug of choice for treating pertussis, diphtheria, and legionnaires' disease. *Clindamycin* is used for treating abscesses and systemic infections caused by gram-negative anaerobic bacteria. Its use, however, may upset the normal anaerobic intestinal flora, thereby promoting the growth of *Clostridium difficile*, an intestinal opportunist that causes severe, sometimes fatal, colitis (inflammation of the bowel). *Chloramphenicol* is an inexpensive broad-spectrum antibiotic that readily diffuses into spinal fluid, the gall bladder, and other body sites that are inaccessible to many other chemotherapeutic agents. Unfortunately, chloramphenicol has caused fatal aplastic anemia, the complete loss of the bone marrow's ability to produce red blood cells. In the United States, chloramphenicol is used only for treating typhoid fever and other life-threatening infectious diseases that fail to respond to safer antibiotics. However, because it can be synthesized inexpensively, chloramphenicol is one of the most commonly used antibiotics in many areas of the world.

■ **Target: Nucleic Acids**

Most chemotherapeutic agents that interfere with nucleic acid synthesis are synthetic compounds and not antibiotics. Only a few agents in this category are selective enough for safely treating infectious diseases.

Rifampin is a semisynthetic derivative of a natural product of *Streptomyces mediterranei*. This drug inhibits transcription of mRNA from DNA by binding to and inactivating bacterial mRNA polymerase. Rifampin is especially effective against *Mycobacterium* species that cause tuberculosis and leprosy. Since rifampin-resistant strains emerge rapidly and chemotherapy continues for long periods of time, it is usually used in combination with other antimycobacterial agents. Unfortunately, rifampin has little therapeutic value in treating the *Mycobacterium avium-intracellulare* infections associated with AIDS patients.

Nalidixic acid belongs to a family of compounds called *quinolines* that inhibit DNA gyrase, an enzyme needed for replication of bacterial chromosomes. High concentrations of nalidixic acid are excreted in the urine, making it useful for treating urinary tract infections. In the serum, however, therapeutic levels are never achieved. It is primarily effective against gram-negative organisms. Newer quinoline derivatives extend the spectrum of activity and are also more widely distributed in tissues.

Flucytosine interferes with nucleic acid replication in yeasts by incorporating into fungal DNA. It does not have a similar effect on humans because only fungi contain the enzyme that activates flucytosine. It is often used in conjunction with amphotericin B for treating systemic *Candida* and *Cryptococcus* infections. *Griseofulvin* also interferes with DNA replication in some fungi. It is used for treating dermatophyte infections of skin, nails, and scalp. The pathogens are not killed by griseofulvin but are ultimately shed with sloughed skin or nails, a process requiring months of antibiotic therapy.

Metronidazole is commonly employed to treat vaginal infections caused by the protozoan *Trichomonas vaginalis* but is also effective against *Giardia* and parasitic amoebas. This synthetic drug is also one of the most active agents against anaerobic bacteria. Almost no aerobic bacteria are affected. Sensitive organisms metabolize metronidazole to a compound that binds to DNA and rapidly kills the cell.

■ Target: Bacterial Metabolism

Several important chemotherapeutic agents called **antimetabolites** competitively inhibit bacterial metabolic reactions. The antimetabolites are usually substrate analogues, compounds with structures that closely resemble the substrate of an enzyme and therefore compete for the enzyme's active site. If the concentration of the inhibitor is high enough, it successfully competes with the substrate and prevents its conversion to products. Inhibitors that interrupt reactions essential to microbial growth but not needed for human metabolism may be used for chemotherapy.

Sulfa drugs, for example, are analogues of *para*-aminobenzoic acid (PABA) (Fig. 16-10a). Normally PABA is enzymatically converted to folic acid, a coenzyme essential for growth. Sulfa drugs react with the enzyme but are not converted to products. They tie up the active site and prevent the production of folic acid (Fig. 16-10b). Bacteria that must synthesize their own folic acid (they cannot absorb folic acid from the medium) are inhibited by the sulfa drugs. Human cells obtain all their folic acid from external sources and are therefore not affected by inhibitors of folic acid synthesis. Consequently the sulfa drugs are selectively toxic for bacteria. Because inhibitor binding is reversible, sulfa drugs are bacteristatic agents.

A variety of folic acid inhibitors have additional advantageous properties. Sulfamethoxazole is more soluble in urine than other sulfa drugs and therefore more effective against urinary tract infections. Increased solubility also reduces the harmful precipitates that may otherwise be deposited in the kidney. *Trimethoprin* blocks a later step in the folic acid pathway. Trimethoprin and sulfa drugs interact synergistically and are often prescribed together. (Cotrimoxazole is a combination of sulfamethoxazole

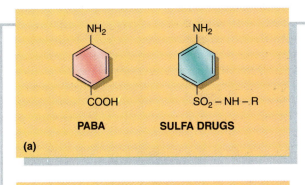

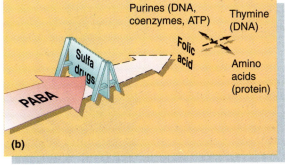

FIGURE 16-10 Metabolic target of sulfa drugs. (*a*) Note the similarity in structure between PABA and sulfa drugs. They are structural analogues of one another. (*b*) Inhibition of folic acid synthesis by sulfa drugs.

and trimethoprin.) Unlike sulfa, trimethoprin interferes with a step that is critical to human metabolism as well as to bacterial metabolism. Fortunately, the affinity of trimethoprin for the critical enzyme is 50,000 times greater in bacteria than in human cells. At therapeutically effective doses, toxicity to the patient is inconsequential.

Another antimetabolite, *isoniazid* (INH), is thought to inhibit mycolic acid synthesis in mycobacteria. It is the most widely used drug for treating tuberculosis. Unlike the sulfa drugs, INH is bactericidal. Since INH resistance is common among mycobacteria, this antimetabolite is usually used in combination with rifampin and other antibiotics.

Chloroquine has been used for over 50 years in treating malaria, but its role as a metabolic inhibitor has only recently been understood. The drug interferes with the enzymatic digestion of hemoglobin during the erythrocytic phase, when *Plasmodium* parasites have invaded red blood cells. The protozoa trap hemoglobin in food vacuoles and degrade the globin protein into amino acids. Normally the protozoa convert the heme portion of the globin molecule to a toxic form, then quickly convert it to a nontoxic pigment, preventing the accumulation of the toxic intermediate. Chloroquine prevents the detoxification of the heme breakdown products, which then kill the protozoa inside the erythrocytes.

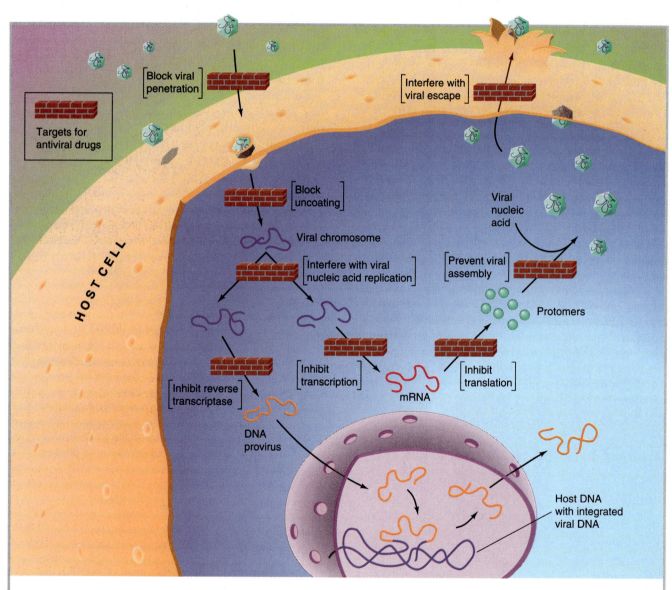

FIGURE 16-11 Viral targets for chemotherapy. The potential targets for antiviral agents are indicated on this hypothetical viral replication cycle, which combines the biological activities of many different types of viruses.

Molecular Decoys— A Cure for the Common Cold?

The approach of the twenty-first century brings with it advanced technologies that promise new drugs to fight disease. We are no longer limited to exhaustive searches for naturally occurring antimicrobial products to protect us from pathogens. Now we can design molecular decoys. The basic strategy of this defense is to create a drug that will inactivate the invading microbe before it can mount an attack. Such drugs can be designed to interfere with specific sites on the pathogen, especially those needed for binding to receptors on a host cell. Molecular decoys may represent some of the most sophisticated weapons in the war against pathogenic microbes.

Two approaches are being used in designing these drugs. One approach requires knowledge of the host cell's receptor for the pathogen; the other requires knowledge of the attachment site on the pathogen itself. The microbe attaches to its host cell by the interaction of these two structures, a process that is usually necessary for initiation and continuation of infection. For example, obligate intracellular parasites such as viruses must attach to their host cell in order to invade. Many extracellular pathogens attach to surfaces to prevent expulsion from the body. Pathogens of the urinary, respiratory, and gastrointestinal tracts bind to host cells during infection.

One of the most elusive diseases to cure has been the common cold, which is usually caused by a member of the rhinoviruses. The cellular receptor for these viruses is a molecule called ICAM-1. Portions of this molecule have been synthesized in a soluble form to which the virus will bind. Once bound to this molecular decoy, the virus cannot bind to the natural receptor on a host cell surface. It is hoped that soluble ICAM-1 incorporated into a nasal spray will have therapeutic value.

Details of the rhinovirus capsid elucidated through X-ray crystallography in 1985 have given scientists the unique opportunity to investigate the construction of the viral attachment site at the molecular level. These observations have been combined with the power of supercomputers to determine drug interactions at this critical site (see computer-generated photo of rhinovirus below). The computers help design low molecular weight compounds that will efficiently and specifically bind to the virus and block its attachment to host receptors. Such drugs may be the future magic bullets not only against colds but also against influenza, AIDS, and other diseases that have thus far proven difficult to treat by conventional therapies.

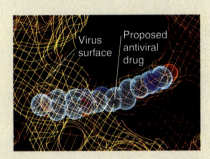

Virus surface   Proposed antiviral drug

CONCEPT CHECK

- Describe the mechanism of selective toxicity of penicillins, sulfa drugs, cephalosporins, aminoglycosides, polyenes, tetracycline, and erythromycin.
- Provide three reasons why cephalosporins and penicillins are the most prescribed antibiotics for treating bacterial infections.
- What are the advantages of semisynthetic drugs over their natural counterparts?

Antiviral Agents

Success at controlling viral infections is due primarily to the development of effective vaccines. Unfortunately, such protection is not available against many viruses, including some that cause life-threatening diseases. For persons infected with these viruses, chemotherapy is needed. Viruses theoretically present a variety of targets for chemotherapy during their replication cycle (Fig. 16-11). For example, drugs may interfere with attachment, virus-specific transcription and translation, replication, or assembly mechanisms. Years of intensive investigations have focused on the unique viral structures and enzymes produced during the infection cycle. Recent developments have been encouraging, indicating that drugs selectively targeting viruses can be designed (see BEYOND THE BASICS: Molecular Decoys—A Cure for the Common Cold?). Currently, however, only a few chemotherapeutic chemicals are licensed for the treatment of a limited number of viral diseases (Table 16-3), and not all of these are both safe and effective in treating existing viral disease.

The drug *amantadine* blocks the uncoating of type A influenza virus and the release of viral RNA into cells but has little or no effect on the virus once it is replicating. Amantadine may prevent influenza infection and can shorten the duration of symptoms by 24 hours if administered within 2 days of disease

TABLE 16-3 SOME DRUGS APPROVED FOR USE AGAINST VIRAL PATHOGENS

Virus	Disease	Drug of Choice
Influenza A	Influenza	Amantadine
Herpes simplex	Herpes keratitis	Trifluridine
	Genital herpes	Acyclovir
	Encephalitis	Acyclovir
Varicella-zoster	Shingles, chickenpox in immunocompromised patients	Acyclovir
Cytomegalovirus	Retinitis	Ganciclovir, foscarnet
	Pneumonia	Ganciclovir
Human immunodeficiency virus	AIDS; AIDS-related complex (ARC)	Zidovudine, ddl, ddC
Respiratory syncytial virus	Pneumonia	Ribavirin
Lassa virus	Lassa fever	Ribavirin
Hepatitis B and C	Chronic hepatitis	Interferon
Human papillomavirus	Genital warts	Interferon

onset. It is most often employed to protect unvaccinated older people and others at risk of contracting fatal infections when epidemic outbreaks are expected.

Acyclovir inhibits DNA synthesis in certain herpesviruses. The drug, an analogue of guanine, must be converted to a triphosphate form to be active. Phosphorylation is performed by a viral enzyme, and only herpes simplex viruses (HSV) or varicella-zoster virus (VZV), the agent of chickenpox and shingles, produce the necessary enzyme. Thus activation of acyclovir's lethal effect occurs only in cells infected with these viruses. When the activated drug becomes incorporated into viral DNA, it causes the termination of DNA synthesis. Acyclovir is used systemically and topically against oral or genital HSV infections to shorten the course of the initial episode. The drug does not prevent recurrences once the virus is established in the nervous system. When administered systemically, it may prevent overwhelming herpesvirus infections in transplant recipients and other immunocompromised patients and is also effective against zoster (shingles) in normal hosts. Acyclovir has few side effects and is the first truly selective antiviral agent. A similar drug, *ganciclovir*, is more readily activated by cellular enzymes, and its spectrum of activity includes cytomegalovirus, which causes life-threatening illness in immunocompromised patients. *Foscarnet* is a structurally unrelated drug that is used to treat infections due to acyclovir-resistant HSV or VZV or ganciclovir-resistant cytomegalovirus.

Idoxuridine (IDU) was the first antiviral agent licensed in the United States. For more than 30 years it has been used topically for treating herpesvirus infections of the eye, but it is too toxic for systemic therapy. Idoxuridine is an inhibitor of DNA replication, and both cellular DNA and viral DNA may be affected. Cells of the cornea are not affected by the drug because they are not actively dividing. Trifluridine and vidarabine (Ara-A) are also used for ocular herpes.

Zidovudine, popularly known as AZT, was the first drug to be approved in the United States for use against HIV, the virus that causes AIDS. A related drug, DDI, has subsequently become available for use in patients who do not respond to AZT, while a third drug, DDC, is licensed for use only in combination with AZT. These drugs are nucleoside analogues that inhibit reverse transcriptase activity. Their action prevents the virus from replicating and spreading to uninfected cells. HIV incorporated into the host genome is not affected, so the drugs do not cure AIDS. Since the integrated virus remains, patients must take medication for the rest of their lives. Side effects, which include headache, fever, and nausea, vary with the stage of the disease. Studies published in 1993 showed that AZT did reduce symptoms of immune deficiency but did not prolong the life of HIV-infected people.

Ribavirin is an analogue of guanine that prevents DNA and RNA synthesis and translation of viral messenger RNA. Ribavirin displays antiviral activity against a variety of viruses in vitro but has not proven as effective in vivo. It is used clinically primarily for treating respiratory syncytial virus in children. It is also active against influenza and a variety of dangerous viruses found mainly in Asia and Africa.

Commercially produced *interferon*, a product of

recombinant DNA technology, is licensed for the treatment of warts and of Kaposi's sarcoma, a tumor that develops in many AIDS patients. It is also approved for use in cases of hepatitis B and hepatitis C. Interferon inhibits viral replication by inducing host cells to produce proteins with antiviral activity ▶ (see *interferon*, Chap. 13, p. 328). Despite interferon's broad antiviral activity, its general use as an antiviral drug is limited because, in most cases, it is more effective in preventing disease than in curing it.

CONCEPT CHECK

- Why are so few drugs effective in treating viral infections?
- What is the basis of the selective toxicity of acyclovir and AZT (zidovudine)?

Antibiotic Resistance

Chemotherapeutic effectiveness depends upon the sensitivity of the pathogen to the agent. Some microbes respond predictably to certain drugs, making selection of treatment easy. Other microbes may vary in their responses, and laboratory tests are usually required to ensure that the selected therapy is appropriate. Antibiotic resistance, however, may develop in microbes within the population. In fact, the history of chemotherapy has been closely paralleled by the history of drug resistance. From the early work of Ehrlich to the recent development of zidovudine, microbes resistant to the agents of their potential destruction have emerged (Table 16-4).

Antibiotics do not create resistant cells or cause mutations that produce resistant organisms. They do, however, selectively favor the survival and proliferation of drug-resistant strains, which otherwise are only a small subpopulation within the vast majority of sensitive cells. Antimicrobial resistance is acquired either by mutation in the pathogen's chromosome or by direct transfer of R-factor plasmids from antibiotic-resistant strains to sensitive recipients.

Random mutations in chromosomal genes may produce a few cells that are resistant to antibiotics. These mutants become significant only when prolonged exposure to the drug favors their survival over that of sensitive cells (Fig. 16-12). In bacteria,

---

TABLE 16-4 TOP TEN DRUG-RESISTANT MICROBES

Microbes	Diseases Caused	Drugs Resisted
1. *Enterobacteriaceae*	Bacteremia; pneumonia; urinary tract, surgical wound infections	Aminoglycosides, beta-lactam antibiotics, chloramphenicol, trimethoprim
2. *Enterococcus*	Bacteremias; urinary tract, surgical wound infections	Aminoglycosides, beta-lactams, erythromycin, vancomycin
3. *Haemophilus influenzae*	Epiglotitis, meningitis, otitis media, pneumonia, sinusitis	Beta-lactams, chloramphenicol, tetracycline, trimethoprim
4. *Mycobacterium tuberculosis*	Tuberculosis	Aminoglycosides, ethambutol, isoniazid, pyrazinamide, rifampin
5. *Neisseria gonorrhoeae*	Gonorrhea	Beta-lactams, spectinomycin, tetracycline
6. *Plasmodium falciparum*	Malaria	Chloroquine
7. *Pseudomonas aeruginosa*	Bacteremia, pneumonia, urinary tract infections	Aminoglycosides, beta-lactams, chloramphenicol, ciprofloxacin, tetracycline, sulfonamides
8. *Shigella dysenteriae*	Severe diarrhea	Ampicillin, trimethoprim-sulfamethoxazole, chloramphenicol, tetracycline
9. *Staphylococcus aureus*	Bacteremia, pneumonias, surgical wound infections	Chloramphenicol, ciprofloxacin, clindamycin, erythromycin, beta-lactams, rifampin, tetracycline, trimethoprim
10. *Streptococcus pneumoniae*	Meningitis, pneumonia	Aminoglycosides, chloramphenicol, erythromycin, penicillin

SOURCE: George Jacoby, *Science*, 257:1036 (1992).

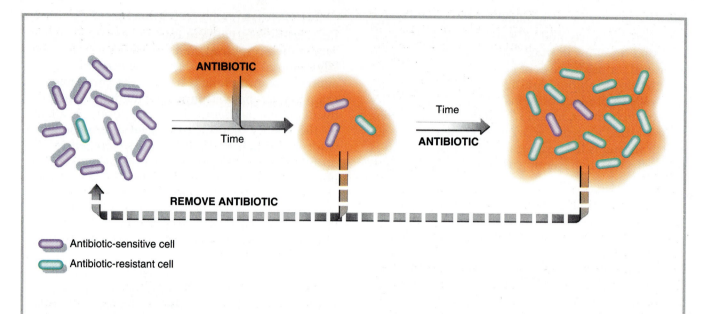

FIGURE 16-12 The emergence of antibiotic resistance by mutation and selection. A population of antibiotic-sensitive bacteria often contains a small number of antibiotic-resistant cells that appear by random mutation. In the presence of antibiotics, most sensitive cells are killed or inhibited; a few resistant cells are uninhibited and continue to grow. Prolonged exposure to the drug prevents sensitive cells from repopulating the area, allowing resistant microbes to become predominant. These antibiotic-resistant cells are not inhibited by the drug, and treatment fails to control the infection. With the removal of the antibiotic, the survival of drug-sensitive bacteria is often favored over that of drug-resistant cells because, all else being the same, sensitive cells waste no resources on unnecessary drug-resistance mechanisms. They would therefore have the competitive advantage as long as the antibiotic remained absent.

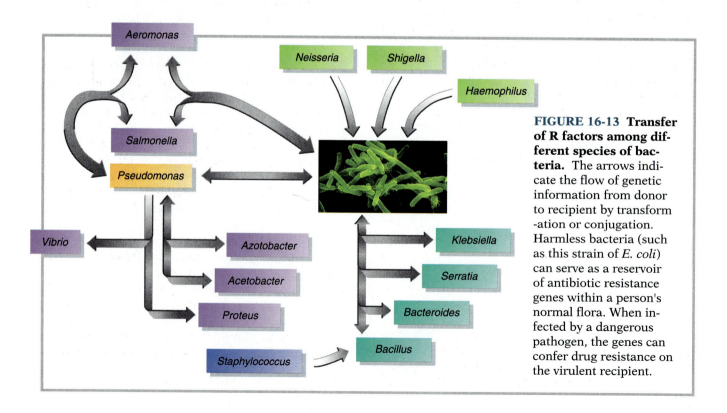

FIGURE 16-13 Transfer of R factors among different species of bacteria. The arrows indicate the flow of genetic information from donor to recipient by transformation or conjugation. Harmless bacteria (such as this strain of *E. coli*) can serve as a reservoir of antibiotic resistance genes within a person's normal flora. When infected by a dangerous pathogen, the genes can confer drug resistance on the virulent recipient.

genetic information for resistance may also be carried by plasmids, which are readily transferred to susceptible cells by conjugation or transformation. Such transmittable resistance is often acquired by the normal flora. The overuse of antibiotics favors these plasmid-carrying strains, establishing a reservoir of R factors in the normal flora. Since plasmids can be transferred among different species, drug-sensitive pathogenic bacteria may acquire R factors from normal flora bacteria, and a patient's disease suddenly becomes untreatable by antibiotics that would have been effective prior to the plasmid transfer (Fig. 16-13). For example, *Escherichia coli*, a harmless intestinal resident, may donate R factors to the bacteria that cause dysentery or typhoid fever or to opportunists such as *Pseudomonas*. Such "plasmid promiscuity" allows for extensive spread of antibiotic resistance throughout a heterogeneous population of bacteria. R factors usually carry genes for multiple resistance, fortifying the bacterial recipient with protection from a number of drugs. Therefore, prolonged exposure to even a single antibiotic may favor the proliferation of bacteria resistant to several drugs.

■ Antibiotics in Animal Feeds

Antibiotic proliferation is a problem that goes beyond indiscriminate use in patients. Antibiotics are used extensively as additives in feeds for pigs, cattle, chickens, and other farm animals to encourage increased meat production. In fact, nearly half the antibiotics produced in the United States are destined for use in farm animals, mostly in sublethal doses too low to kill microorganisms. Unfortunately, this practice selects for the survival of antibiotic-resistant microorganisms, many of which contain R factors. Federal statutes prevent the sale of meat that contains residual antibiotics. However, antibiotic use in livestock feeds poses another potential danger that is not controlled by these statutes. The microorganisms shed by the living animal and during slaughter often contain R factors that may transfer antibiotic resistance to human pathogens and normal flora. Some antibiotic-resistant pathogens, notably *Salmonella* and *Listeria*, can be transferred directly from livestock animals to humans, a problem that has induced several countries to pass laws banning the use of antibiotics that are used to treat humans as growth-boosting supplements in animal feed. After such a law was enacted in The Netherlands, the incidence of tetracycline-resistant *Salmonella* infections in people dropped to less than one-fourth the rate before the supplements were outlawed. The Food and Drug Administration is arguing in favor of similar legislation in the United States, prompted partly by the landmark CDC[1] study correlating the practice of putting antibiotics in livestock feed with an increase in drug resistance in *Salmonella*.

■ Mechanisms of Antibiotic Resistance

Some microorganisms are naturally resistant to certain antibiotics because they lack the target that the antibiotic affects or because the drug cannot reach its site of action. Fungi, protozoa, and viruses, for example, contain no peptidoglycan and are naturally resistant to penicillin and other inhibitors of bacterial cell wall synthesis. Sensitive microbes, on the other hand, may become resistant to a drug by gaining the ability to

- Inactivate or destroy the antibiotic
- Alter their own membranes so they are no longer permeable to the agent
- Alter the target site so it is no longer affected by the drug, or
- Develop a mechanism to bypass the target metabolic reaction.

● INACTIVATION OF ANTIBIOTICS. Many microorganisms produce extracellular enzymes that destroy an antibiotic's activity. Penicillinases, for example, are produced by many bacterial species, including *Staphylococcus*, *Neisseria*, *Pseudomonas*, *Proteus*, *Mycobacteria*, *Yersinia*, *Salmonella*, and *Shigella*. Other bacterial enzymes chemically modify antibiotics to forms that are poorly absorbed by the microbe.

● DECREASED ANTIBIOTIC UPTAKE. Most chemotherapeutic agents must be able to penetrate the cell wall and plasma membrane to achieve effective concentrations at an internal target site. A modification in the plasma membrane may reduce its permeability to the drug, thereby increasing the microbe's resistance. Altered membrane permeability, however, does not confer resistance against penicillin and cephalosporins, as these antibiotics block extracellular assembly of peptidoglycan.

● ALTERATION OF THE TARGET SITE. Microorganisms commonly acquire resistance when a structure or enzyme that is normally impaired by an antibiotic is modified and is no longer recognized by the drug. Bacteria resistant to streptomycin, for example, produce modified ribosomes to which the antibiotic cannot bind. Protein synthesis continues unimpaired in these bacteria even in high concen-

[1]CDC = Centers for Disease Control and Prevention.

Pathogen	Disease	Drug of Choice
Shigella	Bacillary dysentery	Fluoroquinoline
Treponema pallidum	Syphilis	Penicillin
Vibrio cholerae	Cholera	Tetracycline†
Yersinia pestis	Plague	Streptomycin
Acid-Fast Bacilli		
Mycobacterium avium-intracellulare	AIDS-related pneumonia	Various drug combinations
Mycobacterium leprae	Leprosy	Dapsone with rifampin
Mycobacterium tuberculosis	Tuberculosis	INH with rifampin
Other Bacteria		
Actinomyces israelii	Actinomycosis	Penicillin
Chlamydia trachomatis	Nongonococcal urethritis, trachoma	Tetracycline
Mycoplasma pneumoniae	Pneumonia	Tetracycline
Nocardia sp.	Nocardiosis	Sulfonamides
Rickettsia spp.	Spotted fevers; typhus	Tetracycline

*In each case, in vitro susceptibility tests should confirm the pathogen's sensitivity to the drug selected.
†Antibiotics are used in severe cases only.

TABLE 16-6   DRUGS RECOMMENDED AGAINST COMMON EUKARYOTIC PATHOGENS

Pathogen	Disease	Drug of Choice
Fungi		
Aspergillus spp.	Aspergillosis	Amphotericin B
Candida albicans	Disseminated candidiasis	Amphotericin B
	Oral candidiasis	Nystatin
	Vaginal candidiasis	Miconazole or clotrimazole
	Esophageal candidiasis*	Fluconazole
	Chronic mucocutaneous candidiasis	Ketoconazole
Coccidoides immitis	Coccidioidomycosis	Amphotericin B
Cryptococcus neoformans	Cryptococcosis	Amphotericin B with flucytosine; fluconazole
Dermatophytes	Tineas (ringworm)	Clotrimazole or miconazole (topical)
Histoplasma capsulatum	Histoplasmosis	Ketoconazole
Algae		
Prototheca	Protothecosis	Amphotericin B
Protozoa		
Entamoeba histolytica	Amebiasis	Metronidazole with or without diiodo-hydroxyquin
Giardia lamblia	Giardiasis	Quinacine
Leishmania sp.	Leishmaniasis	Stibogluconate
Naegleria fowlerii	Amebic encephalitis	Amphotericin B
Plasmodium sp.	Malaria	Chloroquine
Pneumocystis carinii	Pneumocystis pneumonia*	Cotrimoxazole
Toxoplasma gondii	Toxoplasmosis	Pyrimethamine with sulfonamide
Trichomonas vaginalis	Trichomoniasis	Metronidazole
Trypanosoma sp.	Chagas's disease	Bayer 2502
	Sleeping sickness	Suramin

*In AIDS patients.

● LIMITATIONS OF ANTIBIOTIC SUSCEPTIBILITY TESTS.

The number of drugs that can be employed against eukaryotic pathogens is quite limited (Table 16-6), and antibiotic susceptibility testing is rarely employed. Fortunately, except for malaria, the development of resistance to these drugs has not been a serious problem. Currently, there are no useful drug susceptibility tests for viruses even though there is a high incidence of drug resistance in many viruses. Because of the lengthy treatment people with AIDS or herpesvirus infections receive, the likelihood is high that resistant viruses will emerge within the patient's body.

Unfortunately, standard antibiotic susceptibility tests are not very useful for *Mycobacterium tuberculosis* and other organisms that grow slowly in the laboratory but rapidly in compromised patients. An ingenious approach, however, promises to use genetic recombination to solve this problem. The new test uses a mycobacteriophage as a vector for transferring a firefly gene into the tubercle bacillus. The gene for luciferase production enables the cell to produce light using its own ATP as the energy source. After treatment with the antibiotic, actively growing *Mycobacteria*, after only 2 days of growth, generate enough ATP to ignite a detectable glowing response, whereas antibiotic-sensitive strains produce no ATP in the presence of the drug and fail to glow (Fig. 16-16).

MYTH: The AIDS virus is the most dangerous microbe on earth.

FACT: *Mycobacterium tuberculosis* is currently the planet's most dangerous pathogen, killing 3 million people each year (AIDS has claimed a total of 1.5 million lives since its discovery in 1981).[2] A full one-third of the earth's people may be harboring the tubercle bacillus. One reason the disease persists as the world's foremost microbial menace is the pathogen's resistance to the drugs most frequently prescribed to fight it.

● MONITORING DRUG LEVELS.

Knowing the concentration of antibiotics in body fluids can help in ascertaining whether therapeutic levels have actually been reached at the site of infection. This is particularly important when chemotherapy fails to promote patient recovery even when the pathogen was shown to be sensitive to the agent used. The drug should have a serum level that is between 2 and 8

[2]1993 data from the World Health Organization.

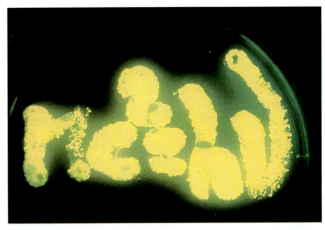

FIGURE 16-16 A glowing report. Because of its slow growth, *M. tuberculosis* typically takes 3 months of culturing before enough bacteria can be obtained to determine antibiotic susceptibility. The patient's tubercle bacteria readily accept the firefly's luciferase gene, enabling it to glow. After being treated with various antimycobacterial agents, the strain is streaked on agar. This glowing strain remains viable enough to produce ATP (needed to fuel the glow) even after exposure to INH. It is therefore resistant to the drug.

times the MIC of the organism to ensure that the tissue concentrations, which are lower than those in the serum, will reach the MIC. Drug concentrations are also monitored to prevent harmful side effects. Patients receiving amphotericin B or aminoglycoside therapy, for example, may develop impaired kidney or liver function if the drug exceeds safe levels.

Antibiotic concentration in tissue fluids is usually monitored by placing a sterile paper disk saturated with the patient's serum or body fluid on a solid medium seeded with the test organism and measuring the zone of inhibition around the disk (Fig. 16-17a). This diameter is compared with zones around disks that contain known amounts of the drug so that the zone size can be translated into actual drug concentrations (Fig. 16-17b).

■ Guidelines for Antimicrobial Drug Use

The most difficult problem associated with antimicrobial chemotherapy is the emergence of antibiotic-resistant microorganisms. The development of new antibiotics that are effective against the resistant organisms provides only temporary solutions. Bacteria resistant to the new drugs quickly appear and may transfer this resistance via plasmids to other bacteria. The rational approach to this problem is to greatly reduce the indiscriminate use of antibiotics. This is the only way to discourage the

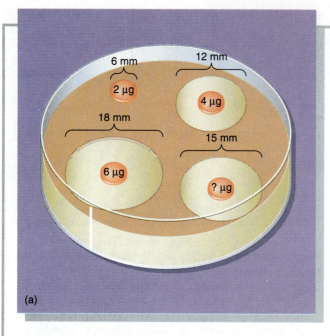
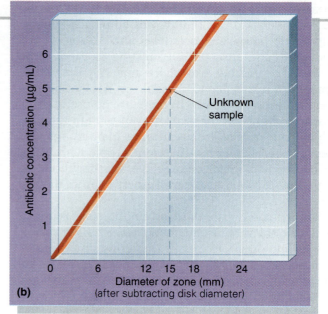

FIGURE 16-17 Measuring drug concentrations in body fluids. (*a*) Paper disks containing known concentrations of antibiotic and one unknown serum sample are placed on a seeded agar plate and incubated until zones of inhibited growth can be measured. The results of the test in this diagram indicate a zone size of 15 mm for the serum sample. (*b*) The data generated by the three known controls are used to generate a standard curve that can be used to determine the unknown concentration in the patient's serum. In this case, the patient's serum contains 5 μg/mL of active antibiotic.

proliferation of antibiotic-resistant microbes and to safeguard the effectiveness of existing antimicrobial drugs. The success of this strategy depends on adoption of the following measures:

- Lower the need for antimicrobial chemotherapy by reducing the transmission of disease.
- Develop antibiotic susceptibility tests that provide accurate results more rapidly so that the use of ineffective or broad-spectrum drugs will be reduced.
- Rely on microbiological laboratory information to reveal which drugs are effective against the isolated pathogen.
- Do not treat viral disease with antibacterial chemotherapeutic drugs; to do so encourages drug-resistant secondary bacterial infections.
- Limit the prophylactic use of antibiotics to those infrequent situations for which their effectiveness has been demonstrated.
- Reserve certain antibiotics for use *only* against infections that fail to respond to the recommended drugs. If the medical community would uniformly agree to restrict the use of these antibiotics, most microbes would rarely be exposed to the reserve drugs, thereby discouraging the proliferation of

resistant pathogens. When needed, these antibiotics would likely be effective.

The overuse of antibiotics has contributed to a profoundly dangerous medical threat to modern health care. In some hospitals, as many as 50 percent of the patients receiving antibiotics have no laboratory cultures. Up to 70 percent of patients receiving antibiotics do not even show evidence of infection. Many hospitals require that all major prescriptions be reviewed and approved by personnel attuned to the problems of antibiotic abuse and aware of the alternatives for treatment. Automatic stop orders ensure that no antimicrobial drug is administered for more than 48 hours without a physician review. If its continued use is truly required, the drug must be reordered.

The public should also be informed of the limitations and hazards of antibiotics. Physicians too often succumb to pressure from patients who expect "wonder drugs" to cure even the mildest ailments. Public awareness may relieve some of this pressure to overprescribe.

If current guidelines governing judicious use of antimicrobial agents were observed, the problem of drug resistance could be controlled. The future effectiveness of these important weapons against infectious disease depends on adoption of a conscientious and responsible approach to antimicrobial chemotherapy.

Antimicrobial Therapy— The Future

In the last few decades, the search for new antibiotics has not yielded many new products. The traditional screening studies have examined soil microbes with such exhaustive investigations that these environments are unlikely sources of future antibiotics. Such investigations, therefore, are being shifted to more promising environments. For example, ecosystems such as tropical rainforests and marine habitats contain plants, animals, and microbes whose physiologies have yet to be explored. Their ecological diversity holds the promise of new antibiotics and other valuable compounds. Plants, for example, synthesize a tremendous variety of organic compounds that deter predatory herbivores or pathogenic microorganisms. Unfortunately, these ecosystems are being disrupted and destroyed at alarming rates. The tropical rainforests are disappearing at the rate of 50 acres per minute, most of them cleared for farm and ranchland. The therapeutic value of tropical rainforest plants is undeniable—they have provided us with more than 40 percent of our medicines. Thousands of pharmaceuticals, from aspirin to lifesaving treatments for leukemia and malaria, come from tropical plants, and we have examined (or even discovered) only a fraction of the species that inhabit this richest of all ecosystems on earth. At the current rate of destruction, however, all the tropical rainforests on earth will be destroyed in the next 50 years, and with them will disappear a treasury of chemotherapeutic agents from plants that grow nowhere else. Cures for cancer and AIDS may be in plants on the brink of extinction. This is only one way in which the destruction of these environments represents a threat to our future.

Scientists are also examining why certain plant and animal materials that are commonly exposed to a microbe-laden environment resist microbial degradation. Why, for example, does a beehive not readily deteriorate? What protects fish and invertebrate eggs discharged into marine environments from microbial attack during their development? Why don't frogs get infections through broken skin from microbes in their murky water habitats? The answers seem to lie in protective antimicrobial substances such as *magainins*, a group of peptides that protect the frog's skin from a wide range of potential pathogens. Dogfish sharks also make an infection-fighting compound, a steroid called *squalamine*, that kills microbes before the fish's immune system even encounters them. Isolation of more of these defensive molecules from animals may lead to a new generation of antimicrobial agents for chemotherapy.

Another new source of potential antimicrobial agents are chemicals released from leukocytes (white blood cells), chemicals called *defensins*. These are natural peptide "antibiotics" from neutrophils, and some scientists believe a whole new generation of chemotherapeutic agents will be patterned on endogenous antimicrobials such as these (Fig. 16-18). These endogenous compounds are broad-spectrum in their killing effectiveness and are even effective in neutralizing enveloped viruses in the body. The amino acid sequences of these mole-

FIGURE 16-18 Crystal of human defensin peptide. This is an endogenous chemical (from leukocytes) with potential therapeutic value as an antimicrobial agent against infectious disease. At least six defensins have been crystallized and found to be effective in killing bacteria, fungi, protozoa, viruses, and tumor cells.

cules have been determined, and their chemical synthesis is under development.

The ability to use available drugs more effectively and with fewer side effects is another area of intense research. Much of this work is directed at new methods of drug delivery. For example, the drugs are packaged into membrane vesicles called **liposomes,** which transport the drug to the pathogen (Fig. 16-19). Liposomes are quickly engulfed by phagocytes and become concentrated in circulating macrophages and in the liver and spleen. They are especially useful vehicles for treating infections of these organs such as those caused by *Histoplasma, Leishmania,* and *Mycobacterium.* Because very little free drug enters the circulation, toxicity is reduced. This means that higher doses of effective but toxic drugs such as amphotericin B may be used in treatment. Liposomes are approved only for investigational purposes in the United States.

Another area of research is controlled release of drugs. Patients who require long-term therapy tend to become discouraged and discontinue their treat-

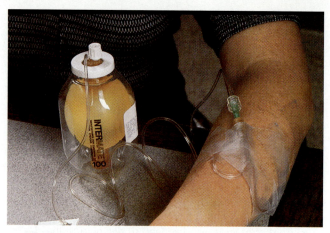

FIGURE 16-20 Home drug delivery system for long-term IV drug use. This commercial system is purchased with the drug in place. The patient need only aseptically attach the tubing to a catheter previously inserted by the physician.

ment. This is especially true if the drug has to be administered in a medical setting, requiring frequent visits to the clinic. Controlled release helps solve these problems. One technique is to incorporate the drug with a polymer into a capsule that is implanted into the patient. The drug dissolves slowly into the bloodstream over an extended period of time. Treatment is also being improved by the introduction of intravenous drug delivery systems that patients can use at home (Fig. 16-20). These disposable devices may come prepackaged with drug and not require professional help to use.

Molecular biology and genetic engineering are also revolutionizing the development of new chemotherapetic agents. As our understanding of pathogens increases, so does our ability to design effective drugs against them. For example, the fungal cell wall is the target of several antifungal agents currently under investigation. Several anti-AIDS drugs are directed against viral protease, an enzyme needed for successful replication. Other potential new treatments include antisense nucleic acids (complementary to viral RNA sequences) and genetically engineered immunopotentiators—peptides and proteins that specifically stimulate the immune system. ▶ (See Antisense Technology, Chap. 9, p. 750; THE EXPLORERS, Chap. 18, p. 470; and Therapeutic Strategies, Chap. 27, p. 750.)

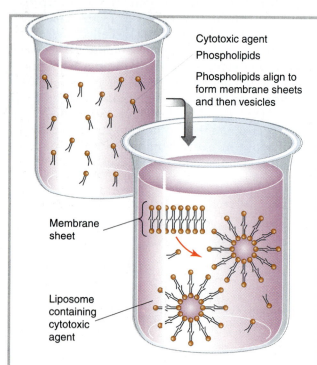

FIGURE 16-19 Liposome drug delivery. Hydrophobic forces align phospholipids into membrane sheets that fold into vesicles. Packaged in these synthetic membrane bags are cytotoxic (cell-killing) agents that will be released only when the liposome membrane fuses with its target cell (such as a fungus, a tumor cell, or a virus-infected cell). In this way the chemical is concentrated on the cells to be killed, sparing the rest of the body exposure to the toxic effects of the agent.

Labels in Figure 16-19:
- Cytotoxic agent
- Phospholipids
- Phospholipids align to form membrane sheets and then vesicles
- Membrane sheet
- Liposome containing cytotoxic agent

CONCEPT CHECK

- What strategies are being used to discover new antimicrobial drugs?
- In what ways is the effectiveness of known chemotherapeutic agents being increased?

Key Terms

Key Facts and Concepts

- **Selective toxicity.** Antibiotics and chemotherapeutic drugs that are selectively toxic to microbes are important agents for treating or preventing infectious disease. They may be used before infection occurs (chemoprophylaxis) or for eliminating a pathogen from the body (chemotherapy).

- **Lethal or inhibiting.** Chemotherapeutic drugs are microbicidal or microbistatic. Microbistatic drugs inhibit microbial growth until host defenses ultimately eliminate the pathogens.

- **Criteria for therapeutic effectiveness.** Chemotherapeutic drugs must not only be selectively toxic against the target microorganism, they must also reach therapeutic concentrations at the site of infection and not be inactivated by other drugs or foods that the patient is consuming.

- **Drug resistance.** Microorganisms may become drug-resistant as a result of a mutation. Bacteria may also receive R factors from a resistant donor. Drug-resistant microbes usually become the dominant organisms in environments that contain the corresponding antimicrobial agent.

- **Guidelines for antimicrobial drug use.** The indiscriminate use of antibiotics encourages the proliferation of drug resistance among pathogens and normal flora. Problems can be minimized by limiting chemoprophylaxis to situations of proven effectiveness, isolating pathogens prior to chemotherapy and identifying their drug sensitivities, and determining the patient's medical history to prevent allergic and toxic reactions.

- **Evaluating therapeutic effectiveness.** It is usually impossible to predict a pathogen's drug sensitivity, so laboratory tests are used to determine which drugs are most effective. Therapeutic effectiveness of several antibiotics can be evaluated by disk diffusion tests or by determining MICs. Similar methods are used to test a patient's body fluids to determine the level of active antibiotic at the site of infection. Automated techniques identify a pathogen's antibiotic sensitivities within 4 to 6 hours of isolation.

- **Choosing the right drug.** In addition to the pathogen's drug sensitivities, other factors to consider when choosing the best agent for antimicrobial chemotherapy include the drug's access to the site of infection, how it interacts with other medications the patient is taking, and the duration of treatment.

Review and Problem-Solving Questions

1. List the major cellular targets of antibiotics and name an antibiotic directed against each target for (a) prokaryotic cells; (b) eukaryotic cells.
2. Distinguish between:
   (a) Broad-spectrum and narrow-spectrum antibiotics
   (b) Natural and semisynthetic penicillins
   (c) Chemotherapy and chemoprophylaxis
   (d) MIC and MBC
3. What are the characteristics of an ideal chemotherapeutic agent? How do each of the following types of drugs conform to your ideal? (a) Penicillins; (b) sulfa drugs; (c) polyenes; (d) aminoglycosides; (e) tetracyclines; (f) acyclovir.
4. Explain why so few antibacterial drugs are targeted against nucleic acids and so many of the antiviral drugs are.
5. Why is a drug such as amphotericin B less selective than tetracycline?
6. Explain why each of the following statements is false.
   (a) The use of antibiotics creates resistant strains of bacteria.
   (b) A zone of inhibition in the Kirby-Bauer test indi-

cates that the antibiotic will be effective.

(c) It is all right to discontinue use of medication as soon as symptoms of a disease have disappeared.

7. Identify three characteristics possessed by penicillin derivatives that make them effective in situations in which penicillin G is ineffective.

8. When is it more appropriate to use a broad-spectrum drug than to use a narrow-spectrum drug?

9. Why is it advisable to select an unrelated antibiotic to treat an infection caused by antibiotic-resistant bacteria?

10. Distinguish between the Kirby-Bauer test and MIC determinations.

11. What is an antibiogram?

12. You are the newly hired infection control officer in Good Care Hospital. After spending 3 weeks reviewing patient and laboratory records, you determine that the number of antibiotic-resistant bacteria isolated from hospital patients is too high. What recommendations would you make for hospital policy to reduce the number of antibiotic-resistant bacteria?

13. As a new physician on a hospital staff, you are encountering situations requiring decisions about whether or not to administer prophylactic antibiotic therapy. For each of the following, explain why you feel the patient does or does not require chemoprophylaxis.

(a) Gastrointestinal surgery

(b) Tonsillectomy on an otherwise healthy child

(c) Healthy patient who is to have a tooth pulled

(d) Peace Corps volunteer going to the tropics

(e) AIDS patient having minor surgery

14. Briefly explain how a host's normal flora can render pathogens resistant to antibiotics.

15. If you were trying to isolate an antibiotic-producing bacterium from a new environment:

(a) Where might you look?

(b) How would you test isolated organisms for antibiotic activity?

(c) How would you determine whether the antibiotic was bactericidal or bacteristatic?

(d) What other factors would determine whether a new antibiotic could be used in humans?

A "NO-WIN" SITUATION FOR HIV?

The development of drug resistance by HIV has proven to be a major problem in long-term therapy of patients. Studies of HIV in cell culture showed that the virus develops mutations that confer resistance to drugs that inhibit reverse transcriptase. Even when the drugs are used in tandem, resistant viruses emerge. Y. K. Chow and associates* reasoned that there were a limited number of changes in amino acid sequence that the reverse transcriptase molecule could undergo and still retain enzymatic activity. Thus, using a combination of three or more drugs, each directed at a different portion of the viral enzyme, they believed they could force the virus into a no-win situation—either succumb to one of the antiviral agents or change the reverse transcriptase so much that it is no longer capable of DNA synthesis. Clinical trials of triple combination therapy against HIV were already in progress by the time their paper was published.

• In your own words explain why it might be better to use three drugs directed at the same target molecule than three drugs directed at different targets.

• Why won't a fatal combination of mutations develop from antibacterial combination therapy using streptomycin, tetracycline, and erythromycin? Using ampicillin, carbenicillin, and amoxicillin?

• What problems that might limit the effectiveness of combination therapy in humans would in vitro tests fail to reveal?

• After publication of their paper, the investigators discovered that the strain of HIV used in the research had a mutation that introduced an uncontrolled variable, making the results questionable. What should they do?

*Y. K. Chow, M. S. Hirsch, D. P. Merrill, L. J. Bechtel, J. J. Eron, J. C. Kaplan, and R. T. D'Aquila, Use of evolutionary limitations of HIV-1 multidrug resistance to optimize therapy, *Nature,* 361: 650–654 (1993).

PART FIVE

Infectious Processes and Host Responses

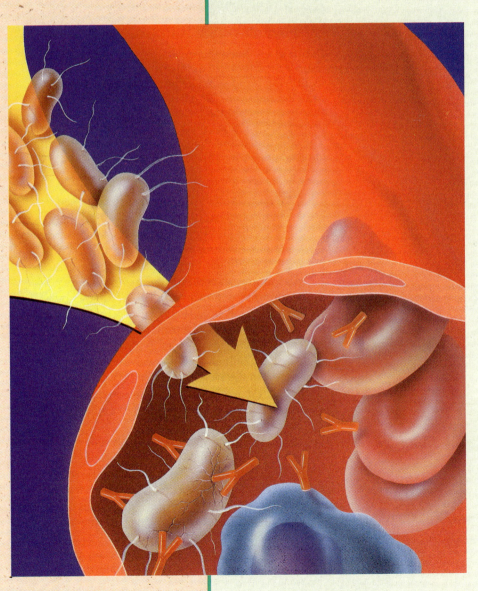

CHAPTER 17

Host–Parasite Interactions

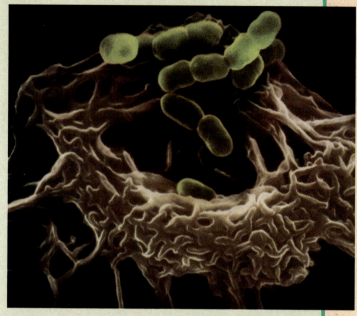

Death grip. Surveillance and attack is the job of many of the protective cells in your body, quickly identifying invaders and eliminating them before they can eliminate you. Here a clutch of doomed bacteria that have entered the tissue fluids are engulfed and destroyed by a macrophage, a type of white blood cell. The effectiveness of this protection is enhanced by another protective weapon, the immune system.

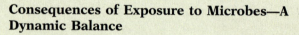

Randy Barlow yelped in pain as a mistake in coordination plunged one of his incisors through the surface of his tongue. Around the campfire, his three friends winced in sympathy and chewed their hamburgers more gingerly, not wanting to compromise the joy of eating by committing a similar painful act. Randy's punctured surface, however, immediately began to heal itself. In fact, the tongue is the second fastest regenerating structure in the human body (the liver is the fastest), quickly resealing the breach in the physical barrier between the inside of the body and the potential pathogens outside. But for Randy Barlow, biting his tongue was a fatal accident.

Streptococcal bacteria that normally reside in the human mouth poured into Randy's bloodstream by the billions through the newly created hole. Randy's backup defensive troops swung into action—his white blood cells and blood proteins rapidly converged on the invaders and destroyed those in the circulation. Unfortunately, Randy had an especially vulnerable heart valve, one that was partially deformed by a bout with rheumatic fever several years earlier. Although his white blood cells and protective proteins successfully fought the invading bacteria in the bloodstream, they could not eliminate the intruders that colonized his compromised heart.

By the third day of his camping trip, Randy was severely sick. His friends carried him out of the mountains and flagged down a passing car, which rushed the critically ill boy to the hospital. Doctors started immediate antibiotic therapy to kill the bacteria, plus emergency measures to rescue his besieged heart. Randy died anyway, a victim of his own normal flora.

We all make the same mistake that killed Randy. We bite our tongues; we step on nails; we cut ourselves—all accidents that damage the "fortress walls" that keep most intruders out of our bodies. Our ability to survive such day-to-day hazards depends on the fight our bodies put up against invading microorganisms. Even as you were reading about Randy's tragic episode, you too were being attacked. Each time you breathe, you inhale thousands of microorganisms suspended in the air, many of which would fatally attack your respiratory tract if you didn't have an arsenal to protect against such invasion. Trillions of microorganisms populate the surface of your body—bacteria and other agents that would invade and use your tissues as their next meal if allowed uncontested entry into the interior of your body. The body harbors 10 times as many bacteria as there are human cells, plus the new microbes you encounter each day. In fact, some microorganisms have very likely entered your bloodstream today, perhaps through cuts in your skin so small you were unaware of them. In your internal tissues and fluids, some microorganism has found a bonanza of nutrients and hospitable environmental conditions. The invasion has begun.

Your body quickly counterattacks. Any intruder faces a battery of internal defenses that quickly neutralize the invaders' chances of success (See opening photo of this chapter). Because these defenses work so well, we usually stay healthy. Even when you do occasionally suffer infectious disease, your immune system and other internal defenses usually come to your rescue. It is a life-and-death struggle, a battle that continues until you die.

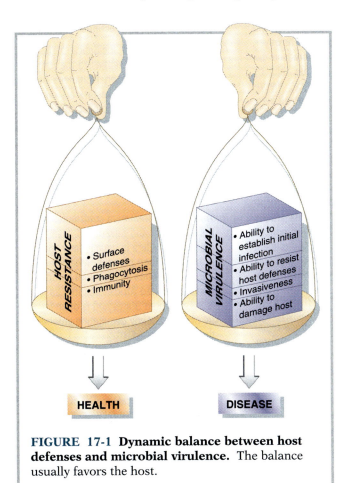

FIGURE 17-1 **Dynamic balance between host defenses and microbial virulence.** The balance usually favors the host.

Consequences of Exposure to Microbes—A Dynamic Balance

Why is it that we generally win the battle against the opportunistic microbes with which we interact so intimately each day? The outcome of exposure to microorganisms depends on a dynamic balance between the relative abilities of the host to resist infection and the invader to cause disease (Fig. 17-1). Many factors influence the balance of this host–parasite relationship. Factors that help a microbe to survive host defenses or to injure the infected host increase its ability to cause infectious disease. Fortunately, normal human defenses usually counteract these microbial factors and we stay healthy. That is not always the case, however, as anyone with influenza, athlete's foot, tuberculosis, or malaria can miserably testify. These people have **infectious diseases,** diseases caused by the growth of microorganisms in or on the body.

With the exception of diseases caused by the inoculation of microbes directly into the bloodstream or other internal tissues, an infectious disease is initiated by **colonization,** the establishment of microbes on the skin or mucous membranes. Microbial colonization of a body site may result in one of

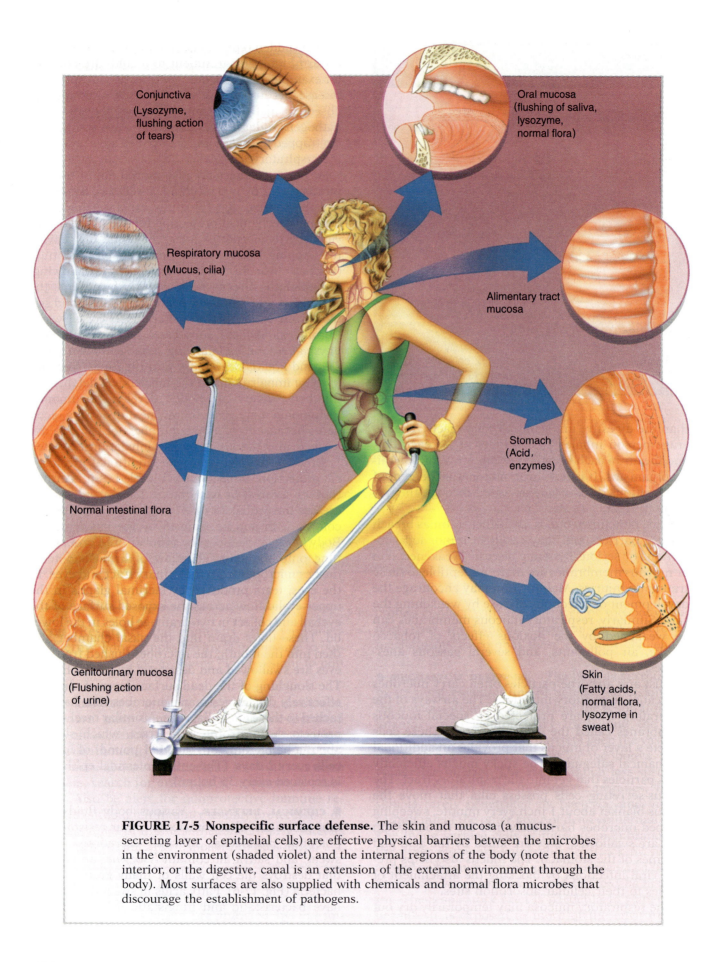

Conjunctiva
(Lysozyme,
flushing action
of tears)

Oral mucosa
(flushing of saliva,
lysozyme,
normal flora)

Respiratory mucosa
(Mucus, cilia)

Alimentary tract
mucosa

Normal intestinal flora

Stomach
(Acid,
enzymes)

Genitourinary mucosa
(Flushing action
of urine)

Skin
(Fatty acids,
normal flora,
lysozyme in
sweat)

FIGURE 17-5 Nonspecific surface defense. The skin and mucosa (a mucus-secreting layer of epithelial cells) are effective physical barriers between the microbes in the environment (shaded violet) and the internal regions of the body (note that the interior, or the digestive, canal is an extension of the external environment through the body). Most surfaces are also supplied with chemicals and normal flora microbes that discourage the establishment of pathogens.

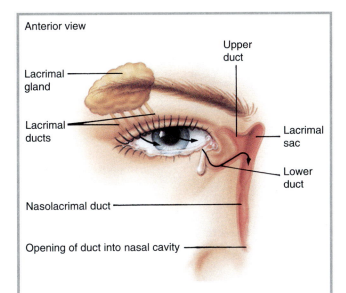

Anterior view

Lacrimal gland

Lacrimal ducts

Nasolacrimal duct

Opening of duct into nasal cavity

Upper duct

Lacrimal sac

Lower duct

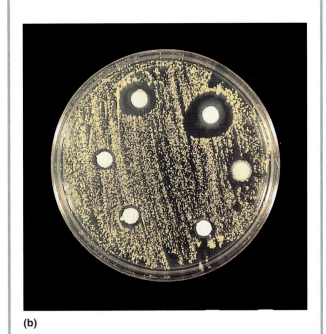

(b)

FIGURE 17-6 Destruction of a gram-positive bacterium by lysozyme in tears. *(a)* The flow of tears across the corneal surface flushes out microbes and also exposes them to antimicrobial chemicals such as lysozome. *(b)* to demonstrate this antimicrobial effect, various dilutions of tears in saline were placed on an agar plate seeded with *Micrococcus*. Growth inhibition is observable where higher concentrations were applied to the plate.

stomach contains a high concentration of hydrochloric acid, which rapidly kills most microbes, and in the small intestines bile destroys many of the survivors. The high acidity of the adult vagina pro-

tects its membranous surfaces against colonization by many types of pathogens. The digestive system produces *mucin,* a glycoprotein that coats microbes and prevents their attachment to intestinal epithelium. Several microbes, however, fight back by secreting enzymes that degrade mucin and inactivate this defense.

Several antimicrobial peptides and proteins have been isolated from cells of the skin and gastrointestinal tract of humans or animals. These molecules are similar in structure and activity to *defensins,* components of phagocytic cells that kill a wide variety of bacteria and yeast. In the host, they likely help protect these vulnerable surfaces from infection.

Interferons are a group of soluble proteins produced by host cells during infection by viruses and some other microbes. Uninfected cells in the area that are exposed to virus-induced interferon become resistant to viral infection ▶ (see *interferons,* Chap. 13, p. 328). Interferons also stimulate the activity of our phagocytic and immunologic defenses. Another protective factor released into the blood is *beta-lysin.* This protein attacks bacterial membranes.

Transferrins are iron-binding proteins that reduce the level of free iron in the blood. *Lactoferrins* perform the same function in milk, respiratory, intestinal, and genital secretions. The presence of free iron is necessary for the growth of most pathogenic microbes. Iron is also required by human cells, but it can be extracted from the iron-binding proteins as it is needed, whereas most bacteria cannot obtain protein-bound iron. Reducing free iron levels in the body therefore helps protect against microbial proliferation. This strategy is enhanced by mononuclear phagocytes that remove even more free iron from the blood during infection.

Interleukin-1 is a fever-inducing protein safely isolated in vacuoles of phagocytic cells, where it has no effect on the host. When stimulated by certain foreign substances such as bacterial endotoxin, phagocytes discharge the contents of these vacuoles, which circulate to the hypothalamic region of the brain and trigger an elevated body temperature. Although fever can itself damage human cells, it has several protective effects: it raises temperatures above that which is optimal for growth of some pathogens; it accelerates the mobilization and protective efficiency of the body's defenses; it stimulates lymphocyte activity in the immune response; and it increases the rate of iron storage, reducing iron availability to pathogens.

Complement is a complex group of serum proteins that act in concert with the immune system or

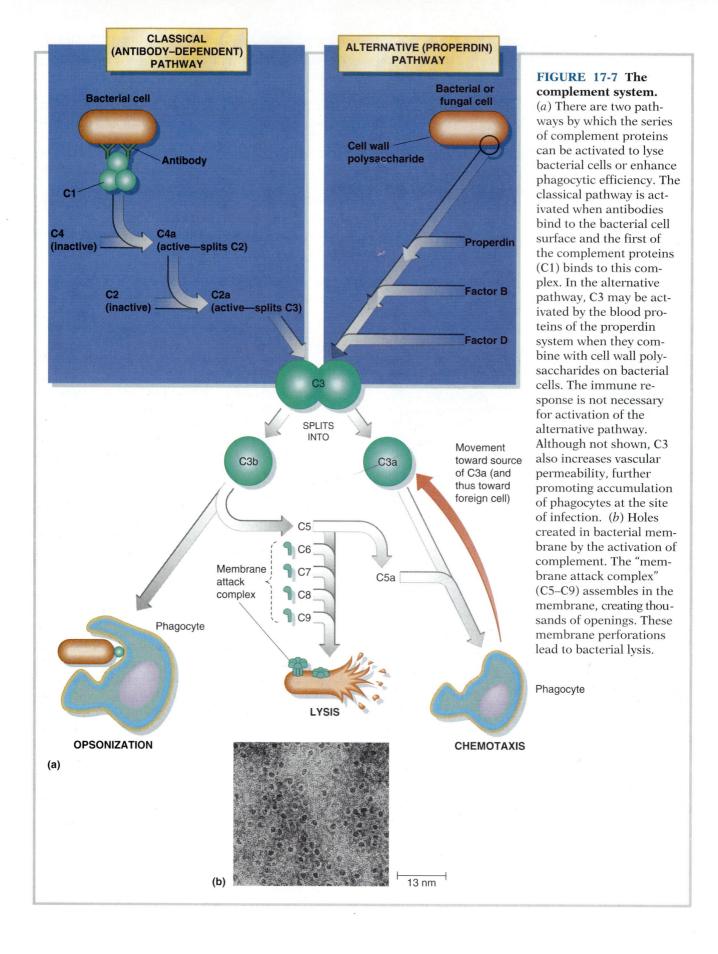

FIGURE 17-7 The complement system. (*a*) There are two pathways by which the series of complement proteins can be activated to lyse bacterial cells or enhance phagocytic efficiency. The classical pathway is activated when antibodies bind to the bacterial cell surface and the first of the complement proteins (C1) binds to this complex. In the alternative pathway, C3 may be activated by the blood proteins of the properdin system when they combine with cell wall polysaccharides on bacterial cells. The immune response is not necessary for activation of the alternative pathway. Although not shown, C3 also increases vascular permeability, further promoting accumulation of phagocytes at the site of infection. (*b*) Holes created in bacterial membrane by the activation of complement. The "membrane attack complex" (C5–C9) assembles in the membrane, creating thousands of openings. These membrane perforations lead to bacterial lysis.

CLASSICAL (ANTIBODY–DEPENDENT) PATHWAY

Bacterial cell

Antibody

C1

C4 (inactive) → C4a (active—splits C2)

C2 (inactive) → C2a (active—splits C3)

ALTERNATIVE (PROPERDIN) PATHWAY

Bacterial or fungal cell

Cell wall polysaccharide

Properdin

Factor B

Factor D

C3

SPLITS INTO

C3b   C3a

Movement toward source of C3a (and thus toward foreign cell)

C5
C6
C7   Membrane attack complex
C8
C9

C5a

Phagocyte

LYSIS

OPSONIZATION

CHEMOTAXIS

Phagocyte

(a)

(b)

13 nm

with other serum proteins to facilitate bacterial lysis and phagocytosis (Fig. 17-7). In the classical pathway, an antibody–pathogen complex activates the complement system. The alternative pathway occurs in the absence of an immune response but requires a blood protein (called *properdin*) to combine with surface molecules of the bacterial cell wall, launching the cascade of events that has the same outcome as the classical pathway—activation of phagocytes and damage to the bacterial membrane. Either pathway can eliminate the foreign cell that triggered the response. (The immune-mediated complement system is discussed in greater detail in Chapter 18.)

● **MICROBIAL DEFENSES.** The body's resident or normal bacterial flora provide another important line of defense. The *normal flora* continually compete with potentially pathogenic microbes for the limited nutrients and space on the body's epithelial surfaces (Fig. 17-8). Because the normal flora are already established, these beneficial microorganisms usually win the competition and prevent either initial infection or the unrestricted growth of pathogenic bacteria and fungi. Resident bacteria also have the ability to alter their local environment so that it is generally unfavorable for the growth of most pathogens. The lactobacilli that normally inhabit the adult human vagina, for example, produce lactic acid as a by-product of growth. This lowers the pH of the vagina to between 4.0 and 4.5, intolerably acidic for most pathogens. Metabolic by-products of skin flora create acidic conditions on the epidermal surfaces (pH of 3.0 to 5.0) that oppose the growth of many potential pathogens.

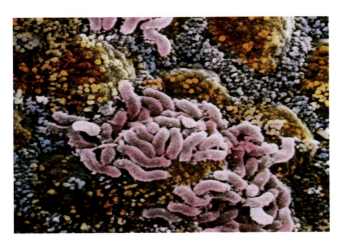

FIGURE 17-8 **Protective covering.** These nonpathogenic *Escherichia coli* normally attach to the intestinal villi in healthy hosts. With attachment sites already occupied by *E. coli,* many intestinal pathogens fail to attach and therefore cannot initiate infection.

MYTH: People cannot survive without their normal flora.

FACT: Normal flora microbes help compete with dangerous pathogens, help maintain vaginal and intestinal balance, and provide some nutrients, but they are not essential for the life of the host. For example, David (often known as "the boy in the bubble") was kept in a sterile environment for all but the last half year of his 12-year life, where he was never in contact with any microbe. Similarly germ-free animals raised for research purposes suffer no adverse effects from having no indigenous microbes. Although desirable, normal flora organisms are not essential.

■ Phagocytic Defenses

When infectious agents penetrate the surface defenses of the body, they are usually devoured by **leukocytes** (white blood cells) or other phagocytic cells. These cells engulf and destroy foreign particles by *phagocytosis* (Fig. 17-9*a*). Most phagocytic cells in the circulation contain large numbers of lysosomes in their cytoplasm. Lysosomes contain a number of digestive enzymes—peptidoglycanases, lipases, and proteases—plus chemicals that oxidize and destroy engulfed cells. As a result of phagocytosis, an invading microbe is trapped within an intracytoplasmic vacuole called a *phagosome.* Lysosomes migrate to the vacuole membrane and discharge their contents into the phagosome (Fig. 17-9*b*). The phenomenon, called *degranulation,* occurs by fusion of lysosomal and vacuole membranes, and the phagocytic vacuole grows into a larger membranous sac called a *phagolysosome.* The chemicals in the phagolysosome kill and digest the engulfed microbe. Human tissue can also be damaged by the lysosomal contents. The vacuole membrane, however, usually prevents tissue injury by safely housing the discharged chemicals, preventing their sudden release into the cell's cytoplasm or into the surrounding tissue.

Phagocytes also kill microbes by converting oxygen to toxic products such as superoxide and hydrogen peroxide (Fig. 17-10). Increased respiration, called the respiratory burst, initiated during the early stages of phagocytosis generates these toxic oxidants.

● **PHAGOCYTIC CELLS.** Lysosomes are characteristic of a group of white blood cells called *granulocytes.* (The cells found in human blood are depicted

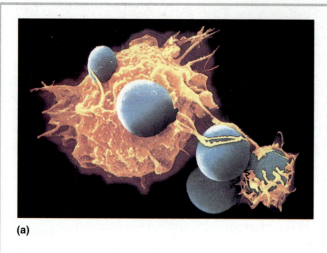

(a)

FIGURE 17-9 Phagocytic defense. (*a*) A human macrophage (a type of phagocytic cell) about to engulf several yeast cells. (*b*) Phagocytic destruction of engulfed particles. Following engulfment, lysosomes discharge their antimicrobial contents into the phagocytic vacuole by membrane fusion. Because the lysosomes (granules) disappear, the process is called degranulation.

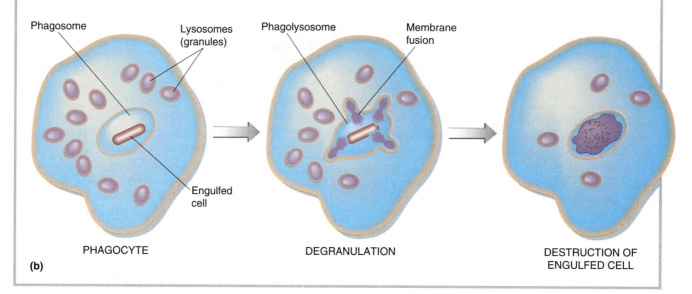

Phagosome   Lysosomes (granules)   Phagolysosome   Membrane fusion

Engulfed cell

(b)   PHAGOCYTE   DEGRANULATION   DESTRUCTION OF ENGULFED CELL

in Fig. 17-11.) Of the three types of granulocytes found in normal blood, **neutrophils** are the most abundant. These cells are also called *polymorphonuclear leukocytes* (PMNs) because of their irregular multilobed nuclei. Neutrophils are actively phagocytic and are usually the first protective cells to arrive at the site of trauma or infection. They do their work rapidly and are fairly short-lived. Two other types of granulocytes—eosinophils and basophils—are not phagocytic but contribute to defense by producing substances that help regulate the host response to injury or infection.

The other group of leukocytes is made up of *agranulocytes*, so called because they have fewer dark-staining lysosomes. Nonetheless, lysosomes are present in numbers great enough to effect the destruction of engulfed particles during phagocytosis. Agranulocytes include *lymphocytes* (immune cells discussed in Chapter 18) and *monocytes*. Mono-

cytes have a single large oval or horseshoe-shaped nucleus. Circulating monocytes migrate into tissue, where they enlarge and differentiate into actively phagocytic **macrophages.** Most macrophages are attached to tissues of the liver, spleen, lymph nodes, and bone marrow and to the walls of blood and lymph vessels. They engulf foreign particles and debris from blood and lymph that flow through these regions. These *fixed macrophages* therefore function as filters to clean debris from the blood and trap potential pathogens. *Wandering macrophages* migrate to the lungs, spleen, and other sites where microbes are likely to be encountered. Macrophages also travel to areas of trauma or infection to participate in the body's overall protective response. In addition, macrophages play an essential role in helping the immune system recognize antigens and respond to these foreign substances (this "processing" of antigens by macrophages is described in Chapter 18).

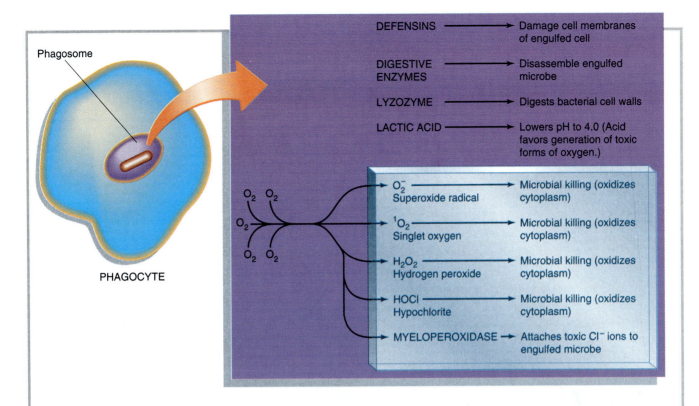

FIGURE 17-10 Respiratory burst and microbial killing. In addition to enzymes that disassemble microbes in phagolysosomes, some toxic effects are associated with a burst of oxygen consumption and an increase in oxidation of glucose. This generates a variety of bactericidal products within the phagocytic cell. These oxygen–dependent killing effects are highlighted in the blue box.

The monocytes and macrophages constitute the body's *mononuclear phagocytic system* (part of the reticuloendothelial system, which consists of all phagocytic cells and tissues to which fixed macrophages attach).

● **INFLAMMATION.** Phagocytosis is an essential component of **inflammation**—a concentrated protective response to infection or tissue injury in which the body attempts to localize and destroy infectious microorganisms and repair damaged tissues. The cells of injured tissues release their contents and increase acidity at the site of injury. This activates local enzymes, called *kinins,* that increase the permeability of capillaries in the immediate region, promoting the escape of plasma from the vessels into the local tissues. Kinins also promote the release of histamine, prostaglandins, and other substances from surrounding cells. These chemicals further increase vascular permeability, so even more plasma accumulates, bringing complement, antibodies, and other protective chemicals to the site of injury. They also stimulate chemotaxis, the migration of white blood cells toward the source of injury or infection. These responses characteristically elicit four symptoms at the site of injury: swelling, pain, redness, and heat (Fig. 17-12). The pain associated with inflammation is also due to the release of irritant chemicals that bind to nerve endings.

MYTH: Inflammation is synonymous with infection.

FACT: Inflammation most often occurs in response to infection and is generally a protective response that fights infection. Sometimes inflammation even occurs in the absence of any microbial attack—following injury, for example.

Inflammation protects the host by promoting phagocytosis of microbes, by localizing infection

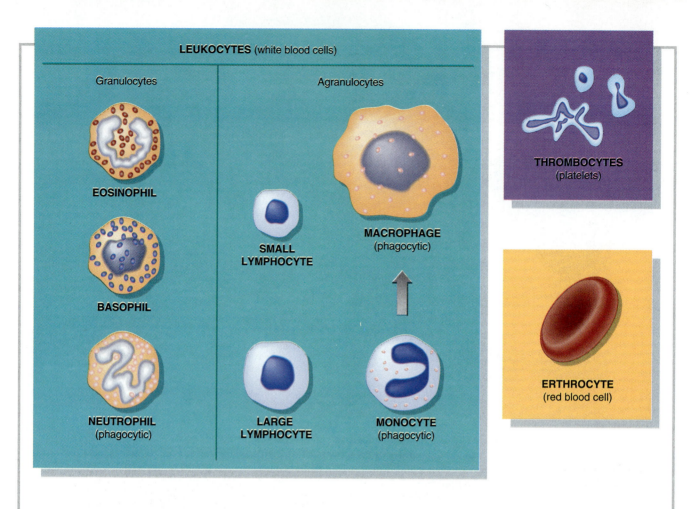

FIGURE 17·11 **Mature formed components in the fluid of human blood.** Erythrocytes (red blood cells), thrombocytes (platelets), and several types of white blood cells (granulocytes and agranulocytes) are all found in normal human blood. The granulocytes are the neutrophils, eosinophils, and basophils. The agranulocytes are the lymphocytes and monocytes (which differentiate into actively phagocytic macrophages). Except for erythrocytes, which carry oxygen, all these cells directly contribute to the body's defense. Thrombocytes aid in blood clot formation, which, in addition to preventing blood loss, helps restrict the spread of microbes. Neutrophils and macrophages are actively phagocytic. Lymphocytes are the body's immune cells, producing antibodies and other factors that contribute to specific acquired immunity.

(microbes are often walled off in capsules of clotted plasma), and by producing an *exudate* (pus), which in some cases allows direct drainage of microbes and dead tissue out of the host's body (as from a pimple, for example). The symptoms of inflammation also alert us to the presence of infection, increasing the likelihood of early diagnoses and treatment.

The pain of inflammation often encourages the use of corticosteroids or other anti-inflammatory agents to reduce discomfort and suffering. Although cortisone therapy is occasionally warranted (for treating chronic inflammation of a noninfectious origin, for example), the body's overall resistance to infectious disease is compromised by such an ap-

proach. Cortisone inhibits mobilization of leukocytes and destruction of pathogens in three ways:

1. By delaying the attachment of leukocytes to the capillary wall, thereby interfering with movement of leukocytes from inside blood vessels to the surrounding tissue
2. By reducing chemotaxis of phagocytes into the area and interfering with phagocytic engulfment
3. By stabilizing lysosomes so they fail to fuse with the vacuole membrane and are therefore unable to release their contents into the phagosome

Although cortisone improves the way an infected person feels by relieving the symptoms of inflam-

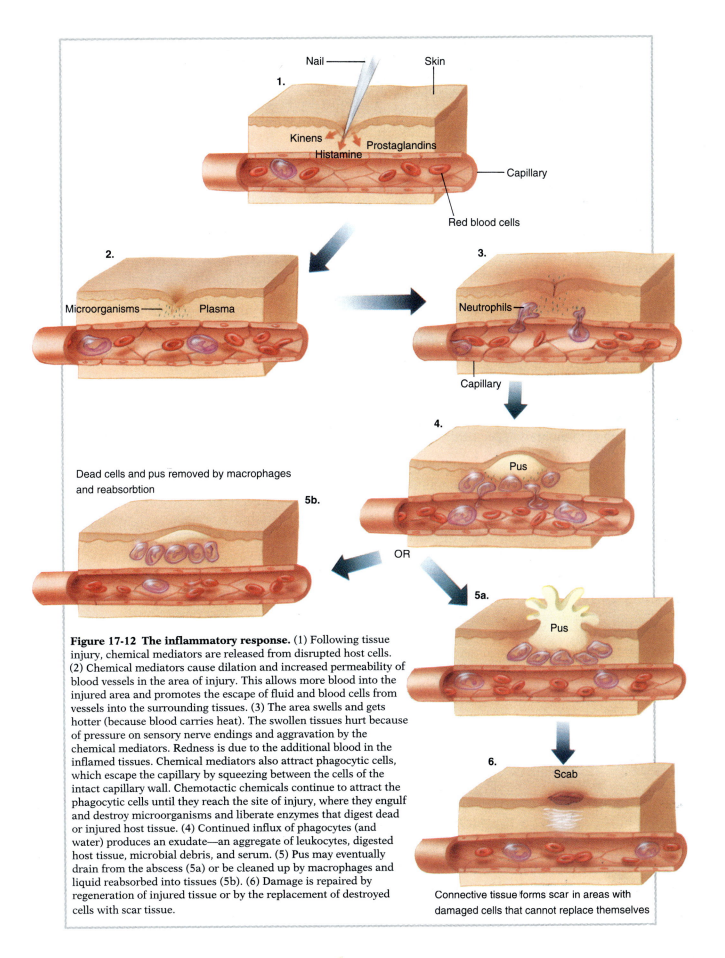

Figure 17-12 The inflammatory response. (1) Following tissue injury, chemical mediators are released from disrupted host cells. (2) Chemical mediators cause dilation and increased permeability of blood vessels in the area of injury. This allows more blood into the injured area and promotes the escape of fluid and blood cells from vessels into the surrounding tissues. (3) The area swells and gets hotter (because blood carries heat). The swollen tissues hurt because of pressure on sensory nerve endings and aggravation by the chemical mediators. Redness is due to the additional blood in the inflamed tissues. Chemical mediators also attract phagocytic cells, which escape the capillary by squeezing between the cells of the intact capillary wall. Chemotactic chemicals continue to attract the phagocytic cells until they reach the site of injury, where they engulf and destroy microorganisms and liberate enzymes that digest dead or injured host tissue. (4) Continued influx of phagocytes (and water) produces an exudate—an aggregate of leukocytes, digested host tissue, microbial debris, and serum. (5) Pus may eventually drain from the abscess (5a) or be cleaned up by macrophages and liquid reabsorbed into tissues (5b). (6) Damage is repaired by regeneration of injured tissue or by the replacement of destroyed cells with scar tissue.

mation, the infection may progressively worsen. Chronic emotional stress can have a similar anti-inflammatory effect due to the long-term release of corticosteroids from the adrenal glands. Similarly, uncontrolled diabetes mellitus reduces resistance to infection by lowering inflammatory protection. Overexposure to X-rays can depress the production of phagocytic leukocytes by the bone marrow and allow the normally harmless nasopharyngeal microbes to cause fatal opportunistic infections of the lower respiratory tract, bloodstream, or central nervous system.

■ Immunity

The body's next line of defense, the immune response, differs from other types of resistance in the following ways:

1. Immunity to a particular disease is acquired during a person's lifetime following encounters with microbes or foreign substances or by the introduction of antibodies or immune cells produced in another individual.
2. The resultant immunity specifically protects against the single type of organism or substance that induced the response. For example, immunity to diphtheria does not confer immunity to unrelated diseases such as measles.

Immunity is discussed in the next two chapters.

■ Compromised Hosts

True pathogens cause disease in otherwise healthy people. Some pathogens, however, cause disease only in those exposed to an overwhelmingly large number of microbes or those who have some **predisposing factor**—some condition that impairs host defenses, rendering them less capable of resisting infection. If a microbe causes disease *only* in people with impaired defenses, it is called an **opportunistic pathogen.** These organisms are usually common environmental contaminants or even members of our normal flora.

Anything that compromises host defenses may alter the outcome of a person's exposure to pathogens by predisposing the individual to the development of infectious disease. Predisposing factors can be categorized as intrinsic or extrinsic. *Intrinsic predisposing factors* are due to naturally occurring malfunctions of the body's defenses, including diabetes, circulatory disturbances, bone marrow failure, disorders of the mononuclear phagocytic system, or any of the other factors listed in Table 17-5. *Extrinsic predisposing factors*, on the other hand, are created by external influences such as trauma, medical procedures for treating other clinical disorders, or unhealthful personal characteristics, notably habitual alcohol consumption, smoking, fatigue, prolonged emotional stress, or malnutrition (Table 17-6).

Repeated infections by organisms of low virulence often indicate serious underlying immunologic deficiencies. For example, children who suffer

TABLE 17-5	SOME INTRINSIC FACTORS THAT PREDISPOSE HUMANS TO DEVELOP INFECTIOUS DISEASE
Predisposing Factor	**Effect**
Hormonal imbalance	Suppresses inflammation and antibody production
Diabetes	Impairs granulocyte function; may prevent protective cells in circulation from reaching local tissues
Extreme youth	Precedes maturation of immune system
Old age	Brings immunologic senescence
Allergy	May impede access of antibodies or protective cells to local tissue
Pregnancy	Rapid hormonal fluctuations may reduce inflammation and antibody production; anatomical changes may predispose for infections of the genitourinary tract
Immune deficiency diseases	Defects in cells of the immune system lead to partial or complete loss of immunity
Rheumatoid arthritis	Reduces chemotaxis of phagocytes, promoting infections of skin and lower respiratory tract
Cystic fibrosis	Depresses phagocyte function; causes poor clearance of mucus from lungs
Renal failure	Depresses immune response, impairs chemotaxis

TABLE 17-6 SOME EXTRINSIC FACTORS THAT PREDISPOSE HUMANS TO DEVELOP INFECTIOUS DISEASE

Predisposing Factor	Effect
Alcoholism	Reduces efficiency of mechanical respiratory defenses; impairs leukocyte function
Smoking	Impairs alveolar phagocytosis; impairs escalator activity of lower respiratory tract, reducing cilia activity by triggering hypersecretion of mucus
Malnutrition	May reduce inflammation, CMI* complement and antibody production; impairs mechanical barriers
Chronic stress	Increases adrenal activity, which may reduce inflammation and antibody formation
Viral disease	May suppress immune response; reduces efficiency of ciliated mucosa
Trauma	Bypasses primary lines of defense, providing access to deeper tissues
Burns	An especially dangerous traumatic injury because normal skin structure is destroyed, not merely disrupted; also impairs phagocytes
Antimicrobial therapy	Alters normal flora; selects for antibiotic-resistant microbes; may encourage growth of opportunistic fungi
Anti-inflammatory therapy with corticosteroids	Depresses phagocytic defenses; depresses interferon function; depresses antibody production by suppressing lymphocyte and monocyte proliferation; promotes diabetic state
Anticancer therapy (ionizing radiation, cytotoxic chemotherapy)	Depresses antibody formation; decreases bone marrow production of leukocytes; depresses phagocyte formation; may compromise mechanical defenses
Immunosuppression to accommodate transplants	Depresses CMI*
Surgical procedure	Provides portal of entry to vulnerable internal tissues; introduces foreign bodies that may be contaminated

*CMI = cell-mediated immunity, an important branch of the immune system that provides essential protection against infectious disease.

repeated episodes of opportunistic infections by *Staphylococcus aureus* or by such common environmental saprophytes as *Serratia* or *Pseudomonas* may have leukemia (which impairs the function of phagocytes) or chronic granulomatous disease (CGD). Children with CGD have defective phagocytes that readily engulf invading microbes but lack one of the critical lysosomal chemicals needed to kill the engulfed bacteria. Phagocyte defects, leukemia, and T-cell deficiencies predispose for systemic candidiasis (Fig 17-13) and other fatal opportunistic infections.

Secondary infections (also called *superinfections*) often occur following reduction of host resistance by a primary infection. Some pathogens paralyze the ciliated mucosa or depress antibody production, thereby increasing the likelihood of secondary infection. In some cases, the secondary infection is more dangerous than the primary one. For example, serious pulmonary disease caused by oppor-tunistic bacteria, most commonly *Streptococcus pneumoniae* and *Haemophilus influenzae*, often fol-

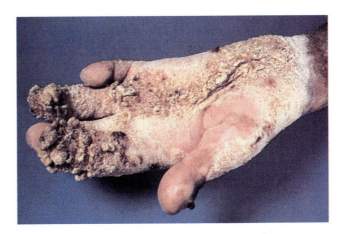

FIGURE 17-13 An opportunistic yeast infection. Mucocutaneous candidiasis in a patient with impaired phagocytic defenses.

lows measles, influenza, or whooping cough. Most AIDS patients ultimately die from fungal or viral pathogens that face no opposition from an immune system destroyed by the human immunodeficiency virus.

Pathogenesis of Infectious Disease

The infectious disease process is considerably more complicated than can be explained by the oversimplified notion that a single pathogen always causes a particular disease and that exposure to the pathogen always has the same outcome. *Staphylococcus aureus*, for example, can produce a broad spectrum of disease, from simple furuncles (boils) to fatal pneumonia, toxic shock, and meningitis. This bacterium is known to produce at least 23 factors that may contribute to its ability to invade virtually every human organ and tissue. No single factor, however, is absolutely necessary for it to elicit disease. *Streptococcus pyogenes* may cause mild pharyngitis (sore throat) in some people and rheumatic heart disease or severe kidney injury in others. On the other hand, a single disease syndrome, such as pneumonia, can be caused by any of a number of bacteria, fungi, viruses, or protozoa.

The actual mechanisms by which a pathogen accomplishes its detrimental effects are thoroughly understood in only a few of the less complex diseases where symptoms are caused almost entirely by the effects of a single exotoxin.

■ Sites of Infection

The affinity most pathogens exhibit for a limited number of preferential tissues is called **tissue tropism.** Tissue tropism determines not only the location of the primary site of colonization (the first habitat infected, sometimes with little or no damage to the area) but also the predilection for specific secondary sites of infection once the initial lines of defense have been breached. Tissues that lack surface receptors complementary to the microbe's attachment sites usually remain uninfected, although they may still be damaged by extracellular toxins or cytocidal substances released either from the microbe or from neighboring injured tissues. To initiate infection, therefore, pathogens must enter the body through a portal of entry that allows them access to tissues displaying the appropriated surface receptors. ▶ (See Portals of Entry and Exit, Chap. 20, p. 507.)

Tissue tropism is also influenced by nutritional and environmental conditions at various body sites. Obligate anaerobes are restricted to areas of low oxygen concentration—for example, the intestinal tract or deep wounds. Some organisms cause only skin lesions because they cannot tolerate temperatures above 33 or 34°C. The dermatophytes' ability to use keratin, a protein found in skin, hair, and nails, permits their growth on these areas of the body.

Tissue tropism is partially determined by inhibitory substances that protect body sites from microorganisms. Gram-negative bacteria, for example, infrequently infect normal skin because of their sensitivity to the high salt and fatty acid content of the epidermis. Neither salt nor fatty acids, on the other hand, inhibit the gram-positive *Staphylococcus aureus,* which causes a variety of skin lesions from boils to impetigo.

■ Spread of Infection

Following infection, diseases develop according to one or more of the following patterns (Fig. 17-14):

1. The organism multiplies locally at the primary site of infection and does not invade surrounding tissues. These organisms are referred to as **noninvasive.** The proliferating pathogen may either elicit local symptoms or, if it is **toxigenic,** it may produce an exotoxin that diffuses into the bloodstream and affects distant tissues.
2. The organism penetrates, multiplies in, and destroys the regional epithelium or other surrounding tissue at the primary site of infection, causing **localized invasive disease.** The characteristic intestinal ulceration of dysentery, for example, is due to invasion of the intestinal epithelium by the pathogen.
3. The pathogen penetrates into deeper tissues and subsequently disseminates to secondary sites

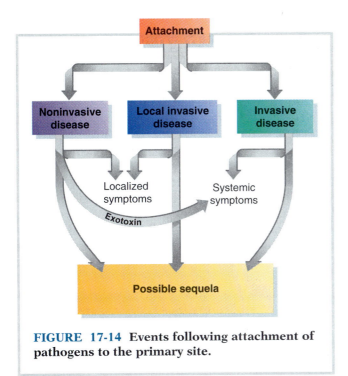

FIGURE 17-14 Events following attachment of pathogens to the primary site.

throughout the body. **Invasive pathogens** usually multiply both at the primary site and in the invaded secondary tissue.

Invasive diseases have much higher mortality rates than localized infections. Invasion of secondary tissues and organs commonly follows *septicemia*, the condition in which microorganisms multiply in the bloodstream. Since all living tissues in the human body are supplied with blood, *hematogenous* spread (*hemato* = blood) may lead to invasion of virtually any organ system of the body. For example, strep throat (pharyngitis caused by *Streptococcus pyogenes*) often leads to severe disease when the pathogen invades the bloodstream and settles in a number of secondary sites. If the lungs are invaded, pneumonia may develop. Skin lesions are common indicators of septicemia. Arthritis signals bone and joint involvement. Visceral complications may injure the heart, kidney, spleen, liver, or any of the internal organs. Meningitis or encephalitis may occur when the streptococci extend from the circulatory system to the central nervous system across the blood-brain barrier. Pathogens can also be spread by the flow of the lymphatic fluid. Since lymph empties into the bloodstream at the thoracic duct, lymphatic routes have dynamics similar to those described for hematogenous dissemination. ▶ (The structure of the lymphatic system is discussed in Chapter 18, p. 443.)

Pathogens may be carried by direct extension to tissues adjacent to the primary site. For example, pneumonia may develop following the transfer of upper respiratory microbes directly into the lower respiratory tract. Pathogens causing pharyngitis may extend to the adjacent auditory canals and cause otitis media (middle-ear infection). From there the infection may further extend to the mastoid sinuses and to the brain, causing encephalitis or meningitis.

Some pathogens cause severe diseases only after invasion of secondary tissue. For example, the primary sites for poliovirus infection are the throat and intestine. In approximately 1 percent of infected people, however, the virus spreads hematogenously from the alimentary tract to motor neurons (those that stimulate movement). The resulting paralysis characteristic of poliomyelitis is due entirely to secondary invasion rather than to primary infection.

■ Mechanisms of Tissue Injury

The human body develops characteristic symptoms in response to infectious disease. Initial indications of infection are usually generalized symptoms: malaise (overall discomfort), headache, listlessness, loss of appetite, weakness, and body aches. As the infection progresses, symptoms become more dramatic. Fever develops, sometimes accompanied by confusion, dehydration, weight loss, skin rash, changes in pulse and blood pressure, and other signs of tissue destruction or metabolic imbalance. People with severe diseases show signs of potentially fatal septic shock, pulmonary distress, coronary dysfunction, brain or neurological impairment, and malfunctions of other critical organ systems. Why does the infected person suffer such damage simply because microorganisms are replicating in the body?

The simplest (but only partially correct) way to explain the tissue damage that accompanies infection is to claim that cytotoxicity (damage or death of host cells) accounts for all the symptoms of infectious disease. Cytotoxicity can be caused by pathogens growing intracellularly or on surface membranes, or it can be caused by direct contact with toxins and destructive enzymes or metabolites (Table 17-7). Although these microbial factors are responsible for many disease symptoms, this oversimplified explanation creates more questions than it can possibly answer. For example, what accounts for the severe, often fatal symptoms evoked by those pathogenic microorganisms that produce no identifiable toxins or factors that directly harm host cells?

In some cases, normally protective responses of the human body can severely harm the host during

and secondary sites of infection. Exotoxin-producing microbes may cause systemic disease even if they don't invade tissues. Many pathogens are both invasive and toxigenic. Adverse host responses that may contribute to pathogenesis include chronic inflammation, hypersensitivity, the release of endogenous substances, and autocytolysis.

- **Sequelae.** Sequelae occasionally develop following recovery from some infectious diseases and bear no symptomatic resemblance to the initial disease. The sequela may be a reactivation of latent infection (although with a different set of symptoms) or a detrimental consequence of the immune response against the pathogen.

- **Clinical diagnosis.** Clinical specimens collected from the appropriate site and examined for host responses and presence of pathogens provide diagnostic information that helps select effective treatment of infectious disease.

Review and Problem-Solving Questions

1. Distinguish between pathogenicity and virulence. What might be the difference between (a) a pathogenic strain and a nonpathogenic strain of the same organism? (b) a highly virulent strain and a strain of low virulence?
2. Describe the events that might occur following exposure to a microbe that causes (a) subclinical infection; (b) acute infection; (c) latent infection.
3. How do the following conditions predispose to infectious disease? (a) Pregnancy; (b) cortisone therapy; (c) cystic fibrosis; (d) diabetes.
4. Fill in the following table comparing endotoxins and exotoxins.

	Endotoxin	Exotoxin
Chemical composition	_____	_____
Produced by	_____	_____
Fever-inducing?	_____	_____
Mechanism of action	_____	_____
Immunogenic (toxoid)	_____	_____
Example of disease	_____	_____

5. Describe three mechanical defenses of the human body and how each protects the host from microbial invasion.
6. How do some of the body's chemical secretions protect against disease?
7. Briefly describe the events of the inflammatory response.
8. How do neutrophils, fixed macrophages, and wandering macrophages differ in function?
9. Discuss the following quotation by Lewis Thomas in *Lives of a Cell:* "Our arsenals for fighting off bacteria are so powerful, and involve so many different defense mechanisms, that we are in more danger from them than from the invaders. We live in a midst of explosive devises; we are mined."
10. What is the importance of biofilms in clinical infections?
11. What are three differences between normal flora *E. coli* and the *E. coli* that cause intestinal disease?
12. Aaron and Christopher are visiting their friend Matthew, who is sick at home with an intestinal infection. Five days later Aaron develops symptoms of a similar infection, but Christopher remains healthy. Provide a possible explanation.
13. Because of an ear infection, Sara's doctor prescribes antibiotics directed against bacteria. As she completes her course of antibiotic therapy, Sara's hearing clears, but she experiences symptoms of an intestinal infection.
    (a) What is a likely etiology of Sara's intestinal infection?
    (b) What should Sara do to cure this infection?
    (c) Should Sara have taken the antibiotics for her ear infection?
14. What are some implications of the finding that the genes for many virulence factors are found on easily transferable plasmids?
15. What factors might affect the length of the incubation period for a given disease?
16. Why is it that only someone who has had chickenpox can get zoster (shingles), and why is shingles rare in children?
17. If you were a bacterial pathogen and could have only three virulence factors, which ones would you choose? Describe your effect on a host.
18. Explain the following quotation. "Only the best-dressed pathogens will gain admittance into the body's most exclusive cells."
19. How can a delay in culturing a specimen affect the accuracy of laboratory results?
20. Most microbial diseases are due to infection by a single pathogen. Why do certain diseases routinely occur as mixed infections?

PLAGUED BY A PLASMID

Yersina pestis, the bacterium responsible for plague, the terrifying Black Death that killed one-fourth of Europe's population, is so closely related to *Yersinia pseudotuberculosis,* an organism that causes a self-limiting mild infection, that it is now considered to be a subspecies of *Y. pseudotuberculosis.* The major genetic difference between these two organisms is the presence of a plasmid in the highly virulent plague pathogen. One of the key virulence factors produced from genetic information on the plasmid is a protease that is located at the outer membrane of this gram-negative bacterium's cell wall. In an effort to determine the role of the protease in pathogenicity, Sodeinde et al.* isolated a mutant of *Y. pestis* with a nonfunctional protease gene and compared it to the parental strain. Both strains reached high concentrations in the liver and spleen of infected mice when injected into the bloodstream, but only the protease-producing strain showed this effect when the bacteria were injected just beneath the skin (subcutaneously).

- Postulate two possible mechanisms by which the protease increases virulence. Would you expect to find bacteria at the site of injection for each of these proposed mechanisms?
- Why do disease symptoms appear only after disseminated infection by this gram-negative bacterium?
- Why is this gene important for the life cycle of *Y. pestis* in nature?
- What problems might occur if the plasmid carrying this gene was transferred to gram-negative bacteria such as *E. coli?*

*O. A. Sodeinde, Y. V. B. K. Subrahmanyam, K. Stark, T. Quan, Y. Bao, and J. D. Goguen, A surface protease and the invasive character of plague, *Science,* 258: 1004–1007 (1992).

CHAPTER 18
Acquired Immunity

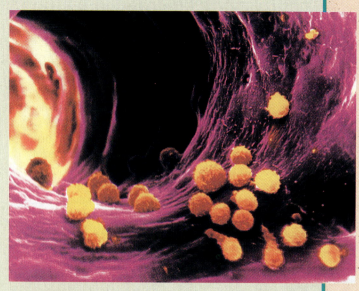

Protective blood cells. Among the blood cells shown in this electron micrograph are lymphocytes non-phagocytic leukocytes that are responsible for the immune response.

Acquired Immunity

It was 1955, the year the fear receded. Across the United States and throughout the world, a killer had claimed millions of casualities. This killer, known to most of us simply as "polio," stole into the body through the digestive tract, spread to the bloodstream, and attacked motor nerves, leaving its victims paralyzed. Many were permanently confined to wheelchairs or iron lungs; others, to a coffin. The year before, there were nearly 28,000 victims in the United States alone.

The schoolchildren standing in line that day understood little of their parents' fear that their family might be polio's next victims. We just knew that we were "polio pioneers" and we were about to participate in a great adventure that made our parents and teachers visibly excited. We would also get a badge signifying our courage and pioneering spirits. I couldn't wait to get mine. Although the line of second and third graders snaked out of the building into the parking lot, it was moving fast; soon I'd take my medicine and get my prize. But entering the building changed everything—I was greeted with an unexpected sound, one that engulfed me with terror. It was the sound of kids crying. Lots of them. I had the sinking feeling that soon I too would be crying. The kids must be getting shots! Oh please, don't let it be a shot!

It was a shot—an injection of killed poliovirus invented by some guy named Jonas Salk. None of the kids cared at that point about the immunologic details of the Salk vaccine. Only that we had to acquire a hole in our arm to get it (and that the painful ordeal would have to be endured twice more for the two dreaded "boosters"). Now the line was moving all too fast, as the nurses efficiently jabbed their needles into unwilling arms. When my turn arrived I started to run. I shouted that I'd rather have polio, and took off. But a firm hand grabbed my biceps, and the needle plunged home. The next minute I was back out in the parking lot, holding a disappointingly flimsy little tin button emblazoned with the words "Polio pioneer." I folded the metal clasp over my shirt pocket and found my friends, who were equally disillusioned by the experience. Two hundred thousand schoolchildren endured the same ordeal, becoming the world's first polio pioneers. And what did we get for our torture? A cheap little pin—and protection against one of the most dangerous diseases in the world. Forty years later, I still have both.

Acquired Immunity

The polio vaccine, indeed all vaccines, provide us with protection by arousing the immune system. We need an active immune system because our nonspecific resistance mechanisms sometimes fail us. Right now in your internal tissues and fluids, some microorganism may have evaded your phagocytic defenses and settled down to a banquet of nutrients and hospitable environmental conditions. The most serious type of invasion has begun.

Fortunately for us, the body can counterattack with its immune system, the final line of defense. Lymphocytes and macrophages team up to respond to the intruder and launch a complex response that eventually generates an effective multidirectional assault on the invading microorganisms. If not for this protective arsenal, a simple accident like biting a hole in your tongue or cheek, inoculating your bloodstream with millions of oral bacteria, could be fatal.

This line of defense, called **acquired immunity,** is a specific type of resistance that differs from the nonspecific lines of defense in the following ways:

1. It is acquired during a person's lifetime as a result of exposure to specific foreign substances.
2. It usually protects against the single type of pathogen or toxin that induced the response.
3. It commonly provides long-term protection against the agent that stimulated the immunity.
4. It is due to the activity of soluble proteins called *antibodies* and nonphagocytic white blood cells, the *lymphocytes,* often with the participation of macrophages ▶ (see Phagocytic Cells, Chap. 17, p. 423).

Acquired immunity also accounts for the success of *vaccination,* the use of nonpathogenic variants of dangerous microbes (or their products) to induce an immune response and stimulate protection against the virulent form of the pathogen in a way that does not expose us to the dangers of disease.

Before examining this protective response, however, let's take one more look at what it takes to mobilize the immune defenses.

Antigens

Any substance that specifically stimulates an immune response when introduced into the body is an **antigen** or immunogen. Several characteristics of antigens are known.

- Antigens are usually composed of protein or polysaccharide or contain these macromolecules as a major constituent. Lipids and nucleic acids are poor antigens unless linked to a protein or polysaccharide.
- Some are soluble; others are particulate. Soluble bacterial exotoxins are antigenic. The structural components of the cell that secreted the toxin are particulate antigens. A bacterium may have several types of particulate antigens (Fig. 18-1).
- In general, larger molecules stimulate a more intense immune response than smaller molecules. Molecules below molecular weight 1000 are very poor antigens.
- Antigens must be recognized by the host as *foreign* before they stimulate an immune response. This is perhaps their most important characteristic, since the system must selectively eliminate foreign susbstances without damaging the host's tissues in the process.
- An antigen contains chemically distinct sites, called **antigenic determinants** (or **epitopes**), that define its specificity (Fig. 18-2). These are the molecular regions against which the immune response is directed. Each site is an antigenic determinant. Antigenic determinants are small, perhaps five or six amino acids or sugars *on the surface of a larger protein or polysaccharide.* Following exposure to an antigen molecule, several

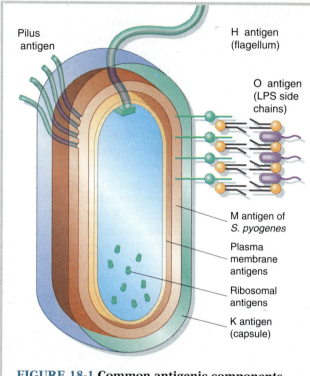

Pilus antigen

H antigen (flagellum)

O antigen (LPS side chains)

M antigen of *S. pyogenes*

Plasma membrane antigens

Ribosomal antigens

K antigen (capsule)

FIGURE 18-1 Common antigenic components of bacterial cells. Antigens (consolidated into a single diagram here) are often identified by assigning each a letter. Antigens are so specific that they can be used to distinguish between strains within a single species. Two different strains of the food-borne pathogen *Salmonella enteriditis*, for example, will have different O antigens.

specificities of antibodies are produced, each specificity directed against one type of antigenic determinant.

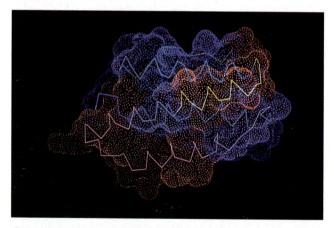

FIGURE 18-2   Antigenic determinants. A single antigenic molecule may contain multiple antigenic determinants and may stimulate the formation of antibodies of many different specificities.

• Small molecules that are nonantigenic by themselves may sometimes become antigenic determinants when coupled to large carrier molecules. These small compounds are called **haptens** (Fig. 18-3). Haptens react with their specific antibodies, but unless they are attached to larger carrier molecules they are too small to stimulate the production of antibodies. For example, the antibiotic penicillin will not stimulate an immune response by itself but becomes immunogenic if it binds to serum proteins.

A substance that is antigenic (recognized as foreign) in one host may be completely tolerated (recognized as "self") in another individual. Transplants, for example, are rejected because the *histocompati-*

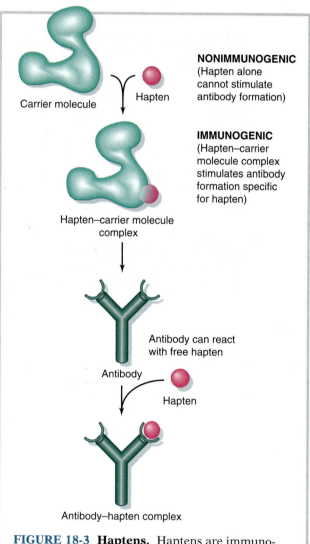

Carrier molecule

Hapten

NONIMMUNOGENIC (Hapten alone cannot stimulate antibody formation)

IMMUNOGENIC (Hapten–carrier molecule complex stimulates antibody formation specific for hapten)

Hapten–carrier molecule complex

Antibody can react with free hapten

Antibody

Hapten

Antibody–hapten complex

FIGURE 18-3   Haptens. Haptens are immunogenic only when they become part of a larger molecule. In their free state, however, they can react with the antibody whose formation was stimulated by the complex.

bility antigens, surface proteins on the cells of the transplanted organ, are foreign to the recipient and are therefore attacked by the immune system. These antigens were recognized as "self" in the donor. Similarly, the *red blood cell antigens* of a donor and recipient may be distinct enough to trigger an immune reaction following transfusion mismatches. Group B blood cells are antigenic in a recipient with group A blood, yet they are immunologically tolerated by people with group B blood.

The mechanism that prevents us from rejecting our own tissue is called **immune tolerance.** In general, those constituents that were present in our bodies during fetal development are tolerated as "self." Most other potentially antigenic substances introduced into the body after birth are recognized as foreign and stimulate an immune response.

Lymphocytes

Special nonphagocytic white blood cells called **lymphocytes** are responsible for immune responses. The various lymphocytes cannot be distinguished microscopically, but each type contains unique markers on its surface. Lymphocytes also contain

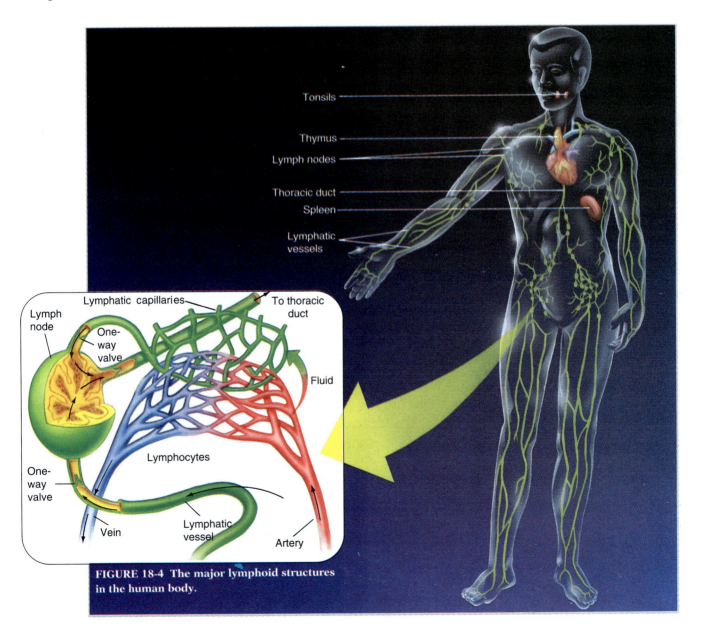

FIGURE 18-4 The major lymphoid structures in the human body.

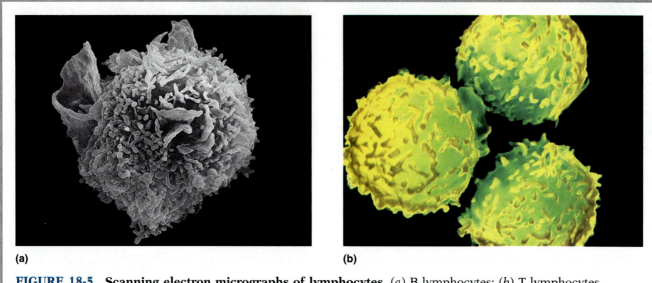

FIGURE 18-5 **Scanning electron micrographs of lymphocytes.** (*a*) B lymphocytes; (*b*) T lymphocytes.

receptors that recognize specific antigenic determinants, and they are activated whenever their specific antigenic determinant is present and binds to their surface. In most cases, lymphocyte activation also requires the participation of phagocytic cells. Interactions between lymphocytes and phagocytes occur in the major lymphoid tissues such as the spleen and lymph nodes (fig. 18-4). Following activation, lymphocytes participate in humoral immunity or cell-mediated immunity, the two major branches of our immune system.

Humoral immunity is the product of lymphocytes that have matured in a special lymphoid organ that programs these cells to produce antibodies in response to specific antigens. This programming of lymphocytes was first discovered in chickens, where it takes place in a structure called the bursa of Fabricius. The bursa-programmed cells are thus called **B lymphocytes,** or simply B cells (Fig. 18-5*a*). The bursa equivalent in humans appears to be lymphatic tissue in the gut, the bone marrow, or perhaps the tonsils. The other immune system branch, **cell-mediated immunity (CMI),** is generated by **T lymphocytes,** cells programmed by passage through the thymus gland (Fig. 18-5*b*). T cells produce no antibodies, but in the presence of antigens they either kill the foreign cells directly or release chemicals that aid in their destruction. Some T cells also help regulate both the humoral and cell-mediated immune responses. With this introduction, we examine each of these two branches of the immune system.

CONCEPT CHECK

- What is the role of the lymphoid system in the immune response?
- Describe two ways in which the humoral and cell-mediated branches of the immune system differ.

Humoral (B-Cell) Immunity

Humoral immunity is due to the production of antibodies, a special class of proteins that are soluble in the body fluids (the "humors"). An **antibody** is produced by a vertebrate host in response to the introduction of an antigen into the body, and it binds specifically with the antigen that stimulated its formation. Antigen–antibody reactions protect the host against many detrimental effects of intruding microbes or other foreign substances.

■ Antibodies

Antibodies are also called **immunoglobulins** because they participate in immune reactions and belong to a class of serum proteins called globulins. Antibodies are *monospecific* molecules—they combine only with the single type of antigenic determinant that stimulated their formation. Most antibodies are also *bivalent*—they possess two identical reactive sites and can couple with two identical antigenic determinants. An antibody does not react with dissimilar antigenic determinants, even if the different determinants are on the same antigen molecule (Fig. 18-6). The reaction between antibody and antigen results in the formation of an *antigen–antibody complex*.

● **ANTIBODY STRUCTURE.**   Although some structural variations exist among antibodies, the typical antibody molecule consists of four protein chains linked together by disulfide bonds in what is usually illustrated as a Y-shaped structure (Fig. 18-7). The two shorter chains, called *light* (L) *chains,* are covalently linked to the branches of the longer *heavy* (H) *chains.* Each chain has variable and constant regions. The specificity of the antibody's combining sites for antigen is determined by the amino acid sequence in the *variable regions* of both the H and L chains. The amino acid sequence in the *constant region* determines other characteristic properties of the antibody, such as its ability to cross certain tissue barriers, to activate the complement system, or to adhere to phagocytic cells. The properties of the tail portion, called the *Fc region,* define the five major classes of immunoglobulins.

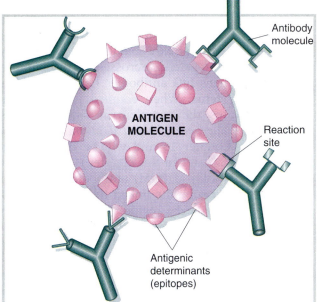

FIGURE 18-6 Antibody specificity. Antigens and antibodies can combine with each other if the reac - tion site of the antibody is complementary to an antigenic determinant on the antigen. Each antibody pictured here has two identical antigen binding sites. The antibody will not react with any other type of antigenic determinant on the antigen molecule.

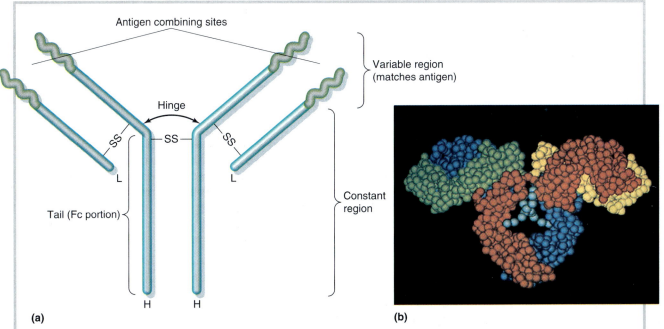

(a)
(b)

FIGURE 18-7 Structure of the most typical antibody, immunoglobulin G (IgG). (*a*) The amino acid sequence of the variable region (shown in green) determines the antigen specificity of the antibody. The rest of the molecule is identical for all IgG produced by each person. The angle of the hinge varies according to antigen size. Large antigens may bend the hinge to an angle of 180 degrees. (—SS— = disulfide bond; H = heavy chain; L = light chain.) (*b*) A more realistic view of IgG antibodies is provided by this computer-generated version (the two variable regions are shown in green and yellow).

Generating Antibody Diversity

The vast array of antigenic determinants that can be recognized by our immune system posed a problem that perplexed scientists for decades: How can each lymphocyte's nucleus house information for antibodies with millions of different specificities? Furthermore, since all cells in the body are descendants of the same zygote, they are genetically identical, so each cell would seemingly have to carry millions of "baggage" genes that would never be used (a muscle or nerve cell, for example, would be encumbered with all the genes for responding to all possible antigens—genetic information it will never need). Encoding all these different antibody specificities would require more DNA than it is possible to fit into the nucleus of a human cell. How is it that we can generate enough antibody diversity to respond to any antigen?

Scientists found that the number of genes for antibody production in a cell is actually relatively small. In fact, the precursor cells that develop into B lymphocytes possess only a few hundred genes for antibody production. During the development of a mature B cell from a pre-

cursor cell, these genes recombine to give rise to the diverse population of B lymphocytes, with more than a million clones of cells, each capable of responding to a different antigen. Here's how it works.

In each person, most of the antibody molecule is constant, that is, identical in all antibodies regardless of their antigenic specificity. Only a few variations punctuate the protein chains, and these lie in the variable region at the ends of the light and heavy chains. This means that a very small number of genes can code for a very large portion of the antibody molecule. Second, a group of genes codes for different portions of the variable regions of light chains. This group of genes is located very near the genes for the constant region of the light chains. These variable genes fall into two clusters called the V (variable) and J (joining) regions.

In the mouse pre-B cell, for example, there are 350 different V-region genes and four different J-region genes. During the development of a B cell, one gene from the V region recombines with a gene from the J region. The remaining DNA is spliced out of the chromosome: This event alone creates a possible 1400 combinations of V and J genes in the mature B-cell population.

Two other factors generate additional variability. Imprecise recombination between the V and J regions creates dif-

ferent nucleotide sequences that ultimately translate into different amino acid sequences in the antibody. In addition, mutations may develop in progeny cells during replication of activated B cells. Information for the constant region is joined to the variable region during the transcription process. The initial RNA product contains an intron separating the V-J transcript from the heavy chain transcript. Splicing combines these two exons to form a functional messenger from which the unique light chain is translated:

A similar series of events controls the development of heavy chains. The heavy-chain variable region, however, is created from three sets of genes called V, D (diversity), and J regions. Using the "Chinese menu" approach to recombination (taking one from column V, one from column D, and one from column J), there are more than 10,000 possible combinations of these genes in mice. Given that any gene for a light chain may occur in the same cell as any gene for a heavy chain, there are more than 14 million (1400 × 10,000) different types of antibodies that may be made, not counting the additional diversity generated by mutation. Such a system guarantees that there will be a clone of antibody-producing cells for virtually any antigen that enters the body without taxing all other cells with millions of unneeded genes.

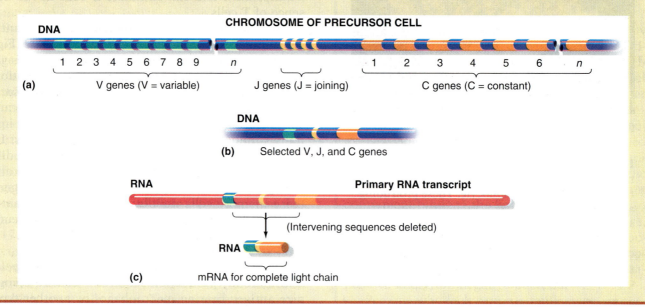

CHROMOSOME OF PRECURSOR CELL

DNA

1 2 3 4 5 6 7 8 9 n

(a)   V genes (V = variable)   J genes (J = joining)   1   2   3   4   5   6 n C genes (C = constant)

DNA

(b)   Selected V, J, and C genes

RNA   **Primary RNA transcript**

(Intervening sequences deleted)

RNA

(c)   mRNA for complete light chain

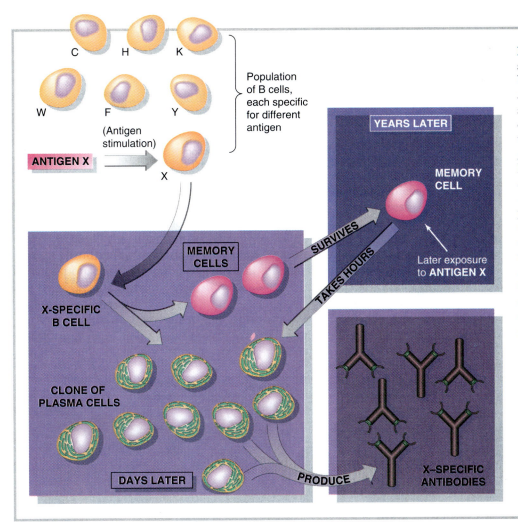

FIGURE 18-11 Clonal selection theory of antibody production. An antigen-specific B cell is selected from other B cells when stimulated by the corresponding antigen (represented here by hypothetical antigen X). If antigen persists, a pool of antibody-producing plasma cells and a few memory cells eventually develop. Later exposure to the same antigen rapidly activates memory cells (shown in red) to differentiate into plasma cells, proliferate, and produce antibodies.

of **antigen-presenting cell (APC)** (Fig. 18-12). These cells then communicate with each other and other nearby leukocytes by secreting proteins called **interleukins.** During this process the following events occur:

1. Macrophages capture and digest foreign particles. The macrophage then transports the antigenic determinant fragments of the digested particles to its own surface, where the epitopes are displayed next to proteins called *class II histocompatibility antigens.* [These are also called human leukocyte antigens (HLAs) or major histocompatibility complex (MHC) proteins.] Class II proteins are surface markers found only on macrophages, B lymphocytes, and a few other antigen-presenting cells.
2. The macrophage-bound antigen is recognized by a subpopulation of T lymphocytes called **helper T cells (T_H).** Helper T cells carry receptors that can recognize a specific antigenic determinant, but only when the antigenic determinant is associated with class II proteins.

3. Antigen-specific B cells in the region also interact with antigen, which may be unbound or presented on the surface of macrophages.
4. The macrophage releases *interleukin-1 (IL-1),* a soluble factor that stimulates bound helper T cells to release another interleukin, *IL-2.*
5. IL-2 then triggers the replication of antigen-activated B cells and causes some of them to differentiate. Ultimately, two populations of B cells are produced: memory cells and antibody-secreting plasma cells.

● **T-INDEPENDENT ANTIBODY PRODUCTION.** A few types of antigens do not require T cells to trigger antibody production. These T-independent antigens are large molecules composed of repeating subunits that act as antigenic determinants. For example, bacterial lipopolysaccharide, capsular polysaccharides, and flagella (polymerized flagellin) are T-cell-independent antigens. Activation may result from cross-linking of receptors on the B-cell surface by copies of the antigenic determinant on the polymer. Antibody production is usually limited to IgM, and

Cell-Mediated (T-Cell) Immunity

Cell-mediated immunity (CMI) is primarily targeted against protozoa, fungi, virus-infected host cells, and bacteria that are intracellular parasites. Resis-

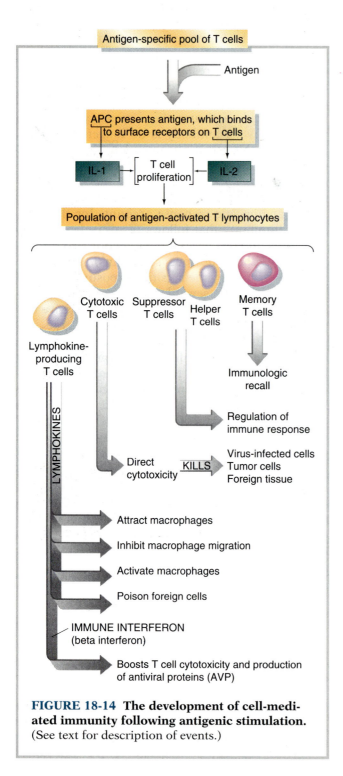

FIGURE 18-14 The development of cell-mediated immunity following antigenic stimulation. (See text for description of events.)

tance and recovery from infectious diseases caused by these agents are provided by the T-cell response. Although these antigens may also induce antibody production, immunoglobulins fail to protect against many of these diseases. CMI also functions as an immune surveillance system that detects and destroys tumor cells that occasionally develop from normal host cells.

The CMI response is triggered by exposure of T lymphocytes to an antigen that has been processed by antigen-presenting cells. Antigen-presenting cells may engulf the whole foreign cell or, in the case of many protozoa and tumors, simply bind to surface antigens shed by the cell. When the antigen-presenting cell displays an antigenic determinant in association with the class II proteins, a specific helper T cell locks onto its surface. This stimulates the release of interleukins by the T_H cell and the transformation of T lymphocytes into the *effector lymphocytes* that provide the actual protection (Fig. 18-14). These effector cells proliferate, forming an expanded clone of immunocompetent lymphocytes in the area of the body where the antigen is concentrated. They protect by directly killing target cells or by releasing soluble proteins called **lymphokines.** Unlike antibodies, lymphokines are not antigen-specific, nor do they combine with antigenic determinants on the target cell. Most of these T-cell-produced proteins attract macrophages or other leukocytes to the site of infection and activate them.

As with humoral immunity, immunologic memory develops following antigen exposure. Some of the activated T lymphocytes differentiate into memory T cells. These quiescent cells remain in the host long after the antigen has been eliminated. Memory cells retain the potential for rapid activation whenever the host is exposed to the corresponding antigen, at which time they proliferate into a pool of effector lymphocytes. This recall response is similar to the anamnestic response of humoral immunity and shortens the latent period (the time between antigen stimulation and development of protective immunity) from about 20 days to 48 hours.

Cell-mediated immunity's effector cells, such as cytotoxic T cells and lymphokine-producing cells, provide the actual protection.

■ Cytotoxic T Cells

Cytotoxic T cells physically attach to their target cells and destroy them by membrane disruption and lysis. Each cytotoxic T cell binds only when it recognizes both the antigenic determinant and *class I histocompatibility proteins*. Class I proteins are found on all nucleated cells. (They are not present on prokaryotic cells or on red blood cells.) Such antigen-specific cytotoxicity enables our immune system to selectively eliminate some of our own host cells that have acquired antigenic determinants that are recognized as foreign, notably tumor cells and virus-infected host cells (Fig. 18-15). In addition, transplanted human organs or skin grafted from one person to another are rejected and destroyed by the recipient's cytotoxic T cells, which become sensitized to the foreign tissue antigens ▶ (see Mechanism of Graft Rejection, Chap. 19, p. 498). Steroids, cyclosporin A, and other drugs that suppress CMI in the transplant recipient can be used to prolong survival of the transplanted organ. Such immunosuppression, however, encourages the development of cancer and fatal infectious disease caused by viruses, fungi, protozoa, and certain bacteria. Cyclosporin A is the most successful of these immunosuppressors because of its specificity in suppressing the cells responsible for transplant rejection with little damage to other immune cells. Although the introduction of the drug greatly increased the survival of transplant recipients, the immune surveillance system that patrols the body for tumor cells is apparently compromised. Cyclosporin recipients run a greater risk of developing cancer. ▶ (The details of intentional immunosuppression are discussed in Chapter 19, p. 491.)

■ Lymphokine Production

Much of the protection afforded by CMI is provided by the action of lymphokines. Certain subpopulations of T cells become **delayed-type hypersensitivity lymphocytes (DTH cells)**, which bind to their antigen and release lymphokines. Although DTH cells do not damage their cellular targets directly, lymphokines hasten the destruction of the foreign cells. Several lymphokines have been discovered; most of them attract macrophages and increase the phagocytic efficiency of these cells. Since they are leukocyte products that affect other leukocytes, most lymphokines are actually interleukins. The following list describes several lymphokines and their protective roles in cell-mediated immunity.

- *Macrophage chemotactic factors* attract macrophages to areas of high lymphocyte concentration (and therefore to the area of high antigen concentration).
- *Migration inhibition factor (MIF)* reduces the mobility of macrophages so that they accumulate around the activated T cells. MIF prevents macrophages from moving away once they reach the site of antigenic stimulation.
- *Macrophage activating factor (MAF)* greatly enhances phagocytosis of any foreign particles in the

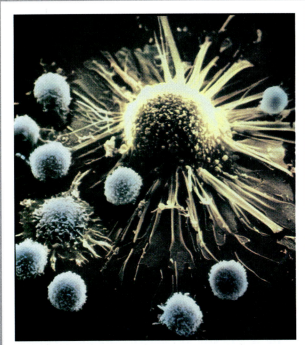

(a)

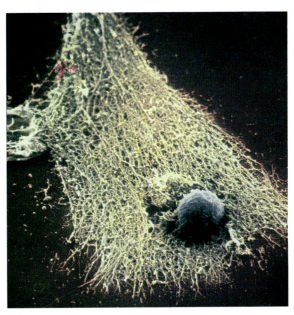

(b)

FIGURE 18-15 Killers disposing of a killer. (*a*) T cells attacking a larger cancer cell elongate as they release a membrane-disrupting complex that lyses the tumor cell. (*b*) All that remains of the destroyed cancer cell is its collapsed cytoskeleton.

area of antigenic stimulation. MAF creates a population of "angry" macrophages with larger, more numerous lysosomal granules. The result is an enhanced intracellular killing capacity following engulfment of any cell, including intracellular parasites.

- *Specific macrophage arming factor (SMAF)* enhances the ability of macrophages to kill the specific antigenic target cell.
- *Lymphotoxin,* or tumor necrosis factor B, nonspecifically destroys potentially pathogenic cells. Unfortunately, host cells may also be damaged by lymphotoxin, which may account for some host tissue damage associated with certain CMI responses ▶ (see Type IV (Delayed-Type) Hypersensitivity, Chap. 19, p. 488).
- *Immune interferon,* produced by antigen-activated lymphocytes, aids the arrest of infections by increasing the cytotoxicity of T cells and by triggering the production of antiviral proteins in neighboring cells ▶ (see *interferons,* Chap. 13, p. 328).

The overall effect of most lymphokines is to enhance the activity of our nonspecific protective mechanisms, especially the phagocytic macrophage system. Activated macrophages are more effectively phagocytic and also release a variety of factors that have immunostimulatory and antitumor effects. In some instances, as with SMAF, these defenses are directed against the specific antigen that induced CMI. Antigenic challenge, however, boosts the CMI defenses in a nonspecific fashion so that an unrelated pathogen may be simultaneously eliminated by the stimulated system.

> **CONCEPT CHECK**
>
> - In what ways is the cell-mediated immune response similar to the humoral immune response? In what ways are they different?
> - How do cytotoxic T cells and lymphokines protect against foreign substances?

Null Cells

Null cells are lymphocytes that are neither T nor B cells. They do not require previous exposure to an antigen but participate in protection from viral infections and tumors, usually in response to lymphokines. Null cells can detect abnormal cells and lyse them by inserting a pore-forming protein into their membrane. There are two kinds of null cells, **natural killer (NK) cells** and **killer (K) cells.** Great numbers of these cells are found in persons infected with viruses such as the mumps virus and herpesviruses. Natural killer cell activity appears to be increased by interferon released from virus-infected cells. The antitumor activity of natural killer cells, which may be important in defending against cancer, is enhanced by lymphokines. Killer cells have receptors for the Fc tail of antibody molecules and react with antibody-coated cells. For example, they bind to antibody extending from the surface of a virus-infected cell, causing it to lyse. This lytic reaction, called *antibody-dependent cell-mediated cytotoxicity,* or ADCC, does not use complement.

> **CONCEPT CHECK**
>
> - What is the relationship between null cells and antibody in providing immune protection? Between null cells and lymphokines?
> - How do null cells differ from cytotoxic T cells?

Regulation of the Immune Response

Activated T lymphocytes include regulatory T cells that help to govern the intensity of the immune response. They also help to ensure that the body normally does not attack itself, and they provide a mechanism for shutting off antibody production and immune cell activation when the antigen has been eliminated. Helper T cells, for example, are necessary to initiate humoral and cell-mediated immune responses. Different subpopulations of helper T cells stimulate antibody production and activation of effector T cells. Cooperation between helper T cells and B cells (or effector T cells) may provide a fail-safe system against the accidental triggering of an immune response. Since two different cells must recognize an antigen as foreign before an immune response is launched, the probability of an accidental attack on "self" tissues is reduced.

The second type of regulatory T cell is the suppressor T cell. **Suppressor T cells** affect immune functions by decreasing the activity of helper T cells. They are activated by antigen at a slower rate than helper or effector T cells. This allows for an effective immune response but helps avoid excessive reactions that might damage the host.

Helper T cells are the target of HIV, the virus that causes AIDS. As the virus destroys these cells, the ratio of helper to suppressor cells decreases, thereby destroying the immune system's ability to function effectively. AIDS and other immunologic dysfunctions are the subject of Chapter 19.

A summary of the cells and soluble factors par-

TABLE 18-2 PRINCIPAL CELLS AND PROTEINS OF THE IMMUNE RESPONSE

Cell	Function
B Lymphocyte	Programmed for a humoral immune response
Plasma cell	Produces antibody
Helper T cell	Activates B and T cells
Macrophage	Presents antigen to helper T cells
Cytotoxic T cell	Lyses target cells on contact
Delayed-type hypersensitivity T cell	Releases lymphokines
Suppressor T cell	Downregulates the immune response
Natural killer (NK) cell	Kills target cells
Killer cell	Lyses antibody-coated cells

Protein Factor	Function
Antibodies	Combine with antigens
Interleukins	Regulate immune system cells
Interferons	Inhibit viral replication; stimulate NK cell activity
Lymphotoxin	Kills target cells

ticipating in the immune response is presented in Table 18-2.

CONCEPT CHECK

- Describe the role of helper T cells in regulating a B-cell response.
- Why is the ratio of suppressor to helper T cells important in maintaining an effective immune response?

Active and Passive Immunity

Immunity is acquired either actively or passively, depending on whether the immune person produces his or her own protective antibodies or T cells or receives presynthesized antibody or sensitized lymphocytes from another individual. Both active and passive immunity can be acquired either naturally or artificially (see Table 18-3).

■ Active Immunity

Active immunity is the production of antibodies or specialized lymphocytes by the host as a result of exposure to a foreign antigen. It is characterized by

(1) a latent period of at least 2 weeks between initial exposure to the antigen and development of protective immunity and (2) an extended duration of immunity that often lasts for years.

Natural active immunity develops following exposure to an antigen as a result of natural infection. Immunity is directed against the specific infectious agent as well as its toxic by-products. Immunologic recall in both cell-mediated and humoral responses maintains protection for months and often for years. The protective efficiency of natural active immunity is demonstrated by the infrequency of second attacks of chickenpox, mumps, and measles. Even in the absence of overt symptoms, these infections may stimulate lifelong immunity.

Artificial active immunity is similar to natural active immunity except for the nature of the antigen and the method of introduction into the host. Instead of natural infection by a potentially virulent microbe, a vaccine is intentionally introduced into the body by a clinical procedure such as injection. A **vaccine** is antigenically similar to a pathogen or to its toxic by-products but has been treated so that it can be administered to people with little danger of disease. The vaccine sensitizes the immune system to the corresponding pathogen, inducing immunity without the danger of infectious disease developing during the lag period. In immunized people, natural exposure to the corresponding virulent pathogen triggers a protective anamnestic response that eradicates the pathogen or neutralizes toxins before symptoms of disease develop.

● **VACCINES.** An effective vaccine retains the corresponding pathogen's antigens but none of the pathogen's ability to damage the host. This is ac-

---

TABLE 18-3 FOUR CATEGORIES OF ACQUIRED HUMORAL IMMUNITY

Active Immunity

Immunized individual produces antibody.
- *Natural active immunity* follows natural infection.
- *Artificial active immunity* follows vaccination (immunization).

Passive Immunity

Immunized individual receives antibody produced by another individual.
- *Natural passive immunity* is due to transfer of maternal antibodies to fetus or suckling newborns.
- *Artificial passive immunity* is induced by injecting presynthesized antibody into recipient.

complished in a number of ways, including the use of killed and attenuated organisms, purified antigen fractions, toxoids, synthetic peptides, and anti-idiotype antibodies.

Vaccines That Are Killed Organisms.

A pure culture of the pathogen is exposed to an agent that will kill the microbe without altering the surface antigens. Injection of these killed organisms into a host provides safe initial exposure to the pathogen and subsequent resistance to disease. Typhoid fever vaccine, for example, is a formalin-killed preparation of the pathogen *Salmonella typhi*. Vaccines containing killed organisms or inactivated viruses have been used to prevent polio (Salk vaccine), cholera, typhoid fever, influenza, whooping cough, rabies, and epidemic typhus. Most of these vaccines require several booster shots before protection is adequate.

MYTH: Rabies vaccine can cure rabies.

FACT: Rabies is an incurable disease; once symptoms occur, mortality is 100%. Rabies vaccines work only if administered before or shortly after exposure (after a bite from a rabid animal). The vaccine is one of the very few examples of vaccination that is effective after the pathogen enters the body. The rabies virus develops so slowly that a protective immune response can be stimulated by the vaccine before the victim's nerves are infected, the resulting antibodies eliminating the virus before the disease can be initiated. An AIDS vaccine, when one is developed, may work in a similar fashion, that is, being administered to persons who are HIV positive before symptoms of AIDS develop.

Attenuated Organisms (Live Vaccines).

The most effective vaccines are those that actually multiply in the host, mimicking the early stages of natural infection. In these vaccines, the pathogens are **attenuated**—alive but weakend in their capacity to cause severe disease. Vaccines containing attenuated organisms are especially effective if they can be introduced by the same routes the corresponding pathogen uses for entering the body. The attenuated microbes harmlessly propagate in the body to provide a greater and more prolonged antigenic stimulation. Furthermore, the attenuated pathogens may be shed by the immunized individuals and vaccinate other susceptible persons.

The attenuated Sabin polio vaccine, for example, is generally superior to the Salk vaccine because it is administered orally and harmlessly replicates in the intestine, similar to the early stages of po-

liomyelitis. Unlike virulent poliovirus, the attenuated variant causes no paralysis. The Sabin vaccine stimulates the formation of secretory IgA, antibodies that neutralize viruses in the intestine before they invade the bloodstream. The inactivated Salk vaccine, on the other hand, must be injected into muscle, inducing little protection in the intestine. In addition, "live" Sabin vaccine virus is shed in the feces of vaccinated persons. Since polioviruses are naturally transmitted by the fecal-oral route, people may be inadvertently vaccinated by ingesting attenuated viruses shed by vaccinated indviduals. Other vaccines that use attenuated versions of pathogens include those that help protect against influenza, measles, mumps, rubella, and tuberculosis.

Unlike killed vaccines, attenuated microbes have been known to regain virulence and have caused serious disease in vaccinated persons. Only those attenuated organisms unlikely to revert can be safely used in vaccines. Another concern about attenuated viral vaccines relates to the fact that viruses are grown in cell culture. The safety of the vaccine is predicated on the cell culture being free of any viruses other than the one being used to make the vaccine preparation.

Purified Antigen Fractions (Subunit Vaccines).

Some vaccines contain only the antigens against which the protective immune reactions are directed. For example, purified extracts of the capsular antigens from virulent *Streptococcus pneumoniae* stimulate the host to produce opsonizing antibodies that effectively protect most immunized persons. This vaccine, called pneumovax, contains capsular antigens of the 14 most common virulent strains of *S. pneumoniae*. About 80 percent of all cases of pneumococcal pneumonia are caused by one of these 14 strains. Antigenic fractions are often prepared by genetic engineering technology (see below). Antigen subunits are also used in vaccines against cholera, plague, and some types of meningitis.

Toxoids.

The dangerous symptoms of tetanus, diphtheria, and a few other diseases are due almost entirely to the effects of soluble exotoxins secreted by the pathogen ▶ (see Toxic Factors, Chap. 17, p. 415). Active immunity against these diseases can be induced by injecting a **toxoid,** an exotoxin that remains antigenically unaltered but has been chemically treated to destroy its poisonous properties (Fig. 18-16). An injection of toxoid stimulates production of antibodies (antitoxins), thereby inducing immunity against the corresponding disease. For example, prior immunization with a toxoid prepared from the toxin that the bacterium *Clostridium tetani*

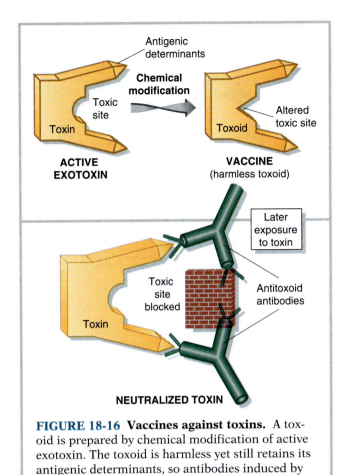

FIGURE 18-16 Vaccines against toxins. A toxoid is prepared by chemical modification of active exotoxin. The toxoid is harmless yet still retains its antigenic determinants, so antibodies induced by immunization with toxoid neutralize the active exotoxin.

Genetically Engineered Vaccines. Several vaccine components are manufactured through genetic engineering using easily cultured bacteria or yeast to synthesize the protein antigen. ▶ (This process is described in Chapter 9, p. 231.) Genetically engineered vaccines against influenza and hepatitis B are cheaper and safer than conventional whole-virus vaccines. Another technological advance in vaccine production is the creation of *piggyback vaccines.* The genes for the desired antigens are inserted into the genome of an infectious, but harmless, virus. The engineered virus is inoculated into the host. As the virus replicates, it produces the "vaccine" protein along with its own products. The host then launches immune responses against the viral products and the extra antigen as well. Vaccinia virus, successfully used in smallpox eradication, is the most commonly used vector for piggyback vaccines.

Anti-idiotype Vaccines. Antibodies themselves can be used as antigens. They stimulate production of other antibodies targeted at their variable regions, the binding sites specific for the corresponding antigenic determinant. Each antibody for a distinct epitope is called an *idiotype.* Anti-antibody antibodies recognize and bind to the antigen binding sites of the antibody idiotype that stimulated their production (Fig. 18-17). These antibodies are aptly known as *anti-idiotype antibodies.* Because the original antigen and the anti-idiotype antibody have identical antigenic determinants, both can bind to the same antibody molecule, so anti-idiotype antibodies can replace a potentially harmful pathogen or toxin as the antigen in a vaccine. The advantage of anti-idiotype vaccines is that there is no pathogen present and therefore no possibility of reactivating the disease. Currently several such vaccines are under investigation.

The various types of vaccines used to prevent infectious diseases among humans are listed in Table 18-4. Unfortunately, many diseases cannot be controlled by vaccination. (See THE MICROBES: Why Not Vaccinate Against All Infectious Diseases?)

releases into the bloodstream from an infected wound site protects against the post-wound disease tetanus. After three boosters, protection against tetanus lasts at least 10 years, explaining why physicians treating injuries always inquire about the patient's last tetanus shot. If it's been more than 10 years, the injury victim gets a booster shot.

Synthetic Peptides. Through the use of sophisticated laboratory technology, the amino acid sequence of purified protein antigens can be precisely determined. Once the sequence is known, it is possible to synthesize peptides of approximately 20 amino acids that may represent antigenic determinants against which protective immunity is directed. Vaccines have been created using peptides of hepatitis B virus surface antigen, *Streptococcus pyogenes* M protein, diphtheria toxin, influenza virus, and a number of other proteins. The advantage of synthetic vaccines is that they ensure exclusion of contaminating materials that might harm the host. The widespread use of synthetic vaccines awaits the development of high-yield production systems.

■ Passive Immunity

Antibodies produced by active immunity in one individual can be transferred in serum to a nonimmune recipient. The antibody recipient is then said to have **passive immunity.** Although these antibodies provide immediate protection, they are eventually depleted and are not replaced by the body. Passive immunization, therefore, provides only temporary, short-term protection, usually lasting no more than a few weeks. Cell-mediated immunity can be passively transferred by immune lymphocytes

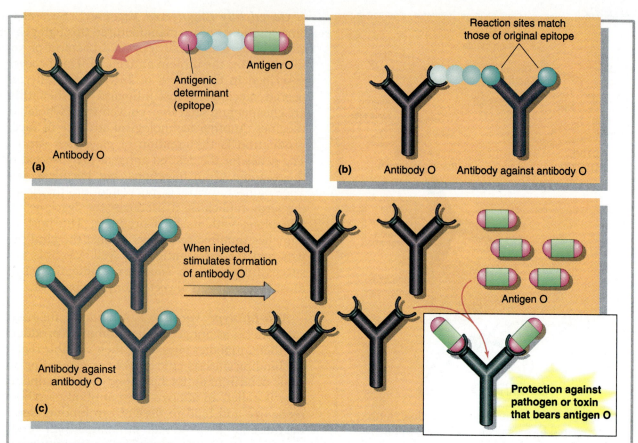

FIGURE 18-17 Anti-idiotype vaccines. (a) The antigenic determinant and the antibody formed against it have complementary structures. (b) Anti-idiotype antibody is also complementary to the antibody in (a). It therefore resembles the antigenic determinant and can be used in its place. (c) Anti-idiotype antibody stimulates synthesis of antibody O without the risk presented by the antigen.

TABLE 18-4	VACCINES USED IN HUMANS TO PREVENT INFECTIOUS DISEASE				
Type of Vaccine	**Disease**	**Etiologic Agent**	**Type of Vaccine**	**Disease**	**Etiologic Agent**
Attenuated (infectious)	Poliomyelitis*	Poliovirus	Antigen fractions	Cholera	*Vibrio cholerae*
	Influenza	Influenza virus		Plague	*Yersinia pestis*
	Measles	Measles virus		Meningitis	*Neisseria meningitidis*
	Mumps	Mumps virus		Meningitis	*Haemophilus influenzae*
	Rubella	Rubella virus		Pneumonia	*Streptococcus pneumoniae*
	Tuberculosis	*Mycobacterium*		Hepatitis	Hepatitis B virus
	Yellow fever	Yellow fever virus			
Killed or inactivated	Whooping cough†	*Bordetella pertussis*	Toxoid	Tetanus†	*Clostridium tetani*
				Diphtheria†	*Corynebacterium diphtheriae*
	Typhoid fever	*Salmonella typhi*			
	Epidemic typhus	*Rickettsia prowazekii*	Genetically engineered vaccines	Influenza	Influenza virus
	Rabies	Rabies virus		Hepatitis B	Hepatitis B virus
	Influenza	Influenza virus			
	Poliomyelitis (Salk)	Poliovirus			

*For Sabin vaccine.
†Routinely administered during the first year of life as the combined vaccine, DPT (diphtheria, pertussis, tetanus).

Why Not Vaccinate Against All Infectious Diseases?

Some early immunologists dreamed of developing a broad complement of vaccines that would make people resistant to all infectious diseases. Yet today we still cannot immunize against many important diseases—gonorrhea, syphilis, AIDS, and even the common cold continue to strike millions of victims each year. The lack of suitable vaccines against many diseases is due partially to the nature of these pathogens and partially to characteristics of the immune system. The following list describes some of the barriers to immunization.

- Some pathogens, such as *Neisseria gonorrhoeae,* are so nonimmunogenic that they fail to trigger protective levels of immunity. Extensive efforts to produce immunogenic vaccines against gonorrhea have, as of 1994, yet to yield a promising product.

- Infections localized on the skin or mucous membranes are not affected by circulating antibodies and are generally controlled by secretory IgA. Unfortunately, immunologic recall is less efficient with these secretory immunoglobulins than with circulating antibodies. Thus uncomplicated gonorrhea and other diseases restricted to the urogenital mucosa stimulate poor natural immunity during infection and are less likely to be controlled by vaccines. In addition, *Neisseria gonorrhoeae* produces a protease that protects the pathogen by enzymatically destroying IgA.

- Some pathogens suppress immunologic response during natural infection. Unfortunately, vaccines prepared against these diseases (malaria, syphilis, leprosy, and AIDS) may similarly depress immunity rather than enhancing it.

- Some viruses become sequestered within host cells where they are protected from the body's immune elements. Herpesviruses, for example, persist in nerve cells following disappearance of the epithelial lesions. Lesions may recur when the virus migrates within infected neurons directly to the skin, even in the presence of specific circulating antibodies.

- Vaccine production usually requires laboratory cultivation of the pathogen to provide a source of antigen. Unfortunately, some pathogens, notably *Treponema pallidum* (syphilis), cannot be readily grown in vitro.

- The common cold viruses, although extremely antigenic, are so numerous it would be virtually impossible to develop vaccines against them all. More than 100 antigenically distinct cold viruses would have to be included in a complete vaccine. Such a vaccine would not only be impractical but would be of little value because poor immunity is induced by simultaneous administration of so many antigens.

- Some pathogens periodically alter their surface antigens. The new antigens are not recognized by immune cells sensitized to the previous determinants. Influenza is the most extensively documented example of a pathogen's tendency toward antigenic shifts. Although influenza vaccines are available, they are type-specific. A new vaccine must be developed whenever a new, antigenically distinct strain appears. Development of a vaccine against AIDS has also been hampered by the ability of the human immunodeficiency virus to change antigenic specificity.

- Vaccines are most successful against diseases controlled by humoral immunity. Few vaccines are available against diseases that are controlled primarily by CMI. For example, we cannot immunize effectively against diseases caused by protozoa or fungi (although some experimental vaccines against malaria use the parasite's antigens to create antibodies that neutralize the sporozoites either before they initially infect a blood cell or when they leave a lysed cell, before they can infect another erythrocyte. (In their intracellular stages, the parasites are protected from antibodies.)

- Some pathogens immunologically camouflage themselves by coating themselves with host blood proteins. In this way the surface antigens of plasmodia (malaria) and trypanosomes (encephalitis) are hidden from immune effectors. Vaccination against these diseases may trigger the formation of antibodies and lymphocytes that cannot locate their camouflaged antigens.

derived from a sensitized donor. Furthermore, a soluble component called *transfer factor* can be extracted from immune lymphocytes and used to specifically sensitize a nonimmune recipient's T cells. Transfer factor has been successfully used to control severe *Candida albicans* infections in persons with apparently deficient CMI against this opportunistic yeast.

Natural passive immunity helps protect newborns from infectious disease. The fetus, which is incapable of producing antibodies, becomes passively immunized by maternal antibodies that cross the placenta from the mother to the fetus. Maternal antibodies persist for many weeks following birth and help protect the baby until he or she can actively produce antibodies. (Immunologic competence begins during the third to sixth month after birth and may not be adequate for 2 or 3 years.) Additional protection is provided by maternal antibodies that are passively transferred in breast milk. Breast milk also contains 25 percent of the mother's daily production of monocytes. Such protection

mediated immunity to eradicate the tumor naturally. This approach is called **immunopotentiation**. Furthermore, while conventional cytotoxic agents for cancer therapy compromise the patient's resistance to pathogens, immunopotentiators would oppose opportunistic infections (see THE EXPLORERS: Immunopotentiators—Priming the Immunologic Pump).

CONCEPT CHECK

- Explain how an antigen–antibody response aids in the diagnosis of disease.
- How do fluorescent antibody tests, ELISA, and RIA differ?
- Describe three ways in which products of the immune system might be used in therapy.

THE EXPLORERS

Immunopotentiators —Priming the Immunologic Pump

In most people with intact immune systems, overwhelming disease is the exception. As your immune system battles potentially pathogenic microbes or cancer cells, it does so with such effectiveness that you will likely never even be aware of the threat it has conquered. But when immunity fails to arrest cancer cells and a person develops the disease, the immune system's activity can sometimes be artificially boosted to eradicate the renegade cells. Scientists observed years ago that some types of microorganisms or their products favorably affect the treatment of cancer by enhancing the natural immune response. The most extensively tested of these immunopotentiators is bacillus Calmette-Guerin (BCG), the vaccine against tuberculosis. BCG is an attenuated strain of the tubercle bacillus that stimulates cell-mediated immunity. The effects of T-cell enhancement are nonspecific; macrophages and NK cells are activated. Not only is tuberculosis prevented but, in many cases, malignancies are also eradicated. The effect is most pronounced against skin cancer and certain leukemias. Pertussis vaccine and nonviable preparations of the bacterium

Corynebacterium parvum also launch an immune-based attack against cancer. In addition to stimulating CMI, these agents may also enhance antibody protection.

Many immunopotentiators are *adjuvants*, substances that enhance the immune response when injected with an antigen. The enhancement is often specific for antigens administered with the adjuvant. For example, killed *Bordetella pertussis* (whooping cough vaccine) enhances immunity against diphtheria and tetanus when administered with the corresponding toxoids. Consequently, the DPT vaccine against diphtheria, pertussis, and tetanus provides better immunity against diphtheria and tetanus than would the toxoids given separately. Unfortunately, there are several dangerous side effects thought to be associated with the bacterial component of the trivalent vaccine, and toxoid of the pertussis toxin is likely to replace the whole cell in the near future.

A variety of molecules that mediate immune cell activity are currently produced by genetic engineering and show promise as immunopotentiating agents. These proteins, which represent our body's natural response modifiers, can now be produced in high yield in cultured cells. They include interferons and molecules that induce interferon production. Interferons stimulate antiviral proteins and increase the effectiveness of the immune system. Some scientists feel that interleukins are the most exciting of the immunopotentiators. For example, inter-

leukin-2 (IL-2), also known as T-cell growth factor, can act directly on macrophages and T cells to generate more T killer cells. Tumor necrosis factor (TNF) activates phagocytic and cytotoxic cells. Colony-stimulating factors (CFSs) promote cell division and maturation of immature immune cells in the bone marrow.

Some recent research has demonstrated the importance of stimulating CMI rather than humoral immunity for protection against some antigens. For example, the disease leishmaniasis (see page 664) is best controlled by CMI but poorly by humoral immunity. In mice given experimental leishmania infections, those that generated T_{H1} cells (helper cells that trigger CMI) lived while those mice that responded with more T_{H2} cells (helpers that promote humoral immunity at the expense of CMI) died. Furthermore, a specific interleukin (called IL-12) promotes the production of CMI-inducing T_{H1} cells, whereas IL-4 encourages humoral responses by activating T_{H2} cells. Using IL-12 with an antigen, designers of vaccines and immunopotentiators can enhance the effectiveness of their product against those diseases that are best controlled by CMI rather than humoral immunity.

The search for nonspecific immunopotentiators opens new doors every year, promising safer, more effective weapons against malignancy and many infectious diseases.

Key Terms

Key Facts and Concepts

- **Acquired protection.** Immunity is acquired during a person's lifetime as the result of exposure to antigens. The resulting immunity specifically protects against pathogens that possess the provoking antigenic determinant. Protection may be due to the production of soluble antibody molecules (humoral immunity) or the activation of T lymphocytes (cell-mediated immunity).

- **Humoral (B-cell) immunity.** Antibodies are subdivided into five immunoglobulin classes: IgG, IgM, IgA, IgE, and IgD. Antibodies specifically react with the antigen that stimulated their formation, often changing it in a manner that aids elimination of the antigen from the host. Such changes include agglutination, precipitation, opsonization, virus or toxin neutralization, and complement fixation. A single antibody molecule may demonstrate several of these activities.

- **Production of antibodies.** Antibodies are synthesized by plasma cells derived from B lymphocytes programmed to recognize specific antigens. Initial exposure to an antigen-presenting cell and a helper T cell triggers the events of the primary response. Antigen-specific B lymphocytes proliferate into clones of antibody-producing plasma cells plus a few memory cells (responsible for immunologic recall). Subsequent antigenic challenges trigger memory cells to quickly differentiate into rapidly proliferating antibody-producing plasma cells, accounting for the events characteristic of the secondary (or anamnestic) immune response. Immunologic memory is responsible for acquired immunity following disease or vaccination. Humoral immunity protects against some bacteria and viruses, toxins, and other extracellular products of pathogens.

- **Cell-mediated (T-cell) immunity.** Cell-mediated immunity also shows immunologic recall, but no antibodies are produced. Instead, T lymphocytes are activated by antigen exposure, generating cytotoxic T cells, which kill their target cell by direct contact, and delayed-type hypersensitivity cells, which produce lymphokines. These soluble chemicals attract and activate phagocytes, inhibit their migration so they accumulate in the region of antigen stimulation, trigger the production of antiviral proteins, or poison target cells. Lymphokines may also activate null cells that further enhance the response. Suppressor T cells protect against immune-mediated tissue damage; helper T cells are required to elicit a humoral immune response from B cells. Memory cells store the information necessary to elicit a rapid response to subsequent antigenic exposure. CMI is usually directed against intracellular parasites, protozoa, fungi, virus-infected host cells, and tumors.

- **Active and passive immunity.** Immunity can be acquired either actively or passively. Natural active immunity often follows infectious disease, whereas vaccination stimulates artificial active immunity. Antibodies are passively acquired transplacentally or in breast milk (natural passive immunity) or transferred from donor to recipient by hypodermic injection (artificial passive immunity). Active immunity provides long-lasting protection after an initial lag period of several days to weeks. Passive immunity confers immediate protection that disappears within a few weeks.

- **The immune response in diagnosis.** Immunologic techniques aid in disease diagnosis. Serology is used to demonstrate elevated titers of pathogen-specific antibodies in the serum of infected patients. Antigen-specific T-cell immunity can be determined by skin testing. Persons who have been exposed to a pathogen usually show a positive skin test against its extracted antigens.

• **Immunotherapy.** Immunologic techniques can also be applied to therapy. Monoclonal antibodies linked to toxins or drugs may be used to deliver cytotoxic chemicals to specific target cells. Genetically engineered immunopotentiators enhance immune activity against foreign antigens.

Review and Problem-Solving Questions

1. How do antibodies differ from enzymes? from antibiotics?
2. How can one antibody molecule (IgG) be both *mono*-specific and *bi*valent? Could the same terms be applied to any of the other four classes of antibodies?
3. Describe a way in which antibodies might protect against each of the following: (a) Tissue damage by diphtheria toxin; (b) establishment of poliovirus in the intestinal tract; (c) multiplication of an encapsulated strain of *S. pneumoniae;* (d) blood infections of gram-negative bacteria.
4. What is the role of memory cells in the immune response?
5. If passive immunity is short-term protection, under what circumstances would it be the preferred method?
6. What is the major role of each of the following cells in the immune response? (a) Cytotoxic T cells; (b) DTH cells; (c) helper T cells; (d) suppressor T cells; (e) plasma cells; (f) natural killer cells.
7. How do lymphokines differ from antibodies? Describe the action of four lymphokines.
8. Distinguish between: (a) Active and passive immunity (b) Natural and artificial immunity (c) T-dependent and T-independent antigens
9. How can a substance that is antigenic in one host be

completely tolerated in another?
10. Why is it recommended that booster shots with tetanus toxoid be obtained every 10 years? Why aren't booster shots given for some other vaccines?
11. Why can a person have antibodies against a pathogen and still lack protection against the corresponding disease?
12. What is the function of the Fc region of an antibody? Diagram the kinds of antigen–antibody reactions you would expect if only individual antigen-binding sites were produced.
13. Explain the processes that generate antibody diversity.
14. What is a hybridoma? Why is the construction of this cell necessary for the formation of monoclonal antibodies?
15. Surface molecules play an important role in the immune response. What is the function of each of the following surface molecules? (a) Type I histocompatibility antigen; (b) type II histocompatibility antigen; (c) receptors on T cells; (d) receptors on B cells.
16. Why are infants first vaccinated against infectious disease when they are 2 months old instead of immediately after birth?

HEADLINES

STRESS AFFECTS DISEASE OUTCOME

Stress may be a universal experience, but scientists are just beginning to define the chemical changes that it induces in the human body.* Stressful conditions prompt the brain to produce factors that increase production of several neuropeptides and factors that stimulate corticosteroid synthesis by the adrenal glands. These chemicals can inhibit the proliferation of lymphocytes in response to antigen and suppress the activity of natural killer cells.

• Design an experiment to test whether stress caused by

final exams increases the susceptibility of students to the common cold. Describe a possible mechanism for increased susceptibility.
• How might stress reduce the effectiveness of vaccines?
• Recurrence of cold sores, herpes simplex lesions, is associated with stress. Hypothesize how stress might affect maintenance of a latent virus in nerve cells.
• Norman Cousins wrote a book for cancer patients that suggested that a positive outlook and a sense of humor helped the body fight tumors. What do you think of this idea?

*B.S. Zwilling, Stress affects disease outcomes, *ASM News,* 58: 23–25 (1992).

CHAPTER 19

Immune Disorders

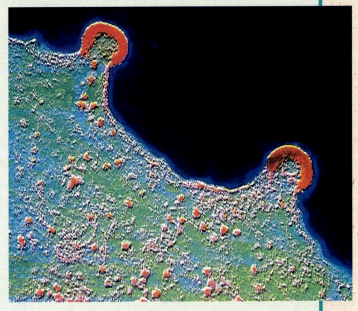

HIV virus particles budding from a "bleb" on the side of a T cell. Acquired immunodeficiency syndrome (AIDS) is one of many immune defects and immune-mediated disorders discussed in this chapter.

The common honeybee is the most dangerous animal in the United States. It kills more people than sharks, bears, bulls, horses, and all other animals combined. This is not the Africanized "killer bee," but the docile version of the common unaggressive honeybee whose sting is as harmless as a pinprick for most people. For a few individuals, however, a bee sting is deadly, as 14-year-old Andrea Morris was about to find out. Strolling shoeless with her mother through a dandelion field, she unwittingly stepped on one of the insects. She instantly recognized the burning pain in her foot as a bee sting, for she had been stung before. Unlike that first encounter, however, the stinging sensation seemed incidental as she began to react violently to the bee's venom. About a minute after the sting, she felt faint and nauseous and was struggling to get a breath. She was going into shock, and her blood pressure plunged. Her mother grabbed Andrea as she lost consciousness and rushed her to the hospital down the street. "Anaphylaxis!" yelled one of the emergency room doctors. "Epinephrine, 10 cc!" He plunged the needle directly into the dying girl's heart, the injected adrenaline circulating throughout the body within a minute. Next he gave her a shot of antihistamines and inserted a respiratory tube to force oxygen into the lungs through Andrea's tightly constricted airways. Slowly, her blood pressure began to climb and her airways opened. Before long she was staring at the doctor, who explained how lucky she was to be alive. A honeybee and an active immune system had almost killed her. Armed with the knowledge that she was allergic to bee venom, she now carries an automatic-injector syringe filled with epinephrine to quell the effects of her hypersensitive immune system should she have another encounter with a bee.

Three floors above the emergency room where Andrea barely escaped death, Hayden Ward stared it in the face. Five years of knowing it was coming didn't make its arrival any more welcome. He remembered his first *Candida* infections, then the recurring cytomegalovirus crises, followed by pneumonia. For the time being, doctors and antimicrobial drugs had saved him from these onslaughts. But now purple cancerous lesions chewed away at the surface of his nose and one foot. He was going to die soon. As he scanned the room, buoyed by the presence of so many people who loved him, he still felt abandoned, forsaken by his internal protector. His immune system, which had fought so gallantly, so victoriously, during his childhood bouts with measles, chickenpox, and even polio, seemed to have just given up, leaving him helpless against many of the opportunists that used to be little more than a day's work for his intact defenses. The system just seemed to cave in under the onslaught of HIV infection. When he was healthy, he hadn't spent much time thinking about the battles his immune system had fought for him, but now he thought about them almost constantly. If only he could rescue his T cells the way they had rescued him so many times before.

As Ward contemplated his crippled immune system, directly across the hall Debra Matthews lay trapped in a body she could no longer control. Debra was also facing death. Like Ward, she dwelled on her immune system, not how it had abandoned her, but how it had turned traitor, attacking her own nervous system (cells it was supposed to protect!), leaving her to die of multiple sclerosis. It seemed like treason, pure and simple—except that there was nothing simple about it. "What a bewildering relationship," she thought, as she sadly pondered being partners with an immune system that betrayed her. "I can't live without it, but I can't live with it either."

Immunity wields a double-edged sword, one that occasionally injures us while it tries to protect us. The cells of the immune system normally protect the body against invasion by foreign agents. As long as the defenses remain intact, people generally remain healthy or recover without consequence from most infections. But when these defenses are either crippled or turned against the very person they are supposed to protect, the consequences can range in severity from uncomfortable to fatal.

Immunity is crippled when immune cells or their products fail either to develop or to function properly, so any infecting agents that circumvent the primary and secondary lines of defense face little opposition within the disarmed host. Alternatively, host tissues can be killed or injured by "friendly fire," the immune system responding to an antigen in a manner that damages the host. For example, the immune system can lose its ability to distinguish host tissue antigens from foreign antigens and will then launch a direct attack on these "self" tissues. Sometimes our immune system overreacts to foreign antigens and injures our own tissues. Dead cells and antibody–antigen complexes generated by immune reactions may impair systems critical to the life of the infected person. This chapter features this dark side of the immune response.

Immunodeficiency

The life-sustaining importance of the immune response is emphasized by the consequences of its loss. **Immunodeficiency** is a deficit in either T-cell or B-cell immunity. Such deficiencies can be either natural (Table 19-1) or induced (to intentionally suppress adverse immunologic reactions, as discussed later in the chapter).

■ Noninfectious Immunodeficiencies

The most severe form of natural immunodeficiency is the failure to develop stem cells in bone marrow (Fig. 19-1). Since stem cells give rise to lymphocytes and phagocytic cells, individuals born with this defect are entirely unprotected from the organisms in their environment and die of overwhelming infections soon after they are born. A defect in the next cell in the chain, the lymphoid precursors, generates *severe combined immunodeficiency* (SCID) of T- and B-cell activity. Although this condition allows phagocytes to develop normally, they get no help

TABLE 19-1 SOME NATURAL IMMUNODEFICIENCIES

Deficiency	Caused by	Effect on Immunity		Increases Susceptibility To
		Humoral	**Cell-Mediated**	
Agammaglobuli- nemia	Genetically acquired B-cell deficiency	Depressed	Normal	Pyogenic bacteria[*]
Hypogammaglobuli- nemia	Dietary protein deficiency	Depressed	Normal	Pyogenic bacteria[*]
Complement deficiency	Lack of C_3'[†]	Normal	Normal	Pyogenic bacteria[*]
T-cell deficiency	Lack of thymus development	Somewhat depressed	Severely depressed	*Candida*, certain viruses, tumors
T-cell deficiency	Certain infections	Usually normal	Depressed	Secondary bacterial infections
Severe combined immunodeficiency (SCID)	Lack of stem cell Lack of interleukin receptors	Depressed Depressed	Depressed Depressed	All the above All the above
T-cell or B-cell deficiency	Lymphoid malignancies	Depressed	Depressed	All the above
Acquired immuno- deficiency syndrome (AIDS)	Destruction of helper T cells	Normal	Depressed	*Pneumocystis* infections; Kaposi's sarcoma, *Candida albicans* and other pathogens listed in Table 19-2

[*]Pyogenic (pus-forming) bacteria include *Staphylococcus aureus, Streptococcus pyogenes, Streptococcus pneumoniae, Neisseria meningitidis,* and *Haemophilis influenzae.*
[†]C_3' is one of the components of the complement system.

from the absent immune system. No one survives SCID. Humoral and cell-mediated immunity are also absent in persons who fail to produce helper T cells. For example, T-cell immunodeficiencies occur in individuals who lack a functional thymus, so lymphocytes cannot be processed to mature T cells. Deficits in cell-mediated immunity (CMI) make these persons extremely susceptible to recurrent infections by fungi, viruses, and intracellular bacteria. They also amplify the danger of developing cancer, rendering these persons a thousand times more likely to suffer from some types of malignant tumors. Furthermore, because helper T cells are needed to induce humoral immunity against most infectious agents, these people have impaired antibody production. Another cause of helper T-cell deficiency (and combined immunodeficiency) is the inheritance of genes defective for the enzyme adenosine deaminase (ADA). This enzyme catalyzes the removal of an amine group from adenine compounds generated in nucleic acid metabolism. In individuals lacking ADA, other enzymes in the T cells process

the adenine intermediates. Unfortunately, the product of this alternative pathway is toxic for T cells. In 1993, the first gene therapy replacement was performed for ADA deficiency. ▶ (See Using Microbes to Transfer Genes into Animal Cells, Chap. 9, p. 230.)

Naturally occurring absence of B-cell immunity is called *agammaglobulinemia* (or *hypogammaglobulinemia* if antibody levels are abnormally low but they are not completely absent). Persons with compromised antibody production are especially prone to opportunistic infections by pyogenic (pus-producing) bacteria, especially *Staphylococcus aureus* and *Streptococcus pyogenes*. Agammaglobulinemia is usually a genetic disorder in which the cells that process precursors to B lymphocytes are absent. It can also be caused by tumors that destroy "processing" cells within the lymphoid organs. Antibody production may also be depressed by protein deficiencies, since amino acids are the building blocks of immunoglobulins. This may occur in malnourished individuals and in burn and trauma patients. Complement deficiencies also compromise immu-

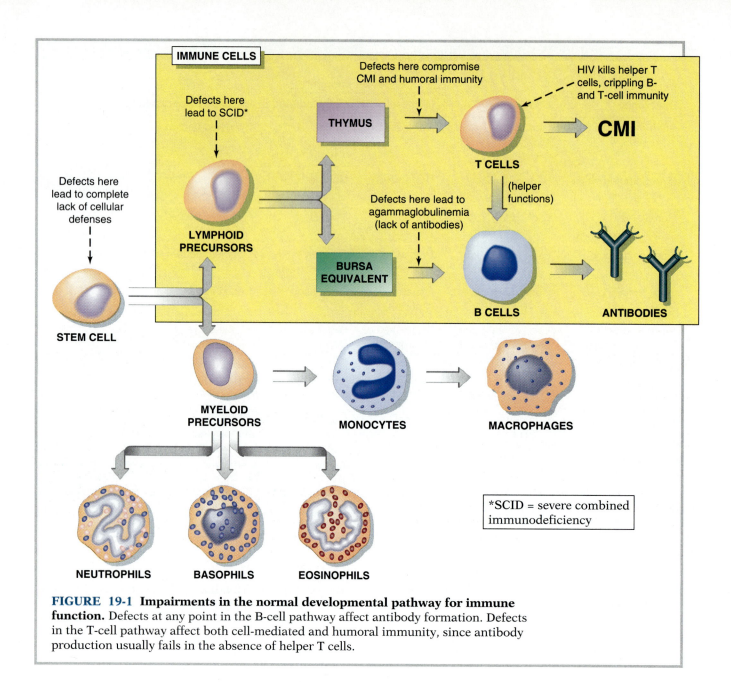

FIGURE 19-1 Impairments in the normal developmental pathway for immune function. Defects at any point in the B-cell pathway affect antibody formation. Defects in the T-cell pathway affect both cell-mediated and humoral immunity, since antibody production usually fails in the absence of helper T cells.

nity because the protective action of many antibodies depends on the lytic and opsonizing action of these proteins.

■ Acquired Immunodeficiency Syndrome

A few pathogens suppress the immune response during the course of an infection. Reduced T-cell immunity is a common feature of syphilis, leprosy, and malaria. Cell-mediated immunity is also crippled by viruses or tumors that destroy the lymphoid organs (for example, herpes and measles viruses or leukemia and Hodgkin's disease). One especially alarming CMI deficiency is **AIDS (acquired immunodeficiency syndrome),** a transmissible form of immune paralysis caused by the **human immunodeficiency virus (HIV).**

HIV might currently be the most famous virus in the world. The World Health Organization estimates that more than 15 million adults will be infected with this virus by 1995. it is also one of the world's most aggressively researched pathogens, a retrovirus that contains two copies of a single-stranded RNA housed in an enveloped virus particle (Fig. 19-2). Reverse transcriptase, an enzyme closely associated with the viral genome, assembles a double-stranded DNA molecule from the virus's RNA

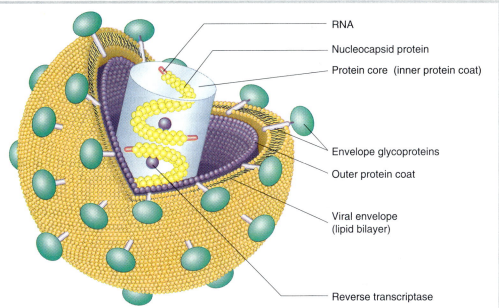

FIGURE 19-2 Human immunodeficiency virus (HIV). The virus contains two copies of its RNA genome in association with the enzyme reverse transcriptase. Surrounding the RNA is a protein capsid (the inner core), which in turn is surrounded by a second protein coat and an outer membrane envelope. Glycoproteins protrude from the surface of the virus. These major envelope proteins are the attachment sites with which the virus binds to host cells.

RNA
Nucleocapsid protein
Protein core (inner protein coat)
Envelope glycoproteins
Outer protein coat
Viral envelope (lipid bilayer)
Reverse transcriptase

genome. ▶ (See THE EXPLORERS: Reversing the Genetic Flow, Chap. 8, p. 197.) This viral DNA intermediate then integrates into the host chromosome. Integration of viral DNA is eventually followed by production of progeny viruses.

HIV preferentially attaches to receptor proteins found on macrophages, helper T cells, and antigen-presenting cells in the mucous membranes, skin, brain, and lymph nodes. A brief period of rapid viral replication generally follows infection, which peaks as immunity develops. Generally the level of circulating virus remains low, and there are no overt symptoms of infection for a period of 1 to 10 years. During this clinically latent state, active viral replication continues; however, it is primarily restricted to the lymph nodes. Eventually the lymph nodes' ability to trap and restrict the spread of progeny virus is overwhelmed, and the peripheral blood vessels once again become viral conduits.

Although the one or more factors that control the rate of viral replication within the lymph nodes remain undiscovered, we do know some ways in which HIV provokes an immunodeficiency. During viral replication, viral proteins become incorporated into the membrane of the infected cell. The immune system recognizes these antigens as "foreign" and launches an attack against the infected T cells and macrophages, an attack that prevents further viral replication in these cells. In doing so, however, the host defenses against other pathogens decline along with the helper T-cell population. As the infection continues, the helper T-cell concentration continues to drop. The host body is especially susceptible to opportunistic infections by such agents as *Candida*

albicans and *Pneumocystis carinii* and to Kaposi's sarcoma, a malignancy of the connective tissue (Table 19-2). When an HIV-infected person develops

TABLE 19-2   CLINICAL CONDITIONS INCLUDED IN THE 1993 AIDS CASE DEFINITION

Candidiasis of esophagus, bronchi, trachea, or lungs
Cervical cancer, invasive
Coccidioidomycosis, disseminated or extrapulmonary
Cryptococcosis, extrapulmonary
Cryptosporidiosis, chronic intestinal (>1 month's duration)
Cytomegalovirus disease (other than liver, spleen, or nodes)
Cytomegalovirus retinitis (with loss of vision)
Encephalopathy, HIV-related
Herpes simplex: chronic ulcer(s); or bronchitis, pneumonitis, or esophagitis
Histoplasmosis, disseminated or extrapulmonary
Isosporiasis, chronic intestinal
Kaposi's sarcoma
Lymphoma: Burkitt's, immunoblastic, or primary of brain
Mycobacterium avium complex, *M. kansasii*, or other species, disseminated or extrapulmonary
Mycobacterium tuberculosis, any site
Pneumocystis carinii pneumonia
Pneumonia, recurrent
Progressive multifocal leukoencephalopathy
Salmonella septicemia, recurrent
Toxoplasmosis of brain
Wasting syndrome due to HIV

SOURCE: *Morbidity and Mortality Weekly Report*, 41(RR-17): 15 (1992).

In the decade since the recognition of the first cases of AIDS, the disease has grown into one of the leading causes of death in the world. Despite rapid progress in isolating and identifying the human immunodeficiency virus (HIV), it is still not clear what causes the collapse of the immune system that makes its victims so vulnerable to opportunistic infections. One thing is clear—although the target for HIV infection is the helper T cell, most helper T cells are *not* infected with the virus. The destruction of these cells as the disease progresses, therefore, must have another cause.

Several proposals have recently been advanced to explain the ultimate destruction of uninfected T cells. One of these proposals suggests that infection with HIV triggers an autoimmune response—that is, the body begins to mount an attack against its own T cells (*auto* = self). Another proposal suggests that HIV acts as a superantigen, stimulating immune cells to respond to the antigen but to respond in a way that ultimately leads to their own death. The HIV antigens might be encouraging immune cells to "commit suicide."

Proponents of the autoimmune hypothesis point to experiments performed in uninfected mice immunized with normal lymphocytes from another strain of *uninfected* mice. After exposure to foreign lymphocytes, the immunized mice produced antibodies to the HIV envelope protein gp 120. This suggests that gp 120 and a lymphocyte antigen share some common properties. Antibodies formed against HIV cross-react with the host's lymphocytes and destroy them. If this is true, it has frightening implications for the hoped-for vaccine against AIDS. A vaccine that stimulates antibody to gp 120 could itself trigger an AIDS-like paralysis of the immune system.

Other researchers have uncovered evidence that HIV can act as a superantigen. Such molecules bind to T-cell receptors in an unusual manner—they need only recognize a portion of the receptor in order to activate the T cells. Superantigens bypass the specificity required of most antigens and activate diverse populations of T cells. Worse, T cells stimulated by superantigens are programmed for death, so the disease is already progressing before the actual loss of T cells is measurable. If the pathology of AIDS is due to a superantigen, then any antiviral therapy would have to be early enough in the infection to prevent exposure of most T cells to the antigen. Like the first proposal, this pathogenic mechanism could seriously impair the development of a suitable vaccine.

If either of these scenarios is true, any superantigens or cross-reacting antigens would have to be identified and meticulously excluded from the vaccine.

Hypersensitivity—Immunity that Backfires

Sensitizing our immune cells to specific antigens provides us with immunologic memory and protection. However, it may also cause tissue damage. This condition, called **hypersensitivity** (or more commonly, *allergy*), affects one of every six Americans. There are four basic types of hypersensitivity.

- *Type I (immediate-type) hypersensitivity* is mediated through chemicals released from mast cells responding to the interaction of antigen and IgE antibody. Sensitized individuals often respond within minutes after exposure to the antigen. Type I hypersensitivity is responsible for hay fever, asthma, drug allergies, and rapidly developing allergic reactions to injected antigens like bee venom.
- *Type II (cytotoxic) hypersensitivity* damages host cells when antibodies react with them and activate complement, phagocytes, or other effector cells. This type of hypersensitivity is not characteristic of typical "allergies" but elicits reactions that generate cell death, such as rejection of transplanted foreign tissue or mismatched blood cells from a donor. Some autoimmune diseases stimulate a type II response when they react with "self" antigens.
- *Type III (immune-complex-mediated) hypersensitivity* injures host tissues by depositing immune complexes—aggregates of antigen and antibody—on capillary, renal, and other vulnerable surfaces.
- *Type IV (delayed-type) hypersensitivity* is associated with sensitized T lymphocytes, which react to antigen by releasing lymphokines. People with type IV allergies have an exaggerated T-cell response, causing symptomatic tissue damage, usually within 24 to 48 hours after exposure (explaining the name "delayed" hypersensitivity). A few autoimmune diseases stimulate a type IV response when the body reacts with "self" antigens. Such reactions also occur in response to foreign tissues in organ transplants.

Table 19-3 compares the important properties of the four types of hypersensitivity.

TABLE 19-3 · COMPARISON OF TYPES OF HYPERSENSITIVITIES

	Type I (Immediate)	Type II (Cytotoxic)	Type III (Immune complex)	Type IV (Delayed)
Mediated by	IgE	IgG, IgM	IgG	Cytotoxic T cells
Symptoms elicited by	Binding to mast cells or basophils and release of soluble factors	Binding to target cells and activation of complement	Binding to antigen, deposition in tissue causing local inflammation	Binding to target cells and release of lymphokines
Reaction time	Minutes	Minutes to days	Days to months	14 to 48 hours
Symptoms	Anaphylaxis; hay fever; asthma; hives	Cell lysis as in transfusion reactions and hemolytic disease of newborns	Glomerulonephritis; hemorrhage; arthritis; serum sickness	Chronic inflammation; allergic contact dermatitis; tissue graft rejection

Several of these hypersensitivity reactions are responsible for tissue rejection. This unwanted response occurs when organs are transplanted into another individual and in autoimmune disorders in which the body no longer tolerates certain self antigens and begins to reject some of its own tissues as foreign.

■ Type I (Immediate-Type) Hypersensitivity

Immediate-type hypersensitivity is due to the production of a unique class of antibodies called immunoglobulin E (IgE). Antibodies in this class are similar to IgG in structure and antigen reactive sites. They are formed in response to an **allergen** (an antigen that elicits an allergic response) and react specifically with the eliciting allergen. The unique property of IgF, however, resides in the antibody's Fc region, the end opposite the antigen-binding sites. This unique immunoglobulin attaches by its Fc region to receptors on a host mast cell or basophil. Once coated with IgE, mast cells are primed for allergic response. (Each sensitized cell may contain 500,000 IgE molecules on its surface.)

Mast cells and basophils contain large numbers of granules (Fig. 19-4). Within these granules are chemicals such as *histamine* that trigger rapid changes in capillaries and smooth muscles. These chemical mediators are normally kept in cellular compartments (the "granules"), and very little of them reaches the surrounding tissues. However, when IgEs on the primed mast cells become crosslinked by reacting with the corresponding allergen, the sensitized cells immediately degranulate, releasing these chemicals into the surrounding tissues, where they trigger the general symptoms of allergy:

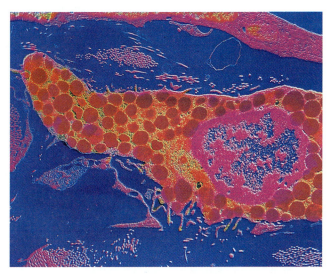

FIGURE 19-4 Mast cells. Electron micrograph of a mast cell shows large granules that contain histamine and other mediators of immediate hypersensitivity. Mast cells are located in tissue. Basophils play a similar role in allergy but circulate in the blood.

itching, edema (accumulation of fluid in tissues), smooth muscle spasms, and vascular dilation. Additional chemicals are synthesized on activation of the mast cell. These include prostaglandins, leukotrienes, and chemotactic factors, chemicals that contribute to the physiological changes associated with allergy. The sequence of events in immediate-type allergy is depicted in Figure 19-5.

The location of allergy symptoms depends on the distribution of the mast cells in host tissue (they are predominantly found in the gastrointestinal and respiratory tracts, connective tissues, and skin) and the route of exposure to the eliciting allergen. Hay fever (allergic rhinitis), for example, is an upper respiratory response to inhaled allergens, often airborne

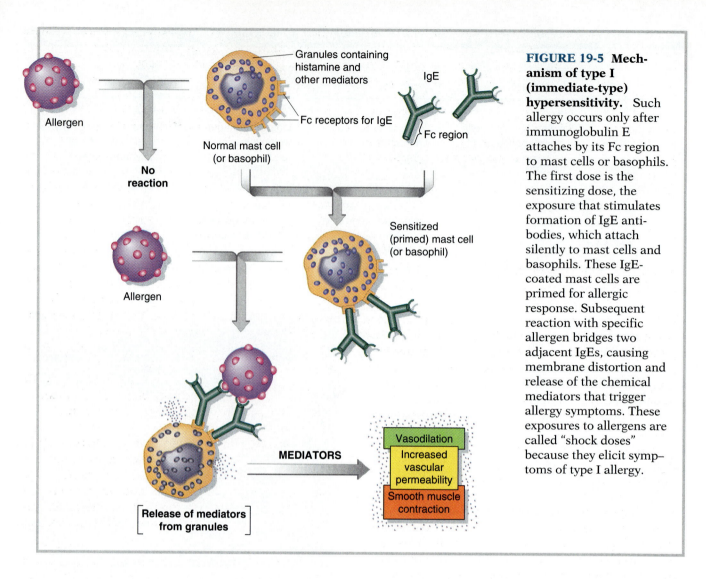

Granules containing histamine and other mediators

Fc receptors for IgE

Normal mast cell (or basophil)

Allergen

No reaction

IgE

Fc region

Allergen

Sensitized (primed) mast cell (or basophil)

MEDIATORS

Vasodilation

Increased vascular permeability

Smooth muscle contraction

Release of mediators from granules

FIGURE 19-5 Mechanism of type I (immediate-type) hypersensitivity. Such allergy occurs only after immunoglobulin E attaches by its Fc region to mast cells or basophils. The first dose is the sensitizing dose, the exposure that stimulates formation of IgE antibodies, which attach silently to mast cells and basophils. These IgE-coated mast cells are primed for allergic response. Subsequent reaction with specific allergen bridges two adjacent IgEs, causing membrane distortion and release of the chemical mediators that trigger allergy symptoms. These exposures to allergens are called "shock doses" because they elicit symptoms of type I allergy.

pollen or fungal spores. The response evokes watery discharges from the eyes and nose because of histamine-induced capillary leakage in these areas. Increased mucus production is stimulated by prostaglandins. Asthma attacks are due to release of chemical mediators other than histamine (explaining why antihistamines fail to relieve asthma symptoms) from mast cells in the bronchioles, where the chemicals trigger smooth muscle spasms that interfere with breathing (sometimes fatally). Hives (urticaria) develop when subdermal mast cells degranulate, causing vascular fluid to leak into skin tissue and form itchy red swellings. Degranulation of intestinal mast cells elicits some of the symptoms of food allergy, although the allergen may be absorbed from the intestine and trigger hives and other generalized allergic symptoms. Most IgE-mediated allergies can be symptomatically treated by using histamine antagonists such as the antihistamines in commercial allergy medications. Antihistamines are not used to treat asthma, the major symptoms of

which are due to release of mediators other than histamine from mast cells. Some symptoms of asthma can be reduced with epinephrine, corticosteroids, or other prostaglandin inhibitors. The best prevention, however, is to avoid contact with the allergen if possible.

● **ANAPHYLAXIS.** One form of type I hypersensitivity can kill a person within 10 minutes of allergen exposure. **Anaphylaxis** occurs when many body sites containing sensitized mast cells and basophils simultaneously release large quantities of chemical mediators into body fluids. The bronchioles contract and trap inhaled air in the lungs. The capillaries dilate, and blood pressure plunges, causing shock. The patient may quickly die of asphyxiation or circulatory failure. Administration of adrenaline (epinephrine) early in the anaphylactic reaction may save the patient's life by counteracting the effects of the chemical mediators of anaphylaxis. Antihistamines are of limited value in systemic anaphylaxis because many of the symptoms are caused by additional substances released along with histamine from degranulating mast cells. Anaphylactic shock is most commonly associated with systemic injection of drugs (notably penicillin), horse serum used for passive immunization, insect venom, and even allergens used for treating allergy (see Allergy Desensitization, below). A patient's history of allergy should always be considered prior to administering antibiotics, horse serum, or any foreign protein, although difficulties in convincing a honeybee to review a potential victim's hypersensitivities before stinging is one of the reasons hundreds of people die each year of bee stings.

● **SENSITIZATION AND SHOCK**   Exposure to an allergen triggers an allergic response only in certain people. These individuals generally produce higher levels of IgE in response to antigens than nonallergic persons, a trait that is genetically inherited. Sensitization occurs when allergen-specific IgE has been produced as a result of *previous* exposure to the antigen and the allergen-specific IgE has attached to host mast cells. Since this sensitized state requires previous exposure to the allergen, the first encounter with any antigen fails to induce allergic symptoms. This initial exposure (the sensitizing dose) is analogous to a primary immune response ▶ (see Primary and Secondary Antibody Responses, Chap. 18, p. 452). Once sensitized to a specific allergen, subsequent exposures (called shocking doses) trigger IgE-primed mast cells to degranulate, inducing the physiological symptoms of hypersensitivity. Subsequent exposures to the allergen act as boosters of the anamnestic response, thereby intensifying the extent of the allergic reaction. Consequently, people who show only mild allergic reactions to penicillin or insect venom may, with subsequent exposures, develop life-threatening anaphylactic reactions.

In clinical tests for allergy, the allergen is introduced by pricking the skin. Allergic individuals will respond within minutes by releasing mediators from mast cells at the test site. The area becomes swollen and red, and it itches (Fig. 19-6). Most symptoms subside within a few hours.

the first or subsequent exposures may stimulate immunity and sensitize the immune system to produce an allergic response to later doses.

● **ALLERGY DESENSITIZATION**   The symptoms of IgE-mediated allergy can be averted by preventing allergens from reacting with IgE-coated cells. This may be accomplished by *allergy desensitization*. Allergic individuals can be desensitized by receiving a series of injections with controlled concentrations of the specific allergen(s) to which they are allergic. These low doses of antigen are believed to stimulate the production of other antibodies, called *blocking antibodies,* that compete with IgE for the allergen. Blocking antibodies are IgG and therefore do not fix to mast cells. They are believed to saturate the system and remove the allergen before IgE can react with it, thereby preventing the development of allergy symptoms. Desensitization may also be assisted by the proliferation of suppressor T cells that inhibit formation of allergen-specific antibody. Results of desensitization, however, are still inconsistent. The technique is effective in only a fraction of the patients treated this way.

It may soon be possible to control allergies by using antibodies to neutralize host factors that contribute to the allergic response. For example, specific interleukins induce IgE and act as mast cell growth factors. Inactivation of these chemicals may help prevent the development of an allergic reaction.

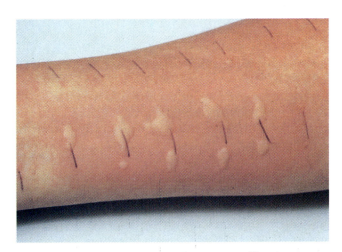

FIGURE 19-6   **Skin test to determine allergic sensitivities.** Minute quantities of many different allergens are scratched under the skin. A half hour later, those allergens to which the patient is allergic collect fluid, itch, and redden. In this way allergists identify which allergens to recommend the patient avoid. These allergens can also be the target of allergy desensitization, described in the next section.

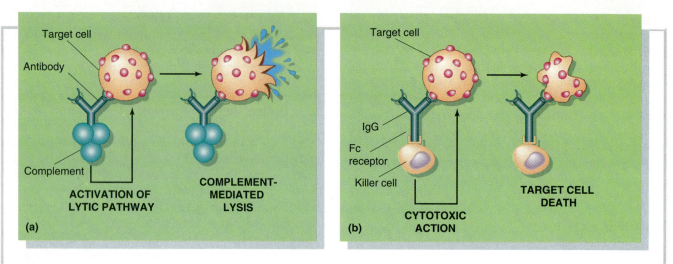

FIGURE 19-7 Mechanism of type II hypersensitivity. Antibodies bind to antigens on their target cell. The Fc receptor extending from the cell reacts with (*a*) complement or (*b*) cells with cytotoxic activity. The target antigen might be a foreign cell (such as a mismatched blood cell), the cells of an organ or tissue transplant, or the cells of the host's own tissues that are no longer tolerated as self (in other words, an autoimmune reaction).

■ Type II (Cytotoxic) Hypersensitivity

The damage associated with type II hypersensitivity results from the interaction of cell-bound antibody with complement or cells such as killer cells or neutrophils. The tail portion of IgG or IgM extending from the target cell is the reactive part of the antibody once the antigen recognition sites are occupied with the target cell. Its interaction with complement or effector cells ultimately leads to lysis of the cell (Fig. 19-7). Thus type II hypersensitivity is also known as **antibody-dependent cell cytotoxicity (ADCC).**

The first complement component, which is an Fc receptor, binds to cell-bound antibody and activates the other complement components in the series of reactions leading to lysis (Fig. 19-7a). Another effector cell with an Fc receptor is the killer cell. Killer cells lyse the antibody-coated cell the moment they bind to it (Fig. 19-7b). Phagocytic cells also contain Fc receptors, but host cells are too large to be engulfed by phagocytes. The "frustrated" phagocytes respond by releasing their lysosomal contents, thereby causing damage to local tissue.

● **TRANSFUSION REACTIONS** The surfaces of blood cells contain many molecules that can act as antigens. When an individual needs a blood transfusion, it is important that the donor blood cells not elicit an immune response in the recipient. To minimize such an occurrence, donor and recipient blood types are determined and only cells with similar antigens are used for transfusions. The ABO blood group and the Rh blood group are the most important antigens to match.

The **ABO blood group** is expressed as a terminal carbohydrate on a cell surface glycoprotein. A single pair of genes control the production of this terminal carbohydrate. There are three possible forms for each gene of this pair: one expresses antigen A (group A); one expresses antigen B (group B); and one expresses no antigen (group O) (Fig. 19-8a). Since each person carries two genes for this carbohydrate (one on each chromosome), six possible genotypes are possible: AA, AO, BB, BO, OO, and AB. Although people tolerate their own blood antigens, they make antibodies, predominantly of the IgM class, to the antigens they lack. For example, the serum of a person with blood group B (genotypes BB or BO) or blood group O (no A or B antigens) will contain anti-A antibodies that destroy any group A blood cells received in a mismatched transfusion. The foreign blood cells are rapidly covered with the A-specific antibodies, leading to agglutination, clearing by phagocytic cells, and hemolysis by complement (Fig. 19-8b). In addition to failing to provide needed erythrocytes, transfusion reactions may cause fever, pain, and circulatory collapse. High concentrations of circulating hemoglobin contribute to the severity of the damage to the kidneys. The desire to avoid such severe transfusion reac-

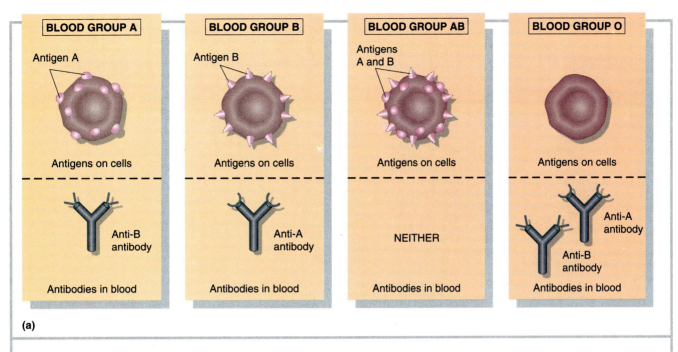

(a)

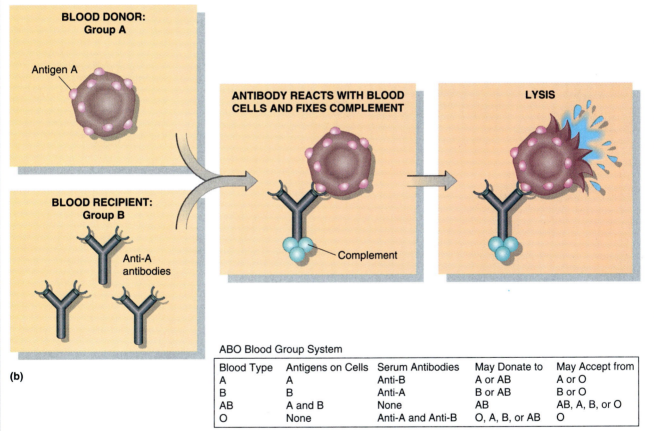

ABO Blood Group System				
Blood Type	Antigens on Cells	Serum Antibodies	May Donate to	May Accept from
A	A	Anti-B	A or AB	A or O
B	B	Anti-A	B or AB	B or O
AB	A and B	None	AB	AB, A, B, or O
O	None	Anti-A and Anti-B	O, A, B, or AB	O

(b)

FIGURE 19-8 ABO blood group reactions. (*a*) Red blood cells may be one of four groups based on the ABO antigens on their surface. Individuals produce antibodies only to those antigens that do not appear on their blood cells (immunologic tolerance prevents synthesis of antibodies to self antigens). An individual may be used as a blood donor only for persons who do not produce antibodies against their blood cells. (*b*) When donor and recipient blood do not match, antibodies in the recipient's serum will bind to donor cells and complement-dependent lysis will occur.

tions is the reason such care is taken to ensure that donors and recipients are compatible for the ABO blood group.

The **Rh** (rhesus) **blood group** is another important surface antigen on red blood cells. Individuals who produce these proteins are designated Rh-pos-

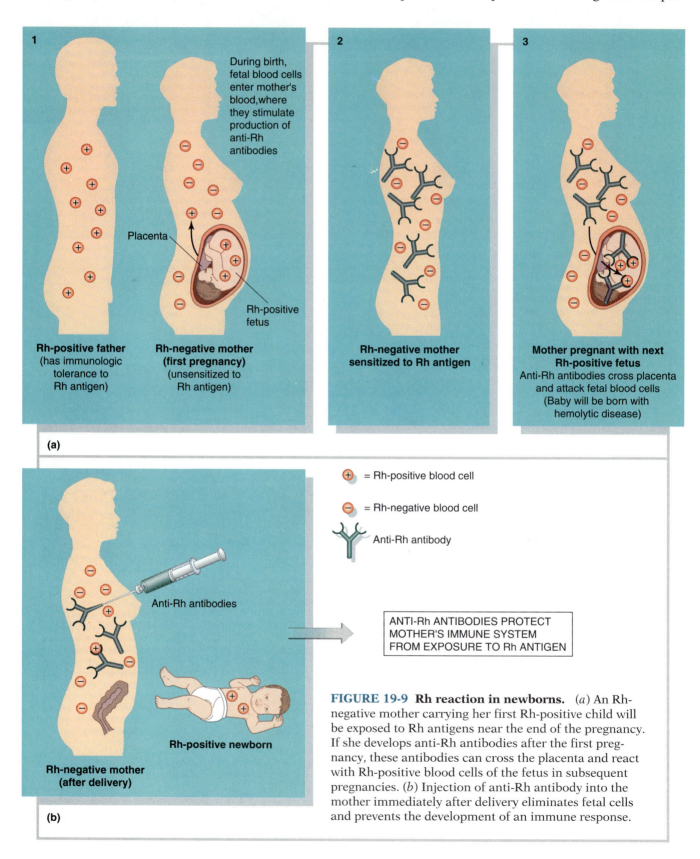

During birth, fetal blood cells enter mother's blood, where they stimulate production of anti-Rh antibodies

Placenta

Rh-positive fetus

Rh-positive father
(has immunologic tolerance to Rh antigen)

Rh-negative mother (first pregnancy)
(unsensitized to Rh antigen)

Rh-negative mother sensitized to Rh antigen

Mother pregnant with next Rh-positive fetus
Anti-Rh antibodies cross placenta and attack fetal blood cells
(Baby will be born with hemolytic disease)

(a)

⊕ = Rh-positive blood cell

⊖ = Rh-negative blood cell

Anti-Rh antibody

Anti-Rh antibodies

ANTI-Rh ANTIBODIES PROTECT MOTHER'S IMMUNE SYSTEM FROM EXPOSURE TO Rh ANTIGEN

Rh-positive newborn

Rh-negative mother (after delivery)

(b)

FIGURE 19-9 Rh reaction in newborns. (*a*) An Rh-negative mother carrying her first Rh-positive child will be exposed to Rh antigens near the end of the pregnancy. If she develops anti-Rh antibodies after the first pregnancy, these antibodies can cross the placenta and react with Rh-positive blood cells of the fetus in subsequent pregnancies. (*b*) Injection of anti-Rh antibody into the mother immediately after delivery eliminates fetal cells and prevents the development of an immune response.

itive; those who do not are Rh-negative. When Rh-negative people receive blood from Rh-positive people, they develop antibodies to the surface proteins. A second transfusion will cause the immediate destruction of the donor blood cells.

● **HEMOLYTIC DISEASE OF NEWBORNS** *Hemolytic disease of the newborn* usually results from incompatibility between an Rh-negative mother and an Rh-positive fetus. IgG antibodies to the Rh antigen cross the placenta from the mother's circulatory system and destroy fetal blood cells. This antibody-mediated death of red blood cells may cause severe anemia or even death if toxic compounds released from cell destruction are deposited in the brain.

Hemolytic disease rarely occurs during a first pregnancy since an Rh-negative mother will develop Rh-specific antibodies only after being exposed to Rh-positive cells (Fig. 19-9*a*). This primary exposure usually occurs during the late stages of the first pregnancy or during delivery (or miscarriage). After her first Rh-positive child, the mother is sensitized to the Rh antigen, so subsequent pregnancies with an Rh-positive fetus elicit an antibody response that is sufficient to damage the fetus (Fig. 19-9*b*). Prior to 1950, hemolytic disease was responsible for almost 1 death per 1000 births. Today almost all cases can be prevented by treating the mother with anti-Rh antibody immediately after her first delivery (Fig. 19-9*b*). The injected antibody reacts with any of the baby's Rh-positive blood cells that entered the mother's circulation, thereby preventing the fetal blood cells from sensitizing the immune system. The antibody must be administered after each pregnancy (even after miscarriage or abortion), since even one exposure of the mother's immune system to Rh-positive blood cells will lead to a rapid production of antibodies against blood cells of any subsequent Rh-positive fetus.

MYTH: An Rh-positive woman who marries an Rh-negative man must also be wary of Rh disease developing in her fetuses.

FACT: Because of immunologic tolerance, an Rh-positive mother will never make antibodies to Rh antigen. Furthermore, if she carries an Rh-negative baby, the lack of Rh antigen on the baby's cells means that the mother's immune system is not exposed to foreign antigen. Only an Rh-negative woman runs the risk of having a baby suffering from Rh disease, and then only if the father of her baby is Rh-positive.

■ Type III (Immune-Complex–Mediated) Hypersensitivity

Immune complex diseases result from the deposition of antigen–antibody complexes on membranes, usually capillary or renal membranes. These immune complexes fix complement, which in turn acts on basophils and platelets, precipitating a local inflammatory response. The damage to host tissue results from the chemical mediators and lysosomal enzymes released by cells drawn to the site of immune complex deposition (Fig. 19-10). Although

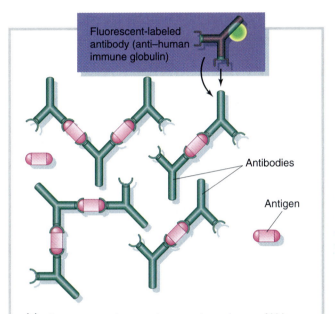

Fluorescent-labeled antibody (anti–human immune globulin)

Antibodies

Antigen

(a) Immune complexes on basement membrane of kidney

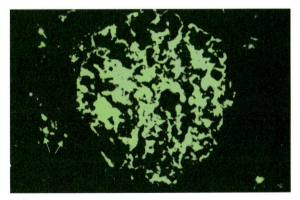

(b)

FIGURE 19-10 Renal membrane of patient with type III hypersensitivity. (*a*) The antigen–antibody complexes deposited in the kidney of a patient with acute glomerulonephritis can be detected with fluorescent antibody that specifically reacts with human antibody in the complex. (*b*) A micrograph of the fluorescent membrane.

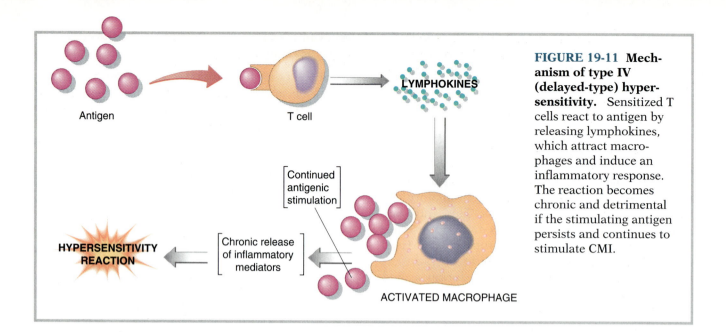

FIGURE 19-11 Mechanism of type IV (delayed-type) hypersensitivity. Sensitized T cells react to antigen by releasing lymphokines, which attract macrophages and induce an inflammatory response. The reaction becomes chronic and detrimental if the stimulating antigen persists and continues to stimulate CMI.

Labels in figure: Antigen; T cell; LYMPHOKINES; Continued antigenic stimulation; Chronic release of inflammatory mediators; HYPERSENSITIVITY REACTION; ACTIVATED MACROPHAGE

antigen–antibody interactions routinely lead to the formation of immune complexes, disease develops only in the presence of large amounts of complex.

Repeated and persistent exposure to an antigen may contribute to the formation of immune complexes. For example, immune complexes localized in the kidney may follow infections by *Streptococcus pyogenes* and some persistent viruses. Acute glomerulonephritis is a kidney disorder that is due to damage caused by deposition of immune complexes on renal membranes. Some other types of immune complex diseases are localized in the lung. These include farmer's lung, pigeon-breeder's disease, mushroom grower's disease, and librarian's lung. Constant exposure to an antigen, such as mold spores in hay or protein in pigeon dung (or in the dust of old library books), stimulates antibody production. Immune complexes form, and an inflammatory response follows. Rheumatoid arthritis is another complication of immune complexes, which elicit an inflammatory response in joints where they have been deposited.

A more generalized distribution of immune complexes often follows passive immunization. Serum sickness results from the overwhelming number of immune complexes that are formed when a person receives horse serum for the second time. Being immunologically sensitized to its antigens, the individual quickly produces antibodies against the foreign horse globulin proteins, generating enormous numbers of antigen–antibody complexes. These immune complexes are too numerous for disposal by phagocytes and are deposited in kidneys, joints, and blood vessels, where they cause inflammation, injury, and sometimes fatal disease.

■ Type IV (Delayed-Type) Hypersensitivity

Delayed-type hypersensitivity is an exaggerated CMI response that damages host tissue. Hypersensitive persons respond to certain pathogens and chemical allergens with a cell-mediated reaction that in itself accounts for many of the symptoms of disease. The common denominator of all these delayed hypersensitivity reactions is the participation of allergen-sensitized T lymphocytes (Fig. 19-11). Antibody and histamine play no role in this type of allergy.

Tuberculosis is the classical example of a disease that induces delayed-type hypersensitivity. In an attempt to eradicate the infection, host T lymphocytes move to the region of infection and release lymphokines, attacting macrophages to the area, where they are activated by macrophage-activating factor (MAF). The intracellular mycobacteria are eliminated by killing the infected host cell, so host protection is inevitably accompanied by some tissue damage. If the pathogen is not destroyed, however, it may continue to replicate within the phagocytes, providing continual and prolonged antigenic stimulation. In addition, sensitized lymphocytes react with the allergen and produce a lymphokine that transforms more lymphocytes into tuberculin-sensitized cells, greatly exaggerating the response. The continual infiltration of sensitized T cells and macrophages into the area and the subsequent release of lysosomal enzymes damage the surrounding tissue. This tissue necrosis in turn further intensifies chronic inflammation, producing a self-perpetuating prolonged allergic disorder. Such situations produce a characteristic lesion called a **granuloma,** so called because of the accumulation of neutrophils and other granulocytes. Symptoms of tuberculosis, therefore, are caused more by the body's "protective" response to infection than by the primary virulence of the bacteria. Immunity is a biological double-edged sword—CMI may kill the pathogen, but delayed hypersensitivity may kill the infected person. Other infectious agents that damage through delayed-hypersensitivity reactions include fungi, protozoa, and a few bacteria.

Delayed hypersensitivity can occur anywhere in the body where T-cell-activating allergens localize. In *sensitized* persons a delay in onset of allergic symptoms, usually 24 to 48 hours, follows exposure to the allergen. This is why a period of 2 days is required for positive tuberculosis skin tests to develop. These skin tests are performed by injecting purified tuberculin extract containing the allergen to which tuberculin-sensitized lymphocytes react. Localized delayed hypersensitivity (redness and swelling) at the site of injection indicates a positive test. A positive tuberculin skin test is due to the presence of T lymphocytes sensitized to the pathogen, suggesting previous exposure to the tubercle bacillus (but not necessarily the presence of active disease). A negative skin test suggests little or no contact with the pathogen; the body's immune system, not having encountered the pathogen, remains unsensitized to the antigen. Tuberculosis skin tests, most commonly the Mantoux test or Tine test, are routinely administered as a screening method to detect possible cases of tuberculosis among college students, health care practitioners, and other large populations. Skin test findings are often the first diagnostic sign suggesting that a person has tuberculosis.

Other allergic reactions that are classified as delayed-type hypersensitivity include **contact dermatitis,** allergic skin reactions to poison ivy plants, cosmetics, soaps, drugs, and certain metals. Most of these antigens are haptens that bind to epidermal cell proteins. The epidermis also contains a population of antigen-presenting cells called Langerhans cells. Lymphocytes migrate from blood vessels in response to the antigen, and macrophages are attracted. The destruction of the epidermal cells gives rise to the red and often blistery symptoms of contact dermatitis (Fig. 19-12).

CONCEPT CHECK

- What is the role of antibody in allergic reactions, transfusion reactions, and immune complex diseases?
- Describe how T cells cause damage in delayed-type hypersensitivity rather than providing protection.

Transplant Reactions

The primary responsibility of the immune system is to prevent foreign substances from becoming established within the body. There are circumstances, however, when it is desirable to permanently implant non-self materials into a person—for example, to replace nonfunctional organs or their products. Heart valves may lose their resiliency, and the heart itself be defective. Kidneys or liver may fail to function. Hip bones may become too fragile to support the body's weight without fracturing. Bone marrow may give rise to leukemic cells or fail to produce red blood cells. Sometimes the damaged part can be replaced with manufactured structures composed of artificial materials—plastics or metals that do not stimulate an immune response and are therefore not rejected. For years, titanium hip joints and silicone heart valves have been successfully implanted in people, and the implants are well tolerated.

Unfortunately, transplanted living tissues (called *grafts*) are usually rejected by the immune system unless they are placed in a "privileged" site, a region of the body that is not accessible to immune lymphocytes. The brain and the cornea are two examples of privileged sites. Both brain and corneal tissues are protected from circulating immune cells and antibodies. The thickened capillaries of the blood-brain barrier prevent the immune system from evoking a damaging response in the central nervous system. Fragments of brain tissue have been successfully transferred from human to human to stimulate the production of neurotransmitters depleted by disease or genetic defects. Corneal tissue is generally well tolerated after transplantation because the cornea is in an area of the body that is not highly vascularized. Corneal transplants in humans have a high success rate without requiring that the donor and recipient be matched. The successful transplantation of other organs, such as kidney, liver, bone marrow, heart, and lung, depends on careful matching of donor and recipient plus medical procedures to prevent a rejection response from the immune system following receipt of the foreign organ.

■ Mechanism of Graft Rejection

Although every cell bears many different antigen groups on its surface, some antigens are more likely than others to elicit a rejection response. Class I and class II histocompatibility antigens on a graft stimulate a strong immune response if they differ from those on the recipient's cells. Donors and recipients

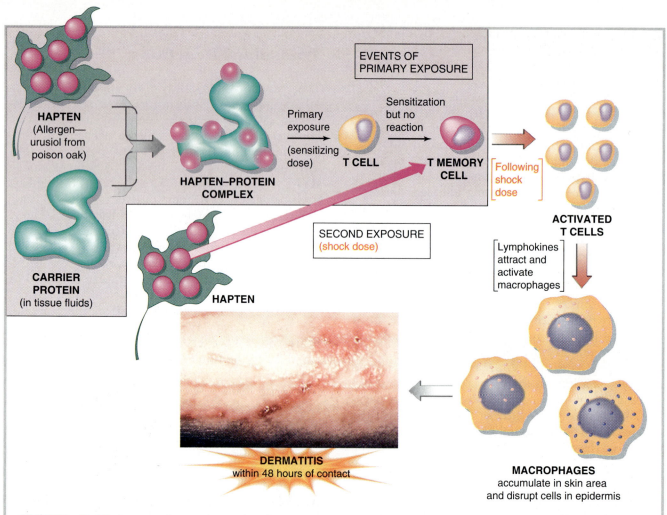

FIGURE 19-12 Contact dermatitis. The allergen (in this case, oil from poison ivy or poison oak) is a hapten that becomes antigenic after combining with a carrier protein. First contact with the allergen is the "sensitizing dose." Once sensitized, any subsequent contact (called a "shock dose") produces characteristic lesions (in photo inset, the lesions have developed in the pattern of contact, producing a replica of the plant's stem and leaf).

must therefore match at these major antigenic sites for the graft to survive. Other sites may also be important, affecting the long-term viability of a graft. It is therefore desirable, although not always possible, to match at these sites as well. (See BEYOND THE BASICS: Are You My Type?) The ideal donor, of course, is an identical twin. Other siblings will have inherited a different combination of parental genes for histocompatibility antigens. Sometimes they are good candidates for donors; sometimes their tissues do not match closely enough.

Antibodies play a role in tissue rejection, particularly in persons previously sensitized to the surface antigens on transplanted tissue. Such individuals have specific circulating antibodies that mediate a type II hypersensitivity reaction, leading to damage and rejection of the tissue within minutes. This rapid rejection develops as complement activation and enzymes released from neutrophils and platelets kill the cells of the foreign organ. Such acute rejections are rare, since tissue typing, which is routinely performed prior to transplantation, alerts practitioners to the presence of circulating antibodies against the graft.

Rejection of most primary grafts is a cell-mediated process that can occur in two ways (Fig. 19-13). Circulating helper T cells activate cytotoxic T cells targeted against the class I antigens on the graft. The rejection response is also facilitated by the production of lymphokines, which attract macrophages to the graft. As with other immune reactions, antigen-specific memory cells are produced. A second graft from the same donor will be rejected more rapidly.

Are You My Type?

Transplant medicine is often thought of as one of modern medicine's more recent advances. But human tissues have been successfully transplanted for hundreds of years. Burn patients, for example, would have skin transferred from healthy areas of their body to the burned region, providing a replacement protective covering over the wound. Such *autografts* (transplants of tissues to different locations in the same person) pose no challenge to the host's immune system (unless the tissue is transplanted from a "privileged site," an area not exposed to the immune system, in which case no tolerance to these tissues has developed). Similarly, *isografts*, transplants between identical twins, provoke no immune response. Most transplant operations, however, are *allografts*, in which the recipient is genetically dissimilar from the donor. The success of these grafts depends on somehow blocking the recipient's immune system from recognizing the donor tissue as foreign. Such immune amnesia is imposed on the recipient by techniques that inhibit immune system function. The extent of immune suppression needed to maintain transplanted tissue depends on how different the donor and recipient tissues truly are. If the match is not close, the prevention of immune rejection may require severe immune suppression, leaving the recipient vulnerable to life-threatening infectious disease.

Immune rejection reactions target proteins on the surface of the foreign cells. The primary targets are surface proteins collectively known as **human leukocyte antigens (HLA)**. These antigens are synthesized under the direction of a cluster of genes (on chromosome 6) called the **major histocompatibility complex (MHC)**. These proteins are identity markers, and each individual has a characteristic set inherited from his or her parents.

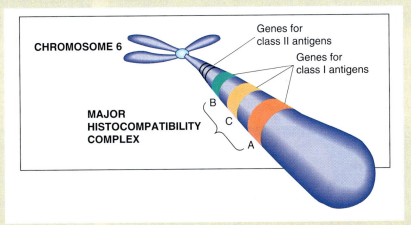

Thousands of different antigenic specificities exist for each of the two genes. With more than 20 types of HLA-A and 50 different HLA-B genes, the chance of two unrelated individuals having the same gene array is very low. Even within a family, the probability of two siblings carrying identical surface proteins is only about 1 in 100 for class I antigen alone (there is a 25 percent chance that they are identical at each of the genes) and 1 in 10,000 for having identity at both class I and class II sites. Identical twins, however, carry identical genetic information and are perfect matches for each other.

Most people do not have an identical twin and must rely on finding a volunteer donor. For bone marrow and kidney transplants, the donor is living. In fact, individuals can be screened and have information on their tissue type available in a registry that facilitates locating them should their marrow be needed. Other living donors are those who donate one of their kidneys, since a single kidney is adequate for survival. Consequently, donors can be selected for their similarity to the recipient and are usually relatives. For essential organs like hearts and lungs, however, selection cannot be so precise. The donor must be dead, so the recipient must take what is available.

The closeness of the match is determined by mixing a blood sample from the donor with antibodies against known HLA specificities. If the blood lymphocytes carry the corresponding specificity of HLA antigen, they will bind with the antibody. Complement is then added to the reaction mixture, and cells that have combined with antibody will lyse. The degree of cell lysis indicates the HLA pattern.

■ Immunosuppressive Therapy

About 16,000 transplant operations are performed each year in the United States. Cell-mediated immunity may be intentionally suppressed in transplant recipients to prolong graft survival. **Immunosuppression** can be induced by drugs that depress lymphocyte proliferation, by irradiating the lymph nodes and the thymus with X-rays that destroy T cells, or by administering antilymphocyte serum (ALS). Lymphocyte-specific antibodies of ALS passively immunize recipients against their own lymphocytes, thereby preventing graft rejection and other T-cell-mediated activities (Fig. 19-14a). To reduce the possibility that ALS will cause serum sickness, monoclonal antibodies directed against anti-

Key Terms

immunodeficiency (p. 474)
acquired immunodeficiency syndrome (AIDS) (p. 476)
human immunodeficiency virus (HIV) (p. 476)
hypersensitivity (p. 480)
immediate-type hypersensitivity (p. 481)
allergen (p. 481)

anaphylaxis (p. 482)
antibody-dependent cell cytotoxicity (p. 484)
ABO blood group (p. 484)
Rh blood group (p. 486)
immune complex diseases (p. 487)
delayed-type hypersensitivity (p. 488)
granuloma (p. 488)

contact dermatitis (p. 489)
human leukocyte antigens (HLAs) (p. 491)
major histocompatibility complex (MHC) (p. 491)
immunosuppression (p. 491)
graft-versus-host disease (p. 493)
autoimmune disease (p. 493)

Key Facts and Concepts

- **Immunodeficiency**. A compromised immune system fails to protect its host from disease and may even be the cause of host injury or death. Disorders of the immune system occur when immune cells or their products are either not produced or respond inappropriately to an antigen. Persons with immunodeficiencies are prone to opportunistic infections, which are often fatal if the deficiency completely paralyzes B-cell- or T-cell-mediated immunity. Immunodeficiency may be genetic in origin or may result from diseases of the lymphoid system.
- **Acquired immunodeficiency syndrome**. AIDS is an immunodeficiency caused by HIV. The viral infection leads to the destruction of helper T cells, which compromises the cell-mediated immune response. There is no vaccine available to protect against HIV infection or drugs that will destroy the virus once it is established in the host. AZT and ddI are two drugs that will reduce viral replication and slow progression of the disease.
- **Hypersensitivity**. Some immune-mediated host damage is caused by hypersensitivity (allergy). Type I (immediate-type) hypersensitivity is mediated by IgE and mast cells. When IgE on the primed mast cells reacts with the corresponding allergen, the sensitized cells immediately degranulate, releasing chemicals that trigger the general symptoms of allergy in the surrounding tissues. Sensitized individuals often develop symptoms within minutes after exposure to the antigen. Type II hypersensitivity is an antibody-dependent cytotoxicity. Host cells are damaged when antibodies react with the cells and activate complement,

phagocytes, or other effector cells. Type III hypersensitivity results from immune complexes deposited in the membranes of capillaries and kidneys, which are then obstructed and damaged by complement fixation. Type IV or delayed-type hypersensitivity is associated with sensitized T lymphocytes. Sensitized T cells react to antigen by releasing lymphokines. People with type IV allergies usually develop symptoms 24 to 48 hours after exposure.
- **Transplant reactions**. Transplanted tissues are usually rejected by the immune system unless they are antigenically similar to the tissues of the recipient, placed in a region of the body that is not accessible to immune lymphocytes, or protected from rejection by suppressing the recipient's immune response. Rejection of most primary grafts is a cell-mediated process. Circulating helper T cells recognize the histocompatibility antigens on the foreign cells and activate cytotoxic T cells while promoting the production of lymphokines that attract macrophages to the graft. To prolong graft survival, cell-mediated immunity can be suppressed with drugs that depress lymphocyte proliferation or interleukin production, by irradiating lymph nodes and the thymus, or with antilymphocyte antibodies. The cost of creating an immunotolerant transplant recipient is increased susceptibility to opportunistic infections.
- **Autoimmune disease**. Autoimmunity follows partial loss of immunologic tolerance, leading to overt disease if the autoantibodies attack and injure host tissues.

Review and Problem-Solving Questions

1. Compare immediate-type hypersensitivity to delayed-type hypersensitivity. What methods can be used to minimize their occurrence?
2. People with genetic deficiencies in complement produce normal levels of antibody and T lymphocytes but are immune-compromised. Why are they more susceptible to infections?
3. Why don't ABO blood group incompatibilities between mother and fetus cause hemolytic disease? (*Hint:* A- and B-specific antibodies are IgM.)
4. Distinguish between:
   (a) Immunodeficiency and immunosuppression
   (b) Autoimmunity and immune tolerance
   (c) Anaphylaxis and inflammation
5. Judy and Rae meet in the endocrinologist's office and find that they are both being treated for autoimmune disease of the thyroid. Judy has Hashimoto's thyroiditis. She is always tired and has gained weight. Rae, who has Graves' disease, is always nervous and is losing weight. Why do these two autoimmune diseases differ in symptoms?
6. How does graft-versus-host disease differ from the host's rejection of a graft?
7. How are symptoms related to the mechanism of damage in each of the following autoimmune diseases? (a) Insulin-dependent diabetes; (b) multiple sclerosis; (c) pernicious anemia.
8. One of the proposed treatments for AIDS and some forms of cancer is interleukin-2 therapy. How would administering this agent possibly help in the treatment of these diseases?
9. Why is it so much more difficult to match organs than to match blood type?
10. How does a superantigen differ from most antigens? How does it cause damage to the immune system?
11. Tuberculosis is usually a chronic, slowly progressing disease. Most people who are infected or who have recovered from the disease will respond to a tuberculin skin test.
    (a) Describe what occurs during infection with *Mycobacterium tuberculosis* in immunocompetent people.
    (b) What reactions lead to a positive tuberculin skin test in cured people?
    (c) Why does tuberculosis progress differently in AIDS patients? Would you expect them to have a positive skin test if they were successfully treated for tuberculosis?
12. Why isn't there an allergic reaction following first exposure to an allergen? Why isn't everyone allergic to the same substances?
13. Why is it inaccurate for someone who does not blister in response to contact with poison ivy to state, "I'm immune to poison ivy"?

HEADLINES

A FAILURE TO COMMUNICATE

Several genetic immune disorders are characterized by a failure to produce protective antibodies in response to invading pathogens. Until recently, all these antibody deficiency disorders were attributed to defects in the immune system's antibody-producing B cells. Now Jean Marx[*] describes results from several laboratories that demonstrate that at least one of these antibody immunodeficiencies is caused by defective helper T cells. The disease is X-linked hyperimmunoglobulin M syndrome (HIM). It is caused by a defective gene on the X chromosome. The gene normally directs formation of a membrane protein by activated helper T cells. This protein recognizes and binds to a B-cell surface protein called CD40. In the presence of certain lymphokines, the bound helper T cell activates the B cell to produce protective antibodies. The helper T cells in patients with HIM, however, cannot communicate the signal needed to stimulate antibody production.

The affected individuals are highly susceptible to bacterial infections and require routine intravenous administration of gamma-globulin to replace the missing antibodies.

- In comparing a "normal" person to one with HIM, explain what differences you might expect in (a) the number of B cells; the number of helper T cells; (b) the ability of their B cells to respond to a normal helper T cell; (c) the ability of their T lymphocytes to respond to their helper T cells.
- Describe three other genetic defects that might lead to antibody deficiency disorders.
- In what ways might this knowledge be used in gene therapy for HIM? in treatment of autoimmune diseases?

[*]Jean Marx, Cell communication failure leads to immune disorder, *Science*, 259:896–897 (1993).

CHAPTER 20

Principles of Epidemiology

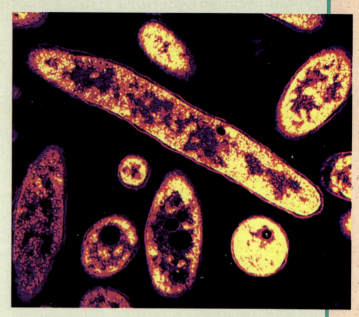

The organism pictured here created a panic in the mid 1970s. This "new" pathogen suddenly swept through a convention site in Philadelphia and killed 29 people. A team of infectious disease detectives mobilized and quickly identified the killer (*Legionella pneumophila*) and its hidden source (the air conditioning system). Before long, the first outbreak of legionnaires' disease was controlled. The detectives responsible for tracking down such microbial killers belong to the field of epidemiology.

Journeying to foreign lands is dangerous. The dangers lurk in the microbial communities there, the pathogens firmly entrenched in the environment or the people of each particular region. These dangers can often be reduced by preparation before and during travel. For example, vaccines stimulate immunity against various exotic pathogens, and antimalarial agents are taken prophylactically while visiting Africa and other areas where malaria is common.

But people don't just travel geographically—some actually journey to different times. These "time travelers" are endangered in the same way as visitors to new geographic locations, but we are less likely to know what pathogens await these voyagers and thus we are less able to protect them. In fact, many "time travelers" to the past have died from the pathogens they unwittingly found there. These endangered time travelers are archaeologists, who transcend time by digging up the past.

When many of the archaeologists and workers who excavated King Tutankhamen's Egyptian tomb died shortly after the dig, the rumor of King Tut's curse was born. It was actually the curse of immunologic naiveté, the explorers encountering old pathogens for the first time. As English archaeologist Howard Carter first drilled a hole through the wall that sealed the burial chamber, he felt air rush onto his face through the hole. He later reported being held spellbound by being the first of his time to breathe this 4000-year-old air. Unfortunately, the chamber contained some pathogens sealed in an environmental reservoir for nearly four millenia. Within 2 months, George Herbert, Carter's partner in the Tut find, was dead of infection contracted from the vault. Subsequent deaths of others who fell victim to the Tut tomb microbes have become a significant deterrent to other archaeologists. Because of the Tut deaths, the Ch'in emperor's tomb still remains sealed, archaeologists fearing the ancient microbes awaiting them. Explorers are considering penetrating the door with a fiber-optics camera instead of opening the tomb, being satisfied with a few photos until procedures for decontaminating the chamber are proven effective in this setting.

Historians are aware of how heavily outbreaks of disease have figured in shaping our history. Hernando Cortez, for example, might have been defeated by Montezuma's superior Aztec army had the Central American natives not been ravaged by smallpox. The virus was inadvertently introduced by the Spanish to a totally vulnerable Aztec population who had never before encountered it. The Europeans, on the other hand, had lived with the virus for centuries. Many had contracted smallpox earlier in life and survived, protected for life by immunity to the disease. Many historians believe that smallpox likely did more to subdue the Aztecs than did the Spanish invaders.

Epidemics—A Continuing Threat

Human history and our personal lives continue to be shaped by **epidemics** (*epidemos* = among the people), sudden outbreaks of infectious disease within a community (Fig. 20-1). Many of the same factors that determine the occurrence of disease in an individual also influence the spread of infectious disease throughout a population. An epidemic occurs when a virulent pathogen is introduced into a population of susceptible people. The affected population may be as small as a single family or as large as the global community. Environmental conditions that bring organisms and hosts together encourage the spread of epidemic disease. These environmental conditions include any physical, chemical, biological, or social factors that are essential to the survival and transmission of the infectious agent. Only when these three major elements—an infectious agent, a susceptible host, and proper environmental conditions—work in concert does infectious disease emerge and persist in a community (Fig. 20-2). The spread of disease can be prevented by interrupting the chain of events leading to infection. This is the ultimate goal of **epidemiology,** the study of disease distribution in populations and of the factors that influence this distribution.

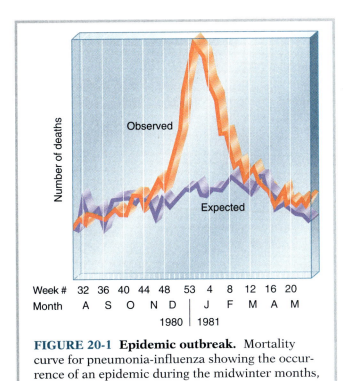

FIGURE 20-1 Epidemic outbreak. Mortality curve for pneumonia-influenza showing the occurrence of an epidemic during the midwinter months, 1980–1981.

CONCEPT CHECK

● What factors determine whether cases of an infectious disease constitute an epidemic?

TABLE 20-4 RECOMMENDATIONS FOR CHILDHOOD IMMUNIZATION

Vaccine	Protects Against (Disease)	Recommended Schedule of Inoculation
DPT (toxoids of diphtheria and tetanus plus pertussis vaccine*)	Diphtheria, pertussis, tetanus	2, 4, 6, and 15 months plus a fifth dose before entering school
Hib vaccine (*Hemophilus influenzae B* polysaccharide conjugated with protein)	Meningitis caused by *Hemophilus influenzae*	2, 4, 6, and 15 months
OPV vaccine (oral poliomyelitis vaccine—trivalent)	All three strains of poliovirus infection and subsequent paralysis	2, 4, and 6 months, plus a fourth dose before entering school
MMR vaccine (attenuated strains of mumps, measles, and rubella viruses)	Mumps, measles, and rubella (German measles)	15 months and again at 5 years
Hep B (hepatitis B antigen)	Hepatitis B	At birth, 1 month, and a third dose between 6 and 18 months
TD vaccine (toxoids of diphtheria and tetanus)	Tetanus and diphtheria	14 to 16 years (and every 10 years thereafter)

*Pertussis portion may be inactivated whole bacteria or acellular bacterial components.
SOURCE: *Morbidity and Mortality Weekly Report,* 43 (RR-1): 9 (1994).

mine the effectiveness of preventive measures that interfere with this chain of events. Epidemiologists collect information that reveals patterns of disease; they determine the frequency of illness, where and how the disease is most often contracted, and the characteristics of those who become ill and those who escape illness. These determinations help epidemiologists identify the factors that contribute to the spread of disease.

■ Quantifying Disease Occurrence

Epidemiologists continually monitor disease **morbidity** (the number of people who become ill with the disease) and **mortality** (the number of people who die of the disease). Morbidity and mortality rates measure the incidence of disease or death in relation to the population size. These values are used to compare the occurrence of disease and death in different populations. If differences exist between populations, epidemiologists try to discover the conditions responsible for low disease rates and attempt to establish those conditions in other populations.

Determining morbidity and mortality requires that accurate data be collected. To assist in this endeavor, some diseases have been declared **notifiable diseases,** those that by law must be reported to local or state public health officials whenever cases are diagnosed (Table 20-5). Most diseases, however, are underreported for a number of reasons (for example, not all affected persons seek medical help).

If the reported information reveals unusual increases in disease incidence, control measures can be initiated to avert an epidemic.

■ Identifying Frequency Patterns

Sporadic diseases are those that occur occasionally without a predictable pattern. Morbidity and mortality of some diseases fluctuate in cyclic variations that are often related to seasonal changes. Seasonal outbreaks of many vector-borne diseases are related to insect life cycles. Respiratory infections generally increase in frequency during cold winter months when people tend to crowd together indoors, encouraging the transmission of these infections. In the past, polio was more prevalent during the summer when swimming is popular. Unchlorinated water was often contaminated with feces from polio-infected persons. A major environmental reservoir of the legionnaires' bacillus is air-conditioning units, which perhaps explains why most cases of legionnaires' disease occur between July and October.

Many diseases are **endemic** in a community; that is, they occur with constant frequency in the population. They persist because of the absence of effective controls against their spread and the presence of susceptible persons in the population. For example, cholera is endemic in areas where malnutrition creates a susceptible population and poor sanitation practices enhance transmission. Coccidioidomycosis is endemic in the southwestern United States where environmental conditions are suitable for proliferation of the fungus in the soil. Histoplasmosis, on the other hand, is endemic to the Mississippi and Missouri river valleys. Some diseases are endemic in hospitals, where many routine procedures may transmit opportunistic infections among highly susceptible hospitalized patients.

A significant increase in the usual number of cases of a disease is classified as an epidemic. The number of cases that constitute an epidemic depends on the normal frequency of the disease within the affected population. In the United States a few cases of cholera or any other rarely occurring disease would be considered an epidemic. In contrast, an increase by a few cases of tuberculosis, an endemic disease, would be epidemiologically insignificant. Occasionally epidemics spread to more than one continent or around the world. Such global outbreaks are called **pandemics.** The most commonly occurring pandemic disease is influenza. The newest and most alarming pandemic disease is AIDS.

Epidemic diseases frequently become endemic.

Organisms that are introduced into a community for the first time in a population of susceptible hosts may cause rapidly spreading epidemics (see THE MICROBES: Deadly Visitors from the Old World). After recovery, most persons have active immunity, and the epidemic runs its course and subsides. Some pathogens, however, may remain and multiply—in humans after recovery from disease, in animals, in soil, or in water. These microbes may attack susceptible individuals who move into or are born into the community or people with impaired defenses. The number of cases will usually become constant, and the disease will become endemic.

■ Identifying Distribution Patterns

Every geographic area is associated with characteristic endemic diseases. Some diseases are geographically restricted because only a few locations can support the growth and survival of the pathogen. Vector-borne diseases occur only in regions where the transmitting arthropod is prevalent. Thus, African sleeping sickness is limited to those areas where there are tsetse flies.

TABLE 20-5 NOTIFIABLE DISEASES — SUMMARY OF REPORTED CASES, UNITED STATES, 1993

Disease	Total	Disease	Total
AIDS	103,691	Meningococcal infections	2,637
Amebiasis	2,970	Mumps	1,692
Anthrax	0	Murine typhus fever	25
Aseptic meningitis	12,848	Pertussis (whooping cough)	6,586
Botulism, total	97	Plague	10
Brucellosis	120	Poliomyelitis, paralytic	3
Chancroid*	1,399	Psittacosis	60
Cholera	18	Rabies, animal	9,377
Diphtheria	0	Rabies, human	3
Encephalitis, primary infections	919	Rheumatic fever, acute	112
Post-infectious	170	Rocky Mountain spotted fever	486
Gonorrhea*	439,673	Rubella (German measles)	192
Granuloma inguinale*	19	Rubella, congenital syndrome	5
Haemophilus influenzae	1,419	Salmonellosis	41,641
Hansen disease (leprosy)	187	Shigellosis	32,198
Hepatitis A	24,238	Syphilis, total all stages*	101,259
Hepatitis B	13,361	Primary and secondary*	26,498
Hepatitis, non-A, non-B	4,786	Congenital <1 year*	3,211
Hepatitis, unspecified	627	Tetanus	48
Legionellosis	1,280	Toxic-shock syndrome	212
Leptospirosis	51	Trichinosis	16
Lyme disease	8,257	Tuberculosis	25,313
Lymphogranuloma venereum*	285	Tularemia	132
Malaria	1,411	Typhoid fever	440
Measles (rubeola)	312	Varicella (chickenpox)	134,722

*Data through February 1994.
SOURCE: *Morbidity and Mortality Weekly Report,* 42:53 (1994).

Deadly Visitors from the Old World

Lahina was about to be visited by a pestilence known as the scourge of the Pacific. On earlier trips this dreadful plague had wiped out more than half the [Hawaiian] population, and now it stood poised in a whaler [ship], prepared to strike once more with demonic force. It was the worst disease of the Pacific: measles.

Men from the infected whaler had moved freely through the community, and on the next morning Dr. Whipple looked out his door and saw a native man, naked, digging himself a shallow grave beside the ocean, where cool water could seep in and fill the sandy rectangle. Rushing to the reef, Whipple called, "Kekuana, what are you doing?" And the Hawaiian, shivering fearfully, replied "I am burning to death and the water will cool me . . . you do not know how terrible the burning [fever] is," and he sank himself in the salt water and within the day he died.

Now all along the beach Hawaiians, spotted with measles, dug themselves holes in the cool wet sand . . . and died. Throughout Lahina, one Hawaiian in three perished.*

Such accounts were all too common among the native inhabitants of the western hemisphere and South Pacific. These native populations had never in their history been exposed to the pathogens common in Europe. They were therefore extraordinarily susceptible to diseases that were relatively harmless to Europeans, who had adapted to the presence of the causative agents by becoming genetically resistant following centuries of exposure. In the Hawaiian Islands, more than 50 percent of the native population died after European visitors and settlers first exposed them to such diseases as measles, influenza, and syphilis.

Similarly, European visitors to the new world were acutely susceptible to the pathogens endemic to these regions. Immigrant settlers in Central and South America, for example, suffered from and died of yellow fever. Fortunately, the settlers who fled the disease could not carry yellow fever home with them because the mosquito vector necessary for transmitting the pathogen fails to survive in temperate European and North American climates.

*Excerpts compiled from James Michener, *Hawaii*, Random House, Inc., New York, 1959.

The distribution of cases often helps to reveal the source of an epidemic. The map that John Snow made during the London cholera epidemic of 1854 shows one of the most famous examples of this approach (Fig. 20-7). Snow observed that almost all the deaths seemed to occur within the vicinity of the Broad Street pump, one of the local sources of water. He had the handle of the pump removed to prevent further consumption of contaminated water. In a more recent example, the 1976 outbreak of legionnaires' disease was traced to the common location—a hotel in Philadelphia that was the site of an American Legion convention. Nearly all infected persons in the outbreak had attended the convention.

Studying disease distribution therefore helps epidemiologists pinpoint the sources of infection. This information is most valuable when the outbreak is a **common-source epidemic,** one that emanates from a single origin. These diseases are either noncommunicable or only poorly spread by person-to-person contact. Examples include food poisoning or any disease for which soil is the sole reservoir of the infectious form of the pathogen. Cases occur only as long as susceptible persons are exposed to the reservoir or contaminated vehicle. Most epidemics, however, are **propagated epidemics** in which any infected host can transmit the disease to other susceptible individuals. The number of cases increases as more infected hosts are produced and as long as they are in contact with a sufficiently susceptible population.

■ Epidemiological Typing

To verify the existence of an epidemic it is necessary to establish that all cases are caused by the same etiologic agent. This generally requires isolation of the pathogen from the majority of infected people. During epidemic outbreaks, specimens from animals or plants or persons other than the patient and samples from the inanimate environment may also provide clues to the source of infection. For example, rodents may be tested for plague, and water may be tested for toxin-producing algae. Isolation of the pathogen is usually accomplished by standard methods for culturing microbes. However, it may be difficult to isolate "new" pathogens during their initial outbreaks if they fail to grow on the more commonly used media or, in the case of viruses, cell cultures.

Identification of the pathogen's genus and species may not be sufficient to distinguish between independent cases caused by different strains of the same organism. For example, an outbreak of pneu-

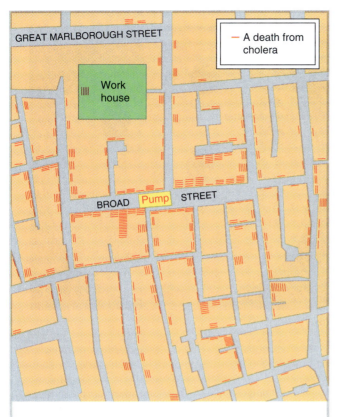

GREAT MARLBOROUGH STREET

— A death from cholera

Work house

BROAD Pump STREET

FIGURE 20-7 Snow's map of the distribution of deaths from cholera during an epidemic in London in 1854. Nearly all the deaths (indicated in red) occurred within a short distance of the Broad Street pump.

monia in a hospital may be an epidemic spread by an infected health care worker or may be due to endogenous infections in a population of compromised patients. Matching the strains of the organisms from various sources will help track down the source of infection. Identifying the individual strains of influenza virus from infected patients is also important. This information helps determine whether cases of influenza are due to a virus that has caused a previous epidemic and left a relatively immune population or is a new strain, the first warning sign of an impending epidemic.

Several techniques for epidemiological typing help establish a microbial "fingerprint" to identify strains within a single species.

- *Biotyping* is based on differences in biochemical properties and is used primarily to distinguish between members of the Enterobacteriaceae.
- *Antibiograms* distinguish between similar organisms by revealing differing patterns of susceptibility to several antibiotics.

- *Serotyping* identifies differences in microbial surface antigens, such as the O (cell wall) antigens, H (flagella) antigens, and K (capsule) antigens. This technique has the broadest range of application and is used for typing virtually all microbes.
- *Bacteriocin typing* is generally available only in research or reference laboratories but is an important tool in distinguishing among strains of *Escherichia coli, Pseudomonas aeruginosa,* and *Serratia marcescens* as well as a few other bacteria. Bacteriocins are substances (usually proteins) produced by bacteria that inhibit or kill closely related bacteria. Strains of bacteria may be differentiated either by the specificity of the bacteriocins they produce or by the spectrum of bacteriocins to which they are sensitive.
- *Bacteriophage typing* distinguishes between strains by revealing differences in susceptibility to infection by a number of phages. It is most useful for typing bacteria, such as *Staphylococcus* and *Salmonella,* that can be lysed by many different types of viruses (Fig. 20-8). Phage typing is often instrumental in locating and controlling sources of staphylococcal infections among hospitalized persons; it is used to identify healthy carriers of bacteria with the same phage type as those causing the nosocomial diseases.
- *Nucleic acid restriction analysis* is one of the most sensitive methods for distinguishing between pathogenic strains of bacteria, fungi, and subgroups of similar viruses. Plasmid or genomic (chromosomal) DNA is isolated from an isolated pathogen and exposed to various restriction endonucleases. (If, like HIV, the organism is not easily propagated, small amounts of the nucleic acid can be amplified through the polymerase chain reaction.) Variations in nucleic acid sequences are revealed by differences in the products of enzymatic digestion (Fig. 20-9). This molecular technique is a powerful tool for investigative epidemiologists and is rarely used for routine typing of pathogens.

> **CONCEPT CHECK**
>
> - How do epidemic diseases become endemic?
> - Why do some diseases fluctuate by season?
> - Compare the spread of common-source epidemics and propagated epidemics. What problems does each present for detection and control?
> - Describe four methods of epidemiological typing of isolated pathogens, and explain what information useful to an epidemiologist can be learned.

Ring-a-ring of roses,
A pocketful of posies.
Atishoo! Atishoo!
We all fall down.

It is ironic that this poem has become a popular children's nursery rhyme, considering the somber origin of its four bitter lines. The poem commemorates the heroic death of 226 residents of Eyam, a small village in England, during an epidemic of bubonic plague more than 300 years ago. The rhyme's roses refer to the rosy hemorrhages that developed on the chest of plague victims. The posies were superstitiously believed to provide some protection from affliction. The ring refers to the unusual way the villagers of Eyam sacrificed themselves to save countless others.

The plague epidemic of 1665 had subsided, leaving 68,500 people dead in London alone. The following year, however, some of the 259 villagers in tiny Eyam were seized by raging fevers, terrible back pains, excruciating buboes (swollen lymph nodes for which bubonic plague is named), and delirium. The plague had again erupted. In terror, the townspeople prepared to evacuate Eyam to escape the source of the pestilence. The town clergyman, however, insightfully realized that such an exodus would spread the disease to neighboring communities and eventually throughout the country. He appealed to the villagers to quarantine themselves in order to contain the disease, and they agreed. A circle around Eyam 1 mile in diameter was marked off with stones. Food and other needed goods were gratefully supplied by the surrounding communities, which remained healthy because of Eyam's sacrifice. Each week the outsiders would leave supplies on the perimeter and anxiously flee.

A widespread wave of devastation may have been avoided because of the self-imposed quarantine ring around Eyam. The plague ravaged the village, however, killing nearly 90 percent of Eyam's population and leaving only 33 survivors. Although the clergyman survived, his wife was not so lucky; she was one of the last to die.

---

ing physical barriers such as protective masks, clothing, and gloves when treating an infected patient, when handling infected materials, or when shedding potential pathogens. Sexually transmitted disease transmission can be interrupted either by abstinence or by the use of a condom. In fact, the condom was first developed as a protective device against venereal disease and not as a contraceptive (hence the common name "prophylactic"—disease-preventing). Any break in these barriers completely negates their protective function.

Isolation and quarantine procedures are used to prevent contact between susceptible persons and those known or suspected to be infectious. In hospitals, patients with highly infectious diseases are usually isolated from susceptible persons. Outside the hospital, a comparable procedure called *quarantine* is used to reduce exposure to persons with cholera, plague, or other dangerous communicable diseases (see THE MICROBES: A Cheerful Rhyme from a Dreadful Event). During illness, voluntary quarantine—staying home from school or work—helps protect susceptible people. Many diseases, however, are most infectious during the incubation period, before symptoms appear. Since isolation usually begins after clinical disease appears, quarantine would be too late to prevent transmission. Persons exposed to someone known to have a dangerous communicable disease should immediately be quarantined until it can be determined that they have not contracted the disease and are not shedding the pathogen.

MYTH: Once symptoms of disease disappear, a person is no longer contagious.

FACT: While true for most diseases, persons with diphtheria, scarlet fever, pertussis (whooping cough), and a few other diseases continue to shed the pathogen for several days to weeks after they appear to have regained health. Furthermore, some diseases, such as typhoid fever and amebic dysentery, may be followed by a permanent "carrier stage" in which there are no symptoms but the person is highly infectious. In the late 1930s, for example, 300 Londoners contracted typhoid (41 died) after a city water worker who was a "healthy" typhoid carrier urinated in the city water supply. The story of Typhoid Mary is an even more famous testimonial to persistent infections without symptoms ▶ (see THE EXPLORERS, Chap. 22, p. 593).

● **IMMUNIZATION OF SUSCEPTIBLE PERSONS.** Active immunization helps to reduce the susceptible population by lowering the number of persons who develop disease and disseminate pathogens. Disease

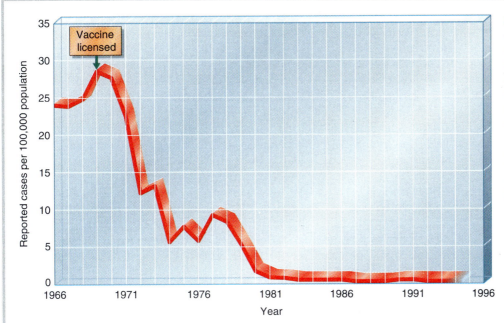

FIGURE 20-10 **Influence of vaccination on the incidence of rubella (German measles).** The number of cases reported in the United States, 1966 to 1993, rose to a high of more than 70,000 (30 per 100,000) in 1969 and then dropped to 255 in 1988. The decline began the year rubella vaccine was licensed.

fails to spread throughout an adequately immunized population. For example, DPT, the triple vaccine against diphtheria, pertussis (whooping cough), and tetanus, has virtually eliminated these diseases as public health problems. The incidence of polio, mumps, rubella, and measles has also been dramatically reduced by vaccines available in developed countries (Fig. 20-10). Worldwide vaccination programs are targeted at providing protection for children who die needlessly of these preventable diseases.

● **EDUCATION OF THE PUBLIC.** The public should be informed of the dangers of uncontrolled disease and of the measures they can take to prevent it. This is a major strategy in rabies control in England (Fig. 20-11). National cooperation has helped prevent the entry of virus-infected people and animals, the sole sources of rabies virus. As a result, the British Isles have been free of rabies for several years, while the European continent has seen a significant increase in the disease. Education is also a critical element in the fight to control the spread of HIV. Without a vaccine or drugs to cure the infection, the only way to reduce the number of newly infected people is to interfere with virus transmission.

■ Eradication of a Disease

In the mid-1960s, the World Health Organization launched a program to eradicate smallpox. This accomplishment required the total elimination of the

FIGURE 20-11 **One element of rabies control in England.** The public is informed about how they can prevent the introduction of the disease into the country. Restriction measures are rigidly enforced. Here a notice aboard a ship warns that the ship's dog must not be taken ashore.

causative organism from all natural sources. The conquest of smallpox was our first (and, to date, only) absolute success against infectious disease. In October 1977, the last case of naturally acquired smallpox was reported (Fig. 20-12). The only smallpox viruses that remain on earth are believed to be those in a small number of laboratory collections. (Unfortunately, some virus from a laboratory in Birmingham, England, infected a medical photographer, who died of the disease in 1978, the last reported case of the disease.) Most diseases are not susceptible to such global eradication. Smallpox was the ideal target for this program because of its unique epidemiological characteristics, which are outlined here.

- Humans are the only known reservoir of smallpox virus. No animal, insect, or environmental reservoirs provide known "hiding places" that could sequester or harbor the virus. Once the pathogen is eliminated from the human population, it is eliminated from the world.
- Smallpox is caused by a virus with a single unchanging antigenic type that induces solid immunity. Thus, persons immune to one smallpox virus are immune to all smallpox viruses. The susceptible population is readily reduced by vaccination, and individuals who have recovered from disease are also immune. Because it is impossible to immunize everyone, however, vaccination of large populations was in itself incapable of totally eliminating the disease.
- Smallpox rarely causes inapparent cases that result in carrier states. Thus, a careful surveillance system identified every case virtually anywhere in the world. These individuals were isolated or quarantined to contain further spread. Because no

viruses are shed during the asymptomatic incubation period, communicability begins only after clinical symptoms appear. Therefore, all persons in contact with known smallpox cases were treated by active and passive immunization and quarantined until it was determined that they were not infectious. This combination of surveillance and containment was successful because, through international cooperation, all cases were quickly reported. In endemic areas, monetary rewards were offered to encourage citizens to report cases of smallpox to the authorities. Local workers and volunteers went to public markets and schools displaying pictures of typical cases and asking if any sick people had been seen. All reports were quickly followed up, and smallpox was eventually eradicated.

Because there is no reservoir of smallpox left on earth, the United States no longer recommends vaccinating anyone against smallpox. America has dropped its requirement for smallpox vaccination of persons entering its borders, regardless of the traveler's origin. WHO, however, still offers a $1000 reward for any confirmed reports of person-to-person transmission of the disease. If some unknown reservoir of infection still remains, this strict surveillance should be an effective mechanism to detect it and prevent the return of the disease. One such reservoir is feared by Canadian health officials to be frozen bodies from early native cultures that have been naturally preserved in ice. The Canadian government has issued guidelines on working with such archaeological specimens to prevent possible infection.

Unfortunately, most common human diseases lack at least one of the epidemiological features that

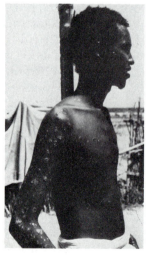 

(a)   (b)

FIGURE 20-12 **The world's last recorded case of endemic smallpox.** (*a*) Onset of rash was on October 26, 1977. (*b*) With the recovery of this patient from Somalia, smallpox was eradicated from the earth.

made smallpox eradication a success. One problem is the lack of vaccines that protect against many of the major killers. Another problem is that some of the available vaccines require refrigeration (unavailable in many areas of the world) or must be given in a series of doses. The failure of many people to complete the series makes it difficult to ensure adequate protection.

One target selected by health authorities for eradication, using many of the techniques successful against smallpox, is measles. Each year, measles kills 1.5 million people, as many as malaria, and perhaps more. In the United States, a dramatic reduction in cases followed the introduction of a vaccine in 1963. At this time, however, the virus has not been eliminated from the population. Outbreaks occur in children who have not been vaccinated or in vaccinated individuals who have not developed a protective immune response or whose immunity has waned in the years since vaccination. Polio has also been targeted for global eradication by WHO based on a program of oral vaccination. In 1993 it was announced that the western hemisphere had become virtually polio-free; there had been no cases of polio associated with wild (nonvaccine) strains for 2 years.

■ The Challenge of AIDS

We are now well into the second decade of AIDS. Since the recognition of the first cases of AIDS in the United States, more than 60,000 people have died of the disease and ten times as many have become infected with HIV, the human immunodeficiency virus. As devastating as these statistics are, they tell only part of the story. AIDS is a disease of global proportions—a pandemic that threatens to destroy large portions of the current generation and the next.

World Health Organization officials estimate that without further intervention, by the year 2000, 40 million people may be infected with HIV. The urgency of the situation emphasizes the need to stop the spread of infection and prevent disease in those already infected. Our current strategies for defeating HIV and AIDS come from our understanding of the epidemiology of the disease.

● **THE RESERVOIR.** The sole reservoir for HIV appears to be humans. This means that control must be directed at treating those already infected or preventing transmission of the virus to susceptible individuals. Unfortunately, there are no drugs available to cure persons carrying HIV. It is hoped, however, that drugs that prevent replication of the virus will reduce infection rates.

Is it possible that reservoirs other than humans harbor HIV? The origin of the HIV within the human population is unclear. There are similarities between the human virus and viruses from other primates ▶ (see BEYOND THE BASICS: The Origin of the AIDS Virus, Chap. 23, p. 621). If these viruses exist and are capable of infecting humans, then it will be necessary to implement primate control programs.

● **MECHANISM OF TRANSMISSION.** Despite its rapid spread, HIV cannot be transmitted casually. Transmission most commonly occurs by direct contact with an infected person or with contaminated body tissues or secretions, usually by one of three primary mechanisms:

1. Through direct sexual contact
2. By inoculation of contaminated blood
3. From infected mother to baby before, during, or after childbirth

A variety of strategies are aimed at interrupting these three modes of transmitting HIV.

Sexual contact is the primary mode of transmission of HIV. At the start of the AIDS epidemic in the United States, transmission was highest among homosexual males. Awareness of the risks within the homosexual community helped promote HIV diagnostic testing and "safer" sex—the use of latex condoms. As a consequence, infection rates among gay men have decreased. Unfortunately, such precautions were taken less frequently among the heterosexual population, and there has been an increase in transmission among heterosexuals. Heterosexual sex has always been the primary mode of transmission of HIV in Africa, a continent that contains over 60 percent of all AIDS cases. In sub-Saharan countries, more than 3 million males and 3 million females are infected. Thirty to forty percent of the children born to HIV-infected mothers are themselves infected with the virus and will die before their fifth birthday. The deaths of HIV-infected parents have orphaned more than 250,000 noninfected children.

The ability to test blood donated for transfusion for the presence of HIV has led to the rapid decline in the number of transfusion-related cases. Nonsexual transmission among adults is now due primarily to the sharing of blood-contaminated needles and syringes. Health care workers are also at risk if they are exposed to the blood or secretions of HIV-infected patients. Hospitals recommend that all blood and body fluids be treated as if they contained HIV and that a series of "universal precautions" be taken. This requires the use of gloves, gowns, and masks and special care to prevent needle sticks or punctures by other sharp instruments. Proper decontamination and disposal of anything contacting blood and tissue is essential.

Sexual transmission can be halted only by ensuring that sexual partners are free of HIV. The HIV status of an individual can be determined by testing a blood sample for antibodies to the virus. The test will usually be positive within 6 months of infection. Next to abstinence, the best precaution against HIV transmission is assurance that the sexual partner is HIV-negative. Although latex condoms are not 100 percent effective in preventing the passage of virus from body secretions, they reduce the rate significantly if used consistently and correctly.

As for many other viral diseases, it is hoped that a safe and effective vaccine will successfully control the spread of AIDS. However, in the absence of drugs to cure infected individuals and until the development of a protective vaccine, the primary aim of control measures must be to educate the susceptible population about the virus and about ways to prevent its transmission.

CONCEPT CHECK

- Why is it important to report cases of notifiable diseases?
- Describe four methods for limiting the occurrence of infectious diseases.
- Compare the factors that led to the successful eradication of smallpox with those that exist for AIDS.

Key Terms

epidemics (p. 501)
epidemiology (p. 501)
reservoir (p. 502)
carrier (p. 502)
incubation period (p. 503)
zoonoses (p. 503)
communicable disease (p. 504)
direct transmission (p. 504)

vertical transmission (p. 504)
horizontal transmission (p. 504)
indirect transmission (p. 504)
vehicle (p. 504)
fomite (p. 504)
vectors (p. 507)
portal of entry (p. 507)
portals of exit (p. 507)

morbidity (p. 510)
mortality (p. 510)
notifiable diseases (p. 510)
sporadic diseases (p. 510)
endemic disease (p. 510)
pandemics (p. 511)
common-source epidemic (p. 512)
propagated epidemics (p. 512)

Key Facts and Concepts

- **Epidemiology.** Epidemiology is the study of factors influencing the morbidity and mortality of diseases in populations. The persistence of disease depends on the continued presence of infectious agents, susceptible hosts, and conditions suitable for pathogen survival and transmission.
- **Disease reservoirs.** The primary reservoirs of human infectious diseases are infected persons, either carriers or persons with overt disease. Animals, soil, and water are also reservoirs of infection.
- **Disease transmission.** Pathogens are transmitted from reservoirs to appropriate portals of entry in a susceptible individual by direct or indirect mechanisms. For diseases with human reservoirs, direct transmission may be vertical (from infected parent to offspring by an infected sperm or egg, during fetal development, or during the birth process) or horizontal (by physical contact with infected persons or their infectious respiratory droplets). Indirect transmission is mediated by vehicles or vectors. Many infectious diseases are geographically limited by the distribution of the reservoir or vector.

- **Endemic, epidemic, pandemic.** Some diseases are endemic within a population, where they have fairly constant morbidity and mortality rates. Others appear as epidemics, often with seasonal patterns. A few diseases, such as influenza, can become pandemics.
- **Monitoring and controlling the spread of disease.** Government agencies and infection control committees in hospitals monitor diseases in order to identify outbreaks and prevent their spread. The methods used to break the chain of infection depend on the disease. They include reducing the number of animal reservoirs, arthropod vectors, and human carriers; preventing microbial contamination of vehicles by adhering to public health and personal hygiene measures; reducing the susceptibility of the human population by immunization; and interrupting direct transmission through isolation and treatment of infected people. These methods, coupled with public awareness and cooperation, help achieve the goal of epidemiology—the eradication or containment of disease.

Review and Problem-Solving Questions

1. Distinguish between:
   (a) A biological vector and a mechanical vector
   (b) A common-source epidemic and a propagated epidemic
   (c) Morbidity and mortality
   (d) Vertical transmission and horizontal transmission
   (e) Communicable disease and infectious disease
2. Identify three communicable and three noncommunicable infectious diseases, their reservoirs, and their modes of transmission.
3. Describe whether each of the following transmits disease directly or indirectly: (a) Respiratory droplet; (b) droplet nucleus; (c) dust; (d) arthropod; (e) contaminated needle; (f) herpes blister.
4. Formulate a job description for an epidemiologist by listing at least five major responsibilities of the job.
5. Describe the basis for differentiating strains in (a) biotyping; (b) serotyping; (c) bacteriophage typing; (d) nucleic acid restriction analysis.
6. A survey conducted each May for the last 5 years indicated that the number of cases of genital herpes increased each year whereas the number of cases of

middle ears are physically connected to the upper respiratory tract by way of the nasolacrimal (tear) ducts and auditory canals and are considered accessory upper respiratory structures because they provide microbes access to the airways. Through these structures, pathogens from the eyes and ears can enter the respiratory system.

The warm, moist surfaces of the human respiratory tract provide ideal conditions for the growth of pathogens. Each person's daily intake of 11,500 liters of air brings numerous airborne pathogens into the lower airways and lungs. The average person inhales at least eight microorganisms each minute—about 10,000 per day. The huge surface area inside the lungs (1000 square feet, about 100 times the surface area of the skin) would be readily colonized and infected if it were unprotected. Fortunately, mechanical defenses of the upper respiratory tract cleanse the air, and the ciliated mucosa of the lower respiratory tract opposes deep microbial penetration. Chemical and cellular defenses further diminish the likelihood of infectious diseases (Table 21-1).

The upper respiratory system is abundantly populated with microorganisms (Table 21-2). Most are harmless members of the normal flora that contribute to the respiratory defenses by bacterial competition. Some of these organisms, however, are opportunistic pathogens that can cause disease if predisposing factors compromise resistance or if they become entrenched in the lower airways. The occasional microorganisms that evade being trapped by the ciliated epithelia and reach the alveoli, the terminal portions of the respiratory tract where gas exchange occurs, are normally disposed of by **alveolar macrophages** (Fig. 21-2). Additional protection is provided by interferon, humoral antibodies, and T lymphocytes. The combined efficiency of respiratory defenses is evidenced by the sterile conditions that exist in healthy human lungs. In persons with compromised defenses, however, these fortifications may be incapable of preventing establishment of pathogens in the lower respiratory tract.

CONCEPT CHECK

- What factors make the human respiratory tract an ideal site for the growth of pathogens?
- In what three ways do the upper and lower respiratory tracts differ?
- Describe five defenses unique to the respiratory tract.

Predisposing Factors

Many factors may predispose a person to develop respiratory disease (Table 21-3). These factors—smoking for example—injure or decrease the effi-

---

TABLE 21-1 RESPIRATORY DEFENSES

Type	Responsible Factors*	Protective Action†
Mechanical	Epiglottis	Prevents particles from entering LRT during swallowing
	Ciliated mucosa	Traps particles in mucus, which is propelled out of the respiratory tract
	Nasal hairs	Tumble inhaled air, increasing chances of trapping particles in mucus
	Neurological sensitivity of nasal mucosa	Discharges particles from URT through sneezing
	Neurological sensitivity of trachea	Discharges particles from LRT through coughing
Chemical	Interferon	Protects uninfected host cells against viral infection
	Lysozyme	Dissolves peptidoglycan in cell walls of many bacteria
Microbial	Normal flora	Competes with pathogens of URT (no normal flora in LRT)
Cellular	Alveolar macrophages (dust cells)	Phagocytize microbes that reach the alveoli
Immunologic	Secretory immunoglobulins	Neutralize pathogens at respiratory surfaces
	T lymphocytes	Enhance killing of pathogens

*See Chapters 17 and 18 for a discussion of each factor.
†LRT = lower respiratory tract; URT = upper respiratory tract.

TABLE 21-2 BACTERIAL FLORA INDIGENOUS TO THE RESPIRATORY TRACT AND CONTIGUOUS STRUCTURES OF HEALTHY HUMANS

Anatomical Site*	Most Common Organisms	Microscopic Characteristics
External ear (URT)†	*Mycobacterium* sp.	Gram-positive acid-fast rods
	Corynebacterium sp. (diphtheroids)	Gram-positive pleomorphic rods
	Staphylococcus aureus	Gram-positive cocci, clusters
	Staphylococcus epidermidis	Gram-positive cocci, clusters
Eye (conjunctiva)† (URT)	*Branhamella catarrhalis*	Gram-negative pleomorphic rods
	Haemophilus sp.	Gram-negative pleomorphic rods
	Corynebacterium sp. (diphtheroids)	Gram-positive pleomorphic rods
	Neisseria sp.	Gram-negative diplococci
	Staphylococcus epidermidis	Gram-positive cocci, clusters
	Viridans streptococci	Gram-positive cocci, chains
Nose and throat (URT)	*Staphylococcus aureus*	Gram-positive cocci, clusters
	Staphylococcus epidermidis	Gram-positive cocci, clusters
	Viridans streptococci	Gram-positive cocci, chains
	Streptococcus pneumoniae	Gram-positive diplococci
	Branhamella catarrhalis	Gram-negative diplococci
	Corynebacterium sp. (diphtheroids)	Gram-positive pleomorphic rods
	Haemophilus sp.	Gram-negative pleomorphic rods
	Actinomyces sp.	Gram-positive rods
	Bacteroides sp.	Gram-negative pleomorphic rods
Trachea, bronchi, bronchioles, and lungs (LRT)	No flora indigenous to this region	

*LRT = lower respiratory tract; URT = upper respiratory tract.
†Although anatomically not respiratory structures, the direct connection of the ear and eye to the upper respiratory tract make them portals of entry into the airways.

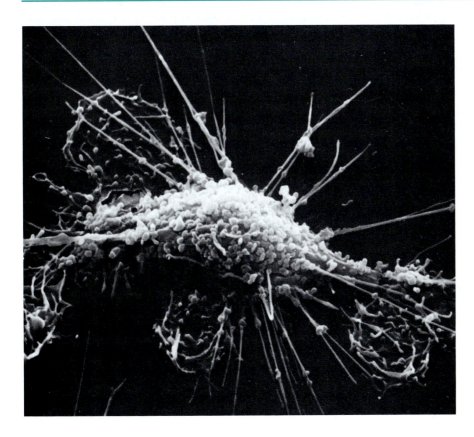

FIGURE 21-2 Alveolar macrophage—a phagocytic defender of the lung. These phagocytes, also called "dust cells," engulf particles that elude the primary respiratory defenses and reach the alveoli.

TABLE 21-3 PREDISPOSING FACTORS THAT ENCOURAGE INFECTIOUS DISEASE OF THE RESPIRATORY TRACT

Factor	Impaired Defense	Site Affected*	Corrective Measure
Tobacco smoking	Ciliated mucosa, phagocytosis	LRT	Stop smoking.
Alcoholism	Cough reflex, phagocytosis, ciliated mucosa	LRT	Reduce alcohol consumption.
Decreased humidity	Ciliated mucosa	URT	Humidify the air.
Chill	Ciliated mucosa	URT	Avoid chilling.
Viral infection	Ciliated mucosa, phagocytosis	URT and LRT	Carefully monitor symptoms; use antibiotics if culture confirms secondary bacterial infection.
Antibiotic therapy	Upper respiratory flora disrupted and bacterial competition reduced	URT	Use narrow-spectrum antibiotics; rely on antibiograms for antibiotic selection.
Certain airborne pollutants such as asbestos and silicon	Phagocytosis	LRT	Change environment.
Allergic rhinitis (hay fever)	Cilitated mucosa	URT	Avoid allergens; desensitize to allergens.
Invasive medical procedures (e.g., respiratory therapy and tracheotomy)	All physical defenses bypassed	LRT	Use sterile equipment and aseptic procedures.
General anesthesia	Ciliated mucosa	URT	Monitor carefully.
Obstructive pulmonary diseases	Airways	LRT	Correct antecedent condition if possible.
Tonsillar inflammation	Drainage of auditory canal	Middle ear	Remove chronically inflamed tonsils.
Aspiration of foreign material	Ciliated mucosa; large inoculum introduced into LRT	LRT	Avoid factors that promote aspiration.
Acquired immunodeficiency syndrome (AIDS)	T lymphocytes	LRT	Avoid potential sources when possible; use chemoprophylaxis with drugs that prevent infections.

*LRT = Lower respiratory tract; URT = upper respiratory tract.

ciency of the host respiratory defenses or bypass them altogether (see BEYOND THE BASICS: Dying for a Cigarette?). Diseases acquired through the respiratory tract are also encouraged by factors that compromise resistance to infectious diseases in general: extremes in age, chronic stress, hormonal imbalances (such as corticosteroid excesses or deficiencies), diabetes mellitus, immunologic deficiencies, disorders of cellular or phagocytic mechanisms, immunosuppressive therapy to accommodate a transplanted organ, anticancer chemotherapy, or severe malnutrition. Individuals with AIDS, for example,

are likely to develop fatal lung infections with *Pneumocystis carinii* as a result of damage to their immune defenses. A person with any of these high-risk factors should be closely attended until corrective measures reverse their predisposition.

Healthy individuals with no predisposing factors can also contract mild or even life-threatening diseases through the respiratory system if the inoculum is large enough or if the pathogen is sufficiently virulent to overwhelm host defenses. *Yersinia pestis* (the agent of pneumonic plague) and measles virus are two examples of virulent pathogens that enter

Dying for a Cigarette?

Cigarette smoking has been unquestionably linked to fatal respiratory disease. Seventy percent of all persons who die of respiratory disease are smokers. Although many of these people die of lung cancer or emphysema, infectious pulmonary disease (especially pneumonia) commonly develops because of the physiological changes that accompany long-term inhalation of smoke. The following cigarette-induced physiological changes contribute to high mortality among smokers and, to some extent, nonsmokers who breathe air contaminated with cigarette smoke.

Chronic cough. Smoking irritates the membranes of the larynx and trachea, triggering vigorous coughing. The violent inhalations that accompany coughing spasms encourage the aspiration of microbe-laden droplets and particles, which may inoculate the lower airways with opportunistic pathogens.

Increased mucus production. Inhaled smoke stimulates the respiratory mucous glands to overproduce mucus. So much mucus accumulates along the respiratory epithelium that it can no longer be elevated by the action of cilia. Instead of being carried out of the respiratory tract, trapped particles descend by gravity deep into the lower airways and lung.

Saturation of dust cells. Smoke is an airborne suspension of carbon particles. Each lungful of inhaled smoke coats the alveoli with these particles, which are phagocytized by protective alveolar macrophages (dust cells). Unfortunately, there are a finite number of dust cells available, and these can be rendered ineffective by diverting their activities to the task of engulfing an overwhelming number of smoke particles. A few opportunistic pathogens inevitably escape destruction by this saturated line of defense.

Paralysis of ciliated mucosa. Nicotine has been associated with temporary loss of ciliary function along the respiratory tract. The mobility of the mucous blanket is eventually recovered in most persons who quit smoking.

AMERICAN CANCER SOCIETY

---

through the healthy respiratory tract and cause diseases in an otherwise healthy person.

CONCEPT CHECK

- Identify three factors that predispose to URT infections and three that predispose to LRT infections. Explain how these factors increase host susceptibility.

Epidemiology and Dynamics of Transmission

Most airborne human pathogens are acquired from infected people or from the inanimate environment. Infected animals are the reservoirs of only a few respiratory-acquired diseases of humans (Fig. 21-3).

■ Human Reservoirs

Person-to-person transmission by contaminated respiratory discharges is the most common mechanism for transmitting infectious disease. Activities that promote the production of airborne respiratory droplets include coughing, sneezing, talking, singing, and spitting. Kissing may also transmit pathogens into the respiratory tract, although no airborne vehicles are involved. Disease transmission primarily occurs indoors, where crowding and close human interaction promote contact with airborne respiratory secretions. Schools, military bases, hospitals, prisons, and other environments and institutions with high indoor population densities have the highest incidence of these diseases.

The most efficient vehicles of transmission are tiny **droplet nuclei** formed from the evaporation of small respiratory droplets before they settle out. Droplet nuclei are more dangerous than the respiratory droplets from which they are formed because desiccation often prolongs microbial viability. Many airborne pathogens survive for longer periods of

Bringing Order to the Chaotic Streptococci

Many species of *Streptococcus* inhabit the bodies of humans and other animals. Some of these, such as the viridans streptococci, are usually harmless members of the normal flora. Others, especially *Streptococcus pyogenes*, are dangerous pathogens. A system of distinguishing between the many groups of streptococci is vital to clinical practitioners and epidemiologists.

The streptococci are divided into three groups according to their lytic effects on red blood cells in blood agar. This effect, called *hemolysis* ▶ (see Chap. 5, p. 116), is due to the action of soluble extracellular substances called hemolysins, which diffuse into the agar and lyse the blood cells. Hemolysis is seen in two characteristic patterns. *Alpha hemolysis* appears as a greenish-brown zone around the colony. This color change is due to alteration of normally red hemoglobin to an unknown product. *Beta hemolysis*, on the other hand, is the complete destruction of red blood cells around the colony, producing a characteristic clear zone with no remaining color. Beta hemolysis is best observed when the blood agar culture is incubated

anaerobically, because one of the chemical hemolysins functions poorly in the presence of oxygen. This hemolysin, called streptolysin O, acts in concert with a second hemolysin, streptolysin S, to produce beta hemolysis. Most normal flora (viridans) streptococci are alpha-hemolytic or produce no hemolysis at all. The dangerous pathogen *S. pyogenes* is beta-hemolytic.

Hemolysis alone, however, is not enough to identify streptococci. For example, not all beta-hemolytic streptococci are *S. pyogenes*. Furthermore, some alpha-hemolytic streptocci (*S. pneumoniae*, for example) are virulent pathogens and must be distinguished from the normal flora. An additional system of classification was devised by Rebecca Lancefield. The Lancefield scheme divides the streptococci into immunologic groups according to different antigenic specificities of the C carbohydrate, a major surface antigen. Thirteen major groups of streptococci—groups A through O (there are no groups I or J)—have been described. Although *S. pneumoniae* and the viridans streptococci possess no group carbohydrate, the Lancefield system is valuable for identifying other important pathogens, including the group A beta-hemolytic streptococci (as *S. pyogenes* is often called). Further serotyping of a surface protein antigen, the M protein, can be used to distinguish strains within this group.

Many clinical laboratories are not

equipped to determine antigenic groups and rely on bacitracin sensitivity to distinguish *S. pyogenes* from other beta-hemolytic streptococci. A paper disk impregnated with the antibiotic bacitracin releases the antibiotic into the surrounding medium when placed on blood agar seeded with the organism. Group A streptococci are sensitive to bacitracin in doses too low to inhibit other streptococcal groups. Thus beta-hemolytic streptococci that fail to grow around the bacitracin disks on blood agar are identified as *S. pyogenes* (see photo).

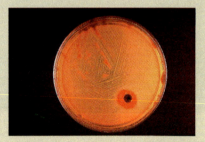

Diagnosis of group A beta-hemolytic streptococcal infection may also be facilitated by detecting antibodies specific for the hemolysins. Streptolysin O is antigenic, and persons with *S. pyogenes* infections develop peak antibody titers against the hemolysin about 2 weeks after initial infection. These antistreptolysin O (ASO) titers are useful in diagnosing poststreptococcal sequelae (rheumatic fever and glomerulonephritis) that develop after the initial streptococcal infection is eradicated.

---

treated for years with large doses of penicillin to prevent the recurrence of streptococcal sore throat that could trigger a fatal attack of the sequela. Glomerulonephritis is an immune-mediated disease, with irreversible damage to the kidney resulting from deposits of circulating antigen–antibody complexes. The gravity of these potential complications emphasizes the need for rapid and accurate identification of *Streptococcus pyogenes* in specimens from infected throats.

Certain lysogenic strains of *S. pyogenes* produce an exotoxin called *erythrogenic* (red-inducing) *toxin* that is responsible for the symptoms of scarlet fever. As its name implies, scarlet fever is characterized by

a red skin rash. The rash is seen most often on the neck and chest. Reddening of the tongue, known as strawberry tongue, is another response to the circulating toxin. Exposure stimulates the development of protective antitoxins. These antibodies, however, still leave the person susceptible to localized infection by toxigenic *S. pyogenes* and to its other sequelae.

■ Nonbacterial Sore Throat

Most sore throats are caused by viruses and are characterized by mild, localized symptoms, general discomfort, and few serious complications. Many of

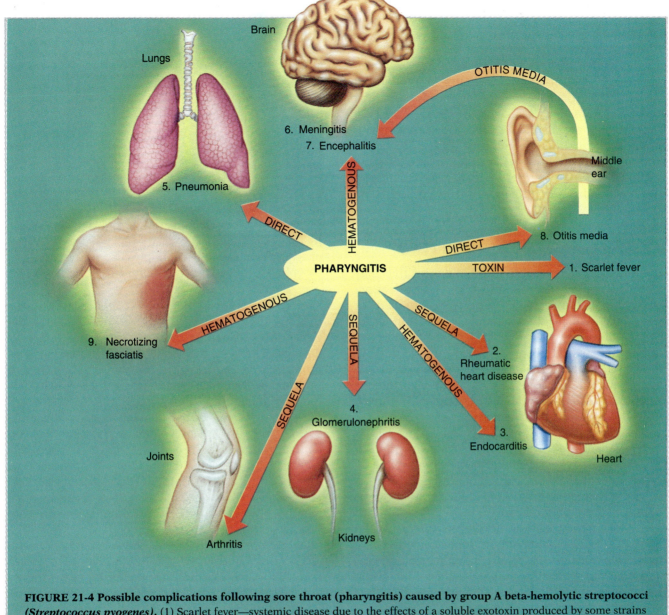

FIGURE 21-4 Possible complications following sore throat (pharyngitis) caused by group A beta-hemolytic streptococci *(Streptococcus pyogenes).* (1) Scarlet fever—systemic disease due to the effects of a soluble exotoxin produced by some strains of *Streptococcus pyogenes.* (2) Rheumatic fever—poststreptococcal sequela that usually damages heart valves and joints (arthritis). (3) Endocarditis—inflammation of the endocardium, the membranous lining of the heart. (4) Glomerulonephritis—poststreptococcal sequela that impairs kidney function. (5) Pneumonia—inflammation of the bronchial or alveolar spaces or supporting interstitial tissue. (6) Meningitis—inflammation of the meninges, the membranous covering of the brain. (7) Encephalitis—inflammation of the brain. (8) Otitis media—inflammation of the middle ear. (9) Necrotizing fasciatis—death of fibrous tissue beneath skin and around muscle.

the viruses infecting the upper respiratory tract, however, can partially paralyze the mucociliary defenses throughout the lower respiratory tract and reduce bronchiolar phagocytic action. Acute pneumonia, therefore, may evolve from a mild viral infection of the upper respiratory tract. In addition, viruses may spread from an infected throat to the epiglottis, nasal sinuses, and the eyes and middle ears.

■ Epiglottitis (Viral)

Acute epiglottitis is characterized by painful swelling and reddening of the epiglottis. This disease is especially dangerous to children, whose small airways are easily closed by a swollen epiglottis. Without immediate measures to provide air to the lungs, a child may asphyxiate. Epiglottitis is usually associated with a *croup syndrome* (acute laryngotracheobronchitis)—inflammation of the larynx, trachea, and bronchi. Croup is commonly caused by a group of RNA viruses called parainfluenza viruses. These agents are also associated with severe pneumonia or bronchitis of children. Parainfluenza infection of adults, however, usually produces mild coldlike symptoms, perhaps because the partial immunity acquired by previous infections limits its severity.

■ Adenoid and Tonsil Infections (Viral)

Adenoids and tonsils are lymphoid tissues that may become enlarged as the result of viral infection. These tissues may remain chronically inflamed for years and predispose for secondary infection of adjacent anatomic sites. Removal of chronically inflamed tonsils and adenoids may reduce the likelihood that these complications will develop but is a controversial approach. Because of their protective function, tonsils and adenoids should be removed only when there is no alternative therapy.

The agents most commonly isolated from inflamed tonsils are the adenoviruses, a group of double-stranded DNA viruses that have been implicated in several upper respiratory and ocular diseases as well as in severe pneumonia in children. Adenovirus infection of the pharynx resembles the symptoms of infectious mononucleosis and strep throat. When blood tests rule out infectious mononucleosis (see discussion later in this chapter) and throat cultures eliminate the possibility of streptococcal infection, the symptoms are likely due to adenoviruses.

■ Diphtheria (Bacterial)

Because of successful immunization programs, diphtheria is not the uncontrolled killer it once was. Each year, however, diphtheria continues to claim some lives in the United States and Great Britain, and it is a more serious problem in developing countries. The disease is a localized throat infection caused by toxigenic strains of *Corynebacterium diphtheriae*, which contain temperate bacteriophages that have genes carrying information for producing an exotoxin called diphtheria toxin. The infected pharynx, tonsils, and nasal region often appear to be covered with a "pseudomembrane" composed of a network of dead tissue, fibrin, bacteria, and white blood cells (Fig. 21-5). This local effect is produced by the action of toxin and the host response to infection.

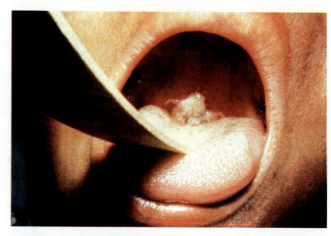

FIGURE 21-5 Pseudomembrane formation in diphtheria. An accumulation of bacteria, dead tissue, and phagocytic cells covers the oropharynx, tonsils, and nasopharynx of patient with diphtheria.

It is usually not severe enough to endanger the patient, although sometimes airway obstruction may develop. The systemic tissue injury characteristic of diphtheria is caused solely by the exotoxin, which inhibits protein synthesis. The targets most profoundly affected by the toxin are the nerves, heart, liver, and kidneys. Subsequent damage to these tissues, if they are left untreated, may be fatal. Toxin neutralization by antibodies (antitoxin) promotes recovery and prevents future cases of diphtheria as long as antibodies or immunologic memory persist.

Corynebacterium diphtheriae is a gram-positive pleomorphic rod that is transmitted by respiratory droplets from its human reservoir of infection. The few cases that occur in the United States are usually among children who failed to receive the trivalent DPT vaccine, which stimulates active immunity against diphtheria exotoxin (D), pertussis (P), and tetanus exotoxin (T). The active disease is treated by passive immunization with antitoxin to neutralize the toxin and with antibiotics such as penicillin or erythromycin to eliminate the infectious agent.

■ Infections that Spread to the Ears or Eyes

The eustachian tubes form a canal between the middle ear and the back of the throat. When functioning properly, the tubes drain the space behind the tympanic membrane (eardrum), preventing fluid from accumulating in the middle ear. When this protective drainage system does not work correctly, the inner auditory canal becomes obstructed. Pharyngeal organisms may ascend the eustachian tubes and proliferate. The result may be infection and inflammation of the middle ear, a condition known as *otitis media*. Accumulation of inflammatory products in the restricted region increases the pressure on sensitive auditory mucosa and distends the tympanic membrane. Accumulation of fluid may be painful and impair hearing. The painful pressure is often relieved by cutting the tympanic membrane before it ruptures, thereby allowing drainage. In the United States, more than half of all children under 5 years old have an ear infection each year. If untreated, permanent hearing loss may occur.

Otitis media is more common in children than in adults, probably because the child's auditory canal is smaller and at an angle that allows microbes to migrate more readily. As children grow older, the angle of the eustachian tubes changes, which allows better drainage and blocks the ascent of microorganisms, thereby reducing the incidence of otitis media. Bacterial pathogens are responsible for more than half of all cases. *Streptococcus pneumoniae*, *Haemophilus influenzae*, and *Streptococcus pyogenes* are the most common bacterial offenders. Infections caused by these pathogens can usually be treated successfully with penicillin or one of its derivatives. The remaining cases of otitis media are caused by viruses and cannot be treated with antibiotics. Surgical implantation of drainage tubes is often used to treat children with recurrent infections. The drainage tubes keep the opening in the tympanic membrane from sealing, thus making reinfection more unlikely. Infection of the tympanic membrane itself, a condition called *myringitis*, is often caused by *Mycoplasma*, which, because it lacks cell walls, is resistant to penicillins and cephalosporins.

Sinusitis, inflammation of the nasal sinus membranes, is an occasional complication of upper respiratory infection. Sinusitis is often caused by the same pathogens that cause otitis media, specifically *S. pneumoniae*, *H. influenzae*, and *S. pyogenes*, as well as many viruses. Strict anaerobes such as *Bacteroides* may also cause sinusitis.

Inflammation of the conjunctiva, mucous membranes that line the eyes and lids, is called *conjunctivitis* ▶ (see Diseases Acquired Through the Eye, Chap. 24, p. 675). Although many pathogens originate from the skin, some may travel to the eye from an infected nasal cavity through the nasolacrimal duct or may enter the eye directly on respiratory secretions from an infected person. *Staphylococcus aureus* and the *aegypticus* biotype of *Haemophilus influenzae* are common causes of conjunctivitis.

CONCEPT CHECK

- Describe four factors that contribute to making the common cold the most prevalent infectious human disease.
- How does *S. pyogenes* cause such different illnesses as pharyngitis, rheumatic fever, scarlet fever, and glomerulonephritis?
- Explain how diphtheria, a localized infection, causes systemic symptoms.
- What factors predispose children to recurrent ear infections?

Diseases of the Lower Respiratory Tract

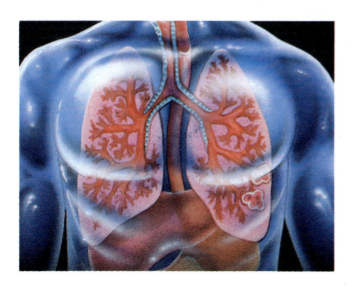

Diseases of the lower respiratory tract include pneumonia, influenza, tuberculosis, pertussis, and several mycoses. Unlike diseases of the upper respiratory tract, these infections of the lungs and lower airways are commonly fatal if untreated. Pneumonia is the fourth leading cause of death worldwide, killing more than 5 million people each year. It is the only infectious disease among the top 10 causes of death in the United States. Throughout history, tuberculosis has been among the most important of human afflictions. Even today it remains unconquered and kills 3 million people annually worldwide.

Diseases of the lower respiratory tract persist in spite of the fact that most of the pathogens causing LRT infections will respond to the antimicrobial effects of appropriate antibiotics. Selection of a chemotherapeutic agent is determined by laboratory analysis of properly collected clinical specimens such as blood samples and secretions from the lower airways. To be effective, treatment must be administered early in the course of disease and continued until the pathogen has been completely eliminated from the body. Many fatalities follow failure to initiate or maintain an adequate therapy regimen.

A summary of infectious diseases of the lower respiratory tract is provided in Table 21-5.

■ Pneumonia

Inflammation of the lungs, **pneumonia**, can be caused by a variety of infectious agents as well as by chemical irritants and allergy (Fig. 21-6). Often the airways of the lower respiratory tract are also inflamed, as in *bronchopneumonia*, for example. Inflammation of the lung's alveolar spaces is called *lobar pneumonia*, whereas inflammation of the interstitial cells that support the alveoli is called

interstitial cell pneumonia. The type of pneumonia depends on the nature of the causative agent. If the site of initial infection is the lung or associated airways, the disease is *primary pneumonia*. Infections that spread from other regions of the body to the lungs are called *secondary pneumonias*. The introduction of pathogens into the lower airways by aspiration of foreign material such as food or gastric contents commonly results in *aspiration pneumonia*. It may be most useful, however, to consider these diseases by etiologic agent.

● **PNEUMOCOCCAL PNEUMONIA (BACTERIAL)**. The most common etiologic agent of bacterial pneumonia is a gram-positive diplococcus, *Streptococcus pneumoniae* (also called the *pneumococcus*). The organism's capsule is its major virulence factor, protecting the pathogen from engulfment by alveolar macrophages. Neutralization of this antiphagocytic property by capsule-specific antibodies eliminates the organism's virulence, is responsible for natural recovery from the infection, and prevents subse-

TABLE 21-5 DISEASES OF THE LOWER RESPIRATORY TRACT

Disease	Pathogen(s)	Incubation Period	Primary Attack Site	Secondary Attack Site	Reservoir
Pneumonia	*Streptococcus pneumoniae*	1–3 days	Lungs	Heart	Infected people
	Gram-negative bacilli	1–3 days	Lungs, bronchi	—	Infected people
Primary atypical pneumonia	*Mycoplasma pneumoniae*	7–14 days	Lungs	—	Infected people
Legionnaires' disease	*Legionella pneumophila*	5–6 days	Lungs	—	Environment
Psittacosis	*Chlamydia psittaci*	4–15 days	Lungs	Blood vessels	Infected birds
Q fever	*Coxiella burnetti*	2–4 weeks	Lungs	Heart, meninges	Infected cattle
Pneumocystis pneumonia	*Pneumocystis carinii*	1–2 months	Lungs	—	Infected people (?)
Tuberculosis	*Mycobacterium tuberculosis, M. bovis*	4–6 weeks	Lungs	GI tract, bones, or any organ	Infected people and cattle
Pertussis	*Bordetella pertussis*	2–5 days	URT	Larynx, trachea, bronchi and bronchioles	Infected people
Influenza	Influenza virus	1–3 days	Membranes of URT and bronchi, bronchioles	Lymphoid tissue	Infected people

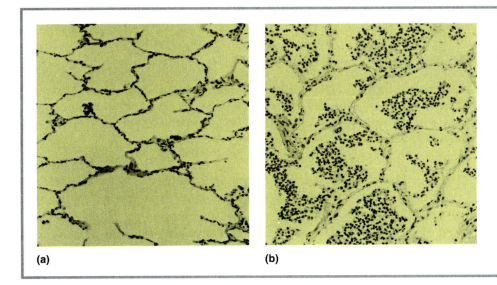

FIGURE 21-6 Bronchial pneumonia infiltrate. Instead of containing air, as seen in the microscopic section of a normal lung (*a*), infiltrated alveoli (*b*) are filled with an inflammatory exudate that contains macrophages and neutrophils.

(a) (b)

quent disease with that serotype. (There are 83 antigenically distinct types of capsules.) Bacterial variants that lack capsules are nonpathogenic.

This bacterium may be part of the normal flora in the oral cavity, restricted from the lower respiratory tract by the mucus and ciliated epithelial de-

Complications or Sequelae	Diagnostic Specimens	Diagnostic Procedures	Treatment
Endocarditis, pericarditis	Sputum, blood	Smears	Penicillin
—	Sputum, blood	Smears, isolation of pathogen	As indicated by antibiograms
—	Sputum, blood	Smears, isolation of pathogen, serology	Tetracycline
—	Sputum, lung biopsy, serum	Smears, isolation of pathogen, injection into guinea pigs, serology	Tetracycline, erythromycin
Vascular damage	Sputum, blood	Isolation of organism, serology	Tetracycline
Endocarditis, meningitis	Sputum, blood, spinal fluid, urine	Isolation of rickettsia in tissue culture, serology	Tetracycline
—	Sputum	Direct microscopic exam	Trimethoprin-sulfamethoxazole
Miliary tuberculosis	Sputum, exudates from local lesions, gastric washings, urine	Acid-fast smears, isolation of pathogen, skin test	Isoniazid with PAS, ethambutol, or rifampin
—	Nasopharyngeal	Smears, isolation of pathogen	Erythromycin, ampicillin (unreliable)
Primary or secondary pneumonia, Reye's syndrome	Blood, throat	Serology, isolation of virus	No specific treatment for active disease (amantadine prophylactically)

fenses. Disease usually develops after influenza or other viral respiratory infections damage these mucociliary defenses. Pneumococcal pneumonia is also prevalent among alcoholics, usually following aspiration of gastric contents vomited while unconscious. Pneumonia compromises pulmonary function by impairment of gas exchange, thereby increasing carbon dioxide concentration in the blood. This triggers increased cardiac activity to accommodate the lack of oxygen. Such cardiopulmonary hyperactivity contributes to the mortality of these respiratory diseases. In addition, the pathogen may spread from the lower respiratory tract and cause septicemia, endocarditis, or meningitis.

The most valuable specimen for diagnosing pneumonia is **sputum**, the secretions of the lower airways. Sputum differs from saliva, the watery secretions of the mouth that contain oral microorganisms. When these URT organisms contaminate the sputum specimen, they may provide false clues that can mislead the diagnostician. The major clue that helps distinguish spit from an acceptable sputum specimen that contains material from deep in the bronchi is the presence of cells sloughed from the oral epithelium (Fig. 21-7). Properly collected sputum specimens from persons with suspected pneumonia should be stained and examined for encapsulated gram-positive diplococci (Fig. 21-8a). Sputum should be cultured on blood agar. The pneumococci and many normal flora streptococci are alpha-hemolytic. Pneumococci, however, are sensitive to the inhibitory effects of the chemical *Optochin*, while the indigenous streptococci are Optochin-resistant. Thus, paper disks that have been impreg-nated with Optochin inhibit the growth of pneumococci on agar plates (Fig. 21-8b). The pathogen can also be identified by the *quellung reaction*. Capsule-specific antibody reacts with encapsulated pneumococci, causing the appearance of capsular swelling when observed with a light microscope. The quellung test, however, is used primarily for epidemiological typing of pneumococci.

Antibiotics, especially penicillin, are usually effective if given early during pneumococcal infections. Penicillin-resistant strains can be treated with erythromycin or other effective drugs. Even with antibiotic therapy, 20,000 to 50,000 people in the United States may die each year of the disease. Most of these are persons under 2 or over 50 years of age. Prevention of pneumococcal pneumonia is complicated by the wide distribution of the pathogen. Healthy carriers that harbor and shed virulent *Streptococcus pneumoniae* continue to be an important source of the pathogens. Development of a universally effective vaccine has been hampered by the existence of approximately 100 types of pneumococci, each with different capsule antigens. Most cases, however, are caused by one of 14 types. A vaccine that contains purified capsular material from all of these pneumococci is now in use. This combined vaccine (called pneumovax) has reduced pneumococcal disease since its introduction.

● **GRAM-NEGATIVE BACTERIAL PNEUMONIA.** Serious pulmonary disease can also be caused by *Klebsiella pneumoniae*, a heavily encapsulated gram-negative bacillus. As with *S. pneumoniae*, the capsule is antiphagocytic. *Klebsiella pneumoniae* is a common

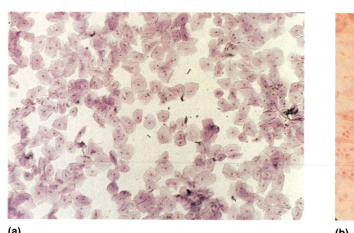

(a)

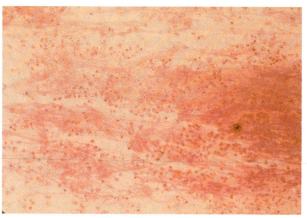

(b)

FIGURE 21-7 **Sputum or spit?** (*a*) The presence of squamous epithelial cells in sputum indicates that the sample is contaminated with saliva. (*b*) Sputum from infected lower respiratory tract contains abundant neutrophils. The absence of squamous cells indicates proper collection.

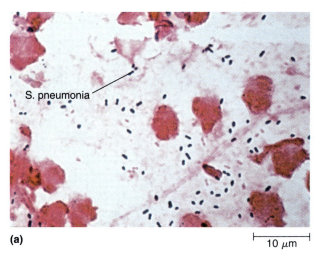

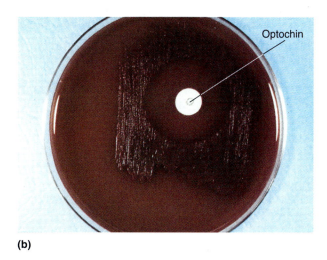

(a)

|— 10 μm —|

(b)

FIGURE 21-8 Diagnosis of pneumococcal pneumonia. (*a*) Gram stain of sputum from patient with pneumococcal pneumonia. Notice the gram-positive cocci in pairs. (*b*) Zone of inhibited growth around the Optochin disk is evidence that this alpha-hemolytic gram-positive diplococcus is *Streptococcus pneumoniae*.

resident of the normal gastrointestinal flora and upper respiratory tract of healthy humans. Most cases of *Klebsiella* pneumonia occur in persons with compromised respiratory defenses; alcoholics are especially vulnerable.

Other gram-negative bacteria that cause human pneumonia, generally among hospitalized patients, include *Escherichia coli*, *Pseudomonas aeruginosa*, *Serratia marcescens*, and various species of *Proteus*. *Haemophilus influenzae* causes secondary pneumonia in persons with influenza. Pneumonia is also part of the complex of clinical illnesses caused by such virulent gram-negative pathogens as *Yersinia pestis* (plague) and *Francisella tularensis* (tularemia). Antibiotic resistance among these gram-negative bacteria is common, and the choice of therapy should be confirmed by laboratory antibiograms on the isolated pathogen. The drug resistance problem is compounded by the ability of these pathogens to transfer plasmids (R factors) that confer multiple antibiotic resistance to recipient bacteria ▶ (see Antibiotic Resistance, Chap. 16, p. 393). In hospitals, *Klebsiella* is believed to be one of the major sources of R factors transferable to other pathogens. Pneumonia caused by *Klebsiella* has a 50 percent mortality rate in spite of supportive therapy and antibiotic treatment. In the United States, even though pneumococcal pneumonia has only a 5 percent mortality rate when treated, *Klebsiella* keeps the overall mortality of pneumonia at 25 percent.

● **MYCOPLASMA PNEUMONIA.** Primary atypical pneumonia, or "walking pneumonia," is a pulmonary infection caused by *Mycoplasma pneumoniae*, a bacterium that lacks a cell wall. This organism is the most common cause of pneumonia among children between 5 and 15 years old. Damage to the ciliated mucosa is mild, and disease runs its course, usually with no dangerous complications (Fig. 21-9). Acute

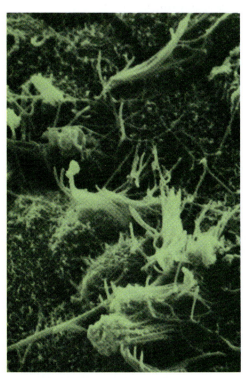

FIGURE 21-9 Ciliated mucosa of the respiratory tract following infection with *Mycoplasma*.

symptoms may be temporarily debilitating, but irreversible tissue damage or fatalities are rare. Most cases are subclinical. Even inapparent infections induce permanent immunity.

Patients with *Mycoplasma* pneumonia develop *cold agglutinins* in their blood. These diagnostically valuable antibodies agglutinate type O red blood cells at 4°C but not at 37°C. DNA probes specific for *M. pneumoniae* are available. Since mycoplasmas lack cell walls, pencillin and cephalosporins are of no therapeutic value. Erythromycin is the drug of choice for treating children. The organism is also sensitive to tetracycline, but this drug is usually reserved for use in adults.

● LEGIONNAIRES' DISEASE (BACTERIAL). In 1976 a severe pneumonia swept through a group of persons attending an American Legion convention in a Philadelphia hotel. Of the 186 persons contracting the disease, 29 died. When all the attempts to isolate and identify the pathogen failed, the malady legionellosis (which acquired the popular name legionnaires' disease) became a national curiosity and a source of potential panic. It forced the closing of one of the nation's largest hotels and triggered speculation of sabotage. The etiologic agent was eventually proved to be a previously undescribed strain of bacterium now called *Legionella pneumophila*. It was growing in the hotel's air-conditioning system and from there was distributed in an airborne form throughout the building.

Since then this organism and several additional *Legionella* species have been implicated in sporadic outbreaks, often associated with their presence in the water-cooling towers of air-conditioning units as well as some potable water sources. Most outbreaks are in hotels, hospitals, or office buildings. It is now known that the organism and the disease occur worldwide. People with compromised immune or respiratory systems are at highest risk for acquiring the disease. Deaths from legionnaires' disease, however, have been reduced by accurate diagnostic procedures and by the discovery that erythromycin is an effective chemotherapeutic agent.

Legionella is a small, gram-negative bacillus (Fig. 21-10) that readily grows when plated on a charcoal–yeast extract agar and incubated in 5 percent CO_2. Laboratory identification of the pathogen may be supplemented by detecting a fourfold rise in blood titers of *Legionella*-specific antibody during the course of disease. The pathogen may also be demonstrated in lung biopsy tissue by staining with fluorescent antibody or reaction with DNA probes.

There is no human reservoir of infection, and the disease is noncommunicable. Prevention depends on detecting the organisms in environmental

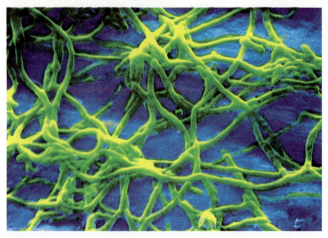

FIGURE 21-10 **Microscopic (SEM) appearance of** *Legionella pneumophila.* In spite of its worldwide distribution and ease of cultivation in the laboratory, this organism remained undiscovered until the 1976 Philadelphia outbreak.

sources that would promote their distribution. Such sources should be routinely disinfected with antimicrobial chemicals.

● ORNITHOSIS (BACTERIAL). Pulmonary infections of *Chlamydia psittaci*, termed *ornithosis*, are transmitted to humans from wild or domestic birds. This disease is an occupational hazard for bird handlers and veterinarians. Although any type of wild or domestic bird may be the source of human infection, birds of the psittacine group (parrots and parakeets) were the first recognized reservoirs of human infection. Thus the disease is often called *psittacosis*. Regardless of the type of bird infected, its feces contain the infectious agent. The etiologic agent belongs to a family of small obligate intracellular bacteria called chlamydias.

Chlamydia[1] forms a highly resistant extracellular particle that facilitates its survival in dried feces outside the host. When inhaled by people, aerosols of dried feces deposit the organism in the lungs, where they replicate and cause pneumonia. Transmission may also occur after contact with infected tissue or feathers. Unless treated, ornithosis pneumonia kills about one of every three persons who develop the advanced symptoms.

Diagnosis requires growth of the organisms in cell culture, identification within a specimen by fluorescent antibody techniques, or serologic tests. Because they are prokaryotes, chlamydia are suscepti-

[1]Chlamydia are true bacteria. These prokaryotes are obligate intracellular parasites of eukaryotic cells.

ble to antibacterial antibiotics that can enter an infected host cell and reach the replicating form. Treatment with tetracyclines usually effects complete recovery. Antibiotic treatment of birds prior to import has helped reduce the incidence of disease.

A related pathogen, *C. pneumoniae*, is a pneumonia-causing *Chlamydia* species with a human reservoir. The infection is usually mild except in persons with pulmonary or immune dysfunction.

● **Q FEVER (BACTERIAL).** *Coxiella burnetii*, a rickettsia[2] that causes pneumonia, is acquired by inhaling aerosols from infected animals, usually cattle. It was initially called "query fever" because the identity of the pathogen was questionable. The name was subsequently shortened to Q fever. Although Q-fever pneumonia is rarely life-threatening, the organism can spread to the meninges or the heart valves and cause fatal meningitis or endocarditis.

The pathogen is spread through its natural animal reservoir by infected ticks, although ticks have never been shown to transmit the disease to humans. Infected livestock animals often show no symptoms, yet shed large numbers of rickettsias in their dander, feces, and urine. High concentrations are also found in placental tissue. *Coxiella* displays a greater resistance to extracellular destruction than other rickettsias, and the organism shed in tick feces and from livestock can retain viability in soil for months. Although inhalation of contaminated dust is the most common mode of transmission, the disease can also be contracted following ingestion of unpasteurized milk from infected cows. Modern pasteurization is designed to destroy any pathogen that may be present in milk. Since *C. burnetii* is the most heat-resistant pathogen in this category, it has replaced *Mycobacterium tuberculosis* as the target of pasteurization.

Q fever is an occupation-related risk for persons who work with livestock or their by-products (hide and meat) and for laboratory workers who handle the organism. A vaccine is available to protect these high-risk individuals. Persons with active Q fever are successfully treated with tetracyclines.

● *PNEUMOCYSTIS* **PNEUMONIA (FUNGAL).** *Pneumocystis carinii* is an opportunistic pathogen that is widely distributed in nature and is also found in the lungs of healthy individuals. It causes a highly contagious interstitial cell pneumonia that is especially severe when it strikes infants or immunocompromised patients. *Pneumocystis* pneumonia is the

[2]Rickettsia are true bacteria. These prokaryotes are obligate intracellular parasites of eukaryotic cells.

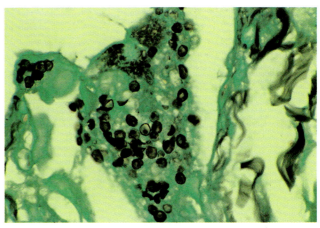

FIGURE 21-11 *Pneumocystis carinii.* This pathogen forms cysts that can be identified in lung tissue.

most frequent cause of death in children with acute lymphoblastic leukemia and in people with AIDS. Before the first AIDS cases appeared in the United States in 1981, *Pneumocystis* pneumonia was a rare disease. Today, it ranks as one of the major health problems in America. In spite of its ubiquity, little is known about its mode of transmission, which is probably by inhalation. Diagnosis depends on microscopic detection of the ovoid or crescent-shaped microbes clustered in packets of eight cells within a membrane (Fig. 21-11). These clusters may be found free or within phagocytes. Trimethoprin-sulfamethoxazole in combination with supportive measures decreases the mortality of this pneumonia. However, if the underlying condition contributing to disease risk cannot be reversed, reinfection is very likely.

● **VIRAL PNEUMONIA.** Pneumonia caused by viruses is less commonly diagnosed than bacterial infections because virological facilities are unavailable in most clinical laboratories. Generally viruses are assumed to be the etiologic agent when no pathogens can be identified in blood or sputum. The most common agents of viral pneumonia are parainfluenza viruses, adenoviruses, and respiratory syncytial viruses. Most cases occur in small children and infants or in older persons with chronic pulmonary disease. Viral pneumonia is also an occasional complication in such LRT diseases as influenza. Antibiotics are useless in treating viral pneumonia.

■ **Influenza (Viral)**

Influenza is a viral disease of the lower respiratory tract caused by an orthomyxovirus, an enveloped RNA virus covered with "spikes" (Fig. 21-12). These

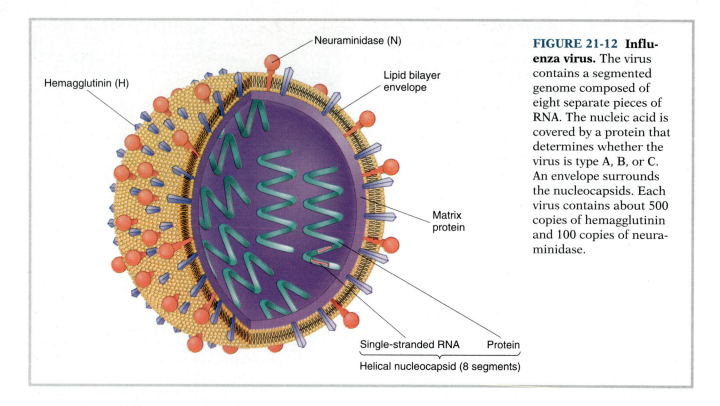

Neuraminidase (N)

Hemagglutinin (H)

Lipid bilayer envelope

Matrix protein

Single-stranded RNA   Protein

Helical nucleocapsid (8 segments)

FIGURE 21-12 Influenza virus. The virus contains a segmented genome composed of eight separate pieces of RNA. The nucleic acid is covered by a protein that determines whether the virus is type A, B, or C. An envelope surrounds the nucleocapsids. Each virus contains about 500 copies of hemagglutinin and 100 copies of neuraminidase.

surface spikes are the *hemagglutinin* (also called the H antigen) and *neuraminidase* (N antigen). The H antigen is the virus's attachment site and is necessary for the initiation of infection. Neuraminidase, on the other hand, functions at the end of the replication cycle in the release of viral progeny as they bud from the infected cell. It plays a role in infection by hydrolyzing mucoproteins in respiratory secretions, thereby destroying the effectiveness of the mucociliary defenses. Neutralization of H antigen by antibodies blocks viral infectivity, whereas N-specific antibodies do not. In addition to H and N antigens, the internal nucleoproteins are antigenic, and their specificities form the basis for classifying the virus into three types—A, B, and C. In humans, type A influenza causes more severe illness than type B. Type C rarely causes clinical disease in people.

Influenza viruses usually affect the bronchioles and bronchi rather than the alveoli and interstitium. Commonly called "flu," the disease ranges in severity from asymptomatic or mild upper respiratory infections to pneumonia and death. Gastrointestinal disease is *not* associated with influenza. The clinical syndrome incorrectly called the "stomach flu" is caused by a different group of etiologic agents ▶ (see *Viral Gastroenteritis*, Chap. 22, p. 579).

The symptoms of influenza are due to viral destruction of the cells lining the upper respiratory tract, trachea, and bronchi. The symptoms begin suddenly and include malaise, fever, chills, headache, muscle aches, and a feeling of weariness.

Acute illness rapidly progresses to a period of prostration lasting 3 to 5 days, characterized by sore throat, nonproductive ("dry") cough, and some nasal obstruction. Uncomplicated influenza is self-limiting, and most people fully recover and are protected by antibody against reinfection by the same strain of virus. Nonetheless, many people die each year of disease complications resulting from damage to the mucociliary defenses. Fatal cases are usually associated with secondary bacterial pneumonia. The predominant secondary invaders are *Staphylococcus aureus*, pneumococci, *Haemophilus influenzae*, and *Streptococcus pyogenes*. Less frequently, influenza viruses attack the alveolar mucosa and cause viral pneumonia. Between 1977 and 1988, more than 10,000 people in the United States died of some complication during each epidemic. In two epidemics, the mortality was greater than 40,000. Most of these people were over 65 years old or suffered from heart or kidney disease, diabetes mellitus, severe anemia, or impaired immunologic defenses. Elderly people and any person having a predisposing factor (as well as health care providers, who can transmit to persons at risk) should receive annual vaccination against the prevalent strains of the influenza virus. The vaccine is up to 90 percent effective in reducing death rates when it is directed against the epidemic strain of virus. The disease can also be prevented by administering amantadine, a chemoprophylactic agent that prevents intracellular development of influenza A virus (but not influenza

B virus). Neither amantadine nor any other chemotherapeutic agent can alter the course of influenza once symptoms appear.

Although influenza outbreaks occur every year, major pandemics sweep the world approximately once every 11 years. The most dramatic influenza pandemic in history occurred in 1918. In less than 2 months, more than 20 million people worldwide died of primary influenza, pneumonia, or secondary bacterial infection. Mortality was not limited to high-risk individuals; many of the deaths occurred in otherwise healthy people between 20 and 30 years of age (Fig. 21-13).

Recovery from one antigenic variety of the flu induces immunity that prevents subsequent infection by the same strain. Over time, however, the virus undergoes changes in its surface determinants. Part of these changes are due to **antigenic drift,** alterations caused by spontaneous mutation, producing minor changes in the antigenic structure of hemagglutinin that reduce the ability of the immunologic memory to recognize the virus. Periodically, however, new strains emerge with unique antigenic determinants for which previously immune persons have no immunologic memory. These new viral strains are therefore not recognized at all by the memory cells of the immune systems and establish a new pandemic. The **antigenic shift** that is responsible for creating these new strains is believed to occur by genetic reassortment of segments of the viral genome when two different strains of influenza viruses coinfect the same cell. The resulting recombinant virus possesses immunologically distinct surface determinants. This coinfection is thought to occur in pigs, ducks, and other animals that serve as reservoirs of the virus.

The etiologic agent causing a new epidemic is named according to the identity of the antigenic variety of the hemagglutinin and neuraminidase, and the geographical location where and year when the virus was initially isolated and identified. This system gives rise to such names as the H1N1 Russian flu (1977), the H3N2 Hong Kong flu (1968), and the H2N2 Asian flu (1957).

■ Hantavirus Infection

In early 1993, the first cases of a new respiratory illness that rapidly produces respiratory failure and death were reported in the southwestern United States. Combined efforts of scientists from local public health services and the CDC quickly identified the likely agent as a hantavirus, a negative-strand RNA virus whose natural reservoir is rodents. Patients demonstrated viral antigens or antibody to hantavirus (or both). Hantaviruses are known pathogens in Asia, where they are associated with kidney failure and hemorrhagic diseases. The new disease, *hantavirus-associated respiratory distress syndrome (HARDS),* results from infection by a unique strain of the virus. Using PCR amplification of hantavirus-specific RNA sequences, researchers have been able to show that the deer mouse carries the same hantaviral strain as humans. Inhalation of aerosolized feces from infected mice appears to be the manner in which humans acquire the disease. Rodent control measures to reduce transmission are combined with infection-reducing education—

FIGURE 21-13 The faces of fear. During the 1918 influenza pandemic, most people took extraordinary precautions to avoid contracting the killer disease. Unlike most epidemics, which kill predominantly elderly and compromised persons, in this case everyone was vulnerable to viral fatalities, even professional athletes.

phocytes and includes a cell-mediated immune response to infected cells and viral antigens. This response produces an abnormally high number of mononuclear white blood cells, hence the name "mononucleosis." T-cell proliferation is also responsible for swelling of lymph nodes, liver, and spleen. Infected patients also produce an unusual antibody that agglutinates sheep red blood cells. This heterophile antibody can be detected by a diagnostic blood test.

The EB virus is also associated with hairy leukoplakia in HIV-infected individuals, as well as two types of tumors, *Burkitt's lymphoma* and nasopharyngeal carcinoma (Fig. 21-18). Hairy leukoplakia are lesions on the tongue associated with immunosuppression. They can be successfully treated with the antiviral drug acyclovir. Burkitt's lymphoma is a malignant tumor of the jaw and occurs primarily among young black Africans. EB virus can be isolated from cells in the tumor. Malaria is also thought to play a role in tumor development. It has been hypothesized that immunosuppression caused by malarial infection allows the EB virus to proliferate within lymphoid cells, explaining why Burkitt's lymphoma is geographically restricted to regions where malaria is endemic. Nasopharyngeal carcinoma is an epithelial cell tumor of adults prevalent in regions of China.

■ **Bacterial Meningitis**

After infecting the nasopharynx, some pathogens enter the bloodstream and from there invade the central nervous system, causing serious infections of the meninges, the membranes that cover the brain. Epidemics of meningitis occur primarily among children between 6 months and 2 years of age. Most of these infections are caused by one of three bacteria, the gram-negative bacillus *Haemophilus influenzae*, the gram-negative diplococcus *Neisseria meningitidis*, and the gram-positive diplococcus *Streptococcus pneumoniae*. Pathogenic strains of each of these bacteria possess an antiphagocytic polysaccharide capsule. Each also possesses a virulence factor (as yet unidentified) that facilitates transfer across the blood-brain barrier.

The bacteria that cause meningitis are commonly found among the normal flora of the upper respiratory tract. Epidemics occur most frequently among children who lack immunity and are in close contact, as in a day care facility. (Adult cases are rare.) The first symptom of meningitis is usually a headache, often accompanied by a fever and vomiting. Diagnosis requires identification of the agent in a cerebrospinal fluid sample. A Gram stain can quickly indicate whether the infection is bacterial, fungal, or viral. (The absence of an identifiable organism suggests that viruses are the cause of the meningitis.) Isolation and identification of the particular bacterium is necessary to ensure that antibiotic treatment is appropriate (Fig. 21-19). Chemotherapy is effective only if the antibiotic can penetrate the "blood-brain barrier" and reach the site of infection. Penicillin is the drug of choice for *Neisseria meningitidis* infections, and cephalosporin capable of penetrating into the CSF is used for *Haemophilus influenzae* meningitis. Untreated meningitis has a mortality rate close to 100 percent and

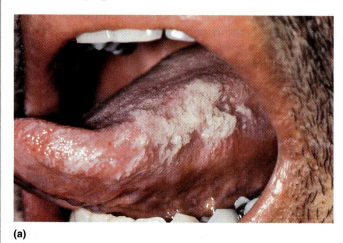
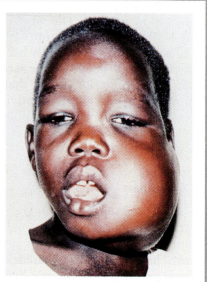

FIGURE 21-18 **Diseases caused by Epstein-Barr virus.** (*a*) The lesions on the tongue of this AIDS patient are characteristic of hairy leukoplakia. (*b*) Burkitt's lymphoma is a tumor of the face.

(a)

(b)

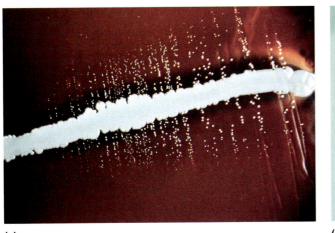

(a)

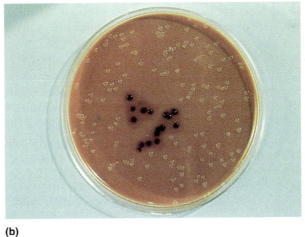

(b)

FIGURE 21-19   Laboratory identification of two important causes of bacterial meningitis. (*a*) Satellite phenomenon for identifying *Haemophilus influenzae*. This bacterium fails to grow on blood agar because the medium lacks two essential nutrients (called X and V factors). *Staphylococcus aureus*, which readily grows on blood agar, produces enough X and V factors to supplement the adjacent medium. Thus, small colonies of *Haemophilus influenzae* develop in areas of *Staphylococcus aureus* growth. This satellite phenomenon classically aids in the laboratory identification of *Haemophilus influenzae*. (*b*) Oxidase test for identifying *Neisseria meningitidis*. *Neisseria* can be distinguished from many other genera of bacteria by their ability to oxidize dimethyl- and tetramethylphenylenediamine hydrochloride (oxidase reagent). When flooded with oxidase reagent, *Neisseria* colonies quickly change color, eventually appearing black. All the members of this genus are oxidase-positive, so the test can-not be used to distinguish between two *Neisseria* species such as *N. meningitidis* and *N. gonorrhoeae*.

may kill a person within 24 hours after onset of symptoms. It is one of the most serious secondary developments following upper respiratory infection. Vaccines composed of capsular polysaccharide material are currently available to protect against bacterial meningitis. Routine immunization is recommended only for children and only against *H. influenzae*. The first dose of the vaccine is given at the age of 2 months to provide protection by the time protective maternal antibodies disappear. The widespread use of this vaccine has dramatically reduced the incidence of disease.

■ Amebic Meningoencephalitis

Several small outbreaks of fatal central nervous system infections have been caused by free-living amoebas that reside in natural bodies of fresh water. Most of these protozoa belong to the genus *Naegleria*. They apparently infect the nasal mucosa of swimmers in contaminated waters, invade the brain and the meninges, and cause meningoencephalitis. Most persons exposed to water containing these parasites fail to develop symptoms. Factors that increase susceptibility to these amebic infections have yet to be identified.

■ Systemic Mycoses

Spores of pathogenic fungi, although readily inhaled, usually cause infections that are mild or inapparent. In some individuals, however, these fungi can cause pulmonary necrosis and disseminate throughout the body, leading to tissue destruction and the death of the infected person.

Two systemic mycoses in the United States are *coccidioidomycosis* (valley fever) and *histoplasmosis*. The fungi that cause these diseases are both dimorphic; they are filamentous in the soil and nonfilamentous in the human body ▶ (see Dimorphism, Chap. 11, p. 263). The filamentous mold produces the infectious spores and grows only in the soil, which is consequently the reservoir of infection. An infected human, on the other hand, harbors only the nonfilamentous form of the organism, a form that damages the body but cannot establish initial infection. The systemic mycoses, therefore, are infectious diseases that are noncommunicable from person to person. Since they are not contracted from other human beings, their prevention depends on reducing exposure to infectious mold spores or changing the environment so that it no longer favors the growth of these fungi. For example, the accumulation of bird or bat droppings enriches the soil with nutrients that encourage growth of infectious *Histoplasma capsulatum*. Deserted houses, chicken coops, or any enclosed structure containing numerous bats or birds are ideal reservoirs of infection for histoplasmosis. Prevention of histoplasmosis is one of the important goals of bird control projects such as Tennessee's effort to reduce the explosive population of starlings in the 1970s. In addition, vaccines are being tested and may soon be used to reduce susceptibility to these diseases.

The agents of these diseases are restricted to characteristic geographical regions (Fig. 21-20). Histoplasmosis is primarily a disease of the eastern American river valleys, where optimal growth conditions exist. Coccidioidomycosis, on the other hand, is restricted to the arid regions of the southwestern states, Mexico, and a small portion of South America. Most individuals residing within these endemic regions have acquired immunity to the corresponding disease, probably because of subclinical infection. Susceptible persons are usually visitors to the regions or new residents who have never been exposed to the endemic pathogens. Immunosuppression or other resistance-crippling factors also create a small population of susceptible persons in the indigenous population. Without appropriate chemotherapy these compromised individuals may eventually die of disseminated infection. Amphotericin B and ketoconazole are used for treatment.

Histoplasmosis is often mistaken for tuberculosis when preliminary diagnosis is based on chest X-rays. Differential diagnosis should be carefully established since tuberculosis (a bacterial disease) and the systemic mycoses are susceptible to different antibiotics. Diagnosis of histoplasmosis and coccidioidomycosis depends on microscopic detection of the fungi in stained tissue specimens as well as identification of the pathogens by laboratory isolation and cultivation ▶ (see Cultivation of the Fungi, Chap. 11, p. 267). In addition, these fungi induce antigen-specific delayed-type hypersensitivity in much the same way as the tubercle bacillus does. Extracts of the fungi (histoplasmin or coccidioidin) can therefore be used for skin testing, although some cross-reactivity exists between the different fungi.

Several other fungi cause serious systemic mycoses when introduced into the lower respiratory tract. The yeast *Cryptococcus neoformans* usually enters the body on dust generated from soil that contains the organism. It cannot be transmitted from an infected person. The pathogen is especially prevalent in environments enriched with pigeon feces, such as bird roosts in buildings and barns. The disease is less frequently acquired through breaks in the skin. When exposed to *Cryptococcus neoformans*, most people develop mild lung infections or escape infection altogether. Some persons, however, fail to eradicate the organism, which may spread to other parts of the body, especially to the central nervous system, where it causes meningitis. Cryptococcal meningitis is distinguished from other meningeal infections by the presence of the encapsulated yeast in the patient's spinal fluid. Specimens are suspended in india ink, which accentuates the large-capsule characteristic of this pathogen. Dissemi-

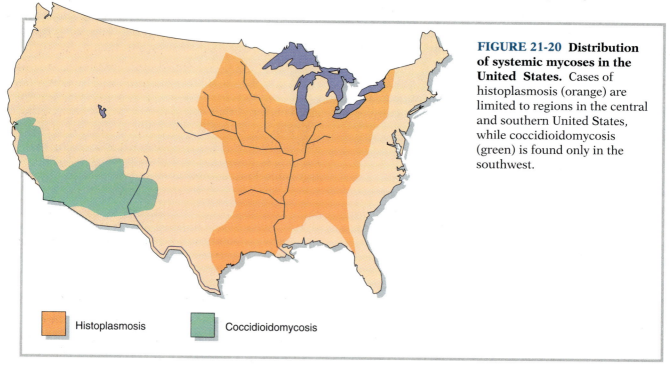

FIGURE 21-20 Distribution of systemic mycoses in the United States. Cases of histoplasmosis (orange) are limited to regions in the central and southern United States, while coccidioidomycosis (green) is found only in the southwest.

Histoplasmosis

Coccidioidomycosis

nated and meningeal cryptococcosis is usually fatal unless treated, usually with amphotericin B in combination with flucytosine.

Two other systemic fungal diseases, *blastomycosis* and *paracoccidioidomycosis*, begin as infections of the respiratory tract and in some persons spread throughout the body, causing fatal disseminated disease. Both pathogens are dimorphic fungi that produce infectious spores while growing in the soil. Blastomycosis occurs primarily in Canada and the United States, whereas paracoccidioidomycosis is endemic in South America.

Immunocompromised persons are susceptible to invasions of the respiratory tract by several opportunistic fungi. These agents are usually harmless contaminants widely distributed in most natural environments. For example, several species of *Aspergillus* (especially *A. fumagatis*) cause invasive pulmonary disease (aspergillosis) in persons with diabetes, cancer, tuberculosis, or AIDS. Similarly, mucormycosis occurs in immunocompromised persons infected with molds of the genera *Mucor* and *Rhizopus*. These fungi grow extremely rapidly and can cause disseminated infections that are often diagnosed after the victim's death. Aspergillosis and mucormycoses are difficult to control because exposure to the pathogens is virtually impossible to avoid. Amphotericin B is the antibiotic most often employed, but many persons with these diseases die in spite of treatment.

CONCEPT CHECK

- Why do measles, mumps, and rubella, usually mild diseases, warrant a national program of control?
- Both varicella-zoster virus and EBV are herpesviruses that cause more than one disease. Compare the symptoms of the diseases caused by each virus.
- Describe the route by which certain respiratory pathogens gain access to the central nervous system.
- Explain why most systemic fungal infections are not contagious.

Prevention and Control

■ Interfering with Disease Transmission

The incidence of respiratory-acquired infections can be decreased by controlling production of and exposure to infectious droplet nuclei. An educated public better knows how to control respiratory discharges. Persons with respiratory diseases should be instructed to stay at home. Sneezes and coughs contained in disposable handkerchiefs shower the air with fewer pathogens. Symptomatic medication can reduce the production of infectious aerosols by suppressing excessive sneezing, coughing, and production of mucous secretions. Frequent and proper *hand washing* is probably the most important intervention in the transmission of many of these diseases.

Additional precautions in the hospital or clinic help prevent the transmission of respiratory-acquired diseases. These include, above all, proper hand washing before and after patient contact. In addition, patients shedding highly contagious, virulent pathogens should be isolated, and all individuals entering the room should wear masks. Other measures include wearing clean gowns to protect clothing; sterilization, disinfection, or careful disposal of all linen, utensils, and equipment (especially equipment to be introduced into the mouth or respiratory tract); careful handling and sterilization of all dishes and trays from potentially infectious patients (dietary personnel should be informed if the utensils are possibly infectious); gentle handling of laundry to minimize turbulence that disseminates airborne pathogens; and thorough disinfection of the room after the infectious patient vacates. Air filtration systems promise to be effective (although expensive) methods for removing infectious droplet nuclei from hospital air.

■ Reducing Population Susceptibility

The most effective method for creating a population of resistant people is active immunization with vaccines. Vaccines are used successfully against the airborne pathogens of diphtheria, pertussis, rubella, measles, mumps, and meningitis and with limited success against influenza and pneumococcal pneumonia. A few diseases are so potentially dangerous that exposure to an infected person warrants prophylactic administration of antibiotics. For example, all close contacts of any person with confirmed tuberculosis or meningococcal meningitis should receive the appropriate regimen of antibiotics. General resistance to infectious respiratory disease can also be promoted by identifying and correcting factors, such as tobacco smoking or alcoholism, that increase susceptibility.

■ Reducing the Reservoirs of Infection

The infected person suffering from overt symptoms should be quickly diagnosed, treated, and isolated until pathogens are no longer being shed. Unrecognized cases of infection represent the greatest infectious hazard to others. Inapparent carriers continually distribute airborne pathogens to susceptible persons, who may acquire overt disease or themselves become carriers. Detection of carriers, as well as of people in the incubation phases of disease, is often impossible in the general community. In a more restricted population, however, such as a military base or a school, bacteriological sampling of all persons usually detects healthy carriers. Such an effort should begin at the first sign of a potential epidemic. Anyone identified as an asymptomatic carrier should receive immediate chemotherapy (if appropriate) to prevent the development of overt disease and to alleviate the carrier state if possible. All persons who have had close contact with a carrier should be notified and, if the disease is extremely dangerous, should also be cultured and treated. Epidemics may be prevented by such prophylactic measures.

Respiratory therapy equipment may be especially dangerous if saprophytes are allowed to proliferate in the moist chambers of the units. Such contaminated equipment serves as a vehicle for introducing these opportunistic pathogens deeply into the susceptible respiratory tract. All instruments and equipment that might produce potentially contaminated aerosols should be carefully monitored and maintained to prevent contamination and growth of microbes.

CONCEPT CHECK

- Describe three actions an infected person can take to reduce the transmission of a pathogen to the respiratory tract of another individual.
- What three protective actions would you suggest be implemented in hospital units housing patients with respiratory infections?

Key Terms

alveolar macrophages (p. 529)	pneumonia (p. 540)	antigenic shift (p. 547)
droplet nuclei (p. 531)	sputum (p. 542)	tubercle (p. 549)
pharyngitis (p. 535)	antigenic drift (p. 547)	Reye's syndrome (p. 555)

Key Facts and Concepts

- **A portal for airborne pathogens.** The respiratory tract is the major portal of entry for human pathogens. Although some virulent pathogens may overwhelm the intact defenses of healthy persons, the anatomy and defenses of the airways usually prevent disease. Many factors reduce respiratory defenses and predispose for subsequent infection.

- **Entry through the URT.** Pathogens entering through the upper respiratory mucosa may cause symptoms limited to these areas (as in the common cold) or may spread to the bloodstream and cause disease in the skin, liver, spleen, bones, or central nervous system. Measles, chickenpox, tuberculosis, and infectious mononucleosis are examples. Some noninvasive pathogens produce toxins that circulate hematogenously and cause systemic damage, as in diphtheria and scarlet fever. The ears may become infected by direct extension of pathogens from the upper respiratory tract. One of the most important URT pathogens is *Strep-*

tococcus pyogenes, which can trigger dangerous sequelae, rheumatic heart disease and acute glomerulonephritis.

- **Extension to the LRT.** The lower respiratory tract may be infected by direct extension from the upper respiratory tract, often by members of upper respiratory flora. Many cases of pneumonia are caused by these opportunistic pathogens. Influenza and pertussis are externally acquired LRT diseases that cause epidemics. Chronic lung infections may follow inhalation of *Mycobacterium tuberculosis* or spores of the fungi that cause systemic mycoses.

- **Diagnosis and treatment.** Accurate diagnosis determines whether the etiologic agent is bacterial or fungal and therefore susceptible to antibiotics. Most pathogens that become established in the upper respiratory tract, however, are viruses that produce self-limiting diseases.

- **Immunization.** Immunization helps prevent many diseases that are acquired through the respiratory tract. People can now be vaccinated against measles, mumps, diphtheria, smallpox (no longer recommended), pertussis, pneumococcal pneumonia, influenza, meningitis, and, in some countries, tuberculosis.

Review and Problem-Solving Questions

1. How does each of the following help the respiratory tract defend against pathogens? (a) Ciliated epithelium; (b) alveolar macrophages; (c) normal flora; (d) IgA.
2. How does each of the following predispose for respiratory infection? (a) Smoking; (b) viral infection; (c) respiratory therapy; (d) alcoholism.
3. Defend the following statement: Frequent hand washing may be the most effective way to prevent acquiring an infection through the respiratory tract.
4. Describe the sequelae associated with recovery from (a) measles; (b) chickenpox; (c) streptococcal pharyngitis.
5. Why is zoster, a disease of the skin and nerves, considered a respiratory-acquired disease?
6. Distinguish between:
   (a) Antigenic drift and antigenic shift; (b) Miliary tuberculosis and reactivation tuberculosis; (c) Primary pneumonia and secondary pneumonia.
7. Discuss five reasons for the current increase in tuberculosis in the United States.
8. What limitations hinder either the effectiveness or development of vaccines against each of the following? (a) Influenza; (b) pneumococcal pneumonia; (c) shingles; (d) the common cold.
9. Discuss the role of CMI in the pathogenesis of tuberculosis.
10. Explain the geographical distribution of cases of histoplasmosis and coccidioidomycosis.
11. Name the disease characterized by each of the following diagnostic findings. (a) Koplik's spots; (b) heterophile antibody; (c) growth requirement for factors X and V; (d) acid-fast bacilli in sputum; (e) spherules in tissue.
12. A young adult comes to the emergency room with a fever, severe cough, and pain in the chest. The physician suspects pneumonia. How can she identify the specific organism that is causing the infection?
13. Compare the microbial mechanism of pathogenesis with the host defense mechanisms for each of the following diseases: (a) Diphtheria; (b) tuberculosis; (c) influenza.
14. It is the beginning of a new semester in the winter and the CDC warns of an impending influenza epidemic. Your instructor wishes to prevent an outbreak among the students in your class. What rules should he impose on the class members, and what precautions might he take in the classroom?
15. What three factors contribute to an increased incidence of respiratory infections during the winter season?

HEADLINES

STORMING THE GATES OF THE BLOOD-BRAIN BARRIER

The bacteria that cause meningitis are among the few microbes that can pass from the blood into the central nervous system by penetrating the boundaries of the blood-brain barrier. Survival of the infected patient depends on the early administration of an appropriate antibiotic. However, even when the bacteria are killed, the patient often dies. Why?

The answer to this puzzle, as described by Elaine Tuomanen[*], lies in an understanding of the pathology of bacterial meningitis. In the absence of antibiotic treatment, the bacteria multiply until they reach a population that triggers an inflammatory response that helps white blood cells squeeze through the blood-brain barrier. This host response contributes to the life-threatening course of the disease. As a result of treatment with penicillins or cephalosporins, bacteria in the CNS die, but in the process they release fragments of

their cell walls. These wall components, unfortunately, are the signals that activate the inflammatory response and mobilize its components into the CNS.

- What are the common agents of meningitis, and how is the cause of a specific case determined?
- What other kinds of drugs might be used in conjunction with antibiotics to prevent white blood cells from having access to the CNS during therapy?
- Reseachers may soon isolate the bacterial wall component that "opens" the blood-brain barrier. How might this molecule be useful in treating brain cancers or diseases such as Alzheimer's?

[*]E. Tuomanen, Breaching the blood-brain barrier, *Scientific American*, 268: 80–84 (1993).

CHAPTER 22

Diseases Acquired Through the Alimentary Tract

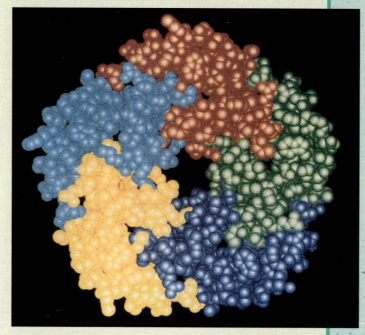

Deadly donut. The toxin that triggers deadly symptoms of cholera, part of which is seen here through the "eyes" of a computer, is produced by the bacterium *Vibrio cholerae* in the digestive tract of persons who have consumed contaminated food or water. Cholera is just one of many microbial challenges to human health that enter through the digestive tract.

President George Bush's 1992 trade mission to Japan was punctuated by a frightening episode. Seated next to the Prime Minister of Japan at a diplomatic dinner, the President slumped in his seat and fell against the Japanese head of state, vomiting on him as the Prime Minister cradled his head. As the President regained consciouness, he quickly quipped, "Just roll me under the table and let me sleep it off." Although the event caused momentary fears of a serious disorder, the news was somewhat mundane—the President was suffering from an illness acquired through the digestive tract. He had what the news media called "food poisoning" and recovered within 48 hours.

Although this may have been the most publicized episode of foodborne illness of the century, other people, even other presidents, have been more seriously affected by pathogens (or their toxins) that entered their bodies through the digestive tract. President Franklin Roosevelt's crippling polio, for example, was contracted from water contaminated with polioviruses. Foodborne and waterborne pathogens have historically been responsible for large outbreaks of serious diseases. In 1900, for example, in the United States diarrheal diseases were responsible for more than 150 deaths per 100,000 population. Now the annual incidence of reportable diseases acquired through the digestive tract averages 7 per 100,000, and these cases are infrequently fatal.

Like the respiratory tract, the digestive tract is constantly exposed to environmental microorganisms. The threat of infection and the seriousness of most of these diseases have been reduced primarily by advances in food preparation and storage and by improvements in personal and environmental sanitation. Although unable to completely eradicate diseases acquired through the digestive tract (*Salmonella*-contaminated poultry and eggs, for example, continue to cause outbreaks of food poisoning in the United States), these procedures have dramatically reduced contamination of ingestible substances and transmission of disease. The effectiveness of this approach is illustrated by the dramatic reduction in the incidence of milkborne diseases following the introduction of pasteurization in 1908. Other foodborne and waterborne diseases, unfortunately, have not been easy to control.

In many countries, poor sanitation, overcrowding, poverty, and malnutrition continue to encourage the transmission of disease through the digestive tract. Cholera has been endemic in India at least as long as recorded history. As it continues to kill tens of thousands each year in India, it has also emerged in South America, where thousands are dying of the disease each year. Infectious diarrhea poses serious problems in scores of developing nations, particularly for their children.

MYTH: Diarrhea is a mild to inconvenient disorder that is rarely life-threatening. Painful stomach cramps are about the most serious consequence the diarrhea sufferer has to face.

FACT: Diarrhea kills more children than any other infectious disease (it is the world's leading cause of death in children under the age of 5). Five to ten million children die each year from the dehydration associated with diarrhea.

Anatomy and Defenses of the Digestive Tract

The digestive tract is essentially a continuous tube lined by mucous membranes that traverses the body from the mouth to the anus (Fig. 22-1). Beginning in the mouth, ingested material passes in sequence through the esophagus, stomach, small intestine, and large intestine. The undigested remains leave the digestive tract through the anus. The primary functions of the digestive system are the disassembly of large food substances into small, water-soluble molecules and the absorption of the digested nutrients into the bloodstream, which distributes the nutrients to all cells of the body. Accessory structures, which include the salivary glands, liver, gallbladder, and pancreas, secrete into the digestive tract substances that are needed for digestion. For this reason, these organs are often considered part of the digestive system. They are also vulnerable to infection by pathogens that enter the body through the digestive tract.

Although the digestive tract is constantly barraged with microbial invaders that enter as contaminants of foods and fingers, the interactions of several natural factors protect the organs of the system from disease (Table 22-1). Microbes in the mouth are exposed to lysozyme in saliva. Those organisms that are not firmly attached to surfaces are swallowed into a chamber few microbes can survive. This chamber, the stomach, contains hydrochloric acid and has a pH of 1.5 to 2.5 (see THE MICROBES: Surviving the Stomach). Many bacteria that successfully pass through the stomach are inhibited by chemicals in bile, secreted from the liver into the small intestine. **Peristalsis,** the rhythmic contraction of the digestive tract's walls, removes microbes that are not securely attached to the mucosal surfaces. These organisms are swept through the digestive system and leave the intestines with the feces. The organisms that remain have a mechanism for firm attachment to the mucosal surfaces. However, the constant shedding and replacement of the intestinal epithelium removes many of these microbes.

Host lymphocytes, macrophages, and phagocytes all participate in the local cell-mediated immune response of the digestive tract. Serum immunoglobulins can enter the intestines but are probably destroyed by digestive enzymes before they can stem a pathogen's attack. On the other hand, secretory IgA produced by local lymphoid tissue plays an important role in resistance to many infections within the oropharynx and intestines. Although the pro-

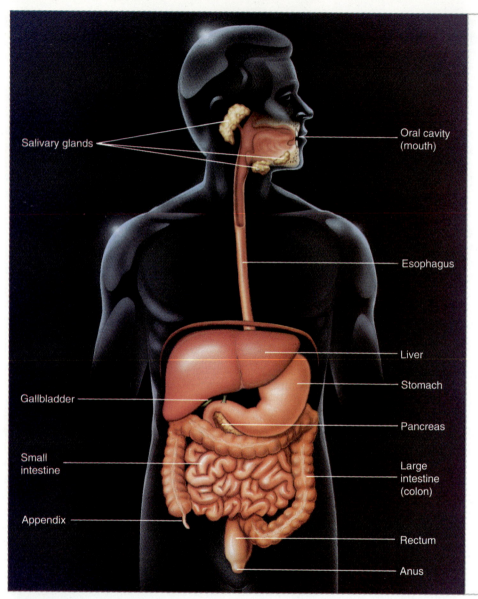

FIGURE 22-1 Anatomy of the human digestive tract. The digestive canal is in orange. The accessory organs are in other colors.

Salivary glands

Oral cavity (mouth)

Esophagus

Liver

Stomach

Gallbladder

Pancreas

Small intestine

Large intestine (colon)

Appendix

Rectum

Anus

Type	Responsible Factor	Protective Action
TABLE 22-1 GASTROINTESTINAL DEFENSES		
Mechanical	Swallowing	Moves microbes into stomach
	Peristalsis	Moves unattached microbes out of the gastrointestinal tract.
	Mucous lining	Prevents microbial attachment to epithelium
	Epithelial sloughing	Sheds epithelial cells to which microbes may have attached.
Chemical	Lysozyme in saliva	Dissolves peptidoglycan in cell walls of some bacteria.
	Gastric juice	Stomach acid (HCl) destroys most microbes. Lytic enzymes in stomach and and intestines disassemble crucial molecules in most microbes.
	Bile salts	Dissolve critical hydrophobic surface components on many microbes.
Microbial	Normal flora	Competes with pathogens; produces substances that inhibit other microbes.
Cellular	Leukocytes	Phagocytize microbes.
Immunologic	Secretory IgA	Neutralizes toxins, viruses, and bacteria.
	T lymphocytes	Enhance killing of pathogens.

Surviving the Stomach

The stomach acts as a decontamination chamber, and its concentrated acid secretions form one of our best barriers against infection. Some factors, however, reduce the effectiveness of the stomach's assault, thereby increasing the likelihood of a microbe's successful passage to the intestines. Pathogens that enter through the digestive tract survive the stomach in one of the following ways.

- Some protozoa form protective cysts that are resistant to acid destruction. Cyst forms of *Entamoeba histolytica*, *Giardia lamblia*, and *Balantidium coli* pass through the stomach unharmed. In the intestines, the fragile trophozoites emerge and reproduce.
- The number of organisms that survive passage through the stomach increases when large doses of a pathogen are ingested within a single inoculum. For most pathogens acquired through the digestive tract, the ingested dose that ensures infection is at least 100 million to 1 billion microbes. Under most circumstances only grossly mishandled materials will contain such high microbial concentrations. For pathogens that are highly virulent, however, only a few organisms have to arrive at the intestines to establish infection leading to disease. As few as 100 *Shigella*, for example, can produce severe dysentery.
- A rapid transit through the stomach shortens the period of exposure to gastric secretions and allows contaminating organisms to escape destruction. In general, fluids and semisolids are transported fastest, while meals of high fat content are held in the stomach longer and increase the likelihood that ingested pathogens will be destroyed.
- The stomach acid may be buffered by the medium in which the microbes are transported. For example, neutralization of the acid with sodium bicarbonate (antacids) can decrease the infectious dose for cholera from 10^9 to 10^5 organisms, a 99.99 percent reduction. Protein in a meal also decreases infectious doses by neutralizing stomach acid while occupying the protein-digesting enzymes.

Some microbes that cause diseases acquired through the digestive tract do not have to survive the stomach's acidity. They induce disease by the effects of toxins they form while growing in the contaminated food prior to ingestion.

---

tective effect of IgA is usually short-lived, it is crucial for preventing the adherence of pathogens to the mucosal surface, for eliminating pathogenic bacteria, and for neutralizing viruses and toxins.

■ Normal Microbial Flora

Although the intestinal tracts of infants are devoid of microorganisms at birth, colonization begins immediately—in fact, with the first cry. Within 2 weeks a complement of indigenous microbes is established. By adulthood, impressive numbers of bacteria, yeasts, and protozoa are harbored as part of the normal flora (Table 22-2). Alpha-hemolytic streptococci reach concentrations greater than 10^9 per milliliter of saliva. The major repository of microorganisms, however, is the colon, crowded with microbes in greater numbers than are obtained in laboratory batch cultures grown under optimal conditions. (The colon contains approximately 10^{11} organisms per gram of intestinal material.) Obligate anaerobes in the colon generally outnumber other microbes by 1000 to 1.

The normal flora helps restrict pathogens from the digestive tract. Some normal flora organisms produce and maintain an anaerobic environment that is inhospitable to aerobic pathogens. Many indigenous organisms produce antibiotics or release ammonia, organic acids, alcohols, or other fermentation by-products that interfere with the survival or multiplication of some intestinal pathogens. Indigenous organisms also occupy the mucosal surfaces and leave few vacant attachment sites for pathogens. Conventional flora forms dense layers that act as remarkably effective barriers to the establishment of most pathogens (see Fig. 17-8, p. 423).

■ Predisposing Factors

Many factors may increase susceptibility to diseases acquired through the digestive tract or may increase the severity of these illnesses (Table 22-3). Malfunctions of the acid-producing machinery of the stomach reduce its effectiveness as a decontamination chamber. Impaired peristalsis in the small intestine encourages local overgrowth of microorganisms, and the microbial population may increase a millionfold as a result. Impaired immunologic systems may fail to restrict proliferation of pathogens. For example, removal of the tonsils, a lymphoid tissue in the oropharynx, predisposes for paralytic polio.

TABLE 22-2 NORMAL FLORA IN THE DIGESTIVE TRACT OF A HEALTHY ADULT

Anatomical Site	Most Common Organisms	Microscopic Characteristics
Mouth		
Saliva and teeth	*Streptococcus* sp.	Gram-positive cocci, chains
	Lactobacilli	Gram-positive rods
	Veillonella sp.	Gram-negative diplococci
	Bacteroides sp.	Gram-negative pleomorphic rods
	Fusobacteria	Gram-negative rods with tapered ends
Oropharynx	*Streptococcus* sp.	Gram-positive cocci, chains
	Branhamella catarrhalis	Gram-positive diplococci
	Corynebacterium sp. (diphtheroids)	Gram-positive pleomorphic rods
	Staphylococcus sp.	Gram-positive cocci, clusters
Stomach	Usually sterile	
Small intestine	Lactobacilli	Gram-positive rods
	Enterococcus faecalis	Gram-positive cocci, chains
	Bacteroides sp.	Gram-negative pleomorphic rods
	Candida albicans	Yeast
Large intestine	*Bacteroides* sp.	Gram-negative pleomorphic rods
	Coliforms	Gram-negative rods
	Enterococcus faecalis	Gram-positive cocci, chains
	Clostridium sp.	Gram-positive rods
	Fusobacteria	Gram-negative rods with tapered ends
	Anaerobic lactobacilli	Gram-positive rods
	Staphylococci	Gram-positive cocci, clusters
	Peptostreptococci	Gram-positive cocci, chains
	Entamoeba coli	Amoeboid trophozoite, cysts with eight or more nuclei
	Trichomonas hominis	Flagellated protozoa

TABLE 22-3 PREDISPOSING FACTORS FOR INFECTIOUS DISEASE OF THE DIGESTIVE TRACT

Factor	Resultant Impairment	Precautionary Measures
Decreased production of stomach acid	Elevates pH of stomach	Minimize exposure to foodborne and water-borne pathogens.
Gastrectomy (removal of stomach)	Deprives system of acid	Minimize exposure to foodborne and water-borne pathogens.
Antacid therapy for ulcers	Neutralizes antimicrobial acidity in stomach	Use lowest effective dose.
Gastrointestinal obstructions; blind loops	Create regions not evacuated by peristalsis	Use surgical procedure and/or antibiotic therapy.
Tonsillectomy	Reduces production of secretory IgA	Remove tonsils only when alternative approaches fail.
Antibiotic therapy	Disrupts normal flora and subsequently reduces bacterial competition	Use narrow-spectrum antibiotics; rely on antibiograms for antibiotic selection.
Malnutrition	Reduces immunologic competence	Provide an adequate diet.
Infancy	Precedes immunologic competence; normal flora not established	Minimize exposure to feces and contaminated foods

Antibiotic therapy disrupts the competitive normal microflora of the intestines and increases susceptibility to doses of pathogens that would be harmless to the undisturbed intestinal ecosystem. Malnutrition may deprive the immune system of the amino acids needed to make immunoglobulins. Proper diet can therefore substantially reduce the morbidity and mortality associated with diarrheal diseases. As with most other infections, diseases acquired through the digestive tract are considerably more prevalent among children than among adults.

CONCEPT CHECK

- Describe four defenses unique to the gastrointestinal tract.
- Name three factors that decrease resistance against gastrointestinal infections, and explain the basis for increased host susceptibility.

Epidemiology and Dynamics of Transmission

Organisms enter the digestive tract through the mouth. These organisms are in the air, foods, water, unwashed hands, and countless numbers of contaminated objects that find their way into the mouth during the course of an average day. For example, as you read this sentence you may have your fingers, a pencil, or a pen placed against your lips or teeth. Fortunately, most diseases acquired through the digestive tract have high infectious doses because few organisms survive the strong acid of the stomach. Usually only heavily contaminated vehicles carry enough pathogens to exceed the infectious dose.

Contamination of vehicles may be due to large numbers of pathogens being deposited directly onto the vehicle, but more commonly, small numbers of organisms are deposited in food or water, where the microbes proliferate and achieve concentrations capable of establishing infection. Direct person-to-person contact is generally not a factor in transmission except when sanitary practices are completely disregarded. Then, however, large quantities of pathogens can be transferred directly from person to person. This pattern of transmission is responsible for outbreaks in day care centers and mental institutions. Direct person-to-person spread is also an important mechanism transmitting diseases that require very small infectious doses, such as bacterial dysentery. However, the general inefficiency of direct person-to-person transmission at least partially accounts for the lower incidence of gastrointestinal diseases compared with the incidence of infections acquired through the respiratory tract.

Lesions on the skin and mucous membranes of food handlers are important sources of foodborne illness. Most pathogens that enter the digestive tract and establish infections, however, are acquired from food or water that has been contaminated with feces. These organisms are spread by untreated water supplies, flies, or fingers. Fecally contaminated fingers may go directly into the mouth or onto other objects placed in the mouth. Several infectious diseases—most notably, hepatitis, shigellosis, amebiasis, and giardiasis—can be transmitted directly in feces during oral-anal activity or by oral-genital contacts following anal intercourse. Some fecally transmitted pathogens are harbored in the intestinal tract of animals. Such microbes contaminate soil (and crops grown in the contaminated soil), water supplies, and foods derived from infected animals. For example, most *Salmonella* infections result from ingesting contaminated meat, poultry, eggs, or unpasteurized milk. These foods may cross-contaminate other foods that are prepared by the same handler or with the same instruments or cutting board.

Although many pathogens found in soil and water are temporary inhabitants shed from human or animal colons in fecal matter, some pathogens exist independently in the environment. *Clostridium perfringens* and *Clostridium botulinum*, which multiply freely in soil, can accidentally contaminate food and cause foodborne illness. *Vibrio parahaemolyticus* and the algal dinoflagellates, normally found only in marine environments, contaminate fish and shellfish. Some fungi are plant pathogens that also cause disease in people who consume food prepared from the infected crops.

The major routes of transmission of diseases acquired through the digestive tract are illustrated in Figure 22-2.

■ Transmission Cycle of Foodborne Diseases

Foodborne diseases are usually associated with failures to prevent either contamination of food or proliferation of microbes in the contaminated foodstuff (Table 22-4). The following conditions are usually associated with outbreaks of these diseases:

1. A vulnerable food is contaminated with a foodborne pathogen. *Vulnerable foods* (also known as "susceptible foods") are those with nutrients, moisture, osmotic pressure, and pH that encourage multiplication of the pathogen. These characteristics of the food determine what types of illness it may transmit. For example, the pathogen of botulism proliferates best in canned foods with low acid content. Eggs and meat are vulnerable to *Salmonella* proliferation.

mediately. Unfortunately, the development of an obnoxious odor or gas depends on the nature of the food or on the strain of the organism, and sometimes there are no telltale signs, even in heavily contaminated food. Since the toxin is so deadly, reliance on a "taste test" to determine edibility could be fatal.

Symptoms of botulism usually begin to appear between 12 and 36 hours after consumption of the toxin. These symptoms are neurological, not gastrointestinal, and include weakness, blurred or double vision, and eventual paralysis. The toxin inhibits the motor neurons' ability to activate their muscle targets, so the ultimate effect is *flaccid paralysis*—the loss of ability to contract skeletal muscles, so they are paralyzed in a state of perpetual relaxation. Without treatment, respiratory or cardiac paralysis and death may occur within 3 to 6 days. Diagnosis is confirmed by immunologic identification of toxin in serum or feces.

The only successful treatment for botulism is antitoxin given early in the course of the disease to neutralize the toxin. Since there are seven distinct serotypes of toxin, each of which is neutralized by a different antitoxin, it is important to identify the specific serotype or to treat with multiple antitoxins. Because the disease is not an infection, antibiotics are of no therapeutic value. Most cases are diagnosed and treated, and only about 5 percent of botulism victims die each year in the United States.

MYTH: Botulinum toxin's deadly effects make it so dangerous that it has no redeeming qualities.

FACT: Because the toxin causes *flaccid* paralysis, it has been used by physicians for treating a number of disorders caused by uncontrolled muscle spasms. In extremely tiny doses, the toxin relieves some cases of severe stuttering, eyelid spasms that keep the lids locked in a closed configuration, rendering the victim functionally blind, and "wry neck" in which muscle spasms permanently tilt the head to one side.

Although botulism is an intoxication rather than an infectious disease, infections by *C. botulinum* can occur. Infections are occasionally acquired by infants through their digestive tract (infant botulism is discussed later in this chapter). *C. botulinum* can also enter the body through a wound, where it establishes a local infection. From this local site, the bacteria release botulinum toxin, which is distributed to all motor neurons through the bloodstream ▶ (see discussion of wound botulism, Chap. 24, p. 668).

■ *Staphylococcus aureus* Food Poisoning

The most common cause of food poisoning in the United States is consumption of an enterotoxin produced by certain strains of *Staphylococcus aureus* growing in vulnerable foods. The symptoms are generally self-limiting, and the disease is of shorter duration than most other food poisonings. Because of this, staphylococcal intoxications are usually not reported to health authorities. However, it is believed that many incidents labeled "ptomaine poisoning"— upset stomach, indigestion, and "stomach flu"—may actually be due to staphylococcus enterotoxin. Thus the definite incidence of this disease is not known.

The staphylococci usually enter the food from a human source, often an infected food handler (Fig. 22-3). They are most commonly shed from nasal secretions or from infected wounds, boils, and abscesses. Toxigenic *S. aureus* have also been isolated from cattle, and the milk from these cows has occasionally been responsible for outbreaks. Protein-rich foods serve as excellent culture media for staphylococci, especially foods rich in eggs or milk—for example, cream-filled pastries, custards, and salad dressings. Meat and poultry also support the growth of *S. aureus*. If contaminated foods are maintained at temperatures between 20 and 35°C for several hours, the pathogens multiply and release enough toxin to elicit symptoms of intoxication. Ham and turkey left to cool at room temperatures or bakery goods stored without refrigeration are particularly vulnerable to *Staphylococcus*. Thanksgiving dinner is often the ideal meal for the staphylococcal agents, particularly when it is prepared by a cook who fails to practice the measures required to safeguard against foodborne disease.

Staphylococcal toxin is referred to as an enterotoxin because the symptoms are primarily related to the gastrointestinal tract. These symptoms include severe nausea, vomiting, cramps, and occasional diarrhea. The toxin produces these symptoms by stimulating the brain's vomit center and by causing local inflammation in the intestine. Generally, the time between consumption of the food and the appearance of symptoms is 2 to 4 hours. The illness rarely lasts more than 1 or 2 days and requires no treatment.

Staphylococcus food poisoning is typically identified by its characteristic symptoms, the nature of the foods involved, the shortness of the incubation period, and the isolation of the bacteria from the implicated food. It is most important to identify the source of the organism so that measures to prevent repeated outbreaks can be implemented. Staphylococci can be isolated on media containing 7.5 per-

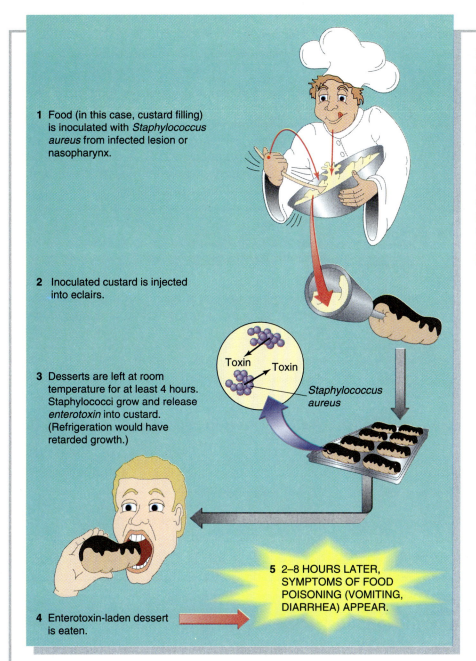

1 Food (in this case, custard filling) is inoculated with *Staphylococcus aureus* from infected lesion or nasopharynx.

2 Inoculated custard is injected into eclairs.

3 Desserts are left at room temperature for at least 4 hours. Staphylococci grow and release *enterotoxin* into custard. (Refrigeration would have retarded growth.)

Toxin   Toxin

Staphylococcus aureus

5 2–8 HOURS LATER, SYMPTOMS OF FOOD POISONING (VOMITING, DIARRHEA) APPEAR.

4 Enterotoxin-laden dessert is eaten.

FIGURE 22-3 Path to painful aftermath. Typically, *Staphylococcus aureus* enters food during preparation from an infected sore or nasopharyngeal discharge of a food preparer. If allowed to incubate in a vulnerable food for more than 4 hours at room temperature, enough enterotoxin accumulates to cause staphylococcal food poisoning. The symptoms are due to ingestion of preformed toxin and not to infection by the bacterium.

cent NaCl. Most toxigenic strains of *S. aureus* can produce coagulase. They are all beta-hemolytic and ferment mannitol. Phage typing ▶ (see Epidemiological Typing, Chap. 20, p. 512) of enterotoxin-producing *S. aureus* isolated from implicated foods provides epidemiological evidence for identifying the person who was the source of the bacteria, thereby avoiding further unfortunate outbreaks from the same source.

Because of its short incubation period, *S. aureus* was capable of causing a rather unique outbreak on an airliner flying from Tokyo to Copenhagen in 1975. During the flight, 277 persons in economy class developed staphylococcus food poisoning after eating contaminated ham. The organism had been shed from a lesion on the hand of a food service employee in Anchorage, an intermediate stop where the food was prepared. The food had been held at room temperature for 6 hours and stored for 14 hours at 10°C prior to delivery to the plane. The ham was held another 8 hours before it was served. Fortunately, the cockpit crew ate first class meals. Because of this incident, regulations were enacted to mandate that pilot and copilot eat from different menus, thereby preventing both from being incapacitated by foodborne illness.

■ Other Bacterial Foodborne Intoxications

Another species of *Clostridium*, *C. perfringens*, causes many cases of food poisoning, although not as severe as those caused by its highly lethal genus mate *C. botulinum*. An enterotoxin released from the organism as it sporulates accumulates in foods under anaerobic conditions and at incubative temperatures. Meats and gravies, if initially undercooked and allowed to sit at room temperature, accumulate enough toxin to cause diarrhea and, less commonly, nausea. *C. perfringens* intoxication lasts about 24 hours and resolves without treatment. The organism, however, can also infect gastrointestinal mucosa and create a condition called *necrotizing enterocolitis* (discussed later in this chapter), which is considerably more severe than intoxication.

Two other toxins, one that causes diarrhea and one that causes vomiting, are produced in starchy foods and dairy products containing the common contaminant *Bacillus cereus*. The effects of the toxins are self-limiting and usually disappear within 24 hours of illness onset. This is not the case for the more dangerous, but less common, food poisonings caused by *Pseudomonas cocovenenans*, so named because of its predilection for coconut-containing dishes. Its frequently fatal toxin strikes most commonly in Polynesia, where many dishes are prepared with coconut. One dish, called bongkrek, is particularly vulnerable to the growth of *P. cocovenenans* and provides the name of the intoxication, which is called bongkrek disease.

MYTH: Dangerously contaminated food can usually be identified by its appearance or smell.

FACT: Even though all agents of foodborne illness can cause detectable spoilage, they also accumulate in food to infectious or even lethal levels before showing any outward evidence of their presence. Grossly contaminated food is usually discarded, so it is the inapparent contaminations that cause most episodes of foodborne illness.

■ Mycotoxicoses

Some fungi produce toxic substances called **mycotoxins** when proliferating in suitable foods ▶ (see Chap. 11, p. 274). For example, many animal feeds and human foods containing peanuts are contaminated with *Aspergillus flavus*, a fungus that produces aflatoxin. In large doses, this toxin causes liver damage. In smaller amounts, aflatoxin can cause cancer (aflatoxins are the most potent natural carcinogenic substances yet discovered). Other mold species, including those from the genera *Claviceps*, *Penicillium*, and *Aspergillus*, produce mycotoxins while growing in stored food products (Fig. 22-4). These toxins can damage the kidney, central nervous system, and gastrointestinal tract.

Molds and mold spores are widely distributed and readily contaminate foodstuffs. Mycotoxin production, however, is enhanced in specific foods—

(a) (b)

FIGURE 22-4 Toxic gallery. Two of the fungi whose toxins cause mycotoxicosis if eaten. (*a*) The characteristic white spots on the brilliant red cap of *Amanita* (death angel mushroom) can be washed away by rain, so their absence should not be used as evidence of nontoxicity. (*b*) *Claviceps purpura* (ergot of rye), which produces a hallucinogenic vasoconstrictive toxin that can cause euphoria as gangrenous appendages, and the person to which they are attached, die.

peanuts, cottonseed, wheat, soybeans, and corn—especially when temperature and humidity are high. Mycotoxicoses are prevalent in tropical areas where peanuts are used as an inexpensive source of protein for malnourished children. In more temperate climates, the toxin is elaborated during storage of contaminated foods in warm, moist silos. In the United States, federal statutes require that animal feeds and many foods for human consumption be surveyed for detectable mycotoxin levels. These statutes prohibit the sale of products that surpass minimum concentrations. Nonetheless, a peanut butter sandwich made with moldy bread and mold-covered jelly can pose a triple threat of cancer.

■ Paralytic Shellfish Poisoning

Several species of dinoflagellates elaborate a lethal neurotoxin for which no known antidote exists. Red tides caused by blooms of these algae contain huge quantities of the toxigenic dinoflagellates ▶ (see Algae and Human Disease, Chap. 12, p. 298). Humans are exposed to the toxin when they eat shellfish that have concentrated the toxin in their tissues while feeding in contaminated waters. The result is paralytic shellfish poisoning (PSP). Another neurotoxin, this one produced by diatoms, can cause loss of short-term memory in sublethal doses. Since neither intoxication can be effectively treated, protection depends on preventing them by monitoring conditions in coastal waters and restricting the consumption of products known or suspected to be contaminated.

The major pathogens responsible for intoxications are summarized in Table 22-5.

CONCEPT CHECK

- How do intoxications differ from infections?
- Compare the conditions that encourage production of botulinum toxin with those that encourage production of *S. aureus* toxin and *C. perfringens* toxin.
- What measures can prevent diseases caused by mycotoxins? by dinoflagellate toxins?

Infections Acquired Through the Digestive Tract

Unlike intoxications, infections are due to the multiplication of viable microbes in the host. Incubation periods are generally longer for infections than for intoxications; usually about 8 to 48 hours elapse before symptoms appear.

Pathogens entering through the digestive tract can cause any of four types of disease processes: (1) infections of the oral cavity; (2) noninvasive infections restricted to the lumen or mucosal surfaces of the stomach or intestines; (3) locally invasive infections of the intestinal mucosa; and (4) invasive infections that cause pathological effects at other body sites.

■ Infections of the Oral Cavity

The mouth is heavily colonized by bacteria, many of which attach to the teeth, gingival (gum) surfaces, cheeks, or tongue, where they resist the flushing action of saliva.

● **DENTAL CARIES.** Microbes on teeth are firmly embedded in a sticky matrix of dextran, a polysaccharide produced by *Streptococcus mutans*, a member of the oral flora ▶ (see Capsule, Slime Layer, and Glycocalyx, Chap. 4, p. 85). Dextran and its embedded microorganisms are called **plaque** (Fig. 22-5). Plaque cannot be removed simply by rinsing with water. Even vigorous toothbrushing leaves a significant amount of material between teeth and at the gingiva. The large populations of bacteria concentrated in this sticky matrix produce acidic metabolic by-products that destroy tooth enamel and cause dental caries (tooth decay), the most costly disease in the United States. Plaque formation is essential to the development of dental caries because it cements microbes to the tooth surface, thereby concentrating the acids at these sites. In the absence of plaque, dental caries fails to develop because saliva dilutes the acid and washes it from the mouth.

The precursor for plaque formation is sucrose, or-

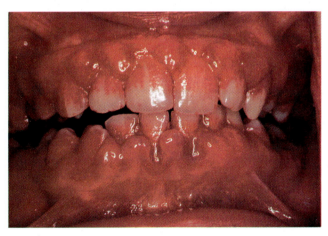

FIGURE 22-5 **Plaque, the source of dental caries and periodontal (gum) disease.** These teeth have been stained to reveal the presence of plaque (stains red).

TABLE 22-5   IMPORTANT MICROBIAL AGENTS OF FOOD INTOXICATIONS

Pathogen	Disease	General Characteristics	Major Vehicle
Staphylococcus aureus	Staphylococcal food poisoning	Gram-positive coccus, heat-stable toxin	Meats, salads containing mayonnaise, custards
Clostridium botulinum	Botulism	Gram-positive rod, strict anaerobe, heat-labile toxin	Home-canned, low-acid foods
Aspergillus flavus	Fungal food poisoning	Aflatoxin	Contaminated nuts and grains
Gonyaulax sp.	Paralytic shellfish poisoning	Dinoflagellate algae, heat-stable toxin	Shellfish
Bacillus cereus	Food poisoning	Gram-positive endospore-forming bacillus, two toxins	Starchy foods, custards, dairy products
Pseudomonas cocovenenans	Bongkrek disease	Gram-negative bacillus	Coconut-containing dishes
Claviceps purpura	Ergotism	Packed mycelia called sclerotia on grain	Foods prepared from moldy grain
Proteus spp.	Scombroid poisoning	Gram-negative bacilli, histamine-like toxin	Tuna and other unrefrigerated vulnerable fish

dinary table sugar. (Other common sugars cannot be used by *S. mutans* to form dextran.) Diets high in sucrose increase the incidence of dental caries by encouraging plaque formation and by providing a sugar substrate for fermentation and acid production (Fig. 22-6). Sucrose can therefore convert a "sweet tooth" to an "acid tooth," one that is prone to caries.

Drinking fluoridated water or taking fluoride tablets during early childhood, when teeth are developing, greatly reduces tooth decay. Fluoride is incorporated as part of the mineralization process that maintains teeth. It "hardens" tooth enamel against the destructive erosion by acids. Fluoride also directly affects cariogenic bacteria in plaque, making them less tolerant of their acid environment.

● **PERIODONTAL DISEASE.** Other bacteria, most notably anaerobes and spirochetes, reside in the spaces between the teeth and gums. These microbes, in conjunction with organisms in plaque, cause **periodontal disease,** an infection of the soft tissues around the teeth. The infection, which begins (and often remains) as *gingivitis,* a mild inflammation of the gums, may progress to ulceration and tissue necrosis, a condition known as Vincent's disease or *trench mouth.*[1] The ultimate consequence of untreated periodontal disease is tooth loss; in fact, this disease is the predominant cause of tooth loss in adults over 35 years of age (Fig. 22-7). Both periodontal disease and caries can best be prevented by daily flossing of the teeth to disrupt bacterial colonies and prevent the accumulation of plaque.

● **INFECTIONS OF THE ORAL MUCOSA.** Type 1 herpesvirus (HSV-1) and *Candida albicans* are the two agents most commonly responsible for infections of the oral mucosa and tongue. Type 1 herpesvirus causes *gingivostomatitis,* which is characterized by tiny vesicles or ulcers on the lips and in the anterior portions of the oral cavity (Fig. 22-8). The lesions probably develop from an inflammatory response to viral replication in the oral mucosa. These herpesviruses are readily spread to other persons ei-

[1]Vincent's disease (acute necrotizing ulcerating gingivitis) is commonly called "trench mouth," since men in the trenches fighting wars have a predilection to develop it, both because of stress and because oral hygiene is a low-priority activity on a list that includes dodging bullets. The disease, which is characterized by severe gum loss and sores, is exacerbated by stress. It can be treated somewhat effectively with antibacterial antibiotics.

Incubation Period	Clinical Picture	Diagnosis	Treatment
2–4 h	Nausea, vomiting, diarrhea	Detection of enterotoxin in food, isolation of organism with same phage type from food, stools, or vomitus of victims and/or skin or nose of food handler	None
12–36 h	Speech difficulty, double vision, dry mouth, nausea, paralysis; death due to cardiac arrest or respiratory failure	Detection of toxin in serum, feces, or food; isolation of organism from food or stools	Antitoxin therapy
Unknown	Liver damage	Identification of toxin	None
30 min–3 h	Numbness of lips, mouth, or face; upper and lower GI symptoms	Detection of toxin in shellfish or of toxin-producing algae in waters from which shellfish were gathered	Induce vomiting
<12 h	Diarrhea	Isolation of organism from stools	None
Hours	Vomiting; spasms (often fatal)	Isolation of organism from suspect food	None
1–2 h	Hallucinations, abdominal pain, gangrene	Clinical picture and history	Vasodilators
Minutes	Cramps, headache, shock	Isolation of organism from fish or assay for amines	Antihistamines

ther directly in saliva or by contaminated fomites. Serological studies indicate that HSV-1 infection is quite common; most persons have herpes-specific antibodies. In fact, most children have already encountered the virus by the age of 5 years, although the vast majority of infections are asymptomatic. Even overt attacks are normally self-limiting, and the lesions disappear within a few days. However, the viruses are not eliminated from the body but remain in a latent state in local nerve cells. A variety of triggering factors, including emotional stress, sunlight, fatigue, and infectious illness, may precipitate the reappearance of lesions on the skin. Because common colds or febrile disease often trigger a recurrence, the lesions are popularly referred to as "fever blisters" or "cold sores." As with most other viral infections, oral herpes is not cured by chemotherapy, although treatment with acyclovir during the primary episode reduces the duration of symptoms and the shedding of the virus and may discourage the establishment of latency.

Oral infections with *Candida albicans* are called *thrush*. The disease often affects infants before they develop their competitive bacterial flora, persons on antibiotic or steroid therapy, persons with diabetes, and persons with AIDS or other immunologically

debilitating diseases. Oral candidiasis is a non-febrile, noninvasive disease characterized by elevated white patches of yeast on the surface of the tongue and mucosa (Fig. 22-9). Topical treatment with the antibiotic nystatin or an imidazole (such as miconazole) effectively eliminates the infection, but recurrences are common as long as the underlying predisposing condition persists.

● **INVASION THROUGH THE ORAL MUCOSA.** Microbes pour into the bloodstream whenever the oral membranes are disrupted by disease, dental manipulations, or minor accidents such as biting one's cheek. Usually these microbial intruders are halted by phagocytic defenses in the blood before systemic invasion can occur. However, the introduction of oral streptococci into the bloodstream of a person with previously damaged heart valves may have serious consequences. The bacteria may colonize abnormal heart valves and cause subacute bacterial endocarditis. Penicillin chemoprophylaxis has proven effective in preventing the development of this potentially fatal complication. Such prophylactic therapy is recommended for all people with defective heart valves when they receive even such routine dental treatment as teeth cleaning.

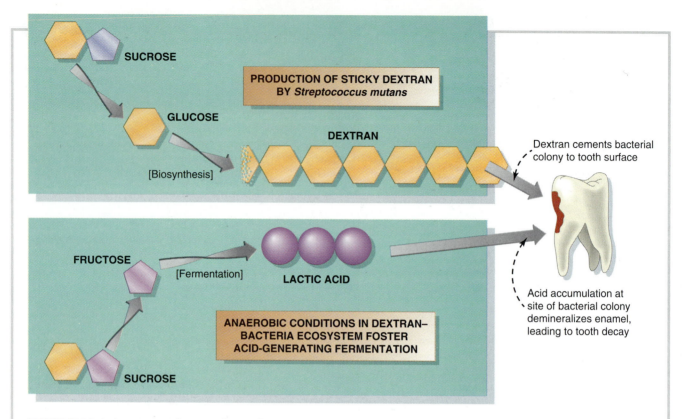

FIGURE 22-6 Sucrose, a key in plaque formation. Acid damage to tooth enamel occurs only when lactic acid-producing bacteria are cemented to the tooth surface, where their acid by-products are concentrated. The cement is dextran, a sticky carbohydrate produced by *Streptococcus mutans* when sucrose is present. (The material holds a complex community of bacteria against the tooth surface.) Dental caries fail to form if either the plaque or the acid is not present. Reducing consumption of sucrose-containing foods and mechanically disrupting plaque deposits are both effective ways to prevent tooth decay.

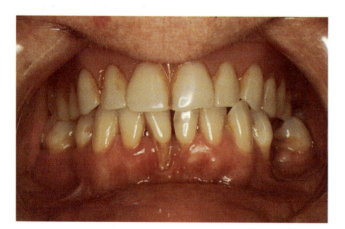

FIGURE 22-7 Impending tooth loss. Chronic periodontal disease is signaled by these inflamed gums in a patient with pronounced periodontal disease. Toxins released from the organisms in plaque initiate periodontal disease by causing gingivitis (gum inflammation). Eventually the chronic release of toxins destroys the tooth-supporting bone, resulting in tooth loss. Notice the loss of supporting gum tissue along the lower row of teeth.

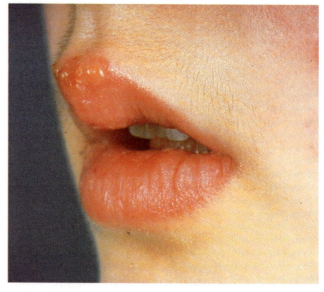

FIGURE 22-8 Herpes gingivostomatitis. This outbreak, also called oral herpes or herpes labialis, is especially severe.

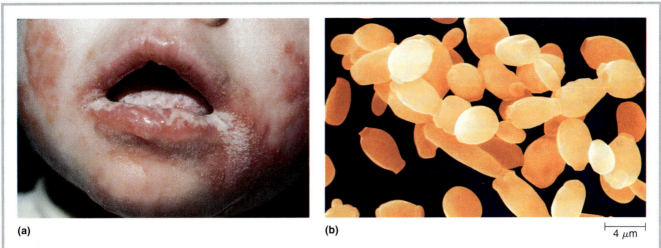

FIGURE 22-9 **Thrush, an oral infection with *Candida albicans*.** (*a*) Victim with thrush. (*b*) *Candida albicans* yeast cells and pseudohyphae stained with fluorescent antibody.

Although its major route of entry into the human body is through parenteral inoculation of skin, hepatitis B virus can invade the bloodstream through cuts and breaks in the oral mucosa and eventually infect the liver. Contaminated dental instruments are occasionally implicated as vehicles of hepatitis infection. The same is true of the human immunodeficiency virus (HIV), with several cases of AIDS being traced back to an infected dentist. In both diseases, however, it appears that dentists more commonly acquire infections from their patients than transmit infections to them.

■ Infections of the Gastrointestinal (GI) Tract

Infections of the gastrointestinal tract can cause mild local damage or severe local inflammation or spread to other body sites through the bloodstream or lymphatic vessels.

● **NONINVASIVE DISEASES.** Most pathogens acquired through the digestive tract are noninvasive and are confined to the intestines, where they cause **gastroenteritis,** a syndrome characterized by diarrhea, nausea, or vomiting. Most noninvasive GI pathogens produce enterotoxins while multiplying in the small intestine. Enterotoxins cause water to be released from the body into the intestines at such a rapid rate that the normal absorptive mechanism of the large intestine is overwhelmed. The result is the production of watery diarrhea. Because these pathogens do not invade the tissues of the intestinal mucosa, they rarely cause fever, dysentery

(bloody stools), or fecal leukocytosis (white blood cells in stools). These microbes do, however, attach to the surface of the intestinal wall and are therefore not readily eliminated with the feces. A summary of pathogens that cause noninvasive infectious diseases of the gastrointestinal tract is provided in Table 22-6.

Cholera. Cholera is caused by the growth of enterotoxigenic *Vibrio cholerae* in the intestines. These comma-shaped, gram-negative bacteria with polar flagella are usually transmitted by water that has been heavily contaminated with feces or vomitus of people suffering from the disease. Food, fingers, and flies sometimes serve as vehicles of cholera transmission but are not important in most outbreaks. The cycle of transmission is therefore easily controlled in areas where sanitary disposal of feces generally prevents the ingestion of fecally contaminated water. Unfortunately, sewage disposal facilities in most of the world's cholera-endemic areas remain inadequate.

Cholera enterotoxin is produced by the pathogen as it grows in the infected gastrointestinal tract. The enterotoxin, which consists of two molecular subunits, attaches to cell membranes of intestinal epithelial cells. The B subunit binds to molecules on the cell's surface and propels the smaller A subunit into the cell, where it stimulates adenylate cyclase, a membrane-associated enzyme that catalyzes the formation of cyclic AMP (cAMP) (Fig. 22-10). Accumulation of cAMP results in massive secretion of salts and water from each affected cell. The rapid loss of water produces a watery diarrhea that may

Pathogen	Disease	General Characteristics	Infectious Dose
Vibro cholerae	Cholera	Gram-negative curved rods, highly motile, enterotoxin producer	10^8
Escherichia coli	Gastroenteritis	Gram-negative rod, enterotoxin producer	10^8
Clostridium perfringens	Necrotizing enterocolitis	Gram-positive rod, strict anaerobe, enterotoxin producer	10^8
Clostridium botulinum	Botulism	Gram-positive rod, strict anaerobe, neurotoxin producer	Unknown
Clostridium difficile	Pseudomembranous colitis	Gram-positive rod, strict anaerobe, toxigenic	High concentration of normal flora
Helicobacter pylori	Gastric (peptic) ulcers	Gram-negative curved rods	Unknown
Giardia lamblia	Enteritis	Trophozoite and cyst forms	10 cysts
Cryptosporidium	Cryptosporidiosis	Protozoan that produces oocysts in intestinal mucosa	Unknown

cause the cholera patient to lose 20 liters of fluid in a single day (doctors have reported people losing more than 70 pounds of body weight in their first day of cholera). Such dramatic water loss leads to severe dehydration, thickening of the blood, a decrease in blood volume, circulatory collapse (shock), and death if not rapidly treated. Some cholera victims die within 6 hours after the onset of symptoms. Half the persons with untreated cholera die of the disease.

Cholera can be preliminarily diagnosed by the characteristic appearance of the watery stools, which are virtually free of feces and contain mostly mucus, epithelial cells, and enormous numbers of *Vibrio cholerae*. These features give the stools a "rice water" appearance. Diagnosis is supported by direct microscopic examination of stool specimens for the presence of the pathogen. Isolation of a pathogen that agglutinates in the presence of *V. cholerae*–specific antisera confirms the diagnosis. Although there are more than 130 different serotypes of *V. cholerae*, most epidemics are associated with a single serotype, *V. cholerae* 01.

In treating cholera, the overriding concern is the continual replacement of fluids and electrolytes until the pathogens are eventually eliminated by the host. In severe cases, intravenous administration is the most rapid method of returning the body fluids to equilibrium. In most cases, fluid and electrolyte balance can be maintained by the oral administration of electrolytes in the presence of glucose.

An oral rehydration solution can be made conveniently and inexpensively from (1) corn syrup (glucose); (2) table salt (sodium chloride); (3) baking soda (sodium bicarbonate); and (4) cream of tartar (potassium tartrate). The glucose is essential because it stimulates uptake of sodium and subsequent osmotic absorption of water. The uptake of water in the absence of glucose is insufficient to replace the fluid loss. This simple treatment reduces the mortality rate of cholera to less than 1 percent. It is also used to decrease mortality from severe diarrhea caused by other diseases.

Outbreaks of cholera have decimated communities since ancient times. The disease continues to be endemic in Asia. Since 1961, however, cholera has been pandemic, steadily spreading to other areas—the Middle East in 1966, eastern Europe in 1970, Africa in 1970, western Europe in 1971, many Pacific islands in 1977, and in 1991 to the western hemisphere, where approximately 2300 South Americans died in the following year.

Incubation Period	Major Vehicle	Clinical Picture	Diagnosis	Treatment
1–3 days	Water	Diarrhea, dehydration	Isolation of *Vibrio* from stool or vomitus	Fluid and electrolyte replacement
12 h	Food, water	Diarrhea, dehydration	Isolation of organism with same serotype in food and stool	Fluid and electrolyte replacement in severe cases
8–12 h	Meat	Diarrhea, cramps	Isolation of organism with same serotype in food and stool or isolation of $>10^5$ organisms per gram in food	None
Unknown	Honey	Constipation, loss of muscle function	Identification of organism or toxin in feces	Supportive
Variable	—	Diarrhea, intestinal necrosis, colitis, formation of pseudomembrane	Identification of organism in feces, detection of toxin in feces	Vancomycin
Unknown	Unknown	Gastric ulcer formation; stomach cancer	Identification of organism	Tetracyclines
1–4 weeks	Water	Diarrhea, dehydration	Identification of organism in feces or duodenal drainage by microscopic examination	Metronidazole or quinacrine
4–14 days	Water, food	Diarrhea, pain, nausea	Identification of cysts in feces	Fluid replacement

The spread of cholera to the western hemisphere was the consequence of a common shipping practice. Large ships take on ballast water when sailing without a cargo. The ballast water is then discharged into the bay of the port where the ship is to pick up its cargo, an epidemiological prescription for international spread of waterborne pathogens. The South American cholera outbreak began when a Chinese freighter discharged its ballast water into a Peruvian harbor, inoculating the bay with the *Vibrio* it had taken on in ballast from the waters of its previous port. The *V. cholerae* discharged by the ship into the bay was concentrated by shellfish, which filter large quantities of water to extract plankton for food. The cholera bacteria were concentrated in the shellfish, which were then eaten raw as ceviche, a popular appetizer. Low levels of chlorination used in the Lima water supply allowed the organisms to multiply and spread. Cases of cholera caused by the same strain are now seen throughout Central America and Mexico.

Approximately 45 countries, mostly in Asia, Africa, and South America, are currently affected by cholera. Unfortunately, the number of cases reported annually has been increasing in those areas where malnutrition is common and where im-provements in sanitation and water purification are neglected. Cholera has been successfully controlled in the United States in spite of the persistence of toxigenic strains of the pathogen. The few sporadic outbreaks are usually caused by consumption of shellfish caught in contaminated coastal waters. Persons with cholera entering the United States account for the other cases. In 1992, for example, 75 people contracted cholera from seafood salad served on an airliner arriving in Los Angeles from South America. After landing in Los Angeles, all sick passengers were hospitalized, and the other passengers were monitored for subsequent signs of cholera. One of the hospitalized patients died.

Prevention of cholera depends on rapidly treating active cases, properly disposing of sewage, chlorinating water supplies, and, when sewage and water treatment facilities are inadequate, boiling water before it is used for drinking, cooking, or washing dishes. In addition, vulnerable foods should not be imported from infected areas. A vaccine against cholera is available, but it is only 50 percent effective and confers immunity for only 6 months. In its current state, the vaccine is useless for preventing disease among residents of endemic areas. It may be beneficial, however, in protecting persons travel-

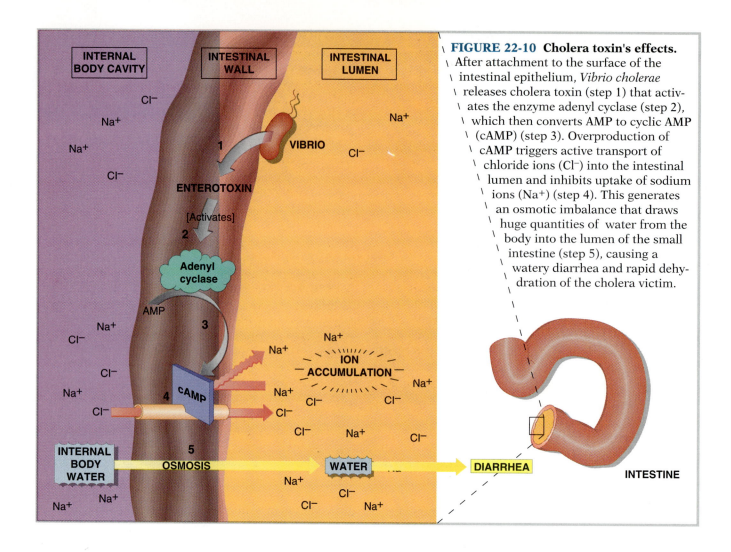

FIGURE 22-10 Cholera toxin's effects. After attachment to the surface of the intestinal epithelium, *Vibrio cholerae* releases cholera toxin (step 1) that activates the enzyme adenyl cyclase (step 2), which then converts AMP to cyclic AMP (cAMP) (step 3). Overproduction of cAMP triggers active transport of chloride ions (Cl⁻) into the intestinal lumen and inhibits uptake of sodium ions (Na⁺) (step 4). This generates an osmotic imbalance that draws huge quantities of water from the body into the lumen of the small intestine (step 5), causing a watery diarrhea and rapid dehydration of the cholera victim.

ing to cholera-endemic regions. Several newer vaccines, some composed of toxin fragments, are being tested for use in endemic areas.

Escherichia coli Gastroenteritis. *E. coli* is a major cause of diarrhea and dysentery-like syndromes that have inspired such imaginative names as "Montezuma's revenge," "turista," "the trots," and "Delhi belly." As members of the normal intestinal flora, most strains of *E. coli* are avirulent and elicit no disease in the gastrointestinal tract. Some strains, however, are pathogenic and cause gastroenteritis. Although a few strains of *E. coli* (called enteroinvasive *E. coli*) can cause local invasive disease of the intestinal epithelium, most pathogenic strains are noninvasive and produce an enterotoxin while growing in the infected person's intestine. These strains are called enterotoxigenic *E. coli* (ETEC). The toxin is physically and antigenically related to cholera toxin and stimulates a similar, although milder, watery diarrhea. Although the two toxins are 80 percent identical, their molecular dif-

ferences are significant—*E. coli* toxin makes people sick, whereas cholera toxin kills.

Unlike cholera, most persons living in endemic regions are resistant to *E. coli* toxin because they have developed antibodies that neutralize it. Endemic regions are usually characterized by poor sanitation, which encourages microbial multiplication in food and water contaminated with feces. The pathogens do not appear to pose a major health threat when they are imported with a returning vacationer, since the unsanitary environment necessary for perpetuating an epidemic does not come home with the traveler. Other strains of the bacterium, called enteropathogenic *E. coli*, cause infectious diarrhea among infants in the United States.

MYTH: If you avoid drinking untreated water in endemic areas, you won't contract Montezuma's revenge or other forms of traveler's diarrhea caused by pathogenic *E. coli*.

FACT: Many persons contaminate their bottled beverages with ice made from tap water (freezing does not kill these pathogens). Even if you don't make this mistake, however, you may still be at risk. Recent evidence indicates that "traveler's diarrhea" is as often acquired from fecally contaminated raw fruits and vegetables as from water and ice. Peeling fruit and avoiding salads are measures that can save your gastrointestinal tract a lot of distress. Almost half of all international travelers (100 million of the 250 million) suffer traveler's diarrhea while visiting a different country.

The large number of bacteria needed to initiate infection plays an important role in the epidemiology of *E. coli* gastroenteritis in the United States, where sanitation facilities usually protect people from exposure to infectious doses. Infected infants, however, shed such high concentrations of the bacteria in their feces that the disease is readily transmitted by improperly washed hands or inadequately decontaminated fomites such as scales and thermometers. Outbreaks in the United States are most commonly associated with inadequate precautions taken by hospital staff in newborn nurseries.

Enterotoxigenic *E. coli* isolated from feces or food are distinguished from normal flora inhabitants by serological tests and by demonstrating enterotoxin production. Because of the short duration of the disease, the etiology of most cases of traveler's diarrhea is not determined, and antimicrobial therapy is of little benefit. Severe cholera-like *E. coli* infections, however, require fluid replacement therapy. Epidemics of diarrhea among infants in hospitals are controlled by isolation of all babies with diarrhea and treatment with antibiotics to which the pathogen is sensitive.

The genetic information for toxin production resides on a plasmid that can apparently be transferred by conjugation in much the same way antibiotic resistance is transmitted. Fortunately, no major population shift to toxigenic strains has occurred among *E. coli* or other gram-negative enteric organisms. However, those strains that acquire the toxin plasmid are also likely to acquire multiple antibiotic resistance plasmids, making severe cases of these infections difficult to treat. Another plasmid that carries information for pilus production is also essential for pathogenicity. Without pili the organisms cannot attach to the surface of the intestinal epithelium and are flushed out of the region by the movement of the intestinal contents.

The most severe cases of *E. coli* infection are caused by enterohemorrhagic strains of the bacteria. These *E. coli* produce a toxin, called *verotoxin*, that causes bloody diarrhea, usually accompanied by severe abdominal cramps. Occasionally, the toxin damages the kidneys, resulting in blood in the urine and a condition known as *hemolytic uremic syndrome* (HUS). HUS is the most common cause of acute renal failure in children, and *E. coli* O157:H7 is the strain that is most often responsible. These bacteria are normal flora in cattle and may contaminate milk or beef during processing. Most outbreaks in the United States are associated with undercooked or raw meat. (If the meat is adequately cooked, the bacteria are killed.) For example, a highly publicized outbreak that began in late 1992 affected over 500 people who consumed undercooked hamburgers at a well known chain of fast food restaurants; four people died. The outbreak prompted even the President to question the routine sanitary precautions used to ensure the safety of meats. ▶ (See THE MICROBES: A Modern Disease for Modern Times, Chap. 15. p. 358).

E. coli O157:H7 can be isolated from bloody stools at a higher rate than most other pathogens. Unfortunately, the pathogen has often not been recognized because its isolation requires use of a modified MacConkey's medium and confirmation by reaction with O157 antiserum, neither of which are routine for examining *E. coli* isolates. Since the 1992 outbreak, however, the CDC recommends that all specimens from persons with bloody diarrhea be cultured for *E. coli* O157:H7.

***Clostridium perfringens* Gastroenteritis.** The versatile pathogen *Clostridium perfringens* is most often associated with gas gangrene, but it is also one of the most prevalent agents of gastroenteritis. In both food and the intestines, this pathogen produces an enterotoxin that causes symptoms much like those of cholera and enterotoxigenic *E. coli* infections, although the disease is usually milder. Diagnosis is rarely attempted, and treatment is unnecessary because the duration of disease is usually 1 day or less.

Some experts believe that *C. perfringens* has the broadest environmental distribution of all pathogenic bacteria (Fig. 22-11). Vegetative cells and endospores are widely distributed in the soil and are normally found in small numbers in the gastrointestinal tract of humans and animals. The disease is acquired from food in which the organism has proliferated to significant numbers. Foods may be contaminated by animal or human feces or by water, soil, or dust containing the organism. Cooking at temperatures above 60°C destroys the vegetative forms, but spores generally survive these temperatures. Heating actually compounds the problem by

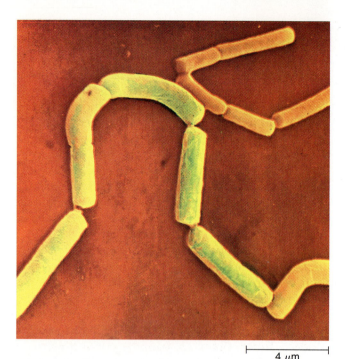

FIGURE 22-11 **The world's most prevalent human pathogen?** Not only does *Clostridium perfringens* cause a diversity of diseases, from gangrene to food poisoning, but its reservoir spans from the inanimate environment to the gastrointestinal tracts of animals.

driving off oxygen, producing the anaerobic conditions that encourage the germination of endospores and growth of the pathogen. The organism proliferates if foods are stored at temperatures between 5 and 60°C or are slowly reheated at temperatures close to 46°C (135°F), the optimum growth temperature for *C. perfringens*. Most outbreaks are traced to cafeterias and restaurants that maintain food on steam tables at temperatures close to optimum for growth of this pathogen.

Meats and poultry prepared in large quantities by fast food establishments or by catering services are common sources of *C. perfringens* gastroenteritis. These outbreaks, which affect large numbers of people, can be avoided if foods are prepared in small portions. Although normal cooking may never reach sporicidal temperatures, the surviving clostridia pose little threat unless allowed to proliferate in food prior to its being eaten. The interior of foods stored in large volumes cools very slowly during refrigeration and becomes an ideal incubator for microbes that survive heating. Food in smaller volumes is more likely to cool thoroughly before significant bacterial growth can occur.

Although prevalence of the organism in the environment makes it impossible to avoid contamination of foods by *C. perfringens*, attention to personal hygiene and proper processing techniques can reduce the levels of contamination and therefore keep the number of organisms below the infectious dose. Meats and other high-protein foods should always be handled with the assumption that spores and vegetative cells are present. Foods should be cooked thoroughly and served immediately or cooled rapidly and kept refrigerated until served or recooked. Foods can be safely reheated or safely kept warm if temperatures are at least 60°C (140°F).

Clostridium botulinum Infections. Infant botulism, a disease first reported in 1976, affects children up to the age of 8 months. The pathogen establishes an infection in the baby's intestines, where it synthesizes the toxin that causes the neurological symptoms of botulism. The resistance of adults to botulism infections is probably due to their normal intestinal flora, which successfully competes with the pathogen. It is presumed that infants younger than 8 months have not yet established the competitive intestinal environment of adults. Epidemiological studies have pinpointed one major source of the pathogen: honey used in infant formulas. Current guidelines recommend that *honey not be fed to children in the first year of life*. It is also believed that 10 percent of all cases of sudden infant death syndrome (SIDS; "crib death") are due to infant botulism.

Pseudomembranous Colitis. An occasional complication of the prolonged oral administration of antibiotics is pseudomembranous colitis, a necrotizing infection of the GI tract caused by toxigenic *Clostridium difficile*. This organism, normally a minor member of the gastrointestinal flora, proliferates to high concentrations in the colon when other normal bacterial inhabitants are inhibited by antibiotic therapy, primarily with clindamycin or ampicillin. (Oddly, *C. difficile* may also be sensitive to these antibiotics—see BEYOND THE BASICS: Antibiotic Boomerang.) *C. difficile* produces a necrotizing exotoxin that may cause massive and potentially fatal loss of fluids. The name of the disease reflects the layer of inflammatory products that accumulates on the surface of the intestinal wall, giving the false appearance of a membrane. This clinical picture, coupled with a history of antibiotic therapy, provides a preliminary diagnosis of pseudomembranous colitis. The high mortality rate (30 to 90 percent) of untreated cases is lowered by simply discontinuing the initial antibiotic. Vancomycin is given to control the population of *C. difficile*.

Helicobacter pylori Infections. *Helicobacter pylori*, a gram-negative spiral bacterium discovered in

Antibiotic Boomerang

It was a simple infection, nothing of serious consequence. Laboratory cultures indicated the causative agent, and the patient received antibiotic therapy. As the infection resolved, however, the patient was stricken with colitis, severe inflammation of the large intestine. An examination of the bowel revealed it to be covered by gray membranous patches, beneath which the tissue was ulcerating and dying. If the patient continued to take the antibiotic, he could soon die of classic antibiotic-induced pseudomembranous colitis.

Although antibiotics precipitated the situation, this person's medical dilemma was caused by *Clostridium difficile,* an anaerobic spore-forming bacillus that normally resides in small numbers in the healthy intestine. Its growth is limited by competition from the other members of the normal intestinal flora. Antibiotic therapy, especially with clindamycin, ampicillin, cephalothin, or metronidazole, which suppress the other members of the indigenous flora, allows *C. difficile* unrestricted growth. The bacterium releases a cytotoxin that induces bowel necrosis and other symptoms of pseudomembranous colitis. In many cases, colitis disappears when the principal antibiotics are withdrawn and the normal flora reestablishes itself. This process is usually helped along by treatment with vancomycin to reduce the pathogenic clostridia population in the colon.

Investigators originally assumed that *C. difficile* causes this disease because of its resistance to antibiotics. In many cases this is accurate. But even antibiotic-sensitive strains of this opportunistic pathogen elicit colitis. In fact, all strains of *C. difficile* are sensitive to ampicillin, one of the drugs that most often induces the disease. Why aren't the clostridia suppressed with the rest of the flora? Recent evidence suggests that another pathogenic mechanism accounts for this paradox. While moderately suppressing drug-sensitive bacteria, ampicillin encourages the growth of other bacteria, those that produce beta-lactamase, an enzyme that inactivates the antibiotic. The clostridia, also protected by the enzyme, overgrow the bowel before the protective members of the normal flora repopulate. It may therefore be possible to prevent ampicillin-induced colitis by administering a beta-lactamase inhibitor along with the antibiotic.

1982, is the only microbe that thrives in the acid environment of the human stomach. It attaches to the surface of the gastric mucosa (Fig. 22-12) and establishes local colonies that are protected from acid attack by ammonia, generated by the prolific amounts of the enzyme urease that these bacteria produce. Ammonia is alkaline and raises the pH around the colony. *H. pylori* causes acute gastritis (inflammation of the stomach) and also the lesions that precede the formation of ulcers in the stomach or small intestine. The bacteria damage the mucus-producing cells to which they attach, creating an area with no effective barrier between the caustic digestive juices of the stomach and the living tissue beneath. These exposed cells may then suffer the enzymatic and acid damage that leads to ulcers. *H. pylori* infections are also implicated as a cause of stomach cancer, which kills 14,000 Americans each year and is epidemic in Asia and South America. Infected people are 6 times as likely as uninfected people to develop stomach cancer.

Antibiotic treatment can eliminate the bacterium and reduce recurrences of ulcers to less than 20 percent. (Conventional antacid therapies have relapse rates as high as 95 percent.) Vaccines that may prevent the bacteria from attaching to the gastric mucosa are being developed by several biotechnology companies.

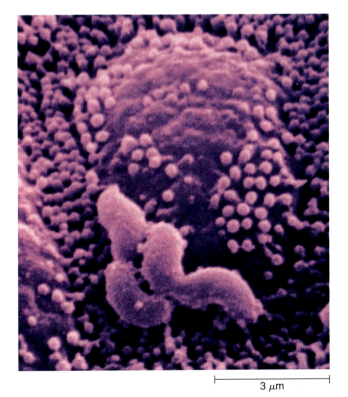

3 μm

FIGURE 22-12 Is an ulcer an infectious disease?
Helicobacter pylori attaches to gastric mucosa, where it damages underlying mucus-producing cells, very likely contributing to the development of ulcers.

dry environments, however, and will not survive on many fomites.

Shigella infections range from asymptomatic cases to life-threatening dysentery. The organisms penetrate and multiply in the superficial epithelium of the colon, where they are believed to release endotoxin that triggers inflammation and causes local damage. When absorbed into the bloodstream, the endotoxin causes the fever associated with the disease. *Shigella* also produces an exotoxin that causes diarrhea. The severity and outcome of the illness are influenced by the status of the patient, the size of the infecting dose, and the virulence of the pathogen. Like most infections, shigellosis is particularly dangerous to the compromised host, one of the reasons for the higher fatality rates among people who acquire the disease while hospitalized.

A preliminary diagnosis of dysentery is provided by microscopic detection of blood cells in feces. *Shigella* can be isolated from fecal specimens or from rectal swabs on selective, differential, or enrichment media that discourage overgrowth by normal flora. Treatment is usually limited to supportive therapy and the replacement of fluids and electrolytes, and the disease is allowed to run its course. In severe cases, antimicrobial drugs, usually ampicillin, tetracycline, or trimethoprim-sulfamethoxazole, are given. Recovery confers serotype-specific immunity.

Salmonellosis. *Salmonella*, another genus of gram-negative bacilli, contains more than 2000 serologically distinct organisms that cause gastroenteritis in humans. These numerous serotypes have been consolidated into three species based on ecological considerations. *Salmonella typhi*, the causative agent of typhoid fever, is a single serotype highly adapted to humans. *Salmonella choleraesuis* is a single serotype commonly found among fowl, cattle, swine, and other nonhuman sources. It is rarely associated with human disease. The third species, *Salmonella enteriditis*, encompasses all the other serotypes, most of which survive equally well in humans and animals. These organisms may cause an acute gastroenteritis of short duration, bacteremia, or a potentially fatal systemic infection called enteric fever.

Salmonella is one of the most commonly reported bacterial agents of gastroenteritis in the United States (it is second only to *Campylobacter jejuni*). These bacteria are shed from the gastrointestinal tract of animals and, less frequently, humans. Salmonellosis is the most widespread of all infectious diseases contracted from animals. The bacteria may be shed in the feces of infected animals or released

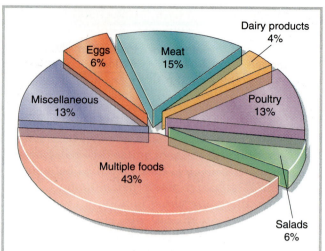

FIGURE 22-15 **Vehicle of transmission in 169 outbreaks of salmonellosis in the United States, 1983–1987.** A miscellaneous category includes vehicles, each of which individually were responsible for less than 3 percent of outbreaks. An additional 173 outbreaks were reported, for which the vehicle was unknown. *MMWR* 39 SS-1, p. 30–39, 1990.

during slaughter and meat processing. Foods obtained from animal sources (eggs, poultry, milk, and sausage) pose high risks for *Salmonella* contamination (Fig. 22-15). Special care in cooking and refrigerating these foods minimizes the potential for *Salmonella* survival and growth prior to ingestion. Pets are also a source of the bacteria and are commonly implicated in fecal-oral spread. Pet turtles have been responsible for several outbreaks of *Salmonella* gastroenteritis, but dogs, cats, birds, reptiles, and fish also harbor and shed the organism (Fig. 22-16). Easter is usually followed by outbreaks of salmonellosis contracted from pet ducks and chickens given to children as Easter gifts. Like humans, animals acquire *Salmonella* in the foods they eat, often from contaminated bonemeal and meat by-products. Federal regulations attempt to prevent the distribution of feeds containing *Salmonella* by prohibiting interstate transport of such materials. Unfortunately, unless entire herds or flocks are free of the organism, a few infected animals can readily transmit the pathogen during transport to the slaughterhouse, during confinement within crowded holding pens, and through contaminated machinery in the slaughterhouse.

Poultry has most recently been the major vehicle of *Salmonella* gastroenteritis. More than 4 billion birds are consumed each year in the United States, and as many of one-third of these are contaminated with *Salmonella* or *Campylobacter* (see next section).

FIGURE 22-16 **Dangerous pet**. Animals, including pets, are frequent sources of *Salmonella*. In the early 1970s, sales of pet turtles were outlawed because 600 cases of human salmonellosis, some fatal, were traced to the turtles. Pet iguanas can also harbor *Salmonella* and have accounted for many documented cases of salmonellosis.

If previous trends continue, 2 percent of the U.S. population will get gastroenteritis from poultry this year. Eggs, another common source of *Salmonella*, become contaminated within the hen's oviduct or by contact with fecal material from infected hens. Organisms on the egg surface may be introduced into food products. The pathogens are usually destroyed by cooking, and outbreaks are often associated with eating raw eggs or salads, pastries, desserts, or drinks that contain raw eggs (Caesar salad, hollandaise sauce, and home-prepared eggnog are common examples). Even undercooked eggs, prepared sunny-side up or soft boiled, are commonly associated with gastroenteritis. (In 1992, New Jersey outlawed the serving of over-easy and sunny-side-up eggs by restaurants.) Outbreaks of salmonellosis have also been caused by marijuana that presumably was contaminated with animal feces during growth or storage.

Because of the widespread distribution of the organisms and the numerous opportunities for spread, outbreaks of *Salmonella* gastroenteritis are common

and difficult to control. Fortunately, *Salmonella* gastroenteritis is usually mild. In fact, most cases are believed to be asymptomatic. The organisms invade the mucosa of the large and small intestines, causing inflammation, moderate fever, diarrhea, abdominal pain, and nausea. The symptoms usually appear within 6 to 36 hours after ingestion of the contaminated food, and most patients recover in 2 to 4 days. In infants, the elderly, and immunocompromised persons, however, the infection is often fatal. The organisms invade the bloodstream and spread to other body sites, killing about 2000 of the estimated 4 million people infected each year in the United States.

Campylobacter Gastroenteritis. *Campylobacter jejuni*, a gram-negative vibrio (Fig. 22-17), is the most common bacterial cause of gastroenteritis in the developed world. First recognized as a human pathogen in 1947, it causes approximately 11 percent of diarrhea cases worldwide. At least one out of every 20 sufferers of diarrhea in the United States is a victim of *C. jejuni* infection. When appropriate diagnostic techniques are employed, this pathogen is isolated more frequently than *Salmonella* and *Shigella* combined.

Campylobacter jejuni is a pathogen of many animals and is transmitted to humans by ingestion of contaminated food, raw milk, or water. The organism can survive in fresh water for up to 5 weeks. It can also be spread to people, usually infants, by direct contact with fecal material from infected pets, including cats and dogs. Gastrointestinal symptoms, usually diarrhea (often with blood in the stool), begin 3 to 5 days after infection. Although most in-

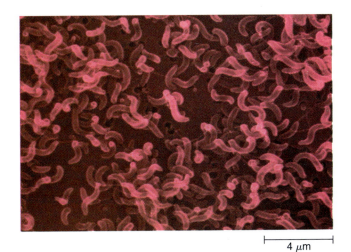

|———————| 4 μm

FIGURE 22-17 **The developed world's most common intestinal invader,** ***Campylobacter jejuni.***

fections are self-limiting, relapses occur in 20 percent of cases.

Healthy persons rarely harbor *C. jejuni*, and isolation of this vibrio from feces is conclusively diagnostic. *C. jejuni* is fastidious and must be grown in special media in a microaerophilic environment. Growth is optimal at 42°C. Severe infections are treated with fluid and electrolyte replacement therapy and with erythromycin.

Enteroinvasive *Escherichia coli* Gastroenteritis.

A few strains of *E. coli* can invade the epithelial cells lining the colon. These enteroinvasive *E. coli* cause fever and a dysentery syndrome that is similar to shigellosis. Clinical manifestations result from bacterial replication in the invaded tissue rather than from enterotoxin production. Enteroinvasive *E. coli* can be identified by their ability to cause conjunctivitis when inoculated in the eyes of guinea pigs. Antibiotics are used only when intestinal disease is severe.

Vibrio parahaemolyticus Gastroenteritis.

Vibrio parahaemolyticus is a halophile most commonly found in marine environments. This bacterium causes gastroenteritis following the ingestion of contaminated seafood. It has been recognized as a major cause of food poisoning outbreaks in Japan since 1956 and has been identified with increasing regularity in the United States in recent years, especially during summers, when the pathogen grows freely in coastal waters and may infect oysters, shrimp, crabs, and fish. During cold seasons, the organisms are submerged in the marine sediment. The disease is associated with eating contaminated raw or inadequately cooked seafood. Cooked foods may be recontaminated if the same surface or utensils are used for cooked and uncooked foods or if the food handler fails to wash hands between manipulations. If recontaminated food is permitted to remain at temperatures between 20 and 35°C, the vibrios can rapidly multiply to infectious concentrations. The disease is rarely transmitted directly from an infected person.

Symptoms of *V. parahaemolyticus* gastroenteritis are similar to those of foodborne illnesses caused by *Salmonella*, *Shigella*, and enteroinvasive *E. coli*. Diagnosis, therefore, depends on isolating the organism from the patient's feces. Standard media used for the identification of enteric organisms will not support the growth of the marine vibrio. Special media containing increased concentrations of sodium chloride are used when *V. parahaemolyticus* is the suspected pathogen. Preventive measures depend on adequately cooking the seafood, storing seafoods at low temperatures, and avoiding the cross-contamination of cooked food by raw seafood. (Cooked and raw food should be handled in separate areas with different utensils, and hands should be washed when going between areas.)

Yersiniosis.

Yersinia enterocolitica is a gram-negative coccobacillus that occasionally causes gastroenteritis in humans following ingestion of feces-contaminated food or water or by direct contact with contaminated feces. The disease is characterized by diarrhea or severe abdominal pain (or both), usually accompanied by a fever. These symptoms often lead to a false diagnosis of appendicitis, and as many as 10 percent of persons with yersiniosis needlessly undergo appendectomies. About 20 percent of the patients infected with *Yersinia enterocolitica* develop disorders believed to be autoimmune. These include arthritis, erythema nodosum, and inflammation of the iris.

Y. enterocolitica grows best between 25 and 30°C on most media used for culturing enteric bacteria (although the organism can also grow in refrigerated foods). The pathogen is more easily isolated if fecal specimens are treated with one of several enrichment techniques prior to plating. Severe infections can be treated with antibiotics.

Amebic Dysentery.

Ingestion of water or food containing cysts of *Entamoeba histolytica* may lead to a severe and sometimes fatal dysentery syndrome. Humans are the sole reservoir of this pathogen. Most infected persons, however, are asymptomatic. The ingested cysts pass through the stomach; the trophozoites emerge in the lower small intestine and multiply in the lumen of the colon, where they engulf red blood cells. Cysts re-form and are shed in feces ▶ (see Fig. 12-4, p. 283, for life cycle of *Entamoeba histolytica*). Approximately 450 million people are believed to be asymptomatically infected, each capable of shedding 3 million cysts daily. In some persons this carrier state continues for years.

Intestinal disease varies from mild diarrhea to severe dysentery, characterized by fever and frequent episodes of bloody, mucus-laden diarrhea. Disease develops when *E. histolytica* invades the mucosa, producing abscesses that ultimately enlarge into ulcers. The infection may spread from the colon to the liver and less often to the lungs and brain. Liver abscesses sometimes occur in the absence of intestinal symptoms. Conditions that predispose to invasion are poorly understood. Infection is most prevalent and severe in areas of the world where crowding and poor sanitary conditions promote fecal-oral spread.

Diagnosis is confirmed by microscopic identification of the trophozoite in stool specimens or abscess aspirates. Trophozoites are fragile, and samples must be examined within an hour of collection or stored in the presence of a preservative. The greatest problem in diagnosis is distinguishing *E. histolytica* from similar amoebas that are part of the normal intestinal flora. The finding of trophozoites containing red blood cells is diagnostic—only invasive *E. histolytica* characteristically ingest erythrocytes. Cysts can be found only in formed stools and are therefore shed only during asymptomatic or mild infections. When cysts can be found, they have one to four nuclei (unlike the normal flora *Entamoeba coli*, which have eight or more nuclei per cyst). Serologic tests are usually used to confirm diagnosis of extraintestinal infections.

Acute dysentery is usually treated with metronidazole in combination with diiodohydroxyquin to ensure the destruction of amoebas in three sites—the bowel lumen, bowel wall, and invaded organs. Control of amebic dysentery depends on sanitary disposal of feces and treatment of carriers. Since persons with acute disease do not shed cysts, it is not necessary to isolate them. Travelers to endemic areas can reduce the likelihood of infection by boiling all suspected water prior to drinking and eating only cooked foods.

Viral Gastroenteritis. The source of the viruses that cause gastroenteritis is usually the infected intestines of humans, and the mode of transmission is primarily fecal-oral, either in food or water or by direct contact with contaminated feces. Gastroenteritis viruses are believed to be responsible for diarrheagenic illnesses that kill as many as 18 million people each year, especially infants and young children in developing countries. Even before their discovery, the existence of such agents was postulated to explain the many cases of gastroenteritis from which no bacterial agents could be cultured. In the early 1970s electron microscopic examinations of stool samples revealed the cause of many of these undiagnosed cases. The Norwalk viruses cause epidemic outbreaks of gastrointestinal illness in schools, families, and communities and may be one of the causes of the erroneously named "stomach flu," a syndrome that is clearly not caused by influenza virus. Epidemic viral gastroenteritis usually is a self-limited illness that lasts 1 or 2 days.

The rotaviruses, unlike most viral pathogens of gastroenteritis, are a major, and perhaps the primary, cause of severe diarrhea among infants and young children. Rotavirus infections often leave children so severely dehydrated that they require

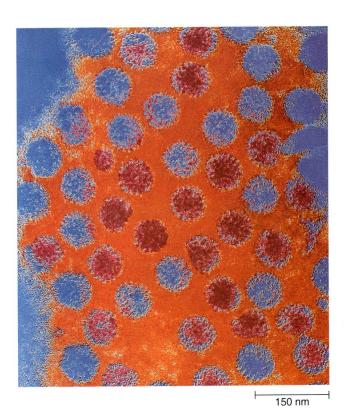

|—————| 150 nm

FIGURE 22-18 *"Wheels of death."* Rotavirus, shown here as isolated from human feces, resembles a wheel, from which its name is derived.

hospitalization to replace the lost body fluids. The pathogenic mechanisms are uncertain, but both types of viruses alter the absorbing surfaces of the small intestine, probably by multiplying in and damaging intestinal epithelial cells. Viral gastroenteritis is diagnosed by serological tests or by using an electron microscope to detect the virus in feces (Fig. 22-18). Several attenuated oral vaccines are undergoing clinical trials.

● **INVASIVE DISEASES.** A few pathogens enter the gastrointestinal tract and subsequently invade tissues in other parts of the body. Although their characteristic symptomology is displayed outside the digestive tract, all of these organisms are excreted within feces and are transmitted primarily through a fecal-oral route. Characteristics of invasive diseases are summarized in Table 22-8.

Enteric Fever. A small number of *Salmonella* strains possess sufficient virulence to invade beyond the intestinal tract and produce a severe systemic infection, characterized by invasion of lymphoid tissue and prolonged fever. This clinical syndrome is called **enteric fever**; its classical form is typhoid fever. The pathogen, *Salmonella typhi*, multiplies

TABLE 22-8 CHARACTERISTICS OF INVASIVE PATHOGENS THAT CAUSE EXTRAINTESTINAL INFECTIONS

Pathogen	Disease	Major Vehicle	Incubation Period	Clinical Picture	Diagnosis	Antimicrobial Treatment
Salmonella typhi	Typhoid	Food, water	1–3 weeks	Fever, malaise, abdominal pain	Positive blood culture during first week, positive stool or urine specimen	Chloramphenicol, ampicillin
Hepatitis A virus	Hepatitis	Food, water	10–45 days	Fatigue, nausea, abdominal pain, jaundice	Clinical and epidemiological findings; RIA of feces or blood	None
Hepatitis E	Hepatitis	Food, water	42–56 days	Fatigue, nausea, abdominal pain, jaundice	Clinical and epidemiologic findings; serological tests	None
Poliovirus	Paralytic polio	Saliva, food, water	7–12 days	Fever, headache, paralysis	Positive throat culture, spinal fluid, or rectal swab	None
Toxoplasma gondii	Toxoplasmosis	Raw meat, cat feces	5–21 days	Neurological and ocular damage in severe cases	Positive serological tests, isolation of pathogen from body fluids	Pyrimethamine plus sulfa drugs in severe cases
Listeria monocytogenes	Listeriosis	Food	4–21 days	Miscarriage, stillbirth, meningitis (fever, headache, nausea)	Growth of pathogen by "cold enrichment" technique	Ampicillin

in the intestinal epithelium, and the progeny are engulfed by macrophages, which fail to kill the bacteria. The parasites are transported in host macrophages to regional lymph nodes, and from there they invade the bloodstream. Circulating bacteria may localize in the liver, spleen, gallbladder, kidneys, bone marrow, heart, lungs, and lymphoid tissue of the gastrointestinal tract. Fever slowly elevates to an average of 104°F, and the skin may become dotted with small hemorrhages called "rose spots." Gastrointestinal symptoms appear late in the course of the disease and are usually characterized by constipation followed by bloody diarrhea. Vomiting and abdominal tenderness may also occur. The disease ranges in severity from asymptomatic or mild to fatal infections. The mortality rate of untreated typhoid fever is 10 to 15 percent, a rate that is reduced by antibiotic therapy. Several *Salmonella enteriditis* serotypes cause a milder form of enteric fever; the disease, called paratyphoid fever, is similar to typhoid but is much less severe.

Enteric fever is transmitted by the fecal-oral route. The disease develops 1 to 3 weeks following the ingestion of the pathogens in food or water that has been contaminated by the feces of overtly sick victims or carriers. During the course of the disease, the infected gallbladder pours pathogens into the feces, seeding the environment with infectious bacteria. The gallbladder often continues to shed the pathogen during convalescence, and it sometimes remains infected after the patient recovers. Such asymptomatic persons become healthy carriers of typhoid fever. Poor personal hygiene habits, especially in food handlers, facilitate the direct inoculation of foods. Improper treatment of public water supplies or inadequate disposal of human wastes can cause outbreaks of typhoid in large populations.

Typhoid fever is diagnosed by identifying the pathogen isolated from blood, feces, or urine. Enrichment media containing selenite or tetrathionate favor the growth of *Salmonella* over that of gram-negative bacteria of the normal flora. These enriched cultures are then plated onto selective differential media to obtain isolated colonies of the pathogen. The bacteria are identified by biochemical tests and agglutination with *Salmonella*-specific antiserum. Rising antibody titers against *S. typhi* can be demonstrated by comparing serum samples drawn during acute and convalescent stages of the disease.

Control of the disease is complicated by the existence of relatively large numbers of chronic carriers. Of those recovering from typhoid fever, 2 to 5 percent become permanent carriers. Controlling the disease depends on identifying these carriers and barring them from occupations that entail food handling or the care of debilitated persons who may be extremely susceptible to infection. If these recommendations are followed, typhoid carriers pose no threat to the general public. The carrier state can also be eliminated by appropriate antibiotic therapy. Although chloramphenicol, despite its potentially severe side effects (including aplastic anemia), is still the most effective drug available for treating active cases of typhoid fever, ampicillin is the antibiotic of choice for treating carriers. Unlike chloramphenicol, ampicillin accumulates in high concentrations in the gallbladder. Antibiotic resistance, however, is becoming a more frequently encountered problem. When drug therapy fails, surgical removal of the gallbladder is the only reliable way to eliminate continued shedding of pathogens (see THE EXPLORERS: Following a Trail of Death).

Listeriosis. *Listeria monocytogenes* is a short gram-positive rod that is primarily an opportunistic pathogen of pregnant women and their fetuses and of immunocompromised individuals. The bacterium is widespread in nature and infects humans who consume foods, particularly those made from vegetables (such as cole slaw), meat (pâté), or milk (soft cheeses), that have been contaminated with the organism. Because the bacterium grows at refrigeration temperatures, infectious doses can be generated if lightly contaminated foods are stored for prolonged periods.

Listeria lives within macrophages of its human host. After the intracellular parasite enters a macrophage, the bacterium secretes enzymes that break down the phagocytic vesicle and escapes into the cytoplasm, where it reproduces. Within hours, the progeny are being propelled through the cell by tails made of polymerized host actin[2] filaments (Fig. 22-19). When the *Listeria* contacts the plasma membrane, a thin projection is formed tightly around the bacterium. The protuberance and its accompanying bacteria are engulfed by a neighboring macrophage, thus facilitating the transfer of the pathogen to a new host cell. The bacterium then secretes enzymes to release itself into the cytoplasm, and the process is repeated.

Listeriosis is usually an asymptomatic infection in healthy adults, but in immunocompromised individuals or cancer patients it may cause meningitis. Infection in pregnant women usually leads to infection of the fetus, resulting in spontaneous abortion, stillbirth, or infection of the newborn. Fatality rates in infants with *Listeria* septicemia or meningitis is high.

Polio. Poliovirus is a small RNA virus often spread by water, fingers, foods, flies, and fomites, the traditional vehicles of fecal contamination. This virus produces mild gastrointestinal infections in more than 90 percent of the persons it infects. It multiplies in lymphoid tissues of the tonsils and in the regional lymphoid tissue of the intestines. Local IgA usually restricts spread of the virus to other body sites. If the virus should spread to the bloodstream, IgG neutralizes it before it can infect other organs.

[2]Actin is a protein found in all eukaryotic cells as a component of their cytoskeleton, a framework that supports the cell and its organelles, and constantly propels the cytoplasm.

THE EXPLORERS

Following a Trail of Death

Successful control of typhoid fever can be a serious problem for public health authorities, as the well-publicized case of Typhoid Mary illustrates. In 1906, Mary Mallon was identified as a carrier of *Salmonella typhi* following several outbreaks of typhoid fever in homes where she had worked as a cook. She changed jobs often, leaving each family with tragedy. One man, health authority Dr. George Soper, became obsessed with her capture. He finally tracked her down, following a pattern of outbreaks in dozens of households and institutions. She was incarcerated for 3 years until she promised not to work in occupations that would expose other people to her infection. Failing to keep her promise, she was responsible for at least 50 more cases of typhoid fever and three deaths before Soper again caught up with her, after she had caused an epidemic in a New York hospital.

Mary Mallon refused to allow surgeons to cure her of the carrier state by removing her gallbladder. For the public's protection she was confined until her death 23 years later.

Fortunately, over the past 70 years, advances in sanitation, antibiotic therapy, and surgical technique make it unlikely that incidents similar to the case of Typhoid Mary will ever be repeated in developed countries.

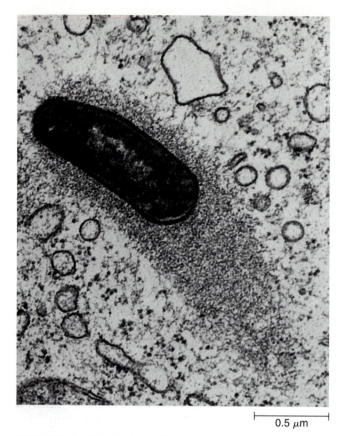

FIGURE 22-19 **Bacterial "comets."** *Listeria monocytogenes* polymerizes an actin tail within its host macrophage, generating its "comet" tail. The tail propels the bacterium to the host cell membrane and facilitates transfer of the pathogen to a new host cell.

In fewer than 1 percent of cases, however, the poliovirus infects the central nervous system. Sometimes this leads to a meningitis that lasts about 10 days and then subsides. A more severe complication evolves when the virus damages motor neurons and causes paralytic poliomyelitis. This form of the disease can be fatal if the muscles controlling respiration and swallowing are paralyzed. Factors that predispose for paralytic complications include tonsillectomy and exposure to an especially large dose of virulent poliovirus. Although polio is characteristically a disease of children, the danger of paralysis is greater in adults. The spread of virus through the infected body is illustrated in Figure 22-20.

Until the introduction of the inactivated poliovirus vaccine (Salk vaccine) in 1955, polio was a major cause of paralysis in the United States. It was not as prevalent in developing countries, where poor sanitation resulted in early exposure to the virus, milder cases, and permanent immunity. In the United States, children were more likely to escape natural exposure to the virus because of better sanitation, producing an older population of suscepti-

ble persons with a greater risk of paralytic sequelae. The Salk vaccine was developed after advances in tissue culture techniques permitted viruses to be propagated in large quantities. In 1961 the introduction of attenuated oral vaccine (Sabin vaccine) further reduced the incidence of disease, so that polio has been effectively eliminated in the United States. The only cases that have occurred since 1979 have been due to an extremely rare phenomenon—the reacquisition of virulence by one of the three viral strains in the Sabin vaccine (which occurs once in about every half million first-time recipients of the vaccine). Although such dangers don't accompany the inactivated vaccine, the advantages of the attenuated vaccine maintain it as the most commonly used polio prophylactic. A new vaccine, called enhanced inactivated polio vaccine, or E-IPV, has been developed for use as the initial injection, which is followed by boosters with the oral vaccine. This approach virtually eliminates the occurrence of reactivation polio.

Although polio remains prevalent in tropical countries, it has been targeted for eradication in Pan American countries. The last nonvaccine case was reported in Peru in 1991, and the World Health Organization projects eradication of the disease from the world by the year 2000.

Polioviruses belong to a group of RNA viruses,

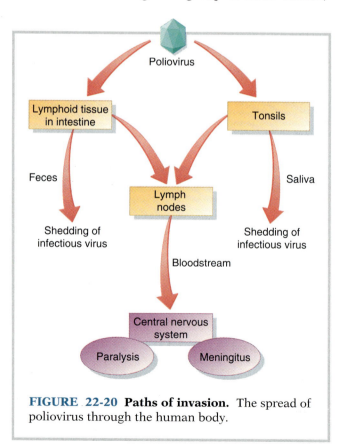

FIGURE 22-20 **Paths of invasion.** The spread of poliovirus through the human body.

called **enteroviruses**, that usually enter the body through the digestive or respiratory tracts and cause asymptomatic infections or mild localized disease in the intestines. There are three categories of viruses in the enterovirus group—Coxsackieviruses, echoviruses, and polioviruses. Although usually inapparent and undiagnosed, infections by these viruses can be systemic or localized in nonintestinal sites such as the meninges. The usual illnesses caused by Coxsackieviruses, for example, are severe sore throats with exudative lesions (herpangia), an influenza-like respiratory disease (summer grippe), and aseptic meningitis. Echoviruses also cause several syndromes, including a diarrhea that is particularly severe in newborns, a common cold-like illness, disseminated infections with skin rash and fever, and aseptic meningitis. Although primarily spread by the fecal-oral route, enteroviruses may also be transmitted by respiratory secretions.

Hepatitis A and Hepatitis E. At least five distinct viruses can cause **hepatitis,** a disease that produces inflammation of the liver and subsequent jaundice (yellowing of the skin and other tissues). These viruses are hepatitis A virus (HAV) (Fig. 22-21), hepatitis B virus (HBV), hepatitis C virus (HCV) (previously referred to as non-A non-B hepatitis virus), hepatitis D virus (HDV or delta virus), and hepatitis E virus (HEV). Although hepatitis B, hepatitis C, and hepatitis D may be acquired through the gastrointestinal tract, all three are primarily acquired by parenteral inoculation. Their prevention and control, therefore, differ from the prevention and control of hepatitis A and E. Because of these epidemiological differences, hepatitis B, C, and D viruses are discussed in Chapter 24.

Hepatitis A and E are transmitted primarily by the fecal-oral route, either in contaminated water and food (especially raw or poorly cooked clams and oysters from contaminated waters) or by direct person-to-person spread. The latter mechanism of transmission is especially prevalent in institutional settings. Because it is transmitted directly, hepatitis A has been popularly called "infectious hepatitis." Hepatitis A is characterized by an incubation period of 2 to 6 weeks during which the virus multiplies in the gastrointestinal tract, spreads to the blood, and invades the liver, where it may proliferate asymptomatically or cause jaundice. Patients usually recover in 6 weeks, acquiring immunologic protection against hepatitis A. There is no permanent carrier state or chronic condition. Difficulty in propagating the virus has hindered studies on replication and pathogenesis, as well as the development of vaccines. Diagnosis by radioimmunoassay (RIA) and ELISA tests immunologically detect the virus in fe-

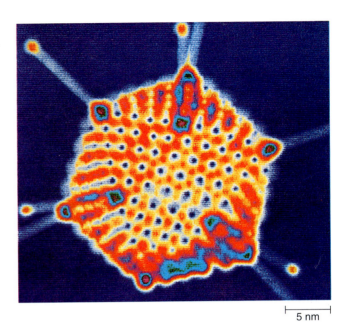

5 nm

FIGURE 22-21 **Hepatitis A virus.** The particle is a nonenveloped virus with a single RNA strand in its nucleic acid core.

ces or blood. Although treatment of hepatitis is limited to supportive therapy, prophylactic administration of immune serum globulin provides temporary protection for persons exposed to the virus. Such passive immunization is recommended for laboratory workers and hospital personnel who will be in contact with infectious patients or clinical specimens. Gowns, gloves, and masks provide added protection.

Although not common in the United States, HEV infections present a global problem, especially in India, parts of Asia, and other areas with poor sanitation. They are usually mild infections but can be severe and even fatal in pregnant woman, especially in the third trimester. As with HAV, the virus is eliminated on recovery and there is no carrier state. Electron microscopy reveals a virus resembling HAV, but immunologic analysis distinguishes the two viruses as having different antigens.

Toxoplasmosis. Although *Toxoplasma gondii* infects the body through the gastrointestinal tract, humans and many animals harbor the asexual cyst stage of this protozoan in various tissues outside the intestine, particularly in muscle and brain. A sexual form, the oocyst, occurs only in cats and is shed in the animals' feces, which can infect adults who clean litter boxes and children who play in sandboxes contaminated with cat feces. Infection is usually acquired by ingestion of cysts in raw or undercooked meats obtained from animals infected by grazing on grass that has been infected with cat feces. Oocysts

5. What is the role of sucrose in the development of dental caries and periodontal disease?

6. Identify the pathogen(s) most likely to be transmitted in each of the following vehicles, and describe how best to prevent associated diseases: (a) Home-canned foods; (b) shellfish; (c) foods with high egg content; (d) untreated mountain springwater; (e) honey.

7. Why are most diseases of the gastrointestinal tract not treated with antibiotics?

8. A patient comes to Dr. Delafield's office on a Saturday in November complaining of a severe diarrhea that started the previous morning. Virtually the whole family (two adults and three children, but not the baby) is sick, and the next-door neighbors (two adults and two children) are also ill. They all ate Thanksgiving dinner together on Thursday afternoon. The menu was traditional—roast turkey with stuffing, cranberry sauce, sweet potato casserole, and custard pie. The adults drank some home-made egg nog, while the children drank milk.

   (a) What pathogens might you suspect as the cause of the diarrhea? Explain your reasoning.

   (b) How would you identify the etiologic agent? the contaminated food?

   (c) What treatment should the doctor recommend?

9. Because symptoms of most gastrointestinal illnesses are so similar, a variety of media and incubation conditions are used for isolation of pathogens from fecal samples. What special considerations would be necessary to identify the following pathogens? (a) *Salmonella;* (b) enterohemorrhagic *E. coli;* (c) *Clostridium perfringens;* (d) *Entamoeba histolytica;* (e) *Vibrio cholerae.*

10. Fill in the following table:

Disease	Pathogens	Symptoms	Mode of Transmission	Control Measures
Botulism				
Shigellosis				
Gastric ulcer				
Hepatitis				

11. Briefly explain the factors that contribute to making diarrheal diseases the major cause of mortality in young children in developing countries. What strategies can be implemented to reduce (a) the death rate and (b) the incidence of disease?

12. Describe how the polymerase chain reaction (PCR) might be used to determine whether shellfish are contaminated with enteric viruses (see p. 232 for PCR description).

13. You are employed in a public health laboratory and have isolated *Salmonella* from both the egg salad and the chicken salad served at a company picnic after which 50 people became ill. How would you show conclusively that one or both of these isolates was the cause of the food poisoning incident?

14. Why is it difficult to achieve active immunity by vaccination against most diseases acquired through the digestive tract?

15. Vegetables are often eaten raw as in salads and thus pose a threat to health if they are contaminated with pathogenic bacteria.

   (a) Design an experiment to test the effect of storage temperature on the viability of *Listeria monocytogenes* and *Shigella sonnei* on lettuce, sliced cucumber, and shredded carrots.

   (b) What are some possible sources of contamination of vegetables with pathogens? What recommendations would you suggest to minimize illness from consuming salads?

16. Why is *E. coli,* but not *Clostridium perfringens,* a major pathogen associated with undercooked hamburgers?

17. Why does drinking directly from a milk carton increase the likelihood of foodborne illness even if you live alone and no one else drinks the milk? (Hint: Consider incubation periods.)

18. A study at Northeastern University revealed that substituting bottled water for tap water actually increases the potential of acquiring waterborne illness in spite of claims of purity on the packaging. Microbes in the water grew to especially high numbers when stored at room temperatures. Design an experiment to evaluate the relative microbiological safety of various bottled waters and tap water, taking into account not only microbial concentrations but also the types of organisms detected.

LEAD-FREE BUT STILL UNSAFE

Sometimes you just can't win. In April 1993, an attempt to reduce lead in the municipal water supply inadvertently resulted in an outbreak of severe diarrhea in more than 350,000 residents of Milwaukee, Wisconsin, almost one-fourth of the city's population. Laboratory tests failed to identify suspected bacterial and viral pathogens. When investigators looked for protozoa, however, they found *Cryptosporidium,* a microorganism with a chlorine-resistant cyst stage.

Diane Edwards[*] describes the circumstances believed to have precipitated the outbreak. Increased water runoff due to thawing snow and spring rains washed manure from surrounding farms into the city's water supply. *Cryptosporidium* is commonly found in the intestinal tract of cattle and other farm animals. At about this time, the chemical routinely used to coagulate particulates in water prior to filtration (which removes the coagulated material) was changed because of concerns that it leached lead from water pipes. Unfortunately, the replacement chemical allowed many of the particles to remain suspended and enter the water supply during March.

Although officials treated the water to reduce any offending color and odor, and chlorination procedures were unchanged, they failed to consider the possible threat of chlorine-resistant pathogens. This mistake became apparent as Milwaukee's citizens began reporting to local physicians.

- What other pathogens can be transmitted in inadequately purified water? Why is it unlikely that bacterial or viral pathogens would have been responsible for this outbreak?
- Describe three things you could have done, had you been living in Milwaukee during this outbreak, to ensure that you would not become infected.
- What recommendations would you make to cities about their water quality programs if you were a member of the Environmental Protection Agency?

[*]D. D. Edwards, Troubled waters in Milwaukee, *ASM News,* 59: 342–345 (1993).

CHAPTER 23

Diseases Acquired Through the Genitourinary Tracts

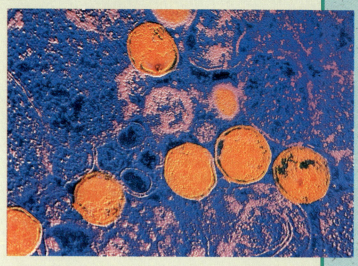

Micrograph of *Chlamydia trachomatis* (in orange), one of the most prevalent sexually transmitted infectious agents in the United States.

Reality collided head-on with Lois and Samuel Feugett's dreams of having children and raising a family. After years of fruitless attempts at pregnancy, Lois and Samuel[1] subjected themselves to the unnerving trial of fertility testing, hoping that they were just "trying too hard." When Samuel's sperm count revealed a healthy 400 million actively swimming normal sperm, Lois reluctantly accepted that some flaw in her reproductive system accounted for their lack of children. They began a long, expensive road of alternatives, beginning with surgical attempts at correcting Lois's condition and ending, after three unsuccessful attempts at in vitro fertilization, with choosing a surrogate mother for their child, another woman who agreed to be impregnated by artificial insemination with Samuel's sperm, then carry the fetus for the Feugetts. After almost 5 years of anxious efforts, they had a daughter, Julia.

Lois and Samuel Feugett's difficult episode actually began 6 years before they even met one another. Lois was 19 years old and quite capable of becoming pregnant. Silently proliferating in her reproductive tract, however, a bacterium called *Chlamydia trachomatis* was beginning to extract its permanent toll. Starting in her vagina, the bacterium invaded the adjacent uterus and, from there, infected the fallopian

---

[1]This fictitious husband and wife share a dilemma suffered by thousands of real couples.

tubes (also called the oviducts), the site where fertilization occurs. The fallopian tubes are also the vessels through which an egg travels to the uterus, where, if fertilized by sperm, it can implant and grow into a bouncing baby. But this bacterium would destroy any possibility of that happening in Lois. Weeks after the uterus and fallopian tubes became infected, Lois began to feel some abdominal pain, which she attributed to menstrual cramps. The pain became intolerable, however, and Lois sought help from her family doctor. He correctly diagnosed her as having pelvic inflammatory disease (PID) and, after finding no microbiological evidence of gonorrhea, prescribed tetracyclines. The antibiotics stopped the bacteria in their tracks, and the pain disappeared. Lois thought nothing of the incident again—until years later when fertility specialists told her that the scars left by the prolonged infection had solidly blocked both her fallopian tubes. Even though every month her competent ovaries released a perfectly good egg, sperm could not reach it, and the egg could not be fertilized. She was sterile.

In the United States, *Chlamydia* infection is the leading cause of preventable sterility in women. Blocked fallopian tubes (in women) or scarred vas deferens (in men) can follow several diseases of the reproductive tract, most of them sexually transmitted, that cause tubal inflammation and eventual scarring (Fig. 23-1). Sterility is one complication of **sexually transmitted diseases (STDs).** Another consequence is death.

Diseases of the reproductive tract are prevalent in many parts of the world, including the United States. Ironically, most developed nations, including Sweden, which is famous for its red light district and its sexual openness, have had decreasing rates of STDs since the 1950s (Sweden's gonorrhea rate declined by 95 percent between 1970 and 1989). Their STD control programs are working. Why does the United States fare so poorly in controlling its devastating STD epidemic? The answer to this question is complicated by new social dynamics (such as the practice of exchanging sex for drugs). But the central answer to this question is simple—it is because the programs have not been implemented and supported by the middle class in the United States.

As you read through this chapter, try to make the connections between the medical aspects of genital and urinary diseases and the complex microbiological, epidemiological, and sociological factors that compound the problem

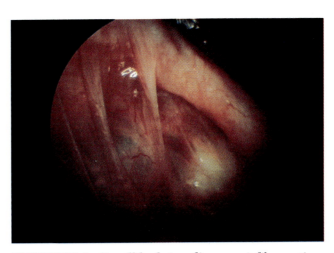

FIGURE 23-1 Roadblock to a lineage. A fiber-optics camera reveals scar tissue blocking the interior of this woman's fallopian tube, which forms a barrier between incoming sperm and the woman's egg. Because an earlier infection left this woman with both fallopian tubes blocked, her chances of normal conception are slim to none.

of bringing these diseases under control in every segment of the global community.

In addition to diseases of the reproductive tract, urinary tract infections are also recurrent problems, bringing discomfort to many people and occasionally threatening the lives of others. Many pathogens of the urinary tract are sufficiently virulent to establish infection in spite of the natural mechanical, chemical, and cellular defenses of the human genitourinary tract, although most require some predisposing factor. Fortunately, most urinary pathogens are susceptible to antimicrobial drugs and, if treated early, can be eliminated without permanent damage to the host. The best strategy for protection from STDs and urinary tract diseases, however, is to understand their transmission well enough to adopt behaviors that prevent infection.

Anatomy and Defenses of the Reproductive Tract

Differences and similarities in the anatomy of the reproductive tracts of males and females are illustrated in Figure 23-2. Each system produces gametes (haploid sex cells) for sexual reproduction. The male gametes, called *sperm*, travel from their site of production in the *testes* through accessory sexual ducts to the *urethra*, which is also part of the urinary tract. The sperm are suspended in a liquid

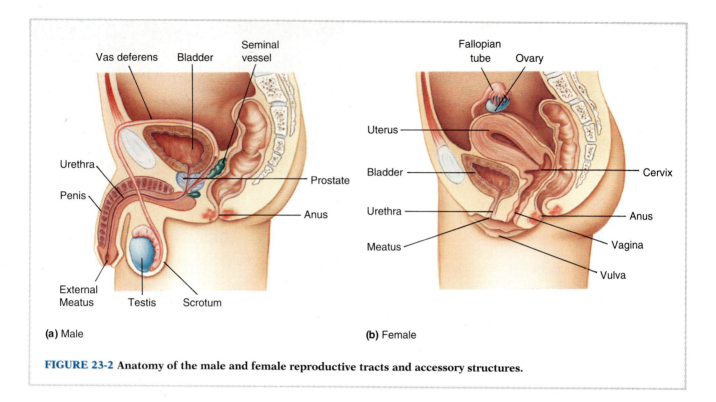

FIGURE 23-2 Anatomy of the male and female reproductive tracts and accessory structures.

produced primarily by the prostate gland and seminal vesicles. This mixture is called *semen*. Hundreds of millions of sperm are normally present in each ejaculate from a fertile man. This system is protected by bactericidal secretions of the seminal vesicles and prostate, by the length of the male urethra, and by the flushing action of urine and semen.

The female gametes, called *ova* or *eggs*, are discharged from the *ovaries* and swept into the *fallopian tubes (oviducts)*, where they may be fertilized by sperm. The egg enters the *uterus*, where, if it is fertilized, it can become implanted and develop into an embryo (which is called a fetus after its first 2 months in utero). Unfertilized eggs continue through the lower portion of the uterus, the *cervix*, into the *vagina* and finally out the external genitals, the *vulva*.

Because of their anatomy, females are more vulnerable than males to many genital infections. The moist, warm environment of the vagina provides a hospitable location for many microorganisms. The vaginal opening, unprotected by a sphincter, leaves the vagina vulnerable to the introduction of potentially pathogenic organisms. Pathogens are often introduced by the insertion of an infected penis. In addition, the close proximity of the anus encourages the accidental transfer of rectal flora to the vulva and into the vagina.

In spite of its vulnerability, the human vagina usually remains healthy. Surface cell sloughing, phago-cytic inflammatory cells, and secretory IgA antibody play important roles in resistance to colonization by pathogens and recovery from infections. The lysozyme content of cervical secretions contributes to the overall bactericidal activity in the vagina and uterus. The healthy state of the vagina can be largely attributed to the protective role of the normal vaginal flora that oppose colonization and proliferation of pathogens.

The defenses of the reproductive tracts are summarized in Table 23-1.

■ Normal Flora

The surface of the penis is colonized by microbes characteristic of the skin. With the exception of the distal portion of the urethra, which is lightly colonized, the internal structures of the healthy male genital tract are free of microorganisms. Similarly, there are no microbes residing in the uterus, fallopian tubes, or ovaries of the healthy female reproductive tract. Many microbes, however, normally inhabit the vagina, where they contribute to host defenses against pathogens. Lactobacilli are predominant bacteria in the vagina of a woman of childbearing age. These important mutualistic bacteria convert glycogen to lactic acid, maintaining the vaginal pH at approximately 4.5, lower than most pathogens can tolerate. In this way, the lactobacilli not only create conditions ideal for their own sur-

TABLE 23-1 DEFENSES OF THE REPRODUCTIVE TRACTS

Type	Responsible Factor	Mechanism of Protection
Mechanical	Flow of urine in males	Flushes away microbes.
	Length of male urethra	Puts distance between internal sexual structures and external source of microbes.
Chemical	Prostatic secretions	Contain bacteriostatic substances.
	Cervical secretions	Contain lysozyme that helps protect vagina and uterus.
Microbial	Normal vaginal flora	Lactobacilli maintain low pH inhospitable to most microbes.
Cellular	Leukocytes	Phagocytose pathogens.
Immunologic	Secretory IgA	Neutralizes pathogens on mucosal surface.

vival and minimize competition from microbes that fare poorly in acid conditions, they also protect the vagina from infection by acid-intolerant pathogens. For example, *Candida albicans* and *Trichomonas vaginalis,* both potential vaginal pathogens, reside in very small numbers in the healthy vagina but produce no symptoms because their growth is held in check by the normal vaginal flora. When an antibiotic or some other factor disrupts the *Lactobacillus,* however, the acidic pH is no longer maintained and these pathogens may flourish. Overgrowth of these organisms leads to vaginal inflammation, or **vaginitis.**

During childbearing years the flora of women is influenced by the action of estrogen and progesterone, hormones that begin to appear shortly before puberty and subside at menopause. These hormones stimulate the vagina to secrete glycogen in concentrations necessary to sustain the growth of the predominant lactobacilli in the vagina. Before puberty and after menopause the vaginal flora is less homogeneous, consisting primarily of coliforms, diphtheroids, and streptococci. The microbes in-

digenous to the normal reproductive tracts are described in Table 23-2.

■ Factors That Lower Reproductive Tract Defenses

A few opportunistic pathogens colonize and injure reproductive systems only when defenses are compromised. Disruption of the normal vaginal flora by broad-spectrum antibiotic therapy or underlying disease such as diabetes mellitus often leads to vaginitis. Vaginitis is also more common during menopause, pregnancy, or oral contraceptive use because of the estrogen and progesterone imbalances associated with these conditions.

Poor personal hygiene also predisposes to infection. In uncircumcised males, for example, bacteria may grow to inflammatory concentrations under the foreskin. Autoinoculation (self-inoculation) due to unsanitary practices poses a frequent infection problem, especially in women. Females may transfer fecal microorganisms from the anus into the vaginal tract, where they can cause vaginitis. Auto-

TABLE 23-2 MICROORGANISMS INDIGENOUS TO THE REPRODUCTIVE TRACTS

Region	Organisms
Urethra, male* (outer third only)	*Streptococcus, Mycobacterium smegmatis, Bacteroides* spp., *Neisseria,* gram-negative enteric bacilli
Vagina (during childbearing years)	*Lactobacillus* (predominant), minor concentrations of other gram-positive bacilli, diphtheroids, gram-negative bacilli, gram-positive cocci, *Candida albicans, Trichomonas vaginalis*
Uterus, fallopian tubes, ovaries, vas deferens, testes	None
Surface of penis	Representative of skin flora

*Also present in the female urethra, which is not part of the reproductive tract.

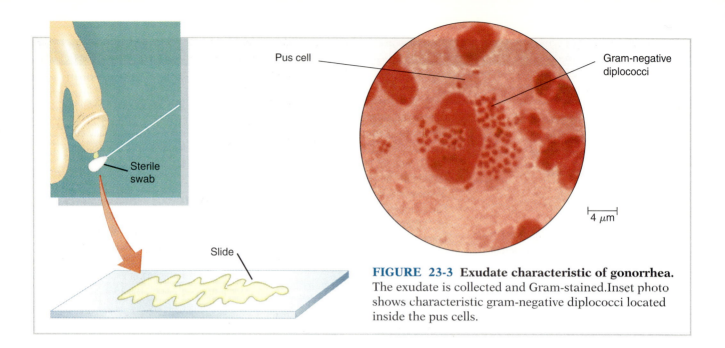

FIGURE 23-3 Exudate characteristic of gonorrhea. The exudate is collected and Gram-stained. Inset photo shows characteristic gram-negative diplococci located inside the pus cells.

treated, the pathogen can produce scarring in the urethra (predisposing for recurrent urinary tract infections) or infect the vas deferens. Scarring in these accessory sex ducts can result in permanent sterility, as does the testicular destruction that occasionally accompanies untreated gonorrhea in men. Gonorrhea may also extend to the prostate.

In women, gonorrhea begins as a mild cervical infection that remains undetected in up to 80 percent of infected women. These patients often report to a physician only when notified by an infected sex partner. Complications are more common in women because they are more likely to be asymptomatic and therefore remain undiagnosed and untreated. In 10 to 20 percent of untreated cases, the pathogen spreads from the cervix to the uterus and finally the fallopian tubes, where painful inflammation, called **pelvic inflammatory disease (PID),** may develop (gonorrhea is one of the major causes of PID). In many cases, such conditions cause sterility by scarring and obstructing the fallopian tubes. Partial scarring can leave a small opening that allows passage of sperm but not the much larger fertilized egg. The embryo cannot reach the uterus and develops in the fallopian tube instead. Tubal or ectopic pregnancies are a dangerous complication of untreated gonorrhea in women.

In both sexes, *N. gonorrhoeae* may invade the bloodstream and cause *disseminated gonococcal infection.* The most common targets for circulating gonococci are the joints (gonorrhea is the most common cause of arthritis in a single joint), the meninges, and the heart. The organism can also be transmitted by fingers from infected genitals to the eyes, where untreated infection can cause corneal ulceration and blindness.

Gonorrhea can be transmitted from an infected mother to her infant during delivery through the infected birth canal. Infected newborns develop a serious infection of the eyes, *ophthalmia neonatorum,* which can lead to blindness. Treating the infected pregnant woman is the best method for preventing infection of the newborn. Because many pregnant women receive no prenatal care, the CDC recommends prophylactic treatment for all newborns. Gonococcal ophthalmia neonatorum can be prevented by instilling drops of a 1 percent silver nitrate solution into the eyes of all newborns. Alternative treatments include using topical erythromycin or tetracyclines. For infants born to mothers with overt gonorrhea, an injection of penicillin prevents neonatal infections at other sites.

● **DIAGNOSIS.** Clinical specimens for diagnoses include the purulent (pus-containing) discharge of males collected on a sterile swab from the opening or outer portion of the urethra, and swabbed material from the cervix of women. The pharynx, rectum, and eyes are sampled whenever evidence suggests that they may be infected. Direct microscopic examination of Gram-stained specimens often reveals gram-negative diplococci inside pus cells (polymorphonuclear leukocytes (Fig. 23-3). The intracellular location aids in distinguishing the gonococcus from nonpathogenic *N. sicca* and *N. subflava,* resident bacteria that are also gram-negative cocci. Identification ultimately depends on culturing the gonococcus on Thayer-Martin agar or chocolate agar and incubating (35°C) in 10 percent CO_2 for 24 to 48 hours. Because the organism dies at low temperatures or if dehydrated, the specimen should be inoculated onto the medium and placed in an incubator immediately after it is collected from the patient. *Neisseria* colonies darken when flooded with oxidase test reagent; in other words, they are oxidase-positive. *N. gonorrhoeae* is distinguished from other oxidase-positive gram-negative diplococci by differential biochemical tests.

Several rapid tests for diagnosing gonorrhea are currently available. These tests rely on the reaction of the gonococcus in clinical specimens with specific antibodies to the bacterium.

● **THERAPY.** Until recently, penicillin was the antibiotic of choice for treating gonorrhea, although the current effective dosage is about 10 times greater than that of 20 years ago. Some strains of *N. gonorrhoeae* are completely resistant to penicillin. Many of these resistant strains were originally isolated from infected military personnel returning to the United States from Vietnam or from sex partners of these persons. In fact, penicillin resistance in *N. gonorrhoeae* is believed to have originated in Vietnam as the result of the widespread availability of penicillin and its use as a prophylactic antibiotic by American soldiers in southeast Asia during the Vietnam war. The incidence of penicillin-resistant gonorrhea has since spread throughout the United States. Antibiotic resistance may be transmitted to the gonococcus by R factors during conjugation with other bacteria, including normal flora *Escherichia coli.*

The current recommended drug of choice for gonorrhea is ceftriaxone, a third-generation cephalosporin. A tetracycline is also used in most cases, since as many as 50 percent of all gonorrhea patients are coinfected with chlamydia. Chlamydia are not affected by cephalosporin; therefore, a tetracycline is administered to ensure the elimination of both organisms. Since syphilis and gonorrhea are often transmitted together, persons diagnosed as having gonorrhea should be serologically tested for syphilis. Penicillin cures syphilis.

Currently there are no vaccines against *N. gonorrhoeae.* Immunity can be developed against the pili, but the pathogen avoids detection and immune attack through genetic variability, switching to genes coding for different pilin subunits. The microbe also produces an enzyme, IgA protease, that destroys secretory antibodies.

■ Chlamydial Infections

Each year an estimated 3 to 5 million Americans become infected with the obligate intracellular bacterium *Chlamydia trachomatis,* making it one of the most common sexually transmitted organisms in the United States. In symptomatic males, *C. trachomatis* causes urethritis similar to the early stages of gonorrhea. Because of its different etiology, however, the disease is termed **nongonococcal urethritis (NGU).** Symptomatic females usually develop urethritis or cervicitis. Most women, however, have asymptomatic infections of the cervix, and men contract the infection primarily during sexual activity with an asymptomatic partner. NGU is a slightly milder disease than gonorrhea, with less purulent discharge. Although several other organisms can also cause NGU, *C. trachomatis* accounts for at least half of all cases.

Like the gonococcus, chlamydia can ascend through the reproductive tract. As a consequence, chlamydia are among the primary causes of sterility attributable to infection in the United States. NGU in males, for example, may progress to the testes, the inflammation causing scarring of the vas deferens and sterility. Cervical infections may extend to the uterus, oviducts, ovaries, and abdominal cavity, causing PID and increasing the risk of problem pregnancies. Unfortunately, asymptomatic infections in both males and females often go untreated, contributing to the spread of the disease and to the incidence of sterility.

Because living cells are required to grow chlamydia, isolation of the organism is difficult and expensive. ▶ (See Animal Parasites, Chap. 14, p. 343.) Diagnosis is most readily achieved by direct detection of the bacteria in a clinical specimen using fluorescent-labeled antibodies (Fig. 23-4). Tetracycline or erythromycin is used to treat infected individuals and any partners.

Chlamydia trachomatis can be transmitted to newborns as they pass through an infected birth

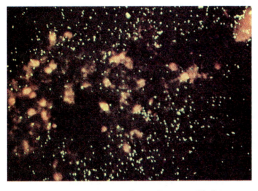

FIGURE 23-4 Detection of an intracellular parasite. The green fluorescent cells in this endocervical specimen stained with fluorescent antibody are *Chlamydia trachomatis.*

canal. Exposure to maternal vaginal infections causes conjunctivitis of newborns. Unlike ophthalmia neonatorum caused by *N. gonorrhoeae*, chlamydial eye infections fail to respond to silver nitrate or topical antibiotics. Because of this, the CDC suggests testing for chlamydia in all cases of conjuctivitis acquired through 30 days of age. Treatment is oral erythromycin for a period of 10 to 14 days.

By far the most common causes of NGU are certain serotypes of *C. trachomatis*. Many other cases of NGU, however, are caused by *Ureaplasma urealyticum* (formerly called T-strain mycoplasma) or, less commonly, by the yeast *Candida albicans* or the protozoan *Trichomonas vaginalis,* often associated with vaginitis in the infected man's sex partner. Because none of these agents of NGU respond to penicillin, Gram-stained smears of exudates are examined to rule out gonorrhea. In most cases of NGU (except those caused by yeast and protozoa), erythromycin or tetracyclines are effective in curing the disorder. Both sex partners should be treated simultaneously to prevent reinfection of the male and to cure asymptomatic (and occasionally symptomatic) infections in the female.

■ Syphilis

Before the antibiotic era in medicine, syphilis (historically referred to as "the great pox" or simply "the pox") was one of the most dreaded diseases. Late stages of syphilis often left its victims mentally degenerated, neurologically incompetent, consumed by destructive lesions, and eventually dead. Although these outcomes are far less common today because of accurate diagnosis and effective therapeutic regimens, people still die of syphilis, and babies continue to be born with tragic congenital anomalies because of exposure during gestation in an infected mother.

Syphilis is caused by an especially fragile spirochete, *Treponema pallidum* (Fig. 23-5). The organism dies quickly outside a living host. Thus direct person-to-person contact is necessary for transmission. Ninety-five percent of all syphilis cases are transmitted by sexual activity. The disease can also be acquired congenitally following transplacental passage to a developing fetus. Nonsexual contact with infectious lesions in which the pathogen enters through breaks in the skin is another nonsexual route, occasionally occurring between medical personnel and patients or between adults and children. Syphilis occurs in four well-defined stages: primary, secondary, and tertiary, plus a latent period that occurs after the secondary stage. The *primary stage* is characterized by the development of a sin-

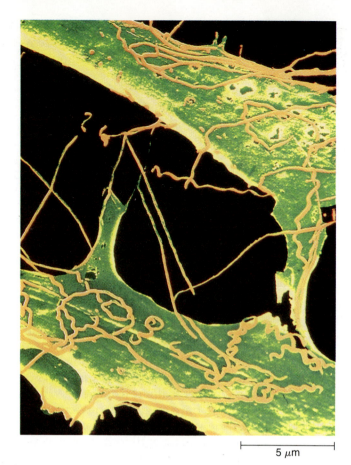

FIGURE 23-5 *Treponema pallidum.* Scanning electron micrograph of the bacterium that causes syphilis.

gle lesion, called a **chancre,** at the site where the pathogen entered the body (Fig. 23-6*a*). Although usually on the genitals, chancres may also be found on the anus, mouth, hand, or breast. The lesion is hard, rubbery, and completely painless and generally appears 7 to 21 days following infection. By this time the spirochetes have spread to the regional lymphatics and invaded the bloodstream. The chancre itself is extremely infectious. Examination of its exudate by darkfield microscopy often reveals swarms of motile, tightly coiled spiral bacteria. Since the chancre is painless, it often escapes detection when located in the vagina or on the cervix. Women may therefore transmit the disease with no knowledge that they are infected. Undetected chancres may also develop in the anal canal, a problem most common among male homosexuals. After several days the chancre spontaneously disappears, often fostering the mistaken belief that the body has "cured itself."

Untreated persons with syphilis enter the *secondary stage* within 3 to 4 months after initial infection. The most common manifestation is a skin rash on the trunk, extremities, and genitals (Fig. 23-6*b*). The rash has a characteristic predilection for

Radical Regimens Against Syphilis

Before it could be cured by chemotherapy, syphilis, also known as the great pox, was one of the world's most feared and tragic diseases. Because no effective treatment existed, one-third of all syphilis-infected adults suffered neurological deterioration and paralysis, horrible skin disfigurements, mental degeneration, insanity, circulatory disturbances, and eventually death. Pregnant females with syphilis commonly gave birth to fatally infected babies. Many remedies were practiced, most without positive effect. Two radical approaches, however, were marginally successful because of the acute heat sensitivity of the pathogenic bacterium.

It had long been observed that some people recovered from syphilis following episodes of high fever. Such observations provided the inspiration for the "hot box." Patients were placed in a heated chamber until their body temperature was elevated well above normal. After 48 hours of such treatment, some patients actually recovered from their infections.* Others died from the treatment.

Another "therapeutic" approach was to intentionally infect a syphilitic person with malaria, a disease that causes periodic episodes of very high fever. Once the induced fever killed the pathogen of syphilis, the malaria was treated with quinine. Although this mode of treatment successfully cured some people of syphilis, many others had both syphilis and malaria for life.†

Although the identities of many of the explorers who created these therapeutic innovations have been lost to time, syphilis attracted the attention of one of the first scientific explorers to successfully create a selectively toxic chemotherapeutic agent against infectious disease. Paul Ehrlich, a German scientist, believed that the ability of certain dyes to selectively stain some cells but not others could be useful in finding ways to *kill* some cells (infectious pathogens) but not others (infected host cells). In the 1890s, he pioneered studies in selective toxicity, invented the term "chemother-apy," and established the first institute for developing antimicrobial drugs, searching for what he called a "magic bullet." Ehrlich's proposed magic bullet was any chemical that, when administered to an infected patient, would kill the pathogen without harming the patient's cells. After a systematic examination of hundreds of chemicals, which he numbered sequentially, compound 606 bore fruit—it was selectively toxic against *T. pallidum*. Salvarsan (the name given to compound 606), an arsenic-containing compound, became the standard treatment for syphilis and remained so for 40 years, saving thousands of lives. It was replaced as the preferred treatment by an even more effective drug, the antibiotic penicillin.

*Heat therapy has been reintroduced as a potential AIDS treatment promoted by Dr. Heimlich (the inventor of the life-saving Heimlich maneuver for dislodging food from an obstructed windpipe).

†Wagner-Jaunegg, the pioneer of this radical therapy, received the 1927 Nobel prize in medicine.

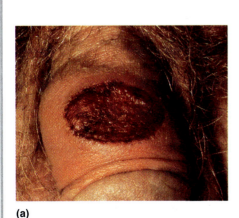

(a)

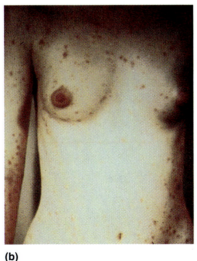

(b)

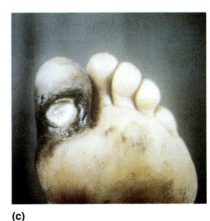

(c)

FIGURE 23-6 **Characteristic manifestations of syphilis.** (*a*) Chancre of primary stage; (*b*) rash of secondary stage; (*c*) gummas of late (tertiary) syphilis.

the palms of the hands and soles of the feet. These disseminated lesions are painless and itchless. Additional lesions, which are highly infectious, may appear on the mucous membranes of the genitals and mouth. Secondary syphilis is therefore more communicable than the primary stage of disease with its single infectious chancre.

Secondary syphilis is often accompanied by generalized symptoms that include low-grade fever, arthritis, aches in the shin, and, less frequently, damage to the nervous system, liver, and spleen. Secondary syphilis spontaneously subsides within a few weeks to several months after onset. In untreated persons, one of three developments follows the secondary stage. About one-third apparently recover from the disease, are no longer infectious, and show no further evidence of syphilis. Another third of untreated persons enter the *latent stage,* an asymptomatic phase that is recognized by the persistence of positive blood tests (see Diagnosis). Relapses of secondary syphilis may occur during latency. *Tertiary syphilis* develops in the other third of untreated cases after the latent period. This may happen 3 years after initial infection or as long as 40 years later. Tertiary syphilis is also known as the destructive phase because of its devastating lesions called **gummas** (Fig. 23-6c). These lesions are hypersensitive immune responses to the presence of residual spirochetes. They develop in virtually any tissue or organ system and may resemble the symptoms of many other diseases. For this reason, late syphilis is known as the "great imitator." If the gummas occur in the nervous system, neurosyphilis develops, causing insanity, deafness, or blindness. Partial paralysis affects the coordinated movements of walking, resulting in a gait sometimes referred to as the "syphilis shuffle." As with cardiovascular syphilis (gummas in the heart or blood vessels), neurosyphilis is potentially fatal.

Although 30,000 to 40,000 cases are reported each year in the United States, not all cases are reported, and the actual number may be as high as 500,000. Today only a small percentage of these cases progress to the tertiary stage. Most cases (and their contacts) are identified and treated in the primary or secondary stages.

Like most STDs, recovery from syphilis does not increase one's resistance to subsequent reinfection ▶ (see THE MICROBES: Why Not Vaccinate Against All Diseases?, Chap. 18, p. 461). Creating a disease-resistant population by artificial immunization has met with little success; all trial vaccines have failed to provide adequate protection. The lack of solid immunity may be due to the pathogen's ability to suppress lymphocyte-mediated cellular immunity to treponemal antigens during the primary stage. Al-though the resultant immune paralysis does not affect humoral antibody production, even high titers of *Treponema pallidum*–specific antibodies are nonprotective against subsequent attack.

● **DIAGNOSIS.** The presence of a chancre, a rash characteristic of secondary syphilis, or a history of sexual contact with an infected person should always be followed by laboratory tests to confirm the diagnosis. A positive test either (1) demonstrates serological reactions to the presence of the pathogen or (2) detects the motile spirochetes in the lesions.

Serological tests detect one of two types of antibodies in the serum of persons with syphilis. *Reagin* is an antibody mixture found in elevated concentrations in persons with syphilis; it is usually absent in uninfected people. Although reagin is not specific for *T. pallidum,* its detection provides the basis for presumptive screening tests. The most common of these tests are the **VDRL (veneral disease research laboratory)** test and the **RPR (rapid plasma reagin)** test. Both procedures rely on precipitation between serum reagin and *cardiolipin,* a soluble antigen extracted from normal beef hearts. This cross-reaction is used to rapidly screen large human populations for syphilis. Sometimes reagin tests produce biologically false positives, especially among persons with malaria, infectious mononucleosis, tuberculosis, hepatitis, leprosy, or any condition that elevates serum reagin levels. Positive reagin results are therefore always followed by tests for specific treponemal antibody.

MYTH: A single blood test can be used to diagnose virtually any sexually transmitted disease.

FACT: The blood test for syphilis provides no information about any disease except syphilis. Each disease requires a separate serological test using antigens that specifically correspond to that pathogen. For some STDs, such as gonorrhea in men, blood tests are diagnostically useless.

The specific treponemal antibody tests are more reliable than determinations of reagin activity because they detect antibodies produced only by people with syphilis. The patient's serum is mixed with laboratory cultures of *T. pallidum.* Positive reactions are usually detected by indirect fluorescence microscopy or by darkfield microscopy. The *fluorescent treponema antibody (FTA-ABS)* test is called an indirect immunologic test because the reaction between antigen and antibody cannot be directly de-

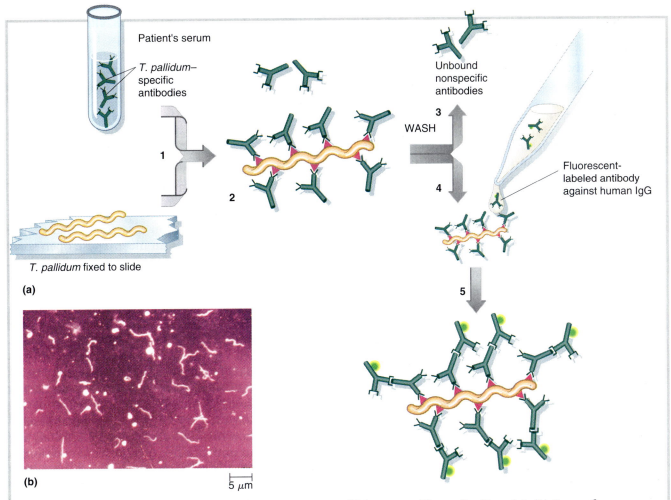

(a)

Patient's serum

T. pallidum–specific antibodies

T. pallidum fixed to slide

Unbound nonspecific antibodies

WASH

Fluorescent-labeled antibody against human IgG

(b)

5 μm

FIGURE 23-7 Positive FTA-ABS test for *Treponema pallidum*–specific antibodies. (*a*) (**1**) Serum from a patient suspected of having syphilis is added to preparation of *T. pallidum* fixed to a microscope slide. (**2**) *T. pallidum*– specific antibodies attach to spirochetes, thereby coating their surface with IgG. Nonspecific antibodies remain unbound. (**3**) The slide is washed to remove unbound nonspecific antibodies. The only human IgG remaining is that attached to the treponemes. (**4**) Fluorescent-labeled antibody specific for human IgG is added to the slide. (**5**) Labeled antibody combines with the treponeme-specific human IgG antibody on the test bacteria, which become coated with fluorescent dye. The spirochetes glow when viewed with the fluorescent microscope. Sera from patients who do not have syphilis contain no *T. pallidum*–specific antibody, so fluorescent-labeled anti-IgG will not coat the test bacteria. Thus in negative tests, the spirochetes do not fluoresce. (*b*) Photomicrograph of positive FTA-ABS test.

tected. The antigen–antibody complex is indirectly detected by adding fluorescent-labeled antibody specific for human IgG (Fig. 23-7).

Living *T. pallidum* grown in rabbit testes are used for detecting serum antibody by the *Treponema pallidum immobilization (TPI) test*. The motility of these living treponemes is observed under darkfield microscopy. Specific antibodies in the sera of patients with syphilis, when combined with living treponemes, immediately immobilize the test bacteria. Sera from healthy persons fail to inhibit motility because they contain no treponema-specific antibodies.

Treponema pallidum cannot be grown readily on inanimate media. Cultivation in rabbit testes is laborious and expensive, making the TPI and FTA-ABS tests uneconomical for screening large populations for syphilis. Thus, specific treponemal tests are usually reserved for use on persons who have positive serum reagin tests or who have clinical signs of syphilis.

● **THERAPY.** Syphilis is uniformly responsive to penicillin, although neurosyphilis requires especially high doses over long periods of time. Antibiotic treatment of infected women early in pregnancy can prevent its spread to the fetus. Antibiotics, however, do not reverse tissue damage that occurs dur-

disease is unpredictable. In many persons it spontaneously subsides in 6 months; in others the lesions persist for years. The cell-mediated immune (CMI) response promotes eventual recovery, but if CMI is depressed, the warts may progress to giant lesions that may destroy genital tissue, especially in males. The lesions also enlarge during pregnancy and may interfere with birth, sometimes necessitating delivery by cesarean section. Secondary bacterial infections complicate some cases of genital warts.

The characteristic appearance of the lesions is often enough to suggest a diagnosis of genital warts, although in some cases the virus may be present without extensive visible symptoms. Occasionally the lesions may be mistaken for lesions of secondary syphilis or, in the tropics, for granuloma inguinale (see p. 625). They are differentiated by lack of positive tests for syphilis and the absence of Donovan bodies, intracellular inclusions typical of granuloma inguinale. As with most viral diseases, no antimicrobial treatment is very satisfactory. Most cases respond to topical applications of podophyllin, an agent that inhibits mitosis and cell proliferation. Because of podophyllin's toxicity and oncogenic potential, many physicians prefer to destroy the lesions by treating them with acid or by electrically burning them, freezing them with liquid nitrogen, or using laser surgery. Podophyllin treatment should be avoided during pregnancy and is not recommended for cervical warts.

An alarming consequence of genital papillomavirus infection in women is an increased likelihood of developing cancer of the cervix, a disease that kills 4500 women annually. Studies show that certain serotypes of papillomavirus, particularly HPV-16 and HPV-18, are present in tumor tissue.

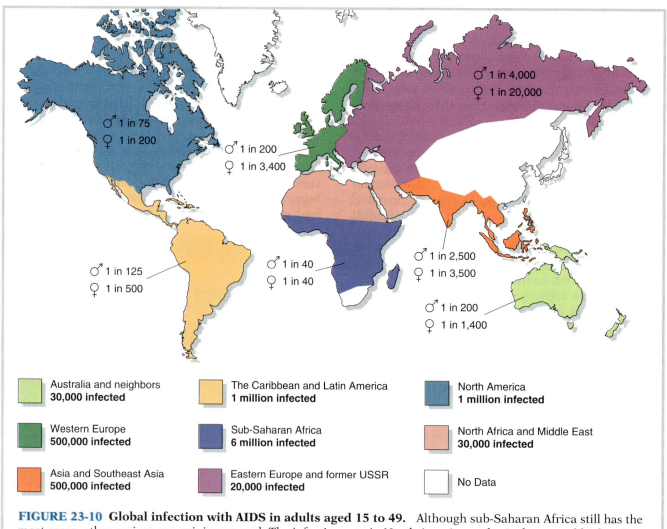

FIGURE 23-10 **Global infection with AIDS in adults aged 15 to 49.** Although sub-Saharan Africa still has the most cases, other regions are gaining ground. The infection rate in North Amerian males is the second highest in the world (*Science* 252: 372).

Because of this epidemiological link it is important that sexually active females have routine examinations for papillomavirus infection and be treated as early as possible to eliminate the virus. It is also sound practice for women who have suffered genital papillomavirus infection to be examined every 6 months for evidence of cervical carcinoma. When diagnosed and treated in its early stages, cervical carcinoma is readily cured. There is also evidence that in males warts may progress to cancer of the penis.

■ Acquired Immunodeficiency Syndrome (AIDS)

This discussion begins with a dangerously prevalent myth.

MYTH: AIDS is primarily a disease of homosexuals and intravenous drug users. If you don't fall into one of these categories, you are not in a group that is "at risk" of acquiring AIDS.

FACT: Belief in this myth can kill you. The overwhelming majority of AIDS cases are sexually transmitted. Furthermore, the World Health Organization projects that by the end of this decade (by 1999) *90 percent of all HIV infections in the world will be transmitted heterosexually.*

Acquired immunodeficiency syndrome (AIDS) is one of the most serious international public health problems, one that generates more than its share of myths and misinformation. As of February 1993, more than 10 million people worldwide—heterosexuals, homosexuals, males, females, adults, and children—were infected with HIV (Fig. 23-10). It is predicted that by the year 2000, pediatric infections alone will reach that magnitude. Transmission of HIV among adults is primarily by sexual contact; the virus enters open wounds or lesions—usually tears in mucous membranes caused by sexual activity. Children who acquire the disease usually do so while gestating in an infected mother when the virus crosses the placenta. Other nonsexual routes of transmission include transfusion with contaminated blood (risk = 1 per 40,000 units of blood) or sharing needles with an infected person, an all-too-common practice among drug abusers. There is no evidence that the disease can be transmitted by casual contact (Fig. 23-11).

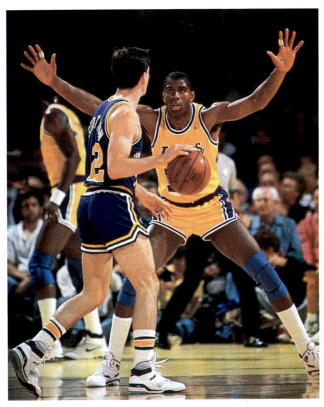

FIGURE 23-11 A myth retires a legend. When Earvin "Magic" Johnson tested positive for HIV, the social impact transcended sports, affecting the whole world. Because physical contact between athletes on a basketball court poses no risk of transmitting the virus, Johnson returned to playing basketball. Reluctantly he permanently retired from the sport in 1992 because some of his competitors feared infection from physical contact with Johnson and stated they would not play as aggressively against him.

MYTH: Men can transmit AIDS to women, but there has never been a documented case of men acquiring AIDS from women.

FACT: Although men are less likely to acquire the disease through heterosexual transmission than are women, about 1 in every 10 male AIDS victims contracted his disease through heterosexual sex with an infected woman.

Two serotypes of human immunodeficiency virus, designated HIV-1 and HIV-2, have been identified, and variants of each type exist. HIV infects cells of the immune system, including macrophages, helper T cells, and certain B lymphocytes. It also can infect glial cells in the brain (Fig. 23-12). Inside infected

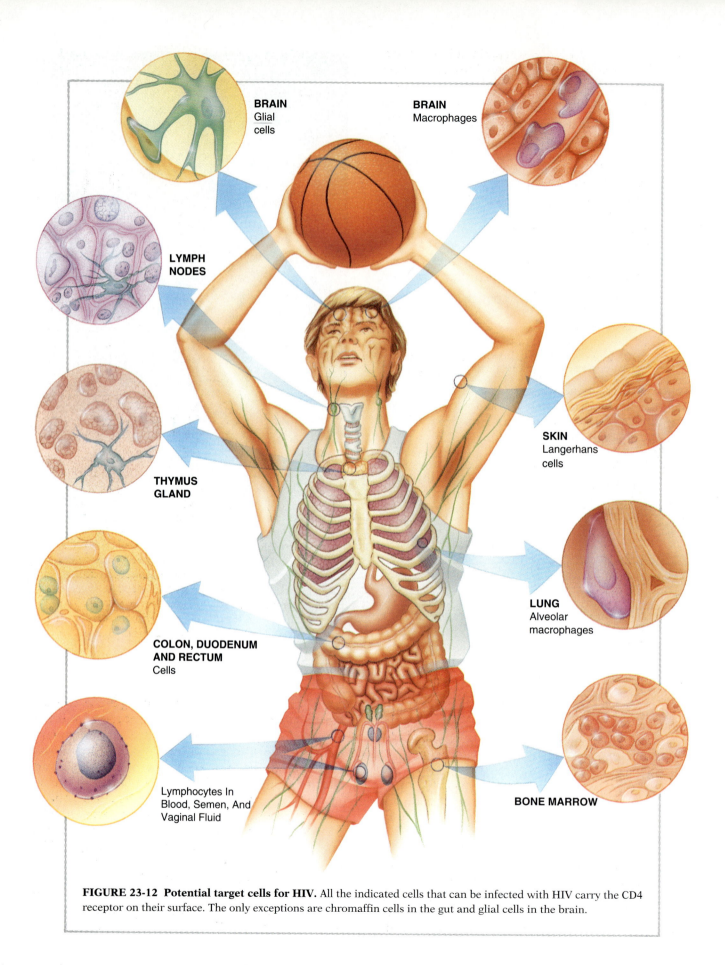

FIGURE 23-12 Potential target cells for HIV. All the indicated cells that can be infected with HIV carry the CD4 receptor on their surface. The only exceptions are chromaffin cells in the gut and glial cells in the brain.

BRAIN
Glial cells

BRAIN
Macrophages

LYMPH NODES

SKIN
Langerhans cells

THYMUS GLAND

LUNG
Alveolar macrophages

COLON, DUODENUM AND RECTUM
Cells

Lymphocytes In Blood, Semen, And Vaginal Fluid

BONE MARROW

cells, the retroviral RNA is transcribed into a molecule of DNA that is then integrated into the host cell chromosome. After a brief period of viral replication, which induces a host antibody response, the virus establishes a slowly progressing infection, with replication occurring primarily in the cells of the lymph nodes. Individuals in this stage of the infection can still transmit the virus in its proviral form. The factors that control the rate of viral replication are unknown; however, some people remain in a disease-free state for more than 10 years. Eventually viral replication spreads from the lymph nodes and is accompanied by a decline in the number of helper T cells (Fig. 23-13). This leads to a serious deficiency in the cell-mediated immune system that leaves the person vulnerable to cancer and pathogens that are normally eliminated by CMI. When a person with HIV develops one of the characteristic opportunistic infections or when the helper T-cell population falls to 200 cells per cubic millimeter, the person is diagnosed with the disease AIDS (concentrations of T cells in healthy persons exceed 800 cells/mm³). *Pneumocystis carinii* pneumonia ▶ (see Acquired Immunodeficiency Syndrome, Chap. 19, p. 476) is by far the most common opportunistic infection, although tuberculosis, candidiasis, toxoplasmosis, and cytomegalovirus (CMV) infections are seen with high frequency. Many individuals also develop AIDS-related dementia, resulting from an encephalitis that appears as the virus replicates in brain cells. About one-third of males contract a formerly rare cancer called Kaposi's sarcoma. These opportunistic diseases are severe and are usually the cause of death in persons with AIDS.

● **DIAGNOSIS AND TREATMENT.** Infection with HIV can usually be diagnosed by an antibody test. There is a period of time, usually 3 months but sometimes more than a year, between infection and the development of an antibody response that is detectable by current diagnostic assays (Fig. 23-14). More expensive tests detect viral proteins or integrated DNA copies of the viral genome in individuals who test negative for antibody. *HIV-positive* individuals are infected with the virus and can transmit it to others. They do not have clinically defined AIDS until the helper T-cell population falls below 200 cells per cubic millimeter of blood or they are diagnosed as having Kaposi's sarcoma or one of the opportunistic infections characteristic of AIDS.

Because there is currently no vaccine and no cure, the threat of AIDS can be controlled only by reducing the transmission of the virus. Nonsexual transmission has been effectively slowed in developed countries by screening blood supplies for the virus. It is not clear whether poorer countries around the

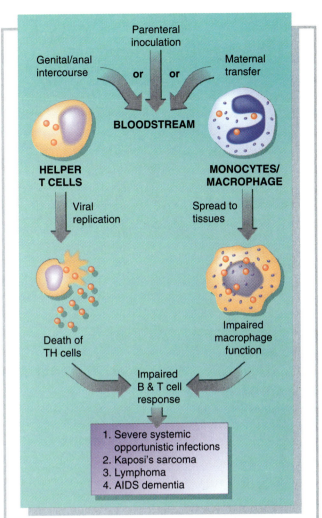

FIGURE 23-13 Pathogenesis of HIV infection. Monocytes (which become macrophages) and helper T cells are targets of HIV infection. Infected macrophages spread the virus throughout the body, including the brain. Viral replication prevents these cells from participating in a normal immune response.

world will be able to afford such vigilance. Reducing the sharing of contaminated needles by intravenous drug abusers will also protect many individuals from infection. Sexual transmission of the virus can be reduced by the use of latex condoms with a spermicide containing nonoxynol-9, which kills HIV as well as sperm. Such safer sex practices, however, are not absolute safeguards because HIV may pass through defects in condoms. The only safe sex is between uninfected partners. Although AIDS is a notifiable disease, results of HIV testing are not reported to public health authorities. Persons who learn that they are HIV-positive, however, should notify all sexual contacts in order to combat the further spread of infection.

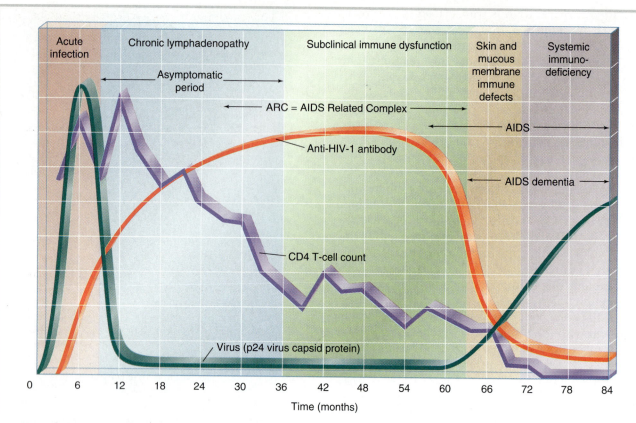

FIGURE 23-14 Stages in disease progression. Antibody can often be detected early in the course of infection as a response to replicating virus. As the asymptomatic period ends, helper T-cell concentration falls, the immune response disappears, and viral concentration increases.

Two drugs, AZT and ddI, are approved by the FDA for use in HIV-positive individuals to slow the progression to AIDS ▶ (see Antiviral Agents, Chap. 16, p. 391). These drugs interfere with reverse transcription and presumably prevent infection of more cells by viral progeny. While successful at delaying clinical progression to AIDS, both of these drugs have side effects, and neither will alter the course of the disease once symptoms develop. Several other compounds discussed in Chapter 16 can block viral replication in vitro and are being tested in clinical studies. These include (1) drugs that inhibit an essential processing enzyme, the viral protease; (2) antisense nucleic acids targeted to critical regions of the virus genome; and (3) genetically engineered soluble receptor molecules that inhibit attachment to cells.

Many research teams are working on a vaccine for prophylaxis or immunotherapy against the virus. These experimental vaccines contain inactivated virus or protein from the envelope or capsid. HIV antigenicity changes through mutations, and there is concern that a vaccine directed to a variable region on the virus would provide only temporary pro-

tection. No vaccine is currently available. Many AIDS researchers believe that conquering this epidemic will require more than vaccines and effective chemotherapy. It will depend on determining the original source that introduced the pathogens into the human population (see BEYOND THE BASICS: The Origin of the AIDS Virus).

■ Candidiasis

One of the most common genital pathogens is the fungus *Candida albicans*, which causes the form of vaginitis commonly referred to as a yeast infection. The infection produces a thick white "cottage cheese" discharge that causes itching and irritation ranging from mild to severe. The usual reservoirs of infection are the colon and the vagina itself, where the opportunistic pathogen resides as a minor component of the normal flora, held in check by the resident lactobacilli. Antibacterial antibiotics, pregnancy, oral contraceptives, or diabetes mellitus can disrupt this balance and predispose for the development of an extremely persistent and painful vaginitis. Recurrent *Candida* vaginal infections may

The Origin of the AIDS Virus

Less than 20 years ago, one of the modern menaces facing human culture was still unknown. It may seem remarkable that a new pathogen could suddenly appear "out of nowhere" and cause such devastation. Where did the human immunodeficiency virus come from? The answer to that question represents another of the mysteries surrounding AIDS. It is also a question that must be answered before we can eradicate this disease from our planet. The epidemic will not be controlled until all reservoirs are eliminated, including any still-viable reservoir that harbored the virus before it entered the human population. Some scientists fear that these reservoirs may reintroduce the virus if it is eliminated from the human population.

Scientific evidence suggests that HIV evolved from similar viruses that infect monkeys and apes, although no virus identical to either of the HIV serotypes has been isolated from either of these primates (see figure). Genetic analysis, however, has shown that HIV-2, the predominant form of the virus in the African population, shares at least 80 percent genetic homology with a virus found in the sooty mangabey, a West African monkey. There are several ways in which the virus may have been transmitted from the monkey to the human population in that area of the world. For example, monkeys are often caught for food and are butchered with few sanitary precautions. The virus could have easily been introduced into the cook who cut himself while preparing the monkey's carcass. Some monkeys are kept as pets or trapped for export. Scratches or bites may have introduced the virus into pet owners.

The reservoir of HIV-1 is less clear. So far the most closely related simian virus has been detected only in chimpanzees, but the two viruses are so dissimilar that it is questionable whether chimps are the recent source of human HIV-1. Perhaps an accidental inoculation occurred years ago and the human and animal viruses evolved along different paths. The original virus, for example, may not have been well adapted for human transmission or disease. There is evidence that immunodeficiency viruses will vary in pathogenicity depending on the species infected. Sooty mangabey virus, for example, does not cause disease in the monkey but does cause disease if transferred to Asian macaques. Some have suggested that simian blood was transferred to humans as part of malaria experiments earlier in the century.[*] Others point to polio vaccines prepared in monkey kidney cell culture (there is no evidence to support this claim). It is also possible that HIV-1 entered both the chimpanzee and human populations from another, as yet unidentified, source.

If the human immunodeficiency viruses are derived from simian viruses, can AIDS be eradicated or controlled even with the development of a vaccine and effective chemotherapy? Some researchers are convinced that until we understand the dynamics of primate-to-human viral transmission, we are seriously handicapped in our ability to stop the introduction of new human pathogens, which may be continuing at a significant rate.

[*]Other hypotheses include sex with monkeys and consumption of uncooked monkey brains.

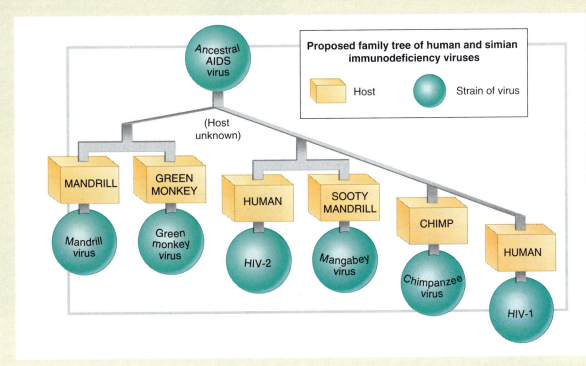

Proposed family tree of human and simian immunodeficiency viruses

Host   Strain of virus

Ancestral AIDS virus

(Host unknown)

MANDRILL   GREEN MONKEY   HUMAN   SOOTY MANDRILL   CHIMP   HUMAN

Mandrill virus   Green monkey virus   HIV-2   Mangabey virus   Chimpanzee virus   HIV-1

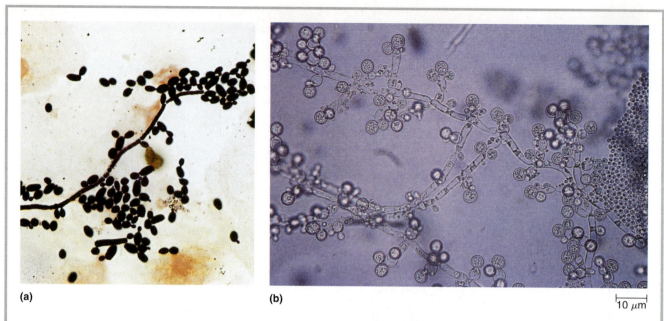

(a)   **(b)**

10 μm

FIGURE 23-15 *Candida albicans.* (*a*) Vaginal exudate showing pseudohyphae and budding yeast cells. (*b*) Typical appearance when cultured on chlamydospore agar, with large chlamydospores, blastospores, and chains of pseudohyphae.

also be associated with occasional transmission by oral-genital or anal-genital sexual activity. Although men rarely develop symptomatic *Candida* genital disease, they may be a source of infection (and re-infection) for women and therefore should be treated as well. Cases of candidiasis that occur in the absence of identifiable predisposing factors indicate that other, poorly defined underlying conditions may also contribute to this disease.

Candidiasis is diagnosed by finding the characteristic fungus in vaginal exudates. Direct microscopic examination reveals pseudohyphae and yeast cells easily distinguished from resident bacteria (Fig. 23-15*a*). The fungus is cultured on chlamydospore agar, a nutritionally deficient medium that stimulates the production of blastospores (budding yeast cells), chains of pseudohyphae, and chlamydospores, the three microscopic hallmarks of *C. albicans* (Fig. 23-15*b*).

Since *Candida* is a fungus, it fails to respond to antibacterial chemotherapy. The most effective agents against *Candida* vaginitis are the azole drugs (butoconazole, miconazole, or clotrimazole) applied topically as a cream or as a vaginal suppository. Several of these drugs can be purchased without prescription. Such over-the-counter preparations should not be used by women unless they have been diagnosed with candidiasis or are experiencing a recurrence of disease.

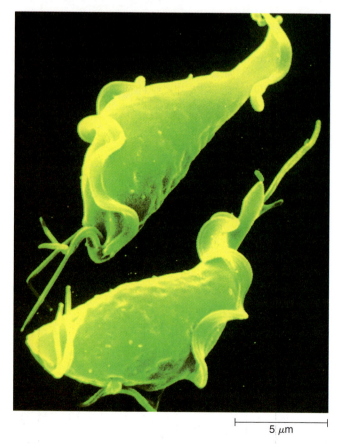

5 μm

FIGURE 23-16 *Trichomonas vaginalis* **(false colored SEM), a frequent cause of vaginitis.**

■ Trichomoniasis

Vaginal infections by the flagellated protozoan *Trichomonas vaginalis* are extremely prevalent, with an estimated 2 to 3 million infections per year in the United States alone. In fact, trichomoniasis may be the most common infection in the world—some 300 million people, males and females, are infected by the protozoan. Fortunately, the disease is not life-threatening nor does it usually have serious sequelae. Trichomoniasis is primarily characterized by vaginal irritation and an unpleasant-smelling yellowish purulent discharge. Although infected males rarely experience symptoms, occasionally they develop urethritis (NGU) or prostatitis. *T. vaginalis* may cause disease when sexually transferred from men to women. The organisms disrupt the local pH balance and elicit a local inflammatory response resulting in vaginitis. Cystitis (urinary bladder infection) may occur in either sex. *T. vaginalis* also poses a danger in pregnant women because it can cause preterm delivery.

Trichomoniasis is diagnosed by microscopic detection of the flagellated protozoa in purulent genital discharges (Fig. 23-16). Metronidazole (Flagyl) is the only effective chemotherapeutic agent. It should be simultaneously administered orally to both sex partners, even though the male appears un-infected; otherwise asymptomatic males reinfect cured females.

■ Bacterial Vaginosis

One form of vaginitis, bacterial vaginosis, is believed to be caused by normal flora bacteria that, for unidentified reasons, proliferate beyond their normal limits. The organisms that replace the normal *Lactobacillus* spp. include obligate anaerobic bacteria (*Bacterioides* and *Mobiluncus*) and *Gardnerella vaginalis*. It is believed that the disease is sexually transmitted since it is rare in women who have never been sexually active. Often, male partners of women with the disease are colonized in their urethras. The thin, milky vaginal discharge of women with bacterial vaginosis usually contains diagnostic "clue cells," epithelial cells with a granular appearance due to the attached *G. vaginalis* (Fig. 23-17). In addition, a fishy odor is produced when the vaginal discharge is mixed with 10 percent potassium hydroxide. Bacterial vaginosis responds quickly to metronidazole therapy.

■ Toxic Shock Syndrome

In 1980, a new and potentially fatal disease exploded into public awareness. The disease, called *toxic*

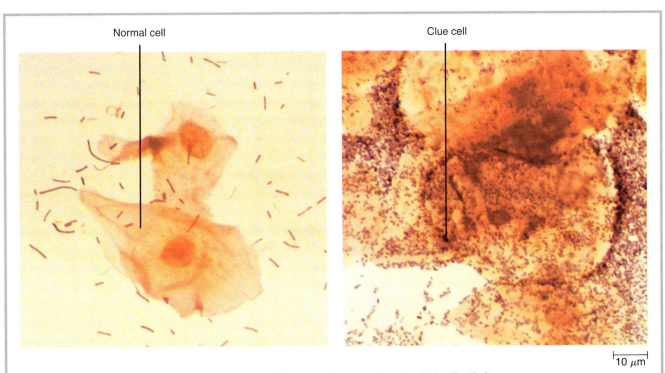

FIGURE 23-17 Clues to the diagnosis. Unlike smooth normal epithelial cells (left), the clue cell, characteristic of *Gardnerella vaginalis* infection, is coated with bacteria, giving it a granular appearance that is most apparent at the edges of the cell (right).

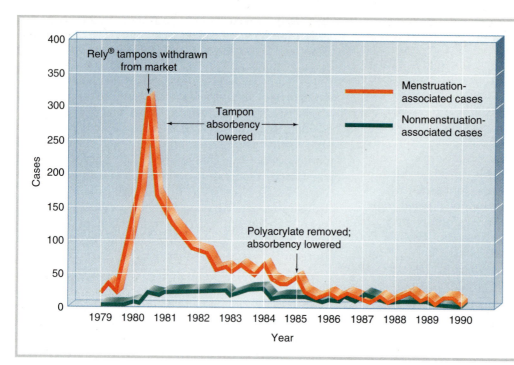

FIGURE 23-18 **A product too effective for its own survival.** The introduction of superabsorbent tampons, especially the brand Rely, was associated with the emergence of the TSS epidemic of 1979. The role of these tampons in the disease is supported by the immediate decline in TSS after the product was voluntarily removed from the market in 1980. Other tampon factors, such as presence of polyacrylate, are also believed to contribute to TSS.

shock syndrome (TSS), had surged in incidence from a few sporadic cases each year to nearly 1000 cases occurring in 1979 and 1980. Ninety-nine percent of these cases were in women, and in 98 percent of these women the symptoms appeared during a menstrual period. Moreover, most of these women were using tampons at the time, generally the superabsorbent products. From 1978 to 1980 the increased sales of the implicated tampons paralleled the increased incidence of TSS. The most popular of the implicated tampons was removed from the market in 1980, leading to an immediate dramatic decline in the incidence of TSS (Fig. 23-18). Toxic shock syndrome killed about 8 percent of the people affected during this period. About half the cases of toxic shock that currently occur result from nongenital *S. aureus* infections ▶ (see Staphylococcal Skin Infections, Chap. 24, p. 642).

The disease strikes with a sudden onset of fever, vomiting, and diarrhea. Blood pressure plunges, and the patient goes into shock. A sunburn-like rash desquamates and peels about 10 days after onset of symptoms. There is usually a purulent discharge from the vagina. The disease is caused by lysogenic strains of *S. aureus* that produce a protein called toxic shock syndrome toxin (TSST-1). These bacteria may reside in the vagina without causing symptoms, but the use of superabsorbent tampons encourages bacterial growth and the elaboration of the exotoxin. In sufficient concentration, TSST-1 causes the symptoms of toxic shock, possibly by increasing

a person's susceptibility to otherwise inconsequential levels of endotoxin released by transient gram-negative bacteria in the circulation. In addition, TSST-1 alters immunity and phagocytic functions, increasing the susceptibility of patients to gram-negative bacterial infections. The extended time that superabsorbent tampons remain in the vagina may allow *S. aureus* to multiply to the higher concentrations necessary for producing enough toxin to elicit symptoms. Women should change tampons at least three or four times daily and wash their hands before inserting a fresh tampon.

Diagnosis of TSS depends on the clinical picture (fever, hypotension, rash, and desquamation) and culturing toxigenic *S. aureus* from the vagina, cervix, urine, feces, or blood of affected patients. Supportive therapy and treatment with penicillinase-resistant antibiotics usually results in complete recovery.

■ Infrequently Encountered Sexually Transmitted Diseases

Three additional diseases transmitted by sexual activity are usually found in residents of tropical areas and occur infrequently in the United States, Canada, and Europe. *Chancroid* is a highly infectious venereal disease caused by a small gram-negative bacillus, *Haemophilus ducreyi*. It is often confused with syphilis because its characteristic lesions, called soft chancres or chancroids (*-oid* = resembling), are similar in appearance to a chancre, the lesion in primary syphilis. Unlike a syphilis chancre, however, these lesions are painful and bleed easily. They appear on the genitals and surrounding areas 3 to 5 days following exposure. The lesions may extend to extragenital sites by autoinoculation with pus from the original ulcers. The organism can be isolated from lesion exudates for diagnosis. Treatment with tetracyclines and sulfonamides has been effective in the past, but antibiotic resistance has made these treatments unreliable. The CDC currently recommends treatment with azithromycin, ceftriaxone, or erythromycin.

Chancroid, which is the major cause of genital sores in Africa, has been rare in the United States since World War II, at least until the mid-1980s. In the 3 years following 1984, the incidence of the disease skyrocketed to 5 times its previous numbers (Fig. 23-19). The increase worries epidemiologists because evidence suggests that chancroid may facilitate the transmission of AIDS by providing open sores through which the virus can enter the bloodstream. In people who are HIV-positive, chancroid is very difficult to eradicate because of a combination of antibiotic resistance of *H. ducreyi* and the immunodeficiency of the patient. The two diseases help the spread of one another, creating an upward spiral that may move chancroid out of the "infrequently encountered disease" category.

Lymphogranuloma venereum is caused by highly invasive serotypes of *Chlamydia trachomatis* that differ from the serotypes that cause NGU. The sexually transmitted disease begins with the formation of a painless genital lesion that showers the bloodstream with pathogens 1 to 2 weeks later. The pathogen can invade the meninges, eyes, and joints and may persist in the lymphatics, where the vessels may become blocked, causing subsequent elephantiasis (swelling due to accumulation of lymphatic fluid). The chlamydia are microscopically visible in inclusions within the cytoplasm of infected cells. Diagnosis can be confirmed by growing the isolated organism in cell culture or embryonated chicken eggs. Treatment requires weeks, sometimes months, of tetracycline therapy, often supplemented by surgical intervention to reduce elephantiasis.

Granuloma inguinale is caused by the gram-negative bacillus *Calymmatobacterium granulomatis*. The disease is characterized by peripherally radiating lesions surrounding the single initial nodule that occurs at the site of entry. Extensive tissue destruction may occur. Although granuloma inguinale is primarily a sexually transmitted disease, it can also be transmitted by sandflies, respiratory secretions, urine, and feces. It is diagnosed by microscopically observing the encapsulated etiologic agent in stained scrapings from lesions. The bacteria often appear as rodlike forms, called *Donovan bodies*, in the cytoplasm of host cells.

■ Diseases Occasionally Transmitted by Sexual Contact

Some diseases that primarily have nonsexual routes of transmission may also be sexually communicable. Viral hepatitis B, for example, is usually transmitted by parenteral routes. However, in the past 10 years over one-third of new cases of hepatitis B have resulted from sexual transmission. Other diseases that fall into the "occasional STD" category are enteric infections with *Shigella*, *Giardia lamblia*, *Enta-*

FIGURE 23-19 Skyrocketing disease. Although chancroid declined to very low levels in the United States, it soared to alarming numbers in just 3 years in the late 1980s. (Adapted from "Sexually Transmitted Diseases in the AIDS Era" by S. A. Aral and K. K. Holmes. Copyright 1991 by Scientific American, Inc. All rights reserved.)

cardiovascular and nervous systems. Immunization of women (before pregnancy) using an attenuated rubella virus vaccine has greatly reduced the incidence of congenital rubella.

Although rare, women with herpes simplex type 1 or type 2 infections may transmit the virus across the placenta to a developing fetus. In these cases disseminated disease usually causes spontaneous abortions or stillbirths due to liver or central nervous system damage.

Approximately 30 percent of children born to HIV-infected mothers are infected in utero. The World Health Organization (WHO) predicts that 3 million children will die of AIDS in the next decade. The factors that determine whether the virus is transmitted to a fetus are not known, but it may be related to the absence of a specific protective antibody. The course of the infection is usually more rapid in infants than in adults, many developing AIDS within the first year. Diagnosis is difficult because detectable antibodies may be of maternal origin. Preventing such maternal–fetal transmission is currently a research priority. One strategy that investigators soon hope to develop is passive immunization of pregnant women with neutralizing monoclonal antibody against HIV.

Women who are infected by the protozoan *Toxoplasma gondii* during pregnancy may transmit the pathogen to their fetuses, resulting in possible mental retardation or injury to several organs later in life ▶ (see Toxoplasmosis, Chap. 22, p. 595). *Toxoplasma* cysts in the tissue are resistant to chemotherapy, so the infection is virtually incurable. Active cases can be arrested, however, by combination pyrimethamine-sulfonamide therapy, which kills the pathogen in the trophozoite stage.

Listeria monocytogenes is an infrequently reported pathogen that causes sporadic outbreaks, usually associated with ingestion of contaminated foods ▶ (see Listeriosis, Chap. 22, p. 593). Unborn infants are highly susceptible to *Listeria* infection, which can lead to abortion or premature delivery. Many infants born alive eventually die of their infections.

The most important infections of the developing fetus are presented in Table 23-5. In addition to these, the agents of mumps, chickenpox, measles, poliomyelitis, influenza, viral hepatitis, and malaria occasionally cause intrauterine infections.

CONCEPT CHECK

- Who is at risk for HIV infection, and how can that risk be reduced? Describe the events that occur in an HIV-infected person that lead to the development of AIDS.
- Compare the possible complications of untreated gonorrhea, chlamydia, syphilis, candidiasis, and papillomavirus infections. What is the appropriate treatment for these infections?
- Why are reinfections common in candidiasis and trichomoniasis? How do these repeated episodes differ from recurrences of genital herpes?
- Name four pathogens transmitted across the placenta from a pregnant woman to her fetus. How can these infections be prevented?

TABLE 23-5 SOME IMPORTANT INTRAUTERINE INFECTIONS OF THE FETUS

Disease	Pathogen	Portal of Entry into Mother	Effect on Fetus
Cytomegalovirus infection	Cytomegalovirus (CMV)	Alimentary tract, genito-urinary tract	Mental retardation; fatal liver, spleen, and neurological damage
Congenital syphilis	*Treponema pallidum*	Genitourinary tract	Abortion or postnatal death; bone deformity; liver and kidney disease; skin and mucous membrane lesions
Toxoplasmosis	*Toxoplasma gondii*	Alimentary tract	Injury to retina or central nervous system; mental retardation
Congenital rubella syndrome	Rubella virus	Respiratory tract	Mental retardation; damage to nerves and cardiovascular system
Congenital herpes infection	Herpes simplex, types 1 and 2	Alimentary tract, genito-urinary tract	Viremia (usually fatal)
AIDS	Human immuno-deficiency virus (HIV)	Genitourinary tract; parenteral	Secondary infections

Anatomy and Defenses of the Urinary Tract

Unlike the reproductive tracts, the urinary tracts of males and females are anatomically similar. The upper urinary tract, in fact, is identical for men and women (Fig. 23-20). In males, however, the urethra is the final segment of both the urinary and reproductive tracts. In women, the urinary tract is physically separate from the reproductive tract.

The upper urinary tract consists of the kidneys and ureters. The kidneys remove metabolic wastes, salts, and water from the blood, forming urine, which is transferred through the ureters into a "holding tank" called the *urinary bladder*. The flow of urine through the ureters flushes out microbes that gain entry from the lower urinary tract. This hydrokinetic washing is assisted by sphincters (circular muscles that close off a tube when contracted). Sphincters at each end of the ureters help prevent *reflux*, a backflow of urine from the bladder that could carry microorganisms into the kidney. Immunologic and cellular defenses also protect the kidneys from pathogens disseminated through the bloodstream.

The lower urinary tract consists of the bladder and the urethra, the tube through which urine flows out of the body during voiding. The external opening of the urethra, called the meatus, is located in the anterior vulva in women and at the distal end of the penis in men. The male urethra is approximately five times longer than the female urethra. The additional length affords men more protection against **cystitis** (infection of the bladder).

The urinary tract of males is also protected by fluid secreted by the prostate that contains spermi-

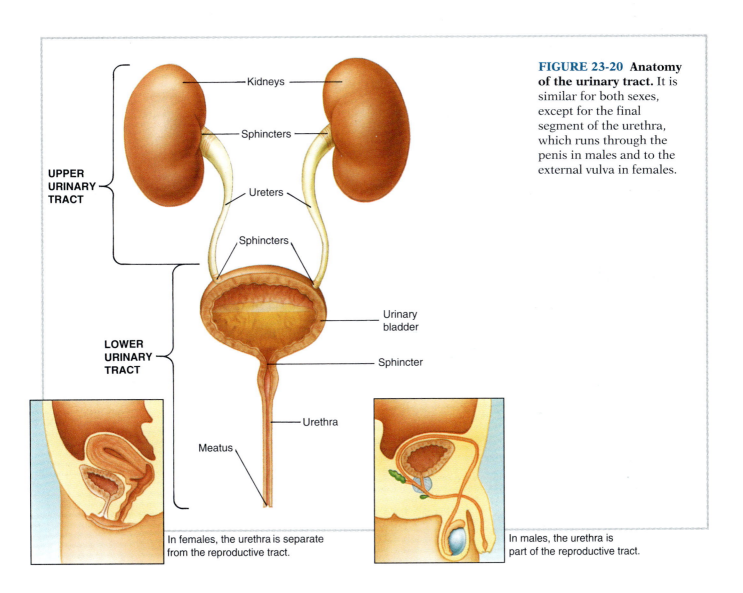

FIGURE 23-20 Anatomy of the urinary tract. It is similar for both sexes, except for the final segment of the urethra, which runs through the penis in males and to the external vulva in females.

UPPER URINARY TRACT

LOWER URINARY TRACT

Kidneys

Sphincters

Ureters

Sphincters

Urinary bladder

Sphincter

Urethra

Meatus

In females, the urethra is separate from the reproductive tract.

In males, the urethra is part of the reproductive tract.

dine, zinc, and other antibacterial chemicals. In females the acidity of vaginal fluids helps protect the region around the urethral opening from colonization by microbes that can cause urinary tract infection.

Physical and chemical properties of urine prevent the growth of some microbes that gain entry into the bladder. Its low pH and high concentrations of dissolved waste products are antibacterial. *Urea*, one of the major urinary constituents, inhibits the growth of some bacteria. However, urine is an excellent culture medium for the growth of some bacteria and supports the growth of a wider range of pathogens when its physical or chemical properties are even slightly altered—by hormonal imbalance, for example. Consequently, pathogens that do manage to invade the urinary tract may find the conditions ideal for growth, producing painful or even dangerous diseases.

The urinary bladder is so well protected that its contents are normally sterile. Microbes are flushed from the urethra by the flow of urine, and reflux is prevented by a set of sphincter muscles. The occasional microbes that reach the bladder will likely become trapped in a layer of sticky mucus that prevents their attachment to the underlying susceptible epithelial cells of the bladder. Normal sloughing of the cells that line the urinary tract sheds attached bacteria before they become entrenched.

Defenses of the urinary tract are summarized in Table 23-6.

■ Factors That Lower Urinary Tract Defenses

Any factor that obstructs the flow of urine or promotes reflux interferes with the hydrodynamic removal of pathogens (Table 23-7). Obstructions may be due to scarring or stricture (abnormal narrowing) in the urethra or ureters, the presence of urinary calculi ("stones"), or the growth of tumors. Habitual overdistention of the bladder forces a backflow of urine into the kidneys and can lead to **pyelonephritis** (inflammation of the kidney). Defective sphincters may also encourage backflow. Urine remaining in the bladder after partial voiding may become a reservoir for organisms that cause cystitis. Incomplete emptying of the bladder may be due to interrupted flow during voiding (some persons habitually stop voiding before their bladder is empty) or to structural abnormalities in the bladder that produce a pool of residual urine.

Females are more affected by these factors because of the short distance between microbes on the external surface and the bladder (approximately 4 cm versus 20 cm in the adult male). Any trauma to the urethral opening in women may facilitate the introduction of microbes into the bladder. For this reason, some women frequently contract bladder infections after sexual activity, a condition referred to as "honeymoon cystitis." Voiding urine immediately after intercourse reduces the likelihood of developing this infection. In men, the most common lower urinary tract infection is urethritis, although most cases of urethritis are acquired through sexual transmission.

Urinary catheters are associated with cystitis in both sexes ▶ (see Mechanisms of Transmission, Chap. 25, p. 684). These instruments often transfer surface microbes into the bladder in sufficient numbers to cause infectious disease. When catheterized individuals are victims of spinal cord injury or other disorders that impair nervous control of sphincters,

TABLE 23-6	DEFENSES OF THE URINARY TRACT	
Type	**Responsible Factor**	**Mechanism of Protection**
Mechanical	Location of kidneys	Isolates kidneys from external source of microbes; connected with external microbes by two well-defended pathways
	Flow of urine	Flushes away microbes from ureters, bladder, and urethra
	Sphincters	Prevent reflux (backflow)
	Length of male urethra	Puts more distance between bladder and external source of microbes, giving men greater protection from cystitis than women
	Mucous lining in bladder	Prevents microbial attachment to bladder epithelium
	Sloughing of epithelium	Sheds attached microbes
Chemical	Low pH of urine	Opposes growth of many microbes
	Dissolved wastes in urine	Osmotic pressure opposes growth of some microbes
	Urea in urine	Is bacteriostatic for some bacteria
	Prostatic secretions	Contain bacteriostatic substances
Cellular	Leukocytes	Phagocytose pathogens
Immunologic	Secretory IgA	Neutralizes pathogens at some urinary surfaces

TABLE 23-7 SOME FACTORS PREDISPOSING FOR THE DEVELOPMENT OF GENITOURINARY TRACT DISEASE

Factor	Site of Disease	Preventive Measure
Overdistention of bladder	Kidney	Prompt voiding of bladder when full
Tumor of lower urinary tract	Bladder, urethra	Excision of tumor
Incomplete emptying of bladder	Bladder	Proper voiding habits, surgical correction of "sump" in bladder
Diminished neurological control of sphincters	Bladder, kidneys	Antimicrobial chemoprophylaxis
Trauma to female urethra	Bladder	Urination after intercourse
Catheterization	Bladder	Proper aseptic precautions
Recurrent lower urinary tract infection	Kidneys	Prompt diagnosis and treatment of cystitis to prevent ascent of infection to kidneys
Obstruction of ureter	Kidneys	Surgical correction, dissolution of calculi (stones)
Obstruction of urethra	Bladder	Surgical correction, dissolution of calculi (stones)
Bacterial prostatitis	Urethra	Early diagnosis and treatment of prostatitis
Hormonal imbalance	Vagina, bladder	Correction of imbalance if possible
Diabetes	Vagina	Insulin therapy
Pregnancy	Vagina, bladder	Hygiene, frequent voiding
Disruption of normal vaginal flora	Vagina	Avoidance of indiscriminate antibiotic therapy

infections of the urinary tract are almost inevitable. Recurrent or prolonged infectious disease of the lower urinary tract predisposes for pyelonephritis by providing a reservoir of pathogens that can ascend from the bladder to the kidneys.

CONCEPT CHECK

- Describe three characteristics of the urinary tract that oppose infection and three factors that reduce the effectiveness of these defenses.
- Why are women more prone to developing cystitis than men?

Dynamics of Transmission of Disease to the Urinary Tract

Many diseases of the urinary tract are transmitted by autoinoculation of normal flora organisms, usually from the intestinal tract. The most common manifestation of this is cystitis in women. Wiping urine from the vulva in a back-to-front motion, for example, carries organisms from the anus to the urethral opening. Cystitis is also common following the insertion of a urinary catheter or *cystoscope* (a fiber-optics tube passed through the urethra to view the bladder's interior). Careless techniques during these procedures encourage the introduction of indigenous or transient microorganisms of the external genitals into the bladder. Contaminants on inadequately "sterilized" equipment or unwashed hands are also important sources of infection. Several other modes of transmission, such as spread of pathogens to the kidneys from the bladder or bloodstream, were discussed in the previous section.

CONCEPT CHECK

- Describe two (nonsexual) routes of transmission by which pathogens gain entry to the urethra.

Urinary Tract Diseases

Infectious diseases of the urinary tract are categorized as either ascending or descending. If a pathogen is introduced through the urethra and extends to the bladder (and perhaps to the kidneys), the infection is **ascending.** Occasionally, pathogens spread through the bloodstream from another infected body site to the kidneys and from there extend downward to the bladder. Such **descending infections** are much less common than ascending disease but are more dangerous to the patient because of the accompanying systemic involvement. They must be differentiated to determine the prognosis and appropriate treatment.

The most common urinary tract pathogen is *Escherichia coli,* which is usually mechanically introduced into the urethra or bladder from its indigenous site in the intestines. *Proteus, Pseudomonas, Enterobacter,* and *Enterococcus* are other intestinal bacteria that frequently cause the same condition. If these organisms bypass the mechanical defenses of the urinary tract, they attach to the bladder mucosa, usually by specific pili, and proliferate.

The most frequent symptom of cystitis is pain and burning during urination. If the disease is not treated at this stage, the microbes may ascend through the ureters to the kidney, and pyelonephritis or other severe renal diseases may develop. Early diagnosis and treatment of lower urinary tract infections is essential to reduce the risk of developing these potentially fatal complications of the upper urinary tract.

Urinary tract infections are generally characterized by **bacteriuria,** bacteria in the urine. Quantitative microbiological assays of urine are necessary to distinguish accidental contamination of the specimen through improper collection from true infection. Contaminating organisms from the distal portions of the urethra and the external genitals may confuse the diagnosis unless bacterial counts are determined using fresh midstream urine specimens collected by a "clean-catch" technique (described in Appendix C). If fewer than 10,000 bacteria per milliliter of urine are detected, it is likely that the source of bacteria is contamination from the external genitals. Detection of more than 100,000 organisms per milliliter is usually indicative of infection (see THE MICROBES: Why Mrs. Baxter's Number Was Up). Further tests are necessary to identify the probable pathogen. This information may indicate whether the infection is ascending or descending. Intestinal bacteria are more likely to be the cause of ascending infection, especially if blood cultures are negative. Descending infections are often caused by pathogens that characteristically infect other areas of the body and spread to the kidneys through the

THE MICROBES

Why Mrs. Baxter's Number Was Up

Evelyn Baxter's number was up. She fell victim to an all-too-common mistake. Because of her pain medication, Mrs. Baxter, the patient in room 415, found it impossible to void urine. The doctor gave her one last opportunity to try to voluntarily empty her bladder. She failed. A urinary catheter was inserted and then withdrawn as soon as the bladder was empty. No further instrumentation was needed. Several days later, however, Mrs. Baxter was preparing to check out of the hospital when she began to develop symptoms of cystitis, a urinary bladder infection. The doctor ordered a urine sample for quantitative bacteriological analysis.

Mrs. Baxter's urine was collected at 9:15 A.M. by the midstream-catch tech-

nique and transported to the laboratory. Mrs. Baxter's room was on the fourth floor and the lab was on the ground floor, and additional specimens were collected from other patients en route. By 1:45 all specimens were delivered together to the laboratory. Quantitative microbiology revealed that Mrs. Baxter's urine contained more than a quarter of a million *Escherichia coli* organisms per milliliter. Because of their great numbers, *E. coli* was presumed to be the etiologic agent, and ampicillin therapy was immediately begun. Antibiograms confirmed the ampicillin sensitivity of Mrs. Baxter's *E. coli.* Nonetheless, the treatment failed. Several days later she still had her painful bladder infection.

Mrs. Baxter was misdiagnosed for two reasons. Her specimen was not promptly transported to the lab, and urine is an excellent culture medium for many microorganisms, including *E. coli.* This bacterium can grow at the rate of one doubling per hour at room temperature. Mrs. Baxter's bladder urine actu-

ally contained the pathogen *Staphylococcus saprophyticus* but no *E. coli.* During voiding, however, her urine was contaminated, as could be expected, with the many microbes on her external genitals. The fresh specimen harbored around 10,000 bacteria per milliliter, a high normal count following clean-catch collection. Had the laboratory processed the urine immediately, *E. coli* would have been dismissed as a probable cause of Mrs. Baxter's infection. During 4½ hours of transport at room temperature, however, these bacteria doubled in number almost 5 times, overgrowing the real pathogen. The concentration of *E. coli* had grown from 10,000/mL to more than 250,000/mL. If Mrs. Baxter's urine had been processed immediately or refrigerated during the delay, she would have been spared the discomfort and danger of misdiagnosis and ineffective therapy. Her coliform number would not have been up.

bloodstream. In these cases, bacteremia can be detected by blood cultures. Any exudate discharged from the urethra should be microscopically examined and cultured *before* collection of urine.

Most cases of cystitis are caused by bacteria that respond to sulfonamides, ampicillin, cephalosporins, or quinolines such as nalidixic acid. Antibiotic resistance is common among urinary tract pathogens, however, emphasizing the need for determining the isolated pathogen's antibiotic susceptibilities. Recurrent infections suggest possible un-

derlying anomalies, the correction of which can reduce the incidence of urinary tract infections.

CONCEPT CHECK

- Why is descending urinary tract disease more dangerous than ascending urinary tract disease?
- How can these diseases be diagnostically distinguished from each other?

Key Terms

sexually transmitted disease (STD) (p. 603)
vaginitis (p. 605)
urethritis (p. 607)
pelvic inflammatory disease (PID) (p. 608)

nongonococcal urethritis (NGU) (p. 609)
chancre (p. 610)
gummas (p. 612)
VDRL (venereal disease research laboratory) test (p. 612)
RPR (rapid plasma reagin) test (p. 612)

intrauterine infections (p. 626)
cystitis (p. 629)
pyelonephritis (p. 630)
ascending infection (p. 631)
descending infection (p. 632)
bacteriuria (p. 632)

Key Facts and Concepts

- **Portal of entry for STDs and urinary tract infections.** The genitourinary tract is a portal of entry for pathogens of the upper and lower urinary tract and for agents of sexually transmitted diseases. Sexually transmitted diseases are mainly contracted by exposure to the pathogen during direct sexual contact.

   Urinary tract disease often correlates with anatomical anomalies, introduction of invasive medical instruments, or other predisposing factors. Lower urinary tract infections can ascend from the bladder to the kidneys, causing fatal pyelonephritis.

- **The consequences of neglect.** Serious consequences may follow unattended cases of diseases acquired through the reproductive tract. Untreated syphilis produces slow neurological degeneration in some of its victims as well as possibly fatal cardiovascular disease. Gonorrhea and chlamydial infections can block reproductive ducts, producing permanent sterility. Congenital disease of a fetus and infections of the newborn present an especially dangerous threat when a pregnant woman has active venereal disease. Whereas adult symptoms of many STDs are localized in the genitals, the fetus or child is vulnerable to deforming or fatal systemic infection. Early VDRL tests for pregnant women, cesarean delivery when the mother has an active case of genital herpes, and instillation of antibiotics into the eyes of newborns have helped to avoid these tragedies.

- **Control of STDs.** The epidemic proportions of the major STDs emphasize the need for improved management. Control measures depend on encouraging people to submit for routine checkups, seek rapid diagnosis at the earliest indications of genitourinary tract disease, obtain prompt therapy, notify contacts, and refrain from sexual relations until the disease is cured. For diseases that are incurable, such as genital herpes and AIDS, use of condoms promotes safer (not foolproof) conditions for sexual interaction. No vaccines are available for most sexually transmitted diseases.

- **Diagnosis of STDs.** Diagnosis of most sexually transmitted diseases requires either staining and culturing the pathogen from the patient's blood or exudates (*N. gonorrhoeae*) or carrying out immunologic tests that show serological reaction to the presence of the pathogen (*T. pallidum*, HIV). Clinical or microscopic signs are sometimes characteristic of a particular disease—for example, a chancre for primary syphilis, itchless rash for secondary syphilis, Donovan's bodies for granuloma inguinale, and multinucleated giant cells or inclusion bodies for herpes.

- **Treatment of STDs.** Most STDs are treated with antibiotics, although AIDS, herpes, and other viral diseases fail to respond to antibiotics. Selection of an effective antibiotic depends on accurate identification of the pathogen. For example, penicillin is usually effective against gonorrhea but not against NGU, which shows the same clinical picture but is usually caused by *Chlamydia*. Treatment should also include additional considerations, such as association of genital warts with late development of cervical cancer. (Patients' cervical cells should be microscopically examined twice a year for early signs of cancer.)

Review and Problem-Solving Questions

1. Distinguish between:
   (a) Chancre and gumma
   (b) Reinfection and recurrence
   (c) Descending infection and ascending infection
   (d) Cystitis and pyelonephritis
2. How do each of the following predispose for infections of the genitourinary tract? (a) Diabetes; (b) antibiotic therapy; (c) urinary catheterization.
3. What is pelvic inflammatory disease? How can it cause sterility?
4. Compare the pathogens and the pathogenesis of gonorrhea, chlamydia, syphilis, and genital herpes.
5. The battle against the spread of AIDS is being fought in several ways. Discuss how each of the following can help reduce the threat of disease: (a) Vaccine; (b) drugs inhibiting reverse transcriptase; (c) mandatory reporting of cases; (d) condoms.
6. Several sexually transmitted infections are asymptomatic in females. If the infection is causing no discomfort, why is it still important to eliminate the pathogen? (Discuss two reasons.)
7. Is urethritis a reproductive tract infection or a urinary tract infection? Explain.
8. Describe the reagin test for syphilis. This test is not very specific to *T. pallidum,* so why is it used?
9. Discuss the biological reasons that explain why *E. coli* is the most common cause of cystitis but is not usually associated with vaginitis.
10. A female has a 50 percent chance of contracting gonorrhea from an infected male, but a male has a 20 percent chance of becoming infected from a female with the disease. What factors might account for this difference in risk?
11. For each of the following diseases describe the pathogen(s) and therapy: (a) Chancroid; (b) nongonococcal urethritis; (c) genital warts; (d) candidiasis; (e) toxic shock syndrome.
12. Dr. Delafield receives the following laboratory reports for her patients. From this information, what diagnosis would you suggest for each patient?

Patient	Specimen	Findings
Adams	Urethral exudate	Gram-negative diplococci
Berens	Lesion scraping	Spirochetes
Jones	Lesion scraping	Donovan bodies
Moore	Urine	*E. coli*, 50,000/mL

13. Explain why HIV is not readily transmitted by casual contact.
14. Fill in the table on the most common types of vaginitis.

Pathogen	Mode of Transmission	Symptoms	Diagnosis	Treatment
Candida albicans				
Gardnerella vaginalis				
Trichomonas vaginalis				

---

the animal reservoir. These include rabies and rat-bite fever, both acquired by being bitten by an infected animal. Tularemia and anthrax enter the human body through minor cuts and abrasions on the skin of a person handling an infected animal or in contact with animal dander. Tularemia and anthrax most commonly afflict hunters, animal breeders, and slaughterhouse workers.

Many diseases are transmitted from animal reservoirs to humans by arthropods. (Vectors also transmit some diseases from person to person.) Most arthropod-borne diseases are noncommunicable, and their continued presence depends on the existence of a population of infected arthropods and the animal reservoir if there is one. Endemic regions are therefore restricted to areas inhabited by the suitable vector and animal reservoirs. Most of these diseases cannot be exported by infected travelers to new locations because of geographical restrictions on where the vectors can survive.

Some microbes that are normally harmless saprophytes in the environment are dangerous when they contaminate objects that cause injury to the skin. Contaminated thorns, nails, hypodermic needles, or other sharp objects simultaneously injure the skin and inoculate the wound with pathogens that thrive in the damaged tissue and cause disease. In causing a wound, a sharp object may accidentally introduce into the body organisms that have been residing on the object itself or on the surface of the skin at the site of injury. Microbes remaining on the skin following improper cleansing and decontamination of puncture sites, for example, readily enter the body as the hypodermic needle is inserted. Wounds can also become contaminated after injury, especially if underlying tissue is exposed to the environment. Burn wounds are among the most susceptible to infection. Open wounds are frequently contaminated by normal flora microbes, either by autoinoculation or through the action of another person. Wounds can also be contaminated by contact with healthy carriers of potential pathogens or with persons transiently colonized by pathogens from other sources. Usually people serve as mechanical vectors of diseases acquired through open wounds—that is, they indirectly transmit microbes from another source. Those few pathogens that infect intact skin are usually spread directly from person to person.

CONCEPT CHECK

- Explain why most infections acquired through the skin are noncommunicable diseases.
- What is the role of people in transmitting pathogens to the skin?

Diseases of the Skin

Infectious disease of the skin may occur in the epidermis **(cutaneous infections)** or in the dermis or fatty tissues **(subcutaneous infections).** Most skin rashes are not considered skin infections because they are cutaneous expressions of systemic diseases acquired by another route. Measles and chickenpox, for example, are acquired through the respiratory tract.

Infections of the epidermal layer generally develop after direct contact with the reservoir. Most pathogens require at least minor breaks in the epidermis to establish an infection. The lesions may be red and flat (a *macule*), elevated (a *papule*), or filled with either clear fluid (a *vesicle*) or pus (a *pustule*). Lesions that have burst (*ulcers*) and necrosed are especially susceptible to secondary infections (Fig. 24-4).

■ Staphylococcal Skin Infections

Some skin infections are characterized by the formation of pus as a product of local inflammation. These **pyogenic** (pus-producing) pathogens are bacteria, most commonly *Staphylococcus aureus* or *Streptococcus pyogenes.* These two bacteria are particularly dangerous for neonates and surgical patients. In hospitals, *S. aureus* is the most frequent cause of skin lesions of newborns. It is difficult to prevent infection because healthy carriers of *S. aureus* are undetected sources of infection. An estimated 40 percent of all healthy persons harbor this bacterium in their nasal passages. Exposure to *S. aureus* is probably universal. About one-third of staphylococcal skin diseases are acquired by autoinoculation, making their control even more difficult. Widespread outbreaks usually do not occur among persons with normal defenses. But when these defenses fail, *S. aureus* can cause a number of cutaneous infections:

- **Folliculitis** is a mild infection of a hair follicle. Bacteria grow in an obstructed hair follicle and elicit an inflammatory response. Folliculitis in the follicle of an eyelash is referred to as a *sty*. The infected site may be walled off with a capsule of clotted fibrin, producing an abscess, called a *furuncle* or boil, that is filled with bacteria and pus. The abscess expands and usually breaks through the skin and drains. Abscesses of the skin of the neck and back tend to spread through the underlying fatty and connective tissues to adjacent follicles, causing a *carbuncle*, an extensive lesion that may develop into a serious, deep-seated infection requiring surgical drainage and antibiotic therapy.

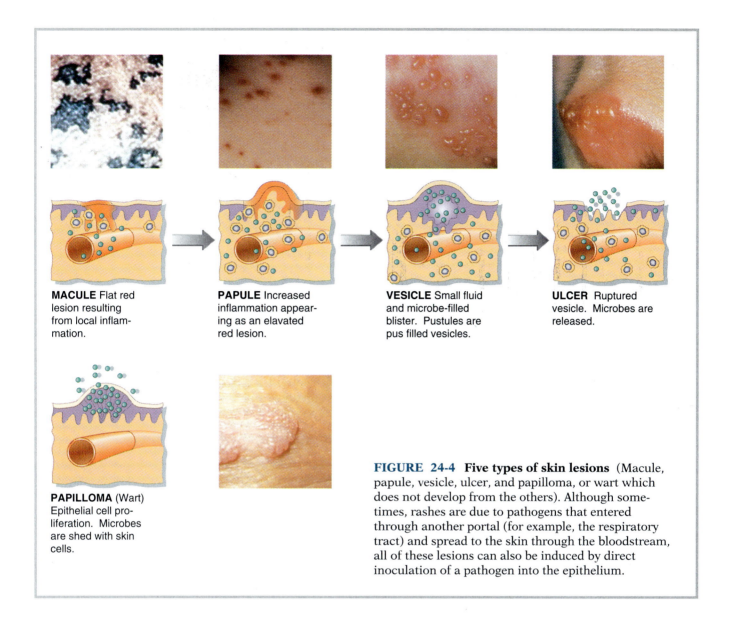

MACULE Flat red lesion resulting from local inflammation.

PAPULE Increased inflammation appearing as an elevated red lesion.

VESICLE Small fluid and microbe-filled blister. Pustules are pus filled vesicles.

ULCER Ruptured vesicle. Microbes are released.

PAPILLOMA (Wart) Epithelial cell proliferation. Microbes are shed with skin cells.

FIGURE 24-4 Five types of skin lesions (Macule, papule, vesicle, ulcer, and papilloma, or wart which does not develop from the others). Although sometimes, rashes are due to pathogens that entered through another portal (for example, the respiratory tract) and spread to the skin through the bloodstream, all of these lesions can also be induced by direct inoculation of a pathogen into the epithelium.

- **Impetigo** is a skin infection that most often occurs in young children, especially on the face and legs. The disease often results from a mixed infection of *S. aureus* and *S. pyogenes*, although either species alone can be the cause. The lesions begin as pustules that rupture and form scabs or crusts (Fig. 24-5a). *S. aureus* is an especially common pathogen in hospital-associated impetigo of the newborn. Bacteria in the discharged pus spread the disease to adjacent areas on the skin and to other persons. Impetigo is a highly communicable disease, often causing epidemics in families, schools, and nurseries.

- **Staphylococcal scalded skin syndrome (SSSS)** is a severe epidermal necrolysis that resembles a burn (Fig. 24-5b). It is caused by strains of *S. aureus* that produce *exfoliatin,* a toxin that causes the epidermis to separate from the dermis, resulting in blisters, reddening, and extensive peeling of the top layers of skin. The disease may be fatal if accompanied by bacteremia. The toxin primarily affects the skin of newborns or older children; adults are usually resistant to SSSS.

- **Cellulitis** is an acute inflammation of the skin that rapidly spreads to the subcutaneous tissues and, if untreated, can lead to bacterial invasion of the bloodstream. *S. aureus* is a major cause of cellulitis following superficial skin infections. (Anaerobic bacteria are more likely to cause cellulitis in traumatized tissues.)

Pus discharged from any of these purulent skin infections is easily obtained for microscopic examination and culture. Pus in a closed lesion is col-

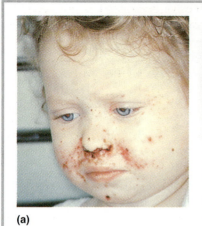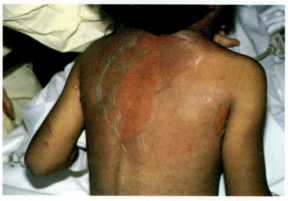

FIGURE 24-5 **Diseases caused by staphylococci.** (*a*) Impetigo. (*b*) Staphylococcal scalded skin syndrome.

lected by aspiration with a needle and syringe. Blood should also be cultured when fever accompanies cutaneous infections. During outbreaks, nasal cultures of suspected carriers help identify the source of the pathogen. Gram stains of pus reveal cocci with cluster arrangements. Laboratory identification, however, is confirmed by biochemical and serological tests that distinguish between pathogens and similar-appearing members of the commensal flora, especially *Staphylococcus epidermidis* or other relatively nonpathogenic staphylococci. The two most commonly used tests for identifying *S. aureus* determine its ability to produce coagulase (an enzyme that coagulates blood plasma) and to generate characteristic colonies on mannitol salts agar. Phage typing of *S. aureus* is often performed to provide a "fingerprint" that allows epidemiologists to determine the source of infection ▶ (see Epidemiological Typing, Chap. 20, p. 512).

Localized cutaneous infections require no systemic antibiotic therapy. Lesions are cleaned and topically treated with antiseptic or antibiotic ointments. In cases of impetigo, the crusts are removed before treatment. The once popular technique of bathing infants in hexachlorophene reduced the incidence of infections, but the routine use of hexachlorophene as an antiseptic for babies was discontinued when it was discovered that it is toxic to the central nervous system if absorbed through the infant's skin. During outbreaks, full-term infants may still be bathed in hexachlorophene, which is subsequently rinsed off completely. Because most strains of *S. aureus* produce penicillinase, antibiotic susceptibility tests are important to determine the best treatment for serious infections. Penicillin-resistant staphylococci can be eliminated by cloxacillin, methicillin, or vancomycin, but the rapid emergence of methicillin-resistant *S. aureus* has reduced the efficacy of this alternative.

Since these infections are usually acquired from carriers or transmitted by autoinoculation, proper and frequent hand washing is necessary to reduce the spread of infection.

■ Streptococcal Skin Infections

Streptococcal skin infections are caused by *S. pyogenes*, also the agent of streptococcal pharyngitis, although different serological types are associated with skin and respiratory infections. Transmission is usually through direct contact with infected individuals or carriers. The most common manifestations of streptococcal skin infections are impetigo and erysipelas. Streptococcal impetigo is most common in children living in hot humid climates. The infection is usually restricted to the skin, but in infants it may seed the bloodstream with bacteria and lead to systemic disease, kidney injury, and death.

Erysipelas is an inflammation that advances through the local lymphatic system (Fig. 24-6). It is caused by *S. pyogenes*, which probably enters the dermis through minute fissures in the skin. (Similar symptoms are caused by a gram-positive bacillus, *Erysipelothrix rhusiopathiae*, common in many wild and domestic animals and acquired by humans from infected animals or contaminated animal products.) Painful red lesions usually spread over the face and legs and may cause bacteremia in debilitated patients. Recurrences are common.

Gram stains of pus from lesions show the characteristic chains of gram-positive cocci. *S. pyogenes* is identified by production of beta hemolysis on blood agar and sensitivity to low-potency (2 µg/mL) bacitracin ▶ (see BEYOND THE BASICS: Bringing Order

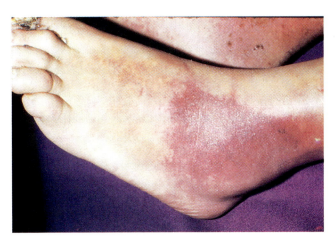

FIGURE 24-6 **Erysipelas.**

to the Chaotic Streptococci, Chap. 21, p. 536). Streptococci are rarely resistant to penicillin, which remains the drug of choice unless the streptococci are accompanied by a coinfecting penicillinase-producing strain of staphylococci.

■ Acne

Acne is a localized inflammation of hair follicles that mainly affects adolescents. The disease occurs in two stages. During the first stage, excessive sebaceous secretions accumulate in hair follicles that have been blocked by keratinized cells. These plugged follicles are called *comedones* (blackheads).

MYTH: Blackheads are pores that have become plugged with dirt.

FACT: The dark color of blackheads is not dirt, but the accumulated skin pigment of keratinized cells.

In the second stage of acne formation, excess sebum is converted to fatty acids by the enzyme lipase produced by *Propionibacterium acnes,* a normally harmless inhabitant of the follicular canal. The fatty acids irritate the skin and cause the inflammation characteristic of acne. The inflamed follicles develop into pustules, which, in severe cases, burst and release bacteria into the surrounding tissue. The lesions may become severe enough to cause permanent pits and scars. Low doses of tetracycline and erythromycin reduce the bacterial population and therefore the production of free fatty acids. This treatment is discontinued when antibiotic-resistant *P. acnes* become predominant.

Penicillin is ineffective because it is not secreted into sebaceous glands. Because of side effects to the fetus, tetracyclines are not prescribed for the control of acne in pregnant women. In 1982, *retinoic acid,* a synthetic derivative of vitamin A, was licensed for treating serious acne that failed to respond to antibiotic therapy. This compound effects a marked improvement or recovery in most persons with acne by suppressing sebum production. This drug is harmful to a fetus and should not be used by pregnant women. Other side effects (dryness, hair loss, and mild muscle pain) also limit its use.

MYTH: Eating chocolate encourages the development of acne.

FACT: It is the oils and fats in many chocolate products, and not chocolate itself, that promote sebum production and subsequent acne. Chocolate in low-fat chocolate milk and in fat-free chocolate candies does not encourage acne. Acne sufferers do not need to give up chocolate; they need to reduce their lipid consumption.

■ Cutaneous Viral Infections

Most viral infections that produce cutaneous lesions do so only as a secondary manifestation of disease that begins in the respiratory or gastrointestinal tract. Only three viruses enter through the epidermis and cause localized lesions in living skin cells.

Wart viruses infect susceptible living cells on the surface of mucous membranes or cells exposed by minor abrasions in the dead strateum corneum of the skin. These viruses stimulate excessive multiplication of infected epithelial cells, producing benign tumors called **papillomas** (warts). Warts may regress spontaneously or persist for many years. They can be physically destroyed by cryocautery (freezing), electrosurgery, or laser surgery.

A second type of viral infection that begins on the body's surface is one caused by herpes simplex viruses, which may cause vesicles (blisters) on skin surrounding the mouth, genitals, or anus ▶ (See *oral herpes,* Chap. 22, p. 577, and Genital Herpes, Chap. 23, p. 614). A third group includes some of the poxviruses that cause pustules at the sites where they enter the skin. These viruses are spread by direct contact with infectious lesions. Humans are the sole reservoir of the poxvirus that causes *molluscum contagiosum.* Lesions may develop on any part of the body, but in adults lesions are most frequent in the genital area. *Orf, cowpox,* and *milker's nodule* are poxvirus diseases transmitted to humans by contact with infected animals. These poxvirus infections usually resolve spontaneously within a few months.

■ Superficial and Cutaneous Mycoses

A variety of fungi can infect intact skin, hair, and nails. Fungi that cause *superficial mycoses* are limited to the outermost dead layers and do not elicit a cellular response from the host. These infections are generally painless cosmetic problems (Fig. 24-7). The agents of *cutaneous mycoses* are also restricted to tissue containing high concentrations of keratin, specifically skin, hair, and nails; however, they elicit an inflammatory response, resulting in more serious skin diseases with uncomfortable, painful, and sometimes even crippling symptoms.

This group of fungi, called *dermatophytes,* cause a variety of infections.

● **DERMATOPHYTE INFECTIONS.** These diseases are commonly known as *ringworm* because the appearance of the lesions led to the erroneous belief that the infected skin harbored worms beneath its surface. Ringworm infections are named "tinea" (a gnawing worm) followed by a second word that designates the infected site. For example, tinea capitis is ringworm of the scalp; tinea corporis is ringworm of the body; tinea pedis is the disease popularly known as athlete's foot; tinea cruris is jock itch; and tinea unguium is ringworm of the nails.

The dermatophytes that cause these diseases belong to three genera of fungi—*Epidermophyton, Microsporum,* and *Trichophyton*—each with an affinity for particular types of keratinized structures.

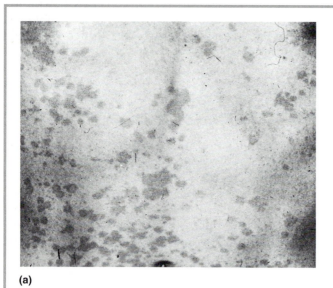

(a)

FIGURE 24-7 **"Cosmetic" diseases.** Superficial mycoses cause observable effects on the infected person but infect no living tissues of the body, so they produce no pain or discomfort. (*a*) Pityriasis versicolor (formerly "tinea versicolor") caused by *Malasezzia furfur* is characterized by abnormal pigmentation in patches of the skin on the torso or arms. (*b*) The agent of tinea nigra, *Exophiala werneckii,* produces a dark-stained lesion. (*c*) Hard nodules develop along the hair shaft of a person with black piedra. The etiologic agent is *Piedraia hortai.*

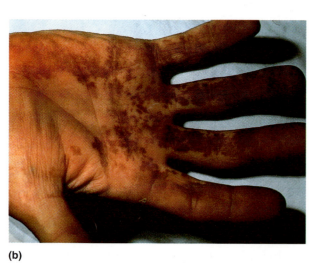

(b)

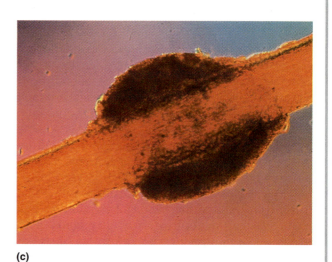

(c)

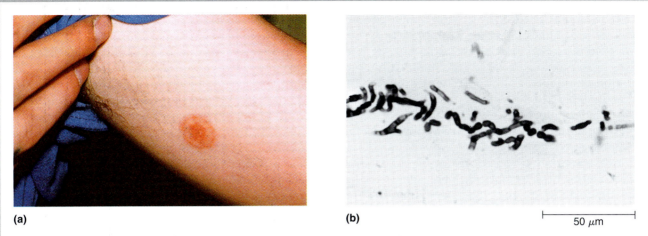

FIGURE 24-8 **Tinea corporis.** (*a*) Circular lesion on the skin and (*b*) microscopic appearance of the dermatophyte (*Trichophyton mentagrophytes*) in tissue.

Epidermophyton infects only skin and nails, *Microsporum* invades only skin and hair, and *Trichophyton* causes disease in all three body sites. None of these organisms can proliferate in living tissue, and all are restricted to the nonliving surface of the body.

The principal dermatophyte reservoirs are infected people or animals, especially dogs, cats, and, in rural areas, cattle, horses, and wild animals. Very few dermatophytes are found primarily in the soil. Transmission is by direct contact with infectious lesions or by contact with infected hair or skin scales that have been shed onto clothing, combs, floors, showers, towels, or other fomites. The soil and animal fungi usually elicit a more marked inflammatory response from the human host than do the organisms that have human reservoirs. The latter appear to be adapted to their host and tend to cause more chronic infections.

In most forms of **tinea capitis** the hair becomes brittle and breaks off. This form of hair loss is temporary, and the hair grows back when the disease is resolved. However, one form of tinea capitis, called *favus,* is characterized by suppurative (pus-discharging) or crusted lesions on the scalp that scar when they heal and cause permanent hair loss.

Children are most susceptible to *Microsporum* infections; adults are virtually resistant to all members of this genus. People of all ages, however, are susceptible to *Trichophyton* species.

Typical ringworm of the smooth (body) skin, **tinea corporis**, is characterized by a flat, spreading, circular lesion representing the continued radial growth of the fungus (Fig. 24-8). Peripheral areas of the lesion contain the living fungus and are often marked by inflammation and pustule forma-

tion, resulting in a red elevated margin (resembling a worm). Alternatively, the lesion may be scaly or crusted. The center of the lesion usually heals and appears like normal skin. Lesions may merge, forming large confluent areas of infection. Infections of the foot (**tinea pedis**) or of other moist surfaces generally lead to extensive scaling of the skin, which may be accompanied by vesicle formation. The appearance of the lesions cannot be used to distinguish between the numerous agents of tinea corporis or tinea pedis.

Tinea unguium occurs in up to one-third of people with tinea corporis, presumably by direct extension of the dermatophyte from skin to nails during scratching of skin lesions. The infection can also be contracted from contaminated fomites or surfaces. Injury to nails increases susceptibility to dermatophyte infection. Any dermatophyte can cause tinea unguium, although *Microsporum* nail infections are rare. Nail dermatophytes may grow as patches on the surface, or they may invade beneath the nail plate, causing an accumulation of keratin debris. Infected nails thicken and become discolored, brittle, and cracked. These infections are usually chronic and rarely resolve without drug treatment.

The characteristic lesions on the infected scalp, skin, or nail are usually distinct enough to suggest dermatophyte infection. Ultraviolet light helps distinguish scalp lesions produced by *Trichophyton* species from those produced by *Microsporum* species. Hair invaded by *Microsporum* fluoresces under ultraviolet light. Clinical suspicion of dermatophyte infection is confirmed by direct microscopic examination of infected hair, nails, or scrapings from the active edge of a skin lesion. Prior to examination, the specimen is heated with 10 percent potassium hydroxide directly on a microscopic slide. This procedure digests the keratin debris that might otherwise obscure the characteristic septate hyphae of the fungus. A variety of media stimulate formation of characteristic spores when the fungus is incubated at 22 to 30°C. These spores aid in identifying the dermatophyte. Dermatophytes are often slow growers and may require 10 days or longer to sporulate.

Many dermatophyte infections can be eradicated by treatment with topical antifungal agents, especially imidazoles (miconazole, ketoconazole, or clotrimazole) or tolnaftate. Griseofulvin is an antibiotic that, when taken orally, inhibits fungal growth while stimulating an increased rate of shedding of the infected keratinized layer. Fungi in skin cells are sloughed and removed during daily bathing. Several months of griseofulvin therapy may be required. Clipping hair and nails short physically removes much of the pathogen. This reduces the probability of the pathogen spreading to other locations from the initial infection site. Since fungi thrive in moist, warm environments, keeping infected regions dry and cool usually shortens the duration of the disease. Dusting powder absorbs excess moisture. Loose-fitting clothing made of natural porous fibers that allow evaporation helps keep lesions dry. Open sandals are recommended for persons with athlete's foot. Bandages increase local humidity, but gauze dressings protect affected skin from friction, irritation, and secondary bacterial infection. The shaving of beards or other infected areas should be discontinued until lesions have healed. Simultaneous treatment of infected pets pre-

vents reinfection. In spite of these measures, some causes of cutaneous mycoses persist.

Infected persons can reduce the likelihood of spreading the disease by washing themselves daily to remove loose hair and skin before it can be shed to other persons; by decontaminating combs, clothing, and other items that contact the lesions; and by avoiding activities that can transmit the disease to others. Sharing towels or clothing encourages transmission of ringworm infections. Floors in dressing rooms, locker rooms, and public showers frequently harbor skin and hair shed from infected persons and should be frequently disinfected with an antifungal agent.

● **CANDIDIASIS.** *Candida albicans* is a yeast that, in addition to its many systemic manifestations, can cause cutaneous mycoses, often on body sites where excessive moisture accumulates. *Candida* infections frequently occur in skin folds of obese individuals, under pendulous breasts, in armpits, in the crotch, and on nails and hands that spend a lot of time immersed in water or covered with nonporous gloves. The yeast multiplies in the cutaneous layers and causes local skin lesions. Persons with impaired immunologic or cellular defenses may develop lesions that cover the entire body surface and the mucous membranes, a disease called *chronic mucocutaneous candidiasis* (CMC) (Fig. 24-9).

The yeast can be isolated from the lesion, grown on Sabouraud's dextrose agar, and identified by biochemical tests. *C. albicans* is unique among yeasts because it possesses the ability to form pseudohyphae (chains of elongated yeast cells), blastospores, and chlamydospores when grown on cornmeal agar;

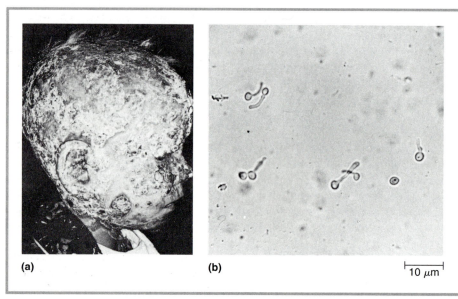

(a) (b) |—10 μm—|

FIGURE 24-9 Chronic mucocutaneous candidiasis. (*a*) A hereditary immunodeficiency in the T-cell response predisposed this child for *Candida* infection of the skin and mucous membranes. (*b*) Characteristic microscopic appearance of germ tubes formed by *Candida albicans* in serum.

TABLE 24-4 SOME CHARACTERISTICS OF INFECTIOUS DISEASES OF THE SKIN

Disease	Pathogen	Reservoir	Diagnostic Procedure	Antimicrobial Treatment
Pityriasis (Tinea) versicolor	*Malassezia furfur*	Soil	Microscopic examination of lesion or culture	Topical fungicides
Cutaneous mycoses	*Microsporum, Trichophyton, Epidermophyton*	Humans, animals, or soil	Microscopic examination of lesion or culture	Topical fungicides, oral griseofulvin
Candidiasis	*Candida albicans*	Humans	Microscopic examination of lesion or culture	Nystatin, imidazoles
Folliculitis	*Staphylococcus aureus*	Humans	Clinical picture	None
Impetigo	*Staphylococcus aureus*	Humans	Isolation and identification of pathogen from lesion	Erythromycin
	Streptococcus pyogenes	Humans	Isolation and identification of pathogen from lesion	Penicillin
Erysipelas	*Streptococcus pyogenes* *Erysipelothrix rhusiopathiae*	Humans Animals	Isolation and identification of pathogen from lesion	Penicillin
Acne	*Propionibacterium acnes*	Humans	Clinical picture	Tetracycline
Molluscum contagiosum	Poxvirus	Humans	Microscopic identification of viral inclusion bodies in cells from lesion	None

it is also the only yeast that forms germ tubes after 2 hours in serum at 37°C (Fig. 24-9*b*).

Cutaneous *Candida* infections are usually treated topically with nystatin, miconazole, halprogen, or clotrimazole. Oral ketoconazole also helps eliminate these infections. As with other opportunistic infections, controlling the conditions that promote proliferation of the pathogen is the best way to prevent *Candida* infections. Unfortunately, many of these predisposing factors cannot be easily identified, corrected, or controlled.

Characteristics of skin infections are summarized in Table 24-4.

CONCEPT CHECK

- Distinguish between cutaneous and subcutaneous infections.
- Why are some staphylococcal skin infections similar to streptococcal skin infections? Why do other infections by the same two organisms present such different symptoms?
- What is the role of *Propionibacterium acnes* in acne infections?
- Why are dermatophyte infections limited to the skin? Describe three ways in which these fungal infections differ from cutaneous candidiasis.

Diseases Acquired Through Animal Bites

A few pathogens, especially those present in animal saliva, are transmitted primarily by the bite of infected animals. Any bite wound, however, even one inflicted by another human, can readily become infected. In industrialized countries, the number of people who acquire infection through animal bites is kept down by most peoples' limited exposure to animals other than domestic pets. Animal handlers, researchers, and other people who have contact with a wide variety of animals, however, are at risk of acquiring such infections.

■ Rabies

Each year more than 30,000 people in the United States suspect they have been bitten by rabies-infected animals. Yet no more than 18 people died of rabies in the United States between 1960 and 1993, and only eight of those cases were due to domestically contracted infection (the rest of the cases were contracted in other countries and imported to the United States). The low rabies death rate is due to an effective (and expensive) policy that includes vaccinating domestic animals and monitoring infection

The Cost of One Rabid Dog

In Yuba County, California, a dog was placed under observation after it bit three people in a parking lot. Because the dog appeared ill, it was killed; tissues were tested and found positive for rabies. The subsequent investigation by Sutter-Yuba County Health Department personnel eventually resulted in the identification of 70 persons with known or probable exposure to the dog, and all 70 subsequently received antirabies prophylaxis. Because investigators found that only 20 percent of the dogs and cats in the area had up-to-date vaccinations, special clinics were held in which 2000 dogs were vaccinated, and more than 300 unclaimed dogs and cats were destroyed.

No persons or other animals were known to develop rabies as a result of this episode. However, the costs generated by this single rabid dog were estimated to be $92,650 for human antirabies treatment, $4190 for animal vaccination and veterinary services, and $8950 for heath department and animal control programs. The total cost of the episode was $105,790 (about $1500 per person treated), not including lost work time, patient travel time, and costs of the 6 months quarantine imposed on animals exposed to the rabid dog.

SOURCE: CDC, *Morbidity and Mortality Weekly Report*, 30:527 (1981).

---

within the animal population (see THE EXPLORERS: The Cost of One Rabid Dog).

Before the development of an effective vaccine, little could be done to save the life of anyone bitten by a rabid animal. Even today, once symptoms appear the disease follows an invariably fatal course (only three persons are on record as having survived rabies). The unrelenting progression of neurological symptoms in rabies is characterized by increased salivation, irrational behavior, convulsions, and hydrophobia (fear of water). The last symptom develops partially as a result of painful spasms in the throat when the victim attempts to swallow and partially as a manifestation of irrational behavior (rabies sufferers often refuse to bathe because of fear of water). Death often occurs within a week after onset of symptoms. In contrast to the low mortality in the United States, worldwide more than 30,000 people die of this disease each year (in 1989, one country, India, lost 25,000 people to rabies).

The rabies virus, a member of the bullet-shaped rhabdoviruses, is readily transmitted in populations of domestic and wild animals. Raccoons are the major reservoir in the United States, followed closely by skunks, but dogs, cats, cattle, bats, and foxes (the major reservoir in Europe) are also potential sources of rabies. Most fatal human cases occur in developing countries as a result of dog bites. In all animals except bats, the virus causes a fatal infection. Prior to the appearance of symptoms, however, the virus is released into salivary secretions. As the infection progresses, many animals become aggressive and tend to bite, thereby promoting the spread of disease. Some animal victims suffer a stuporlike state rather than agitation. Vicious attacks are not necessary for disease transmission, however, because even licking an abrasion on the skin or mucous membrane may be sufficient to inoculate a lethal dose of virus.

The viral particles replicate in muscle tissue at the site of inoculation, invade the regional nerve endings, and eventually reach the brain (Fig. 24-10). From the central nervous system the virus migrates through nerves to other parts of the body, including the eyes and salivary glands. The incubation period varies between 1 week and 1 year (most often 4 to 10 weeks), depending on the site of inoculation and the size of the infecting dose. Symptoms develop more rapidly when the site of entry is close to the brain.

Transmission infrequently occurs by routes other than direct inoculation from an infected animal. Rabies may follow inhalation of dust heavily contaminated with rabies virus. Cave explorers run a high risk of contracting airborne rabies by inhaling dried feces of rabid bats. The disease has also been transmitted by surgical transplants of infected corneas obtained from persons with undiagnosed rabies at the time of death.

The appearance of typical neurological symptoms and the history of an animal bite suggest a presumptive diagnosis of rabies. The virus can sometimes be isolated by cell culture or by inoculating mice with saliva, but these attempts are often unsuccessful. Rabies-specific fluorescent antibody detects virus in frozen specimens of skin, corneal impressions, or mucosal scrapings. When there is no evidence of animal exposure or when symptoms are atypical, the disease is usually diagnosed following autopsy, by the microscopic examination of the victim's brain cells, where cytoplasmic inclusion bodies, called *negri bodies*, have developed (Fig. 24-11).

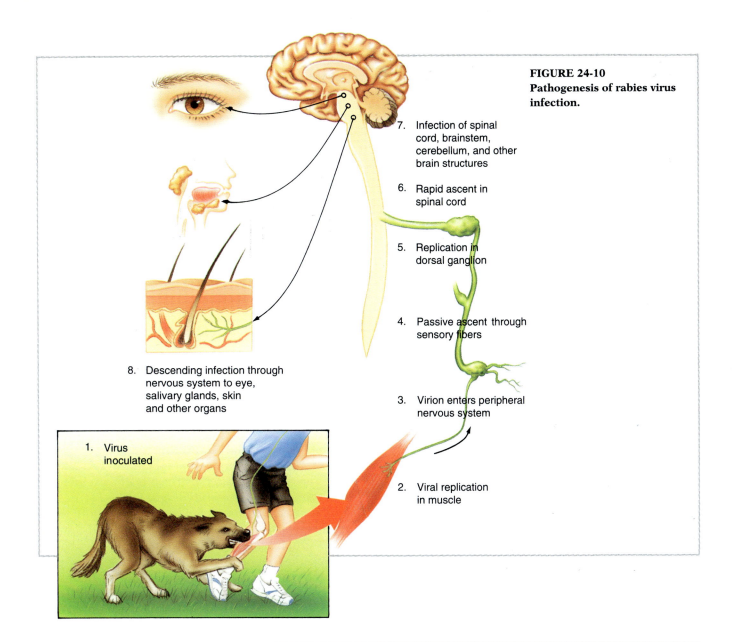

FIGURE 24-10
Pathogenesis of rabies virus infection.

7. Infection of spinal cord, brainstem, cerebellum, and other brain structures

6. Rapid ascent in spinal cord

5. Replication in dorsal ganglion

4. Passive ascent through sensory fibers

3. Virion enters peripheral nervous system

8. Descending infection through nervous system to eye, salivary glands, skin and other organs

2. Viral replication in muscle

1. Virus inoculated

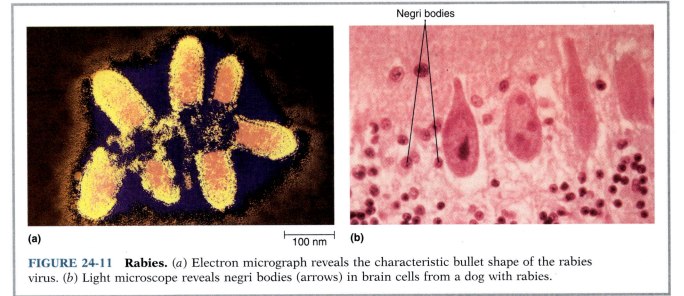

Negri bodies

(a)   100 nm   (b)

FIGURE 24-11   **Rabies.** (a) Electron micrograph reveals the characteristic bullet shape of the rabies virus. (b) Light microscope reveals negri bodies (arrows) in brain cells from a dog with rabies.

Anyone exposed to rabies should be treated to prevent the disease, since a fatal outcome is practically guaranteed once symptoms develop. Immediate action is critical if clinical disease is to be avoided. The inoculation site should be thoroughly cleansed with soap and water, detergent, alcohol, iodine, or quaternary ammonium compounds to physically remove and inactivate any virus remaining in the wound. Passive immunization with hyperimmune antirabies serum provides immediate protection, especially if injected directly around the wound as well as intramuscularly. In addition, a series of daily injections of rabies vaccine builds active immunity before the virus attacks the nerve cells. The original rabies vaccine developed by Louis Pasteur was obtained from homogenized brain tissue of experimentally infected rabbits (passage through rabbits attenuated the virus). The vaccine was administered in 14 to 21 painful inoculations under the abdominal skin. A newer, more potent vaccine uses viruses propagated in tissue culture, chemically inactivated, and injected intramuscularly in only four to six shots. Because there is a small risk of encephalitis following vaccination, only those persons in danger of developing rabies receive the vaccine. This includes anyone bitten by an animal that either was examined and found to have rabies or is not available for examination.

The following guidelines are recommended to help reduce the danger of human rabies by reducing the number of rabies-infected animals:

- Routinely vaccinating pets protects them from acquiring and transmitting the disease.
- Confining pets to the home reduces the likelihood of their exposure to rabid animals.
- Reporting to local authorities any animal suspected of having rabies better enables the authorities to capture, examine, and, if necessary, destroy these sources. Any animal behaving unnaturally should be suspected of having the disease. (The behavior need not be aggressive; a nocturnal animal such as a raccoon or skunk walking the streets at noon is likely to have rabies.)
- Isolating persons with suspected rabies helps protect others from their infectious secretions. The disease can be transmitted by materials soiled with salivary secretions, tears, urine, or spinal fluid of an infected individual. The virus is not present in the blood.
- Vaccination of the wild animal population can be partially accomplished by distributing bait that contains the vaccine. The animals consume the food and vaccine and develop immunity to the virus. This decreases the size of the reservoir of

infection, thereby reducing the likelihood of the disease spreading to humans or other animals.
- Monitoring the population of wild animals alerts public health officials to naturally occurring epizootics (epidemics among animals) and provides an indicator of any increase in the risk of people acquiring rabies.
- International travel restrictions requiring animals to be quarantined or vaccinated before they are admitted to areas that are free of rabies help prevent the emergence of the disease in these regions. Approximately 30 countries have maintained a rabies-free animal population by rigid enforcement of such regulations.

In addition to these measures, people whose occupations increase the risk of contracting rabies should be actively immunized with the vaccine. Persons at risk include veterinarians, wildlife conservation personnel, and laboratory or kennel workers who may handle rabid animals.

■ Rat-Bite Fever

Rat-bite fever is uncommon in the western hemisphere. The rare cases that occur in the United States are usually caused by the bite of a rat infected with *Streptobacillus moniliformis*. In the Far East, a similar disease is due to *Spirillum minor*. Both these organisms may be members of the normal flora of a rat's oropharynx. In the rat-bite victim, the bacteria invade the bloodstream, producing a disease characterized by chills and fever and a disseminated rash. Approximately 10 percent of untreated rat-bite victims die of the disease. *Streptobacillus moniliformis* can be isolated on artificial media and identified for diagnosis. *Spirillum minor* can be grown in animals inoculated with materials from lesions, blood, or lymph of an infected person. Chemotherapy with penicillins or tetracyclines is usually effective. Prevention is best accomplished by controlling the rat population (see Plague, p. 654).

■ Cat-Scratch Disease

The bite or scratch of a cat, and occasionally that of a dog, may transmit a microbe that causes a mild local wound that extends to the regional lymph nodes. The first known case, described in 1931, was that of a 10-year-old boy who slept with cats and had many scratches on his arms. Over 20,000 cases occur in the United States annually, most often in young children. The disease rarely becomes systemic and is usually self-limited. About 1 percent of

lapses are due to the pathogen's ability to repeatedly change the antigenic structure of its surface as the patient acquires immunity against the existing antigenic determinants. The death rate in untreated cases is 2 to 10 percent. The disease is diagnosed by microscopic detection of the spirochete in the blood and is treated with tetracyclines. Preventive measures are similar to those discussed for tick- and louse-borne typhus (discussed below).

■ Rickettsial Diseases

Ticks, lice, fleas, and mites are vectors of a variety of diseases caused by rickettsias (Table 24-6). With the exception of epidemic typhus, all these diseases have nonhuman reservoirs, and humans are incidental hosts. Because they all require vectors for transmission, none of these diseases is directly communicable from person to person.

Rickettsias have had an almost unparalleled impact on world history, decimating armies and causing widespread famine. Epidemic typhus is believed to have contributed to Napoleon's defeat by Russia, and it significantly increased the mortality rates among soldiers in World Wars I and II.

Spotted fevers are a group of clinically similar diseases, worldwide in distribution, with the specific etiologic agent varying with geographic region. **Rocky Mountain spotted fever** is the most common and severe rickettsial disease in the United States, where it is harbored in dogs and wild rodents and transmitted by ticks. The etiologic agent, *Rickettsia rickettsii*, multiplies in the tick's salivary gland and readily enters the bloodstream of a mammal bitten by the infected arthropod. Ticks can also acquire the rickettsia by transovarian transmission directly from their infected mother. In humans, the disease begins with a high fever (105°F), headache, and muscle pain, followed by a rash that first appears on the ankles and wrists, then spreads over most of the body (Fig. 24-14). Rickettsias multiply in the endothelial cells lining blood vessels, causing capillary

TABLE 24-6 ARTHROPOD-BORNE RICKETTSIAL DISEASES

Disease	Etiologic Agent	Vector	Reservoir	Geographic Distribution	Positive Weil-Felix Agglutination*	Treatment
Spotted Fevers						
Rocky Mountain spotted fever	*Rickettsia rickettsii*	Tick	Wild rodents, dogs, ticks	Western hemisphere	OX-19 and OX-2	Tetracycline, chloramphenicol
Boutonneuse fever	*Rickettsia conorii*	Tick	Wild rodents, dogs	Africa, Europe, Middle East, India	OX-19 and OX-2	Tetracycline, chloramphenicol
Rickettsialpox	*Rickettsia akari*	Mite	Mites, mice	United States, Russia, Africa, Korea	None	Tetracycline, chloramphenicol
Typhus						
Epidemic typhus	*Rickettsia prowazekii*	Body louse	Humans	Worldwide	OX-19	Tetracycline, chloramphenicol
Brill-Zinsser disease	*Rickettsia prowazekii*	[Recurrence of epidemic typhus]		North America, Europe	Usually none	Tetracycline, chloramphenicol
Endemic typhus	*Rickettsia typhi*	Flea	Rats	Worldwide	OX-19	Tetracycline, chloramphenicol
Scrub typhus	*Rickettsia tsutsugamushi*	Mite	Mites, rats	Southeast Asia, southwest Pacific	OX-K	Tetracycline, chloramphenicol

*Strains of *Proteus* OX agglutinated with sera from persons with the disease.

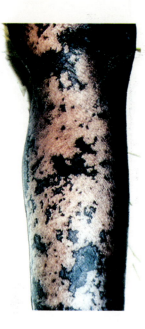

FIGURE 24-14 **Rocky Mountain spotted fever.** The rash usually appears first on the extremities.

hemorrhage responsible for the purple-black rash. If untreated, fatal disseminated intravascular coagulation frequently develops. Currently, most cases of Rocky Mountain spotted fever are found in the eastern United States, although a few years ago it was a serious problem in the Rocky Mountain states. Several other tick-borne spotted fevers occur in different parts of the world. These include boutonneuse fever (Africa and India), Queensland tick typhus (Australia), and North Asian tick fever. All these diseases are caused by species antigenically similar to *Rickettsia rickettsii.* **Rickettsialpox** is a mild spotted fever last reported in the United States in 1969. It is transmitted from infected mice to humans by mites.

Epidemic typhus is the prototype disease in the typhus group. Unlike other rickettsial diseases, however, humans are the sole reservoir of infection. Epidemic typhus is transmitted from person to person by body lice. The louse becomes infected when it feeds on a victim of the disease. During subsequent meals it releases its rickettsial inhabitants by defecating onto human skin. People inadvertently force the rickettsias into the bite wound by rubbing or scratching the area. The clinical symptoms of epidemic typhus are similar to those of Rocky Mountain spotted fever—high temperature, rash, headache, and muscle pains. The spleen, liver, myocardium, and nervous system may be injured in severe cases. Most persons who recover from epidemic typhus are permanently immune to reinfection. In some people, however, epidemic typhus may recur years later in a milder form known as *Brill-Zinsser disease,* which is caused by reactivation of latent infection in the lymph nodes. The pathogen responsible for epidemic typhus is *Rickettsia prowazekii,* named after Howard Ricketts and Stanislaus von Prowazek, who both died as a result of accidental infections while researching the disease. The last outbreak in the United States occurred in 1921.

Endemic typhus (murine or flea-borne typhus) is generally a milder disease than epidemic typhus. The pathogen, *Rickettsia typhi,* is harbored in infected rats, where it causes an inapparent, long-lasting infection. The organism is transmitted among rats by fleas, which also transmit the disease to humans. Once infected, the flea remains an active vector for life, which may be as long as a year. Usually, fewer than 50 cases are reported in the United States each year.

Scrub typhus, transmitted by the bite of a mite infected with *Rickettsia tsutsugamushi* (*tsutsugamushi* = dangerous bug) is endemic in central, eastern, and southeastern Asia. These regions are usually characterized by scrub overgrowth on terrain that serves as a natural habitat for the wild rodents that are hosts to mites. Scrub typhus was almost unknown to the western world until World War II when Allied troops entered areas endemic for the infected mites. Similar outbreaks occurred among soldiers in Vietnam.

Most clinical laboratories are not equipped with the facilities or personnel necessary for isolating rickettsias. Preliminary diagnosis depends on the **Weil-Felix reaction,** a serological test that takes advantage of an immunologic coincidence—most rickettsial infections stimulate formation of cross-reacting antibodies that agglutinate variants of *Proteus* OX, a strain of *Proteus vulgaris* (Table 24-6) (see THE MICROBES: A Vaccine Against Nazis). Blood specimens can be mailed to special laboratories, where diagnosis can be confirmed by growing the pathogens in host cells. The intracellular parasites are usually identified by fluorescent antibody techniques.

Rickettsial diseases are treated with tetracyclines and chloramphenicol, antibiotics that reach the pathogen's intracellular locations. Control measures are directed at preventing vector bites and reducing reservoirs of infection. These measures include eliminating rodent harborages and reducing the vector population by a combination of good sanitation and insecticides.

■ Viral Diseases

More than 500 arthropod-borne viruses, collectively called "arboviruses" (all are togaviruses or flaviviruses), cause disease in vertebrates, including humans. Most of these cause zoonoses, with humans as incidental hosts. They are transmitted from an animal reservoir, usually birds or small mammals, to humans by the bite of a vector. Mosquitoes are the principal vectors, although ticks transmit several arboviral diseases (Table 24-7). In most cases, the vectors cannot become infected by biting infected humans. There is no direct person-to-person spread of any of these diseases. These viruses cause one of three clinical syndromes: (1) benign fever often accompanied by a rash but with a short course and few complications; (2) hemorrhagic fever, characterized by a purplish rash due to ruptured capillaries in the skin; or (3) disease of the central nervous system, usually encephalitis.

The benign viral fevers usually last a week or less and resolve without serious consequences, although hemorrhagic fever or encephalitis may be a rare complication. Most arbovirus-induced benign fevers are restricted to rural or jungle areas where the vectors and natural animal reservoirs are prevalent. These benign fevers occur primarily in tropical areas and include dengue fever, bunyamwera, chikungunya, O'nyong-nyong, Rift Valley, West Nile, and sandfly fevers. In the United States, Colorado tick fever is a similar but rarely reported disease. Although clinically similar, these geographically restricted diseases are caused by different viruses. Because infection normally confers permanent immunity, children are more likely to contract these diseases than adults.

The best known arboviral hemorrhagic fevers are **yellow fever** and **dengue hemorrhagic fever.** Each

TABLE 24-7 IMPORTANT ARTHROPOD-BORNE VIRAL (ARBOVIRAL) DISEASES

Disease	Vector	Reservoir	Mortality Rate, %	Geographic Distribution
Dengue fever	Mosquitoes	Humans	Low	Tropics and subtropics
Yellow fever	Mosquitoes	Monkeys	5–50	Tropics
Dengue hemorrhagic fever	Mosquitoes	Humans	5–10	Southeast Asia
St. Louis encephalitis	Mosquitoes	Birds	5–10	Southern, central, and western United States
Eastern equine encephalitis	Mosquitoes	Birds, horses	50–70	Eastern United States
Western equine encephalitis	Mosquitoes	Birds, horses	3	Western United States
Colorado tick fever	Ticks	Small mammals	Low	Rocky Mountain states
California encephalitis	Mosquitoes	Rabbits, rodents, squirrels	Low	North, central, and southeast United States
Venezuelan equine encephalitis	Mosquitoes	Rodents	Low	Northern South America, Central America, and the United States

disease ranges in severity from asymptomatic to fatal. Hemorrhagic fevers are characterized by bleeding of the gums and gastrointestinal tract, purpura (hemorrhage into the skin), and thrombocytopenia (decreased number of platelets in the blood).

Yellow fever virus is restricted primarily to the jungles of Central and South America and Africa. In the jungle, yellow fever virus is naturally propagated in monkeys and transmitted between monkeys by mosquitoes. Infected mosquitoes also spread the disease to people in the endemic regions. This is the cycle for *jungle* yellow fever. If the disease spreads to urban centers, it is transmitted among infected persons by domestic *Aedes* mosquitoes. Thus, *urban*

yellow fever is an epidemic, and humans rather than monkeys are the reservoir.

Because of vector control measures and an effective vaccine, urban yellow fever no longer decimates large human populations (see THE EXPLORERS: Microbiology and the Panama Canal). *Aedes aegypti* mosquitoes still exist in many areas, including the United States, where, fortunately, there are no reservoirs of yellow fever. In many tropical regions, however, the disease remains endemic.

Dengue hemorrhagic fever is a cause of disease and death primarily among children in tropical Asia. Only people with antibody to dengue virus develop this presumably immune-mediated disease. There-

---

THE EXPLORERS

Microbiology and the Panama Canal

Although the economic and political advantages were immeasurable, the French had given up in defeat. Panama had simply claimed too many lives. Too many of the engineers sent there had developed fever and muscle pains, begun vomiting, and died within a week. There was no treatment and no known means of prevention. The waterway connecting the oceans would have to wait until yellow fever, known in many areas as the "black vomit," could be conquered.

In the late 1890s the United States began where the French had quit. A yel-

low fever commission, led by Major Walter Reed, was dispatched to Havana, Cuba, where the disease had killed more American soldiers than the Spanish-American War. The commission's major task was to determine how yellow fever was transmitted. Most researchers believed that yellow fever was a "filth disease" similar to typhoid fever—a disease spread by human feces. Improving sanitary conditions, however, failed to reduce its incidence in Cuba. Clean Havana was as disease-ridden as filthy Havana. The disease spread in a curious pattern, sometimes striking one side of a street but not the other. It was more prevalent in low wetlands than nearby highlands and was usually spread in the direction of the prevailing wind. The disease flourished in hot weather and disappeared with cold and frost.

These observations led Reed to investigate an unpopular theory proposed 20 years earlier by the ridiculed "mosquito doctor," Carlos Finlay. Finlay believed that the disease was acquired by inoculation with a "living miniature hypodermic needle"—an infected mosquito. But because he was unable to demonstrate mosquito-borne transfer of the disease, his proposal was not considered seriously by his contemporaries. Reed's group discovered the reason for Finlay's failures. Finlay had mosquitoes bite yellow fever patients and then let these freshly charged insects bite healthy persons. Unfortunately for his hypothe-

sis, these persons remained healthy. With carefully timed experiments and a handful of courageous volunteers, Reed showed that mosquitoes become infected only when they draw blood from a person during the first 4 days of the fever. In addition, the infected mosquito's bite can transmit the disease only after the pathogen has incubated in the insect for 7 to 10 days.

Reed also ruled out other possible modes of transmission. Healthy volunteers were confined for 3 weeks in special mosquito-proof houses where they slept on bed sheets and blankets taken from yellow fever wards and wore pajamas removed from the bodies of yellow fever victims. All these volunteers remained healthy. Reed wrote, "The bubble of the belief that clothing can transmit yellow fever was pricked by the first touch of human experimentation." To show that these men were not immune to the disease, two of them volunteered to be intentionally exposed to infected mosquitoes, and both contracted the disease.

With experiments like these, Walter Reed conclusively proved that the mosquito *Aedes aegypti* is the vector of yellow fever. This discovery provided the first successful tactic for controlling the disease. Mosquito-reduction programs allowed construction of the Panama Canal to begin in 1904, 4 years after Walter Reed's yellow fever team arrived in Havana.

---

fore, only individuals who have recovered from dengue fever or infants with maternal antibody are susceptible. Some of these individuals fail to develop neutralizing antibody but possess opsonizing antibody. On a second exposure to the virus, these antibodies facilitate viral entry into macrophages where the viruses replicate and stimulate the release of soluble host factors. The initial symptoms resemble benign dengue fever and appear about 5 days after inoculation of dengue virus by the *Aedes* mosquito. As the fever subsides, however, the patient, in response to the host factors released from macrophages, may hemorrhage and, in severe cases, develop vascular defects that cause fluid loss. Unless the fluid is replaced, death from shock may follow within hours.

In the United States, the most common arthropod-borne disease is arbovirus encephalitis. The neurotropic arboviruses all cause similar diseases, with symptoms ranging from mild fever to convulsions, paralysis, coma, and death. Disease severity varies among the different agents, but they can all cause irreversible neurological damage that leaves survivors permanently retarded or paralyzed. The most common neurotropic arboviral disease in the United States is St. Louis encephalitis (SLE), which causes the majority of deaths. Eastern equine encephalitis and western equine encephalitis cause fatal infections in horses as well as humans. In spite of its name, California encephalitis is most prevalent in the Mississippi and Ohio river valleys. Venezuelan equine encephalitis was originally a disease of South and Central America but has spread to the southern United States. Most outbreaks of arbovirus encephalitis occur in summer and fall, and the disease is conspicuously absent during winter months when mosquitoes are dormant. Mosquitoes are believed to acquire the infection from sources other than infected people, because the low-level viremia that occurs in humans is not a rich supply of viruses. Birds, horses, rodents, squirrels, and rabbits develop high-grade viremias, and their blood provides mosquitoes a stock of readily accessible viruses. Early detection of these viruses in their natural animal reservoirs may alert health officials to an impending epidemic before a single person contracts the disease. Chickens have been used as sentinels to monitor for St. Louis encephalitis in Florida since 1977. These domestic birds are periodically examined for evidence of infection by SLE virus. Mosquito control programs and public announcements to avoid the vector have prevented many cases of arbovirus encephalitis.

Arboviruses stimulate formation of virus-specific antibodies. These diseases are most often diagnosed by detecting rising antibody titers. The viruses that cause benign and hemorrhagic fevers can be isolated from the blood of infected persons by propagation in cell culture. Recovery of viruses from the blood of patients with arbovirus encephalitis is difficult because viremia is mild and short-lived.

Although no specific antimicrobial agent is available for treating patients with arbovirus infections, supportive therapy to reduce symptoms can minimize disease severity. Replacement of intravascular fluids caused by hemorrhagic fevers helps prevent fatal shock.

Although vaccines are available against yellow fever, Japanese B encephalitis, and the equine encephalitides, vector control is the most important preventive measure. Eliminating mosquito breeding grounds gives long-term protection, while spraying human habitats with insecticide reduces spread during epidemic outbreaks. Personal protection is also afforded by the use of repellents, protective clothing, and mosquito nets.

■ Protozoan Infections

The major arthropod-borne diseases caused by protozoa are malaria, leishmaniasis, and trypanosomiasis. The pathogens causing these diseases have proven difficult to control, particularly because they possess effective mechanisms to elude immune response (see BEYOND THE BASICS: Deceptive and Deadly Protozoa). These diseases are summarized in Table 24-8.

● **MALARIA.** Malaria is one of the great human scourges of all time and is still one of the most common fatal infectious diseases. According to WHO, 270 million people currently have malaria, and each year 1 to 2 million of these infected people die of the disease. The disease is naturally transmitted by the *Anopheles* mosquito. Although the vector has been eliminated from most areas in the United States, it is still common in the southwestern and southeastern parts of the country. Malaria, however, is not, probably because the reservoir of infection (people with malaria) is small to nonexistent. In California's San Diego County, however, malaria has been a problem since 1986 because migrant workers from malaria-endemic countries provide a temporary reservoir of infection. In addition, malaria continues to be transmitted through imported cases (soldiers, immigrants, and travelers) and by alternative means—for example, by transfusion of contaminated blood. Drug users often acquire the disease by sharing contaminated needles. A fetus may acquire malaria by transplacental passage from an infected mother. In 1990, approximately 2000 cases of malaria were reported in the United States.

Deceptive and Deadly Protozoa

Protozoa that cause blood infections employ some of the microbial world's "cleverest" tricks for eluding an arsenal of host defenses, a defensive battery usually very effective at defeating microscopic invaders. Most of these strategies rely on deceit and timing. Some protozoa hide their foreign antigens by covering themselves with host proteins so they are not recognized as invaders. Others periodically shed these foreign molecules and synthesize new, antigenically distinct proteins. Just as the body's immune response builds to an effective level, it is rendered ineffective by an inability to recognize the new proteins, so it must start again against the new antigens.

Occupying a host cell as an intracellular parasite is another strategy that protects a protozoan from recognition by the immune system. *Leishmania* performs what might seem like a reckless and potentially self-destructive move—it invades the macrophage itself, penetrating into the core of one of the body's most potent weapons against invaders. It disarms the macrophage's antimicrobial defenses by neutralizing lysosomal enzymes and detoxifying oxidative metabolites as they are dumped into the phagocytic vacuole in which *Leishmania* replicates. To protect against the acidic environment of the vacuole, these intracellular parasites differentiate into acid-stable forms. Another virulence factor that accounts for *Leishmania*'s success as a pathogen is its ability to synthesize surface glycoproteins that mimic those of their host. Molecular mimicry hides the pathogen from the host immune system, disguising it as "self." A combination of antigenic deceit and changeability contributes immeasurably to *Leishmania*'s success as a human pathogen.

Trypanosomes come well equipped to escape immune surveillance. They contain over 1000 genes for unique proteins that exhibit little or no cross-reactivity with one another. The periodic switch in gene expression creates antigenically distinct pathogens within the infected host. As the host's immune system eventually responds to the pathogens during the course of infection, many of them escape detection because they no longer match the antigenic specificity against which the immune response is targeted. The antibodies fail to react with the new glycoprotein antigens on the surface of these variant trypanosomes. This immunologic game of "hide and seek" creates cycles of heightened parasitemia (parasites in the bloodstream). One form of the trypanosome multiplies until the host responds and destroys the predominant form. Another antigenic variant then emerges, and the cycle is repeated.

Plasmodia that cause malaria protect themselves from the host defenses by a combination of methods, including becoming intracellular parasites and changing forms during infection. The highly immunogenic sporozoite phase, which would be easily eliminated by the immune response, is present in the blood for only a brief period of time before it disappears into the liver. The merozoites released from infected liver cells quickly invade red blood cells, again becoming invisible to the immune system within seconds. As the immunologically distinct gametocytes (the forms picked up by the mosquito vector during its blood meal) are produced, again there is no immune retaliation. To halt the progress of malaria, immunologic protection against all three forms will likely be necessary.

Unfortunately, these mechanisms not only challenge the immune system but are also obstacles to the development of effective vaccines. This is one of the reasons protozoan diseases still play such an important role in developing countries.

TABLE 24-8 ARTHROPOD-BORNE PROTOZOAN DISEASES

Disease	Pathogen	Vector	Reservoir	Incubation Period
Leishmaniasis				
Cutaneous	*Leishmania tropica*	Sandfly	Rodents, canines	Days to months
Mucocutaneous	*L. braziliensis*	Sandfly	Rodents, canines	Months
Visceral	*L. donovani*	Sandfly	Humans, rodents, canines	2–4 months
Trypanosomiasis				
African	*Trypanosoma rhodesiense*	Tsetse fly	Wild game, cattle	2–3 weeks
	T. gambiense	Tsetse fly	Humans	Months to years
American	*T. cruzi*	Reduviid bug	Humans, many animals	5–14 days
Babesiosis	*Babesia microti*	Tick	Mice, cattle	10–20 days
Malaria	*Plasmodium vivax*	*Anopheles* mosquito	Humans	14 days
	P. malariae	*Anopheles* mosquito	Humans	30 days
	P. falciparum	*Anopheles* mosquito	Humans	12 days
	P. ovale	*Anopheles* mosquito	Humans	14 days

Four species of *Plasmodium* cause malaria in humans—*P. malariae, P. vivax, P. ovale,* and *P. falciparum.* The protozoa are injected into the bloodstream from the saliva of the female *Anopheles* mosquito during a blood meal. The infectious forms of the protozoa, called *sporozoites,* invade the liver, where they divide by multiple fission into cells called *merozoites.* A single sporozoite can produce as many as 40,000 merozoites. These cells are released into the bloodstream, invade erythrocytes, and undergo a series of morphological changes into ring-shaped *trophozoites.* The trophozoites reproduce inside the red blood cells, forming more merozoites. Ultimately the red blood cell is destroyed by lysis, and the released merozoites are free to infect more red blood cells. Rupture of the erythrocytes occurs in a synchronized periodic manner—every 48 to 72 hours, depending on the multiplication rate of the particular *Plasmodium* species. Rupture is accompanied by the release of toxic materials into the blood, producing the cycles of chills and fever that are typical of malaria.

Some cells released from the erythrocytes differentiate into male and female *gametocytes.* These cells do not mature in the human but must be ingested by the female *Anopheles* mosquito to complete their development. Gametocytes fuse in the insect, and the resulting zygotes divide into asexual sporozoites, which then make their way to the insect's salivary glands. The sporozoites are introduced into a human when the mosquito feeds. Additional details of the *Plasmodium* life cycle are presented in Figure 24-15.

The most characteristic symptoms of malaria are the periodic cycles of chills, fever, and sweating accompanied by headache and nausea. Additional manifestations may be serious, especially those caused by *P. falciparum* (jaundice, anemia, gastrointestinal disturbances, coagulation disorders, shock, and coma). As many as 60 percent of the circulating erythrocytes may be parasitized. Falciparum malaria is fatal in more than 10 percent of untreated cases. The other types of malaria are rarely life-threatening, but without appropriate treatment the parasites persist in the liver and may cause relapses. Relapses never occur when the disease has been acquired from contaminated blood or instruments because only the sporozoite, which develops only in the mosquito, can infect the liver. Laboratory diagnosis of malaria is established by detecting the trophozoite in red blood cells.

Most cases of malaria respond to therapy with chloroquine or related compounds, although drug resistance in *P. falciparum* is becoming more common. These resistant strains can be treated with quinine. Malaria relapses are prevented by combining chloroquine, which kills the circulating form of the pathogen, and primaquine, which destroys the parasites in the liver. Prophylactic drug therapy is recommended for persons traveling to areas at high risk for malaria. Since protection is virtually ensured, it is imperative that travelers be informed of the serious consequences of disease and the overwhelming success of preventive measures.

In the United States, infection by transfusion is prevented by screening blood donors and not accepting blood from persons who have returned from an area endemic for malaria until they are proven

Clinical Specimen	Diagnostic Feature	Treatment
Tissue	Ovoid, nonflagellated trophozoite	Antimonials
Tissue	Ovoid, nonflagellated trophozoite	Antimonials
Tissue	Ovoid, nonflagellated trophozoite	Antimonials
Blood, lymph, CSF	Flagellated trophozoite	Suramin, melarsoprol
Blood, lymph, CSF	Flagellated trophozoite	Pentamidine, melarsoprol
Blood	Flagellated trophozoite	Bayer 2502
Tissue	Nonflagellated trophozoite	
Blood	Erythrocytes containing tear-shaped trophozoites	Chloroquine, quinine
Blood	Enlarged erythrocytes with stippled cytoplasm and amoeboid trophophozoites	Chloroquine
Blood	Normal-sized erythrocytes; band-formed trophozoites	Chloroquine
Blood	Normal-sized erythrocytes	Chloroquine, quinine for resistant strains
Blood	Enlarged erythrocytes; ring-shaped trophozites	Chloroquine

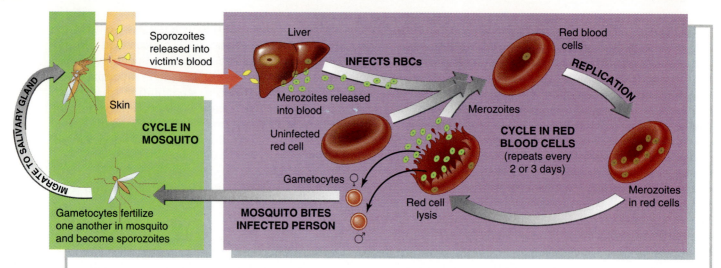

FIGURE 24-15 Life cycle of the malaria parasite. Plasmodia require two hosts, a human and a mosquito. When an infected mosquito bites a person, it releases sporozoites into the blood. The parasite replicates in cells of the liver. (This is the pre-erythrocyte phase of malaria.) Days later, the bloodstream is showered with merozoites released from the liver. (This may also occur months after symptoms have disappeared and may cause a relapse of malaria.) The merozoites infect red blood cells and begin the second (erythrocytic) phase of malaria. Some merozoites in red blood cells mature into gametocytes. When a mosquito draws blood from an infected person, these gametocytes fertilize each other in the new host, where they complete the pathogen's life cycle. Infectious sporozoites are shed in the mosquito's saliva.

uninfected by remaining free of symptoms for at least 6 months.

The failure of vector elimination to control malaria plus the emergence of drug-resistant parasites emphasizes the need for an effective vaccine. In March 1993, the first successful clinical trial of a malaria vaccine was reported. The study enrolled 1500 volunteers in areas of South America endemic for malaria, half of whom received the vaccine and half a placebo. The vaccine reduced the incidence of malaria by 40 percent. The vaccine, composed of chemically synthesized proteins similar to those manufactured by the protozoan in the blood, provided measurable protection by stimulating antibody production. Efforts to develop more effective vaccines continue.

● **LEISHMANIASIS.** The flagellated protozoa that cause leishmaniasis are usually transmitted from infected humans, canines, or rodents through the bite of the sandfly. The protozoa are lodged in and block the insect's proboscis, making feeding impossible. During its numerous attempts to feed, the fly inoculates the parasites into the host. The parasites are then engulfed by macrophages, lose their flagella, and multiply in the macrophages by a series of binary fissions, ultimately rupturing the parasitized cell and releasing progeny, which continue to invade new white blood cells (Fig. 24-16). The infected white cells may remain localized in the skin at the

site of inoculation or may carry the pathogen to mucosal surfaces of the nose and mouth or to various internal organs (spleen, lymph nodes, or liver) where the protozoa can multiply and be released into the bloodstream. Thus, three forms of leishmaniasis occur: cutaneous (oriental sore), mucocutaneous (espundia or American leishmaniasis), and visceral (kala azar). The species of *Leishmania* determines which of these diseases will develop. Geographic distribution is dictated by the occurrence of the corresponding vector.

Cutaneous leishmaniasis is usually manifested as an ulcerated lesion that heals spontaneously. Infections that spread to mucous membranes of the nasopharynx, however, may be fatal. When the blood is infected, the disease is characterized by fever and enlarged lymphoid organs. The victim becomes weak and will usually die if no treatment is provided.

Leishmaniasis is diagnosed by microscopic observation of the nonflagellated form of the protozoan in appropriate specimens (lesions, bone marrow, spleen, liver, lymph nodes, or blood) depending on the form of the disease. Sodium antimony gluconate, available only from the CDC, is recommended for treating the infection. Other antimonials and amphotericin B are sometimes effective. Control measures rely on interrupting the transmission cycle. Controlling the canine population, for example, decreases the number of human cases when dogs are the major reservoir. Insecticide spraying programs, eliminating fly breeding habitats, avoiding sandfly-infested areas, and wearing protective clothing all help control disease spread.

● **AFRICAN TRYPANOSOMIASIS.** Like *Leishmania,* trypanosomes are flagellated protozoa that invade the bloodstream after the bite of an arthropod vector. These parasites are pleomorphic, varying from long and slender to short and stumpy during their life cycle. Trypanosomes live in blood plasma and do not infect blood cells. Three species cause human disease, and many others infect and kill domestic animal species, thereby destroying cattle and similar food sources in affected areas. One such parasite prevented the successful introduction of horses for transportation in Africa, significantly reducing the development of the continent.

African trypanosomiasis (African sleeping sickness), caused by *Trypanosoma gambiense* or *T. rhodesiense,* is transmitted by the bite of the tsetse fly ▶ (see Diseases Caused by Protozoa, Chap. 12, p. 291). Since tsetse flies are found only in Africa and southern Arabia, the disease is restricted to these regions. Despite this narrow geographical range, the disease kills about 20,000 people annually.

Fever, headache, and enlarged lymph nodes are the earliest signs of disease. When the organisms spread to the central nervous system, victims exhibit lethargy, severe weight loss due to anorexia, mental deterioration, and somnolence (the daytime sleeping from which the disease derives its name). If untreated, most cases are fatal.

Trypanosoma gambiense has a human reservoir, whereas *T. rhodesiense* lives in wild animals. The gambian disease runs a prolonged course. This in-creases the period of shedding, allowing the parasite ample time to infect new hosts before killing the infected person. *T. rhodesiense* is more virulent for humans and usually results in death in less than 1 year of infection.

Diagnosis is confirmed by observing trypanosomes in lymph, blood, or, in later stages, cerebrospinal fluid. Suramin is effective for treating early infections, and pentamidine is an alternative. When either protozoan has spread to the central nervous system, melarsoprol is used. In the United States, these drugs are available only from the CDC. Prophylactic administration of pentamidine confers protection against *T. gambiense* for 3 to 6 months. However, because this drug is potentially toxic and because drug-resistant strains are becoming prevalent, pentamidine is used only by persons at high risk of being exposed to the pathogen.

● **AMERICAN TRYPANOSOMIASIS.** American trypanosomiasis, also known as **Chagas's disease,** is caused by *Trypanosoma cruzi,* a protozoan found among domestic and wild animals from South America to Mexico and many southern states of the United States. The disease affects 16 to 18 million people and is responsible for 30 percent of deaths in Brazil. The vectors are blood-sucking insects called reduviid bugs. These bugs commonly seek a blood meal by biting the person's lips, accounting for their nickname, the kissing bug. When an infected bug feeds, it deposits contaminated feces on the skin surface. The protozoa usually enter the body by being rubbed into the bite wound, where they multiply and frequently cause a swollen skin lesion called a chagoma. As the infection progresses,

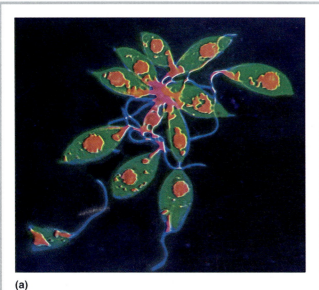

(a)

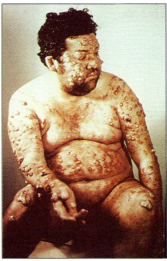

(b)

FIGURE 24-16 Two views of leishmaniasis. (*a*) Colorized light micrograph of *Leishmania* protozoa reproducing within its insect host. (*b*) Clinical appearance of victim of leishmaniasis.

fever develops, local lymph glands enlarge, and the parasites disseminate to the bloodstream and tissues. Acute disease is most common in children and may be fatal if severe myocarditis or meningoencephalitis develops. Most infections, however, are chronic and are not fatal. Some chronic infections are asymptomatic; the others are usually characterized by various degrees of myocardial damage.

Chagas's disease is diagnosed by demonstrating the pathogen's presence in the blood, by direct microscopic examination, by culturing the organisms, by inoculation into mice, or by allowing an uninfected vector to feed on the patient and after a few weeks examining the bug's feces for trypanosomes. There is no treatment that is both safe and effective against American trypanosomiasis. Bayer 2502 (nifurtimox), available from the CDC, is useful in some cases of acute infection.

CONCEPT CHECK

- Describe two advantages that arthropod transmission has for a pathogen. What are two disadvantages?
- Why is plague considered a zoonosis, but malaria is not? How does this difference affect methods of control against each disease?
- Compare the transmission cycles for urban yellow fever and equine encephalitis.
- How do spotted fevers, rickettsialpox, and typhus infections differ?
- Why do most cases of Lyme disease occur during the summer months? What factors determine the geographic distribution of the disease?
- Compare the vectors, mode of transmission, and clinical features of trypanosomiasis and leishmaniasis.

Wound Infections

Pathogens that enter through wounds may cause infections at their inoculation sites, or they may invade surrounding tissue, enter the lymph or blood systems, and cause systemic disease.

■ Tetanus

Endospores of *Clostridium tetani* are common contaminants of soil, dust, or feces and are frequently found on objects that cause accidental injuries. These spores germinate in the oxygen-starved environment provided by the necrotic tissues of deep

wounds. Vegetative cells growing in the injured tissue produce a potent neurotoxin called *tetanospasmin* that is spread throughout the body by the bloodstream. The toxin blocks inhibitory motor neurons that prevent overstimulation of voluntary muscles. Without inhibitory impulses, voluntary muscles become paralyzed in a state of tetanic contraction. *Lockjaw*, the common name for the disease, refers to the toxin's predilection for jaw and neck muscles, which become rigidly locked. The toxin eventually affects most of the voluntary muscles in the body, causing rigid, uncontrolled contractions that are sometimes strong enough to break the patient's spine (Fig. 24-17a). Without treatment, the fatality rate may reach 70 percent, usually from respiratory paralysis.

Laboratory diagnosis confirms the clinical suspicion of tetanus. The pathogen is isolated only from the local lesion and only under anaerobic transport and culture conditions. *Clostridium tetani* is identified by its characteristic terminal, swollen endospores (Fig. 24-17b). Because of the need for immediate treatment, therapy should not be delayed while waiting for laboratory confirmation when clinical symptoms suggest tetanus. Patients with histories of tetanus immunization are treated with tetanus toxoid to quickly boost their immunity against the effects of the toxin. This is usually sufficient to prevent any further manifestations of disease. Patients who have never been immunized should receive passive immunization with tetanus antitoxin to provide immediate protection. These measures neutralize the toxin before it attaches to nerve cells. Once the toxin is bound to neurons, its effects cannot be reversed by immunotherapy. Paralysis persists until the toxin is slowly metabolized by the patient. Penicillin is administered to kill the pathogens in the wound. Sedatives, muscle relaxants, and respiratory support systems help maintain the patient until neurological functions normalize.

The best protection against tetanus is routine immunization with tetanus toxoid. A series of four DPT (diphtheria, pertussis, and tetanus) shots is recommended for all children at 2, 4, 6, and 18 months of age. Booster shots at 10-year intervals maintain immunity. Because of the danger of hypersensitivity to the antigens in the vaccine, toxoid boosters are no longer given as a part of routine wound treatment unless either 10 years has elapsed since the patient last received toxoid or there are signs of developing tetanus. In areas of the world where immunization is not common and sanitation is poor, tetanus causes 700,000 deaths a year. In the United States the total number of reported cases of tetanus annually is less than 100. Most cases are in newborns or in per-

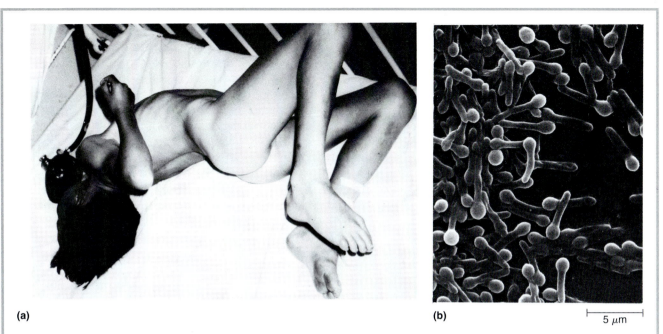

(a)   **(b)**   ⊢——⊣ 5 μm

FIGURE 24-17 Tetanus. (*a*) Tetanic paralysis characteristic of advanced disease.
(*b*) Scanning electron micrograph of sporulating *Clostridium tetani* cells showing
the swollen terminal endospores.

sons over 60 years old who presumably failed to maintain adequate immunizations.

MYTH: Rust causes tetanus if introduced into a wound—for example, by stepping on a rusty nail.

FACT: Rusty nails are more likely to be contaminated with tetanus endospores because they have been exposed to soil and dust longer than new, rust-free nails. Any object that causes a wound, rusty or not, can inoculate the tissue with the bacterial endospores of *C. tetani*. Rust itself neither causes tetanus nor makes it worse.

■ Gas Gangrene

Gas gangrene is a rapidly advancing muscle necrosis caused by several species of *Clostridium*, most notably *C. perfringens*, *C. novyi*, and *C. septicum*. These clostridia grow especially well in anaerobic tissues of severe traumatic injuries such as gunshot wounds, compound fractures, septic abortions, and surgical incisions. Postsurgical gas gangrene is an occasional complication when, because of poor asepsis, surgical wounds are contaminated with endospores during the operation or during subsequent

therapy. Tissues that are insufficiently supplied with blood—for example, the poorly oxygenated tissues in the extremities of diabetic individuals—are also prone to infection. The clostridia release toxins and histolytic enzymes that kill the tissue surrounding the infected site. As the area of necrosis advances, enzymes dissolve muscle and connective tissue, facilitating the spread of infection into the desolidified areas and accounting for most of the clinical symptoms (see Fig. 15-1). The enzymes produced by *C. perfringens*, the most common agent of gas gangrene, include collagenase, protease, and lecithinase (also called alpha toxin). Lecithinase is an enzyme that dissolves the cell membranes of muscle cells. It is the same poison that is found in rattlesnake venom and appears to be the most important virulence factor in gas gangrene. The pathogens also produce carbon dioxide and hydrogen gases as metabolic by-products. These gases can distend tissues until blood flow through the area is obstructed, thus enhancing tissue necrosis. If not halted within hours, the disease may claim an entire limb or even the patient's life. Gas gangrene of the thoracic or abdominal region is particularly life-threatening.

Gas gangrene often requires immediate therapeutic action to prevent irreversible tissue damage or death. A Gram stain of infected tissue provides a rapid preliminary diagnosis by differentiating between the major gram-positive and gram-negative agents of anaerobic wound infections (the clostridia are gram-positive rods). This helps in selecting the therapy most likely to arrest microbial growth and halt further damage. An important step in treating gas gangrene is the removal of necrotic tissue, a procedure called *debridement.* Sometimes debridement requires surgical removal of gangrenous muscle or even amputation of a limb. Simultaneous treatment with penicillin helps control the growth of pathogens not removed by debridement. This combination of treatments has reduced death rates to about 20 percent. *Hyperbaric oxygen treatment* also appears to reduce tissue loss and mortality rates. For this therapy, the patient is placed in a high-pressure (3 atmospheres) oxygen-filled chamber that forces elevated concentrations of oxygen into the tissues. This increased oxygen concentration is lethal to many obligate anaerobic bacteria, which may account for the beneficial effects of hyperbaric oxygen in treating gas gangrene.

■ Other Anaerobic Wound Infections

In addition to gas gangrene, many other anaerobic infections are caused by microbes introduced into wounds. These infections are usually caused by a mixture of bacteria that may include anaerobic streptococci and species of *Bacteroides, Fusobacterium,* or *Clostridium.* Although most of these organisms are sensitive to antibiotics, the lack of blood flow to the necrotic region may prevent effective concentrations of the drug from reaching the infected site. Thorough wound cleansing and prompt surgical debridement are important therapeutic measures.

Wound botulism is a rare infection in which *Clostridium botulinum* multiplies in the anaerobic surroundings of a deep wound. The pathogen releases *botulin,* a toxin that, when absorbed, causes classical symptoms of botulism (flaccid paralysis).

■ Burn Wound Infections

The dead or dying tissues of a burn wound provide an ideal environment for microbial growth as well as a portal of entry for invasion of the bloodstream by pathogens. Burns also impair many major defenses of the immune response. Circulating IgG and complement are reduced, T-cell functions are altered, and chemotaxis of phagocytic white blood cells is inhibited. Each year in the United States between 10 and 30 percent of the approximately 60,000 people hospitalized for burn wounds die of infection, which is the major cause of death in persons hospitalized for burn wounds who have survived the first 48 hours.

Because the integrity of the skin is breached, colonization and subsequent infection of the wound are inevitable. Organisms are usually transmitted to the wound by autoinoculation with the patient's normal flora or by contact with colonized personnel or contaminated fomites. Air is generally not a major route of transmission. The most prevalent microbes in burn wounds are *Pseudomonas aeruginosa, S. aureus,* and *S. pyogenes.* The presence of *Pseudomonas* is often recognizable by the pigment it produces, which imparts a green color to the wound and dressings (see Fig. 25-8), accompanied by a foul smell. This bacterium produces a variety of toxins and enzymes that help it invade the wound site, spread to adjacent tissues, and harm the host. These include elastase and protease (which destroy host tissue), hemolysins and leukocidins (which kill red and white blood cells), and an exotoxin resembling diphtheria toxin. Before the availability of antibiotics, *S. pyogenes* was the most frequent cause of invasive burn wound infections. This bacterium could spread so rapidly that bacteremia could be fatal within a few hours. Currently, *S. pyogenes* is cultured from only 5 percent of infected burn wounds. Enterobacteria and the yeast *Candida albicans* are other common burn wound contaminants, although any pathogen can probably infect extensively burned areas.

There is no such thing as a sterile wound. However, the danger of invasion increases as the types and numbers of microbes in the wound increase. Thus, persons with extensive burns are usually kept in areas where access to the patient is restricted and where rigorous precautions are practiced, particularly the wearing of masks and gowns and mandatory washing of hands. Topical antimicrobial agents, most commonly silver sulfadiazene, mafenide, or povidine-iodine, are used to inhibit microbial growth. Unfortunately, some of these drugs also delay epithelial growth and wound repair. In addition, they are sometimes ineffective because of the emergence of resistant pathogens. Appropriate therapy depends on determining antimicrobial sensitivities and using drugs that are effective. Debridement or surgical excision of the nonliving tissue decreases infection, especially if the wound is covered with a skin graft.

■ Viral Hepatitis

At least five viruses can cause hepatitis, an inflammation of the liver. The clinical manifestations of the diseases are similar despite the fact that the agents are different. Together these viruses are among the most commonly reported agents of infectious disease and are believed to infect 300,000 people in the United States each year. Hepatitis A and hepatitis E viruses are acquired through the alimentary tract ▶ (see Chap. 22, p. 595). Hepatitis B, hepatitis C, and hepatitis D (the delta hepatitis agent) are usually acquired by parenteral inoculation with blood products or instruments (needles, syringes, or hemodialysis equipment) contaminated with the virus. The disease may also be spread in saliva and semen by oral or sexual contact and is highly prevalent among male homosexuals. Women who acquire the disease during pregnancy often transmit the virus to the fetus, usually at birth. In certain areas of the world, this is a major mode of spread.

Worldwide, **hepatitis B virus (HBV)** infects several hundred million people each year. The clinical course of the disease ranges from asymptomatic to fatal. In general, after an average incubation period of 90 days, the virus begins to cause nausea, vomiting, fever, and abdominal pain. Jaundice is a common symptom that usually disappears within 6 weeks. In the United States, hepatitis B is most common in persons aged 15 to 29, especially among drug addicts and workers in blood banks, renal dialysis units, and medical or dental laboratories. Mortality is approximately 1 percent, and recovery from infection provides permanent immunity to the virus. However, approximately 5 to 10 percent of the individuals who recover from hepatitis B become chronic carriers. The carrier rate among perinatally infected infants is 90 percent. More than 120 million people in the world currently are carriers and continue as potential disseminators of disease. In many of these individuals, the infection eventually develops into cirrhosis of the liver or hepatocarcinoma (liver cancer). The loss of life due to these sequelae makes hepatitis B infection the ninth leading cause of death in the world.

Hepatitis B is caused by a DNA virus. Electron microscopy of serum containing hepatitis B virus shows the complete virus, which is called the *Dane particle*, and two other particles that are composed solely of viral surface antigen (Fig. 24-18). The infection is diagnosed by immunoprecipitation, immunoelectrophoresis, or radioimmunoassay to detect viral antigens circulating in the patient's serum. There is no specific antiviral treatment for hepatitis B, although interferon is being used to treat chronic infection. Passive immunization with sera from immune patients may prevent disease if administered shortly after exposure to the virus. Prevention of hepatitis depends largely on avoiding exposure to contaminated products. Blood products and donors are screened for hepatitis B antigen to reduce transmission of virus by transfusion. The use of disposable needles and syringes further reduces the danger. Hepatitis B virus is especially resistant

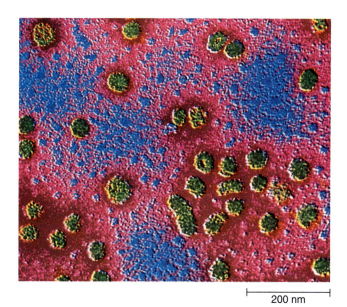

FIGURE 24-18 **Hepatitis B viruses.** The infectious spherical viruses are often known as Dane particles.

200 nm

to many chemical disinfectants but can be destroyed by solutions of sodium hypochlorite (household bleach), formaldehyde, or glutaraldehyde. In 1982 a vaccine containing viral surface antigen was licensed for use in the United States and is recommended for protection of persons with high risk of exposure to the hepatitis B virus. Persons who work in blood banks, medical laboratories, and other high-risk areas should be well informed of the mode of transmission and methods for preventing infection. Due to increases in the population of carriers in the United States, the use of hepatitis B vaccine for routine infant immunization is recommended by the CDC.

Another particle, the **delta hepatitis agent** (also called **hepatitis D virus**) often accompanies hepatitis B infections. The genome of this particle is a single-stranded circular RNA that is similar to many plant viroids. The nucleic acid is packaged in a protein capsid (the delta antigen), which is covered by the hepatitis B surface antigen. Without this scavenged envelope, the delta hepatitis agent cannot infect target liver cells. Thus it is a defective virus, dependent on coinfection with hepatitis B virus. The delta hepatitis agent increases the severity of HBV infections and increases mortality. Chronic infection with HBV continues to support the growth of its dependent partner. Delta agent can be detected by serological tests that measure delta antigen. Elimination of HBV gets rid of the delta agent, and recovery from infection or vaccine-induced immunity to HBV will also provide protection against the delta hepatitis agent.

TABLE 24-9 CHARACTERISTIC FEATURES OF HEPATITIS VIRUSES

Feature	Hepatitis B	Hepatitis A	Hepatitis C	Hepatitis D	Hepatitis E
Viral nucleic acid	DNA	RNA	RNA	RNA	RNA
Average incubation period	90 days	30 days	50 days	90 days	50 days
Common vehicles of transmission	Blood products, contaminated needles and syringes	Food, water, shellfish	Blood products	Same as HVB	Water
Major portal of entry	Parenteral; sexual	Oral	Parenteral	Parenteral	Oral
Major portal of exit	Wound, usually created by a needle	Fecal	Same as hepatitis B	Same as HVB	Fecal
Carriers	Yes	No	Yes	No	No
Mortality in untreated cases	1%	<0.1%	<1%	<1%	1%*

*20 percent in pregnant women.

Screening donors and monitoring blood for the presence of hepatitis B virus has lowered the incidence of posttransfusion hepatitis. Many transfusion recipients, however, have acquired classical viral hepatitis from blood that was verified free of hepatitis B virus. Many of these cases of *non-A, non-B (NANB) hepatitis* are now known to be caused by another parenterally transmitted hepatitis virus designated **hepatitis C virus (HCV)**. The agent is a positive-strand RNA virus antigenically distinct from hepatitis A or hepatitis B virus. Hepatitis C has an incubation period of 50 days, intermediate between that of hepatitis A and hepatitis B. Although acute infection with HCV is usually mild (70 percent have no symptoms), up to half of infected individuals will become chronically infected and have an increased risk of liver damage. A screening test for this virus is now available. Despite all the screening tests, some cases of transfusion hepatitis still occur, suggesting that other agents have yet to be identified.

Major differences that exist between HBV and other hepatitis viruses are detailed in Table 24-9.

■ Acquired Immunodeficiency Syndrome (AIDS)

Although primarily transmitted by sexual intercourse, AIDS can be transmitted by inoculation with blood contaminated with HIV. Blood donations are not accepted from HIV-positive individuals, and blood supplies for transfusions are routinely screened for the virus. The current screening procedures identify HIV-contaminated blood by the presence of antibodies. It is estimated that the risk of infection by transfusion is 1 in 40,000 units. Most bloodborne transmission occurs among intravenous drug abusers sharing contaminated needles. ▶ (Details on the replication and pathogenesis of HIV can be found in Chap. 23, p. 617.)

■ Fungal Infections

Many opportunistic fungi in soil or on vegetation can be introduced through the skin by piercing wounds. Spores of these fungi are found on thorns, barbs, and splinters. **Sporotrichosis,** the most common parenterally acquired mycosis in the United States, begins as a local subcutaneous lesion where the organism enters the body ▶ (see subcutaneous mycoses, Chap. 11, p. 272). The entry site is often a wound made by the accidental prick of a plant, thorn, or barb contaminated with the pathogen *Sporothrix schenckii*. The disease is an occupational hazard of gardeners, farmers, and horticulturists (hence the common name "rose gardener's disease"). The initial lesion often progresses from the site of inoculation to the regional lymph nodes, forming a series of enlarged nodules that extend up the arm or leg. The nodules may become necrotic and ulcerate. Although invasion of lungs and other organs is an occasional severe complication, spreading beyond the local lymph nodes is rare.

Mycetomas are common parenterally acquired mycoses in tropical and subtropical regions, especially among persons who seldom wear shoes. The initial lesion develops at the site of entry, most often on the feet (Fig. 24-19a). A variety of environmental organisms may produce the swelling and the local suppurative lesions of the skin and subcutaneous tissues that are characteristic of mycetoma. The organisms include *Pseudoallescheria boydii* (the most common agent of mycetoma in the United States), *Madurella mycetomatis*, and numerous other molds. Lesions usually remain localized but can

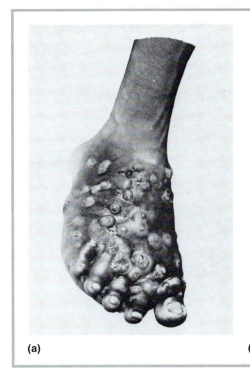
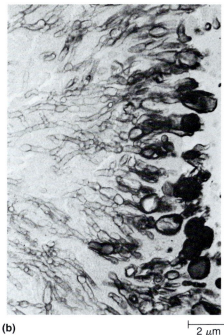

FIGURE 24-19 **Mycetoma.**
(*a*) Clinical appearance and
(*b*) microscopic appearance of
the fungus *Madurella myceto-matis* in tissue.

2 μm

(a)

(b)

cause crippling disfigurement unless they are surgically removed. Another mycosis acquired by puncture wounds to the extremities is **chromomycosis,** caused by fungi from the genera *Phialophora, Fonsecaea,* and *Cladosporium.* Lesions slowly develop into a cauliflowerlike mass on the inoculated limb. Rarely, the pathogens invade the bloodstream and brain.

Diagnosis of the parenterally acquired mycoses depends on microscopic observation of fungi in tissue preparations (Fig. 24-20*b*) and isolation of the pathogen. Materials aspirated from the lesions are the best specimens for examination and culture. The lesions of mycetoma characteristically produce granules of packed hyphae that are readily observed microscopically. Pigmented segmented spores called *fission bodies,* or sclerotic bodies, are indicative of chromomycosis. *Sporothrix schenckii,* however, is rarely detected by examining tissue. Positive identification of this dimorphic fungus depends on growing the mycelial phases at 25°C and the yeast phase at 37°C.

Sporotrichosis is treated by oral administration of potassium iodide or intravenous therapy with amphotericin B. Chromomycoses also respond to amphotericin B or to flucytosine. No drugs are effective against mycotic mycetomas, and surgical removal of lesions remains the best therapy.

The characteristics of diseases acquired by wound contamination are summarized in Table 24-10.

CONCEPT CHECK

- Why are wounds, especially burns, so susceptible to infections?
- Discuss at least four examples that show how the nature of the wound influences which pathogen(s) infect the wound.
- Describe the bacterial virulence factors that contribute to the pathogenesis of each of the following: tetanus; gas gangrene; *Pseudomonas aeruginosa* burn wound infections.
- Explain why delta hepatitis virus infects only in the presence of hepatitis B virus.

Diseases Acquired Through Superficial Breaks in the Skin

Minor breaks in the integrity of the skin may allow entry of microbes. Such inoculation probably requires direct contact between the open lesion and infected tissues.

■ Leprosy (Hansen's Disease)

Throughout history leprosy has been an especially alarming disease, slowly disfiguring its victims. In

Disease	Pathogen	Reservoir	Incubation Period	Diagnosis	Antimicrobial Treatment
Tetanus	*Clostridium tetani*	Humans, animals, spores in soil	4–21 days	History of wound and clinical picture	Penicillin
Gas gangrene	*Clostridium perfringens*, other clostridia	Humans, animals, spores in soil	1–3 days	Isolation and identification of pathogen from wound	Penicillin
Sporotrichosis	*Sporothrix schenckii*	Soil, vegetation, wood	1 week to 3 months	Cultivation and identification of fungus from lesion	Iodides
Mycetoma	*Madurella mycetromi*, *Pseudoallescheria boydii*	Soil, decaying vegetation	Months	Microscopic identification of fungus in lesion or isolation in culture	None
Hepatitis B	Hepatitis B virus	Humans	60–90 days	Serological identification of viral antigens in blood	None
Hepatitis C	Hepatitis C virus	Humans	40–60 days	Serological identification of viral antigens in blood	None

the past, people with leprosy were often confined to remote asylums or made to carry conspicuous alarms, such as bells and lanterns, to warn others of their approach. These abuses supposedly helped protect uninfected people from contracting the disease. Today we know that leprosy is one of the *least* communicable of all infectious diseases. Most people can safely live in the same household with a leprosy patient. For unidentified reasons, however, a few people are susceptible to contracting the disease, perhaps by direct intimate contact with leprosy lesions. The details of transmission are still poorly understood, but the pathogen may enter the body through otherwise inconsequential skin abrasions. Inhalation of contaminated nasal secretions is another possible route of infection.

Leprosy has been an especially difficult disease to study because the causative agent, *Mycobacterium leprae*, cannot be grown on laboratory media. *M. leprae* can be cultured only in the nine-banded armadillo or in the foot pads of mice, where the pathogen grows extremely slowly, about one generation every 12 days. (Compare this to *Escherichia coli*'s generation time of 20 minutes.) Each experiment may require months or years to obtain even the earliest results. This slow growth probably accounts for leprosy's long incubation period of 2 to 10 years and the slowly progressive nature of the disease.

The tissue affinity of *Mycobacterium leprae* for pe-

ripheral nerve cells triggers chronic neurological inflammation, often causing loss of feeling in the region. This anesthetizing effect promotes physical injury and secondary infections of hands, fingers, and other numb extremities. The primary symptoms, however, are largely due to a delayed hypersensitivity response to the pathogen, which ultimately produces the characteristic cutaneous lesions and disabling neurological inflammation. This form of the disease, called *tuberculoid leprosy*, is infrequently fatal, and patients may recover without treatment. In persons with defective cell-mediated immunity, on the other hand, the pathogen may invade every organ in the body. If untreated, this progressive form, called *lepromatous leprosy*, is usually fatal. Its victims may die because of secondary bacterial infection, often tuberculosis, or because of waxy deposits that accumulate in and injure the liver, kidney, and spleen.

In some persons symptoms alone are characteristic enough to diagnose leprosy. Otherwise the disease can be diagnosed by microbiological findings such as demonstrating acid-fast bacilli in cells from the lesions. Skin tests, using an extract of lepromatous tissue called lepromin, are also used. Because the disease is poorly transmitted, isolation of leprosy patients is unnecessary. Most patients eventually recover after years of treatment with sulfone drugs (dapsone) usually with rifampin and/or clofazimine (Fig. 24-20). An estimated 12 million persons are believed to currently have leprosy. Most of these are in Africa, Asia, and Latin America. There are about 250 cases of leprosy reported annually in the United States.

■ Yaws and Pinta

Yaws and pinta are treponemal infections that are primarily transmitted by nonvenereal routes among persons living in the tropics. These diseases are usually contracted by skin-to-skin contact among children or by flies that feed on the open lesions. *Treponema pertenue*, the etiologic agent of *yaws*, causes a primary skin lesion, called a "mother yaw," that releases bacteria into the bloodstream. If untreated, the pathogen spreads to bone, lymph, and skin sites over the body. The skin lesions are infectious if they become ulcerated. *Pinta* is caused by *T. carateum*. The hyperpigmented lesions later become nonpigmented and hyperkeratotic (the skin thickens). They usually appear on the hands, feet, and scalp.

The microscopic appearance and serologic responses of *T. pertenue* and *T. carateum* are identical to those of *T. pallidum*. The three agents are so similar that persons who naturally recover from yaws and pinta acquire some cross-protection against syphilis. Penicillin is effective in treating these diseases.

■ Anthrax

Anthrax is a disease of animals that spreads to humans primarily through minor breaks in the skin or mucous membranes. Spores of *Bacillus anthracis* can survive in soil or on articles for years. Livestock are often infected while feeding on contaminated soil or contaminated feed products. Infected animals develop a severe, usually fatal disease. Most human infections occur in workers whose occupations expose them to infected animals or animal products (hides, wool, or hair). In addition to direct inoculation through broken skin (cutaneous anthrax), the bacterium can be acquired by inhalation or, more rarely, by ingestion of contaminated, insufficiently cooked meat.

Cutaneous anthrax first appears as a papule that develops into a vesicle and in 2 to 6 days into a *black eschar* (a hard crust or scab) that is surrounded by small vesicles. If untreated, *B. anthracis* may spread through the lymphatic system to the bloodstream. The bacillus produces a potentially lethal toxin, and between 5 and 20 percent of untreated cases are fatal.

Diagnosis is confirmed by direct microscopic identification of the gram-positive pathogen or by isolation of *B. anthracis* from lesions or blood. The infection responds to penicillin and probably induces immunity. The incidence of anthrax is minimized by immunizing susceptible animals and persons most likely to be exposed to the pathogen. Additional control measures depend on disinfection and sterilization of animal products as well as on the destruction and safe disposal (burning or burial) of infected animals.

Anthrax finds an unfortunate place in history as one of the diseases studied for use in biological war-

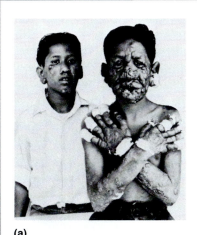 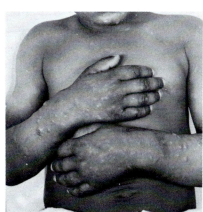 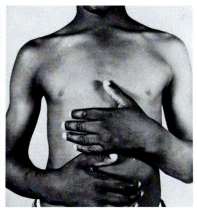

(a) (b)

FIGURE 24-20 The effect of sulfone on the severity of leprosy. (*a*) Before sulfone was available, the disease in this patient followed a relentless course, within 2 years producing the results shown at the right. (*b*) Four years of sulfone treatment halted the progress of the disease in this patient.

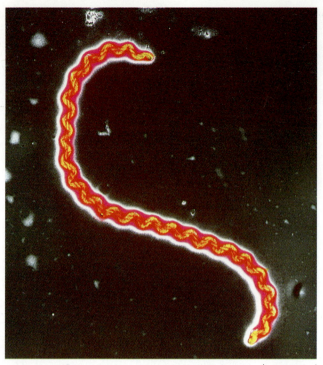

FIGURE 24-21 *Leptospira interrogans.* The shape of this curved spirochete accounts for its name (*interrogans* = question mark).

fare. The ability of the spores to survive in soil left some testing sites uninhabitable until measures were implemented to eradicate the spores ▶ (see THE MICROBES: Imortal Seeds of Death, Chap. 4, p. 94).

■ Tularemia

Many wild animals, domestic animals, and various ticks harbor *Francisella tularensis*, which causes *tularemia* in humans. Disease symptoms vary with the route of infection of this small gram-negative bacterium. It is most commonly acquired either by inoculation of minor skin wounds while handling infected animals or by the bite of an infected tick. An ulcer usually forms at the inoculation site, and regional lymph nodes become enlarged. Hematogenous spread may produce infection of other organs and lead to meningitis or pneumonia. Occasionally the disease is acquired by eating undercooked meat from an infected animal, resulting in gastrointestinal symptoms. When inhaled in contaminated dust, the bacteria may cause pneumonia.

Tularemia was first observed in Tulare County in California, but the disease occurs throughout the United States, Canada, the European continent, Russia, and Japan. It is most prevalent among hunters who fail to wear gloves when skinning or handling animals. The major animal reservoir in the United States is rabbits. No direct person-to-person transmission occurs.

Diagnosis usually depends on demonstrating increased antibody titers during the infection. Isolating and identifying the organism from lesions is difficult, because the microbe stains poorly in tissue and has fastidious growth requirements. Growth on specialized laboratory media is slow, and biohazard precautions are necessary to prevent infection of laboratory personnel. Most cases respond to streptomycin. An attenuated vaccine gives some protection and may be administered to high-risk individuals.

■ Leptospirosis

Animals that are infected with *Leptospira interrogans* shed the spirochetes in their urine, thereby contaminating water or soil. Humans contacting infected animals or bathing in contaminated water may become infected, especially if their skin is abraded. The disease can also be acquired by ingesting food or water contaminated with the urine of an infected animal. Most cases of leptospirosis result from recreational or occupational exposures. Rice and sugar cane field workers, farmers, sewer workers, veterinarians, and slaughterhouse workers are at high risk of exposure.

The infection may be asymptomatic or mild and flulike, or it may progress to severe disease. Symptoms usually appear within 10 days and characteristically include fever, severe headache, chills, and vomiting. Often the liver is affected and jaundice develops. Occasionally a rash, skin hemorrhage, or meningitis occurs.

Diagnosis is confirmed by isolating and identifying the curved spirochete in blood or urine or by immunologic tests. Penicillin and tetracycline are effective if administered early. In untreated patients with jaundice the mortality rate is about 20 percent.

Table 24-11 presents the major features of diseases acquired by inoculation through superficial breaks in the skin.

> **CONCEPT CHECK**
>
> - Compare the methods of diagnosis and treatment for leprosy, yaws, and tularemia.
> - Describe the transmission cycle to humans for the zoonotic pathogens of anthrax, leptospirosis, and tularemia.

Disease	Pathogen	Characteristic of Pathogen	Reservoir	Incubation Period	Diagnosis	Antimicrobial Treatment
Leprosy	*Mycobacterium leprae*	Acid-fast rod	Humans	3–6 years	Microscopic identification of pathogen in lesion	Dapsone + rifampin
Yaws	*Treponema pertenue*	Spirochete	Humans	14–90 days	Microscopic identification of pathogen in lesion	Penicillin
Pinta	*Treponema carateum*	Spirochete	Humans	3–60 days	Microscopic identification of pathogen in lesion	Penicillin
Anthrax	*Bacillus anthracis*	Gram-positive spore-forming rod	Domestic farm animals, spores in soil, animal fur, wool, and fur products	2–5 days	Demonstration of pathogen in blood or lesions by serology or culture	Penicillin
Tularemia	*Francisella tularensis*	Gram-negative rod	Rabbits, muskrats, ticks	2–10 days	Isolation and identification of pathogen from lesion or blood	Streptomycin
Lepto-spirosis	*Leptospira interrogans*	Spirochete	Farm and pet animals, rodents	4–19 days	Serologic tests or isolation of pathogen from blood or urine	Penicillin

Diseases Acquired Through the Eye

Like the skin, the eye is constantly exposed to microorganisms. The eyelids provide mechanical protection from infection, as does the flushing action of tears. Tears also contain lysozyme and other antimicrobial substances. A few pathogens readily infect the conjunctiva, the mucous membranes that line the eyes and eyelids. Inflammation of the conjunctiva is called **conjunctivitis.** Several pathogens establish themselves in eye tissue that has been damaged, for example, by the prolonged use of contact lenses. If microbes are not eliminated, blindness may result.

■ Trachoma

Chlamydia trachomatis causes *trachoma,* a conjunctivitis that blinds more people in the world than any other cause (Fig. 24-22). The microbe possesses surface molecules that bind to epithelial cells. It enters the host cell and thus avoids being rinsed away by

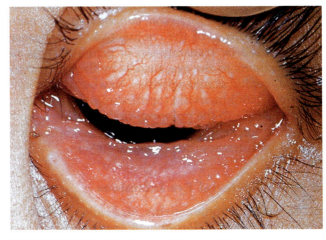

FIGURE 24-22 **Trachoma.** The world's leading cause of blindness is an eye infection by *C. trachomatis.*

Key Terms

parenteral inoculation (p. 637)
epidermis (p. 637)
dermis (p. 637)

cutaneous infections (p. 642)
subcutaneous infections (p. 642)
pyogenic (p. 642)

papillomas (p. 645)
buboes (p. 654)
Weil-Felix reaction (p. 658)

Key Facts and Concepts

- **Surface defenses and vulnerabilities.** Intact skin is an inhospitable environment for most microbes and effectively prevents their access to underlying tissues. Only a few pathogens can multiply on the surface or in the follicles of undamaged skin. Cutaneous infections are most likely to occur in areas of the skin where moisture accumulates or in persons with decreased surface shedding. Underlying diseases can also predispose for cutaneous infections. Most infections, however, are acquired through breaks in the skin's physical integrity and are not confined to superficial surfaces.

- **Breaching the wall.** Abrasions, nicks, and cuts that would otherwise be inconsequential and often unnoticeable are portals of entry for several pathogens. Transmission of these microbes is usually by autoinoculation or by direct contact with an infected reservoir or contaminated vehicle.

- **Inoculation through the skin.** Wounds that penetrate beyond the superficial layers of skin are portals of entry for some dangerous pathogens, most of which gain access to the subcutaneous tissues at the time the wound is inflicted. The microbes may then be disseminated by the blood, lymph, or nervous system. Rabies and rat-bite fever are transmitted by the bite of an infected animal. Arthropods are vectors of diseases caused by bacteria (plague, relapsing fever, spotted fevers, and typhus), viruses (benign and hemorrhagic fevers and encephalitis), and protozoa (leishmaniasis, trypanosomiasis, and malaria). Controlling these diseases depends on reducing the animal reservoir and arthropod vector populations.

- **Wounds that become infected.** Extensive wounds and burns expose underlying tissue to pathogens in the environment. The threat of infection can often be minimized by proper cleansing and care of the wound. *Clostridia* and other anaerobic pathogens survive in the oxygen-poor environment of deep wounds. *Clostridium tetani* releases a neurotoxin from its localized infection site in deep wounds. *Clostridium perfringens,* on the other hand, releases enzymes that destroy surrounding host tissues and facilitate bacterial spread and the development of gas gangrene.

- **Infection of mucosa of the eyes.** The eye is protected from invasion by mechanical and chemical defenses. *Chlamydia trachomatis* avoids these defenses by its intracellular existence and causes trachoma or inclusion conjunctivitis, depending on the serotype. Other bacterial and viral pathogens also infect the conjunctiva or cornea. Contact lenses predispose the wearer to eye infections by *Pseudomonas aeruginosa* and *Acanthamoeba*.

Review and Problem-Solving Questions

1. What distinguishes *S. aureus* from the usually non-pathogenic strains of staphylococci?
2. What factors predispose for each of the following? (a) Acne; (b) dermatophyte infections; (c) cutaneous viral infection; (d) candidiasis; (e) hepatitis B; (f) *Acanthamoeba* eye infections.
3. How are the following used in disease diagnosis? (a) Negri bodies; (b) Weil-Felix reaction; (c) hepatitis B surface antigens.
4. What evidence suggests that humans are probably not the reservoirs of infection for viral encephalitis?
5. Describe three routes by which malaria can be transmitted among humans. In each case, how can transmission best be interrupted?
6. Compare the pathogenic agents, modes of transmission, and methods of treatment and control of rabies, plague, tetanus, and leprosy.
7. Discuss the role of enzymes in the development of gas

gangrene. Why is this disease difficult to cure with antibiotics alone?

8. Distinguish between the following:
   (a) Bubonic and pneumonic plague
   (b) Jungle and urban yellow fever
   (c) Benign and hemorrhagic fever
   (d) Tinea unguium and tinea capitis
   (e) African and American trypanosomiasis
   (f) Hepatitis B and hepatitis C
   (g) Conjunctivitis and keratitis

9. Lyme disease and hepatitis C are relatively new diseases. What factors may account for their recent appearance?

10. Discuss why the following quote is probably true: "Reducing populations of mosquitoes and other blood-sucking pests has saved more human lives than all antibiotics combined."

11. Why aren't all bloodborne pathogens, such as HIV, spread by blood-sucking arthropods?

12. Historically, individuals with leprosy have been isolated from the community even though it is not a highly communicable disease, while people with tuberculosis, a more contagious mycobacterial infection, have not. Why would such an irrational situation exist?

13. Four patients arrive at the local emergency room with wounds to the leg. Arlene has stepped on a nail while shopping in a hardware store. Dave has also stepped on a nail; however, this occurred when he was repairing a fence. Aaron, who is only 5 years old, scraped himself after falling from his bike, and his mother Judy ran into a rose bush as she went to help her son. What would be your concerns for each patient? How should they be treated?

14. Explain the following findings.
   (a) Among adults the risk of tetanus in those over 80 is 10 times the risk of those between 20 and 29 years old.
   (b) The use of chickens as sentinels has reduced the number of cases of viral encephalitis.
   (c) Although rabies is common among wildlife species, dogs are responsible for more than 90 percent of human cases worldwide.

15. The following data come from a study designed to determine factors associated with increased risk for malaria among U.S. personnel living in endemic areas of Africa. Each person was asked about their behaviors related to malaria prevention. What conclusions do you draw from the data, and what precautions would you take for preventing malaria?

Preventive Behavior	Percent of Indicated Group Who Contracted Malaria	
	Persons Using Behavior	Persons Not Using Behavior
Chemoprophylactic medicines	9	24
Mosquito nets around beds	7	15
Insect repellent	0	15
Protective clothing	20	22

16. Design an experiment to determine how many and what types of microbes are present in used mascara wands.

HEADLINES

BLOOD-SUCKING ARTHROPODS: NEW IMPROVED VARIETIES

Arthropods serve as vectors for transmission of several of the deadliest diseases of humans, and their elimination has been viewed as the only way to eradicate these diseases. The appearance of insecticide-resistant arthropods, however, has frustrated attempts at control. Furthermore, the emergence of drug-resistant pathogens and the failure to develop effective vaccines indicate that new strategies against these diseases must be developed. C. B. Beard and coworkers[*] suggest a novel approach—making the vector inhospitable for the pathogen, thereby cutting off the pathogen's access to humans. They suggest that foreign genes that are toxic to the parasite or block its development be introduced into the vector by mutualistic bacteria that are associated with most arthropods, often in obligate relationships. The genetically engineered vectors would then be released into the environment to dilute the species with the pathogen-resistant versions.

- What kind of gene functions might be appropriate to be transferred, and where might one find the genes?
- Outline a procedure for creating modified symbiotic bacteria.
- How might these symbionts be spread into the insect population?
- What diseases might be controlled by this method?

[*]C. B. Beard, S. L. O'Neill, R. B. Tesh, F. F. Richards, and S. Aksoy, Modification of arthropod vector competence via symbiotic bacteria, *Parasitology Today*, 9: 179–183 (1993).

CHAPTER 25

Nosocomial Infections

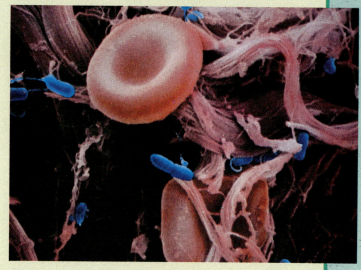

Burn wound colonized by *Pseudomonas aeruginosa*. The bacteria are blue in this false-colored SEM; one disk-shaped erythrocyte is visible among the debris. *Pseudomonas* and other hospital-acquired pathogens cause especially troublesome problems. This chapter discusses the reasons why the epidemiology of hospital-acquired infections is so different from that of diseases acquired in the general community.

With explosive fury, tragedy strikes a 13-year-old American and his Brazilian cousin as they burn the family trash. Gathering pressure within the flames, an aerosol paint can sets off a chain reaction of events that will claim lives thousands of miles away. The can explodes in the fire, splattering the American with flaming embers. Seconds later he rolls on the ground to extinguish his burning clothes. Although his quick thinking prevents him from burning to death, it does not save his life.

As he lies in his hospital bed with second and third degree burns, the event that will kill him and so many others occurs without notice. The boy coughs while his dressings are being changed, discharging several dozen types of microorganisms into the air, some of which settle on his burned flesh. One of these microbes is the species *Providencia stuartii*. These bacteria have been living on the boy for most of his life without causing disease, but in the burn wounds they are no longer held in check by the skin's natural barrier or phagocytic cells. Growing uncontested in the nutrient-rich burn sites, the microbe soon reveals its presence to the attending staff—it is clear that the boy is suffering from infection. He is treated with antibiotics, but his infection worsens, invading his bloodstream. The boy is flown to a hospital in the United States that specializes in burn treatment. The staff prepares for his arrival.

Specimens are quickly collected for microbiological examination, and he begins a regime of several antibiotics, one of which will hopefully knock out his infection. But his condition deteriorates. Lab reports reveal the invader to be *Providencia stuartii*, but antibiotic sensitivity tests fail to find any chemotherapeutic agent that will stop its growth. Soon four other patients in the hospital, all severely immunocompromised by their injuries or illnesses, acquire the organism. By the time the boy dies of his infection, others are struggling for their lives against the bacterium. Then things grow much worse. Several other types of bacteria begin causing new infections in the hospital, all resistant to every antibiotic that the doctors can throw at them. Laboratory analysis reveals that the original *Providencia* harbors a transmissible plasmid, an R factor that makes it virtually invulnerable. This new plasmid carries genes that impart resistance to every antibiotic commonly used in the hospital plus many others. *Providencia* has conjugated with other genera of opportunistic gram-negative bacteria, which are now causing equally tough infections in other patients. A dozen victims die before the hospital epidemic is controlled by isolating the infected patients and assigning separate personnel to treat only them. The isolated patients continue to die, but new patients in the hospital remain free of the plasmid-bearing bacteria.

This composite of several actual episodes reveals how today's health care system can be a double-edged sword. The unprecedented battery of life-saving resources are themselves instrumental in creating a hospital environment in which the threat of infection is far more serious than in the general community. This is, in fact, one of the major problems in contemporary medicine: the alarming numbers of people who contract infectious diseases as a direct result of exposure to the hospital environment. These hospital-acquired diseases are called **nosocomial infections** (*nosokomeion* = hospital).

Each year in the United States about 2 million people reportedly contract an infectious disease during their hospital stay. This reported nosocomial infection rate of 5 percent is considered by many experts to be lower than the actual incidence. Nosocomial rates in some American hospitals may run as high as 15 percent, usually depending on the type of patients present in a particular hospital. In other words, 1 in every 7 patients may acquire an additional infectious disease after admission to the hospital. On average, the victim of nosocomial disease stays in the hospital an additional 4 days. At a cost of $600 per day, nosocomial diseases add an annual $4 billion to hospitalization expenses. The cost in terms of human lives and suffering is even more important. More than half of all infectious diseases seen in hospitals in industrialized countries are nosocomial. Each year in the United States at least 20,000 patients die as a direct result of infections they acquire while hospitalized for other causes.

Epidemiology of Nosocomial Infections

Patients who develop a disease within their first few days in the hospital may have been in the incubation phase of a *community-acquired infection* (CAI) at the time they were admitted. Nosocomial infections, however, represent diseases for which there is no evidence that infection was present or incubating at the time of hospital admission. Infections in newborns that are acquired during passage through the birth canal are also considered nosocomial (those acquired during pregnancy by transplacental passage are not). With nosocomial infections, exposure to the pathogens occurred during and as a direct result of hospitalization.

Hospital-acquired infections also pose a potential hazard to the general community since symptoms often do not develop until after the patient is discharged, and many of these patients harbor virulent pathogens. Infectious disease introduced into the community may be difficult to control as individuals freely and continuously enter and exit.

Prevention of hospital-acquired infections depends on cooperation among hospital personnel, especially those in contact with patients. It requires an understanding of the special factors in the hospital that contribute to the problem. The types of reservoirs, the unusual opportunities for transmission, and the nature of the susceptible population combine to create a unique epidemiological setting. Because of these special features, the types and frequencies of infections contracted in the hospital are different from those of infections acquired within the community outside the hospital (Fig. 25-1).

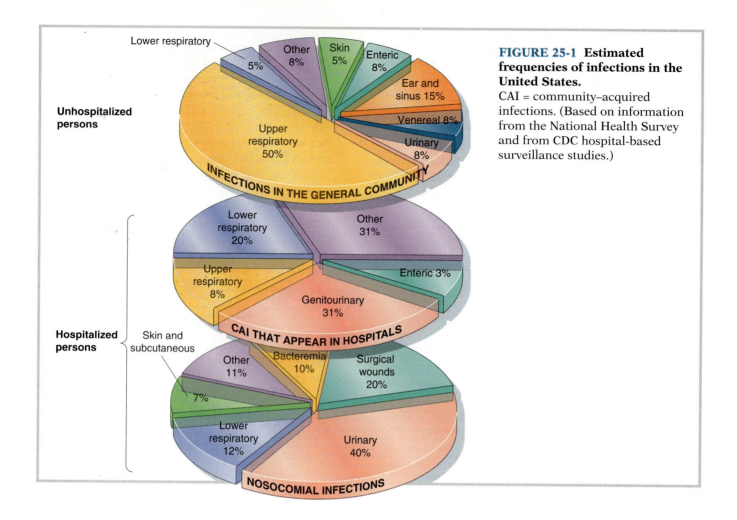

FIGURE 25-1 **Estimated frequencies of infections in the United States.**
CAI = community–acquired infections. (Based on information from the National Health Survey and from CDC hospital-based surveillance studies.)

Unhospitalized persons

Lower respiratory 5%
Other 8%
Skin 5%
Enteric 8%
Ear and sinus 15%
Venereal 8%
Upper respiratory 50%
Urinary 8%

INFECTIONS IN THE GENERAL COMMUNITY

Hospitalized persons

Lower respiratory 20%
Other 31%
Upper respiratory 8%
Enteric 3%
Genitourinary 31%

CAI THAT APPEAR IN HOSPITALS

Skin and subcutaneous
Other 11%
Bacteremia 10%
Surgical wounds 20%
7%
Lower respiratory 12%
Urinary 40%

NOSOCOMIAL INFECTIONS

■ Characteristics of the Susceptible Host

The hospital contains a group of **compromised patients** who are unusually susceptible to infectious disease because of illness and medical procedures that reduce host resistance. Most infectious diseases acquired in the hospital are caused by opportunistic pathogens that pose no danger to normally healthy persons or to persons not receiving treatments that increase the risk of infection. The following factors commonly found in hospitalized patients create an especially susceptible population:

- Some infectious diseases for which patients may require hospitalization reduce resistance and predispose for secondary infection. Viral respiratory tract infections, for example, may be followed by bacterial infection. Bacterial pneumonia is responsible for most fatal cases of influenza.
- The generous use of antibiotics disrupts the protective normal flora, leaving the patient susceptible to colonization by hospital strains of microorganisms resistant to the antibiotic used.

- Large wounds, especially burns and surgical wounds, provide access to tissues that are ideal for bacterial growth.
- Invasive procedures may introduce hospital strains of pathogens directly into the target area, bypassing normal host defenses.
- Uncontrolled diabetes and other constitutional diseases increase susceptibility to infection by the opportunistic hospital strains.
- Newborn infants and the elderly are especially susceptible to serious infections by organisms of low virulence. More than 50 percent of all nosocomial infections occur in people older than 65.
- Immunosuppression to accommodate organ transplants leaves a patient vulnerable to infectious disease by most microorganisms in the hospital. Infections in such patients are commonly fatal.
- Treatment of cancer patients with cytostatic or cytotoxic drugs kills not only tumor cells but also the lymphatic cells needed for phagocytosis and acquired immunity. Cancer chemotherapy thus lowers the patient's ability to fight infectious disease.

- AIDS patients among the adult and pediatric population represent a growing proportion of hospitalized patients. Infection control is aimed at preventing transmission of HIV to hospital personnel and reducing the risk of infection to the immunocompromised individuals.

Because so many hospitalized patients are suffering from at least one of these predisposing conditions, the population most susceptible to serious infections by opportunistic pathogens is concentrated in hospitals.

■ Reservoirs of Infection

Hospitals are unavoidably reservoirs of virulent and opportunistic pathogens. Some nosocomial diseases are *exogenous infections*—that is, the microbe is transmitted from another source to the patient. They may be acquired as **cross infections** from other infected patients, visitors, or hospital personnel or as environmental infections from contaminated fomites, food, water, or air. People with active disease shed infectious agents that, unless confined, may contaminate the environment or directly infect other patients, employees, or visitors. Healthy carriers also shed pathogens, usually with no knowledge of their asymptomatic infections, an especially dangerous situation if the carrier treats patients. Most of these carriers probably acquired the pathogens from infected patients, usually because of failure to wash hands after contact with the patient, inadequate aseptic technique during treatment, or improper isolation of infectious patients. Persons who fail to take these precautions need not become infected to transfer organisms from an infected patient to other susceptible persons. Hands that harbor transient contaminants are common mechanical vectors of infection in hospitals. Surveys show that one in five medical professionals carries potentially pathogenic antibiotic-resistant pathogens on his or her hands. Hand washing by medical professionals in some hospitals occurs at only 30 percent of the ideal rate. Infected or colonized medical professionals (including physicians) are prevalent sources of epidemics within the hospital (Fig. 25-2). *Failure to wash one's hands before and after each patient contact is probably the most important contributor to the spread of these infections.*

Nosocomial infections are also acquired by contact with contaminated inanimate materials. For example, in one nationwide outbreak, 400 cases of *Enterobacter* bacteremia followed intravenous therapy with contaminated fluid distributed by a large manufacturer. Fifty-two of these patients died before the

FIGURE 25-2 Bacteria grown from a hand. This imprint was made by a nurse who had just engaged in cardiopulmonary resuscitation. Many of these organisms are opportunistic pathogens that could infect other patients should the nurse fail to wash his or her hands before treating them.

source of the infection was identified. Contaminated respiratory equipment has been responsible for epidemics of severe lower respiratory tract infections. Intestinal infections have been caused by contaminated endoscopes (used for examining internal mucous membranes), and infant diarrhea is often due to feeding contaminated bottled milk. Some opportunistic pathogens grow in soaps or lotions. In some instances, the disinfectants and antiseptics employed to reduce microbial contamination become heavily populated with resistant bacteria. Outbreaks have been associated with mycobacteria growing in presumably sterile gentian violet solutions (used to mark incision sites prior to surgery). Flower vases are no longer allowed in surgical wards and burn units because of their potential for harboring dangerous bacteria. Some opportunistic pathogens proliferate in the ventilation systems and are subsequently dispersed throughout the hospital.

Occasionally the source of nosocomial pathogens is the patient's own normal flora. These *endogenous infections* usually occur when the indigenous flora are transferred to susceptible body sites by invasive procedures or when a predisposing factor compromises the patient's normal defense mechanisms. Often infections with endogenous organisms cannot be prevented despite careful precautions.

The epidemiology of nosocomial infections is depicted in Figure 25-3.

use of antibiotics in the hospital. With the discovery of penicillin and sulfa, a new era of medicine and medical education began, an era that we associate with significant reduction in the fear of infectious disease. As powerful and important as antibiotics are, they quickly acquired a reputation as a virtual panacea against infection. The consequences of such unrealistic expectations being placed on antibiotics by people in the general community and professionals in medicine have been severe.

MYTH: Antibiotics and chemotherapeutic agents are capable of preventing infection when proper asepsis is breached. (This myth is still prevalent in the medical community.)

FACT: While the positive attributes of antibiotics should not be underrated, reliance on drugs rather than on medical asepsis has led to relaxed aseptic technique and has contributed to an alarming increase in nosocomial infections.

The overuse of antibiotics in the hospital has created another serious problem—a preponderance of drug-resistant pathogens (Fig. 25-7). While antibiotics reduced the prevalence of many formerly common pathogens in hospitals, the "ecological vacuum" left by these organisms has been rapidly filled by microbes that survive the most frequently used chemotherapeutic agents. These new breeds of hospital pathogens often cause infections that are difficult to cure. As a consequence, many patients die before an effective chemotherapeutic regimen can be initiated. Furthermore, multiple-drug-resistant plasmids can be readily transferred from drug-resistant strains to antibiotic-susceptible bacteria. ▶ (See Genetic Transfer in Bacteria, Chap. 8, p. 205.) The number of drug-resistant organisms therefore increases as antibiotics are used more frequently and less judiciously. Laboratory reports and antibiograms reduce reliance on the shotgun approach to chemotherapy and consequently discourage the emergence of drug-resistant pathogens. Although new antibiotics may provide successful chemotherapy against multiple-drug-resistant pathogens, bacterial populations quickly become resistant to these drugs when they are used excessively. ▶ (See Antibiotic Resistance, Chap. 16, p. 393.)

Concern about the problems associated with antibiotics has prompted hospital personnel to examine the reasons for prescribing antibiotic therapy. Data obtained from several hospitals on antibiotic use have indicated that more than 50 percent of the patients received unnecessary or inappropriate antimicrobial therapy. For example, although antibiotics have been proven to be of no value in preventing catheter-associated urinary tract infections or bacterial infections secondary to influenza and common colds, they continue to be frequently prescribed for these purposes. Such unjustifiable use of antibiotics not only increases the risk of nosocomial infections but substantially adds to the cost of medical care. It is a task of the hospital infection control committee to establish and enforce prudent guidelines for antibiotic use.

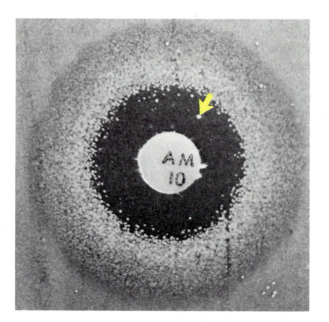

FIGURE 25-7 Emerging antibiotic-resistant microbe. The colony of ampicillin-resistant *Haemophilus influenzae* (arrow) growing in the usual zone of inhibition reveals a reservoir of resistant bacteria that may become predominant in an infected patient treated with ampicillin.

CONCEPT CHECK

- Why aren't all infections that emerge within the hospital considered nosocomial? Why don't all nosocomial infections emerge within the hospital?
- Explain why many hospitalized patients are at increased risk for opportunistic infections.
- Describe four factors that contribute to the high incidence of nosocomial urinary tract infections.
- Compare mechanisms of transmission for endogenous microbes to mechanisms of transmission from an exogenous source.
- Why can't the problem of nosocomial infection be eliminated by placing *all* hospitalized patients on broad-spectrum antibiotics to prevent new infections?

Etiology of Nosocomial Infections

Hospitals are no longer pesthouses where people are constantly exposed to other patients with highly infectious diseases. Patients known to be shedding highly virulent pathogens are kept well isolated from susceptible patients. It is not the highly virulent pathogens that present the major nosocomial problems in modern hospitals, but microorganisms of surprisingly low virulence, those to which we are all frequently exposed. Many are free-living saprophytes.

Nearly any strain of bacterium residing in the hospital can cause nosocomial infection, but the most common of these opportunistic pathogens are among the enteric bacteria (Table 25-2). These and several other genera of gram-negative rods have replaced the staphylococci as the prevalent nosocomial offenders. This change may be partially due to increased recognition that the "harmless" saprophytes are not merely contaminants but cause serious disease.

■ Staphylococci

These gram-positive cocci have historically been a major nosocomial problem, primarily because of the prevalence of healthy carriers. At least 30 percent of the adult population harbors *Staphylococcus aureus* in the nasopharynx or on the skin. *S. aureus* is among the pathogens most commonly isolated from nosocomial infections of adults and is the most common cause of newborn-associated nosocomial infections. In newborns, this organism is usually associated with cutaneous or ocular infections; however, it has been isolated from virtually every site. Most of the strains within the hospital are resistant to antibiotics. One particularly dangerous group is *methicillin-resistant S. aureus (MRSA)*. These organisms are resistant to a number of antibiotics and are often difficult to treat. The major danger areas for severe staphylococcal infections are the nursery, the delivery room, the operating room, and the burn unit, although transfer of infections among patients in hospital wards is not uncommon. Patients with open wounds infected by staphylococci should be separated from highly susceptible individuals such as postsurgical patients. The major mode of transmission is on colonized hands of health care workers who have failed to properly wash or change gloves between patient contacts.

Infection with *S. epidermidis* is frequently associated with intravenous catheterization, implanted devices, and central nervous system shunts. These coagulase-negative staphylococci enter the bloodstream and account for over one-fourth of nosocomial bacteremias. The microbes originate from the

TABLE 25-2 COMMON AGENTS OF NOSOCOMIAL INFECTIONS		
Organism	**Percentage of Nosocomial Infections**	**Common Sites from Which the Pathogen Is Isolated**
Gram-Negative Bacteria		
Escherichia coli	16	Urinary tract, surgical wounds, blood, lower respiratory tract
Pseudomonas sp.	12	Urinary tract, lower respiratory tract, burn wounds
Enterobacter sp.	7	Lower respiratory tract, surgical wounds, urinary tract
Klebsiella sp.	5	Lower respiratory tract, urinary tract, surgical wounds
Proteus sp.	4	Urinary tract, surgical wounds
Serratia sp.	2	Lower respiratory tract, urinary tract, surgical wounds
Bacteroides fragilis	1	Surgical wounds, bacteremia
Gram-Positive Bacteria		
Enterococcus faecalis	11	Urinary tract, surgical wounds, blood
Staphylococcus aureus	10	Skin, surgical wounds, blood, lower respiratory tract
Staphylococcus epidermidis	9	Skin, surgical wounds, blood
Streptococci	1	Blood, upper respiratory tract, skin
Fungi		
Candida albicans	2	Urinary tract, blood
Other fungi	1	Urinary tract, blood, lower respiratory tract

endogenous flora of the patient or health care worker.

■ Streptococci

Streptococcus agalactiae (group B streptococcus) is responsible for infections of the uterus following childbirth and for septicemia and meningitis in newborns. *S. agalactiae* is a normal vaginal resident in 20 to 30 percent of pregnant women. Although most infants escape these infections, a few develop neonatal disease, usually within the first week of life. About half of these infected children die, usually from septicemia, pulmonary invasion, or meningitis. Half the survivors of meningitis are blind, deaf, mentally retarded, or stricken with epilepsy or cerebral palsy. Premature infants are at greatest risk. The risk of neonatal group B streptococcal disease can be reduced by administering penicillin at the onset of labor to women whose vaginas are colonized by this bacterium. Group B streptococcal infections are generally treated with penicillin in combination with gentamicin.

Group A streptococci *(S. pyogenes)* and the pneumococci *(S. pneumoniae)* are infrequent causes of nosocomial infections, although they are prevalent pathogens of community-acquired infections.

■ Enterococci

Enterococci, formerly classified as group D streptococci, are members of the normal bowel flora. *Enterococcus faecalis* causes the greatest number of enterococcal infections in the hospital. This organism is second only to *Escherichia coli* as an etiologic agent of nosocomial urinary tract infections. *E. faecalis* is also one of the common causes of surgical wound infections. In fact, enterococci are the third most common cause of all nosocomial infections in the United States. They are readily transmitted from patient to patient whenever hand washing and hygienic procedures are poorly practiced. In addition to causing urinary tract and postsurgical infections, enterococci frequently invade heart valves, intraabdominal locations, and prosthetic devices of hospitalized patients. Patients with heart defects should be given prophylactic antibiotics before surgery on the gut. Treatment usually includes a combination of penicillin and aminoglycosides.

■ Enterobacteria

Before 1960, disease caused by *Serratia marcescens*, a common saprophyte, was unusual. Now this "harmless" bacterium, *Klebsiella*, *Escherichia*, and *Proteus* are among the most common pathogens in the hospital. Each causes disease that often cannot be cured by the commonly used antibiotics. The mortality rate from bacteremia with any of these gram-negative organisms is 20 to 50 percent among hospitalized patients. *Escherichia coli* is the hospital's most troublesome pathogen, causing nearly twice as many nosocomial infections as any other microbe. It is the most common agent of bacteremia and causes one-third of all hospital-acquired urinary tract infections. This microbe, as well as other intestinal bacteria *(Klebsiella, Enterobacter,* and *Serratia)*, readily grows in glucose solutions used for intravenous therapy. (Most bacteria either die or do not proliferate in these solutions.) It readily colonizes the oropharynx of chronically and severely ill persons, who may develop pneumonia following direct spread of bacteria to the lungs.

■ Pseudomonads

Pseudomonas is the prototype of saprophyte turned pathogen and is a major cause of nosocomial infections in American hospitals. A particularly hazardous hospital pathogen, *P. aeruginosa* is a free-living gram-negative bacterium found in the bowels of 5 percent of healthy persons. In hospitals, this figure increases to 40 percent. *Pseudomonas* is also the most frequent cause of pneumonia acquired in the hospital. Overall, it is the third most frequent cause of all nosocomial infections, partially because of its widespread distribution within hospitals.

Pseudomonas aeruginosa is one of a group of gram-negative bacilli collectively called "water bacteria" that present a special medical problem—their ability to grow in distilled and deionized water allows them to proliferate in the moist chambers of respirators, mist therapy units, renal dialysis units, and improperly dried plastic tubing used for intravenous therapy. Microbes in this group also include other species of *Pseudomonas, Acinetobacter, Flavobacterium,* and *Aeromonas*. When contaminated items are used in a patient, high numbers of bacteria are introduced directly into the bloodstream or respiratory system, and severe disease is virtually inescapable. The number of *Pseudomonas* organisms is often not enough to make water appear turbid, so contamination cannot be visually detected. Contaminated water looks as clear as sterile water.

P. aeruginosa is especially problematic in burn wounds, where the organism becomes easily established (Fig. 25-8). From an infected burn wound, the bacteria may cause septicemia, pneumonia, meningitis, and several other serious diseases. Infected burns often infect other burn patients who are present in the same ward. Burned persons isolated in

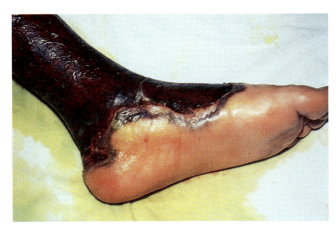

FIGURE 25-8 *Pseudomonas* **infection of a burn wound.** Such pigmented lesions are characteristic of *Pseudomonas*.

burn units or located with patients who are not burned are less likely to contract serious infections than those confined to burn wards. Sometimes the source of infection is the patient's own gastrointestinal tract, especially following treatment with antibiotics to which the strain is resistant. These antibiotics may actually encourage growth of *Pseudomonas* in the bowel by reducing competition from other members of the normal microbial flora.

In other patients, the organism is often introduced into body orifices on contaminated catheters, cystoscopes, or other invasive instruments. One startling source of infection is contaminated disinfectants and medications used to prevent infection. Medications packaged in single doses are less likely to be contaminated, especially if vials are dated and sterilized after preparation. Poorly designed wash basins that are difficult to disinfect are a common source of *Pseudomonas* infections in newborns.

■ *Mycobacterium tuberculosis*

Since 1985, the incidence of tuberculosis has dramatically increased within the United States. ▶ (See Tuberculosis, Chap. 21, p. 548.) The highest infection rates occur among immigrants from Asia, Africa, and Latin America, among low income populations with poor medical care, among residents of long-term care facilities, and among HIV-infected individuals. The reemergence of this disease in the community has been accompanied by an increase in the number of nosocomial cases. This hospital-based epidemic has been noticed primarily because the incubation period for tuberculosis among AIDS patients and other immunosuppressed persons may be quite short—less than 1 month. Spread to other patients, health care workers, and visitors occurs through an airborne route. Early diagnosis and prompt drug treatment will reduce coughing and shedding of pathogens within days. Unfortunately, recently emerging strains of *Mycobacterium tuberculosis* often display resistance to isoniazid and rifampin, the major drugs used in treatment, thus increasing the period of infectiousness. Nosocomial tuberculosis poses a real hazard for the community. Discharged patients often fail to continue the necessary therapy for the 9 months needed to cure the disease, thus becoming transmitters of the pathogen into the community.

■ Fungi

In addition to opportunistic bacteria, molds and yeasts of normally low virulence also can establish nosocomial infections (Fig. 25-9). Antibiotic therapy or severe immunosuppression often precedes extensive invasion by *Candida albicans* and fatal lung infections with either *Aspergillus* or molds of the order Mucorales. Sixty percent of patients dying of acute leukemia have fungal infections, as do 45 percent of those who die after renal transplant. Infections with fungi are also common among burn patients, bone marrow and heart transplant recipients, and recipients of cardiac valve prostheses.

■ Viruses

Because of the difficulty in detecting and propagating viruses, knowledge of their role as agents of nosocomial infection is very limited. Indeed, we have yet to determine which viruses can normally be expected to be isolated from healthy humans. It is estimated, however, that about one-third of pediatric nosocomial infections are of viral origin. We

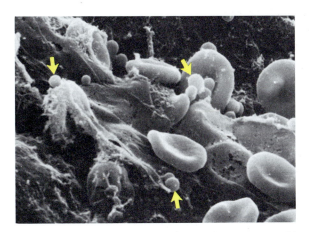

FIGURE 25-9 *Candida albicans* **colonizing a fibrin clot.** This yeast (indicated by yellow arrows) causes opportunistic infections following cardiac surgery, often colonizing prosthetic heart valves and catheters.

also know that patients treated with immunosuppressive agents to prevent rejection of organ transplants often die of overwhelming herpesvirus infection (cytomegalovirus, herpes simplex, or varicella-zoster). The HIV and hepatitis viruses, on the other hand, pose a threat to all people in the hospital, regardless of immunologic competence. These viruses are usually spread by the parenteral route through contaminated blood or invasive equipment. Inapparent carriers of the virus increase the danger of hospital-acquired disease, especially for surgical personnel, people performing renal dialysis, and personnel in blood banks. Strict adherence to guidelines for patient care (see Lowering the Infection Rate) has been effective in preventing HIV infection in virtually all health care workers.

Newborns are often at risk of developing viral infection, for example, following vaginal delivery if the mother has an active genital herpes infection. Perinatal transmission of HIV and hepatitis B virus can also occur. Pregnant health care workers may put their fetuses at risk of death or damage from rubella or human parvovirus B19.

Transmission of respiratory viruses is another problem in hospitals. Outbreaks of epidemic influenza can affect susceptible patients and personnel. Respiratory syncytial virus, chickenpox, and measles are highly communicable among the pediatric population. Infected children who are admitted should be placed in rooms by themselves or with other children with a similar infection.

■ Biofilms—A Stronghold for Pathogens

Many of the etiologic agents are more difficult to eradicate from surfaces of tubes, needles, respirators, catheters, and prosthetic devices if they form biofilms, aggregates of microbes that attach to surfaces. ▶ (Review Microbial Aggregates and Biofilms, Chap. 14, p. 337.) It is a thousand times more difficult for the body's natural defenses or antimicrobial therapy to eradicate bacteria that have established themselves in biofilms (Fig. 25-10). The infected human body can eliminate large numbers (over a million) of free-floating bacteria introduced into sterile organs such as the lungs or heart. The same microorganism, if attached to the surface of these organs, causes persistent infection when fewer than 1000 cells are introduced. Because of the difficulty in eliminating these biofilm organisms with conventional antibiotic therapy (even if the individual cells are sensitive to the antibiotic), foreign bodies, such as prosthetic devices, infected with attached bacteria often have to be removed before the infection can be cured with chemotherapy. Not only do antibiotic-resistant biofilms cause chronic infec-

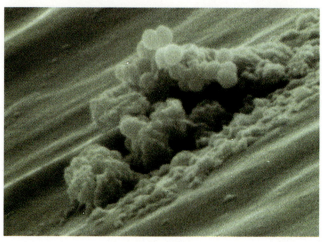

⊢————⊣ 2 μm

FIGURE 25-10 **Biofilm aggregate colonizing a cardiac pacemaker.** Attached to this medical implant is a microcolony of *Staphylococcus aureus*. The individual cells don't look spherical because they have covered themselves with a protective layer of secreted polysaccharide, a glycocalyx. Even after 6 weeks of intensive antibiotic therapy, the bacteria in these biofilms remain alive, necessitating the removal of the device. Even when antibiotic therapy apparently resolves the infection, the infection will be reestablished by surviving cells in the biofilm if the device is not removed.

tions, they are also a source of bacteria that break away from the colony and establish another locus of infection elsewhere, sometimes breaking away in large enough clumps to cause strokes.[1] Postsurgical strokes plagued the Jarvik artificial heart implant when bacterial clumps in biofilms broke free from the implant and lodged in the capillaries of the brain.

MYTH: Antibiotic sensitivity tests are almost always accurate indicators of which antibiotic is best for use in the patient infected with the tested pathogen.

FACT: Biofilm formation on the internal surfaces of patients, especially on medical devices, renders the information obtained from standard Kirby-Bauer tests an unreliable indicator of whether the drug can actually cure the patient.

Biofilm researchers have suggested a strategic course for dealing with biofilms in the medical en-

[1]A stroke occurs when the blood supply to part of the brain is blocked.

vironment. These strategies include the following:

- Ensure that air conditioners, mist units, and respiratory therapy apparatuses are free of biofilm-forming organisms and their residues.
- Ensure that an instrument or object introduced into a patient is not only sterile but also thoroughly clean. (Even sterile materials introduced into patients' bodies can promote biofilm formation if they contain the residues of biofilm-forming bacteria.)
- Base antibiotic dosage on the amount needed to eradicate bacteria in biofilms rather than on conventional tests that measure the sensitivity of unprotected cells.
- Accelerate efforts by the pharmacological industry to develop therapeutically safe penetrating agents to be administered with antibiotics that enhance the drug's effectiveness against pathogens protected within aggregates.

CONCEPT CHECK

- Discuss three factors that contribute to maintaining *E. coli* as the major cause of nosocomial infections.
- What are the sources of other pathogens that frequently cause nosocomial infections, and how do these pathogens gain entry into the body?
- Why are biofilm-forming bacteria more of a problem in the hospital than are bacteria that form no such aggregates? Answer the question from the "before infection" and "after infection" point of view.

Lowering the Infection Rate

Even under ideal conditions, some nosocomial infections are inevitable. For example, severely compromised patients cannot be entirely protected from endogenous infections caused by their own normal flora microbes. The present nosocomial infection rate, however, can be significantly reduced.

■ Infection Control Committees

Hospital surveillance has shown that most nosocomial infections result from failure to adhere to established procedures for handling patients and infectious materials. Many of these procedures are established by an infection control committee, which, since 1971, is required in every accredited hospital in the United States.

The primary responsibility of the hospital's **infection control committee** is to identify and control hospital-acquired infections. The committee is usually composed of representatives from each major clinical department, the laboratory, hospital administration, and nursing. Housekeeping, pharmacy, dietary, and blood bank personnel may also sit on the committee. The major responsibilities for surveillance are assigned to the chief infection control practitioner (ICP), usually a nurse epidemiologist. This individual collects data on all infections that occur in the hospital. These data include the body site, causative agents, a record of the responsible pathogen's antibiotic susceptibilities, and any host factors that may contribute to development of disease. This information alerts the committee when an increase in reported infections suggests the possibility of an epidemic. The data are also used to assess the effectiveness of measures implemented to prevent nosocomial infections.

The infection control committee does not require routine microbiological surveillance of possible environmental reservoirs of infection, although some critical items are monitored periodically. The effectiveness of autoclaves and other sterilization systems is assessed using biological indicators to ensure that instruments, equipment, and fluids for direct use on or in patients are effectively treated and are sterile. Hospital-prepared injectable fluids or infant formulas are monitored for the presence of microbial contaminants. Dialysis fluid may be tested periodically. The ICP also establishes policy for employee monitoring. Potential employees are screened to detect conditions that might endanger patients, especially if they are to work with children or immunocompromised patients. Although not routinely monitored, all personnel must notify their supervisors whenever they have infectious disease. If necessary, they are treated or restricted from patient contact until the incubation period is over. Ideally, only those employees who are immune should be exposed to patients with highly communicable diseases. Immunization of appropriate personnel against hepatitis B, rubella, and influenza helps reduce their infection rates and the risk of spread to other patients.

During an outbreak, appropriate environmental and personnel monitoring is implemented to determine the source of infection. Often epidemiological typing must be performed to establish the similarity between isolates from infected patients and suspected sources. ▶ (See Epidemiological Typing, Chap. 20, p. 512.) Remedies are implemented on the basis of laboratory findings.

Most nonhospital medical care facilities have no infection control committees (see BEYOND THE BASICS:

TABLE 25-3 ISOLATION PROCEDURES

Strict Isolation

1. Hands must be washed thoroughly on entering and before leaving room.
2. Gowns, gloves, and masks must be worn by all persons entering room and discarded before leaving room.
3. Articles must be discarded or washed and wrapped before being taken from room for disinfection or sterilization.
4. Confinement to a private room (or with others who have same proven diagnosis) is necessary.

Recommended for patients with anthrax (inhalation); plague (pneumonic); smallpox; systemic vaccinia infections; major burn, skin, or wound infections with *Staphylococcus aureus* or *Streptococcus pyogenes;* pneumonia (*S. aureus* or *S. pyogenes);* diphtheria; chickenpox; disseminated herpes zoster; congenital rubella syndrome; disseminated neonatal herpes simplex; rabies

Respiratory Isolation

1. Hands must be washed thoroughly on entering and before leaving room.
2. Masks must be worn by all susceptible persons entering room.
3. Articles coming into contact with patient's secretions must be discarded or washed and wrapped before being taken from room for disinfection or sterilization.
4. A private room is necessary.

Recommended for patients with tuberculosis; meningococcal meningitis; meningococcemia; measles; mumps; rubella; pertussis

Enteric Isolation

1. Hands must be washed thoroughly on entering and before leaving room.
2. Gowns and gloves must be worn by persons having direct contact with excretions of patient.
3. Articles coming into contact with patient's excretions must be discarded or washed and wrapped before being taken from room for disinfection or sterilization.
4. A private room is desirable, especially for children.

Recommended for patients with cholera; *Escherichia coli* gastroenteritis; salmonellosis; shigellosis; *Yersinia entero-colytica* gastroenteritis; typhoid fever; acute diarrhea with suspected infectious cause; viral hepatitis

Wound and Skin Isolation

1. Hands must be washed thoroughly on entering and before leaving room.
2. Gowns and gloves must be worn by persons having direct contact with patient and must be changed between patients.
3. Articles must be discarded or washed and wrapped before being taken from room for disinfection or sterilization.
4. A private room is desirable.

Recommended for patients with *Clostridium perfringens* gas gangrene; localized herpes zoster; limited burn, skin, or wound infections; plague (bubonic)

Reverse (Protective) Isolation

1. Hands must be washed thoroughly *before* entering room.
2. Gowns, masks, and gloves must be worn by all persons entering room and discarded outside of room.
3. A private room with special ventilation is necessary.

Recommended for patients with agranulocytosis; severe and extensive noninfected vesicular, bullous, or eczematous dermatitis; certain patients receiving immunosuppressive therapy; certain patients with lymphomas and leukemias

stead they should be placed directly into a puncture-resistant container (Fig. 25-12).

A rational approach to the use of antibiotics will reduce the likelihood of infection as well as the emergence of antibiotic-resistant pathogens. Chemoprophylaxis reduces the incidence of surgical wound infection following *some* types of surgery, particularly of the gastrointestinal and genitouri-nary tracts. (These drugs should be discontinued by the end of the operative day.) The effectiveness of a prescribed antibiotic depends on determining if the pathogen is sensitive, if the drug interferes with other medications the patient is taking, and if the antibiotic is truly needed. ▶ (See Guidelines for Antimicrobial Drug Use, Chap. 16, p. 401.)

When noninvasive alternatives are unavailable,

TABLE 25-4 BLOOD AND BODY FLUID PRECAUTIONS FOR CONTAINMENT OF HIV INFECTIONS IN CLINICAL SETTINGS

1. Wash hands before and after direct patient contact.

2. Use gloves if handling blood, drainage, secretions, or excretions. Wash hands after removing gloves, and change gloves between patients.

3. Use masks and protective eyewear during procedures that can generate droplets of blood or other body fluids to prevent exposure of mucous membranes of the mouth, nose, and eyes.

4. Wear gowns if soiling of clothes is likely.

5. Bag all contaminated articles before removing them from room, and clearly label them for infectious precautions.

6. Disinfect surfaces with household bleach diluted 1:10 in water.

7. Place needles and disposable sharp instruments in a puncture-resistant container containing disinfectant.

invasive therapies may be necessary. Invasive equipment should be monitored for contamination, changed at defined intervals, and removed from the patient's body as soon as possible.

The control of nosocomial infections also depends on educating all hospital personnel about the importance of excellent medical asepsis, the potential sources of infection, and how infection can be controlled (Table 25-5). Each health care professional has a personal responsibility for adhering to infection control policies and exercising adequate precautions, the details of which are discussed in Chapter 15. For example, if there is doubt about the sterility of any object, it must be treated as if it were unsterile. Any sterile object that has contacted an unsterile object (even the outside surface of its package) is contaminated and should be discarded. If the package is paper or cardboard, it must be dry to prevent contamination of the enclosed object. Objects in moist paper packaging cannot be considered sterile. Many hospitals are reprocessing disposable patient care items in an attempt to cut costs. These items are purchased sterile as warranted by the manufacturer. In-house resterilization processes

THE MICROBES

Infectious Wastes and Public Safety

In 1990, the American public recoiled from news reports that syringes and other "medical wastes" were washing onto the beaches of New Jersey. Subsequent investigations targeted hospital disposal systems and small clinics. Hospitals generate large amounts of wastes that must be discarded. If these materials contain pathogens in sufficient numbers to cause disease they are considered **infectious wastes.** About 1000 tons of infectious hospital wastes is generated each day in the United States. These wastes include patient specimens, laboratory cultures, microbiologically contaminated supplies and equipment, blood products or blood-contaminated items, and contaminated sharps. These items

may pose a threat to public health unless they are appropriately decontaminated or destroyed. The recommended disposal and treatment methods for infectious wastes include incineration and/or steam sterilization. Appropriate chemical inactivation of cultures or blood may also be acceptable.

Simply observing universal precautions and confining wastes to waterproof bags and puncture-proof boxes reduces the likelihood that pathogens will cause infections prior to disposal. In fact, the only serious hazard to hospital personnel is accidental inoculation with contaminated sharps. Special impervious containers for the storage of used sharps until decontamination are now recommended in all facilities.

Not all hospital wastes are decontaminated. Unless classified as infectious wastes, these items are not considered to be any different from normal household wastes that are loaded with bacteria. Several studies have even shown that refuse from private households contains

up to 100 times as many microbes with pathogenic potential as are in hospital wastes. The sources of these household pathogens include facial tissues, disposable diapers, pet litter, and spoiled foods. There has never been a documented case of infection being transmitted to the general community or even to waste industry workers by current disposal schemes.

But what about the public safety when syringes appeared on public beaches? Extensive investigations of several occurrences revealed that most of the syringes contained insulin or cocaine and came from home disposal by diabetics or were discarded by intravenous drug abusers. In the rare instances where medical wastes washed onto beaches, investigators found that malfunctions in municipal sewage treatment or uncontrollable changes in weather patterns were responsible. Although aesthetically unappealing, health risks to the public were virtually nonexistent from these materials.

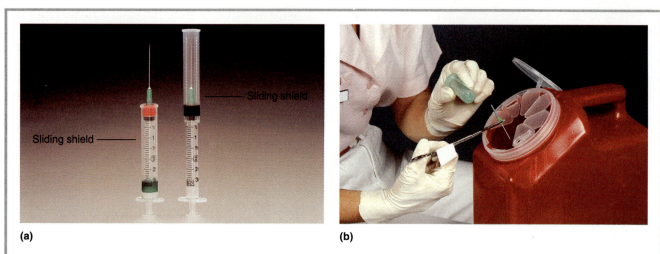

(a)   **(b)**

FIGURE 25-12 **Precautions against accidental needle sticks.** These include (*a*) using injectors that automatically cover the needle with a guard as it is removed from the patient and (*b*) placing used needles in safe containers.

TABLE 25-5   INANIMATE SOURCES AND VEHICLES OF NOSOCOMIAL INFECTIONS

Source or Vehicle	Precautions*
Water	Use sterile water for nebulizers and humidifiers. Replace drinking water carafes daily.
Mops	Do not allow mops to stand in water. Use mophead once (at most for 1 day), then autoclave or wash at 70°C.
Respiratory therapy equipment	Sterilize or properly disinfect before use, and change every 24 hours when in continuous use.
Renal dialysis machine	Disinfect distribution system immediately before use; disinfect machine overnight; treat water by reverse osmosis.
Anesthesia equipment	Disinfect artificial airways and breathing bags before each use.
Instruments and fluids for injection into patients	Use disposable needles and syringes; open packages immediately before using; prepare skin properly; use single-dose vials when possible; store multiple-dose vials in refrigerator.
Thermometers	Use disposable thermometers; otherwise, sterilize instruments with ethylene oxide between patients and store in fresh 70% ethyl alcohol.
Ventilation systems and air conditioners	Disinfect water reservoirs and monitor for contamination.
Linen	Gather soiled linen with minimum amount of turbulence; place immediately in bags; wash linen (and bag) *before* sorting; use sterile (autoclaved) linen for compromised patients and in operating and delivery rooms.
Patient rooms	Wet-vacuum floors after flooding with water containing germicide; regularly wash all surfaces (walls, ceilings, etc.); disinfect following discharge of patient with a communicable disease; wash furniture, lamps, and other fixtures daily; wash plumbing fixtures twice daily with a detergent-germicide; scrub tubs and showers immediately after use; change curtains after patient leaves.

*Precautions may vary according to the hospital.

must be monitored to ensure that each product is safe and maintains its effectiveness in use.

Aseptic techniques should be strictly adhered to at all times. This is especially true for physicians, since many nurses and therapists model their behavior on that displayed by the medical doctor. Personnel must never consider themselves "sterile," and their behavior should reflect awareness that potential pathogens are constantly being shed from their bodies. This danger is best combated by maintaining excellent hygiene of the total body (hands, hair, and clothing).

Ignaz Semmelweiss developed a system of disinfecting instruments and hands that dramatically reduced the infection rate following childbirth. Yet he was chastised by his colleagues for his radical ideas, and they resisted the implementation of these practices. One hundred years later, many medical professionals still fail to perform the simple lifesaving task of washing their hands before and after contact with each patient. Studies by infection control researchers reveal that although most of these persons are aware of the importance of hand washing in preventing nosocomial disease, they are deterred by the "inconvenience" of the procedure. Some persons neglect to wash their hands frequently because they believe it will have a detrimental effect on their skin. Neither of these reasons is justified.

The alarming rate of nosocomial infections has helped maintain community mistrust of the hospital environment. The persisting stigma of the "pesthouse" will disappear only after practice of asepsis by medical personnel is elevated to a level required to prevent the transmission of infectious disease among patients and staff alike.

CONCEPT CHECK

- Why is proper hand washing considered the most important element in controlling nosocomial infections?
- What is the role of the infection control committee in reducing hospital infection rates?
- Why are universal precautions applied to all hospital patients, even though most are not infectious?

Key Terms

nosocomial infection (p. 681)
compromised patient (p. 682)
cross infections (p. 683)
infection control committee (p. 693)

strict isolation (p. 695)
reverse (protective) isolation (p. 695)
universal blood and body fluid precautions (p. 695)

sharps (p. 695)
infectious wastes (p. 697)

Key Facts and Concepts

- **Hospital-acquired diseases.** The prevalence of nosocomial infections is due to the concentration of highly susceptible persons in the hospital, the ubiquity of opportunistic pathogens, medical procedures that encourage infection, and breaches in sound aseptic practices.
- **Susceptible hosts.** The compromised patient is the most common victim of nosocomial infection. Illness and medical procedures combine to increase susceptibility to the many opportunistic pathogens in the hospital. Identifying these susceptible persons is one step in controlling nosocomial infections so that additional precautions can be employed to protect them.
- **Causative agents.** The microorganisms that most commonly cause nosocomial infections cause few infections in the general community. These opportunistic pathogens, such as *Escherichia coli*, *Enterococcus faecalis*, and *Pseudomonas aeruginosa*, are common residents of the human colon and skin surrounding the anus. The source of these opportunists may be the patient, the health care provider, or inanimate vehicles. The tendency of many pathogens to form biofilms complicates the prevention and treatment of disease.

- **Drug-resistant pathogens.** Hospital-acquired infections are often difficult to treat because widespread use of antimicrobial chemotherapy in the hospital encourages the emergence of drug-resistant microbes. Many of these strains possess R factors that confer resistance to more than one antibiotic. Drug-resistant hospital strains colonize the hospital staff, who then transmit them to patients.
- **The infection control committee.** Controlling hospital-acquired infections is the task of all personnel but is coordinated by the hospital's infection control committee. This committee reviews cases of nosocomial infection and collects data on their occurrence. The committee also conducts educational programs to encourage awareness and practice of aseptic procedures by the hospital staff, monitors the use of antimicrobial chemotherapy, and coordinates surveillance of possible reservoirs.
- **Wash your hands.** Hand washing before and after each patient contact is the single most important precaution for lowering the incidence of nosocomial infections.

Review and Problem-Solving Questions

1. Distinguish between:
   (a) Community-acquired infections and nosocomial infections
   (b) Exogenous infections and endogenous infections
   (c) Strict isolation and reverse isolation
2. Explain why the most common community-acquired infections differ from the most common hospital-acquired infections.
3. Why does each of the following increase the risk of infection for hospitalized patients? (a) Antibiotic therapy; (b) respiratory therapy; (c) cancer and its treatment; (d) urinary catheterization.
4. Explain why each of the following statements is false.
   (a) Aseptic techniques can be safely relaxed when treating patients who are taking antibiotics.

   (b) Nosocomial infections are usually caused by highly virulent pathogens.
   (c) All patients should be separated into wards according to their disease.
5. You have been hired as chief infection control practitioner at University Hospital. Under what circumstances would you place work restrictions on members of the hospital staff who come into direct contact with patients? on kitchen staff who prepare meals?
6. Describe two techniques that can be used to reduce the rate of nosocomial infection of (a) the urinary tract; (b) surgical wounds; (c) neonatal skin.
7. What are the specific risks of infection for patients and personnel associated with dental care, and how can these be avoided?

8. Dr. Adler heads the laboratory in University Hospital, where there is an outbreak of wound infections. Laboratory personnel have isolated *S. aureus* from 12 specimens. Nasopharyngeal cultures from the nurses and physicians have yielded another eight *S. aureus* isolates. Dr. Adler wishes to know if any member of the staff is transmitting the pathogen to the patients. What tests should the laboratory personnel be instructed to perform?

9. The following data come from a study to determine the efficacy of dry cleaning to disinfect fabrics of uniforms, soft toys, and rugs contaminated with viruses or bacteria. What recommendation would you make based on the data?

Organism	Percent Infectious Particles Remaining
Poliovirus	80
Herpes simplex virus	0
Influenza virus	0
E. coli	1
P. aeruginosa	1
S. aureus	1

10. A manufacturer of intravenous catheters wishes to test the ability of catheters coated with various antibiotics to reduce the rate of catheter-related *S. aureus* infections. He has asked you to evaluate the catheters. Design a study to compare the efficacy of antibiotic-coated catheters in preventing infections to that of uncoated catheters.

11. The following data were collected by the infection control personnel to determine if antibiotic prophylaxis was warranted for several types of surgeries. How might you account for the different rates of infection? What recommendations would you make based on these findings?

Surgery	Percent Infections with No Prophylaxis
Hysterectomy	34
Appendectomy	27
Hip replacement	5
Heart transplant	85

HEADLINES

ARE GLOVES EFFECTIVE BARRIERS?

The hands of health care workers are responsible for much of the transmission of microorganisms in the hospital. Hand washing and the routine use of disposable gloves whenever there is contact with mucous membranes and broken skin has helped to reduce infection rates. R. J. Olsen and colleagues* wanted to determine the efficacy of vinyl and latex gloves alone as barriers to hand contamination. They cultured the exterior surface of 135 gloves for gram-negative rods after patient contact. Eighty-six gloves were positive—42 of the vinyl gloves and 44 of the latex gloves. In 11 cases, the same gram-negative organism as that found on the glove was cultured from the hands of the health care workers. These gloves were also tested for leaks. Their findings are presented in the table below.

*R. J. Olsen, P. Lynch, M. B. Coyle, J. Cummings, T. Bokete, and W. E. Stamm, Examination gloves as barriers to hand contamination in clinical practice, *Journal of the American Medical Association,* 270: 350–353 (1993).

Glove Type	Leak-Test Result (Did Glove Leak?)	Microorganism	Colony Count on Gloves, CFU	Colony Count on Hands, CFU
Vinyl	Yes	*Enterobacter cloacae*	2.0×10^3	1.0×10^1
Vinyl	Yes	*Acinetobacter calcoaceticus*	1.2×10^5	4.0×10^1
Vinyl	Yes	*A. calcoaceticus*	6.5×10^2	5.0×10^0
Vinyl	No	*A. calcoaceticus*	3.0×10^5	2.5×10^2
Vinyl	Yes	*A. calcoaceticus*	4.2×10^4	1.0×10^1
Vinyl	Yes	*A. calcoaceticus, Enterobacter aerogenes*	—†	—†
Vinyl	Yes	*A. calcoaceticus*	5.2×10^3	9.0×10^1
Vinyl	No	*Pseudomonas aeruginosa*	2.1×10^3	2.0×10^1
Vinyl	No	*Escherichia coli*	2.0×10^6	2.0×10^1
Vinyl	No	*P. aeruginosa*	1.3×10^4	2.0×10^1
Latex	No	*A. calcoaceticus*	1.5×10^4	1.0×10^1

CFU = colony-forming units.
†No data available.

Continued on page 702

Continued from page 701

- Summarize the results.
- How might you explain the contaminated hands in the absence of a leak in the glove?
- What conclusions would you draw regarding the effectiveness of vinyl versus latex gloves in preventing hand contamination?

- What other factors besides glove type might influence the likelihood of hand contamination after glove use?
- From these data, would you recommend that hand washing is not necessary if personnel use gloves?

PART SEVEN

Applied and Environmental Microbiology

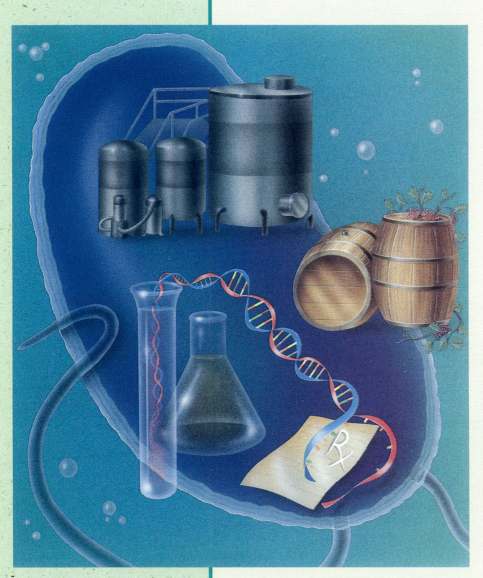

CHAPTER 26

Microbes and Planet Earth: Environmental, Soil, and Water Microbiology

The envelope of life. Life is restricted to a thin layer of air, water, and land that envelopes the earth, a zone we call the *biosphere*. This thin film where life thrives is a mere 14 miles thick. Compared to the size of the earth, the biosphere is about as thick as the skin of an apple. Without microorganisms, however, its thickness would be zero. The biosphere would cease to exist.

In our efforts to solve problems, we often find that our solutions create unanticipated complications. Malaria has been endemic for centuries in many parts of the world. In Bolivia, officials unleashed DDT against the mosquitoes that carried the pathogen. The incidence of malaria indeed diminished, but so did the cat population. The poisoned cats were no longer a factor in controlling the rodent population, which quickly exploded into enormous numbers. Although rats in the area had always carried the arenavirus that causes human hemorrhagic fever, the disease was never a problem because cats had kept their numbers low. In the aftermath of DDT spraying, however, the ensuing epidemic of hemorrhagic fever left countless people dead.

Larger scale problems emerged from a technological breakthrough, the 1964 invention of phosphate-containing biodegradable laundry detergents. Synthetic detergents had already become vastly more popular than laundry soap, and the addition of phosphates improved both their cleaning efficiency and their commercial success. The consequences of this success story, however, were far more significant than whiter whites. Wastewater that contained the phosphate-rich detergents inevitably ended up in inland waterways and lakes. Enriched by the phosphates, many of these lakes began to support huge blooms of algae and cyanobacteria whose growth had previously been limited by naturally low concentrations of phosphate. The photosynthetic microbes blanketed many lakes in thick mats, then died. Bacteria in the lake decomposed the dead mats, consuming all the oxygen in the process. Fish and all other oxygen-requiring organisms suffocated, killing virtually all the desirable organisms in the lake. Perhaps the largest and most famous of the lakes that nearly "died" is Lake Erie. Legislation outlawing phosphates in detergents, as well as other measures that reduce the phosphate load in our waterways, has helped dramatically in reviving these aquatic habitats.

These stories join countless others that reveal a common theme that continues to be repeated in similar episodes. That theme is the inevitability of ecological chain reactions, ripple effects in which one change generates greatly magnified and often unanticipated consequences. Our growing awareness of the ways in which environments operate contributes to our ability to predict and thus avoid such consequences, but the pattern continues to be repeated, often because of a failure to consider the most subtle level of environmental operations—the activities of microorganisms and their interactions with the nonliving chemicals in the environment. This chapter takes you on a tour of these subtle environments and the roles that microbes play in sustaining, or in some cases disrupting, habitat stability. In addition, because people acquire their water and food from the environment, discussions of soil and water microbiology are included in this chapter to reveal the close links between human well-being and microbes in the environment.

Microbial Ecology

There are places on earth teeming with life where humans have never set foot, and if they did they would find it virtually impossible to survive. In many of these regions, conditions are so severe and so inhospitable that not a single plant or animal can be found. Only microorganisms find such places inhabitable. Of course, microorganisms also abound in more hospitable environments, where animals and plants also live. Anywhere there is life, there are microbes.

Microorganisms grow abundantly in most soils, oceans, and bodies of fresh water. They also populate the atmosphere close to the earth's surface. When they share their environment with other microbes or multicellular organisms, crucial interactions inevitably influence the lives of all. Some of these are cooperative interactions; others are battles for survival; still others have no direct effect on one another. These living individuals make up a **community**—all the organisms that reside in a particular area. These organisms not only interact with one another but are also inseparably dependent on water, air, sunlight, warmth, and other nonliving aspects of the environment. Together, the living community plus the nonliving physical and chemical properties of a particular environment form an **ecosystem** (Fig. 26-1). Understanding ecosystems requires knowledge not only of how the organisms in a community interact with each other but also of how they affect and are affected by their nonliving surroundings. This knowledge comes from **ecology**, the study of interactions of organisms with each other and with their environment—in other words, the study of ecosystems ▶ (see BEYOND THE BASICS: Collecting Environmental Specimens).

The environment for any individual microbe is very small. The conditions in such a small area, or **microenvironment**, may differ dramatically from those of adjacent environments. The lawn in front of your science building, for example, represents many different microenvironments. Certain regions may contain trees and shrubs that influence the soil around them, perhaps by producing inhibitory compounds or rapidly absorbing water and minerals. Other areas may have metal pipes that leach metallic ions or other chemicals into the ground. A leaking sprinkler concentrates water in one spot while prolonged sun exposure may create drier conditions in another area of the lawn. The gardeners may have poured fertilizers around a newly planted rose bush and toxic pesticides where snails and slugs are a problem. Critical differences even exist between distances as small as the diameter of an individual soil particle, where the surface is aerobic and the interior is anaerobic. A microbe might find a grain of sand and a crumb of fertile soil to be as different as the moon is to the earth in its ability to support its life.

In most habitats, microorganisms grow fixed to

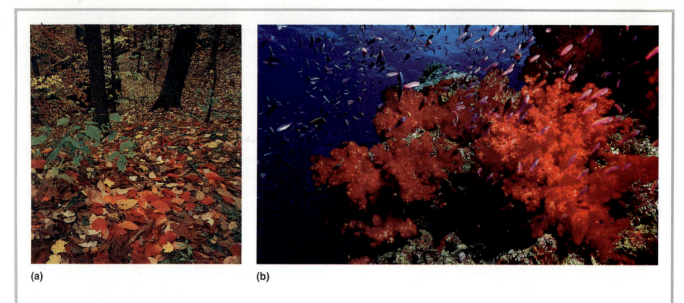

FIGURE 26-1 **Some typical ecosystems.** (*a*) Microorganisms in this litter on the forest floor decompose the organic debris and recycle the nutrients. (*b*) The coral and other animals in this marine habitat all depend on algae to provide food to the community.

surfaces, forming microcolonies, or adhere to other organisms, forming a *biofilm* (Fig. 26-2). ▶ (See Microbial Aggregates and Biofilms, Chap. 14, p. 337.) The biofilm matrix protects the interior microbes from harmful chemicals. It also allows them to share nutrients, with end products released from one organism being used as substrates for other organisms within the tight-knit community. This cooperation may produce an aggregate with a metabolic potential that is greater than the sum of those of the individual organisms.

Microbes play various roles within ecosystems, some harmful and others indispensable to the continued life of the ecosystem. Some microorganisms are ultimately responsible for providing energy or nutrients to all other organisms within the ecosystem. These microbes are essential for the maintenance of life. Most types of microbes have a less crucial influence on an ecosystem—their elimination will change the balance of organisms but will not lead to the loss of all life in the system.

■ Food Webs

Every organism within an ecosystem requires energy and organic compounds for growth. Whereas *heterotrophs* must take in preformed organic material to fulfill these needs, *autotrophs* convert carbon dioxide (and water) into organic compounds using energy derived from either sunlight or inorganic compounds. ▶ (See Energy and Carbon Sources, Chap. 5, p. 105.) In other words, autotrophs are the

FIGURE 26-2 **Life in a biofilm.** A series of organisms become embedded in an organic polymer produced by the microbes. The organisms on the margins of the aggregate have access to more oxygen, nutrients, and water and consequently grow faster. The interior organisms are protected from toxins and ultraviolet light and form a reservoir of microbes that quickly reseed an environment. (See Fig. 14-4, page 337, for a microscopic view of a biofilm.)

Collecting Environmental Specimens

Specimens from natural environments, soil, sediments, fresh water, and marine waters are usually collected in order to study the organisms within these habitats or to test for the presence of pathogens. The goal may be to understand the natural ecology, to monitor the fate of genetically engineered organisms released into the environment, to monitor the safety of drinking and recreational waters, or to search for isolates that can potentially be exploited in the biotechnological revolution. Among the extraordinary numbers of microbes found within these environments, it is likely that most have yet to be identified and their natural activities explored. Thus microbiologists hope to discover organisms that can be used for degrading pollutants, for producing new antibiotic and antitumor drugs, or for enhancing the world's food supply.

No single sampling technique is adequate to ensure the survival and subsequent isolation of every type of organism from these diverse environments. Many different methods of sampling are used in order to match the microbe's microenvironment, which maximizes the likelihood of success. Although this often requires custom design of collection techniques for microbes from extreme environments, such as deep-sea thermal vents, or microbes with fastidious metabolic requirements, standardized procedures are used for obtaining most types of specimens.

Public health officials collect samples of potable waters after processing and waters discharged directly from storm drains. They fill bottles submerged about 1 foot beneath the surface of lakes, rivers, and oceans in areas where people are likely to swim. These samples are tested for indicator organisms such as fecal coliforms and enterococci. Deeper ocean sampling requires equipment that will collect at a specific depth without contamination from the waters above (see Fig. 1). Some sampling devices are designed to remain closed until the desired depth is reached. They are then opened, filled, and sealed. Other samplers are open as long as the device is descending. Flaps close and trap water as the device is brought up to the surface. Remote-controlled submersible vessels can also be used to sample within the deeper parts of the ocean (see Fig. 2).

Figure 2 Launch of remote controlled submersible *Odyssey* for collecting microbial specimens from deep in the Antarctic Ocean.

Sediments at the bottom of oceans and freshwater bodies can be acquired using remote sampling devices that dig into the sediment and snap shut around the specimen. When the floor is not too deep, divers can descend and obtain a sample. Submersible vessels can also remove sediment samples.

Surface soil can be readily retrieved with a spatula, scoop, or shovel. Core samplers are used to collect soil from specified depths. With any of these terrestrial samples, moisture is retained by storing the specimens in sealed bags until processing. Baiting techniques are used to attract specific microbes. Coated slides, for example, can be buried and removed at different time intervals.

Figure 1 Photo of deep water sampling device.

producers, the organisms that generate all the usable energy that supports every organism in the ecosystem. Producers synthesize organic carbon compounds and incorporate them into living tissue, and they serve as food for heterotrophs. Because heterotrophs rely on a diet derived from other organisms, they are categorized as **consumers**. Producers (autotrophs) are the basic suppliers of food for all consumers (heterotrophs) in an ecosystem. Producers form the base of all **food webs**, the feeding relationships among the organisms in an ecosystem (Fig. 26-3).

MYTH: All life on earth depends on photosynthesis.

FACT: Although virtually all the world's organisms are either photosynthetic or consumers in a food web fueled by photosynthetic producers, a few producers use chemical energy released when they ox-

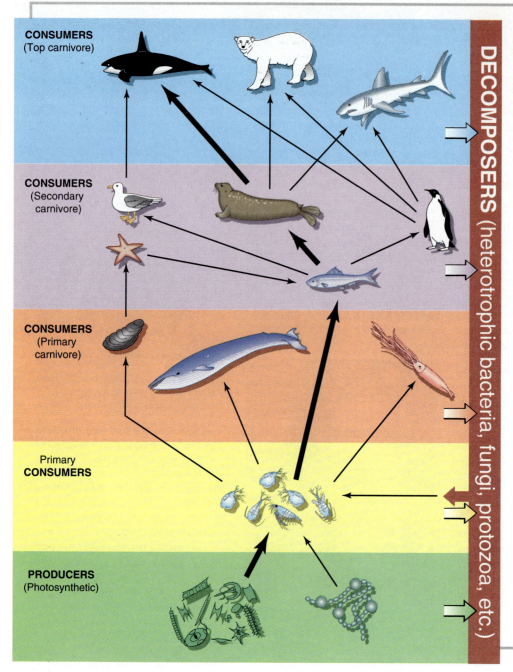

CONSUMERS (Top carnivore)

CONSUMERS (Secondary carnivore)

CONSUMERS (Primary carnivore)

Primary **CONSUMERS**

PRODUCERS (Photosynthetic)

DECOMPOSERS (heterotrophic bacteria, fungi, protozoa, etc.)

FIGURE 26-3 **Energy and nutrient flow in a food web.** All energy in virtually all ecosystems is supplied by the sun. Photosynthetic organisms (in this case algae and cyano-bacteria) harvest this energy. All organisms in the ecosystem depend either directly or indirectly on these primary producers for their energy. (The broad arrows follow a "typical" food chain embedded in the food web.) Note that decomposers use nutrients from organisms at every trophic level and are directly consumed by zooplankton.

idize inorganic compounds instead of sunlight. These chemoautotrophs are found in diverse habitats.

The producers in most ecosystems are photosynthetic organisms: cyanobacteria, algae, and plants. In aquatic systems exposed to sunlight, photosynthetic microbes are common producers. Oceans and lakes, for example, are populated by animals that depend on energy and nutrients introduced by algae and cyanobacteria. Even heterotrophic bacteria,

which serve as the base of the food web for many animals, depend on chemical energy stored in organic carbon compounds of animal feces and dead organisms as well as dissolved molecules released from phototrophs. In most terrestrial environments, plants form the base of the food web. But what about environments that are perpetually dark and isolated from photosynthetically supported organisms (for example, deep trenches miles beneath the surface of the ocean)? In these and other extreme habitats (such as the bottom of Yellowstone Lake, where the water is too hot for photosynthetic microbes to live), *chemoautotrophs* produce the food

(a)

(b)

FIGURE 26-4 **Volcanic "oasis."** (*a*) This picturesque bay lies in a volcanic crater whose water is heated to almost boiling by underlying magma. The heated water is saturated with sulfur, which is oxidized by chemoautotrophic bacteria. (*b*) The chemoautotrophs form dense mats (in combination with photosynthetic algae) that feed about 150 species of invertebrate animals, such as these anemones. As sunlight, temperature, and chemical composition of the water vary, the dominant primary producers shift between the photosynthesizers and chemosynthesizers.

that drives the ecosystem. These bacteria acquire energy by oxidizing inorganic substrates such as hydrogen, iron, sulfur, or ammonia that stream from hot water fissures in the lake bottom. In some areas, the presence of chemoautotrophs and photoautotrophs together maximizes the food supply as conditions vary (Fig. 26-4).

■ Nutrient Cycling

In every environment on earth, microbes are required for the **biogeochemical cycles** that recycle chemicals between living organisms and the physical environment. When producers and consumers die, carbon, nitrogen, and other elements are left trapped within their bodies. Bacteria and fungi ultimately decompose these dead organisms, freeing their molecular constituents to be reintroduced into the food web. These microbial **decomposers** break down complex organic material into simpler compounds. When the end products are inorganic molecules, the decomposition process is termed **mineralization**. Without such activity, supplies of carbon, nitrogen, sulfur, and phosphorus, essential to the lives of all organisms, would be quickly exhausted. This recycling process influences the composition of soils, waters, and the atmosphere. Mi-

crobes also play an important role in the flow of iron, manganese, calcium, silicon, and other elements needed by all organisms.

● **CARBON CYCLE.** Carbon is brought into the food chain when photosynthetic (or chemoautotrophic) organisms convert carbon dioxide to organic carbon compounds, a process called *carbon dioxide fixation*. Carbon dioxide is returned to the atmosphere, and relatively constant levels are maintained primarily as a result of respiration by heterotrophic organisms. This movement of carbon between inorganic and organic forms is called the **carbon cycle** (Fig. 26-5). Through the carbon cycle, almost 80 billion tons of inorganic carbon are converted into organic material annually. Photosynthetic microbes in the ocean are responsible for slightly more than half of this activity. Terrestrial photosynthesis accounts for most of the rest. Two possible fates may await the producer and the carbon trapped within its organic molecules. It may die, with its remains then broken down by decomposers (an outcome typical in forest habitats), or it may be eaten by animals or other consumers such as protozoa that ingest algae and cyanobacteria. In both cases, most of the organic carbon is oxidized by the process of respiration (by the decomposer or consumer), releasing CO_2 in the

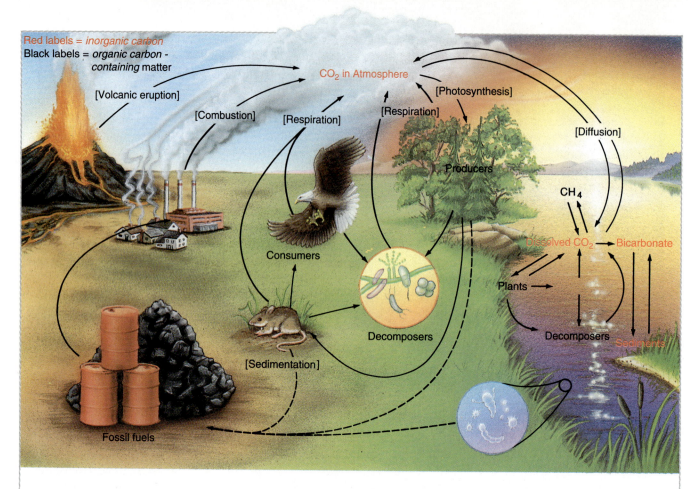

Red labels = *inorganic carbon*
Black labels = *organic carbon -
     containing* matter

[Volcanic eruption]

CO₂ in Atmosphere

[Photosynthesis]

[Combustion]

[Respiration]

[Respiration]

[Diffusion]

Producers

CH₄

Consumers

Dissolved CO₂ → Bicarbonate

Plants

Decomposers

Decomposers

Sediments

[Sedimentation]

Fossil fuels

FIGURE 26-5 Carbon cycle. Carbon is constantly converted between its inorganic forms and organic forms, creating the carbon cycle. Without inorganic carbon to provide CO₂ for autotrophy, life on earth would cease.

process. ▶ (See Respiration, Chap. 7, p. 155.) Some of the carbon, however, is incorporated into the tissues of the consumer. The carbon that enters consumers ultimately ends up in their excretory products, in their dead remains, or in the body of a higher order consumer. Ultimately, decomposers use the dead organism's organic nutrients for their growth, and through respiration CO₂ returns to the atmosphere.

Certain archaeobacteria, called methanogens, produce methane (CH₄) from CO₂ or other simple organic compounds such as formic acid or acetate. These anaerobic cells use an alternative pathway for carbon metabolism. Methanogens are common in sediments where decaying plant material is deposited (Fig. 26-6); they are also found in the digestive systems of animals such as cows. Methane cannot be metabolized by most microorganisms, and were it not for a group of methane-oxidizing bacteria it would rapidly accumulate in the atmosphere. Methane oxidizers create CO₂ from methane,

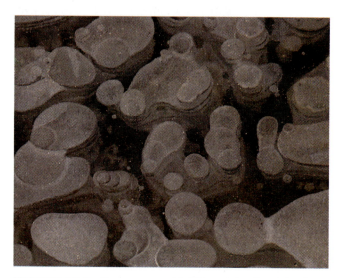

FIGURE 26-6 Swamp gas. Bubbles rising from the bottom of stagnant water pools and swamps contain methane gas generated by the methanogens living in the anaerobic sediments. (The bubbles in this photo have been trapped beneath the surface ice of a frozen swamp in winter.)

thereby completing this side loop of the carbon cycle.

Some carbon does end up in a biological "dead end," inaccessible to decomposers. Fossil fuels such as coal and petroleum are deposited below the earth's surface with no water and are therefore incapable of supporting life. As humans remove these resources from their burial places and burn them as fossil fuels, the carbon is released as carbon dioxide to the atmosphere. Increased CO_2 concentrations could have a dramatic effect on the environment.

Relatively little carbon dioxide is fixed by chemolithotrophs, since most of the world's autotrophs are photosynthesizers. Thermal vent ecosystems, however, rely on these chemoautotrophic bacteria to support the intricate invertebrate communities of sponges, anemones, worms, and crustaceans.

● **NITROGEN CYCLE.** Molecular nitrogen (N_2) accounts for 79 percent of the atmosphere. Although all living organisms require nitrogen, only a few can use the abundant nitrogen gas in the air. Instead, molecular nitrogen must be converted into biologically usable compounds, such as ammonia, before it can be incorporated by plants. This process, called **nitrogen fixation**, is the first step in the **nitrogen cycle** (Fig. 26-7). In addition to nitrogen fixation, three more reactions are critical to the cycle: nitrification, ammonification, and denitrification.

Nitrogen fixation is carried out only by a few prokaryotes. In aquatic environments, the process is performed by free-living bacteria and cyanobacteria. Free-living bacteria and symbiotic bacteria in the root nodules of certain plants fix most of the nitrogen in the soil. The ammonia produced by symbiotic bacteria is immediately available to the plant for assimilation into organic matter. In contrast, am-

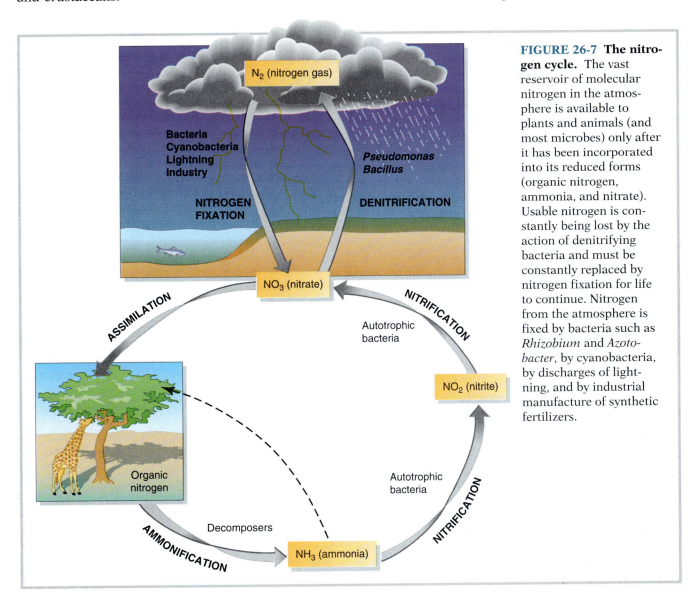

FIGURE 26-7 The nitrogen cycle. The vast reservoir of molecular nitrogen in the atmosphere is available to plants and animals (and most microbes) only after it has been incorporated into its reduced forms (organic nitrogen, ammonia, and nitrate). Usable nitrogen is constantly being lost by the action of denitrifying bacteria and must be constantly replaced by nitrogen fixation for life to continue. Nitrogen from the atmosphere is fixed by bacteria such as *Rhizobium* and *Azotobacter*, by cyanobacteria, by discharges of lightning, and by industrial manufacture of synthetic fertilizers.

monia produced by free-living bacteria is unavailable for absorption by plant roots. It dissolves in the soil and combines with hydrogen to form positively charged ammonium ions, which quickly combine with negatively charged soil particles. The ammonium ions would remain in the soil were it not for autotrophic bacteria that convert them into negatively charged nitrites and nitrates. Each step in this process, called **nitrification**, is performed by a different group of nitrifying bacteria. Both nitrification processes are accomplished by bacteria that usually live close together, sometimes even sharing the same soil particle.

The negatively charged nitrate ions are easily absorbed by plants and assimilated into organic compounds. Herbivores acquire these nutrients by consuming the plants; the nitrogen-containing nutrients pass through the food web as higher order consumers feed on lower order consumers. Microbes ultimately decompose dead plants and animals, releasing some of the organic nitrogen once again as ammonia, a process called **ammonification**. Nitrifying bacteria once again convert ammonia to nitrate, and much of the nitrogen is recycled into plants.

Unlike nitrogen-fixing and nitrifying bacteria, which introduce nitrogen into the food web, denitrifying bacteria return nitrogen to the atmosphere.

Through a series of reactions called **denitrification**, these bacteria transform nitrates into molecular nitrogen. Denitrifying bacteria, therefore, reduce the amount of nitrogen available for metabolism by other microbes and plants. The process costs farmers dearly, since much of the nitrate in expensive fertilizers is converted to useless nitrogen gas, thereby reducing crop yields. In natural ecosystems, however, nitrogen-fixing bacteria reintroduce the nitrogen gas into the food web, and the cycle of life continues.

● **SULFUR AND PHOSPHORUS CYCLES.** Phosphorus and sulfur are among the other elements that microbes help recycle. Phosphorus is incorporated into nucleic acids, energy storage compounds, and membranes. Sulfur is found in the sulfhydryl groups (—SH) of certain amino acids. It also is essential for energy generation for a number of autotrophic bacteria. Sulfur metabolism is particularly important in the ecology of deep-sea vent communities (sulfate-rich water from geothermal vents provides the inorganic energy sources for many of these chemoautotrophic producers), drainage of minerals from soils, and formation of acid rain.

Phosphates are usually present in relatively low concentrations in most environments. In fact, it is

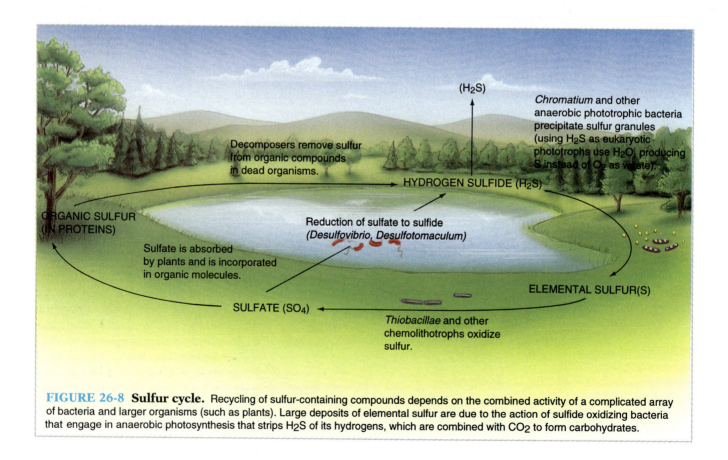

FIGURE 26-8 **Sulfur cycle.** Recycling of sulfur-containing compounds depends on the combined activity of a complicated array of bacteria and larger organisms (such as plants). Large deposits of elemental sulfur are due to the action of sulfide oxidizing bacteria that engage in anaerobic photosynthesis that strips H_2S of its hydrogens, which are combined with CO_2 to form carbohydrates.

often the concentration of phosphorus that limits the growth of producers and thus the number of microbes within an ecosystem. Bacteria improve the productivity of an ecosystem by increasing the amount of phosphorus available. Some do this by producing nitric or sulfuric acids, which in turn dissolve inorganic phosphorus-containing rocks, releasing the trapped phosphates. Other microbes produce enzymes that hydrolyze organic phosphate molecules. Plants and microorganisms then absorb the soluble inorganic phosphates and incorporate them into organic molecules such as nucleic acids and ATP or use them to drive glucose metabolism. These decomposers return the phosphates in dead organisms to the soil or aquatic habitat.

Plants, algae, and many microbes absorb sulfate ions for assimilatory metabolism, the incorporation of sulfur into organic material (Fig. 26-8). The biologically transformed sulfur supplies a nutritional need for consumers in the food web. When these organisms are decomposed, the sulfur is released as hydrogen sulfide (H_2S). Sulfates are also converted directly into H_2S when used as final electron acceptors for anaerobic sulfate-reducing bacteria. The H_2S produces the characteristic odor of rotten eggs. It is also extremely toxic for most organisms, including humans. Fortunately, H_2S is usually metabolized by bacteria, and toxic sulfide levels accumulate only where oxygen supplies are scarce—for example, in sewers, in deep sediment, or in aquatic systems covered with a dense mat of algae.

Bacteria dispose of H_2S in several ways. Under anaerobic conditions, photosynthetic sulfur bacteria convert H_2S into sulfur and occasionally into sulfate. Several chemoautotrophic bacteria oxidize sulfide into sulfur under aerobic conditions. They store the sulfur within the cells, and when H_2S supplies become limited, the sulfur reserves are further oxidized to sulfate.

From food webs to nutrient recycling, microbial ecology is central to the earth's ecology. There are microbiologists, however, who are interested in the possible ecology of extraterrestrial environments. As we leap further into space, scientists are exploring the ecosystems that we might encounter or create as the result of our endeavors (see THE EXPLORERS: Microbiologists Investigate the "Final Frontier").

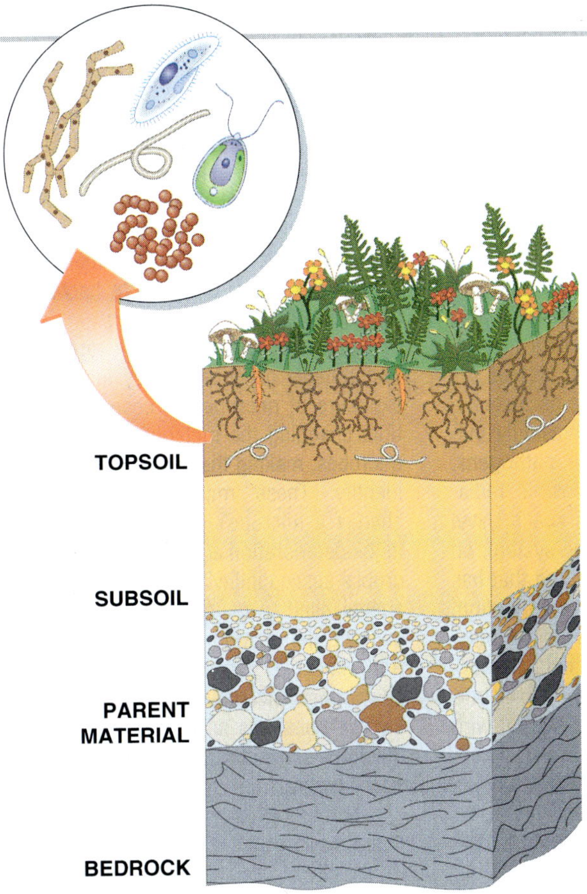

FIGURE 26-9 Soil profile. Four distinct layers (called horizons) comprise typical soil. Virually all soil organisms reside in the topsoil.

TOPSOIL

SUBSOIL

PARENT MATERIAL

BEDROCK

CONCEPT CHECK

- Distinguish between a microbe's microenvironment and the ecosystem to which it belongs.
- In what ways does a food web differ from a biogeochemical cycle? What roles do microbes play in each?
- Describe how decomposers mineralize organic carbon compounds and free the nitrogen trapped in organic compounds.
- Why are the sulfur and phosphorus cycles important?

Terrestrial Microbiology

Soil is composed of all three phases of matter—solid, liquid, and gas. The solid phase is a complex of inorganic and organic compounds. Inorganic compounds arise from the hard mineral layers that lie below the earth's surface. As the mineral layer gets closer to the surface, it is acted on by physical, chemical, and biological processes that break down the rocks. Inorganic matter mixes with organic matter, primarily from decomposing plants, to form topsoil, the fertile zone that supports terrestrial life (Fig. 26-9). The organic matter, called *humus*, soft-

Water is often badly polluted by the time it collects in reservoirs. It may contain human and animal excrement, industrial wastes, and a variety of potential pathogens and noxious chemicals. Suspended particulate materials tend to increase water turbidity (cloudiness). The larger particles settle out when water is allowed to sit undisturbed in a holding reservoir. The colloidal solids, however, remain suspended in the water unless precipitated with a chemical, such as alum (aluminum potassium sulfate), that complexes the particles into *floccules*, large aggregates that quickly settle to the bottom of the tank. Flocculation also removes many viruses and bacteria. The clarified water is then *filtered* through sand or activated charcoal, a process that physically removes most bacteria, protozoa, and other cellular microbes as well as many viruses. Poliovirus and other small viruses, however, may escape filtration. The addition of *chlorine* is the final step in ensuring a safe, *potable* (drinkable) water supply. (This process is illustrated on p. 727 in Figure 26-18, left side.) Even in low concentrations, chlorine inactivates most of the pathogens in water. (Some protozoan cysts survive chlorination, but they are readily removed from water by filtration.) Although in higher concentrations chlorine and its by-products are deleterious to people and the environment, the amounts used to render public water safe do not exceed the danger level.

MYTH: Drinking water that comes from municipal water departments has been treated and is free of all microbes as it leaves your tap.

FACT: Although it has been treated, drinking water contains harmless bacteria when it leaves the treatment plant. Additional microbes may be picked up along the route to your tap. As long as your water is free of fecal coliforms or pathogens, however, it is considered safe to drink.

● **EVALUATING WATER QUALITY.** Water that has been contaminated by feces may contain pathogens and transmit serious diseases. Fecal contamination is most commonly detected by assaying water for the presence of fecal coliforms, predominantly *Escherichia coli*. The only natural source of these bacteria is the intestines of humans and other mammals. Although several strains of *E. coli* are pathogenic, coliforms are usually not pathogenic when ingested by healthy people. Their presence in water, however, indicates fecal contamination and thus the possible presence of waterborne pathogens. In the United States, water is considered safe for drinking only if it contains no more than one fecal coliform per 100 mL, as determined by the membrane filter method (Fig. 26-16).

Coliforms can also be detected by a multiple-tube fermentation technique called the **most probable number (MPN)** test. In this test, the bacterial sample is diluted into a medium that specifically supports the growth of coliforms. If the sample is diluted sufficiently, a point will be reached at which there are no coliforms in the diluted sample. This endpoint can be used to estimate the original population of coliforms in the sample.

A more modern and more sensitive method for detecting *E. coli* employs the polymerase chain reaction. ▶ (See Chap. 9, p. 232.) Bacteria collected from filtered water are broken open and incubated with primers specific for *E. coli* genes necessary for lactose utilization. (DNAs from other coliforms do not respond to the primers.) Within a few hours, this technique amplifies genetic information sufficiently to allow the detection of even a single organism in the original sample (Fig. 26-17). Similar methods are being developed for monitoring pathogenic enteric viruses in water.

Water to be used for injectable pharmaceuticals must be subjected to purification to remove endotoxin, the remnants of the cell wall of gram-negative bacteria. The inoculation of endotoxin into the circulatory system may cause fever, vascular collapse, and even death. ▶ (See *Endotoxin*, Chap. 17, p. 415.) The Food and Drug Administration (FDA) requires that all injectable solutions be free of endotoxin. Two tests are commonly used to evaluate endotoxin concentrations in water for injection. ▶ (See THE MICROBES: Dangerous Ghosts of Dead Bacteria, Chap. 4, p. 84.) The *pyrogen test* measures the induction of fever in rabbits that have been injected with a sample of the test material. The *limulus amoebocyte lysate (LAL)* test detects endotoxin by its ability to cause the formation of a clot in a lysate from the blood cells of the horseshoe crab (*Limulus polyphemus*).

■ Sewage Disposal

There are several approaches to disposal of domestic wastewater. One is to return the raw sewage directly to the environment, burying it or dumping it into oceans, lakes, or rivers. Microorganisms, the decomposers, eventually digest the organic load and convert it to the inorganic state. This is simply completion of the biogeochemical cycle. Decomposers also eliminate the pathogens in sewage by outcom-

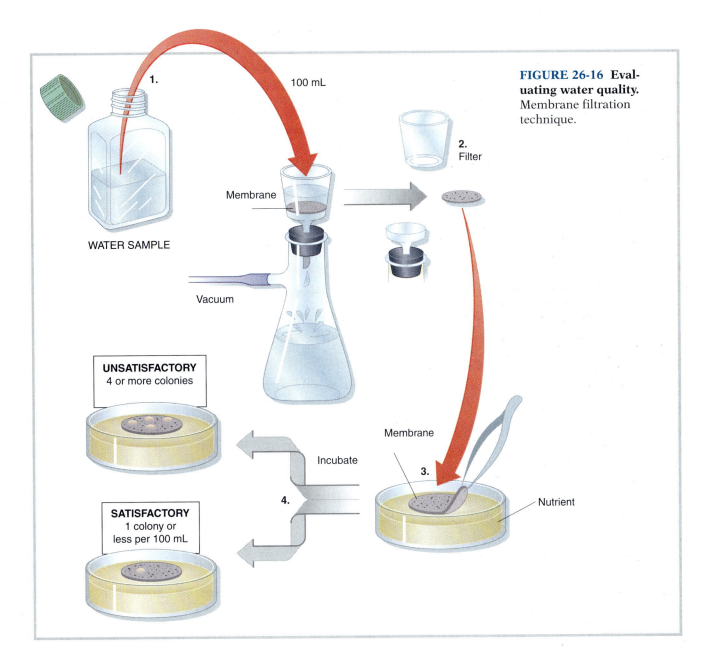

1.

100 mL

WATER SAMPLE

Membrane

Vacuum

2.
Filter

Membrane

3.

Nutrient

Incubate

4.

UNSATISFACTORY
4 or more colonies

SATISFACTORY
1 colony or
less per 100 mL

FIGURE 26-16 Evaluating water quality. Membrane filtration technique.

peting them, rendering the material noninfective. After a period of time, the waste is safely free of pathogens, noxious chemicals, and offensive odors. Unfortunately, this process requires so much time that sewage-contaminated water will likely be reused before the pathogens are eliminated, resulting in epidemics of serious waterborne disease. The natural decomposition cycle is not rapid enough to keep up with the volumes of sewage generated by communities, especially large municipalities.

Families may safely dispose of their sewage in *septic tanks*. These are containers that store sewage while solids settle out and decomposers break down organic compounds. Complete breakdown, however, occurs only in the presence of molecular oxygen,

and the biological activity in the tank quickly consumes oxygen and produces anaerobic conditions. Thus, decomposition is incomplete, and the liquid discharged from septic tanks contains solubilized organic compounds. In most systems, the liquid leaving the tank trickles along pipes or troughs lined with gravel. In this way, the **effluent** (discharged liquid) is aerated, and breakdown to inorganic substances is completed by the decomposers attached to the surfaces over which the liquid flows. The treated effluent is discharged into the surrounding soil.

Larger communities and cities in the United States rely on large-scale sewage treatment plants for disposal of wastewater (Fig. 26-18, right-hand side). In these systems, removal of solids and microbial decomposition are accelerated to keep pace with the huge volumes of wastewater generated. Sewage is first subjected to *primary treatment*. Large-particle solids are screened or skimmed off for burning and burying. The liquid is then transferred to a sedimentation tank where more solids settle out to form *sludge*. Primary treatment is therefore more a process of physical separation than one of microbial decomposition.

Both the effluent from the primary treatment and the sludge contain potentially dangerous pathogens and a heavy load of organic compounds that give them a high **biochemical oxygen demand (BOD)**. The BOD is a way of expressing the amount of organic compounds in sewage as measured by the volume of oxygen required by bacteria to metabolize these compounds. The more dissolved organic matter there is in the sewage, the more oxygen will be required by bacteria to decompose it. Dumping sewage that has a high BOD elevates the concentration of soluble organic compounds in the environment where it is discarded. Digestion of these organic compounds in natural ecosystems such as

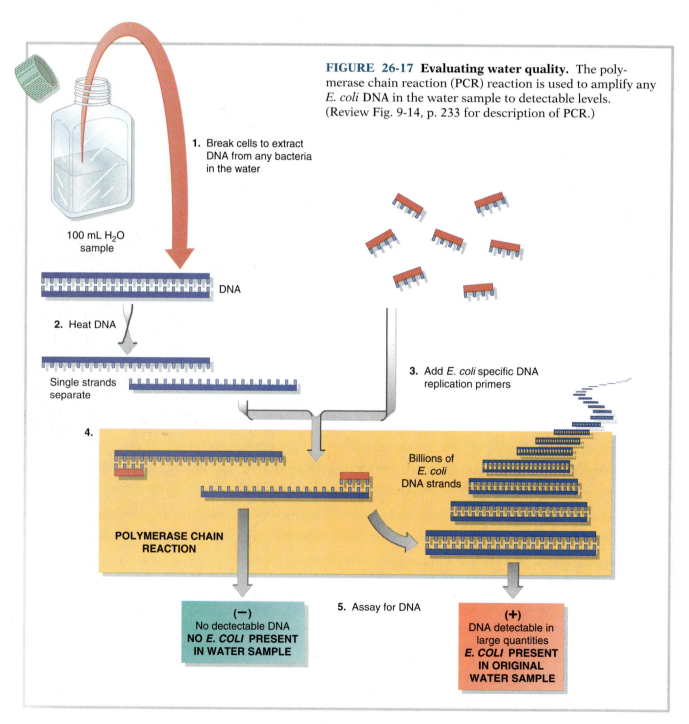

FIGURE 26-17 Evaluating water quality. The polymerase chain reaction (PCR) reaction is used to amplify any *E. coli* DNA in the water sample to detectable levels. (Review Fig. 9-14, p. 233 for description of PCR.)

1. Break cells to extract DNA from any bacteria in the water

100 mL H₂O sample

DNA

2. Heat DNA

Single strands separate

3. Add *E. coli* specific DNA replication primers

4.

Billions of *E. coli* DNA strands

POLYMERASE CHAIN REACTION

5. Assay for DNA

(−)
No dectectable DNA
NO *E. COLI* PRESENT IN WATER SAMPLE

(+)
DNA detectable in large quantities
***E. COLI* PRESENT IN ORIGINAL WATER SAMPLE**

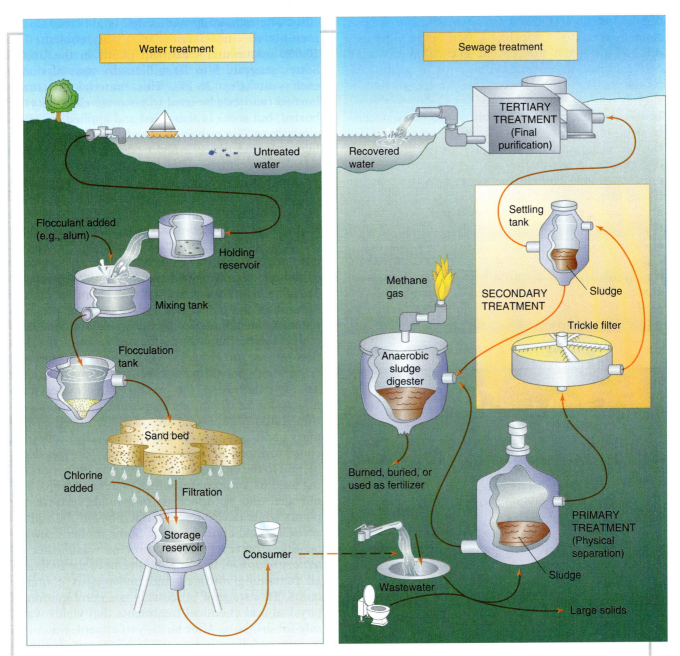

FIGURE 26-18 **Water treatment and sewage disposal—the two facets of maintaining water safety.** Treatment of municipal water supply is shown on the left, the steps in wastewater (sewage) treatment on the right. The screened background represents the relative level of water contamination (darker background = higher level of chemical or microbial contamination). Although tertiary sewage treatment is ideal, its cost is prohibitive in most municipalities, which release treated wastewater after secondary treatment.

McKane/Kandel • Fig. 26-18

lakes can deplete available oxygen and result in asphyxiation of fish. This type of environmental disruption is minimized by further treating sewage to reduce the BOD prior to its return to the environment. Primary treatment removes approximately one-fourth of the BOD from sewage.

Secondary treatment is mostly a biological process that depends on complete aeration of the system.

The presence of excess oxygen during secondary treatment reduces the BOD of sewage by another 80 to 90 percent. One of two systems is usually employed to saturate the sewage with oxygen. In *trickle filtration*, slowly moving sprinkler arms trickle the effluent from the settling tank over a bed of gravel. As the sewage seeps through the gravel, microbes attached to the rocks rapidly oxidize the dissolved

Key Terms

community (p. 705)
ecosystem (p. 705)
ecology (p. 705)
microenvironment (p. 705)
producers (p. 707)
consumers (p. 707)
food webs (p. 707)
biogeochemical cycles (p. 709)
decomposers (p. 709)

mineralization (p. 709)
carbon cycle (p. 709)
nitrogen fixation (p. 711)
nitrogen cycle (p. 711)
nitrification (p. 712)
ammonification (p. 712)
denitrification (p. 712)
siderophores (p. 718)
greenhouse effect (p. 723)

most probable number
  (MPN) (p. 724)
effluent (p. 725)
biochemical oxygen de-
  mand (BOD) (p. 726)
bioremediation (p. 729)
recalcitrance (p. 729)
bioaugmentation (p. 730)

Key Facts and Concepts

- **What is ecology?** Ecology is the study of ecosystems, the living and nonliving components of an environment. Ecologists examine the interactions among organisms in an ecosystem plus the interrelationship of the organisms and the nonliving factors in the environment such as water, temperature, oxygen, and pH.

- **Food webs**. Food webs depend on primary producers, autotrophs that capture energy (from sunlight or oxidation of inorganic chemicals) and synthesize energy-rich organic molecules. The chemical energy is harvested by consumers that eat the producers or other consumers, and by decomposers, which break down dead organisms and wastes, using some of the nutrients for growth and returning some of the chemicals to the environment in mineralized form.

- **The critical role of microbes**. Every ecosystem depends on microorganisms to maintain life. Decomposers recycle nutrients, providing the resources to create new organisms. In fact, all biogeochemical cycles depend on prokaryotes.

- **Nitrogen's role**. The nitrogen cycle is driven by nitrogen fixation, the conversion of molecular nitrogen to a biologically accessible form such as ammonia. Nitrogen fixation is performed by some cyanobacteria and some free-living bacteria, but most nitrogen is fixed by *Rhizobium* in root nodules of legumes. Nitrifying bacteria convert ammonia to nitrate, the form in which nitrogen must be before it can be used by most plants. Nitrogen is returned to the atmosphere by denitrifying bacteria that convert nitrate to molecular nitrogen or nitrous oxide, both of which are gases. Denitrification reduces soil fertility and wastes much of the fertilizer applied to crops.

- **Sulfur's role**. The sulfur cycle is also propelled by prokaryotes that convert elemental sulfur or hydrogen sulfide to sulfate, the form needed for assimilation into proteins of plants, animals, and other organisms. Decomposers return the sulfur in proteins to hydrogen sulfide, which can then be oxidized to elemental sulfur or sulfate.

- **Terrestrial microbiology**. The ecology of soil includes solid elements and minerals, water, and air plus humus, the organic material that accumulates from dead organisms, primarily plants. Decomposers improve soil fertility by releasing nutrients from dead organisms for use by plants and by processing humus to improve the soil's ability to retain water and other physical properties. Nitrogen-fixing bacteria improve soil fertility. Soils containing fungi that form mycorrhizae on roots of crop plants are also conducive to crop production, as are soils that contain strains of *Pseudomonas* that produce iron-binding siderophores.

- **Marine microbiology**. Microorganisms contribute to the ecology of aquatic ecosystems in several ways in addition to decomposition and nutrient recycling. In marine environments, viruses may play an important role in preventing the overgrowth of bacteria. Marine waters contain photosynthetic microbes at the surface, heterotrophs below that, and decomposers in the sediment on the seafloor. One exception to this is the thermal vent ecosystem, which depends on chemoautotrophic bacteria as the producers that support a unique community of organisms near geothermal vents.

- **Freshwater microbiology**. Freshwater ecosystems are less stable than oceans and seas and are more susceptible to fluctuations in temperature, pH, and oxygen concentrations. They are also more susceptible than oceans to the immediate effects of pollution.

- **Microbes as pollutants**. Microorganisms play several roles in causing pollution and aggravating its negative impact. Water pollution includes contamination with untreated sewage (which can generate outbreaks of disease), increased nitrate and phosphate content (which accelerates eutrophication, algal blooms, and subsequent death of oxygen-requiring organisms), and introduction of toxins that poison plant and animal life. Microbes corrode metal and concrete, increase the likelihood of heavy metal poisoning, and release greenhouse gases that may contribute to global warming.
- **Water treatment**. Before water can be safely used, it must be subjected to a battery of treatments to remove potential pathogens and impurities. Clarification of water by mechanical treatment (skimming and sedimentation to remove large solids) is followed by filtration to remove impurities. Chlorine is added to the water in hold-

ing facilities to kill any microbes that were not mechanically removed and to prevent microbial growth in water awaiting use. The purity of drinking water is assessed by determining the coliform count as evidence of fecal contamination.

- **Sewage disposal**. Sewage (wastewater) is rendered safe for disposal by using a combination of physical removal of particulates, microbial digestion of organic compounds, and microbial elimination of pathogens by bacterial competition. In complete sewage treatment, BOD is completely removed by microbial decomposition of organic compounds. Sewage treatment is accomplished in septic tanks or in large-scale municipal treatment plants.
- **Bioremediation**. Microbes are also natural "antipolluters" that metabolize toxic compounds and remove recalcitrant pollutants. Oil spills and other types of environmental pollutants can be eliminated by inoculating the area with bacteria or by adding nutrients that encourage the growth of indigenous microbes that can consume the contaminating substance.

Review and Problem-Solving Questions

1. Discuss the relationship among microbes in (a) ecosystems; (b) biogeochemical cycles; (c) bioremediation.
2. Distinguish between:
   (a) Mycorrhizae and root nodules
   (b) Consumers and decomposers
   (c) Colonies and biofilms
3. In what ways do the types and numbers of microbes in soil differ from those found in ocean waters? What factors contribute to these differences?
4. (a) Diagram the nitrogen cycle.
   (b) Are nitrifying bacteria necessary in fields that contain symbiotic nitrogen-fixing bacteria? (Explain why or why not.)
   (c) Why are *Rhizobium* able to form root nodules only with leguminous plants?
   (d) What are the possible consequences of significantly increasing the amount of artificially fixed nitrogen?
5. Discuss the following statement: "Because of microbes, the molecules that comprise your body have a timeless history and future."
6. Cellulose is an abundant substrate for microbial growth, but it is not a source of nitrogen. It has been suggested that a cellulose-degrading bacterium that also had the ability to fix nitrogen would be a useful organism.

   (a) How would you isolate a nitrogen-fixing cellulolytic bacterium from soil?
   (b) How might this organism be adapted to provide the needed environment to maintain activity of nitrogenase?
7. How could industrial pollution of rivers lead to the death of a lake?
8. Herbicides are among the most recalcitrant chemicals. A new herbicide called Weedstompa has just been produced. The toxic component is a chemical created in the laboratory that resembles chlorinated phenols.
   (a) Describe how you might develop a microbe that would degrade this herbicide.
   (b) Under what circumstances might you need to use such a microbe?
9. What is the role of biofilms in corrosion?
10. At least 27 major cities dump their untreated sewage directly into the Mediterranean Sea. The sea is an excellent dispersal system because of its size and constant wave action. Discuss whether you believe this practice is adequate for sewage disposal in these areas. Describe two of the possible adverse consequences.
11. A terrorist group plans to destroy the earth's reserves of petroleum by inoculating oil-digesting bacteria used for bioremediation down the shafts of active oil wells. Why would this plot fail to achieve the objec-

tive even if every oilfield's reserves were inoculated?

12. A terrorist group hatched a plot to create mass chaos by dumping dangerous botulism toxin (a protein) into the pretreated water reservoirs of large cities. Botulism toxin is the most poisonous substance on earth. Would this sinister plan accomplish its objective? Why or why not?

13. Discuss the implications of the following observations.
   (a) Many human pathogens released in sewage retain viability in ocean waters.
   (b) Genetically engineered microorganisms used in bioremediation are introduced live into soils.

HEADLINES

THE RECOVERY OF SPIRIT LAKE

In May 1980, Mount St. Helens erupted, covering the landscape with downed trees and volcanic debris. One of the casualties of this natural calamity was Spirit Lake, a pristine body of water nestled in the mountains of Washington. The clarity of the lake prior to the volcanic blast reflected the limited life in its waters, likely due to the extremely low concentrations of organic and inorganic substances needed to sustain organisms. For 5 years after the blast, volcanic deposits blocked the lake's only outlet, virtually turning the lake into a large batch culture (see Chap. 5). Biological activity in the lake increased until the microbial populations approached stationary phase. This changed the conditions in the lake, allowing new microbial communities to replace the stationary ones. In 1985, a tunnel was opened as an outlet for water to prevent flooding. This renewed flow of water eventually led to the restoration of the lake to its previous clarity and limited microbial growth.

- Why would chemotrophic bacteria repopulate the lake more quickly than photosynthetic organisms?
- What effect would limited oxygen supplies have on the composition of the microbial population?
- What other factors might determine the kinds of microbes that appeared in the lake after the eruption?
- Where might the microbes repopulating the lake come from? Do you think that the ecology of the lake after recovery is identical to its pre-eruption ecology?
- What do you think would have occurred if the water outlet tunnel had not been built and the lake waters had been allowed to stagnate?

Source: D. Larson, The recovery of Spirit Lake, *American Scientist*, 81: 166–177 (1993).

CHAPTER 27

Food and Industrial Microbiology

Industrial microbiologists in action. This model of a bioreactor will be scaled up to create the giant fermentation tanks used in microbe-driven industrial processes. Although they are minuscule in size, bacteria and fungi produce immense quantities of foods and chemicals that vastly improve the quality of our lives.

History abounds with stories of people fascinated with the prospects of turning lead into gold. Alchemy, as such pursuits are called, has sent many talented people on misguided quests, not unlike the search for the mythical fountain of youth. Never in history, however, have we come so close to accomplishing something that approaches alchemy, the transformation of substances of little worth into products of great value. In these instances the "alchemy" does not require magic, but genetically engineered microorganisms. Using such organisms we are developing techniques for turning a persistent nuisance into a great treasure. The nuisance is human garbage; the treasure is food and fuel. Food and fuel are just two of the fruits of **microbial biotechnology**, the manipulation of microorganisms to increase their practical benefit.

Biotechnology is just one of the rewards of applying the lessons of microbiology. Many applications have been discussed throughout this book, and the previous chapter specifically introduced you to the applied principles of soil, sewage, water, and waste management microbiology. In this chapter, we consider microbial biotechnology as a modern adjunct to the more traditional applications of food and industrial microbiology.

People first started putting microbes to work long before microorganisms were first observed. Processes discovered by accident thousands of years ago are still used today to produce bread, beer, wine, cheeses, yogurt, and many oriental foods. Sophisticated techniques are providing new and exciting ways of tapping the technological potential of microorganisms. This technology is already a well-established factor in the world economy. In the United States alone, microorganisms add tens of billions of dollars to our annual income.

Some of the more advanced technologies used today would have astonished many early microbiologists. So far, however, the applied fields of microbiology are still dominated by traditional applications of microbial activities—industrial fermentations, treatment of wastewater, and production of alcoholic beverages, foods, antibiotics, and vaccines. One of the goals of the new biotechnology is to improve the efficiency of the microorganisms performing these established processes or to develop new strains of organisms that can perform desirable tasks not accomplished by their naturally occurring microbial counterparts. Another objective is to develop strains of organisms that will readily provide resources that either are unavailable or are difficult to obtain from conventional sources.

MYTH: "Biotechnology" is synonymous with "genetic engineering" (recombinant DNA technology or "gene splicing").

FACT: Biotechnology not only includes genetic "reconstruction" of organisms by gene splicing, it also embodies hybridoma technology (production and use of monoclonal antibodies), new substance-delivery systems (using liposomes, for example), and manipulation of a microbe's environment to maximize a desired activity or to force accelerated evolution. Biotechnology is the use of technology to enhance a desirable characteristic of an organism such as its ability to produce a marketable commodity.

Microbes and Food Production

Microorganisms continue to be used in the food industry primarily in three ways.

1. Specific metabolic activities, usually fermentation reactions, generate organic compounds that accumulate and transform some marginally edible substances into foods with more desirable characteristics. These altered properties usually help preserve foods, and they often enhance their flavor, texture, or digestibility. Dairy products, breads, soy sauce, pickled vegetables, alcoholic beverages, and vinegar are just a few of the edibles produced by microbial fermentation.
2. Microbial cells, cultured in large quantities, are used as protein supplements in feed for livestock. Such microbe-generated proteins have also been suggested as alternative food resources for humans.
3. Certain microbes produce metabolic by-products that have nutritional or flavor-enhancing properties when added to foods and feeds. Enzymes isolated from microorganisms are also instrumental in food production.

■ Dairy Products

Milk is fermented to cheese, yogurt, sour cream, or buttermilk by lactic acid–producing bacteria (*Lactobacillus*, *Leuconostoc*, and *Streptococcus* species). Fermented milk products taste different, usually have a lower moisture content, and are more resistant to spoilage than milk. Some cheeses, for example, can be stored for months at room temperature, whereas whole milk spoils quickly even when pasteurized and refrigerated. Dairy fermentations can best be controlled if milk is pasteurized to inactivate contaminating organisms and enzymes before processing. Starter cultures of desired microbes are then added. The final product depends on the kinds of microbes inoculated, the type of milk (cow's, sheep's, or goat's), and the conditions of processing.

Cheese production (Fig. 27-1) begins with *curdling*, or *souring*, the bacterial conversion of lactose (the sugar in milk) to lactic acid. The resulting drop in pH causes protein in the soured milk to coagu-

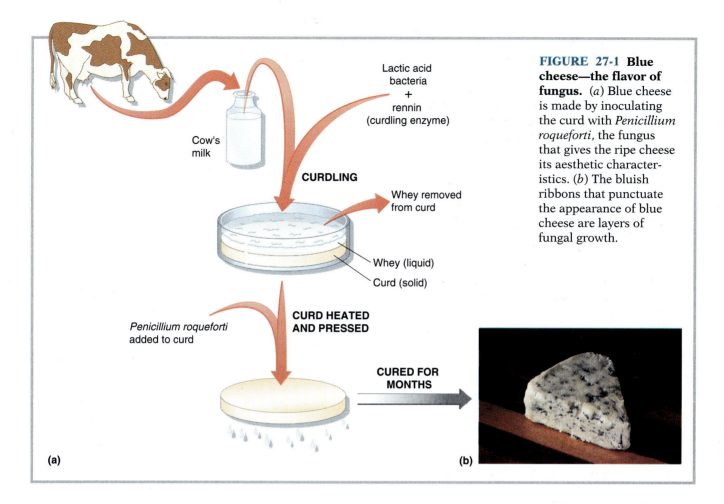

FIGURE 27-1 Blue cheese—the flavor of fungus. (a) Blue cheese is made by inoculating the curd with *Penicillium roqueforti*, the fungus that gives the ripe cheese its aesthetic characteristics. (b) The bluish ribbons that punctuate the appearance of blue cheese are layers of fungal growth.

Labels in figure (a):
Cow's milk
Lactic acid bacteria + rennin (curdling enzyme)
CURDLING
Whey removed from curd
Whey (liquid)
Curd (solid)
Penicillium roqueforti added to curd
CURD HEATED AND PRESSED
CURED FOR MONTHS
(a)
(b)

late. This process may be facilitated by the addition of the milk-curdling enzyme *rennin*. The coagulated solids form a curd, which, when separated from *whey* (the watery liquid), becomes cottage cheese. When cream rather than milk is curdled, the solid product is cream cheese. The curd can also be used to produce other types of cheese. It is first heated to reduce its moisture content and then allowed to ripen through additional microbial activity. Depending on the residual moisture content, cheeses are classified as soft, semisoft, or hard. The microbes used for ripening break down proteins and fats or produce fermentation products that generate the characteristic flavors, aromas, textures, and appearances of cheeses. Brie and Camembert are soft cheeses ripened by fungi growing on the surface of the pressed curd. The semisoft Roquefort cheese, on the other hand, is ripened by a blue-green mold that can be seen growing throughout the curd (Fig. 27-1b). Bacteria are responsible for ripening most hard cheeses. For example, propionibacteria transform a curd to Swiss cheese by producing propionic acid (which imparts flavor) and carbon dioxide, the gas that accumulates as large bubbles that form the holes of Swiss cheese (Fig. 27-2).

Milk can also be acidified and flavored by con-

trolled fermentation to produce yogurt. Milk solids are added before fermentation, and two bacteria, *Lactobacillus bulgaricus* and *Streptococcus thermophilus*, convert lactose to lactic acid and acetaldehyde, which impart the characteristic flavor to yogurt. Kefir, a dairy drink from eastern Europe that is gaining popularity in the United States, contains

FIGURE 27-2 The hole story. Holes in Swiss cheese are bubbles of carbon dioxide generated by propionibacteria growing in the curd. The bubbles remain suspended as the curd hardens.

about 1 percent alcohol. It is formed by the combined fermentations of lactic acid–producing bacteria and yeast that convert lactose in milk to alcohol. Buttermilk is usually made by fermenting pasteurized skim milk (originally it was the liquid remaining after the fat was removed in churning cream to butter). Sour cream, as the name implies, is fermented cream. Lactic acid bacteria in cream produce a chemical called *diacetyl* that is responsible for the characteristic flavor of the butter churned from the cream. Although butter is not a product of fermentation, its flavor is.

■ Nondairy Foods

Vegetables, particularly cucumbers, cabbages, and olives, can be preserved by the fermenting activities of the lactic acid bacteria and yeast that naturally reside on their surfaces. The growth of these microbes is selectively encouraged by placing the vegetables in salt solutions called *brines*. Lactic acid production usually continues until no fermentable carbohydrates remain. The salt, low pH, and absence of carbohydrates so effectively prevent the growth of spoilage organisms that pickles, sauerkraut, and olives can usually be kept indefinitely.

Soy sauce originated in China centuries ago. Its production requires two microbial transformations. Cooked soybeans and crushed wheat are first incubated with the mold *Aspergillus oryzae*, which converts much of the starch to fermentable sugars. The mash is then fermented for 6 to 12 months with a mixture of lactic acid bacteria and yeast. The resulting liquid is soy sauce; the remaining solids are used as livestock feed.

Bread production by fermentation is another ancient process. Leavened breads are made by adding baker's yeast, *Saccharomyces cerevisiae*, to dough. The yeast metabolizes sugar in the dough aerobically. The oxygen minimizes alcohol production and maximizes carbon dioxide. Bubbles of carbon dioxide gas become trapped in the dough, which rises—literally inflates—and acquires a light texture. The small amount of alcohol that is produced evaporates during baking, so you can eat bread without fear (or hope) of inebriation.

Sourdough and various other breads acquire their characteristic flavors from additional microbial processes. In sourdough production, two additional organisms produce the acid by-products responsible for the bread's tanginess. One organism, *Saccharomyces exiguus*, leavens the dough, while the bacterium *Lactobacterium sanfrancisco* generates the lactic and acetic acids that impart that elusive taste once believed achievable only if the bread were baked in San Francisco.

■ Alcoholic Beverages

The term *fermentation*, from the Latin word for boiling, was originally applied to the production of alcoholic beverages because the generation of carbon dioxide gas gives the fermenting liquid a frothy, boiling appearance. Most alcoholic beverages are manufactured by the metabolic activity of yeasts in the genus *Saccharomyces*. Wines are prepared by the direct action of *S. cerevisiae* or *S. ellipsoideus* on sugars in fruit juices. Beers, on the other hand, are made from the starch in barley or other grain extracts (Fig. 27-3). Starch is a nonfermentable substrate, so beer production requires that the starch be hydrolyzed to fermentable sugars, a process that begins with *malting*. During malting the barley germinates and produces starch-digesting enzymes. In the next step, called *mashing*, the crushed malted grain is mixed with water, at which time the enzymes digest starch to form the fermentable sugars glucose and maltose. These sugars are dissolved in the aqueous extract of the malted barley, called *wort*, which is then boiled with hops, a flower with bacteriostatic properties that imparts a characteristic bitter flavor to the brew while preventing unwanted bacterial growth. The boiling also denatures the starch-digesting enzymes, halting the enzymatic release of sugars. Brewers prepare mash in concentrations that provide only enough sugar to produce between 3 and 6 percent ethanol (drinking alcohol). After boiling, the hops is removed and yeasts are added to ferment the sugars to alcohol and flavor-enhancing compounds. Because the undigested starch in the wort cannot be fermented by brewer's yeasts, most beers have a high carbohydrate content, containing about 4 percent unfermented starch (each gram of which adds about 4 calories to the beverage). Yeasts that digest this starch have been mated with brewing yeasts to yield hybrid organisms that produce beers containing less than 1.5 percent undigested carbohydrate. These yeasts are used to manufacture some brands of low-calorie ("light") beer. Other light beers depend on predigestion of starch in the wort by adding enzymes instead of whole organisms.

MYTH: Bakers use yeasts completely different from those used by most producers of alcoholic beverages.

FACT: *Saccharomyces cerevisiae* is often used for both processes. In fact, until the mid-1800s, commercial bakeries used yeast left over from the brewing of beer to leaven their pastries. Modern bakeries

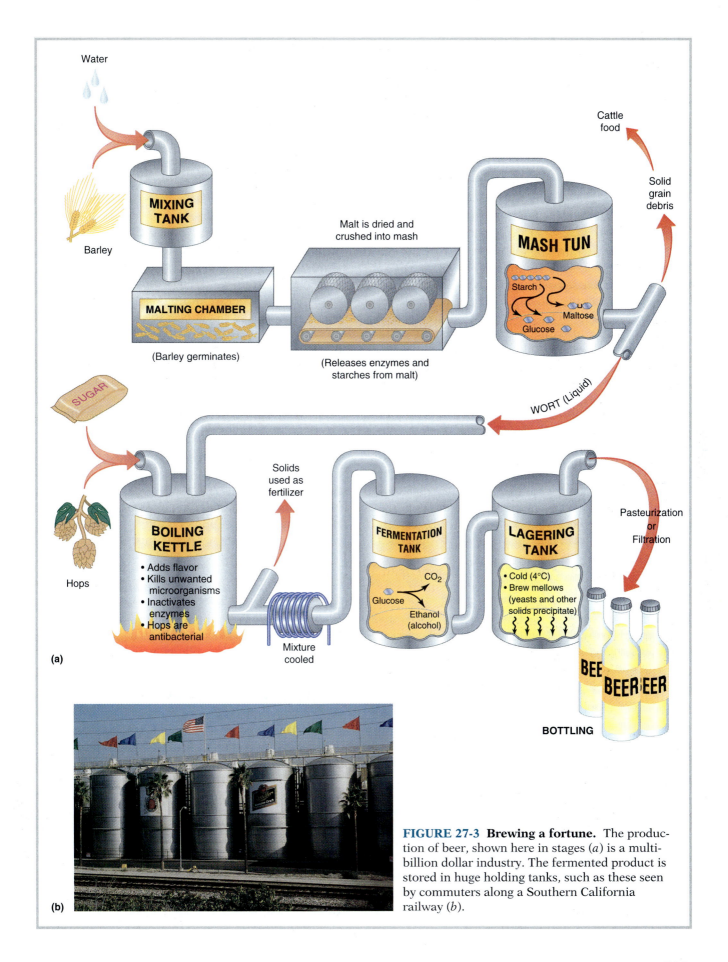

Water

MIXING TANK

Barley

MALTING CHAMBER

(Barley germinates)

Malt is dried and
crushed into mash

(Releases enzymes and
starches from malt)

MASH TUN

Starch
Maltose
Glucose

Cattle
food

Solid
grain
debris

WORT (Liquid)

SUGAR

Solids
used as
fertilizer

Pasteurization
or
Filtration

BOILING KETTLE

• Adds flavor
• Kills unwanted
  microorganisms
• Inactivates
  enzymes
• Hops are
  antibacterial

Hops

Mixture
cooled

FERMENTATION TANK

CO$_2$

Glucose

Ethanol
(alcohol)

LAGERING TANK

• Cold (4°C)
• Brew mellows
  (yeasts and other
  solids precipitate)

BEER BEER BEER

BOTTLING

(a)

(b)

FIGURE 27-3 Brewing a fortune. The production of beer, shown here in stages (*a*) is a multi-billion dollar industry. The fermented product is stored in huge holding tanks, such as these seen by commuters along a Southern California railway (*b*).

use different strains of *S. cerevisiae* than those used by producers of alcoholic beverages, which have their own strains of the fungus that are often closely guarded industrial secrets. Although some breweries use a different species of *Saccharomyces* (*S. carlsbergensis* for some beers; *S. ellipsoides* for some wines), the fermentation process is the same in both bread leavening and alcoholic beverage production.

The starch in rice used to make sake must also be broken down into sugars before it can be fermented. This process is accomplished by *Aspergillus oryzae* inoculated into steamed rice. The resulting sugar is fermented to alcohol by yeast (*S. cerevisiae* and *S. sake*) and to lactic acid by bacteria. Most distilled spirits are also made from fermented grain products in a process similar to beer brewing. The malted mash, however, is not boiled, and enzymes continue to hydrolyze starch to sugar during the fermentation process, resulting in higher alcohol production. The amount of alcohol produced is limited by the inability of yeast to thrive in ethanol concentrations higher than 14 to 18 percent. Thus, products of natural fermentation never exceed these concentrations of alcohol. Brandies, whiskeys, and other spirits have higher alcoholic contents as a result of *distillation*—the process of vaporizing the volatile alcohol by heating the brew (leaving the water behind) and collecting the vapor as it condenses back to liquid.

Grapes and other fruits used to make wine naturally contain large numbers of yeasts on their surface (up to 10^5 per grape). Originally, the sugar in crushed grapes was fermented to wine solely by these yeasts. However, a more consistent product is obtained when the natural flora is inhibited, after which a starter culture of *S. cerevisiae* is added to the grape mash. The resulting wine usually contains about 12 percent alcohol. Red wines are fermented from crushed red grapes. The entire grape, including juice, pulp, skin, and seeds, is present during fermentation. Pigment in the grape's skin provides the wine's red color. Juice that is separated from the solids after 1 day and then fermented becomes rosé wine. White wines are fermented from either red or white grapes; however, only the juice is used. Fortified wines such as sherries, ports, and vermouths contain 15 to 20 percent alcohol and are produced by adding brandy to wine. The increased alcohol content of fortified wines inhibits microbial growth, thereby increasing the stability of the wines.

Vinegar is literally sour wine, originally manufactured by intentionally exposing wine to airborne bacteria. The bacteria then grow in the wine and convert ethanol to acetic acid, the chemical that gives vinegar its characteristic odor and taste. The modern method of vinegar production is not very different from the original "let-alone" process. It consists of inoculating wine or hard (alcoholic) apple cider with *Acetobacter*, which, under highly aerated conditions, converts the ethanol to acetic acid.

Some foods traditionally produced by microbes are summarized in Table 27-1.

■ Single-Cell Protein

Most microbes reproduce faster than either plants or animals. Many bacteria, for example, double their numbers in less than 1 hour in a favorable environment. The generation time of yeasts is usually 1 to 3 hours, that of algae is 2 to 6 hours, and even the more slowly growing molds double their mass in 4 to 12 hours. Microbial cells are generally rich in protein; some are as much as 80 percent protein. In reproducing themselves, these cells can quickly generate huge quantities of proteins. While a 1000-pound steer produces 1 pound of new protein daily, and 1000 pounds of soybean plants manufacture 80 pounds of protein daily, 1000 pounds of yeast may yield 50 tons of protein per day. Such rapid protein production makes microbes attractive as sources of **single-cell protein (SCP)**, alternative or supplementary protein sources to relieve shortages caused by the limited production of plants and animals. Because protein deficiency is the most common form of malnutrition worldwide, SCP becomes even more attractive.

The most useful single-cell proteins supply all the essential amino acids. Several bacteria and yeasts, for example, have protein values comparable to those of animal products and in many cases are better protein sources than plants, which tend to be deficient in the amino acid lysine. Often these single-cell proteins can be produced inexpensively by growing microbes on inorganic nitrogen (ammonium salts or nitrates), using cheap sources of carbon and energy. Petroleum hydrocarbons were among the first and most efficient substrates for producing single-cell proteins but are no longer used because of their current demand as fuel. Agricultural and industrial waste products provide several inexpensive substrates, including molasses, fruit pulp, milk whey, sulfite liquor (wastes from the paper industry), and methane (or methanol) generated from sewage treatment. When grown with these substrates, microbes not only supply a valuable resource but simultaneously help dispose of huge amounts of garbage.

Algae and photosynthetic prokaryotes use light

TABLE 27-1   SOME COMMON FERMENTED FOOD PRODUCTS AND FERMENTING MICROORGANISMS

Fermented Product	Starting Material	Main Fermenting Microorganisms
Dairy Foods		
Cheese	Milk	*Lactobacillus* sp. (early reactions)
Yogurt	Milk	*Streptococcus thermophilus, Lactobacillus bulgaricus*
Kefir	Milk	*Lactobacillus* sp., *Saccharomyces* sp.
Acidophilus milk	Milk	*Lactobacillus acidophilus*
Buttermilk	Milk	*Lactobacillus lactis, Leuconostoc cremoris*
Vegetables		
Pickles	Cucumbers	*Pediococcus cerevisiae, Lactobacillus plantarum*
Olives	Green olives	*Leuconostoc mesenteroides, Lactobacillus plantarum*
Sauerkraut	Cabbage	*Leuconostoc mesenteroides, Lactobacillus plantarum*
Kimchi	Cabbage and other vegetables	*Lactobacillus plantarum*
Beverages and Other Liquids		
Coffee	Coffee beans	*Erwinia dissolvens, Saccharomyces* sp.
Beer	Barley wort	*Saccharomyces carlsbergensis*
Wine	Grapes	*Saccharomyces ellipsoideus*
Sake	Rice	*Aspergillus oryzae, Saccharomyces sake*
Soy sauce	Soybeans	*Aspergillus oryzae, Saccharomyces rouxii, Lactobacillus delbrueckii*
Vinegar	Ethanol (wine)	*Acetobacter orleanensis*
Other Foods		
Bread	Wheat flour	*Saccharomyces cerevisiae*
Sourdough bread	Wheat flour	*Saccharomyces exiguus, Lactobacillus sanfrancisco*
Country-cured ham	Pork	*Aspergillus, Penicillium* sp.
Dry sausage	Pork, beef	*Pediococcus cerevisiae*

energy and carbon dioxide for growth. Because they need not be supplied with a prepared carbon and energy source, they are potentially inexpensive sources of SCP. Unfortunately, these organisms are difficult to culture. They must be grown in shallow ponds or lagoons and will produce high yields only if they are provided adequate sunlight and controlled incubation temperatures. Often it is impossible to prevent contamination, and it is difficult to harvest the organisms, particularly the unicellular forms. Despite these problems, some algae, notably *Spirulina*, are being used in dried form as human food.

Although dried yeasts have supplemented human diets for generations, most microbes cultivated as food sources are used primarily as animal feeds. For example, over 150,000 tons of yeast is harvested annually as fodder for the world's livestock. Microbes have yet to gain widespread acceptance as dietary supplements for humans. Many organisms are difficult to digest because of their cell walls or have a disagreeable appearance, smell, or taste. Some microbes produce endotoxins or other products that make them unsafe for use as food. Furthermore, most rapidly dividing cells contain high nucleic acid concentrations, which, when ingested in amounts that the body cannot effectively excrete, pose a threat to human health. Degradation of excess nucleic acids generates uric acid, a compound that forms kidney stones and accumulates in joints and tissues, causing symptoms of gout. Unfortunately, the cost of removing nucleic acids reduces the economic feasibility of these products. Nonetheless, several countries, including France and Japan, are safely culturing bacteria on petroleum wastes and using the cells as a human protein supplement.

■ Food Additives

Many vitamins, amino acids, nucleotides, and enzymes that are commercially valuable to the food industry are obtained in high concentrations from

TABLE 27-3　PRESERVATIVES COMMONLY USED IN FOODS

Preservative	Protected Foods
Sorbic acid/sorbates	Salad dressing
Benzoic acid/benzoates	Carbonated beverages, jams, relishes
Citric acid	Carbonated beverages
Propionic acid/propionates	Breads, cakes
Sulfites/sulfur dioxide	Wine, dried fruits, molasses, soft berries
Nitrites	Cured meats

CONCEPT CHECK

- What factors determine whether microbial growth produces edible foods, a spoilage process, or foods capable of causing illness?
- Why aren't all foods sterilized? What techniques other than heat sterilization are used to control microbial growth? What advantages do these methods have over heat sterilization?
- What is the role of lactic acid–producing bacteria in the production of cheese? of pickled vegetables? How does the metabolic activity of *Saccharomyces* produce leavened bread and beer?
- List four advantages of using single-cell protein or genetically modified foods rather than traditional food sources. Discuss four problems that currently limit their use.

Industrial Microbiology

Most industrially important microbes are used as metabolic factories that can manufacture many commercially important substances that are not readily or economically obtained from other sources. Three categories of natural microbial products are of industrial importance: (1) primary metabolites, (2) secondary metabolites, and (3) enzymes. **Primary metabolites** are compounds that are either intermediate or end products of biochemical pathways essential for the growth of the microbe. Sugars, amino acids, vitamins, nucleotides, and some organic acids and alcohols are examples of primary metabolites. They are usually produced in greatest abundance during logarithmic growth. **Secondary metabolites,** on the other hand, are not essential for growth, and in many cases their natural role in the microbe's physiology is unknown. Most secondary metabolites are low molecular weight compounds that accumulate during the stationary phase. Many antibiotics are secondary metabolites.

In addition to yielding metabolic products and by-products, industrial microbiological processes sometimes give intact living cells as final products. Baker's yeast, frost-preventing bacteria, and microbes used for bioremediation would fall into this category, as would living microbial pesticides and powdered *Lactobacillus* used for fortifying or reestablishing normal flora bacteria. In addition, any product that is a "starter culture" contains viable microbes. In all these cases, living microbes are harvested, lyophilized (freeze-dried), and packaged.

■ Industrial and Pharmaceutical Products

Some primary metabolites and enzymes of industrial importance that are produced by microbes are summarized in Table 27-4. The organic solvents glycerol, ethanol, acetone, and butanol were previously derived from microbes, but they are currently produced more economically by chemical synthesis from petroleum products. However, microbes are becoming economically competitive with these chemical processes as the cost of petroleum substrates increases and as recombinant DNA technology produces microbes that can use cheap petroleum hydrocarbon wastes as substrates for solvent production.

Microbial fermentations manufacture more than 175,000 tons of organic acids annually throughout the world. Much of this is in the form of citric acid, a metabolic intermediate that accumulates when an early step in the TCA cycle is blocked and the cycle cannot be completed ▶ (see Tricarboxylic Acid Cycle, Chap. 7, p. 159). Citric acid is the principal acid added to jams, candies, fruit drinks, and other foods. It is also used to adjust the pH of cosmetic lotions and pharmaceutical products. Citric acid has metal-binding properties that make it useful in shampoos, electroplating, and leather tanning and in clearing metal-clogged pipes in the oil industry. Itaconic acid is another compound derived by bioconversion of a TCA cycle intermediate. Its major application is in the manufacture of plastics and synthetic fibers. Other organic acids are employed in the preparation of perfumes, plastics, dyes, and pharmaceuticals.

Dextrans and xanthans, the principal polysaccharides of industrial importance, are produced by bacteria as components of their capsules or slime layers. *Dextrans* have adsorbent properties that aid in chemical separation and purification procedures. They are also used to expand the volume of blood

TABLE 27-4 SOME INDUSTRIAL PRODUCTS AND ENZYMES OBTAINED FROM MICROBES

Product	Major Source	Application
Citric acid	*Aspergillus niger*	Food acid, cosmetics
Itaconic acid	*Aspergillus terreus*	Plastics
Dextrans	*Leuconostoc mesenteroides*	Blood plasma expander, adsorbent
Xanthans	*Xanthomonas campestrans*	Drilling muds, stabilizers, and emulsifiers
Cellulose	*Acetobacter xylinum*	Filters, fiber production
Amino acids	*Corynebacterium glutamicum*	Food supplement and additive
Amylase	*Bacillus*	Textile industry, detergents
Proteases	*Bacillus, Streptomyces*	Detergents
Lipase	*Rhizopus, Saccharomycopsis*	Degreasing wool, digestive aid
Streptokinase	*Streptococcus*	Dissolution of blood clots
Uric oxidase	*Aspergillus*	Treatment of gout
Penicillin acylase	*Escherichia coli*	Production of semisynthetic penicillin

plasma for transfusion. *Xanthans* are the most widely used microbial polysaccharides and are the only bacterial exopolysaccharides regarded by the Food and Drug Administration as safe food additives. They are produced by *Xanthomonas campestrans*, a bacterium originally isolated from a rutabaga plant. The gelling and emulsifying properties of xanthans make them ideal solidifying agents in foods and pharmaceutical and cosmetic products as well as stabilizers for suspending insoluble components of dairy products, salad dressings, and fruit drinks. Xanthans are also emulsifiers in paints and ceramic glazes and thickeners in sauces and syrups, cosmetics, pharmaceuticals, textiles, and glue. When dissolved, xanthans produce highly viscous solutions that are used in the petroleum industry to lubricate drills, to counterbalance the pressure of oil in the well, and to recover otherwise unobtainable oil.

Some microbial enzymes that can be harvested in high concentrations have industrial applications, predominantly in the food industry. Enzymes are also included in some commercial laundry detergents. These "cleaning enzymes" are proteinases that dissolve blood, mucus, chocolate, and other proteinaceous stains. Enzymes are being isolated from thermophilic microorganisms to be used in the high temperatures of hot water washes. Proteinases also soften and prepare hides for tanning. Lipid-splitting enzymes (lipases) are used to degrease wool, and starch-digesting amylases remove starch sizing from textiles. Amylases and lipases also serve as digestive aids in preparing foods for people who produce deficient amounts of these enzymes. Another enzyme, a temperature-stable DNA polymerase, has revolutionized biology and biotechnology. This enzyme, which comes from the thermophile *Thermus aquaticus,* made possible the poly-

merase chain reaction that allows the production of unlimited amounts of identical DNA from a single molecule ▶ (see Gene Amplification—Polymerase Chain Reaction, Chap. 9, p. 232).

Enzymes are used to catalyze the production of several substances of industrial or medical value, thereby eliminating the need for whole cells to accomplish these reactions. This approach has several advantages. It eliminates the need for growth medium, thereby reducing the likelihood of contaminants growing in the nutritionally deficient system. It also reduces the number of subsequent steps needed for purification of the product. The efficiency of these processes can be augmented by attaching the enzyme to a solid carrier. **Immobilized enzymes** remain in the reaction vessel as the medium flows through, carrying the substrate in and the products out (Fig. 27-5). This approach is especially useful when the enzyme is very expensive or is available only in very low quantities. Immobilized enzymes, which accounted for 20 percent of the total enzyme market in 1990, are routinely used to convert glucose to fructose in the production of high-fructose syrups, to hydrolyze lactose to form low-lactose milk, and to catalyze the formation of various derivatives of amino acids and antibiotics. Annual sales of these enzymes, currently at $600 million, are expected to rise dramatically with the commercial availability of new fixed-enzyme systems. For example, enzymes must be fixed to a surface to manufacture extremely sensitive chemical-detection devices, called biosensors ▶ (see p. 752).

Many microorganisms provide useful products for the pharmaceutical industry. Each year over 100,000 tons of antibiotics—primarily penicillins, cephalosporins, and tetracyclines—are produced in the world. These drugs are secondary metabolites, or their derivatives, that are synthesized by bacteria

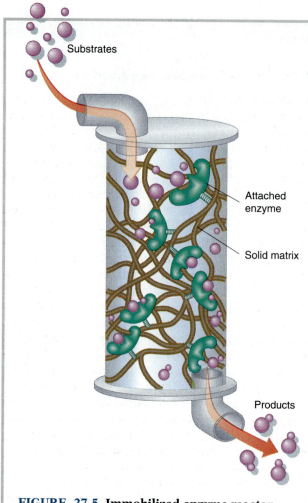

FIGURE 27-5 Immobilized enzyme reactor.
Enzymes attached to a solid matrix, such as cellulose fibers, catalyze formation of products in a continuous-flow reaction vessel. The active enzymes remain attached to the cellulose fibers, preventing their loss in the effluent.

ple, the use of microbes to synthesize cortisone reduces the number of chemical reactions from the 37 needed for chemical synthesis to 11, some chemical and some microbial, reducing the cost of cortisone 400-fold. Other steroids produced by microbial transformation include hydrocortisone and prednisone.

Vaccine production is entirely dependent on microbial growth. The manufacturer must culture the target pathogen (or a genetically engineered organism that produces the target antigen) to obtain high yields of the microbe itself or of specific antigens—surface antigens, toxins, or capsular material. The development of vaccines against many pathogens has been hampered by an inability to culture the pathogen or by the expense of growing certain or-

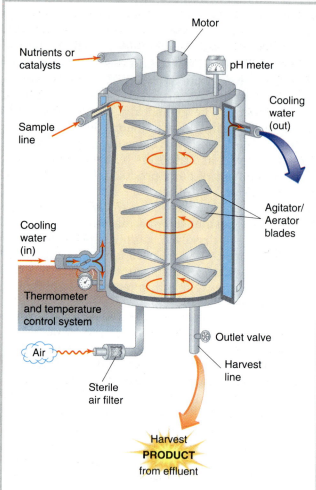

FIGURE 27-6 More than a giant container.
The anatomy of fermentation tanks and bioreactors reveals the spinning aerator/agitator blades that keep conditions uniform throughout the tank as well as ports for introducing substances and monitoring conditions such as pH.

or fungi during the stationary phase of growth. (Chapter 16 details the source organisms for the major antibiotics.) Antibiotic-producing microbes are usually grown for several days in aerated batch cultures in huge tanks, some with volumes in excess of 100,000 liters (Fig. 27-6). Under these conditions, penicillin production takes approximately 8 days. Conversion of penicillin G to semisynthetic derivatives usually requires a second microbial process. Penicillin G is incubated with bacteria that produce an enzyme that hydrolyzes the antibiotic to 6-aminopenicillanic acid, the core molecule onto which chemical groups are attached to produce semisynthetic penicillin derivatives.

Microbial bioconversions are important in the manufacture of some steroid hormones. For exam-

ganisms. New vaccines that contain only the antigenic component(s) against which immunity is directed are being developed by genetic engineering of microbes (see p. 755).

Techniques for Maximizing Production

Industrial microbiologists maximize the harvest of microbially produced materials by using high-yield strains and then manipulating their growth environment—for example, by including in the medium substrates that induce the formation of the desired enzymes. Feedback inhibition presents another obstacle to maximum production—most cells turn off production of a product after it has reached a particular concentration. Feedback inhibition can be overcome by promoting rapid excretion of a product before it can accumulate inside the cell and shut down its own production. In this way, detergents and other substances that cause nonlethal leaks in the cell membrane promote high yields of end products by reducing their intracellular concentrations.

High-yield strains can be obtained using a number of strategies. Ultraviolet irradiation and chemical mutagenesis help generate mutant strains that overproduce desired materials. The loss of regulation may be at the level of enzyme activity or enzyme synthesis (Figs. 27-7 and 27-8). The combination of organism selection, genetic alteration, and manipulation of cultural conditions to improve product yield has been exploited with great success, as evidenced by the 10,000-fold increase in today's penicillin production over that of Fleming's original strain.

Gene splicing technology has created new frontiers in industrial microbiology. Recombination can occur following *fusion* of genetically distinct protoplasts to form a single hybrid cell with combined characteristics. This technique is valuable for incorporating into a single organism desirable traits that ordinarily occur in separate cells—for example, fusing a cell that produces high yields with one that grows rapidly. Recombinant DNA technology has unleashed the potential for engineering organisms with even greater production capabilities. One such process, *gene amplification*, increases the number of copies of a gene in the cell, thereby increasing the amount of the corresponding product. Productivity is also increased by manipulating the operon that regulates a cluster of genes—for example, inserting protein-coding regions and a promoter (the site to which messenger RNA polymerase binds) directly adjacent to each other in a plasmid. With the operator region eliminated, coding regions cannot be re-

pressed, and huge quantities of enzymes are produced.

Products of Genetic Engineering

In addition to creating "improved" microorganisms that are highly efficient at their natural metabolic activities, recombinant DNA technology has also produced "superbugs" that are capable of doing things that neither they nor any other microorganism has done before. (This technology is discussed in Chapter 9.) Such superbugs are employed in the production of some vaccines. A gene that directs the synthesis of a surface antigen (the capsid of a virus, for example) can be extracted from virulent pathogens and transferred to harmless bacteria. The newly received gene would instruct the microbe to manufacture large quantities of a single viral antigen. Pure preparations of the immunogen can be readily harvested from the growth medium for use as a vaccine that carries no risk of infection by viable pathogens. In addition, such genetically engineered vaccines could eliminate the risk of deleterious side effects from additional antigens, endotoxins, or other materials from the pathogen.

In addition to producing pure antigens for vaccines, bacteria and yeasts have been genetically manipulated to produce human growth hormones, interferon, and insulin. Practical quantities of these products are virtually unattainable from their natural sources. For example, it would take 100 sheep brains to obtain the amount of somatostatin (a growth-regulating hormone) produced in a 2-liter culture of the *Escherichia coli* strain engineered with the human somatostatin gene. Large-scale production of interferon for clinical studies is also made possible by genetically manipulated microbes that have been given human interferon genes. Products of microbial origin are often safer than those derived from traditional sources. Human insulin produced by microbes, for example, does not cause the allergic reaction associated with commerical insulin obtained from its traditional sources, cows and pigs. Recent studies comparing the clinical effectiveness of biotechnological drugs to conventional drugs show that the biotechnological products have a higher rate of success.

Application of recombinant DNA techniques is limited by the accessibility of genes. Considering the tremendous amount of DNA in a human cell (1000 times that found in *E. coli*), the likelihood of being able to isolate any given gene seems remote at best. The use of synthetic genes may soon overcome this limitation. The nucleotide sequence of a gene can be deduced by determining the amino acid sequence

Tobacco—A Killer Turned Hero?

Tobacco may soon have its nasty reputation replaced with a more positive image—that of producer of health-promoting products. This crop may help keep the population healthy. Genetic engineers have successfully introduced foreign genes into a variety of plants, including tobacco, using plant viruses as vectors. By placing the desired gene adjacent to an active viral promoter, the plant is tricked into producing large concentrations of the foreign protein. Plants thus have become another vessel for synthesizing foreign gene products along with bacteria, yeasts, and mammalian cells in culture.

Among the successful demonstrations of this technique have been the production of an anti-AIDS drug and an antihypertensive protein by tobacco plants infected with a modified tobacco mosaic virus. The future is likely to bring other therapeutic proteins, vaccine components, and industrial enzymes.

Studies indicate several advantages that plants have over the conventional unicellular systems. They produce large volumes of protein that are more readily extracted and ultimately cheaper. These preparations are also free of many contaminating materials that pose problems in bacteria-produced products (such as endotoxin) that might be harmful to humans. In addition to being safe, plant-derived vaccines have often proved to be more effective than bacteria-derived vaccines because of the way the antigen is assembled.

Plants may become a cost-effective way to produce human or animal vaccines and may eventually provide a painless route of administration. Scientists are already trying to develop edible vaccines. By eating an antigen-containing fruit or vegetable, children might someday be inoculated against the various pathogens that threaten their health. Alfalfa, lettuce, and tomatoes are already producing vaccine components against cholera toxin and hepatitis B virus. The challenge now is to figure out ways to ensure that these proteins retain their antigenicity as they travel through the digestive tract and that their antigenic determinants can stimulate adequate immunity.

Glowing tobacco plant is the product of genetic engineering, having grown from a tobacco cell that received a copy of the firefly gene that enables the insect to illuminate its abdomen. Just as the firefly gene is expressed when incorporated into this tobacco plant, genes for other products, more useful to us than a glowing plant, are expressed, making the plant a factory for the desired product.

---

patients who would be falsely identified as negative by other diagnostic procedures. In addition to amplifying DNA from pathogens and genetically defective human cells, PCR can amplify the DNA from a single hair left by a criminal at the scene of a crime, thereby aiding in the "diagnosis" of the perpetrator.

■ Therapeutic Strategies

Several technologies provide new methods for treating disease. These include the destruction of unwanted cells with monoclonal antibodies, repression of harmful genes by antisense nucleic acids, insertion of active genes to replace missing functions in gene therapy, and computer-aided design of antimicrobial drugs.

Monoclonal antibodies recognize and bind to specific targets that possess the one antigenic determinant against which the antibodies are directed. This selectivity is exploited in therapy—for example, when toxic molecules are attached to the antibodies the result is an *immunotoxin* that concentrates the lethal chemical only on the target cells. Some monoclonal antibodies do not need to be attached to toxins; they inactivate their targets directly. Cancer cells, virus-infected cells, or endotoxin are the primary targets of most monoclonal antibodies in development. One major obstacle to their success thus far is the body's reaction to the administration of nonhuman antibodies. However, new technologies are designed to develop "human-friendly" monoclonal antibodies.

Many inherited diseases are caused by defective genes, ones that fail to produce a functional protein essential for a normal cell. Gene therapy would replace these faulty genes with functional versions, perhaps even during embryological development,

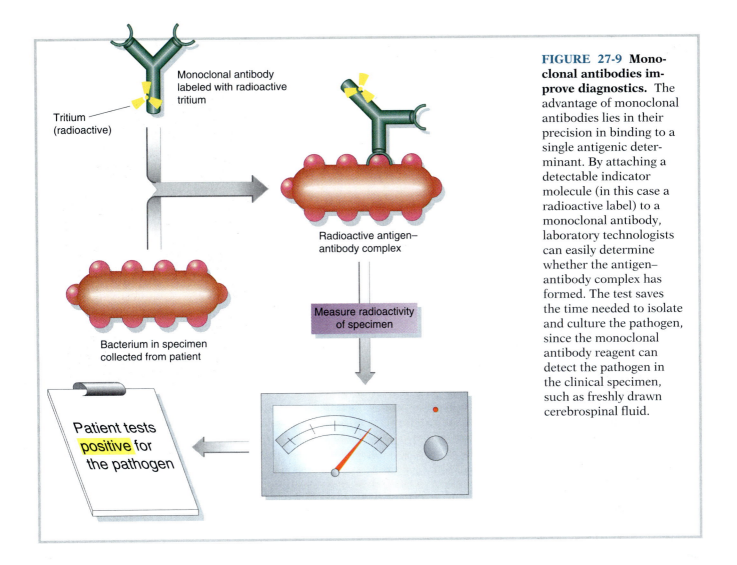

Tritium (radioactive)

Monoclonal antibody labeled with radioactive tritium

Radioactive antigen–antibody complex

Bacterium in specimen collected from patient

Measure radioactivity of specimen

Patient tests positive for the pathogen

FIGURE 27-9 Monoclonal antibodies improve diagnostics. The advantage of monoclonal antibodies lies in their precision in binding to a single antigenic determinant. By attaching a detectable indicator molecule (in this case a radioactive label) to a monoclonal antibody, laboratory technologists can easily determine whether the antigen–antibody complex has formed. The test saves the time needed to isolate and culture the pathogen, since the monoclonal antibody reagent can detect the pathogen in the clinical specimen, such as freshly drawn cerebrospinal fluid.

providing the necessary genetic information and preventing the development of these often tragic illnesses. People with immunodeficiencies, muscular dystrophy, insulin-dependent diabetes, and hundreds of other disorders may someday find their diseases curable by gene replacement. The potential is staggering. In 1993, gene therapy was used successfully in several children with ADA, an immunodeficiency described in Chapter 19.

Microbes, especially viruses, play a significant role in such gene therapy. For example, in some cases the gene itself is unavailable but RNA transcribed from the desired gene can be obtained. Reverse transcriptase (obtained from retroviruses) can then be used to make a DNA copy of the RNA information. Furthermore, the maintenance of the desired gene within cells is greater if the DNA is incorporated into the host genome. Once again retroviruses provide the necessary properties. The new gene is inserted into the viral genome, which makes a DNA copy of itself and then integrates this

copy of virus (and the desired gene) into host DNA.

Based on successful animal studies, viruses are also being used to deliver lethal instructions to tumor cells. For example, retroviruses carrying herpes simplex virus thymidine kinase (TK) genes have been inoculated into brain tumors. The patient is then treated with ganciclovir, a drug that is activated by the TK gene product and subsequently kills the infected tumor cell.

Often it is not practical to kill a cell that is producing harmful products or reproducing out of control. The goal of antisense technology in human therapy is to shut off unwanted gene expression in these cells. Retroviruses can be used to deliver into a cell DNA that encodes an antisense message that blocks the harmful gene's expression. There are several clinical trials testing this technology against HIV and tumors.

Computers are also being used to create drugs and improve on natural antimicrobials. The computer can assist in rational drug design when the target molecule has been well described at a molecular level (Fig. 27-10). Computer-aided drug design relies on detailed information about some essential site on the pathogen or toxin. For example, the capsid protein on rhinoviruses (which cause the common cold) is necessary for their attachment to their target cells. The amino acid sequence and the three-dimensional structure of this protein have been determined. The region participating in recognition and attachment appears as a pocket or canyon on the surface of the protein. Using this information, computers generate a polypeptide sequence that is small enough to have access to this site and will bind and block viral attachment to host cells. Such drugs have proven efficacy in vitro but must be tested for safety and efficacy in humans before they will be made available to the medical community. Commercial success will also depend on the cost of the final products.

■ Biosensors

Biosensors are devices that contain biological material (live organisms or cell products) that respond to specific chemicals. The response is linked to a sensor system and monitored by an electronic readout. Biosensors are used in environmental monitoring and in medicine. For example, metabolic poisons in water supplies can be detected by measuring light generation in luminescent bacterial cells. Light

generation by *Photobacterium phosphoreum* requires ATP, so substances that inhibit energy-generating metabolism (and thus ATP production) will reduce the amount of light emitted by the cell. These bacteria are incorporated as part of a device along with a photodetector and a visual readout of light intensity. Other possible applications include measurement of glucose in the bloodstream (Fig. 27-11), fermentation products in industry, and atmospheric pollutants, and battlefield detection of chemical warfare agents such as neurotoxins.

■ Microbial Pesticides

The cost of insect damage to crop plants and orchards is staggering. Insects substantially increase world hunger by annually consuming nearly one-third of the human food supplies on earth. These pests have traditionally been fought with synthetic organic chemical insecticides. Many of these insecticides have long-term toxicity, pollute the environment, and accumulate in the food chain. Their lethal effect is also fairly nonspecific, killing beneficial insects and spiders, predators that naturally control the pest population, along with the target populations. In a predator-free environment, pesticide-resistant insects quickly repopulate to overwhelming proportions that result in even greater crop damage than would have occurred without the use of the pesticide.

Several insect-specific pathogens are being produced commercially for use as **microbial pesticides**. Microbial pesticides have several advantages over chemical insecticides. They are specific for a small number of insect species, sparing plants, animals, and many beneficial arthropods. They are also much cheaper than organic pesticides, which rely on petroleum for their production. In addition, some of the microbial pesticides are self-perpetuating in the field. *Baculoviruses*, for example, are sprayed on a crop and ingested by the target insect pests. The virus does not kill immediately. Infected insects have time to mate and inoculate others with the virus. In 4 to 8 days, all members of the insect population have received a lethal dose of baculovirus. Unfortunately, during this time the insects generally increase their food intake, and large-scale crop damage may occur. To make the viruses more effective, genes for insect-specific toxins have been added to the baculovirus genome. These toxins kill infected insects more quickly.

Cultures of sporulated *Bacillus thuringiensis* have been used worldwide to control damage to crops, trees, and ornamental plants. During endospore formation, this bacterium produces toxic crystals (Bt toxin) that make it a good pesticide. Each strain pro-

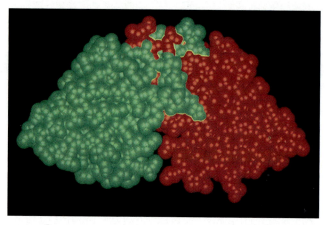

FIGURE 27-10 Computer-designed drugs. This computer-generated model depicts an enzyme crucial to infection by Rous sarcoma virus, a retrovirus that infects chickens. By creating molecular mimics of the enzyme's substrate that are slightly different from the natural substrate, scientists can design a drug that will inactivate the enzyme and block productive viral infection. A similar enzyme is present in HIV, so it may soon be possible to prevent the development of AIDS by injecting a substitute substrate that prevents the replication of the virus.

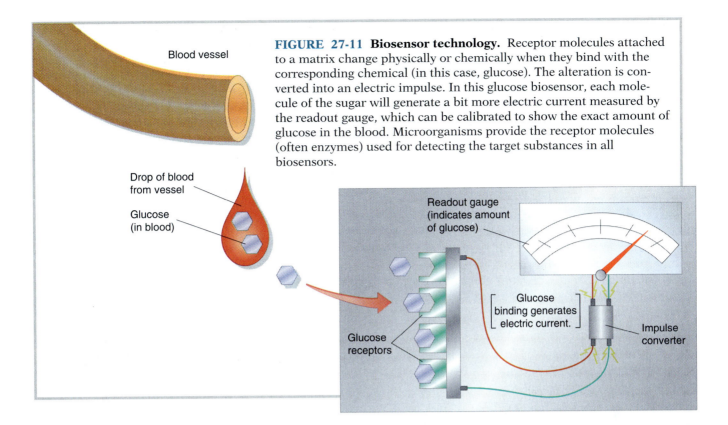

FIGURE 27-11 **Biosensor technology.** Receptor molecules attached to a matrix change physically or chemically when they bind with the corresponding chemical (in this case, glucose). The alteration is converted into an electric impulse. In this glucose biosensor, each molecule of the sugar will generate a bit more electric current measured by the readout gauge, which can be calibrated to show the exact amount of glucose in the blood. Microorganisms provide the receptor molecules (often enzymes) used for detecting the target substances in all biosensors.

Blood vessel

Drop of blood from vessel

Glucose (in blood)

Readout gauge (indicates amount of glucose)

Glucose binding generates electric current.

Impulse converter

Glucose receptors

duces a single type of protein that targets specific insects. The bacteria or purified toxin are dried and dusted onto plants. When ingested by gypsy moth larvae, cabbage worms, or tent caterpillars, the toxic protein lethally damages the insects' intestinal tracts (Fig. 27-12). The bacteria make a nontoxic precur-

(a) (b)

FIGURE 27-12 **Deadly yet desirable.** (*a*) After ingesting toxin produced by *Bacillus thuringiensis*, this cabbage looper larva shows signs of the poisoning that will soon kill it. Crystals of the fatal poison (*b*) produced by the bacterium during sporulation may save farmers millions of dollars with the development of genetically engineered plants that contain the bacterial toxin gene.

sor that is broken down into toxic subunits by enzymes in the insect gut. Insects, plants, and animals lacking this enzyme cannot convert the protein to its toxic form and are unaffected by the pesticide. A major problem in using *B. thuringiensis* is that it is not self-perpetuating and toxin preparations rapidly deteriorate in the field. Thus, it is necessary to frequently reapply the spores or toxic crystals to crops. To overcome this problem, Bt toxin has been packaged into dead bacterial cells that break down slowly, extending the pesticide's field life. The bacterium's toxin has also been engineered into various plants in an attempt to create insect-resistant plant tissues (see p. 755). Another problem, development

of resistance to the toxin by insects, remains a threat to the continued effectiveness of this approach.

Some fungi attack insects and are used to prevent damage to crops, especially in humid environments. In most cases, fungal spores attach to the insect's surface cuticle, germinate (if there is enough moisture), and penetrate to the interior of the insect. So far, the commercial development of fungal insecticides has been limited by the need for moisture and regular reapplication, but these limitations may eventually be overcome.

Now that the safety of many of these agents, as well as their profit potential, has been demonstrated, such easy-to-produce pesticides may reduce

A Multimillion Dollar Fungus

From killer bees to kudzu plants that overgrow houses and whole forests, we continue to bear a legacy of foreign species unleashed into new habitats where they have flourished at the expense of native species and the environment. A classic example stemmed from one man's obsession with breeding a better silkworm. Leopold Trouvelot transported gypsy moth larvae from France to Massachusetts in 1889, convinced that the caterpillar would improve the species. His experiment failed, and the moth soon escaped from Trouvelot's home and has wreaked destruction on America's woodlots and forests for the last 100 years. Each gypsy moth larva can eat a full square meter of foliage from oak, aspen, and numerous other valuable trees. In one year, 13 million acres were defoliated in 17 northeastern states. The destruction costs millions of dollars in lost recreational income and destroyed wood resources for the furniture industry. Millions more have been spent battling the pest (between 1979 and 1983, $24 million went into the war against the moth, the second most expensive entomological war in history). And we are still losing.

Rescue might come in an unexpected form—a fungus that suddenly appeared in 1989 and 1990 during a mysterious

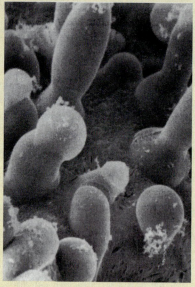

The destructive gypsy moth and *Entomophaga maimaigma*, the fungus that kills it.

die-off of gypsy moths in a four-state area. Ann Hajek of Cornell University discovered that the blackened carcasses of the larvae were packed with hyphae of *Entomophaga maimaigma*, a fungus that had twice been tried as a biological control agent against the gypsy moth. Both attempts had apparently failed. The first trial was conducted more than 80 years ago by Harvard researchers who propagated the fungus from infected caterpillars in Japan, then released the fungal spores around the Boston, Massachusetts, area. Disappointing results discouraged the scientists, who abandoned the project. Hajek attempted a repeat of the experiment in 1985 and 1986 in New York State and Virginia, but the anticipated die-offs of the moth larvae in the experimental regions never materialized. So where did the fungus come from? Hajek believes the fungus originated from the 1910 inoculations. Conditions in 1989 and 1990 were obviously good for the growth of the fungus, although precise factors that fostered the fungal growth are still undefined. The multimillion dollar question now is, What factors are necessary to trigger such massive infections of gypsy moth caterpillars? Perhaps an even more important question is, Can the fungus be engineered so it is not so responsive to fluctuations in environmental conditions? If the answer is yes, then we might finally win this 100-year war.

TABLE 27-5 EXAMPLES OF MICROBIAL PESTICIDES

Pesticide Producer	Target Pest	Target Environment
Bacteria		
Bacillus thuringiensis	Caterpillars Mosquitoes	Fields, forests Water, soil
Bacillus sphaericus	Mosquitoes	Water, soil
Bacillus popillae	Japanese beetle	Fields
Viruses		
Baculoviruses	Cotton bollworm	Cotton, corn crops
	Codling moth	Orchards
	Sawfly	Forests
Fungi		
Arthrobotrys sp.	Nematodes	Soil
Beauveria bassiana	Colorado beetle	Potato crops
Hirsutellat hompsonii	Rust mites	Citrus crops
Lagenidium giganteum	Mosquito larvae	Water
Metarhizium anisopliae	Beetles	Sugar cane
Verticillium lecanii	Aphids	Greenhouses

world starvation and help preserve our forests. In fact, in 1992, a huge area of Washington State was sprayed with a microbial pesticide to combat the newly arrived Asian gypsy moth. The quick action and effectiveness of the program prevented an estimated $55 billion worth of forest destruction. Another strain of the gypsy moth, one that has been killing U.S. trees since the mid-1800s, may also be on the brink of succumbing to a microbial pesticide (see THE EXPLORERS: A Multimillion Dollar Fungus).

Some microbial candidates for controlling pests are described in Table 27-5.

■ Harnessed Decomposition

Microbes cultured and marketed for environmental and municipal cleanup jobs are already proving to be lucrative industrial products. One market for these bacteria is removal of pollutants such as recalcitrant contaminants and petroleum from specific environments. Although genetically engineered bacteria may someday do most of these jobs, currently most of these cleanup microbes are either un-

modified bacteria isolated from environments contaminated with the target substance or products of "forced" evolution. ▶ (See Bioremediation, Chap. 26, p. 728.)

Microbial biodegradation is also being used for many tasks that are traditionally done mechanically. For example, municipal sewer lines eventually become clogged with a buildup of insoluble materials on the inside of the pipes. In Washington, D.C., 3000 miles of sewer pipe was effectively cleared by drip-feeding commercial bacterial cultures into manholes at the extremes of the system. The project cost about one-third that of mechanically removing the buildup, saving the city nearly $3 million. The same process is being used on a smaller scale to clean drainpipes, grease traps, and septic tanks.

■ Frost-Free Plants

Crop damage occurs when frost forms on plants; as the water freezes and expands, it bursts plant cells and reduces food yields. When temperatures drop below 0°C, water freezes where proteins form a seed for ice crystallization to begin. In most cases, these ice-forming molecules are produced by *ice-nucleating bacteria* that colonize the plant surface. One of the earliest successes in genetic engineering was the development of an "ice minus" strain of the bacterium *Pseudomonas syringiae*. When sprayed on plants, the ice minus bacteria outcompete other bacteria, dramatically reducing the number of ice-nucleating bacteria on the plant. The absence of ice nucleation protein in this bacterium prevents damage from frost (Fig. 27-13).

■ Genetically Engineered Plants

Several groups of scientists have transferred foreign genes into commercially valuable plants that supply us with fabrics, food, and wood. As previously described, this transfer may convert plants into sources of foreign proteins for pharmaceutical use. Much of the work, however, is aimed at transferring genes that increase the productivity and enhance the quality of crop, forest, or decorative plants (Table 27-6). Transfer of genes from *B. thuringiensis*, for example, has already created plants with toxin-laden leaves that kill insects that dine on them, conferring resistance to such insects directly to the plant itself. Plants have also been made herbicide-tolerant and disease-resistant by introducing microbial genes. Herbicide-tolerant plants survive treatment with chemicals designed to destroy overgrowing weeds. Some herbicide tolerance is controlled by a bacterial enzyme, expressed by the plant, that inactivates the herbicide. Several crop plants, such as alfalfa,

FIGURE 27-13 **The first authorized release of genetically engineered organisms into the environment.** This scientist in the "space suit" is spraying ice minus bacteria on a test plot of strawberries. (The suit prevents contaminating the plants with microbes from the body of the person applying the spray, contaminants that could affect the results of this controlled experiment.) Frost damage in the experimental plot was virtually nil, whereas the plants in the plot that received no ice minus cells suffered significant frost damage.

potatoes, and tomatoes, have been modified by the addition of the gene coding for the coat protein of a disease-causing virus. Expression of high levels of the coat protein inhibits the replication of infecting viral nucleic acid. Such "molecular breeding" to increase crop yields has led to diverse improvements in plants. In another example, incorporation of an *E. coli* gene into potatoes increased starch concentrations. Genes that delay ripening or increase sweetness have also been engineered into plant cells.

Foreign genes are more readily transferred into plant cells when the DNA is incorporated in a vehicle that facilitates the integration of the newly acquired genes into the chromosome. One such method exploits the natural infection process of the plant pathogen *Agrobacterium tumefaciens*. ▶ (Review Fig. 9-11, p. 229.) The bacterium contains a plasmid, of which a portion called *transfer DNA* (*tDNA*) is inserted into the infected plant's chromosome. Using recombinant DNA technology, foreign genes are inserted into the tDNA of *A. tumefaciens*. To ensure expression of the genes within the plant cell, they are placed next to a recognizable strong promoter, such as one derived from cauliflower mosaic virus. The resulting recombinant plasmid is incorporated into the cells of plants once they have been infected with *A. tumefaciens* carrying the modified tDNA. Cells containing the integrated plasmid and foreign genes are isolated and cultured. An entire plant can be regenerated from one of these genetically modified cells. Succeeding generations would acquire the new trait by normal inheritance.

CONCEPT CHECK

- How do primary metabolites of industrial importance differ from secondary metabolites? Give three examples of each, and describe how they are used. Provide four examples in which whole bacterial cells (rather than metabolites) are the marketable products.
- Describe two advantages of using enzymes in an immobilized form.
- How can the production of industrially valuable metabolic by-products be increased without genetic manipulation of the microbe? How can it be increased by regulating feedback inhibition or end-product repression?
- Compare the ways in which monoclonal antibodies and DNA probes are used to diagnose disease. How are monoclonal antibodies used in therapy?
- What roles may retroviruses play in human therapy?
- How do microbes in biosensors detect substances?
- What advantages do microbial pesticides have over chemical pesticides? Compare the modes of action of baculoviruses and *B. thuringiensis*.

Microbes as Suppliers of Natural Resources

■ Fuel Production

Dependence on petroleum and other nonrenewable energy sources will lead to a fuel crisis as the world's finite reserves become depleted. Available oil sup-

TABLE 27-6 MICROBE-MEDIATED BIOTECHNOLOGY APPLICATIONS IN AGRICULTURAL FOOD PRODUCTION

Product/Food	Benefit	Technology
Apples	Insect resistance	A bacterially derived insect resistance gene is inserted.
Bananas	Early detection of Sigatoka fungus (black streak) as part of integrated pest management	Technology is available to produce detection kits.
Cabbage, mustard, rapeseed	Crops without the damage caused by Coleopteran and Lepidopteran insects	A natural insecticide protein produced by *Bacillus thuringiensis* kills the insects (see Potatoes, Tomatoes).
Corn	Insect resistance	*Bacillus thuringiensis* genes are inserted into corn.
Cottonseed	Herbicide resistance	Gene for resistance to glyphosate (Roundup) is inserted.
Lettuce, tomato	Lepidoptera larvae are selectively infected and killed, preventing major crop damage	Insecticidal viral products (baculovirus) are sprayed on plants.
Potato (Mexican)	Virus resistance	Viral coat protein genes are introduced into potato.
Potatoes	Potato chips with less fat	Starch production gene from *E. coli* was introduced, and the genetically altered potatoes will have 30–60% higher starch and lower moisture; they will absorb less fat on frying.
Potatoes	Resistance to Coleopteran and Lepidopteran insect pests; reduced chemical treatments	Expression of genes derived from *B. thuringiensis* disrupts ion transport of cell membranes in the insect gut.
Potatoes	Sweeter tasting	Antisense gene that blocks ADP-glucose pyrophosphorylase was inserted so sucrose is not converted to starch.
Soybean/legumes	Productivity increased with limited fertilizer application	Improved strains of *Rhizobium* used in the soil to enhance nitrogen fixation.
Soybeans	Herbicide resistance	Gene for resistance to glyphosate inserted.
Strawberries	Resistance to viral diseases; increased yields with less chemical treatment	Expression of genes that produce viral coat proteins.
Tomatoes	Slow ripening of tomatoes that stay green, firm, and unflavored until exposed to exogenous ethylene.	Expression of antisense RNA for ACC synthase that blocks ethylene production.
Tomatoes	Slow softening of tomatoes that can vine ripen and stay firmer in shipment or in home storage	Expression of antisense RNA for polygalacturonase, which would otherwise degrade pectin.
Tomatoes	Resistance to post-harvest rot; resistance to freezing	Tomatoes produce the enzyme chitinase, which protects against fungi. Gene inserted from winter flounder.
Wheat	Herbicide resistance	Gene for glutofinate resistance inserted.

plies will probably be exhausted within the next 50 years. Microbes may provide a partial solution to this problem by converting waste materials and sewage into usable fuel. Already methane gas released as a by-product of sewage treatment is harvested and used to generate electricity. Most of the possibilities for using microorganisms in future energy technology, however, are yet unexplored. Biotransformation of sunlight into usable fuel is perhaps the most exciting, since the source is

Monitoring the Microbes—Quality Control

Industrial microbiologists responsible for **quality control** routinely monitor components critical to food production, industrial and pharmaceutical processes, and municipal water supplies for public consumption in order to prevent product contamination and costly microbial damage or the distribution of dangerous pathogens. In industry, specimens are routinely collected to determine whether harmful contaminants threaten commercial processes. Quality control monitoring focuses primarily on the microbiology of the raw materials, process water, equipment, and air as well as the final product. It is not necessary in most manufacturing processes for the starting materials to be sterile, even if the final product must be. However, starting materials cannot be grossly contaminated or contain organisms that will harm the production process (or personnel).

The acceptable levels or types of microbes in the final product depend on the particular product and its use. Microbial contamination of water used in manufacturing processes is a particular concern, especially in the production of injectable fluids or devices that will be implanted into humans. Any gram-negative bacteria or their cell walls present in the final product could cause endotoxic shock in the recipient. The best guarantee for the manufacturer is to use water that is free of such contaminants. Process water is sampled routinely, usually by collecting a portion of the liquid through a collection port in the line of flow.

Air sampling is performed to assess the level of contamination from the environment. Critical processes are carried out in special hoods or clean rooms that are protected from microbes by HEPA filters. ▶ (See Chap. 15, p. 360.) Air within these environments is usually tested by *settling plates*, open petri dishes that contain solid medium exposed to the air for a specific period of time. In contrast, special equipment has been developed to sample air within larger, open environments. In these situations, known volumes of air are forced through these devices and the microbes are deposited directly onto a nutritional medium. The results of testing air with these devices provide more precise quantification of airborne contaminants than does the use of settling plates.

Like many foods, certain natural products contain preservatives to reduce biodeterioration by environmental microbes. For example, paints, textiles, fuels, and cosmetics are subject to such destruction. One of the roles of the microbiologist in charge of quality control is to monitor the effectiveness of the preservative added to the product, thus securing the integrity of the product.

Actually, much of this book has been devoted to quality control. Familiarity with microorganisms, knowing which ones should cause no alarm and which are harbingers of dangerous or harmful situations, is one of the most important principles providing insight into not just monitoring the microbes but also interpreting the results. Louis Pasteur's insight averted the decline of the French wine industry in the late 1800s, one of the first successes of systematic microbial quality control. We hope this journey through the world of microorganisms has provided you with some similar insight. As we move into the next century, with its new medical and industrial technologies, there will be an ever-increasing need for inventors and practitioners of novel ways to monitor microbes. Microbiological knowledge and insight will never become obsolete.

CONCEPT CHECK

- Why would moderate quantities of harmless gram-negative bacteria be of little concern when detected in drinking water but a serious concern in water used for making injectable solutions, even if the solutions were to be autoclaved after packaging?
- Invent a technique that uses monoclonal antibodies in microbial quality control. Repeat the exercise using biosensors; settling plates; differential medium.

Key Terms

microbial biotechnology (p. 736)
single-cell protein (SCP) (p. 740)
primary metabolites (p. 744)
secondary metabolites (p. 744)
immobilized enzymes (p. 745)
biosensors (p. 752)
microbial pesticides (p. 752)
quality control (p. 760)

Key Facts and Concepts

- **Traditional food microbiology.** Microbes are used to protect food from spoilage organisms and to impart aesthetic characteristics to edible substances. Milk and cream are converted into cheese, yogurt, sour cream, or buttermilk by bacteria and fungi. Vegetables are soaked in brine solutions, then fermented to produce ripe olives, sauerkraut, pickles, and other foods that are much more resistant to spoilage than the original, untreated vegetables. Bread is leavened when carbon dioxide is released by *Saccharomyces cerevisiae*, which metabolizes glucose in the dough. This yeast, and others closely related to it, also ferment sugars in fruit juices and grain mash, producing wine, beer, and other alcoholic beverages.

- **Single-cell protein (SCP).** Microbes are potential sources of nutrition for protein-starved people in developing nations. Much of the world's biomass ends up as wastes that could be used as nutrients for growing huge quantities of SCP. Currently SCP is used primarily as a livestock feed to increase meat and dairy production.

- **Genetically engineered foods.** Some fruits come from plants whose genes for ripening have been suppressed to prevent overripening before they reach the shelves. Other plants have been engineered with genes to produce resistance to plant viruses. Yet others have received a toxin gene from *Bacillus thuringiensis* so they can produce their own pesticides (after eating from a leaf, the insect dies).

- **Food supplements.** Microbes are sources of vitamins, amino acids, nucleotides, and other food supplements as well as of enzymes used for a variety of purposes in food preparation.

- **Industrial microbiology.** Microorganisms synthesize many industrially valuable products. Production of these compounds is increased by selecting deregulated mutants that overproduce the desired product. Recombinant DNA technology generates strains that give greater yields or that manufacture useful substances not normally synthesized by microbes. The products of industrial fermentation are either primary metabolites such as vitamins, organic acids, and alcohols or secondary metabolites such as antibiotics.

- **Pharmaceutical microbiology.** In addition to antibiotics, the pharmaceutical industry depends on microbes for the manufacture of vaccines and some steroid hormones. Improved vaccines that contain only those antigens against which immunity is directed may soon be constructed by genetically engineered bacteria, yeast, or the vaccinia virus. Genetic engineering has already yielded microbes that synthesize several medically valuable substances that are difficult to obtain from traditional sources. The use of synthetic genes manufactured in vitro promises to contribute to the potential of genetic engineering.

- **Fuel production microbiology.** Gene-splicing technology has generated new strains of bacteria that may help alleviate the energy shortage by producing usable fuel. The most attractive of these technologies is one that converts industrial waste and garbage into combustible hydrocarbons. Bacteria already aid in recovering hard-to-obtain petroleum deposits trapped in shale.

- **Microbial miners.** Lithotrophic bacteria are used to extract copper and uranium from low-grade ore. In leaching solutions, these bacteria dissolve the metal from the ore, after which the solution is treated to release the metal.

Review and Problem-Solving Questions

1. Describe how microbes participate in the production of (a) Swiss cheese; (b) soy sauce; (c) butter; (d) sauerkraut.
2. Compare the processes of wine and beer production. Why must grains, but not fruits, be hydrolyzed prior to fermentation?
3. How does the production of red wine differ from the production of white wine? How is the alcoholic content of wine controlled? How are the higher alcohol contents of whiskey and brandy attained?
4. List five enzymes of microbial origin and their industrial applications.
5. Distinguish between:
   (a) Industrial microbiology and microbial biotechnology
   (b) Gene amplification and gene deregulation
   (c) Gene splicing for antisense technology and gene splicing for new protein production
6. Describe three ways in which genetic engineering contributes to new uses of microorganisms.
7. What are the industrial or biotechnological applications of each of the following microbes?
   (a) *Thiobacillus ferrooxidans*
   (b) *Bacillus thuringiensis*
   (c) *Pseudomonas syringiae*
   (d) *Xanthomonas*
   (e) *Thermus aquaticus*
8. Why aren't antibiotics used as food preservatives? Describe three foods that are protected from spoilage because of a property of the food.
9. Why is water monitoring so important in industry?
10. Why aren't humans affected by Bt toxin if they consume foods that have been sprayed with endospores of *B. thuringiensis*?
11. Describe two ways plants can be genetically engineered to make them resistant to pathogens.
12. The mold *Acremonium chrysogenum* produces cephalosporin C, the precursor of cephalosporin antibiotics. The reaction depends on oxygen. A strain of *Acremonium chrysogenum* was transformed with (a) a plasmid containing a bacterial gene for a hemoglobin-like molecule or (b) a plasmid containing a human gene for hemoglobin. The following results were obtained.

Transforming Plasmid	Cephalosporin C Produced (g/L)
None	1
Bacterial "hemoglobin"	4
Human hemoglobin	0.8

Give a possible explanation for these results.

13. You have discovered a new species of *Streptomyces* that produces an antibiotic that kills *Pseudomonas aeruginosa*, but it is produced only in low concentrations. Outline at least three methods you might use to increase the yield of the antibiotic.
14. Describe four ways to determine whether a food is spoiled. What specific microbial activities are responsible for most of these changes?
15. What are the commercial uses of each of the following microbial products? (a) Lysine; (b) glucose isomerase; (c) amylase; (d) dextran; (e) citric acid.
16. Indicate which of the following graphs would represent the production of ethanol and which graph would represent the production of penicillin. Explain your answer.

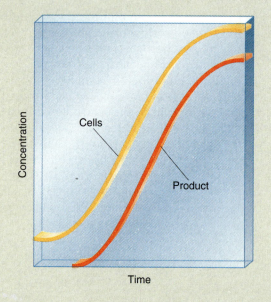

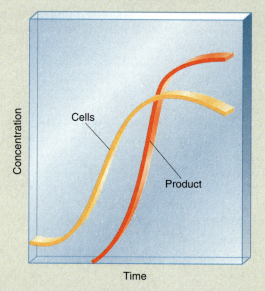

AN INSIDE ADVANTAGE

Antibodies are among the most important natural defenses of the body. These secreted proteins circulate throughout the bloodstream, protecting mucous membrane surfaces and keeping pathogens from entering tissues. In the earliest examples of "biotechnology," scientists produced vaccines capable of coaxing the cells of the immune system to produce and release these protective molecules before an individual is exposed to disease. Now, as revealed by John Travis,[*] researchers have gone even further. They have developed techniques to make cells other than immune cells synthesize antibodies that remain inside the cell and block the activity of harmful proteins. For example, such intracellular antibodies can react with virus components or tumor-inducing proteins. This approach has been successful in preventing HIV repli-cation in cell culture and perhaps in the future may become a therapy for HIV-infected people.

- Outline the procedures you would follow to engineer antibody-producing genes into a mammalian cell in culture.
- Against what proteins produced during HIV replication might you direct an intracellular antibody? (Refer to Fig. 16-11.) How would the antibody protect the cell?
- In order to use this approach as therapy in HIV-infected people, what other experiments must the researchers do?
- What is an advantage of using an intracellular antibody rather than inducing extracellular antibody production?

[*] J. Travis, Putting antibodies to work inside cells, *Science* 261: 1114 (1993).

EPILOGUE

Microbiology in the 21st Century

As a new millenium dawns, microbiologists are poised on the brink of discoveries and developments expected to generate new dimensions in understanding and applying the principles of life. Some will lead to disappointment; others could exceed the most optimistic projections. One thing seems certain, however—microbiologists will continue to make scientific history.

Making predictions is always a risky business, especially when it's about the future.

Niels Bohr

Your introductory journey through the microbial world is not quite complete. In fact, you may never finish the voyage, which will continue as long as you remain alert to the consequences of microbes in all aspects of life. For example, as you read the news this week, you will probably come across a story or two about some promising advance in technology or basic science that has its roots in microbiology. Some of these advances will solve problems we never knew we had; others might help us blunt the thrust of grave dilemmas that could affect all of Planet Earth's occupants. As we enter the 21st century, if we fail to solve these difficult problems they could create a chain reaction of disasters. Human overpopulation, global warming, solid waste accumulation, environmental pollution, and new diseases that are resistant to our most effective medical weapons threaten the quality of life on our

planet. Fortunately, we have unprecedented scientific knowledge—perhaps our greatest ally—in our corner. Microbiology is among the most significant of these sciences, one that will continue to change life on earth.

One of the most stimulating aspects of any science is projecting what these changes will be. Throughout this book, we have tried to draw attention to future developments that may emerge from today's microbiological knowledge. Here are a few more of the developments that many microbiologists foresee happening as this century gives way to a new millennium.[1]

[1]By the time you read this, some of these projections may have already borne fruit or been abandoned. New promises will undoubtedly develop as fresh discoveries open doors to unexplored territory.

Pollution Reduction

In addition to using microbes for bioremediation of toxic or environmentally detrimental chemicals (see Chap. 27), bacteria may help solve problems of acid rain, pesticide contamination of groundwater, and countless other problems caused by environmental pollutants. Acid rain, for example, showers parts of the earth with dilute solutions of sulfuric acid and nitric acid, killing fish, forests, and whole ecosystems (Germany's famous Black Forest, for example, is rapidly dying from the effects of acid rain). When high-sulfur fuels are burned in factories, power plants, or vehicles, they emit sulfur dioxide in the smoke. Sulfur dioxide accumulates in the atmosphere, eventually dissolving in airborne water droplets, forming sulfuric acid (H_2SO_4). This falls to earth as acid rain. One solution to this problem is to remove the sulfur from the fuel before burning it. Conventional methods of extracting sulfur from crude oil, however, will require a $25 billion initial investment and a sustained outlay of $6 billion each year (all of which will raise the price of petroleum products).

A recent bacterial isolate, called *Rhodococcus rhodocrous,* can remove sulfur from crude oil without digesting the combustible hydrocarbons. The enzymes responsible for sulfur removal by *R. rhodocrous* may soon be isolated and immobilized, producing a continuous-flow biocatalytic reactor. Sulfur-laden crude oil would go into the reactor, and sulfur-free crude would flow out. Sulfur-oxidizing bacteria may also provide an economically feasible method of removing sulfur from coal, the major producer of sulfur emissions. The same approach may be applied to the removal of nitric acid–producing nitrogen compounds from crude oil, releasing the nitrogen in a form that can be used to make fertilizer. Microbiologists may help prevent severe acid rain damage while saving consumers billions of dollars in energy costs.

Resource Enhancement

Natural resources such as fuel and food will be dramatically enhanced by many of the techniques discussed in Chapter 27 (such as using microorganisms to convert garbage into food or fuel). In addition, microbes may soon be helping us explore for oil and perhaps even gold. Volatile hydrocarbon vapors permeate the soil overlying some oil deposits. Bacteria

that specifically metabolize these hydrocarbons can be used as bioindicators in exploring for oil. When inoculated into media mixed with a soil sample, the bacteria will grow only if the soil contains the hydrocarbons that the microbe uses as its sole carbon source—no petroleum, no growth. A more ambitious approach would be to isolate the hydrocarbon receptors from the membranes of these bacteria and use the proteins to build oil-finding biosensors (see p. 752). Microbiologists might also be in demand by gold prospectors, who will use them to determine the concentration of *Bacillus cereus* in soil samples. These bacteria have been found to be twice as numerous as usual in soil near gold deposits.

Another exciting and largely unexplored resource is submerged beneath the seas that cover four-fifths of the earth's surface (Fig. 1). Marine microbiology and biotechnology are brimming with opportunities. At least two compounds from deep-sea microorganisms, for example, show anti-HIV activity in vitro (but successfully using these microbes to produce a promising AIDS fighter has so far been hampered by difficulties encountered in cultivating either organism in the laboratory). Aquatic organisms are also abundant reservoirs of food, as commercially successful aquaculture ventures testify. Marine and freshwater aquaculturists (fish growers) are working with microbial geneticists to introduce into fish the gene for a growth hormone that will

FIGURE 1 Underwater "airplane." Engineered to "fly" beneath the waves, the *Deep Flight* is one of a growing list of craft used for exploring submarine environments. All these submersible vehicles will have the capacity to explore marine microenvironments, collecting samples for microbiological analysis. With 99 percent of the world's microbes still undiscovered, marine specimens promise to substantially increase the number of known species, promoting new microbial applications and adding to our understanding of life.

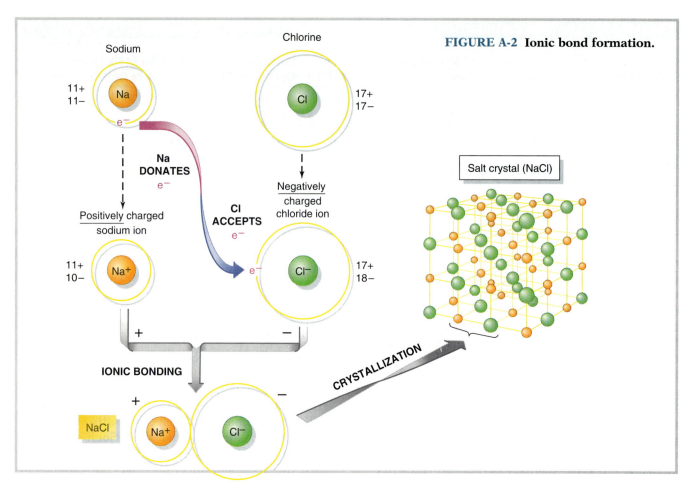

FIGURE A-2 **Ionic bond formation.**

trates triple bond formation, each nitrogen in the duplex donating three electrons to be shared with the other atom of nitrogen. No atom can form a quadruple bond with another atom; three is the maximum number of linkages between any two atoms. Carbon, with its surplus (or deficit) of four outer-shell electrons, must combine with at least two other atoms to achieve its ideal state. Methane, for example, is the product of such multiple covalent bonding, with carbon eliciting the cooperation of four hydrogens to generate saturated shells for all five participating atoms. (Methane is natural gas, a flammable organic compound often used for heating and cooking.)

■ Covalent Bonding and Organic Molecules

Carbon's ability to form four covalent bonds imparts special importance to its role in the chemistry of life. It enables carbon to form long chains by linking with itself. These carbon chains provide the chemical backbone of the enormously complex molecules needed to organize and direct the life processes, as well as those that make up the fabric of life itself. These carbon backbones may be linear, branched, or cyclic. Equally important, every carbon in the chain has at least two free bonding sites to which other elements and chemical groups can attach. These accessory chemicals influence the property of the compound. The presence of carbon (and usually hydrogen) distinguishes a compound as **organic.** In some organic compounds, carbon forms double or triple bonds with other carbon atoms.

■ Hydrogen Bonds

Another type of adhesive force between molecules is formed when two molecules share an atom of hydrogen. This occurs only between **polar** molecules. Polar molecules have their electrons unequally distributed and therefore have slight positive and negative charges concentrated in different parts of the molecule. The single oxygen atom in water, for example, tends to attract electrons more efficiently than do its two hydrogens. The shared electrons therefore spend more time at the oxygen end of the molecule. This imparts an unequal electron distribution to the molecule. The oxygen end becomes slightly more negatively charged than the two hydrogen ends, which are left with a small positive

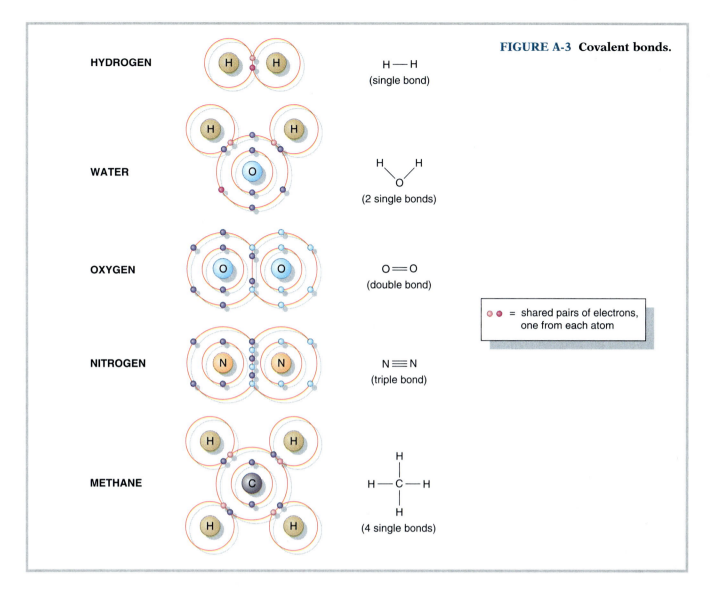

FIGURE A-3 Covalent bonds.

HYDROGEN

H — H
(single bond)

WATER

H   H
 \\ /
  O
(2 single bonds)

○ ● = shared pairs of electrons, one from each atom

OXYGEN

O ═══ O
(double bond)

NITROGEN

N ≡≡≡ N
(triple bond)

METHANE

```
    H
    |
H — C — H
    |
    H
```
(4 single bonds)

charge. As a result, water molecules are drawn together, each positively charged end of the molecule attracted to the negatively charged portion of another water molecule. (The attraction is similar to that between the opposite poles of a magnet.) In effect, the oxygen atoms of two water molecules share a common hydrogen atom (see Fig. A-4). The sharing is not equal, of course, since the shared hydrogen is covalently bonded to one of the oxygens; nonetheless, the effect is to generate a weak association between the two participating molecules, an attraction called a **hydrogen bond.** The strength of these bonds keeps water molecules closely associated as a liquid rather than breaking apart and vaporizing (unless the water is heated to the boiling point).

Hydrogen bonding is not limited to water but is common to many polar molecules. The double strands of DNA are held together by hydrogen bonds. Some large polar molecules even form hy-

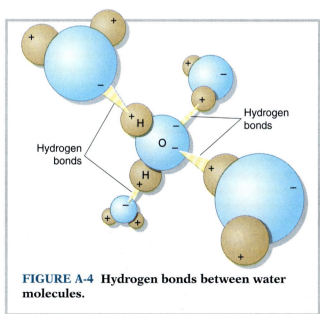

FIGURE A-4 **Hydrogen bonds between water molecules.**

drogen bonds with other atoms within the same molecule. For example, parts of a single protein molecule form hydrogen bonds with each other, bonds that help stabilize the giant molecule. Although a hydrogen bond is only about 1/20 as strong as a typical covalent bond, a large molecule may contain several hundred, or even thousands, of these polar linkages, creating considerable combined strength.

Life-Supporting Properties of Water, Solvents, and Solutes

Life as we know it on earth depends on water. About 70 to 80 percent of an organism's living material is composed of water. In addition to being a reactant in or product of some essential chemical reactions, water provides the matrix in which nearly all of life's biochemical processes occur. The peculiar properties of water that make it invaluable to life are largely the product of its polar nature.

■ Molecular Structure

Water's single oxygen tends to attract electrons more efficiently than do its two hydrogens. Such polarity creates interactions between individual water molecules that dictate its properties. These interactions result from the attraction between the oppositely charged ends of adjacent molecules, weakly pulling water molecules together by hydrogen bonding. Each molecule of water can form hydrogen bonds with four other water molecules (review Fig. A-4).

The polarity of water makes water earth's most efficient solvent, gives it surface tension, propels it upward against the pull of gravity (capillary action), and makes it ideal for temperature regulation in organisms. Each of these special properties of water is discussed in the following sections.

■ Water as a Solvent

More types of substances dissolve in water than in any other medium. Many biologically active molecules, large and small, perform their functions only when dissolved in water. Virtually any substance that is either ionic (a salt, for example) or polar can dissolve in water, forming a solution.

■ Solutions and Suspensions

When a substance dissolves into individual molecules in a fluid, the resulting mixture is called a **solution.** The fluid matrix in which the substance is dissolved is the **solvent;** the material that goes into solution is the **solute.** Prior to being dissolved, a solute can be a solid, liquid, or gas. Water is such an efficient solvent that rainwater becomes a solution containing the dissolved gases oxygen, nitrogen, and carbon dioxide by the time it reaches the ground.

Some substances that don't dissolve into small molecules become suspended as particles in a fluid matrix, forming a **suspension.** Particles that are 10^{-4} μm or larger in diameter eventually settle out of suspension, collecting at the bottom of the fluid. If the particles are small enough, however, they resist gravity's pull and remain in suspension, forming a **colloidal suspension.** Common examples of colloids include milk, Jell-O, and fog (water droplets suspended in air). Cytoplasm is also a colloidal suspension, a semifluid substance in which most of the life processes take place.

■ Surface Tension

The attraction between the water molecules imparts to the liquid a high surface tension, that is, a tendency of the surface to resist being easily broken. Position a leaf broadside on top of a bowlful of water and see what happens. It not only floats, it seems to be supported by a membrane across the water's surface (you can even see indentations in the water's surface around the edge of the leaf). This "membrane" is the product of the cohesive force generated by hydrogen bonding between adjacent water molecules. It takes considerable force to overcome the attraction and break the water's surface. The same effect is demonstrated by filling a glass with water just beyond the top of the container. The water bulges above the top of the glass rather than spilling out because of hydrogen bonding with the underlying water molecules.

Water molecules are attracted not only to each other, but also to any polar (hydrophilic) substance. Many surfaces, such as glass, are hydrophilic and attract water molecules with enough force to overcome the effects of gravity. The results of this type of attraction are apparent when one dips a narrow glass tube into water. The water seems to crawl up the sides of the tube, pulled along by the attraction to the hydrophilic (wettable) surface (Fig. A-5). The water molecules adjacent to the surface pull nearby molecules up by the strength of hydrogen bonds. This tendency of water to be pulled into the tube is called *capillary action.*

■ Temperature Regulation

Most organisms are remarkably flexible in their ability to withstand changes in the amount of heat to which they are exposed. This flexibility is largely the product of the thermal stability of water—its ability to absorb or lose large amounts of heat without changing temperature.

■ Ice Formation

As the temperature of water decreases, the molecules tend to slide in closer to one another. In other words, as water gets colder it becomes heavier, at least until it reaches 4°C. At this temperature water is at its heaviest state. As the temperature approaches 0°C, however, this trend reverses itself, and the molecules orient themselves in an arrangement that takes full advantage of hydrogen bonding. Below 0°C, water freezes to a solid, forming a rigid molecular lattice in which each molecule is firmly hydrogen-bonded to four other water molecules. Since this configuration prevents the molecules from slipping into closer proximity with each other, the intramolecular space is greater in ice than in liquid water. Consequently, water expands as it freezes (this explains why ice is lighter than liquid water).

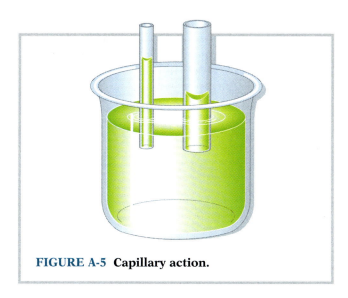

FIGURE A-5 Capillary action.

Freezing is fatal to many organisms because the expansion of water tends to rupture cell membranes and tissues, especially if large crystals with sharp edges form. Rapid freezing with extremely cold temperatures tends to reduce the formation of such destructive crystals.

Bergey's Manuals—Identification of an Unknown Bacterium

Bergey's Manual of Determinative Bacteriology[1]

This single-volume paperback ninth edition, published in 1994, is a classification of bacteria that contains descriptions of all prokaryotic species that had been described and cultured at the time of its publication. It is designed to serve as an inexpensive, practical guide for the identification of bacteria. The bacteria are divided into 35 groups, and each group is assigned to one of four major categories:

I. Gram-negative eubacteria that have cell walls (groups 1–16)
II. Gram-positive eubacteria that have cell walls (groups 17–29)
III. Eubacteria lacking cell walls (group 30)
IV. Archaeobacteria (groups 31–35)

Assignment to a specific group within each category is based primarily on phenotypic characteristics such as (1) morphology, (2) energy and carbon source, (3) mode of locomotion, (4) gaseous requirements, and (5) ability to produce endospores. The distinguishing features of each group are described in Chapter 10, Table 10-1. ▶ (See p. 247.) Within each group, the manual contains keys and tables describing additional criteria for identification of genus and species.

[1]J. G. Holt, N. R. Krieg, P. H. A. Sneath, J. T. Staley, and S. T. Williams (Editors), *Bergey's Manual of Determinative Bacteriology*, 9th ed., Williams and Wilkins, Baltimore, Maryland, 1994.

■ Using *Bergey's Manual*

When an unknown bacterium is isolated in *pure culture* in the laboratory, it is usually identified by a combination of information from microscopic observations, cultural (growth) characteristics on agar and in broth, and biochemical tests. The following information describing an unknown bacterium could be readily obtained in most laboratories.

Morphological
  Shape: straight rod
  Arrangement: single or in pairs
  Gram stain: negative
  Buds or sheaths: negative
  Motility: motile

Cultural
  Nutrition: heterotroph
  Agar colonies: white
  Oxygen tolerance: facultative anaerobe

Biochemical Tests
  Oxidase: negative
  Catalase: positive
  Indole: positive
  Methyl red: positive
  Voges-Proskauer test: negative
  Citrate: negative
  Hydrogen sulfide: negative
  Nitrate: reduced
  Urea hydrolysis: negative
  Lipase: negative
  Acid aerobically from: arabinose, glucose, lactose, mannitol, xylose

We can now identify the unknown bacterium using *Bergey's Manual*. Groups 16 through 35 can be immediately eliminated because the unknown is a gram-negative eubacterium and thus belongs to category I. Within this category, we can narrow the choices to group 5 based on the following:

1. The organism is a straight rod (eliminate groups 1, 2, 3, 8, 13, 14, 15, and 16).
2. The organism is a facultative anaerobe (eliminate groups 2, 4, 6, 7, 8, 9, 10, 11, 12, 14, and 16).
3. It is heterotrophic, not lithotrophic, phototrophic, or an obligate intracellular parasite (eliminate groups 9, 10, 11, and 12).

Thus the organism must belong to group 5—Faculatively anaerobic gram-negative rods.

Group 5 is further differentiated into three families (Enterobacteriaceae, Vibrionaceae, and Pasteurellaceae) and five individual genera, all but one of which can be eliminated by the results of the oxidase, catalase, indole, and motility tests. Thus, the unknown is one of the more than 100 organisms within the family Enterobacteriaceae. Based on the combined findings of the biochemical tests, the unknown is identified as *Escherichia coli*.

A CD-ROM version of *Bergey's Manual* is currently in development. This computerized format can inform the investigator which organisms within the database match the input information (test results) about the unknown. Additional tests can then be performed to distinguish among the remaining possibilities.

Bergey's Manual of Systematic Bacteriology

Bergey's Manual of Systematic Bacteriology is a four-volume work, published between 1984 and 1989, that can also be used for identification. In it, prokaryotes are divided into 33 sections based upon the best available genetic information. The manual contains detailed descriptions of the characteristics of organisms within each section, as well as discussions of media and tests used to grow and identify these organisms.

Volume I (1984)

Section 1 The Spirochaetes
Section 2 Aerobic/Microaerophilic, Motile, Helical/Vibrioid Gram-Negative Bacteria
Section 3 Non-Motile (or Rarely Motile) Gram-Negative Curved Bacteria
Section 4 Aerobic Gram-Negative Rods and Cocci
Section 5 Facultatively Anaerobic Gram-Negative Rods
Section 6 Anaerobic Gram-Negative Straight, Curved, and Helical Rods
Section 7 Dissimilatory Sulfate- or Sulfur-Reducing Bacteria
Section 8 Anaerobic Gram-Negative Cocci
Section 9 The Rickettsias and Chlamydias
Section 10 The Mycoplasmas
Section 11 Endosymbionts

Volume II (1986)

Section 12 Gram-Positive Cocci
Section 13 Endospore-Forming Gram-Positive Rods and Cocci
Section 14 Regular, Non-sporing, Gram-Positive Rods
Section 15 Irregular, Non-sporing, Gram-Positive Rods
Section 16 Mycobacteria
Section 17 Nocardioforms

Volume III (1989)

Section 18 Anoxygenic Photosynthetic Purple Bacteria
Section 19 Oxygenic Photosynthetic Bacteria
Section 20 Aerobic Chemolithotrophic Bacteria and Associated Organisms
Section 21 Budding and/or Appendaged Bacteria
Section 22 Sheathed Bacteria
Section 23 Nonphotosynthetic, Nonfruiting, Gliding Bacteria
Section 24 Gliding, Fruiting Bacteria
Section 25 Archaeobacteria

Volume IV (1989)

Section 26 Nocardioform Actinomycetes
Section 27 Actinomycetes with Multilocular Sporangia
Section 28 Actinoplanetes
Section 29 Streptomycetes with Related Genera
Section 30 Maduromycetes
Section 31 Thermomonospora and Related Genera
Section 32 Thermoactinomycetes
Section 33 Other Genera

APPENDIX C

Collection and Examination of Clinical Specimens

Microbiological laboratory findings provide the attending physician with information needed for a complete and correct diagnosis of infectious disease, which in turn dictates the most effective antimicrobial therapy if any is appropriate. Information revealed by examining properly collected specimens also alerts medical staff to any potential epidemic dangers to the hospital staff or the general community.

To supplement the discussion provided in Chapter 3, this appendix supplies details for collecting and examining specific clinical specimens.

Collection of Clinical Specimens

■ Throat and Nasopharyngeal Specimens

Throat or nasopharyngeal specimens aid in the diagnosis of infectious disease of the upper respiratory tract. They are usually obtained on a sterile swab tipped with a cottonlike material such as calcium alginate wool. Because many pathogens will not survive dehydration, drying of the specimen must be prevented by immediate inoculation of culture media or by placing the swab in a transport medium and taking it immediately to the laboratory.

Obtaining the optimum throat culture requires care in avoiding the slightest contact with the teeth, gums, cheeks, tongue, and other regions heavily populated with normal flora. Such precautions will prevent the throat culture from becoming a mouth culture, which is of no diagnostic value.

Nasopharyngeal cultures are obtained by using a sterile flexible swab (usually a cotton-tipped wire) that may be passed up behind the uvula or down through the nasal passages and gently rotated.

■ Specimens from Sputum

Although the lower respiratory tract is usually sterile, its secretions, called **sputum,** may become contaminated by oral microflora during the process of expectoration. Sputum differs from saliva, the watery secretions of the mouth that contain oral microorganisms. When these organisms contaminate the sputum specimen, they may provide false clues that can mislead the diagnostician.

Sputum may be collected by one of three methods: expectoration, nasotracheal or orotracheal suction, or transtracheal aspiration. Ideally, expectorated sputum is collected in the morning immediately after the patient awakens. This is the time when the most sputum is available and the patient is strongest for coughing it up. The procedure should be carefully explained to the patient, and the importance of elevating the material from deep in the bronchi should be emphasized. Oral flora contamination can be reduced by washing the mouth before coughing. If coughing is difficult, the process can be assisted by *nebulization*, exposure to a mist of hypertonic saline solution to loosen and liquefy bronchial secretions while stimulating a cough reflex. Collecting sputum in a large-mouthed sterile cup with a tight-fitting lid prevents spillage and contamination of the specimen.

Patients incapable of voluntary coughing and expectoration may require *tracheal suction* of sputum with a device that is passed through the mouth or nose into the lower respiratory tract. This technique does *not* reduce the problem of contamination by the normal oral or nasal flora. *Transtracheal aspiration,* on the other hand, eliminates the presence of upper respiratory organisms in the specimen. In this procedure, a sterile large-bore needle is passed through the decontaminated skin of the neck and into the trachea. Sputum is then aspirated into a syringe.

■ Ear and Eye Specimens

Although the human eye possesses a characteristic microflora, bacterial numbers are relatively small because of the continual flushing activity of tears and the action of antibacterial substances such as lysozyme. These natural inhibitors may also destroy bacteria in a specimen that is not immediately processed. Specimens are obtained by retracting the eyelid and stroking the infected regions with a sterile swab. Caution should be exercised to avoid touching the sensitive iris. The swab may then be placed in a sterile test tube containing just enough transport medium to keep the material moist. The specimen is immediately inoculated onto growth medium if *Neisseria gonorrhoeae* infection of the eye is suspected.

Prior to collecting a specimen from the ear, the external surfaces should be decontaminated with alcohol or an iodophore. The clinical material can be removed by sterile swab, or pus can be aspirated into a sterile syringe.

■ Stool Specimens and Rectal Swabs

Most pathogens that cause gastrointestinal illness reside in the lumen as well as in or on the walls of the intestine. Since these organisms are shed in the feces, it is not necessary to obtain scrapings of intestinal walls to recover them in culture. Like sputum, the optimal fecal sample is available in the morning, the first stool of the day. These specimens are most valuable when obtained early in the course of the disease. The material can be collected by scraping a small amount of feces from a fresh stool left in a bedpan or in an infant's diaper. Care should be exercised to avoid contamination of feces with urine. Unlike other specimens, stool specimens are collected in unsterile containers.

Many intestinal pathogens are sensitive to desiccation and die unless immediately inoculated or placed in a transport medium. Intestinal protozoa alter their characteristic forms in specimens that are not examined soon after collection.

A few pathogens are found in rectal ulcers and are not readily isolated from feces. Others are fastidious organisms that may lose viability quickly in feces. In these cases, rectal specimens taken on sterile swabs may prove more beneficial than a stool specimen. Intestinal lesions should be visually located, using a proctoscope, and directly swabbed. Fastidious microbes can be removed by swabbing feces in anal crypts approximately 1 inch beyond the anal sphincter.

Isolation of pathogenic viruses from feces usually requires shipping the specimen to the nearest viral laboratory. The feces should be frozen quickly at −40°C until processing can occur.

■ Genital Tract Specimens

Venereal diseases and genital infections are the most commonly reported infectious diseases in the United States. Cervical-vaginal swabs, urethral exudates, and fluid from genital lesions are among the most common types of specimens collected. Unless protected, many venereal pathogens quickly die when removed from their genital sanctuaries.

Specimens from women with vaginitis are best obtained by swabbing the vaginal mucosa at the place where the discharge is most plentiful. The close proximity of vagina and rectum necessitates care to avoid contamination with fecal organisms. The swabs should then be placed in tubes of sterile saline. If trichomoniasis is suspected, the specimen should be microscopically examined within 15 minutes to locate the characteristic trichomonads, which quickly lose motility and become increasingly difficult to identify. These specimens are also inoculated into appropriate media to help detect asymptomatic trichomonas carriers.

Blood is the predominant specimen employed in the diagnosis of HIV infection and primary syphilis. Specimens for microscopic detection of *Treponema pallidum* may also be recovered from fluid expressed from the characteristic genital lesions (chancres).

■ Urine Collection

Bladder urine is normally sterile and contains microorganisms only when bladder, ureters, or kidneys are infected. Even healthy persons, however, void urine that becomes heavily contaminated with the microbial residents of the urethra and external genitalia. Since these resident organisms may also cause urinary tract infection, quantitative determinations are necessary to distinguish whether these microbes are causing disease or are simply contaminants in urine specimens. For example, *Escherichia coli*, a common contaminant of urine voided from all persons, causes 80 percent of all uncomplicated urinary tract infections. Because contamination of properly collected urine is inevitable, fewer than 10,000 bacteria per milliliter is considered normal. When the concentration of enteric gram-negative bacilli exceeds 100,000 cells/mL, they are declared the etiologic agent. If the count is between 10,000 and 100,000 per milliliter, new specimens are collected and processed. These numbers apply only to specimens collected by the midstream-catch technique (described below).

Any procedures that alter the microbial proportions in the sample invalidate the laboratory results and confuse the diagnosis. Since urine itself is an excellent culture medium, specimens allowed to stand at room temperature for more than 30 minutes become overgrown and may lead to misdiagnosis. If delay is unavoidable, specimens should be stored at 4°C, and even then for no longer than 6 hours before cultivation.

To reduce the extent of contamination during collection, voided urine should be gathered by the "clean-catch" or **midstream-catch** technique. This requires thorough cleansing of the external genitalia. Uncircumcised males should retract the foreskin, and females should clean from front to back to avoid transferring rectal flora to the vagina. The patient begins voiding and, without stopping the stream, collects the midstream portion in a sterile cup until the container is no more than half full.

Alternatives to the clean-catch method are occasionally employed to reduce contamination. The once-common practice of inserting a catheter up the urethra and into the bladder does reduce microbial contamination of the sample but is discouraged because of the danger of urinary tract infections due to this procedure. Urine specimens from patients with existing catheters, however, may be collected with a sterile syringe either from the port of the catheter or through the tubing. Both sites need to be decontaminated prior to insertion of the needle. Because the genital flora are avoided by collecting from existing catheters, any gram-negative enteric bacterium that exceeds concentrations of 100/mL in catheter urine is considered the etiological agent of infectious urinary tract disease. Specimens should *never be obtained from the urine "incubating" in the catheterized person's collecting bag.*

■ Specimens from Skin Lesions

Draining rashes or purulent (pus-producing) lesions are swabbed after being cleaned with alcohol or io-dine to remove the pus. Swabs that accidentally contact the surrounding skin and its indigenous flora are discarded. Closed lesions are not exposed to surface flora, and their exudates are excellent specimens when aspirated directly into sterile syringes.

Dermatophytes from cutaneous lesions cannot be obtained on a swab. These fungi are usually collected by obtaining a portion of the infected tissue or aspirating any exudate that may be present.

■ Specimens from Wounds

Uninfected wounds are free of established microorganisms. Many transient residents of the skin, however, may cause serious infections in wounds. Contaminants introduced into the specimen during collection may therefore confuse the diagnosis.

Infected superficial wounds usually contain aerobic organisms. Deep wounds or abscesses often support the growth of obligate anaerobes, and specimens should be cultured anaerobically, especially if the area is characterized by a foul odor or by necrotic or gangrenous tissue. The same is true of any infection of a deep sinus, the pleural cavity, or joints.

■ Specimens from Biopsy Material

Biopsy specimens are commonly fixed in formalin before they are microscopically examined. Since formalin kills the microorganisms in the tissue, fixed biopsy specimens are useless for culture. The fresh specimen should be divided in half and the unfixed portion sent to the laboratory for microbiological analysis.

Biopsy material may have to be aseptically ground prior to cultivation to release pathogens hidden in the tissue. Because aerosols are inevitably created by this vigorous procedure, tissues should be ground under a contagion hood. Biopsy material should be cultured both aerobically and anaerobically.

■ Cerebrospinal Fluid (CSF) Specimens

Infectious disease of the central nervous system often causes rapidly progressing irreversible damage or death. Because of the urgency of this situation, delays in microscopic and cultural analysis of CSF specimens should be avoided so that the most effective antimicrobial agents can be determined as quickly as possible.

Normal CSF is sterile and not likely to be contaminated by surface organisms if the skin is decontaminated with an iodophor and if the fluid is properly collected. If Gram-stained smears of the fluid are microscopically examined immediately after collection, a complete morphological description of any microbes in the CSF smears can be returned to the physician within minutes. This information often suggests which antibiotics will most likely inhibit the microscopically observed organisms.

■ Blood specimens

Septicemia, bacterial endocarditis, and most systemic infectious diseases can be diagnosed by isolating and identifying the pathogens from the blood. The value of the specimen, however, can be destroyed by inadequate preparation of the skin at the venipuncture site, since many organisms residing on the skin are opportunistic pathogens. Introduction of contaminants into blood cultures produces a microbiological "smoke screen" that diverts diagnostic attention away from the actual pathogens.

Venipuncture sites are most commonly decontaminated by using a combination of alcohol (or acetone) and iodine. Each of these antiseptics should be applied with a circular motion in an outward spiral to move residual microorganisms away from the proposed site of puncture. After applying a tourniquet, the specimen is drawn according to the aseptic procedures—for example, using sterile needles and careful preparation of the skin to protect both the specimen and the patient from being inoculated with surface microorganisms.

Enough blood should be withdrawn to inoculate both aerobic and anaerobic culture media. Before inoculating each medium, a fresh sterile needle should replace the one used to penetrate the skin, since it is likely contaminated with residual skin flora. Additional blood samples may be collected for quantitative plate counts, microscopic examination for pathogens, and differential characterization of stained blood cells. Blood specimens are also collected for immunological analysis to detect microbe-specific antibodies (see Chap. 18).

Laboratory Examination of Clinical Specimens

Several procedures in the microbiology laboratory are critical to the diagnosis of infectious disease:

Microscopy Preliminary microscopic examination of stained smears or wet mounts of clinical material can provide early indications of the types of pathogens that may be causing the patient's disease. Gram's stains of sputum, CSF, or blood can provide evidence to guide the physician in early selection of antimicrobial agents in urgent situations that require antibiotic therapy soon after the specimen is collected. Gram's stains of bacteria in CSF, for example, can indicate bacterial etiology of suspected meningitis or encephalitis and suggest which antibacterial agents would be most effective. Acid-fast smears can alert the staff to a potential case of tuberculosis. Direct dark-field examination of fluid from suspected syphilis chancres can often reveal the presence of motile spirochetes.

Culture Diagnosis may be confirmed by isolating, culturing, and identifying suspected pathogens. Quantitative cultures help determine the causative role of the isolated species. Some specimens may yield positive cultural data only after substantial delay. Blood cultures, for example, should be maintained for at least 3 weeks before being discarded as negative.

Antibiograms The determination of the antibiotic susceptibilities of the isolated microorganisms is as important as species identification of pathogens. The resultant antibiograms are used as guidelines to the selection of effective antibiotic therapy (see Chap. 16).

Serology Serological tests on patients' sera provide diagnostic information revealing the presence of antibodies that react with the antigens of specific pathogens. This information offers indirect indications of the etiology of disease, often before the pathogen can be isolated and identified. In cases where cultivation of fastidious or delicate pathogens is difficult or impossible (as with the agent of syphilis, for example), serological findings become increasingly important. Serological tests are also used to identify antigens of unknown specificity (see Chap. 18).

DNA tests Like serology, genetic analysis may be used to identify microorganisms without culturing them. As such, it is a rapid method for analyzing clinical specimens, especially for organisms that are difficult to grow. Detection of a specific microorganism relies on reacting the clinical specimen (or isolated organisms) with labeled probes consisting of DNA sequences unique to the microbe of interest. Genetic tests are replacing other rapid tests as DNA probes for many microbial pathogens become commercially available. When the population of the pathogen within a specimen is believed to be too low to be detected directly, it is often possible to amplify its DNA using the polymerase chain reaction prior to reaction with a DNA probe.

alternative pathway. Antibody-independent pathway for complement activation.

alveolar macrophage. Phagocytic cell in the lungs; also called "dust cell."

alveolus. Terminal portion of the respiratory tract where gas exchange occurs.

amantadine. Antiviral agent used to prevent influenza.

amensalism. Antagonistic relationship in which one organism does not benefit from the harm it does to the other.

Ames test. Laboratory procedure using bacteria to screen compounds for mutagenic activity.

amination. Addition of an amino group (NH_2) to a compound.

amino acid. Organic building block of protein. It contains an amino group (NH_2) and a carboxyl group (COOH).

aminoglycoside. Any of a group of related antibiotics that attach to ribosomes and interfere with protein synthesis. Streptomycin, gentamicin, and neomycin are examples.

amino group. —NH_2.

ammonification. Chemical release of ammonia (NH_3) from nitrogen-containing organic compounds.

amoeba. Protozoan in the class Sarcodina that moves and feeds by pseudopods.

amphitrichous. Having polar flagellation with tufts of flagella at each end of the cell.

amphotericin B. Polyene antibiotic used for treating systemic fungal infections.

amylase. Enzyme that hydrolyzes starch.

anabolism. Biosynthesis of complex molecules from simple compounds.

anaerobe. Organism that grows in the absence of molecular oxygen.

anaerobic glove box. Enclosed unit used for manipulations requiring an oxygen-free atmosphere.

anaerobic respiration. Energy production that depends on an inorganic molecule other than oxygen as the terminal electron acceptor.

anamnestic response. Immunological memory that accounts for rapid immune reactions following antigen exposure in sensitized individuals.

anaphylaxis. Generalized IgE-mediated allergic response; may be fatal.

anecdotal evidence. Nonscientific information based solely on personal experience.

anemia. Deficiency in erythrocytes or hemoglobin (or both).

angstrom. Unit of measure equal to 1×10^{-10} meter.

anionic. Possessing a negative charge.

anoxygenic photosynthesis. Photosynthesis in which oxygen is not produced.

antagonism. (1) Reduction in the effectiveness of drugs when used in combination; (2) in ecology, relationship in which one organism harms another.

antibiogram. Spectrum of antibiotic susceptibility of a microbe.

antibiotic. Chemical produced by microorganisms that in low concentrations selectively inhibits or kills other microorganisms.

antibiotic resistance. Ability of a microbe to survive in the presence of an antibiotic.

antibody. Protein produced in the body in response to an antigen and capable of specifically binding with the antigen that stimulated its formation.

antibody-dependent cell-mediated cytotoxicity (ADCC). Damage resulting from the interaction of cell-bound antibody with complement, phagocytes, or other effector cells.

antibody titer. Concentration of an antibody in the blood.

anticodon. Three-nucleotide region on transfer RNA that binds to the condon of messenger RNA.

antigen. Substance that stimulates an immune response when introduced into the body.

antigenic determinant. Small site on the antigen that determines the specificity of the immune response.

antigenic drift. Minor structural changes in the viral surface due to mutation in the viral genome.

antigenic shift. Sudden major structural change in the viral surface resulting in immunologically distinct strains; usually due to recombination.

antigen-presenting cell. Macrophage or other host cell that processes antigen and displays fragments bound to MHC to other cells of the immune system.

antihistamine. Chemical that inhibits the effects of histamine in the body.

anti-idiotype antibody. Antibody that recognizes and binds to the antigen-binding site of another antibody.

anti-inflammatory agent. Substance that reduces the intensity of the inflammatory response.

antimetabolite. Chemical that competitively inhibits specific microbial metabolic reactions.

antimicrobial agent. Substance that kills microbes or inhibits their growth.

antiparallel. Orientation in DNA in which one strand is $5' \rightarrow 3'$ and the other is $3' \rightarrow 5'$.

antisense nucleic acid. Single-stranded nucleic acid complementary to portions of another nucleic acid, capable of binding to it and interfering with its activity.

antisepsis. Inhibition or destruction of microorganisms on the surface of living tissue.

antiseptic. Chemical used to destroy or inhibit microbes on body surfaces.

antiserum. Blood serum that contains antibodies of known specificity.

antitoxin. Antibody that neutralizes a toxin.

aphotic zone. Region in an aquatic environment where light penetration is insufficient to support photosynthesis.

Archaea. Domain containing prokaryotes characterized as archaeobacteria.

archaeobacteria. Prokaryotes that have certain chemical and metabolic properties not characteristic of eubacteria.

arthrospore. Fungal spore formed by segmentation of hyphae into thick-walled cells.

artifact. Artificial characteristic that develops in a specimen as a result of its manipulation.

artificial active immunity. Production of antibodies or lymphocytes by the host in response to vaccination.

artificial passive immunity. Immunity acquired by injecting presynthesized antibody into a nonimmune host.

Ascomycetes. Class of fungi characterized by production of sexual ascospores.

ascospore. Sexual fungal spore produced in an ascus.

ascus. Sac that contains ascospores.

aseptic technique. Procedure that prevents the introduction of microbial contaminants.

asexual reproduction. Reproduction without fusion of haploid sex cells.

asexual spore. Spore formed without fusion of nuclear material.

asthma. Allergic response characterized by narrowing of the trachea or bronchi.

asymptomatic. Exhibiting no symptoms.

atom. Smallest unit of an element that retains the characteristic properties of that element; it is composed of protons, neutrons, and electrons.

attachment site. Location on the viral surface of the molecules that adhere to the cell's receptor site.

attenuation. (1) Reduction in a pathogen's capacity to cause disease; (2) regulation of gene expression by premature termination of transcription.

autoclave. Instrument that sterilizes with steam under pressure.

autocytolysis. Killing of an infected host cell by its own lytic enzymes.

autoimmune disease. Disease caused by an immune response against one's body tissues.

autoinoculation. Transfer of microbes indigenous to one area of the body to another area where they may cause disease.

autolysin. Bacterial enzyme that degrades peptidoglycan.

autotroph. Organism that uses carbon dioxide as the sole source of carbon.

auxotroph. Mutant that has acquired an additional nutritional requirement.

avirulent. Lacking the capacity to produce disease.

axial filament. Structure responsible for spirochete motility.

bacillus. Rod-shaped bacterium.

bacitracin. Antibiotic active against the cell wall of many gram-positive bacteria.

bacteremia. Presence of bacteria in the bloodstream.

Bacteria. Domain containing prokaryotes characterized as eubacteria.

bactericidal agent. Substance or process that is able to kill bacteria.

bacteriochlorophyll. Light-absorbing pigment in photosynthetic bacteria.

bacteriocin. Substance produced by some bacteria that inhibits or kills closely related bacteria.

bacteriological membrane filter. Filter that contains pores small enough to prevent the passage of bacteria.

bacteriophage. Virus that infects bacteria.

bacteriorhodopsin. Light-absorbing pigment in membrane of certain halophilic archaeobacteria.

bacteriostatic agent. Substance or process that is able to inhibit the growth of bacteria.

bacterium. Prokaryotic cell.

bacteriuria. Presence of bacteria in the urine.

bacteroid. Modified form of nitrogen-fixing bacteria within the root nodules of legumes.

baking. Hot-air sterilization process requiring exposure to temperatures between 150 and 180°C.

barophile. Organism requiring high pressure to grow.

barotolerant. Organism able to grow at high pressures but not requiring them.

basal body. Structure at base of flagellum that anchors it to cell.

base. Substance with a pH above 7.0.

base analogue. Chemical that resembles one of the purine or pyrimidine bases normally found in nucleic acids; some may cause mutations by being incorporated into DNA in place of the normal base.

base pairing. Mutual attraction between adenine and thymine (or uracil) and between guanine and cytosine.

Basidiomycetes. Class of fungi characterized by production of sexual basidiospores.

basidiospore. Sexual fungal spore produced by a basidium.

basidium. Structure that produces basidiospores.

basophil. Nonphagocytic granulocyte that stains with basic dyes.

batch culture. Culture system containing a medium to which nutrients are neither added nor removed during the growth of the culture.

BCG vaccine. Vaccine to provide immunity against tuberculosis; made from attenuated bovine tubercle bacillus.

Bergey's Manual. A published classification of bacteria used as a practical aid for their identification.

beta hemolysis. Zone of complete clearing around colonies of certain microbes growing on blood agar.

beta lysin. Soluble blood chemical that disrupts bacterial membranes.

bidirectional replication. Replication of circular DNA at two forks that travel away from each other.

binary fission. Process of asexual reproduction whereby one cell divides into two cells of equal size.

binomial nomenclature. System of naming organisms with two terms, the first representing the genus, the second the species.

bioassay. Method of measuring the concentration or activity of a substance by determining its effect on a living organism.

bioaugmentation. Introduction of microbes into a contaminated environment for the purpose of bioremediation.

biochemical. Compound that is a component of living organisms.

biochemical oxygen demand (BOD). Amount of oxygen consumed by aerobic organisms; used to determine the concentration of biodegradable organic matter in water.

biodegradation. Deterioration of material resulting from microbial activity.

biofertilization. Nutrient enrichment of soil as a result of microbial activity.

biofilm. Surface aggregate composed of layers of bacteria embedded in a polysaccharide matrix.

biogenesis. Doctrine stating that living organisms arise only from other living organisms.

biogeochemical cycle. Series of conversions by which nutrients are recycled in nature.

biological indicator. Living organism whose viability or activity is used as a measure of a specific process.

biological vector. A type of vector in which the pathogen proliferates, as opposed to a mechanical vector, which is a carrier only.

biomass. Weight of all organisms that inhabit a defined area.

biomining. Use of microorganisms to recover metals from ores.

biopsy. Surgical removal of tissues for diagnosis.

bioremediation. Microbial transformation of harmful materials into harmless molecules.

biosensor. Device that contains live organisms or their products that respond to specific chemicals, linked to a sensor system that can monitor the response.

biosynthesis. Construction of chemical components in the growing cell.

biotechnology. The use of technology to enhance a desirable characteristic of an organism.

biotype. Subgroup of organisms within a species that possess biochemical properties distinguishing them from other members of the species.

bivalent. Property of an immunoglobulin that enables it to combine with two identical antigens at the same time.

blastospore. Asexual fungal spore produced by budding.

blocking antibody. Antibody able to bind with an allergen and prevent an allergic response.

blood agar. Solid culture medium containing whole blood.

blood-brain barrier. Barrier that determines which substances will pass between brain capillaries and brain tissue.

bloom. Visible accumulation of algae or cyanobacteria in a body of water.

B lymphocyte. Cell that has been programmed to participate in the humoral immune response; when stimulated by antigen it proliferates and differentiates into antibody-producing plasma cells and memory cells.

booster. Immunization given to enhance the memory response to an antigen.

brightfield microscope. Microscope that uses a direct light source for illumination.

brine. Salt solution used to preserve foods by restricting the growth of spoilage microbes.

broad-spectrum antibiotic. Antibiotic affecting a wide variety of microorganisms.

bubo. Swollen lymph node due to inflammation.

bubonic plague. Form of plague transmitted from rats by fleas and characterized by the development of buboes.

budding. Asexual reproduction in which progeny arise from protuberances on the surface of a parent cell; in virology, the release of an enveloped virus from the surface of an infected cell.

buffer. Chemical that helps resist changes in the pH of a solution.

burst size. Number of viruses produced by an infected cell.

Calvin cycle. Carbon dioxide fixation pathway commonly used by autotrophs.

cancer. Uncontrolled proliferation of abnormal host cells.

candle jar. Container used to create microaerophilic or capneic conditions.

capnophile. Microorganism requiring increased levels of CO_2.

capsid. Protein coat surrounding the nucleic acid of a virus.

capsomere. Repeating protein unit of which a capsid is composed.

capsule. Structure composed of polysaccharide or polypeptide that surrounds the cell wall in some bacteria and fungi.

carbohydrate. Organic molecule that contains carbon, hydrogen, and oxygen, with a hydrogen-to-oxygen ratio of 2:1.

carbon cycle. Series of reactions in nature in which carbon dioxide is converted to organic carbon and ultimately metabolized back to carbon dioxide.

carboxyl group. An organic acid group; —COOH.

carboxysome. Cytoplasmic inclusion that contains the enzymes for carbon dioxide fixation found in autotrophic bacteria.

carbuncle. Extensive suppurative inflammation of skin and underlying tissues caused by bacterial infection.

carcinogen. Any agent that causes cancer.

cardiolipin. Component of beef heart extract used in rapid screening tests for syphilis.

carrier. Person with an inapparent infection who sheds pathogens and can therefore transmit disease.

caseation. Dead tissue with a cheeselike appearance, typically seen in tuberculosis.

catabolism. Degradation of complex molecules to simpler molecules.

catabolite repression. Inhibited production of certain catabolic enzymes in the presence of glucose.

catalase. Enzyme that decomposes hydrogen peroxide to water and oxygen.

catalyst. Substance that increases the rate of a chemical reaction without being consumed in the reaction.

catheter. Hollow tube used to gain access to narrow body canals in order to drain fluids from body cavities or to introduce material into the body.

cationic. Having a positive charge.

cDNA. DNA complementary to RNA usually synthesized with reverse transcriptase.

cell. Structural and functional unit of all living organisms.

cell culture. Growth of eukaryotic cells in vitro.

cell-mediated immunity (CMI). Immune response mediated by T lymphocytes.

cell membrane. *See* plasma membrane.

cellulose. Polysaccharide that makes up the cell wall of plants and most algae.

cell wall. Rigid structure surrounding the cell membrane in most plant, algal, fungal, and bacterial cells.

Centers for Disease Control and Prevention (CDC). Federal agency that gathers and analyzes data on notifiable diseases.

cephalosporin. Any of a group produced by *Cephalosporium* or a semisynthetic derivative; inhibits cell wall synthesis in gram-positive bacteria.

chancre. Ulcer with a hard, rubbery base; the initial lesion of syphilis.

chemical bond. Force that holds atoms together in a molecule.

chemical energy. Energy stored within a chemical bond and released when that bond is broken.

chemically defined medium. *See* synthetic medium.

chemiosmosis. Use of a proton gradient across a membrane (generated by electron transport) to drive the formation of ATP.

chemoautotroph. Organism that uses inorganic chemicals as a source of energy and carbon dioxide as the sole source of carbon.

chemoheterotroph. Organism that uses organic chemicals as a source of energy and carbon.

chemolithotroph. *See* chemoautotroph.

chemoprophylaxis. Administration of drugs to prevent infectious disease.

chemostat. Device used to maintain continuous cultures.

chemotactic factor. Substance that attracts phagocytic white blood cells.

chemotaxis. Movement of an organism toward or away from a chemical stimulus.

chemotherapeutic agent. Chemical administered in treatment of disease.

chemotherapy. Use of chemicals in the body to treat disease.

chemotroph. Organism that uses chemicals to secure energy.

chitin. Structural polysaccharide of fungal cell walls.

chlamydium. Obligate intracellular parasitic bacterium that lacks a functional ATP-generating system.

chlamydospore. Thick-walled asexual fungal spore formed within a vegetative hypha.

chloramphenicol. Antibacterial chemotherapeutic agent that inhibits protein synthesis.

chlorine. Halogen used in water purification and as a disinfectant.

chlorophyll. Light-absorbing pigment used by cyanobacteria, algae, and plants for photosynthesis.

chloroplast. Cellular site of eukaryotic photosynthesis.

chloroquine. Antimalarial drug.

chocolate agar. Solid medium containing blood that has been heated.

chromatophore. Invagination of the cell membrane in photosynthetic bacteria that contains enzymes and pigments for photosynthesis.

chromosome. Structure that carries genes and is responsible for heredity.

chronic disease. Slowly progressing disease of long and often indeterminable duration.

cilia. Short, hairlike processes on the eukaryotic cell surface that provide motility or allow the cell to move surrounding liquids.

ciliated mucosa. Lining of the respiratory tract that traps microbes and other particles in a layer of mucus and propels them toward the pharynx.

Ciliophora. Class of protozoa that possess cilia.

cirri. Tufts of fused cilia.

citric acid cycle. *See* tricarboxylic acid cycle.

classical pathway. Antibody-dependent pathway of complement activation.

classification. Categorization of organisms according to evolutionary relatedness (phylogenetic) or phenotypic similarity (determinative).

clean room. Enclosed area maintained free of contamination and used for manipulating sterile materials.

clinical specimen. Fluid or tissue removed from the human body for laboratory analysis.

clonal section theory. Theory explaining how antigen stimulates a single B or T cell to form a large population of cells with identical antigen specificity.

clone. A group of genetically identical cells that have descended from a common parent cell, or a member of such a group.

cloning vector. DNA used as a carrier of foreign genes in genetic engineering.

CoA. *See* coenzyme A.

coagulase. Enzyme produced by staphylococci that causes the clotting of blood plasma.

coagulation. Formation of a blood clot.

coccobacillus. Oval bacterium.

coccus. Spherical bacterium.

codon. Triplet sequence of nucleotides in messenger RNA that specifies a particular amino acid to be inserted in the protein chain.

coenzyme. Small organic molecule required by some enzymes to function.

coenzyme A (CoA). Coenzyme that transfers acetic groups.

cofactor. Metallic ion required by some enzymes to function.

coinfection. Infection caused by the presence of more than one kind of pathogen.

cold agglutinin. Antibody formed in response to some infections that clumps type O erythrocytes at 4°C but not at 37°C.

cold sterilization. Sterilization method that does not rely on heat for killing.

coliforms. Colon bacilli; *Escherichia coli* and similar lactose-fermenting

gram-negative rods that normally inhabit the colon.

collagenase. Enzyme that hydrolyzes collagen.

colonization. Establishment of microbes on the skin or mucous membranes.

colony. Visible aggregate of cells on a solid medium, all descended from the same parent organism.

colony-forming unit. Single cell or cluster of cells that generates an isolated colony on a solid medium.

commensalism. Association between organisms in which one is benefited and the other is neither benefited or harmed.

common-source epidemic. Outbreak in which the agent is acquired from a contaminated source rather than from an infected individual.

communicable disease. Disease that may be transmitted from an infected host to uninfected individuals.

community. All organisms living within a given area.

community-acquired infection. Infection that develops as a result of exposure to a pathogen outside a hospital environment.

compatible solute. Compound that accumulates in cytoplasm to balance osmotic environment for cells growing in high concentrations of salt or sugar.

competitive inhibition. Inhibition of enzyme activity due to competition from a chemical that is a structural analogue of the substrate.

complement. Group of serum proteins that facilitate bacterial phagocytosis and lysis.

complementary base pair. Nucleotides united through hydrogen bonds between their bases.

complement fixation. Reaction of complement with an antigen-antibody complex.

complex medium. Medium in which the exact identity and concentration of each chemical compound is unknown.

complex viral morphology. Capsid shape that is neither helical or icosahedral.

compound. Chemical composed of two or more elements bound to one another.

compound microscope. Microscope containing at least two lenses.

compromised patient. Person who is highly susceptible to infectious disease because of illness or injuries that reduce host resistance.

concentration gradient. Existence of a higher concentration of a substance in one region of a system than in another.

condenser. Part of the microscope that focuses light on the specimen.

congenital infection. Infection acquired in utero and present at birth.

conidium. Asexual fungal spore formed at the tip of aerial hypha.

conjugation. Transfer of genetic material requiring cell-to-cell contact.

conjunctivitis. Inflammation of the conjunctiva.

constitutive protein. Protein whose production is not sensitive to regulation.

consumer. Heterotroph in an ecosystem that feeds on other organisms or their wastes.

contact dermatitis. Delayed-type allergic reaction resulting from skin contact with the allergen.

contagious disease. Infectious disease that can be readily transmitted from person to person.

continuous culture. System in which microbial populations are maintained in exponential growth.

contractile vacuole. Osmoregulatory organelle in protists that contracts and expels excess water from a cell.

control. Material subjected to conditions identical to experimental materials except for the absence of the variable that is proposed to cause the result.

controlled experiment. An experiment performed with the appropriate controls.

convalescence. Recovery from disease.

core. Nucleocapsid of an enveloped virus.

corepressor. Molecule that activates the repressor in a repressible enzyme system.

coupled reactions. Chemical reactions that are linked to one another by a common intermediate.

covalent bond. Chemical bond in which atoms share electrons.

cresol. Any of a group of disinfectants derived from phenol.

cross infection. Infection acquired from an infected person.

cross reaction. Reaction between an antibody and a different antigen than that which stimulated its formation; this phenomenon is due to similarities between the two antigens.

cross resistance. Insensitivity of a microbe to several related antibiotics.

cubic symmetry. Virus morphology that resembles a spherical structure.

culture. Microorganisms grown in an artificial medium.

culture medium. Nutrient material that supports cell growth.

curd. Solid portion of milk after it coagulates.

cutaneous mycosis. Fungal infection of the skin, nails, or hair.

cyanobacterium. Prokaryote that contains chlorophyll a and performs O_2-generating photosynthesis.

cyclic AMP (cAMP). Regulatory molecule synthesized in cells from ATP.

cyst. Protective, dormant structure formed by some protozoa.

cystitis. Inflammation of the urinary bladder.

cytochrome. Iron-containing compound that functions in electron transport in respiration and photosynthesis.

cytopathic effect (CPE). Virus-induced change in cell cultures.

cytoplasm. Internal contents of the cell other than the nucleus.

cytoplasmic inclusion. Intracellular storage granule in a prokaryote; examples are metachromatic granules and poly-β-hydroxybutyric acid granules.

cytoplasmic streaming. Constant motion of cytoplasm in the eukaryotic cell.

cytosine. Pyrimidine nucleotide base found in nucleic acids.

cytoskeleton. Network of filaments and tubules that give shape to the cytoplasm of a eukaryotic cell.

cytostome. Oral opening of some protozoa.

cytotoxicity. Death of cells.

cytotoxic T cell. T cell that physically attaches to cells and destroys them by membrane disruption and lysis.

Dane particle. Infectious form of hepatitis B virus in serum.

darkfield microscope. Compound microscope in which specimens appear light against a dark background.

dead-end infection. Infection that is not normally transmissible to another host.

deamination. Removal of an amino group (NH_2) from a compound.

death phase. Stage in the growth curve when organisms are dying at an exponential rate.

debridement. Removal of necrotic tissue from a wound.

decarboxylation. Removal of carbon dioxide from an organic molecule.

decimal reduction time. *See* D value.

decomposer. Microorganism that degrades organic material into inorganic molecules.

decontamination. Process of removing or destroying harmful microorganisms or their toxic products.

defective virus. Virus capable of replicating only in the presence of another helper virus.

defensin. Antimicrobial peptides present in certain white blood cells.

degradation. Conversion of an organic compound to a smaller organic compound.

degranulation. Discharge of lysosomal contents into the phagocytic vacuole.

dehydration. Removal of water.

delayed hypersensitivity (DHS) lymphocyte. T cell that enhances immunity by secreting lymphokines.

delayed-type hypersensitivity. Allergic response produced by the cell-mediated immune system.

deletion mutation. Loss of a section of DNA; may be loss of a single nucleotide or of entire genes.

denaturation. Change in the configuration of a macromolecule, often resulting in a loss of function.

denitrification. Metabolic conversion of nitrate to nitrite or nitrogen gas.

dental caries. Localized disintegration of teeth due to acids produced by oral bacteria adhering to the tooth surface.

dental plaque. Combination of bacteria and dextran adhering to teeth.

deoxyribonucleic acid (DNA). Macromolecule that encodes the genetic information of the cell.

dermatophyte. Fungus that invades superficial keratinized areas of the body such as skin, hair, and nails.

dermis. Layer of skin below the epidermis.

detergent. Surface-acting agent that often has antibacterial activity.

Deuteromycetes. Class of fungi characterized by the absence of sexual reproduction.

dextran. Type of polysaccharide composed of glucose.

diacetyl. Bacterial fermentation product responsible for the characteristic flavor of butter.

diapedesis. Process by which white blood cells pass through blood vessels into tissues.

diarrhea. Increased frequency of movement and fluid consistency of stools.

diatom. Unicellular alga having a cell wall composed of silica.

diatomaceous earth. Sediment composed of the silica remnants of the cell walls of diatoms.

differential blood count. Percentage of each kind of leukocyte in a blood sample.

differential medium. Culture medium used to distinguish microorganisms on basis of their characteristic appearance.

differential stain. Stain used to differentiate bacteria, their structures, or their tissue components.

diffusion. Tendency of molecules to move from areas where they are in higher concentration to areas of lower concentration.

dikaryon. Stage in replication of some fungi when cell contains two haploid nuclei.

dimorphic fungus. Fungus that exists as either a yeast or a mold, depending on environmental conditions.

dimorphism. The characteristic of existing in two structurally distinct forms.

dinoflagellate. Alga with two flagella. Some are associated with toxic algal blooms.

diphtheroid. Any of a group of bacteria morphologically resembling the diphtheria bacillus.

diplobacilli. Bacilli that occur in pairs.

diplococci. Cocci that occur in pairs.

diploid. Containing two copies of each type of chromosome.

direct count. Determination of the number of microorganisms by microscopic observation.

direct transmission. Immediate transfer of an infectious agent from the reservoir to a host with no intervening intermediate.

disaccharide. Sugar consisting of two monosaccharides.

discontinuous synthesis. Synthesis of DNA on antiparallel strand in small fragments that must be joined by ligase.

disease. Injury or damage to the body.

disinfectant. Agent used on inanimate objects that destroys or inhibits pathogenic microbes.

disinfection. Elimination from an inanimate object of the vegetative forms of most pathogenic organisms.

disk diffusion method. Method for determining the antibiotic sensitivity of a microbe; performed by placing filter-paper disks impregnated with the drugs onto a plate seeded with the test organism and measuring any zone of inhibition that develops.

DNA gyrase. Enzyme that unwinds double helix of DNA.

DNA homology. The degree of similarity between the nucleotide sequences in different DNAs.

DNA library. Collection of cells into which have been cloned foreign DNA fragments that together make up the entire genome of an organism.

DNA polymerase. Enzyme that synthesizes molecules of DNA.

domain. Highest phylogenetic level in evolutionary classification scheme.

Donovan body. Intracellular inclusion found in cells of persons infected with *Calymmatobacterium granulomatis*.

doubling time. *See* generation time.

downstream processing. Procedures used to purify a product of genetic engineering.

DPT vaccine. An immunizing mixture against diphtheria, tetanus, and pertussis.

droplet. Expelled particle of moisture that can transmit microbes between hosts.

droplet nucleus. Dried droplet containing microorganisms; it is smaller than a droplet and may remain suspended in the air; if inhaled it may enter the lower respiratory tract.

D value. Decimal reduction time. Time required to reduce a population to one-tenth its original size.

dysentery. Inflammation of the intestine characterized by diarrhea with blood and mucus.

early protein. Protein synthesized by a virus during the early period of replication; most early proteins are enzymes.

eclipse phase. Time during a viral infection of a host cell when mature virions cannot be found.

ecology. Study of ecosystems.

ecosystem. The organisms and the physical and chemical components within a defined environment.

edema. Accumulation of fluid in tissues.

effector. Compound that inhibits an allosteric enzyme.

effector cell. T lymphocyte that provides protection from an antigen.

effluent. Fluid discharged for sewage treatment.

electron. Negatively charged subatomic particle.

electron microscope. Instrument that uses electron beams instead of light to produce images.

electron transport system (ETS). Series of compounds that generate energy by the transfer of electrons.

electroporation. Use of electric current to increase uptake of DNA into cell.

element. Chemical composed of only one kind of atom.

elementary body. Extracellular, infectious form of chlamydia.

ELISA (enzyme-linked immunosorbent assay). Immunological test that uses an enzyme-mediated reaction as an indicator of an antigen-antibody reaction.

Embden-Meyerhof pathway. Glycolytic pathway.

encephalitis. Inflammation of the brain.

encystment. Process of cyst formation.

endemic. Occurring with a constant frequency in the population.

endergonic. Pertaining to a chemical reaction that consumes energy.

endocarditis. Inflammation of the lining of the heart.

forms the basis for dividing bacteria into two groups, gram-positive and gram-negative.

granulocyte. White blood cell characterized by granules in its cytoplasm. Eosinophils, basophils, and neutrophils are examples.

granuloma. Lesion characterized by the accumulation of granulocytes.

greenhouse effect. Change in atmospheric temperature resulting from accumulation of certain gases such as methane and carbon dioxide.

griseofulvin. Fungistatic antibiotic used to treat dermatophytic infections.

group translocation. An active transport process in which carrier proteins chemically modify molecules as they cross the membrane.

growth curve. Graphic expression of the changes in microbial population size in a batch culture.

growth factor. Organic substance required for growth.

growth rate. Number of doublings per hour.

guanine. Purine nucleotide base found in nucleic acids.

gumma. Rubbery necrotic lesion characteristic of tertiary syphilis.

halogen. Any of the elements chlorine, iodine, bromine, or fluorine.

halophile. Organism that requires a high concentration of salt for growth.

hand washing. A mechanical method extremely successful in reducing the population of pathogens that can be transmitted by the hands.

haploid. Containing a single copy of each type of chromosome.

hapten. Small molecule that reacts with a specific antibody but is non-antigenic unless coupled to a large carrier molecule.

hay fever. Allergic rhinitis; antibody-mediated upper respiratory response to inhaled allergens.

heavy metal. High molecular weight metal that kills cells by denaturing proteins.

helical symmetry. Viral morphology that resembles a long rod.

helper T cell. T lymphocyte that cooperates with B cells to initiate an antibody response.

hemadsorption. Attachment of red blood cells to the surface of virus-infected cells.

hemagglutination. Clumping of red blood cells caused by antibodies or by some enveloped viruses.

hemagglutination inhibition. One type of test to determine the amount of antiviral antibody in a person's blood.

hemagglutinin. Substance that causes red blood cells to clump.

hematogenous. Relating to blood; disseminated by the bloodstream.

hemolysin. Substance that destroys red blood cells.

hemolytic. Having the ability to lyse red blood cells. (*See also* alpha hemolysis *and* beta hemolysis.)

hemorrhagic fever. Acute febrile disease caused by certain viruses and characterized by fever and hemorrhages into the skin and internal organs.

HEPA filter. High efficiency filter that removes more than 99.97% of particles in the respirable range.

hepatitis. Inflammation of the liver.

herd immunity. Protection of individuals in a group from infection by virtue of extensive acquired immunity that decreases the number of infected individuals and therefore the likelihood of exposure of nonimmune persons to the pathogen.

heterocyst. Structure in some cyanobacteria specialized for nitrogen fixation.

heterophil antigen. Antigen from plant or animal that coincidentally cross-reacts with antibody to specific microbes.

heterotroph. Organism that requires carbon in the form of organic molecules.

hexachlorophene. An antiseptic derivative of phenol.

Hfr cell. Donor bacterium containing an F factor integrated into the chromosome; it can transfer chromosomal information to the F$^-$ cell by conjugation.

high-level germicide. Chemical that can be used for sterilization.

histamine. Chemical released from tissue cells that triggers changes in capillaries and smooth muscles.

histiocyte. Fixed macrophage; a macrophage attached to tissue.

histone. Protein specifically associated with eukaryotic DNA.

HLA. Human leukocyte antigen; marker on human cells that is recognized on transplanted tissue by the immune system.

horizontal transmission. Spread of disease from person to person within a group.

host. The organism on or in which a parasite lives.

human immunodeficiency virus (HIV). Retrovirus that causes AIDS.

humoral immunity. Immunity due to the production of antibodies.

humus. Organic matter in soil from decayed plants.

hyaluronidase. Enzyme that hydrolyzes hyaluronic acid in connective tissue, facilitating the spread of pathogens in tissues.

hybridization. Formation of bonds between complementary nucleic acids.

hybridoma. Cell resulting from the fusion of a tumor cell with an antibody-producing cell; it produces abundant quantities of monoclonal antibody.

hydrogen bond. Weak attraction between a positively charged hydrogen of one compound and a portion of another molecule or a different part of the same molecule that has a negative charge.

hydrogen ion. Positively charged nucleus of a hydrogen atom (a proton).

hydrolysis. Breaking of a molecular bond by the addition of water.

hydrophilic. Having an affinity for water molecules.

hydrophobic. Lacking an affinity for water molecules.

hyperhidrosis. Excessive sweating.

hyperimmune serum. Serum that contains specific antibodies from persons who have been immunized with a vaccine.

hypersensitivity (allergy). Immune-mediated response to antigen that injures the host.

hypertonic. Having a solute concentration greater than another solution against which it is compared.

hypha. Filament of a mold.

hypochlorite. Chlorine-containing compound with disinfectant and bleaching properties.

hypogammaglobulinemia. Abnormally low amounts of antibody in the blood.

hypothesis. Proposed explanation for observed phenomena whose validity can by verified by testing.

hypotonic. Having a solute concentration lower than another solution against which it is compared.

icosahedral. Viral morphology resembling a spherical structure with 20 triangular sides.

ID$_{50}$. Number of microbes required to cause disease in 50 percent of laboratory animals experimentally infected with the pathogen.

idiotype. Group of antibodies all of which recognize the identical epitope in their variable region.

IgA. Principal antibody found in body secretions.

IgD. Immunoglobulin found on the surface of B cells; may function during fetal development of the immune system.

IgE. Antibody associated with allergic reactions.

IgG. Most abundant antibody in blood.

IgM. First antibody to appear after initial exposure to an antigen.

imidazole. Antifungal drug that acts on the cell membrane.

immediate-type hypersensitivity. Allergic response mediated by IgE antibodies; symptoms appear within minutes of exposure to antigen.

immobilized enzymes. Enzymes attached to a solid carrier.

immune-complex disease. Damage caused by the deposit of antigen-antibody complexes on capillary or renal membranes.

immune interferon. Interferon produced by DHS cells.

immune tolerance. Lack of immunological activity against certain antigens such as the ability of the body to recognize and not reject "self."

immunity. Body's resistance to invasion by microorganisms and damage by foreign substances.

immunization. Process rendering a host immune.

immunodeficiency. Deficit in either T-cell or B-cell immunity.

immunodiffusion test. Laboratory technique that uses immunoprecipitation in agar to indicate the presence of a specific antigen or antibody.

immunogen. An antigen.

immunoglobulin (Ig). An antibody.

immunology. Study of antibodies and immune cells and their interactions with antigens.

immunopotentiation. Enhancing the natural immune response.

immunosuppressant. Substance that inhibits the immune response.

immunotoxin. Tissue-specific antibody linked to a cytotoxic chemotherapeutic agent.

inapparent infection. Infection that causes no clinically apparent symptoms.

incidence. Rate of occurrence of a disease within a defined population.

incidental host. Host that is not essential to the life cycle of a parasite.

incineration. Sterilization procedure by which contaminated materials are burned to ash.

inclusion body. Microscopically visible site within the cell that contains aggregates of developing viruses.

incubation period. Time interval between exposure to a pathogen and appearance of disease symptoms.

indicator organism. Organism whose presence provides indirect information on the condition of a specimen.

indigenous. Belonging to a certain location or environment.

indirect transmission. Transfer of an infectious agent from an infected host by an intermediate, either a vector or a vehicle.

inducer. Substance that stimulates synthesis of an enzyme.

inducible enzyme. Enzyme that is synthesized only when the appropriate inducer is present.

induction. In virology, the onset of virus production (lytic cycle) in a lysogenic cell. (*See also* enzyme induction.)

infant botulism. A disease caused by *Clostridium botulinum* infection in children less than 8 months old.

infection. Growth of microbes in the body or on its surface.

infection control committee. Epidemiological unit in the hospital responsible for monitoring nosocomial infections.

infectious disease. Injury of host tissue due to infection.

infectious dose. Number of organisms needed to initiate infection in a host.

infectious waste. Discarded material containing sufficient numbers of pathogens to cause disease.

infectivity. Ability of a microbe to infect a host.

inflammation. Nonspecific host response to tissue injury characterized by swelling, pain, redness, and heat.

initiator codon. Nucleotide triplet on messenger RNA that specifies the site where protein synthesis begins.

inoculum. Microorganisms introduced to culture medium or into a host.

inorganic compounds. Most compounds that do not contain carbon.

insertional inactivation. Disruption of gene activity by insertion of a segment of DNA.

insertion sequence. Transposable element that consists of no more than the genes to transpose itself.

integration. Incorporation of a segment of DNA into a larger DNA molecule.

intercalating agent. Chemical that inserts into DNA helix, causing distortions that may result in mutations.

interferon. Group of proteins released by animal cells in response to antigens and certain other triggers. Some interferons induce antiviral activity; others enhance the immune response.

interleukins. Proteins produced by white blood cells that regulate other leukocytes during an immune response.

intermediate-level germicide. Chemical that kills most vegetative cells and viruses but not endospores.

intoxication. Food poisoning following consumption of foods containing preformed microbial toxins.

intracellular pathogen. Microbe that invades and multiplies within the cells of the host.

intrauterine infection. Infection acquired by the fetus as it develops within the uterus.

intron. Coding sequences in eukaryotic DNA that are removed from the primary transcript and do not end up in mRNA.

in-use test. Evaluation of disinfectants under actual conditions of use.

invasive pathogen. Microbe capable of penetrating into deep tissues and disseminating to secondary sites in the body.

invasive procedure. Any technique that introduces an instrument into the body.

in vitro. (Latin, "on glass") Occurring in a test tube or culture dish.

in vivo. (Latin, "in life") Occurring in a living organism.

iodine. Halogen used as an antiseptic.

iodophor. Antiseptic composed of a complex of iodine and surfactant that slowly releases the iodine.

ion. Atom or group of atoms having either a positive or negative charge.

ionic bond. Bond formed when electrons are transferred from one atom to another.

ionizing radiation. High-energy radiation that causes molecules to be ionized.

isolation. Separation of an individual from the community to prevent the transfer of infectious disease.

isoniazid (INH). Tuberculostatic drug.

jaundice. Yellowing of the skin due to an increase of bile pigments in the blood; a common symptom of hepatitis.

keratin. Protein found in skin, hair, and nails.

keratitis. Inflammation of the cornea.

killer cell. Protective cell that reacts with the Fc tail of antibodies on antibody-coated cells.

kinetic energy. Energy doing work (in motion).

Kirby-Bauer technique. Disk diffusion method for determining antibiotic susceptibility of bacteria.

Koch's postulates. Criteria used to establish the etiology of an infectious disease.

Koplik's spots. Oral rash characteristic of measles.

Krebs cycle. *See* tricarboxylic acid cycle.

lag phase. Initial stage in the bacterial growth curve. The microbes may be metabolically active but are not increasing in number.

laminar flow hood. Large box for performing microbial manipulations with minimum chance of contamination; sterile air is circulated through the hood and passed through a series of filters.

Lancefield group. Classification of streptococci into groups A through O, based on antigenic differences in a surface carbohydrate.

latent infection. Stage of an infection in which a pathogen remains in a host for long periods of time without producing disease.

late protein. Protein synthesized by a virus during the late period of replication; most are structural components of the virion.

leader region. Sequence at beginning of mRNA to which ribosomes bind to initiate translation.

lecithinase. Enzyme that degrades lecithin (in cell membranes).

lectin. Surface protein of plant cells to which microbes, such as nitrogen-fixing bacteria, can attach.

legume. Plant with root nodules containing nitrogen-fixing bacteria.

leukemia. Uncontrolled proliferation of leukocytes and their immature precursors.

leukocidin. Substance that kills white blood cells.

leukocyte. White blood cell.

leukocytosis. Increase in the number of circulating leukocytes.

leukopenia. Decrease in the number of circulating leukocytes.

L form. Bacterium that has lost the ability to synthesize peptidoglycan and therefore lacks a cell wall.

lichen. Symbiotic association between a fungus and an alga or cyanobacterium.

ligase. Enzyme that seals gap in the sugar-phosphate backbone of DNA.

limulus amoebocyte lysate assay (LAL). In vitro test for endotoxin; also employed as a rapid screening test for diagnosing gonorrhea.

lipase. Enzyme that hydrolyzes lipids.

lipid. Organic molecule characterized by water insolubility. Lipids include fats, oils, waxes, phospholipids, and steroids.

lipid A. Component of lipopolysaccharide in the gram-negative bacterial cell wall that possesses endotoxin activity.

lipopolysaccharide. Compound composed of lipid and polysaccharide; an important component of the outer membrane of gram-negative bacteria.

lipoprotein. Compound composed of lipid and protein.

liposome. Artificial membrane vesicle.

local infection. Infection limited to a single body site.

localized invasive disease. Disease in which pathogens invade tissues in one body site but do not disseminate to other body sites.

logarithmic (log) phase. Stage in the bacterial growth curve characterized by increases in cell numbers at an exponential rate.

lophotrichous. Having polar flagella arranged in tufts at one end of a cell.

low-level germicide. Chemical that kills only a few types of bacteria, fungi, and viruses.

lymphocyte. Type of agranulocyte; B cells and T cells are examples.

lymphokine. Soluble substance released by T cells in response to antigenic stimulation; lymphokines include lymphotoxin, macrophage-activating factor (MAF), specific-macrophage-arming factor (SMAF), migration-inhibition factor (MIF), and immune interferon.

lymphotoxin. Lymphokine that damages or lyses many types of cells.

lyophilization. Freeze-drying; rapid dehydration of organisms while they are in a frozen state.

lysis. Bursting of a cell; also splitting of a chemical bond, for example, proteolysis.

lysogenic cell. Bacterium that contains a prophage.

lysogenic conversion. Acquisition of new properties by a cell after lysogeny.

lysogeny. Integration of the DNA of a temperate bacteriophage into the bacterial chromosome.

lysosome. Intracellular vesicle containing hydrolytic enzymes and some powerful oxidizing chemicals.

lysozyme. Enzyme that degrades bacterial cell walls by disassembling peptidoglycan; found in many types of body secretions.

lytic cycle. Viral infection that results in cell lysis.

macromolecule. Protein, polysaccharide, nucleic acid, or other large molecule composed of subunits assembled in polymeric chains.

macronucleus. The large nucleus in some protozoa that controls vegetative, but not sexual, reproduction.

macrophage. Active phagocytic white blood cell that develops from a monocyte.

macrophage-activating factor (MAF). Lymphokine that enhances intracellular killing efficiency of macrophages.

macule. Small red spot on the skin.

magnetosome. Cytoplasmic deposits of iron oxide that allow the bacterium to align along the earth's magnetic field.

magnification. Enlargement of an object's image.

maintenance medium. Medium used to store microbes in which they remain viable but grow very slowly.

malignancy. Progressively growing tumor that tends to metastasize.

malting. Digestion of starch in grain to fermentable sugars.

marker gene. Gene encoding an easily monitored phenotypic trait.

mast cell. One type of cell that binds IgE and, in the presence of specific allergen, releases histamines and other chemicals that elicit symptoms of allergy.

Mastigophora. Class of protozoa that use flagella for motility.

MBC. *See* minimum bactericidal concentration.

mechanical vector. Living organism, especially an arthropod, that transfers pathogens from one host to another but is not required for the multiplication or development of the microorganism.

meiosis. Process by which the number of chromosomes in a diploid nucleus is reduced by half.

membrane attack complex. Pore created by complement components that causes lysis of target cell.

memory cell. Antigen-sensitive B or T lymphocyte that persists in the host long after the immunological response to that antigen has waned. Subsequent exposure to the antigen triggers the cell to differentiate into immunologically active T cells or plasma cells.

meningitis. Inflammation of the membranes of the brain or spinal cord.

mercurial. Compound that contains mercury or its salt, often used as an antiseptic.

merozoite. Product of asexual reproduction of *Plasmodium* in humans.

mesophile. Organism whose optimal growth temperature is between 20 and 40°C.

mesosome. Cytoplasmic invagination of the cell membrane in prokaryotes.

messenger RNA (mRNA). Single-stranded RNA that determines the sequence in which amino acids are assembled during protein synthesis.

metabolic intermediate. Compound in a metabolic pathway that is neither the initial substrate nor the final product of that pathway.

metabolic pathway. Series of reactions that sequentially alter a starting compound to a final end product.

metabolism. The sum of all the chemical reactions performed by an organism.

metachromatic granule. Storage form of phosphate in some bacteria. (*See also* volutin granule.)

metastasis. Spread of disease from its primary site to another part of the body.

metronidazole. Antiprotozoal chemotherapeutic agent; also used to treat infections caused by anaerobic bacteria.

MHC. Major histocompatibility complex; a cluster of genes that determine the unique surface antigens that mark an individual's tissues as well as surface proteins that help regulate the immune response.

MIC. *See* minimum inhibitory concentration.

microaerophile. Organism that requires molecular oxygen at concentrations less than that in normal air for growth.

microbial biotechnology. The use of technology to develop microorganisms of greater practical benefit.

microbial pesticide. Microorganism used to control the growth of an insect by virtue of its specific pathogenicity for the undesirable arthropod.

microbicidal. Capable of killing microbes.

microbistatic. Capable of inhibiting microbial growth.

microenvironment. Area immediately surrounding a microbial cell.

micrometer. 10^{-6} meter.

micronucleus. Smaller of two nuclei in some ciliated protozoa; required for sexual reproduction but has no role in vegetative reproduction.

microorganism. Life form too small to be seen with the naked eye.

microtubule. Hollow protein filament in eukaryotic flagella and cilia.

midstream-catch technique. Procedure for obtaining a urine sample that reduces contamination from resident microorganisms on the genitals.

migration-inhibition factor (MIF). Lymphokine that retains macrophages at the site of activated T cells.

mineralization. Conversion of organic material to an inorganic state.

minimum bactericidal concentration (MBC). Smallest concentration of a drug that kills the pathogen against which the drug is to be used.

minimum inhibitory concentration (MIC). The smallest concentration of a drug that inhibits the multiplication of the pathogen against which the drug is to be used.

minimum medium. Medium that supplies only a source of carbon, nitrogen, inorganic salts, and energy.

missense mutation. Nucleotide change that results in replacement of one amino acid with another.

mitochondrion. Organelle in eukaryotic cells that generates ATP by respiration.

mitosis. Duplication of chromosomes and division of the nucleus in eukaryotic cells; usually accompanied by cell division.

mixed culture. Culture containing more than one species of microorganism.

mixed infection. Infection with more than one type of organism.

mold. Fungus that forms large, multicellular aggregates of long, branching filaments.

molecule. Smallest quantity of a compound that retains its characteristic properties.

Monera. Kingdom of prokaryotes; one of the five kingdoms defined by Whittaker; the other kingdoms are those of protists, fungi, plants, and animals.

monoclonal antibody. Immunoglobulin, specific for a single antigenic determinant, produced in vitro by hybridomas (lymphocytes fused with tumor cells).

monocyte. Mononuclear phagocyte from which macrophages differentiate.

monogenic. Possessing messenger RNA that carries the code for a single protein.

monolayer. Uniform single layer of cells growing on the surface of a culture container.

monomer. Subunit from which a polymer is made.

mononuclear phagocytic system. Defense system that consists of monocytes and macrophages.

monosaccharide. Sugar that cannot be hydrolyzed to a simpler sugar.

monospecific. Capable of reacting with only one kind of antigenic determinant.

monotrichous. Having a single polar flagellum at the end of the cell.

morbidity. Number of cases of a disease in a population.

Morbidity and Mortality Weekly Report (MMWR). Publication from the Centers for Disease Control and Prevention that reports statistics on notifiable diseases and current information on public health issues.

mordant. Substance that forms an insoluble complex with a dye in the cell, increasing the intensity or tenacity of staining.

morphology. Physical form of an organism or its parts.

mortality. Number of people in a defined population who die of a specific disease.

most probable number (MPN). Statistical estimate of microbial population determined by the dilution end point for growth in liquid medium.

motility. Ability to move by oneself.

M protein. Antiphagocytic antigen on the surface of *Streptococcus pyogenes*.

mucormycosis. Infection, usually of a compromised host and usually severe, caused by fungi of the group Mucorales.

multiple fission. Asexual reproduction characterized by a series of nuclear divisions followed by cytoplasmic division into as many parts as there are nuclei.

mutagen. Any physical or chemical agent that increases the rate of mutation.

mutant. Organism in which a mutation has occurred.

mutation. Alteration in the nucleotide sequence in DNA.

mutualism. Symbiotic relationship that benefits both members of the relationship.

mycelium. Filaments of fungi forming an interwoven mass.

mycology. Study of fungi.

mycoplasma. Group of eubacteria that lack cell walls.

mycorrhiza. Growth of fungi on the roots of plants in a relationship that benefits both the plant and the fungus.

mycosis. Disease caused by a fungus.

mycotoxicosis. Disease caused by ingestion of mycotoxin.

mycotoxin. Poisonous substance produced by a fungus.

myxamoeba. Amoeba-like feeding stage of slime molds.

myxomycete. Plasmodial slime mold.

NAD (nicotinamide adenine dinucleotide). Coenzyme that transfers hydrogen atoms between metabolic reactions.

NADP (nicotinamide adenine dinucleotide phosphate). Coenzyme that transfers hydrogen atoms between metabolic reactions.

nanometer. 10^{-9} meter.

narrow-spectrum antibiotic. Drug with high degree of specificity against a small number of microbial types.

natural active immunity. Active immunity induced by natural infection.

natural killer (NK) cell. Protective cell that lyses virus-infected cells and tumor cells.

natural passive immunity. Immunity acquired by transfer of maternal antibodies across the placenta to the fetus or in breast milk to an infant.

nebulization. Dispersal of a liquid in a fine spray; used to loosen and liquefy respiratory secretions.

necrosis. Localized tissue death that occurs in response to infection or injury.

negative stain. Procedure that stains the background and leaves the cells colorless.

negative-strand virus. Virus whose single-strand RNA genome is the complement of messenger RNA.

negri body. Cytoplasmic inclusion body that develops in nerve cells infected with rabies virus.

neuraminidase. Enzyme that digests neuraminic acid, a component of mucus and cell membranes.

neurotoxin. Toxin that damages nerve tissue.

neutralizing antibody. Antibody that inactivates the harmful effects of a virus or toxin.

neutron. An electrically neutral subatomic particle.

neutrophil. Highly phagocytic granulocyte that does not stain with either basic or acid dyes.

neutrophile. Microorganism that grows best between pH 5.5 and 8.5.

nitrification. Conversion of nitrogen in ammonia to nitrites and nitrates.

nitrogenase. Enzyme that converts nitrogen (N_2) to ammonia (NH_3).

nitrogen cycle. Series of reactions in nature in which atmospheric nitrogen (N_2) and inorganic nitrogen are converted to organic nitrogen and ultimately degraded back to inorganic and atmospheric forms.

nitrogen fixation. Conversion of molecular nitrogen (N_2) into a biologically useful form.

nodule. *See* root nodule.

nongonococcal urethritis (NGU). Inflammation of the urethra caused by an organism other than *Neisseria gonorrhoeae*, most often a chlamydia or mycoplasma.

noninvasive disease. Infectious disease restricted to the primary site of infection.

nonsense codon. Codon for which no amino acid is specified; it signals protein chain termination.

nonseptate hypha. Hypha that lacks partitions.

normal flora. Those microorganisms that live on the body surfaces in a harmonious relationship with their host.

nosocomial infection. Infection caused by exposure to a pathogen within the hospital environment.

notifiable disease. Disease that by law must be reported to public health officials whenever diagnosed.

nuclear membrane. Membrane separating the nucleus from the cytoplasm in eukaryotes.

nucleic acid. Macromolecule composed of nucleotides; RNA and DNA.

nucleic acid probe. Unique DNA segment used to identify organisms or gene sequences.

nucleocapsid. Viral structure composed of the capsid and the enclosed nucleic acid.

nucleoid. Region in a prokaryotic cell that contains the chromosome.

nucleolus. Structure within the cell nucleus where ribosomal RNA is synthesized.

nucleotide. Monomer from which nucleic acids are constructed; composed of a purine or pyrimidine base, a five-carbon sugar, and a phosphate group.

nucleus. Membrane-bound structure in eukaryotes that contains the genetic material.

numerical aperture. In a microscope, a property of the objective lens that determines how much of the light that has passed through the specimen enters.

numerical taxonomy. Comparison of organisms based on testing a large number of characteristics, each given equal weight in the comparison.

nutrient agar. Nutrient broth that has been solidified by the addition of agar.

nutrient broth. Complex medium containing beef extract and peptone.

nystatin. Polyene antibiotic used topically for treating localized yeast infections.

O antigen. A surface antigen that is a component of lipopolysaccharide in gram-negative bacteria and is useful for their identification.

objective lens. Lens located closest to the specimen in the compound microscope.

obligate aerobe. Organism that grows only in the presence of oxygen.

obligate anaerobe. Organism that grows only in the absence of oxygen.

obligate intracellular parasite. Microorganism that grows only in the cytoplasm or nucleoplasm of a host cell.

ocular lens. The lens in the eyepiece through which the image is viewed.

oligodynamic action. Antimicrobial activity of low concentrations of heavy metals.

oncogene. Gene responsible for transforming normal cells to malignant cells.

oncogenesis. Formation of a tumor.

oncogenic virus. Virus that can induce tumor production.

one-step growth experiment. Experiment that measures viral replication after simultaneously infecting every cell in a culture with virus.

oocyst. Sexual stage of many protozoa, especially Sporozoa.

oomycete. Fungus characterized by production of asexual zoospores.

oospore. Sexual fungal spore produced by many Zygomycetes.

operator. Region in the operon that can bind a specific repressor protein that prevents transcription of the operon's structural genes.

operon. Genetic unit consisting of an operator site and the adjacent structural genes, which are regulated as a unit.

ophthalmia neonatorum. Eye infection acquired by newborns during passage through an infected birth canal.

opportunistic pathogen. Microbe that can cause disease only in an injured, debilitated, or immunologically compromised host.

opsonin. Serum protein, such as an antibody, that promotes phagocytosis of the antigen with which it reacts.

opsonization. Enhancement of phagocytosis by coating a cell with opsonin.

optical density. *See* turbidity.

optimal growth. Multiplication at the maximum growth rate (shortest generation time).

optocin. Chemical used to distinguish *Streptococcus pneumoniae* from other alpha-hemolytic streptococci by virtue of its lethal effect on the pathogen.

organelle. Specialized structure bound by membranes in eukaryotic cells.

organic compound. Compound that contains carbon.

osmophile. Organism capable of growing in media with high solute concentrations.

osmosis. Movement of a solvent across a semipermeable membrane from a dilute solution into a more concentrated one.

osmotic pressure. Pressure exerted against a membrane when solutions of different concentrations are present on opposite sides of the membrane.

osmotroph. Organism that absorbs dissolved nutrients directly through its cell membrane.

outer membrane. Lipid-protein bilayer that surrounds the peptidoglycan layer in the cell walls of gram-negative bacteria.

oxidation. Loss of electrons.

oxidative phosphorylation. Production of ATP using energy provided by an electron transport chain.

oxygenic photosynthesis. Photosynthesis in which oxygen is released from water, characteristic of cyanobacteria and photosynthetic eukaryotes.

palisade arrangement. Orientation of bacilli in parallel rows resembling a picket fence.

pandemic. Epidemic that occurs on several continents.

papilloma. Benign cutaneous tumor. Warts are papillomas.

papule. Small, elevated lesion of the skin.

para-aminosalicylic acid (PAS). Anti-tuberculosis chemotherapeutic agent.

parasite. Organism that lives in or on a host organism from which it secures some advantage.

parasitism. Relationship in which a small organism invades and adversely affects a larger one.

parenteral. Entering the body by means other than through the alimentary canal.

passive immunity. Transfer of antibodies from an immune host to another individual.

Pasteur effect. Preference of facultative anaerobes to use aerobic pathways when molecular oxygen is available.

pasteurization. Mild heating process used to destroy spoilage microorganisms and most types of pathogens.

pathogen. Disease-causing microbe.

pathogenesis. Sequence of events during the development of disease and mechanism(s) by which tissues are injured.

pathogenicity. Ability to cause disease.

PCR. *See* polymerase chain reaction.

pelvic inflammatory disease (PID). Infection of the female genital tract, often accompanying the spread of gonorrhea or chlamydia to the fallopian tubes.

penicillin. Any of a group of antibiotics produced by fungi of the genus *Penicillium* that inhibit bacterial peptidoglycan synthesis. The natural antibiotic is often chemically modified to generate semisynthetic penicillins.

penicillinase. Enzyme that hydrolyzes penicillin to an inactive form.

pentose phosphate pathway. Alternative pathway for glucose oxidation that yields a variety of carbon intermediates including pentoses.

peptide. Short chain of two or more amino acids.

peptide bond. Chemical linkage between amino acids, forming the fundamental backbone of proteins.

peptidoglycan. Rigid matrix of eubacterial cell walls; consists of *N*-acetyl glucosamine and *N*-acetyl muramic acid arranged in strands crosslinked to one another by amino acid side chains.

peptone. Mixture of short amino acid chains produced by partially hydrolyzing protein; used as a nutrient supplement in culture media.

percent G+C. Amount of guanine and cytosine present in an organism's DNA.

periodontal disease. Disease of the tissues surrounding the teeth.

periplasm. The region between the plasma membrane and cell wall. Also called periplasmic space.

peristalsis. Rhythmic muscular contractions that force substances to move through a tube, such as food through the digestive tract.

peritrichous. Having flagella characterized by even distribution of the appendages over the entire surface of the bacterium.

permease. Protein receptors that transport specific molecules across the plasma membrane.

peroxidase. Enzyme that converts hydrogen peroxide to water.

petri dish. Shallow container used for solid culture media.

pH. Measure of the relative hydrogen-ion concentration of a solution. pH values below 7.0 indicate acidic solutions, and those above 7.0 indicate basic (alkaline) conditions.

phage. *See* bacteriophage.

phage typing. Identification of bacterial strains using bacteriophage susceptibility as the specific indicator.

phagocyte. White blood cell that is actively phagocytic.

phagocytosis. Cellular engulfment of solids.

phagolysosome. Vacuole formed within the phagocyte by the fusion of a phagocytic vacuole and a lysosome.

phagotroph. Organism that feeds by engulfing food, forming vesicles within the cell.

phase-contrast microscope. Compound microscope with optics that detect density differences and highlight details of structures within a cell.

phenol (carbolic acid). Caustic chemical used as an antimicrobial agent.

phenol coefficient. Number that describes a chemical's antimicrobial effectiveness relative to that of phenol.

phenotype. Total of the genetically controlled characteristics that are expressed at a given time.

phospholipid. Complex lipid containing two fatty acids and phosphate attached to a single glycerol molecule.

phosphorylation. Coupling of inorganic phosphate (PO_4) to an organic molecule; a term often used to describe ATP production.

photic zone. Region in aquatic environments penetrated by enough sunlight to support growth of photosynthetic organisms.

photoautotroph. Organism that uses light as its energy source and CO_2 as its principal carbon source.

photoheterotroph. Organism that uses light as its energy source and organic molecules as its principal carbon source.

photolysis. Use of light energy to split water into hydrogen and oxygen.

photophosphorylation. Photosynthetic transformation of light energy into chemical energy of ATP.

photosynthesis. Process by which cells convert energy of sunlight into chemical energy.

phototroph. An organism that relies on sunlight as its principal energy source.

phycology. Study of algae.

phytoplankton. Photosynthetic aquatic microbes residing in the photic zone.

pilosebaceous unit. Complex of a sebaceous gland and a hair follicle.

pilus. Appendage extending from the surface of some bacteria that is used for attachment and conjugation.

pinocytosis. Endocytic engulfment of liquids by a cell.

placebo. Substitute for the experimental variable that is known to have no effect.

plaque. Zone of clearing in a layer of susceptible cells due to viral replication.

plasma cell. Antibody-producing cell derived from a B lymphocyte after antigen activation.

plasma membrane. Selectively permeable structure enclosing the cytoplasmic contents of a cell; composed of phospholipids and proteins.

plasmid. Small circular piece of extrachromosomal DNA in bacteria. Some plasmids can integrate into a bacterial chromosome.

plasmodium. Multinucleate cytoplasmic mass formed by myxomycetes.

plate count. *See* viable count.

pleomorphic. Exhibiting several different shapes.

pneumonia. Inflammation of the lung.

pneumonic plague. Form of plague that is transmitted directly from person to person by respiratory droplets from someone infected with *Yersinia pestis*.

pneumovax. Multivalent vaccine that contains capsule antigens of the 14 most common strains of *Streptococcus pneumoniae*.

point mutation. Change in DNA sequence by substituting one nucleotide for another.

polar flagellum. Flagellum located on one or both ends of a bacillus or vibrio.

poly-β-hydroxybutyric acid. Storage granule for lipids found in some bacteria.

polyene. Any of a group of antibiotics that react with sterols in eukaryotic membranes.

polygenic. Messenger RNA that carries the code for more than one protein.

polymer. Large molecule composed of simpler molecules (monomers) repeating in a linear or branched arrangement.

polymerase chain reaction. In vitro method for the rapid amplification of DNA.

polymorphism. Changes in form associated with different stages of a complex life cycle.

polymorphonuclear leukocyte (PMN). Phagocytic white blood cell with a multilobed nucleus. (*See also* neutrophil.)

polymyxin. Antibiotic that alters the permeability of the cell membrane by disrupting phospholipid.

polysaccharide. Macromolecule composed of repeating sugar subunits.

polysome. Chain of ribosomes held together by mRNA; the functional protein synthesis complex.

porins. Protein channels in the outer membrane of gram-negative bacteria that allow for entry of small hydrophilic molecules.

portal of entry. Site in the host that provides a microorganism access to tissues in which initial infection is established.

portal of exit. Site from which pathogens are shed from an infected individual.

positive-strand virus. Virus whose RNA genome serves as a messenger RNA.

potable. Suitable for drinking.

potential energy. Stored energy, in a state of readiness for doing work.

pour plate. Bacterial culture prepared by suspending organisms in melted agar medium and poured into a petri dish to solidify.

precipitation. Solidification of soluble substances, often used to detect antigen-antibody reactions.

precipitin. Antibody that reacts with soluble antigens and converts them to a solid precipitate.

predation. The act of preying; generally a large organism attacks and eats a smaller one.

predisposing factor. Condition that reduces a person's resistance to infectious disease.

preservation. Prevention of microbial growth to retard spoilage of prepared products or to maintain viability of laboratory cultures.

preservative. Compound that retards microbial proliferation in prepared products.

prevalence. Number of persons with a specific disease present in a defined population at a given time.

primary immune response. Initial production of antibodies following the first exposure to a specific antigen.

primary infection. Acute infection that causes the initial illness in a complex disease process.

primary metabolite. Intermediate or end product of a biochemical pathway essential for growth of the microbe.

primary stain. Initial dye used in a differential staining process.

primary treatment. Mechanical removal of most solids from sewage during the first stage of its treatment.

primer. Nucleotide sequence bound to a template that is used by DNA polymerase to initiate DNA replication.

prions. Infectious protein complexes that appear to lack nucleic acid.

probenecid. Adjunct that, when injected with penicillin or cephalosporins, prolongs therapeutic effectiveness by delaying excretion of the drug.

procaine. Adjunct that is often injected with penicillin to delay absorption of the antibiotic and prolong therapeutic effectiveness.

prodromal period. Earliest stage of a developing condition or disease.

producer. Autotrophic organism that generates energy to support heterotrophs within an ecosystem.

product. Substance formed by a chemical reaction.

productive infection. Viral infection resulting in the release of viral progeny.

prokaryote. Cell whose genetic material is not surrounded by a nuclear membrane.

promoter. Site on DNA to which messenger RNA polymerase binds.

propagated epidemic. Outbreak in which the infectious agent can be directly transmitted by infected individuals rather than through a common source.

prophage. DNA of a temperate bacteriophage that has integrated into a bacterial chromosome and established lysogeny.

prosthetic group. Nonprotein portion of an enzyme.

protein. Macromolecule composed of a linear sequence of amino acids that folds into a specific shape.

proteinase. Enzyme that hydrolyzes protein to peptides or amino acids.

protists. Kingdom of single-cell eukaryotic organisms; one of the five kingdoms defined by Whittaker; in an older, three-kingdom classification scheme, the kingdom of all bacteria, algae, fungi, and protozoa.

protomer. Repeating structural unit in a virus capsid.

proton. Positively charged subatomic particle.

protonmotive force. Proton gradient established across a membrane through electron transport that represents a reservoir of energy.

protoplast. Gram-positive bacterium from which the cell wall has been completely removed.

protozoan. Unicellular, nonphotosynthetic eukaryote that lacks polysaccharide cell walls.

pseudohypha. Chain of elongated yeast cells that resembles a mold filament.

pseudopeptidoglycan. In certain archaeobacteria, a modified peptidoglycan lacking NAM and D-amino acids.

pseudopod. Extension of an amoeboid cell's surface for motility and phagocytosis.

psychrophile. Organism that grows optimally at temperatures below 20°C.

pure culture. Culture that contains a single species of microorganism.

purified protein derivative (PPD). Purified protein derived from *Mycobacterium tuberculosis* cultures and used for skin testing.

purine. Category of nucleotide bases. Adenine and guanine are purines.

purulent. Pus-producing.

pus. Accumulation of dead white blood cells, bacteria, and serous fluid in tissues.

pustule. Elevated surface lesion that contains pus.

pyelonephritis. Inflammation of the kidney.

pyrimidine. Category of nucleotide bases. Thymine, cytosine, and uracil are pyrimidines.

pyrogen. Fever-inducing substance.

pyruvic acid. A three-carbon compound that is the end product of glycolysis.

quadrant streak. Method for obtaining isolated colonies in which sample is streaked successively through a portion of the previously streaked area.

slant. Solid culture medium that solidifies in a tilted test tube to produce a slanted surface for microbial inoculation.

slime layer. Loosely adhering surface layer surrounding some types of bacteria.

slime mold. Organism that has an amoeba-like feeding stage and that develops a multicellular spore-bearing structure for reproduction.

sludge. Solid matter that settles during sewage treatment.

smear. Material on a glass slide for microscopic examination.

solubility. Substance's tendency to dissolve in a solute, usually water.

SOS response. Inducible response to DNA damage in bacteria.

specialized transduction. Transfer of bacterial genes to a recipient by a temperate bacteriophage; the transferred genes lie adjacent to the prophage integration site.

species. An organism's ultimate designation in the taxonomic hierarchy; members of the same species share common characteristics that distinguish them from members of other species.

specific-macrophage-arming factor (SMAF). Lymphokine that enhances the ability of macrophages to kill specific antigenic target cells.

spectrophotometer. Instrument that measures the amount of light transmitted through a sample.

spike. Protein projection on the surface of some enveloped viruses.

spirillum. A rigid spiral or corkscrew-shaped bacterium.

spirochete. A motile, corkscrew-shaped bacterium that possesses an axial filament.

splicing. Removal of introns from primary eukaryotic transcript to form messenger RNA.

spontaneous generation. Theory that nonliving substances can be converted into living organisms in the absence of preexisting cells.

sporadic disease. Disease that occurs occasionally within the population.

sporangiospore. Asexual fungal spore contained in a sporangium.

sporangium. Sac containing one or more sporangiospores borne on the tip of an aerial hypha.

spore. Reproductive structure produced by some bacteria, fungi, and a few primitive plants; some spores are more resistant to adverse conditions than vegetative cells are.

spore strip. Filter paper impregnated with bacterial endospores and used to determine the effectiveness of sterilization processes.

sporicidal. Lethal to spores.

sporophore. Sporulating structure formed by plasmodial slime molds.

Sporozoa. Class of protozoa characterized by the absence of motility in the adult forms.

sporozoite. Infectious stage in the complex life cycle of the malaria parasite (*Plasmodium* sp.). These trophozoites develop in the mosquito and are transmitted by its bite.

sporulation. Production of one or more spores by vegetative cells.

spread-plate method. Technique of obtaining isolated colonies by spreading microbes in a liquid sample on the surface of a solid medium.

sputum. Secretion of the lower respiratory tract.

standard curve. Graphic plot that shows the relationship between two characteristics, such as bacterial concentration and turbidity.

starch. Linear polysaccharide composed of glucose subunits linked together by alpha-glycosidic bonds.

starter culture. Inoculum for initiating fermentations or other useful microbial processes.

stationary phase. Stage in the bacterial growth curve when total bacterial numbers are neither increasing nor decreasing.

sterile. Free of all living microorganisms.

sterilization. Elimination of all microbial forms of life, including spores and viruses.

steroids. Group of lipids characterized by a four-ring structure.

sterol. Type of lipid found in the membranes of eukaryotes and mycoplasma. Cholesterol and ergosterol are examples.

sticky ends. Single-stranded ends of a linear piece of double-stranded DNA; they base-pair with complementary sticky ends.

stock culture. Stored culture from which working cultures are prepared.

strain. Subgroup of individuals within a single species that possess several properties that distinguish them from the other members of the species.

streak-plate method. Technique for obtaining isolated colonies by inoculating a sample over a large area of the surface of the solid culture medium.

streptobacilli. Rod-shaped bacteria that form chains.

streptococci. Spherical bacteria that form chains.

strict isolation. Restriction of an infectious patient to a private hospital room that no one enters without wearing a mask, gown, and gloves.

structural analogue. Chemical that closely resembles another compound in molecular structure.

structural gene. Gene that specifies the production of an enzyme.

sty. Infected eyelash follicle.

subacute sclerosing panencephalitis (SSPE). Rare form of brain inflammation that occurs in children, usually with fatal consequences; a sequela of measles.

subclinical infection. *See* inapparent infection.

subculture. Microbial culture that results from the transfer of microbes from one culture to a fresh medium.

subcutaneous. In the tissues beneath the skin.

subcutaneous mycosis. Fungal infection usually restricted to the skin, the tissue beneath the skin, and the lymphatics.

substrate. Reactant in an enzyme-mediated reaction.

substrate-level phosphorylation. Production of ATP by direct transfer of a high-energy phosphate group from an organic substrate to ADP.

sulfa drugs. Synthetic sulfur-containing antimicrobial compounds that are structural analogues of para-aminobenzoic acid. (*See also* sulfonamide.)

sulfonamide. Synthetic chemotherapeutic agent that competitively inhibits microbial metabolism by interfering with the synthesis of folic acid from para-aminobenzoic acid.

supercoil. Twisted form of circular DNA.

superficial mycosis. Fungal infection that affects only the outer dead layers of the skin.

superinfection. Secondary infection that develops during the course of chemotherapy against another infectious disease.

superoxide dismutase. Enzyme

that degrades the superoxide (O^-) radical.

surfactant (surface-active agent). Wetting agent, detergent, or any other compound that interferes with the interaction between the cell's surface and its aqueous environment.

susceptibility. Lack of resistance to a disease.

swarm cell. Flagellated feeding stage of myxomycetes.

sylvatic plague. Bubonic plague acquired from a wild animal reservoir rather than from domestic rats.

symbiosis. Close association of two dissimilar organisms living together.

symptom. Disease-induced change in condition that is perceived by the person suffering from the disease.

syndrome. Complex of signs and symptoms that accompany a specific disease.

synergy. Enhancement of effectiveness of chemicals when used in combination. The increased activity is greater than the sum of the activities of the two agents used alone.

synthetic medium. Nutrient growth medium in which each chemical compound is added separately so that the exact chemical nature of the solution is known.

systemic infection. Active proliferation of microorganisms throughout the body.

systemic mycosis. Fungal disease that may affect the brain, bone, viscera, skin, or any area of the body; also called deep mycosis.

---

target organ. Body site most commonly attacked by a particular pathogen or its by-products.

taxonomy. Science of classifying organisms into categories that contain individuals with similar characteristics.

teichoic acid. Macromolecules composed of sugars and phosphates found in cell walls of gram-negative bacteria.

temperate bacteriophage. Virus that establishes lysogeny with its host bacterium; its genome integrates into the host chromosome and replicates with the bacterial DNA.

tertiary treatment. Final treatment of sewage; use of chemical and physical means to remove BOD, nitrogen, and phosphorus from wastewater, rendering it suitable for drinking.

tetanospasm. Neurotoxin, produced by *Clostridium tetani,* responsible for the paralytic symptoms of the disease tetanus.

tetracycline. Broad-spectrum antibiotic isolated from *Streptomyces* and used for treatment of bacterial and amoebic infections. Tetracyclines interfere with protein synthesis.

tetrads. Arrangement of cocci in packets of four.

thermodynamics. Study of energy transformations.

thermophile. Organism that grows best at temperatures above 40°C.

thrombocyte (platelet). Smallest of the cells in the blood; essential for coagulation.

thylakoid. Folded membrane structure that contains the photosynthetic pigments of cyanobacteria and chloroplasts.

thymine. Pyrimidine nucleotide base that is found in DNA but not in RNA.

thymine dimer. Adjacent thymines joined by abnormal covalent bonds that may result from exposure to ultraviolet light.

thymus. Organ that programs lymphocytes to become T cells.

tincture. Solution that uses alcohol or a water-alcohol combination as the solvent.

tinea. Cutaneous infection by dermatophytes; ringworm.

Ti plasmid. Plasmid in *Agrobacterium tumefaciens* that contains transfer DNA and is used to transfer foreign genes into plant cells.

tissue tropism. Affinity of a pathogen for a particular tissue.

T lymphocyte. Lymphocyte that has been programmed by the thymus for participation in cell-mediated immunity.

toxemia. Presence of toxins in the blood.

toxic shock syndrome. Severe acute disease caused by strains of *Staphylococcus aureus,* phage group I, that produce an exotoxin that is distributed by the blood from a site of local infection; occurs most often in menstruating women using highly absorbent tampons.

toxigenic. Exotoxin-producing.

toxoid. Inactivated form of toxin that is antigenically identical to the active toxin; used for immunization and skin testing.

trace element. Element essential for growth but in extremely small amounts.

tracheostomy. Surgical opening into the trachea to maintain an airway to the lungs.

transcriptase. Enzyme responsible for transcribing mRNA from DNA; DNA-dependent mRNA polymerase.

transcription. Process of assembling a molecule of messenger RNA with a nucleotide sequence complementary to a corresponding segment of DNA.

transduction. Bacteriophage-mediated gene transfer from one bacterium to another.

transfection. Uptake by a cell of free viral nucleic acid.

transfer DNA (tDNA). Plasmid segment transferred from the bacterium *Agrobacterium tumefaciens* to infected plant cells, where it integrates into the eukaryote's chromosome.

transfer factor. Soluble component extracted from immune T lymphocytes that can sensitize other T cells to an antigen.

transferrins. Iron-binding proteins that reduce free-iron levels in the body.

transfer RNA (tRNA). Class of RNA molecules that carry specific amino acids and insert them into growing protein chains.

transformation. Transfer of genetic information by free DNA released from disrupted bacteria.

transgenic animal. Animal containing genes from more than one genus.

transient flora. Microbes that temporarily colonize the surface of a person or animal.

translation. Process of protein synthesis using the nucleotide sequence in mRNA to determine the amino acid sequence in the corresponding protein.

translocation. During protein synthesis, the shift of the ribosome-mRNA-tRNA complex bringing the next codon into position for accepting its specific tRNA.

transmission electron microscope. Microscope that transmits an electron beam through thin sections of a specimen.

transport medium. Medium used to maintain the viability of microorganisms in a specimen being transported to the laboratory.

transposable element. DNA sequences able to move around the chromosome.

transposon. Transposable element that contains genetic information that it also moves.

transtracheal aspiration. Suction of sputum into a sterile needle passed through the skin into the trachea.

***Treponema pallidum* immobilization (TPI) test.** Diagnostic test for syphilis that uses darkfield microscopy to observe whether spirochetes from a lesion are immobilized by *T. pallidum*-specific antibodies.

tricarboxylic acid (TCA) cycle. Metabolic pathway that oxidizes two-carbon compounds to CO_2 and water, with the release of energy-rich electrons that yield ATP when processed by the electron transport system. Also called Krebs cycle or citric acid cycle.

trickle filter. Apparatus for secondary treatment of sewage. Wastewater is sprayed over a gravel bed seeded with microbes that digest the organic compounds in sewage.

trophozoite. Vegetative state of a protozoan.

tubercle. Characteristic lesion produced by *Mycobacterium tuberculosis* infection.

turbidity. Cloudiness of a liquid.

tyndallization. *See* fractional sterilization.

type strain. Strain whose characteristics are used to define the species.

---

ultramicrotome. Knife that cuts ultrathin sections of a specimen, usually in preparation for transmission electron microscopy.

ultraviolet (UV) radiation. Radiation at wavelengths between 180 and 390 nm.

uncoating. Release of viral nucleic acid from its capsid.

undulating membrane. Flexible sheet of material that joins the flagella of certain protozoa to the cell surface.

unicellular. Composed of a single cell.

universal precautions. Guidelines to reduce the risk of infection to persons exposed to contaminated blood or body secretions.

uracil. Pyrimidine nucleotide base found in RNA but not in DNA.

---

urethritis. Inflammation of the urethra.

use-dilution method. Laboratory evaluation of a disinfectant; determines the highest dilution that kills a standard number of test bacteria.

---

vaccination. Induction of active immunity against disease by introducing a vaccine into the individual; artificial active immunity.

vaccine. Preparation of microbes or toxoid that can no longer induce severe disease but can still stimulate immunity against the corresponding pathogen or toxin.

vacuole. Membrane-bound intracellular inclusion in eukaryotic cells; a hollow, gas-filled chamber in some prokaryotes.

vaginitis. Inflammation of the vagina.

variable. Changing condition that, in an experiment, might be responsible for the result.

VDRL (venereal disease research laboratory) test. Rapid screening test for syphilis that detects the presence of reagin as an indicator of *Treponema pallidum* infection.

vector. Living organism, such as an arthropod, that transmits disease from one individual to another. (*See also* biological vector *and* mechanical vector.)

vegetative cell. Actively growing, feeding, and proliferating stage of an organism, as opposed to dormant forms such as spores or cysts.

vehicle. Nonliving material or object that can transmit infectious disease.

venereal disease. Infectious disease acquired by sexual intercourse or genital contact.

venipuncture. Insertion of a needle into a vein.

vertical transmission. Spread of disease from parent to offspring by an infected sperm or egg, by passage across the placenta, or during the birth process.

vesicle. Fluid-filled blister.

viable count. Laboratory process of determining the microbial concentration in a sample by plating samples on (or in) solid media and counting the resulting colonies.

vibrio. Comma-shaped bacterium.

viremia. Presence of viruses in the blood.

---

virion. Infectious virus particle.

viroid. One of a group of small RNA molecules that cause some infectious diseases of plants.

virology. Study of viruses.

viropexis. Engulfment of viruses by a host cell using the process of phagocytosis.

virucidal. Capable of destroying viruses.

virulence. Degree of pathogenicity of a microorganism.

virulence factors. Microbial attributes that increase either infectivity or severity of disease.

virus. Submicroscopic, noncellular infectious entity that is an obligate intracellular parasite; consists of nucleic acid surrounded by a protein coat and sometimes an envelope.

virustatic. Capable of inhibiting viral proliferation without destroying the virus.

vitamin. Nutrient essential in small quantities as a coenzyme or its precursor.

volutin granule. Intracellular reservoir of phosphate characteristic of some bacteria; the granules have a marked affinity for basic dyes. (*See also* metachromatic granule.)

vulnerable food. Food that can support the proliferation of microbes that cause foodborne illness.

---

wandering macrophage. Actively phagocytic cell that can travel to sites of infection or inflammation.

water activity. Measure of available water.

Weil-Felix reaction. Serological test for diagnosing rickettsial disease; uses strains of *Proteus* OX as the indicator of positive reactions.

whey. Fluid portion of milk after curdling.

wild type. Organism with the nonmutant genotype.

working culture. Culture used as source of routine inoculations.

wort. Aqueous extract of malted barley produced during beer production.

---

xanthan. Microbial exopolysaccharide used as a solidifying and stabilizing agent in cosmetics and many other commercial products.

yeast. Unicellular fungus cell.

zone of inhibition. Clear area within a lawn of bacteria in which growth fails to occur because of the presence of an inhibitor.

zoonosis. Disease transmitted from a vertebrate animal to a human.

zooplankton. In aquatic environments, protozoa and small invertebrates that feed on phytoplankton and each other.

zoospore. Flagellated asexual spore characteristic of oomycetes.

Zygomycetes. Class of fungi characterized by aseptate hyphae, asexual sporangiospores, and sexual zygospores.

zygospore. Sexual spore produced by Zygomycetes fungi.

Photo Credits

N. Furjanick. FIGURE 5-17: a & b, Herb Charles Ohlmeyer, prepared by Dr. Philip Tierno/Fran Heyl Associates. FIGURE 5-18: G. W. Willis, MD/Biological Photo Service. FIGURE 5-20: K. Talaro/Visuals Unlimited. PAGE 108: D. Foster/Visuals Unlimited.

CHAPTER 6
OPENER: NASA/Science Source/Photo Researchers. FIGURE 6-1: Andy Levin/Photo Researchers; inset, Herb Charles Ohlmeyer/Fran Heyl Associates. FIGURE 6-3: NASA/Science Source/Photo Researchers. FIGURE 6-13: Paul W. Johnson & John Sieburth University of Rhode Island/Biological Photo Service. FIGURE 6-18b: Marie Green, Graphics & Visual Imaging Laboratory, Computing & Communications, University of California, Riverside. PAGE 146: Photo Researchers.

CHAPTER 7
OPENER: James King-Holmes/Photo Researchers. FIGURE 7-1: Dr. Jason C. H. Shih. FIGURE 7-5: Leon J. Le Beau, Ph.D, Indianhead Park, IL. FIGURE 7-11: Sharon Gerig/Tom Stack & Associates. FIGURE 7-13: WHOI, D. Foster/Visuals Unlimited. FIGURE 7-14: a, C. C. Remsen, S. W. Watson, J. N. Waterbury, and H. G. Truper, *J. Bacteriol* 95:2374 (1968); b, G. A. Peters and R. A. Cellarius, *Bioenergetics* 3:345 (1972).

CHAPTER 8
OPENER: Longridge/Dan McCoy/Rainbow. FIGURE 8-1: a, CNRI/SPL/Photo Researchers; b, David Scharf/Peter Arnold; c, N. Furjanick; d, Patricia L. Grilione, Ph.D., San Jose State University. FIGURE 8-2: a & b, Dr. Rasika Harshey, University of Texas at Austin. FIGURE 8-3: Science Photo Library/Photo Researchers. FIGURE 8-5: Nelson L. Max, Lawrence Livermore National Laboratory. FIGURE 8-16: Professor Oscar Miller/SPL/Photo Researchers. FIGURE 8-27: Dr. Dennis Kunkel/Phototake.

CHAPTER 9
OPENER: Hank Morgan/Rainbow. FIGURE 9-1: B. N. Ames, University of CA, Berkeley J. McCann, and E. Yamasaki, *Mutation Res.* 31:347 (1975). FIGURE 9-2: Courtesy Dr. Daniel C. Williamson and the Lilly Microscope Laboratory. FIGURE 9-3: Perkin Elmer Corporation, Applied Biosystems Division. FIGURE 9-5b: Professor Stanley Coren/SPL/Photo Researchers. FIGURE 9-10: Herb Charles Ohlmeyer/Prepared by Dr. Darzins/Fran Heyl Associates. FIGURE 9-11a: Dan McCoy/Rainbow. PAGE 234: Courtesy Perkin Elmer Corporation.

CHAPTER 10
OPENER: Stephen Frink/Stock Market. FIGURE 10-1: John Walsh/SPL/Photo Researchers; inset, David M. Phillips/Visuals Unlimited. FIGURE 10-6: a, Patricia L. Grilione/Phototake; b, Courtesy Dr. John Smit; c, Science VU-EPA/Visuals Unlimited. FIGURE 10-7: Leon J. LeBeau, PhD, Indianhead Park, IL.

CHAPTER 11
OPENER: David M. Philips/Visuals Unlimited. FIGURE 11-1: a, James W. Richards/Visuals Unlimited; b, Biological Photo Service. FIGURE 11-2: a, Michael A. McClure, Ph.D/Phototake; b, Cabisco/Visuals Unlimited. FIGURE 11-5: a, Dr. Dennis Kunkel/Phototake; b, Herb Charles Ohlmeyer/Fran Heyl Associates; c, David Scarf/Peter Arnold; d, Garry T. Cole, Ph.D/Biological Photo Service; e, SEM by M. Giles; computer enhanced by Pix-Elation/Fran Heyl Associates. FIGURE 11-7b: Michael Fogden/Earth Scenes/Animals Animals. FIGURE 11-8: Ed Reschke/Peter Arnold. FIGURE 11-9: M. Giles, computer enhanced by Pix-Elation/Fran Heyl Associates. FIGURE 11-10: Manfred Kage/Peter Arnold. FIGURE 11-11: a, Herb Charles Ohlmeyer, prepared by Dr. Michael McGinnis/Fran Heyl Associates; b, Dr. Michael McGinnis/Fran Heyl Associates. FIGURE 11-12: Jeff Lepore/Photo Researchers. FIGURE 11-13: Dr. Jeremy Burgess/Science Photo Library/Photo Researchers. FIGURE 11-14: Fran Heyl Associates. FIGURE 11-15: Dr. G. L. Barron, University of Guelph, Ontario. FIGURE 11-16: EXXON, vol. 3, 3rd Quarter, 1985.

FIGURE 11-17: a, Ken Greer/Visuals Unlimited; b, Herb Charles Ohlmeyer/Fran Heyl Associates. FIGURE 11-18: a, Courtesy Armed Forces Institute of Pathology, Neg. No. 57-6860; b, Daniel E. Snyder/Visuals Unlimited. FIGURE 11-19: a, Ken Greer/Visuals Unlimited; b, E. Gueho/CNRI/Science Photo Library/Photo Researchers. FIGURE 11-20: a, Courtesy G. P. Segal; b & c, Courtesy S. H. Sun. FIGURE 11-21: a & b, Larry Tackett/Tom Stack & Associates. FIGURE 11-22: Dr. Jeremy Burgess/Science Photo Library/Photo Researchers. FIGURE 11-23: a, Bruce Iverson; b, Dr. E. R. Degginger. PAGE 259: Left, G. L. Barron, University of Guelph/Biological Photo Service; right, Herb Charles Ohlmeyer/Fran Heyl Associates.

CHAPTER 12
OPENER: Kevin Schaefer/Tom Stack & Associates; inset, Phillip A. Harrington/Fran Heyl Associates. FIGURE 12-1: Cecil Hobbs & JoAnn Burkholder, North Carolina State University. FIGURE 12-2: Manfred Kage/Peter Arnold. FIGURE 12-3: a, Larry Jensen/Visuals Unlimited; b, M. Abbey/Visuals Unlimited. FIGURE 12-5b: Michael Abby/Science Source/Photo Researchers. FIGURE 12-6: a & b, Thomas Eisner, Ithaca, NY. FIGURE 12-8: Courtesy Professor Romano Dallai, University of Siena, Italy. FIGURE 12-9: Eugene Small, Donald Marszalek & Gregory Antipa, courtesy Gregory Antipa, San Francisco State University. FIGURE 12-11: M. Abbey/Visuals Unlimited. FIGURE 12-13: inset l, Martin Dohrn/Science Photo Library/Photo Researchers; inset 2, Eric Gravé/Photo Researchers; inset 3, Professors Gilles and Peters, Mosby-Wolfe Publishing '92/Fran Heyl Associates. FIGURE 12-14: Courtesy World Health Organization. FIGURE 12-16: Pollack/Biological Photo Service; inset, S. Elems/Visuals Unlimited. FIGURE 12-18: Volker Steger/Peter Arnold. FIGURE 12-19: David Hall. FIGURE 12-20: Biophoto Associates/Science Source/Photo Researchers. FIGURE 12-21: D. W. Schindler, 1974. *Science* 184:897-899. FIGURE 12-22: Courtesy R. R. Davies and the Liverpool School of Tropical Medicine. FIGURE 12-23: a, David M. Phillips/Visuals Unlimited; b, Kevin Schafer/Peter Arnold; c, C. C. Lockwood. FIGURE 12-24b: Dr. E. R. Degginger. FIGURE 12-25b: Carolina Biological Supply Company/Phototake. PAGE 291: Jerome Paulin/Visuals Unlimited.

CHAPTER 13
OPENER: Dan McCoy and Richard Feldmann/Rainbow. FIGURE 13-1: Lee D. Simon/Science Source/Photo Researchers. FIGURE 13-2: R. L. Steere and T. L. Schaffer, *Biochem. Biophys. Acta* 28:241 (1958); inset, CNRI/Science Photo Library/Photo Researchers. FIGURE 13-4: a, Hans Gelderblom/Visuals Unlimited; b, CDC/Science Source/Photo Researchers; c, Dr. O. Bradfute/Peter Arnold; d, CNRI/Phototake; e, K. G. Murti/Visuals Unlimited. FIGURE 13-10: Visuals Unlimited. FIGURE 13-14: Dr. Karl Maramorosch, Dept. of Entomology, Rutgers University. FIGURE 13-15: Bruce Iverson. FIGURE 13-16: a & b, Herb Charles Ohlmeyer/Fran Heyl Associates.

CHAPTER 14
OPENER: Institute of Oceanographic Sciences, Deacon Laboratory. FIGURE 14-1: Larry Jensen/Visuals Unlimited. FIGURE 14-2: a, Steven C. Wilson/Entheos; b, A. Ahmadjian/Visuals Unlimited; c, Sharon Cummings/Dembinsky Photo Associates. FIGURE 14-3: M. Abbey/Visuals Unlimited. FIGURE 14-4: a, BIOS-F & J. L. Zeigler/Peter Arnold; b, Dr. T.J. Beveridge & S. Shultze/Biological Photo Service. FIGURE 14-5: Manfred Kage/Peter Arnold. FIGURE 14-6: a, B. Ben Bohlool, NiftAL Project, University of Hawaii; b, Dr. J. Burgess/Science Photo Library/Photo Researchers; c, C. P. Vance/Visuals Unlimited. FIGURE 14-7: G. I. Bernard/Oxford Scientific Films/Animals Animals. FIGURE 14-8: a & b, Courtesy Jeffrey Burnham, BioCheck Laboratories, Inc. Toledo, Ohio. FIGURE 14-9: a & b, M. R. Gambrill and C. Wisseman, Infec. Immun. 8:519 (1973). FIGURE 14-10: a, J. W. Moulder, in D. Schlessinger (Ed.), *Microbiology* 1979, American Society for Microbiology, 1979; b, Courtesy J. W. Moulder.

CHAPTER 15
OPENER: Dr. Jeremy Burgess/Science Photo Library/Photo

Index

Note: Page numbers in italic type indicate figures and tables; page numbers followed by n indicate footnotes.

819

820